T0145104

Lecture Notes in Computer Science 13946

Founding Editors

Gerhard Goos
Juris Hartmanis

The series Lecture Notes in Computer Science (LNCS), including its subseries Lecture Notes in Artificial Intelligence (LNAI) and Lecture Notes in Bioinformatics (LNBI), has established itself as a medium for the publication of new developments in computer science and information technology research, teaching, and education.

LNCS enjoys close cooperation with the computer science R & D community, the series counts many renowned academics among its volume editors and paper authors, and collaborates with prestigious societies. Its mission is to serve this international community by providing an invaluable service, mainly focused on the publication of conference and workshop proceedings and postproceedings. LNCS commenced publication in 1973.

Xin Wang · Maria Luisa Sapino ·
Wook-Shin Han · Amr El Abbadi · Gill Dobbie ·
Zhiyong Feng · Yingxiao Shao · Hongzhi Yin
Editors

Database Systems
for Advanced Applications

28th International Conference, DASFAA 2023
Tianjin, China, April 17–20, 2023
Proceedings, Part IV

Springer

Editors
Xin Wang ⓘ
Tianjin University
Tianjin, China

Maria Luisa Sapino ⓘ
University of Torino
Turin, Italy

Wook-Shin Han
POSTECH
Pohang, Korea (Republic of)

Amr El Abbadi
University of California Santa Barbara
Santa Barbara, CA, USA

Gill Dobbie ⓘ
University of Auckland
Auckland, New Zealand

Zhiyong Feng
Tianjin University
Tianjin, China

Yingxiao Shao ⓘ
Beijing University of Posts
and Telecommunications
Beijing, China

Hongzhi Yin ⓘ
The University of Queensland
Brisbane, QLD, Australia

ISSN 0302-9743   ISSN 1611-3349 (electronic)
Lecture Notes in Computer Science
ISBN 978-3-031-30677-8   ISBN 978-3-031-30678-5 (eBook)
https://doi.org/10.1007/978-3-031-30678-5

This Springer imprint is published by the registered company Springer Nature Switzerland AG
The registered company address is: Gewerbestrasse 11, 6330 Cham, Switzerland

Preface

It is our great pleasure to present the proceedings of the 28th International Conference on Database Systems for Advanced Applications (DASFAA 2023), organized by Tianjin University and held during April 17–20, 2023, in Tianjin, China. DASFAA is an annual international database conference which showcases state-of-the-art R&D activities in database systems and advanced applications. It provides a premier international forum for technical presentations and discussions among database researchers, developers, and users from both academia and industry.

This year we received a record high number of 652 research paper submissions. We conducted a double-blind review following the tradition of DASFAA, and constructed a large committee consisting of 31 Senior Program Committee (SPC) members and 254 Program Committee (PC) members. Each valid submission was reviewed by at least three PC members and meta-reviewed by one SPC member, who also led the discussion with the PC members. We, the PC co-chairs, considered the recommendations from the SPC members and investigated each submission as well as its reviews to make the final decisions. As a result, 125 full papers (acceptance ratio of 19.2%) and 66 short papers (acceptance ratio of 29.3%) were accepted. The review process was supported by the Microsoft CMT system. During the three main conference days, these 191 papers were presented in 17 research sessions. The dominant keywords for the accepted papers included model, graph, learning, performance, knowledge, time, recommendation, representation, attention, prediction, and network. In addition, we included 15 industry papers, 15 demo papers, 5 PhD consortium papers, and 7 tutorials in the program. Finally, to shed light on the direction in which the database field is headed, the conference program included four invited keynote presentations by Sihem Amer-Yahia (CNRS, France), Kyuseok Shim (Seoul National University, South Korea), Angela Bonifati (Lyon 1 University, France), and Jianliang Xu (Hong Kong Baptist University, China).

Four workshops were selected by the workshop co-chairs to be held in conjunction with DASFAA 2023, which were the 9th International Workshop on Big Data Management and Service (BDMS 2023), the 8th International Workshop on Big Data Quality Management (BDQM 2023), the 7th International Workshop on Graph Data Management and Analysis (GDMA 2023), and the 1st International Workshop on Bundle-based Recommendation Systems (BundleRS 2023). The workshop papers are included in a separate volume of the proceedings also published by Springer in its Lecture Notes in Computer Science series.

We are grateful to the general chairs, Amr El Abbadi, UCSB, USA, Gill Dobbie, University of Auckland, New Zealand, and Zhiyong Feng, Tianjin University, China, all SPC members, PC members, and external reviewers who contributed their time and expertise to the DASFAA 2023 paper-reviewing process. We would like to thank all the members of the Organizing Committee, and the many volunteers, for their great support in the conference organization. Lastly, many thanks to the authors who submitted their papers to the conference.

March 2023

Xin Wang
Maria Luisa Sapino
Wook-Shin Han

Organization

Steering Committee Members

Chair

Lei Chen — Hong Kong University of Science and Technology (Guangzhou), China

Vice Chair

Stéphane Bressan — National University of Singapore, Singapore

Treasurer

Yasushi Sakurai — Osaka University, Japan

Secretary

Kyuseok Shim — Seoul National University, South Korea

Members

Zhiyong Peng	Wuhan University of China, China
Zhanhuai Li	Northwestern Polytechnical University, China
Krishna Reddy	IIIT Hyderabad, India
Yunmook Nah	DKU, South Korea
Wenjia Zhang	University of New South Wales, Australia
Zi Huang	University of Queensland, Australia
Guoliang Li	Tsinghua University, China
Sourav Bhowmick	Nanyang Technological University, Singapore
Atsuyuki Morishima	University of Tsukuba, Japan
Sang-Won Lee	SKKU, South Korea
Yang-Sae Moon	Kangwon National University, South Korea

Organizing Committee

Honorary Chairs

Christian S. Jensen	Aalborg University, Denmark
Keqiu Li	Tianjin University, China

General Chairs

Amr El Abbadi	UCSB, USA
Gill Dobbie	University of Auckland, New Zealand
Zhiyong Feng	Tianjin University, China

Program Committee Chairs

Xin Wang	Tianjin University, China
Maria Luisa Sapino	University of Torino, Italy
Wook-Shin Han	POSTECH, South Korea

Industry Program Chairs

Jiannan Wang	Simon Fraser University, Canada
Jinwei Zhu	Huawei, China

Tutorial Chairs

Jianxin Li	Deakin University, Australia
Herodotos Herodotou	Cyprus University of Technology, Cyprus

Demo Chairs

Zhifeng Bao	RMIT, Australia
Yasushi Sakurai	Osaka University, Japan
Xiaoli Wang	Xiamen University, China

Workshop Chairs

Lu Chen	Zhejiang University, China
Xiaohui Tao	University of Southern Queensland, Australia

Panel Chairs

Lei Chen   Hong Kong University of Science and
                                Technology (Guangzhou), China
Xiaochun Yang   Northeastern University, China

PhD Consortium Chairs

Leong Hou U.   University of Macau, China
Panagiotis Karras   Aarhus University, Denmark

Publicity Chairs

Yueguo Chen   Renmin University of China, China
Kyuseok Shim   Seoul National University, South Korea
Yoshiharu Ishikawa   Nagoya University, Japan
Arnab Bhattacharya   IIT Kanpur, India

Publication Chairs

Yingxiao Shao   Beijing University of Posts and
                                Telecommunications, China
Hongzhi Yin   University of Queensland, Australia

DASFAA Steering Committee Liaison

Lei Chen   Hong Kong University of Science and
                                Technology (Guangzhou), China

Local Arrangement Committee

Xiaowang Zhang   Tianjin University, China
Guozheng Rao   Tianjin University, China
Yajun Yang   Tianjin University, China
Shizhan Chen   Tianjin University, China
Xueli Liu   Tianjin University, China
Xiaofei Wang   Tianjin University, China
Chao Qiu   Tianjin University, China
Dong Han   Tianjin Academy of Fine Arts, China
Ying Guo   Tianjin University, China

Hui Jiang   Tianjin Ren'ai College, China
Kun Liang   Tianjin University of Science and Technology,
                                China

Web Master

Zirui Chen   Tianjin University, China

Program Committee Chairs

Xin Wang   Tianjin University, China
Maria Luisa Sapino   University of Torino, Italy
Wook-Shin Han   POSTECH, South Korea

Senior Program Committee (SPC) Members

Baihua Zheng   Singapore Management University, Singapore
Bin Cui   Peking University, China
Bingsheng He   National University of Singapore, Singapore
Chee-Yong Chan   National University of Singapore, Singapore
Chengfei Liu   Swinburne University of Technology, Australia
Haofen Wang   Tongji University, China
Hong Gao   Harbin Institute of Technology, China
Hongzhi Yin   University of Queensland, Australia
Jiaheng Lu   University of Helsinki, Finland
Jianliang Xu   Hong Kong Baptist University, China
Jianyong Wang   Tsinghua University, China
K. Selçuk Candan   Arizona State University, USA
Kyuseok Shim   Seoul National University, South Korea
Lei Li   Hong Kong University of Science and
                                Technology (Guangzhou), China
Lina Yao   University of New South Wales, Australia
Ling Liu   Georgia Institute of Technology, USA
Nikos Bikakis   Athena Research Center, Greece
Qiang Zhu   University of Michigan-Dearborn, USA
Reynold Cheng   University of Hong Kong, China
Ronghua Li   Beijing Institute of Technology, China
Vana Kalogeraki   Athens University of Economics and Business,
                                Greece
Vincent Tseng   National Yang Ming Chiao Tung University,
                                Taiwan
Wang-Chien Lee   Pennsylvania State University, USA

Xiang Zhao	National University of Defense Technology, China
Xiaoyong Du	Renmin University of China, China
Ye Yuan	Beijing Institute of Technology, China
Yongxin Tong	Beihang University, China
Yoshiharu Ishikawa	Nagoya University, Japan
Yufei Tao	Chinese University of Hong Kong, China
Yunjun Gao	Zhejiang University, China
Zhiyong Peng	Wuhan University, China

Program Committee (PC) Members

Alexander Zhou	Hong Kong University of Science and Technology, China
Alkis Simitsis	Athena Research Center, Greece
Amr Ebaid	Google, USA
An Liu	Soochow University, China
Anne Laurent	University of Montpellier, France
Antonio Corral	University of Almería, Spain
Baoning Niu	Taiyuan University of Technology, China
Barbara Catania	University of Genoa, Italy
Bin Cui	Peking University, China
Bin Wang	Northeastern University, China
Bing Li	Institute of High Performance Computing, Singapore
Bohan Li	Nanjing University of Aeronautics and Astronautics, China
Changdong Wang	SYSU, China
Chao Huang	University of Notre Dame, USA
Chao Zhang	Tsinghua University, China
Chaokun Wang	Tsinghua University, China
Chenyang Wang	Aalto University, Finland
Cheqing Jin	East China Normal University, China
Chih-Ya Shen	National Tsing Hua University, Taiwan
Christos Doulkeridis	University of Pireaus, Greece
Chuan Ma	Zhejiang Lab, China
Chuan Xiao	Osaka University and Nagoya University, Japan
Chuanyu Zong	Shenyang Aerospace University, China
Chunbin Lin	Amazon AWS, USA
Cindy Chen	UMass Lowell, USA
Claudio Schifanella	University of Torino, Italy
Cuiping Li	Renmin University of China, China

Damiani Ernesto	University of Milan, Italy
Dan He	University of Queensland, Australia
De-Nian Yang	Academia Sinica, Taiwan
Derong Shen	Northeastern University, China
Dhaval Patel	IBM Research, USA
Dian Ouyang	Guangzhou University, China
Dieter Pfoser	George Mason University, USA
Dimitris Kotzinos	ETIS, France
Dong Wen	University of New South Wales, Australia
Dongxiang Zhang	Zhejiang University, China
Dongxiao He	Tianjin University, China
Faming Li	Northeastern University, USA
Ge Yu	Northeastern University, China
Goce Trajcevski	Iowa State University, USA
Gong Cheng	Nanjing University, China
Guandong Xu	University of Technology Sydney, Australia
Guanhua Ye	University of Queensland, Australia
Guoliang Li	Tsinghua University, China
Haida Zhang	WorldQuant, USA
Hailong Liu	Northwestern Polytechnical University, China
Haiwei Zhang	Nankai University, China
Hantao Zhao	ETH, Switzerland
Hao Peng	Beihang University, China
Hiroaki Shiokawa	University of Tsukuba, Japan
Hongbin Pei	Xi'an Jiaotong University, China
Hongxu Chen	Commonwealth Bank of Australia, Australia
Hongzhi Wang	Harbin Institute of Technology, China
Hongzhi Yin	University of Queensland, Australia
Huaijie Zhu	Sun Yat-sen University, China
Hui Li	Xidian University, China
Huiqi Hu	East China Normal University, China
Hye-Young Paik	University of New South Wales, Australia
Ioannis Konstantinou	University of Thessaly, Greece
Ismail Hakki Toroslu	METU, Turkey
Jagat Sesh Challa	BITS Pilani, India
Ji Zhang	University of Southern Queensland, Australia
Jia Xu	Guangxi University, China
Jiali Mao	East China Normal University, China
Jianbin Qin	Shenzhen Institute of Computing Sciences, China
Jianmin Wang	Tsinghua University, China
Jianqiu Xu	Nanjing University of Aeronautics and Astronautics, China

Jianxin Li	Deakin University, Australia
Jianye Yang	Guangzhou University, China
Jiawei Jiang	Wuhan University, China
Jie Shao	University of Electronic Science and Technology of China, China
Jilian Zhang	Jinan University, China
Jilin Hu	Aalborg University, Denmark
Jin Wang	Megagon Labs, USA
Jing Tang	Hong Kong University of Science and Technology, China
Jithin Vachery	NUS, Singapore
Jongik Kim	Chungnam National University, South Korea
Ju Fan	Renmin University of China, China
Jun Gao	Peking University, China
Jun Miyazaki	Tokyo Institute of Technology, Japan
Junhu Wang	Griffith University, Australia
Junhua Zhang	University of New South Wales, Australia
Junliang Yu	The University of Queensland, Australia
Kai Wang	Shanghai Jiao Tong University, China
Kai Zheng	University of Electronic Science and Technology of China, China
Kangfei Zhao	The Chinese University of Hong Kong, China
Kesheng Wu	LBNL, USA
Kristian Torp	Aalborg University, Denmark
Kun Yue	School of Information and Engineering, China
Kyoung-Sook Kim	National Institute of Advanced Industrial Science and Technology, Japan
Ladjel Bellatreche	ISAE-ENSMA, France
Latifur Khan	University of Texas at Dallas, USA
Lei Cao	MIT, USA
Lei Duan	Sichuan University, China
Lei Guo	Shandong Normal University, China
Leong Hou U.	University of Macau, China
Liang Hong	Wuhan University, China
Libin Zheng	Sun Yat-sen University, China
Lidan Shou	Zhejiang University, China
Lijun Chang	University of Sydney, Australia
Lin Li	Wuhan University of Technology, China
Lizhen Cui	Shandong University, China
Long Yuan	Nanjing University of Science and Technology, China
Lu Chen	Swinburne University of Technology, Australia

Lu Chen	Zhejiang University, China
Makoto Onizuka	Osaka University, Japan
Manish Kesarwani	IBM Research, India
Manolis Koubarakis	University of Athens, Greece
Markus Schneider	University of Florida, USA
Meihui Zhang	Beijing Institute of Technology, China
Meng Wang	Southeast University, China
Meng-Fen Chiang	University of Auckland, New Zealand
Ming Zhong	Wuhan University, China
Minghe Yu	Northeastern University, China
Mizuho Iwaihara	Waseda University, Japan
Mo Li	Liaoning University, China
Ning Wang	Beijing Jiaotong University, China
Ningning Cui	Anhui University, China
Norio Katayama	National Institute of Informatics, Japan
Noseong Park	George Mason University, USA
Panagiotis Bouros	Johannes Gutenberg University Mainz, Germany
Peiquan Jin	University of Science and Technology of China, China
Peng Cheng	East China Normal University, China
Peng Peng	Hunan University, China
Pengpeng Zhao	Soochow University, China
Ping Lu	Beihang University, China
Pinghui Wang	Xi'an Jiaotong University, China
Qiang Yin	Shanghai Jiao Tong University, China
Qianzhen Zhang	National University of Defense Technology, China
Qing Liao	Harbin Institute of Technology (Shenzhen), China
Qing Liu	CSIRO, Australia
Qingpeng Zhang	City University of Hong Kong, China
Qingqing Ye	Hong Kong Polytechnic University, China
Quanqing Xu	A*STAR, Singapore
Rong Zhu	Alibaba Group, China
Rui Zhou	Swinburne University of Technology, Australia
Rui Zhu	Shenyang Aerospace University, China
Ruihong Qiu	University of Queensland, Australia
Ruixuan Li	Huazhong University of Science and Technology, China
Ruiyuan Li	Chongqing University, China
Sai Wu	Zhejiang University, China
Sanghyun Park	Yonsei University, South Korea

Sanjay Kumar Madria	Missouri University of Science & Technology, USA
Sebastian Link	University of Auckland, New Zealand
Sen Wang	University of Queensland, Australia
Shaoxu Song	Tsinghua University, China
Sheng Wang	Wuhan University, China
Shijie Zhang	Tencent, China
Shiyu Yang	Guangzhou University, China
Shuhao Zhang	Singapore University of Technology and Design, Singapore
Shuiqiao Yang	UNSW, Australia
Shuyuan Li	Beihang University, China
Sibo Wang	Chinese University of Hong Kong, China
Silvestro Roberto Poccia	University of Turin, Italy
Tao Qiu	Shenyang Aerospace University, China
Tao Zhao	National University of Defense Technology, China
Taotao Cai	Macquarie University, Australia
Thanh Tam Nguyen	Griffith University, Australia
Theodoros Chondrogiannis	University of Konstanz, Germany
Tieke He	State Key Laboratory for Novel Software Technology, China
Tieyun Qian	Wuhan University, China
Tiezheng Nie	Northeastern University, China
Tsz Nam (Edison) Chan	Hong Kong Baptist University, China
Uday Kiran Rage	University of Aizu, Japan
Verena Kantere	National Technical University of Athens, Greece
Wei Hu	Nanjing University, China
Wei Li	Harbin Engineering University, China
Wei Lu	RUC, China
Wei Shen	Nankai University, China
Wei Song	Wuhan University, China
Wei Wang	Hong Kong University of Science and Technology (Guangzhou), China
Wei Zhang	ECNU, China
Wei Emma Zhang	The University of Adelaide, Australia
Weiguo Zheng	Fudan University, China
Weijun Wang	University of Göttingen, Germany
Weiren Yu	University of Warwick, UK
Weitong Chen	Adelaide University, Australia
Weiwei Sun	Fudan University, China
Weixiong Rao	Tongji University, China

Wen Hua	Hong Kong Polytechnic University, China
Wenchao Zhou	Georgetown University, USA
Wentao Li	University of Technology Sydney, Australia
Wentao Zhang	Mila, Canada
Werner Nutt	Free University of Bozen-Bolzano, Italy
Wolf-Tilo Balke	TU Braunschweig, Germany
Wookey Lee	Inha University, South Korea
Xi Guo	University of Science and Technology Beijing, China
Xiang Ao	Institute of Computing Technology, CAS, China
Xiang Lian	Kent State University, USA
Xiang Zhao	National University of Defense Technology, China
Xiangguo Sun	Chinese University of Hong Kong, China
Xiangmin Zhou	RMIT University, Australia
Xiangyu Song	Swinburne University of Technology, Australia
Xiao Pan	Shijiazhuang Tiedao University, China
Xiao Fan Liu	City University of Hong Kong, China
Xiaochun Yang	Northeastern University, China
Xiaofeng Gao	Shanghai Jiaotong University, China
Xiaoling Wang	East China Normal University, China
Xiaowang Zhang	Tianjin University, China
Xiaoyang Wang	University of New South Wales, Australia
Ximing Li	Jilin University, China
Xin Cao	University of New South Wales, Australia
Xin Huang	Hong Kong Baptist University, China
Xin Wang	Southwest Petroleum University, China
Xinqiang Xie	Neusoft, China
Xiuhua Li	Chongqing University, China
Xiulong Liu	Tianjin University, China
Xu Zhou	Hunan University, China
Xuequn Shang	Northwestern Polytechnical University, China
Xupeng Miao	Carnegie Mellon University, USA
Xuyun Zhang	Macquarie University, Australia
Yajun Yang	Tianjin University, China
Yan Zhang	Peking University, China
Yanfeng Zhang	Northeastern University, China, and Macquarie University, Australia
Yang Cao	Hokkaido University, Japan
Yang Chen	Fudan University, China
Yang-Sae Moon	Kangwon National University, South Korea
Yanjie Fu	University of Central Florida, USA

Yanlong Wen	Nankai University, China
Ye Yuan	Beijing Institute of Technology, China
Yexuan Shi	Beihang University, China
Yi Cai	South China University of Technology, China
Ying Zhang	Nankai University, China
Yingxia Shao	BUPT, China
Yiru Chen	Columbia University, USA
Yixiang Fang	Chinese University of Hong Kong, Shenzhen, China
Yong Tang	South China Normal University, China
Yong Zhang	Tsinghua University, China
Yongchao Liu	Ant Group, China
Yongpan Sheng	Southwest University, China
Yongxin Tong	Beihang University, China
You Peng	University of New South Wales, Australia
Yu Gu	Northeastern University, China
Yu Yang	Hong Kong Polytechnic University, China
Yu Yang	City University of Hong Kong, China
Yuanyuan Zhu	Wuhan University, China
Yue Kou	Northeastern University, China
Yunpeng Chai	Renmin University of China, China
Yunyan Guo	Tsinghua University, China
Yunzhang Huo	Hong Kong Polytechnic University, China
Yurong Cheng	Beijing Institute of Technology, China
Yuxiang Zeng	Hong Kong University of Science and Technology, China
Zeke Wang	Zhejiang University, China
Zhaojing Luo	National University of Singapore, Singapore
Zhaonian Zou	Harbin Institute of Technology, China
Zheng Liu	Nanjing University of Posts and Telecommunications, China
Zhengyi Yang	University of New South Wales, Australia
Zhenya Huang	University of Science and Technology of China, China
Zhenying He	Fudan University, China
Zhipeng Zhang	Alibaba, China
Zhiwei Zhang	Beijing Institute of Technology, China
Zhixu Li	Fudan University, China
Zhongnan Zhang	Xiamen University, China

Industry Program Chairs

Jiannan Wang	Simon Fraser University, Canada
Jinwei Zhu	Huawei, China

Industry Program Committee Members

Bohan Li	Nanjing University of Aeronautics and Astronautics, China
Changbo Qu	Simon Fraser University, Canada
Chengliang Chai	Tsinghua University, China
Denis Ponomaryov	The Institute of Informatics Systems of the Siberian Division of Russian Academy of Sciences, Russia
Hongzhi Wang	Harbin Institute of Technology, China
Jianhua Yin	Shandong University, China
Jiannan Wang	Simon Fraser University, Canada
Jinglin Peng	Simon Fraser University, Canada
Jinwei Zhu	Huawei Technologies Co. Ltd., China
Ju Fan	Renmin University of China, China
Minghe Yu	Northeastern University, China
Nikos Ntarmos	Huawei Technologies R&D (UK) Ltd., UK
Sheng Wang	Alibaba Group, China
Wei Zhang	East China Normal University, China
Weiyuan Wu	Simon Fraser University, Canada
Xiang Li	East China Normal University, China
Xiaofeng Gao	Shanghai Jiaotong University, China
Xiaoou Ding	Harbin Institute of Technology, China
Yang Ren	Huawei, China
Yinan Mei	Tsinghua University, China
Yongxin Tong	Beihang University, China

Demo Track Program Chairs

Zhifeng Bao	RMIT, Australia
Yasushi Sakurai	Osaka University, Japan
Xiaoli Wang	Xiamen University, China

Demo Track Program Committee Members

Benyou Wang	Chinese University of Hong Kong, Shenzhen, China
Changchang Sun	Illinois Institute of Technology, USA
Chen Lin	Xiamen University, China
Chengliang Chai	Tsinghua University, China
Chenhao Ma	Chinese University of Hong Kong, Shenzhen, China
Dario Garigliotti	Aalborg University, Denmark
Ergute Bao	National University of Singapore, Singapore
Jianzhong Qi	The University of Melbourne, Australia
Jiayuan He	RMIT University, Australia
Kaiping Zheng	National University of Singapore, Singapore
Kajal Kansal	NUS, Singapore
Lei Cao	MIT, USA
Liang Zhang	WPI, USA
Lu Chen	Swinburne University of Technology, Australia
Meihui Zhang	Beijing Institute of Technology, China
Mengfan Tang	University of California, Irvine, USA
Na Zheng	National University of Singapore, Singapore
Pinghui Wang	Xi'an Jiaotong University, China
Qing Xie	Wuhan University of Technology, China
Ruihong Qiu	University of Queensland, Australia
Tong Chen	University of Queensland, Australia
Yile Chen	Nanyang Technological University, Singapore
Yuya Sasaki	Osaka University, Japan
Yuyu Luo	Tsinghua University, China
Zhanhao Zhao	Renmin University of China, China
Zheng Wang	Huawei Singapore Research Center, Singapore
Zhuo Zhang	University of Melbourne, Australia

PhD Consortium Track Program Chairs

Leong Hou U.	University of Macau, China
Panagiotis Karras	Aarhus University, Denmark

PhD Consortium Track Program Committee Members

Anton Tsitsulin	Google, USA
Bo Tang	Southern University of Science and Technology, China

Hao Wang   Wuhan University, China
Jieming Shi   Hong Kong Polytechnic University, China
Tsz Nam (Edison) Chan   Hong Kong Baptist University, China
Xiaowei Wu   State Key Lab of IoT for Smart City, University of
                                  Macau, China

Contents – Part IV

Industry Papers

Demo Papers

PhD Consortium

Applications of Machine Learning

Event Relation Extraction Using Type-Guided Attentive Graph Convolutional Networks

Ling Zhuang[1,2,3], Po Hu[1,2,3(✉)], and Weizhong Zhao[1,2,3(✉)]

[1] Hubei Provincial Key Laboratory of Artificial Intelligence and Smart Learning,
Central China Normal University, Wuhan, Hubei, China
`zl_sy@mails.ccnu.edu.cn, phu@mail.ccnu.edu.cn, wzzhao@ccnu.edu.cn`
[2] School of Computer Science, Central China Normal University,
Wuhan, Hubei, China
[3] National Language Resources Monitoring and Research Center for Network Media,
Central China Normal University, Wuhan, Hubei, China

Abstract. Event relation extraction is a fundamental task in text mining, which has wide applications in event-centric natural language processing. However, most of the existing approaches can hardly model complicated contexts since they fail to use dependency-type knowledge in texts to assist in identifying implicit clues to event relations, leading to the suboptimal performance on this task. To this end, we propose a novel type-guided attentive graph convolutional network for event relation extraction. Specifically, given the input text, the event-specific syntactic dependency graph is first constructed, from which both the local and global dependency knowledge related to events are derived. Then, a dependency-type guided attentive graph convolutional network is designed for learning representations of events, in which the local and global dependency information are utilized to effectively aggregate semantic context among texts. Finally, the event relation is predicted based on the representations of event pair and the representation of the whole text, completing the task of event relation extraction. The experimental results on multiple datasets show that our method significantly outperforms the state-of-the-art baselines. Moreover, further analysis reveals that our method can effectively capture complex relationships between long-distance events, improving accordingly the performance of event relation extraction.

Keywords: Subevent Relation Extraction · Event Temporal Relation Extraction · Attentive Graph Convolutional Networks · Dependency Type Augmentation

1 Introduction

Understanding the complex event structure contained in texts is fundamental for semantic-based text mining and knowledge discovery. Generally, event structure contained in texts is expressed by the event relation in the field of natural language processing. As two typical event relationships, temporal relation

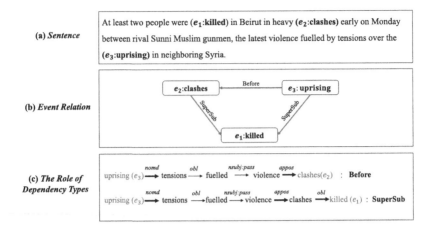

Fig. 1. An example for illustrating TRE and SRE. (a) The text in which each event is annotated in bold (i.e., e_1, e_2, and e_3). (b) Results of TRE and SRE on the text in (a). Note that the blue arrows indicate subevent relationships, and the red arrows indicate temporal relationships. (c) Syntactic dependencies between events. The labels on the arrow are the dependency types, and thick arrows indicate more contributions of corresponding dependency types to task. (Color figure online)

extraction and subevent relation extraction have a widely application in information retrieval and information extraction, including news timeline generation [7], event evolution analysis [12,13], event knowledge graph construction [8,14], and machine reading comprehension [20,24]. Therefore, the two types of event relations, i.e., temporal relation and subevent relation, are the focus of this study.

Given each text, the task of event temporal relation extraction (TRE) aims to determine the temporal order (e.g., "before", "after", "equal", etc.) between each pair of events, while subevent relation extraction (SRE) aims to identify whether an event is a subevent of another (e.g., "SubSuper" (an event is a subevent of another event), "SuperSub (an event contains another event)", "Coref" (coreference event), etc.). For example, given the sentence in Fig. 1 (a), in which the tokens corresponding to three events are marked in bold, and the results of TRE and SRE are presented in Fig. 1 (b), i.e., the event e_2(clashes) contains the subevent e_1(killed), while the event e_3(uprising) occurs before the event e_2(clashes).

Identifying event temporal and subevent relationships from texts is very challenging because events often have diverse expressions and there are even no explicit clues indicating their relations in many cases. Therefore, effective modeling the context between events is essential to improve the performance of event relation extraction. Over the last few years, in order to take advantage of the contextual information of events, some dependency-based methods have been proposed to enhance both TRE and SRE. For example, Phung et al. (2019) [23] adopt a hierarchical graph convolutional network to encode the dependency information between events to learn a more effective context representation. Zhang et al. (2022) [34] propose to encode syntactic dependencies using graph transformer to establish a connection between two events.

Although existing dependency-based models can improve the performance of TRE and SRE in some scenarios, they generally suffer from the following two drawbacks. First, the vast majority of previous studies focused more on syntactic dependency arcs, while ignoring dependency type information which is critical for capturing the semantic context between events. For example, as shown in Fig. 1 (c), *"clashes"* (e_2) and *"violence"* are appositions (*appos*) to each other; *"violence"* is the noun subject (*nsubj:pass*) of the passive modifier *"fuelled"*; *"tensions"* is the oblique noun (*obl*) of *"fuelled"*, which is an additional dative; *"uprising"* (e_3) is the noun modifier (*nomd*) of *"tensions"*. With the meanings contained in the dependency types (i.e., *appos*, *nsubj:pass*, etc.), it is relatively simple to infer that *"uprising"* occurred before *"tensions"*, successfully extracting the temporal relation *"uprising* $\xrightarrow{Before}$ *clashes"* in the example. The subevent relation can be extracted accurately as well with the guidance of dependency types, which is shown in Fig. 1(c). According to the observation, we argue that the types of syntactic dependencies can provide important clues for TRE and SRE.

In addition, most existing dependency-based methods usually treat different types of dependencies equally when modeling the semantic context among events, inducing the deteriorated performance of event relation extraction. However, different types of dependencies might contribute differently, treating them distinctively is the key to extract event relations effectively, especially when two events are distant from each other or contain complex contexts. As shown in Fig. 1 (c), the two dependency types *"violence* $\xrightarrow{appos}$ *clashes"* and *"uprising* $\xrightarrow{nomd}$ *tensions"* (highlighted with thick arrows) are more helpful to identify the temporal relationship between the e_2(*clashes*) and e_3(*uprsing*). Therefore, the effectiveness of different dependencies on event relation extraction needs to be considered.

Based on the above discussion, we proposes a new Type-Guided Attentive Graph Convolutional Network (TGAGCN) for TRE and SRE, which fully utilizes the dependency type information for modeling the context between events. Specifically, we first construct an event-specific syntactic dependency graph that contains both local dependencies (expressed by one-hop connections with events) and global dependencies (expressed by the shortest path connecting a pair of events). Second, we use dependency types as the guide to encode event-specific syntactic dependency graphs by using a multi-layer enhanced attentive graph convolutional network to derive a richer representation of events. This enables the model to effectively capture key information between remote events while isolating weakly correlated information. Finally, we take the event pair representation and the sentence representation as the input of the event relation predictor to identify the relationships between events. Experimental results on both event temporal relation extraction and subevent relation extraction show that TGAGCN significantly outperforms the state-of-the-art baselines. Further analysis also reveals that our method can effectively capture complex relationships between long-distance events.

To sum up, our main contributions are listed as follows:

– To guide the representation learning of graph for event relation extraction, we propose a graph neural network model called TGAGCN that fully utilizes the knowledge of dependency types.

– We design a new mechanism to aggregate critical information in context to event-related key nodes by an enhanced attentive graph convolutional network, enabling the model to discriminate the importance of different dependencies.
– We conduct extensive experiments, the results of which demonstrate the effectiveness of the proposed method.

2   Related Work

In this subsection, we review the related work to TRE and SRE, including the traditional models, deep learning-based models, and syntactic dependency-based models.

Traditional TRE and SRE methods are based on hand-crafted features [6,9, 33]. For example, Bethard et al. (2007) [3] use syntactic features of verb clause to build a time structure tree. However, such methods are labor-intensive.

Recently, many researchers have adopted deep learning approaches for TRE and SRE. Some researchers exploit complementarity between multiple tasks to improve performance [32]. For example, Zhou et al. (2020) [35] propose to apply duration prediction to assist SRE. Moreover, several new strategies have been applied. Cao et al. (2021) [4] perform semi-supervised learning using an uncertainty estimation algorithm for data augmentation. Man et al. (2022) [15] design a reinforcement learning mechanism to select critical contexts from document. Wang et al. (2020) [30] and Hwang et al. (2022) [11] propose different constraints to extract temporal and subevent relation.

Since the rich linguistic knowledge is contained in syntactic dependencies, many syntactic dependency-based methods have been proposed. Aldawsari et al. (2019) [1] use an ancestor event of another in a dependency tree while fusing discourse and narrative features to obtain richer event representation. Meng et al. (2017) [17] use a sequence encoder to encode the shortest dependency path between events to identify their relationship. Since the sequence encoder cannot learn the structural information well, some graph-based methods are proposed to be applied to this task. Zhang et al. (2022) [34] propose to use a graph transformer to capture TRE-related knowledge in the syntactic graph. Mathur et al. (2021) [16] and Tran et al. (2021) [27] use syntactic dependency while introducing other knowledge, such as rhetoric, discourse, and semantics, to enrich the representation of events through the interaction between nodes in different graphs.

However, these approaches do not take full advantage of dependency-type information. In addition, when there are complicated contexts among events or they are distant from each other, treating different dependencies discriminately is particularly important to capture critical information in the context. Therefore, we propose to use dependency types as guiding signals to learn the dependency arc weights to enhance the effectiveness of graph-based event relation extraction.

3 Methodology

3.1 Problem Formulation

Given an input text with two event mentions, the goal of event relation extraction is to predict the type of relationships between the two events. Formally, the text is a sequence of s tokens, which is denoted by $S = \{w_1, \cdots, w_s\}$. Without loss of generality, the i-th token and j-th token are two event mentions of interest, which are denoted by e_i and e_j, respectively. The expected output can be expressed as a triplet $<e_i, r, e_j>$, in which r indicates the relationship between e_i and e_j. In this study, for TRE, four types of relationships are involved, including "Before", "After", "Equal" and "Vague", while another four types of relationships are investigated for SRE, including "SubSuper", "SuperSub", "Coref" and "NoRel".

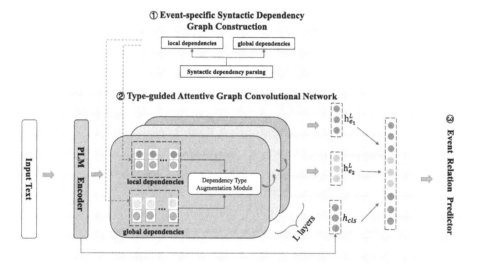

Fig. 2. The overall architecture of TGAGCN.

3.2 Model Architecture

TRE and SRE can be regarded as typical classification tasks. Figure 2 is the overall architecture of TGAGCN. It mainly consists of three parts: (1) *Event-specific Syntactic Dependency Graph Construction*, in which both local and global dependencies as well as their type information are derived; (2) *Type-guided Attentive Graph Convolutional Network*, which can aggregate critical information in context to event-related key nodes; (3) *Event Relation Predictor*, which stacked two-layer feedforward neural networks to fuse event representations and text representation for relation prediction.

Event-Specific Syntactic Dependency Graph Construction. In order to capture event-related syntactic dependency information comprehensively, we construct an event-specific syntactic dependency graph. The detailed construction process is shown in Fig. 3.

First, we use an off-the-shelf syntactic parser to parse out the syntactic dependency trees of the input text S. The derived dependency tree is represented by a set of N triplets, denoted by $D = \{(w_{i_1}, a_1, w_{j_1}), \cdots, (w_{i_n}, a_n, w_{j_n}), \cdots, (w_{i_N}, a_N, w_{j_N})\}$. Note that in the n-th triples (w_{i_n}, a_n, w_{j_n}), $i_n \in \{1, \cdots, s\}$ and $j_n \in \{1, \cdots, s\}$ are the indexes of the two involved tokens, and a_n is the annotation type of dependency. In this study, the dependencies immediately adjacent to each event are viewed as local semantic context, which is defined as the set of triples connecting with it. Formally, the local semantic context of e_i and e_j are denoted by Eq. (1) and Eq. (2), respectively.

$$\mathbf{L_{e_i}} = \{(w_{i_n}, a_n, w_{j_n}) \mid w_{i_n} == e_i \ or \ w_{j_n} == e_i\} \tag{1}$$

$$\mathbf{L_{e_j}} = \{(w_{i_n}, a_n, w_{j_n}) \mid w_{i_n} == e_j \ or \ w_{j_n} == e_j\} \tag{2}$$

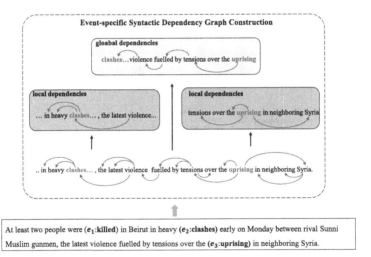

Fig. 3. Event-specific Syntactic Dependency Graph Construction

Moreover, in the dependency tree, the shortest dependency path between a given event pair is considered as the global semantic context, consisting of a set of triples on the path. Formally, the global semantic context between e_i and e_j is denoted by G_{e_i,e_j}.

Then, an event-specific syntactic dependency graph is constructed according to the triples in the union of L_{e_i}, L_{e_j}, and G_{e_i,e_j}, in which the nodes are the involved tokens and the edges (with types) are the syntactic dependency triples covering both local and global connections among events.

Finally, we formalize the event-specific syntactic dependency graph by an adjacency matrix **A**. For the entry A_{ij}, the value is 1 if $i == j$ or there is a dependency connection between w_i and w_j; the value is 0, otherwise.

Type-Guided Attentive Graph Convolutional Network. Graph convolutional network [31] (GCN) can consider the features contained in each node and its neighbors in the graph. It is often used to encode nonlinear structures and enhance expressiveness. Many studies have demonstrated its effectiveness to encode contextual information [10,26]. Therefore, in this paper, we encode syntactic information using GCNs.

First, we employ a sequence encoder to extract initial contextual features, which take the whole text as input. We use Roberta as the initial encoder, which is based on the Transformer architecture [29] and has yielded state-of-the-art results on this task. Specifically, for a given text S, after encoding with the sequence encoder, the initial embedding representation of each token can be obtained as the following,

$$\mathbf{H} = \mathbf{PrLM}\left\{h_{[CLS]}, h_1, h_2, ..., h_i, ..., h_j, ..., h_s\right\} \tag{3}$$

where H is the hidden embedding of S; The i-th token's embedding vector in S is h_i; $h_{[CLS]}$ is the representation of the whole text.

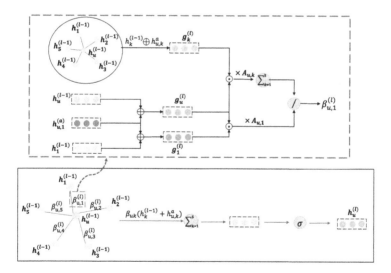

Fig. 4. Dependency Type Augmentation Module. In the figure, $h_k^{(l-1)}$ is the vector representation of k-th token at the $(l-1)$-th layer GCN (here $k \in [1,5]$); $h_{u,k}^{(a)}$ denotes the dependency type representation between u-th token and k-th token; The weight value on the link between the u-th and k-th tokens on the $(l-1)$-th layer graph network is referred to as $\beta_{u,k}^{(l)}$; The adjacency matrix **A**'s value in the u-th row and k-th column is indicated by $A_{u,k}$.

Second, due to the diversity of forms in which event relationships are expressed and the lack of explicit indicative clues in most of them, the rich structural and semantic information in dependency types is important. To fully use dependency-type information and capture critical information more accurately, we design a dependency-type augmentation module at each layer of GCN. The module can distinguish the importance of different dependencies by using information from the dependency types. Its detailed structure is shown in Fig. 4. Specifically, we use the dependency types as guidance to optimize the weights of different dependency arcs so that the information flow on the paths of long-distance event connections can be effectively fused. First, we use a new type-driven weight optimization approach to learn how to distinguish important contexts. Suppose that the $(l-1)$-th layer representation of the w_u is currently obtained through the first $(l-1)$ layers of TGAGCN denoting $h_u^{(l-1)}$, and the triple (w_u, a, w_v) is known from the event-specific syntactic dependency graph construction, where $u \in (1, s)$ and $v \in (1, s)$, a is the dependency type between w_u and w_v. Then the importance score $\beta_{u,v}^{(l)}$ of the dependency arc between w_u and w_v in the l-th layer of TGAGCN is calculated according to the Eq. (4) and (5):

$$\mathbf{g}_u^{(l)} = \mathbf{h}_u^{(l-1)} \oplus \mathbf{h}_{u,v}^a \tag{4}$$

$$\beta_{u,v}^{(l)} = \frac{A_{u,v} \cdot \exp\left(\mathbf{W}_0^{(l)} \mathbf{g}_u^{(l)} \cdot \mathbf{W}_0^{(l)} \mathbf{g}_v^{(l)}\right)}{\sum_{k=1}^s A_{u,k} \cdot \exp\left(\mathbf{W}_0^{(l)} \mathbf{g}_u^{(l)} \cdot \mathbf{W}_0^{(l)} \mathbf{g}_k^{(l)}\right)} \tag{5}$$

where $\mathbf{W}_0^{(l)}$ denotes weight matrix; "$\oplus$" denotes concatenation operation; $h_{u,v}^a$ is the representation of the dependency type a between w_u and w_v. Then, we use the optimized weights to integrate the linguistic knowledge contained in syntactic dependencies, which is shown in the Eq. (6) and (7).

$$\mathbf{h}_u^{(l)} = \mathbf{Relu}\left(\sum_{k=1}^s \beta_{u,k}^{(l)}\left(\mathbf{W_1}^{(l)} \cdot \hat{\mathbf{h}}_k^{(l-1)} + \mathbf{b_1}^{(l)}\right)\right) \tag{6}$$

$$\hat{\mathbf{h}}_k^{(l-1)} = \mathbf{W}_2^{(l)} \cdot \mathbf{h}_k^{(l-1)} + (\mathbf{W}_3^{(l)} \cdot \mathbf{h}_{u,k}^a + \mathbf{b}_2^{(l)}) \tag{7}$$

where $\mathbf{W}_1^{(l)}$, $\mathbf{W}_2^{(l)}$ and $\mathbf{W}_3^{(l)}$ represent the weight matrix and $\mathbf{b}_1^{(l)}$ and $\mathbf{b}_2^{(l)}$ represent the bias functions; $Relu$ is the activation function; $h_{u,k}^a$ is the representation of the dependency type between w_u and w_k.

All in all, our approach makes full use of the guidance of dependency types in obtaining representations, allowing the model to give more attention to task-aware dependency information, thereby improving the performance for event relation extraction.

Event Relation Predictor. After passing through the L-layer attentive GCNs with a dependency type augmentation module, we can obtain an enhanced representation of each token: $\hat{H} = \left\{h_1^{(L)}, h_2^{(L)}, ..., h_s^{(L)}\right\}$. The representations of the event pair is $h_i^{(L)}$, $h_j^{(L)}$, respectively. Additionally, we also use the $[CLS]$ representation obtained by the sequence encoder. Next, we adopt a similar approach in [32] to fuse the two representations, which consists of a two-layer feedforward neural network with a *tanh* activation function and a softmax layer. It can transform the representation into a probability distribution, as shown in Eq. (8).

$$a_{i,j} = \text{FFN}_b \left(tanh \left(\text{FFN}_a \left(\left[h_i^{(L)}; h_j^{(L)}; h_{[CLS]} \right] \right) \right) \right) \qquad (8)$$

where $\text{FFN}_a(x) = W_a x + b_a$; $\text{FFN}_b(x) = W_b x + b_b$; W_a and W_b mean the weight matrix; b_a and b_b mean the bias functions. We take cross-entropy as the optimization objective for event temporal relation and subevent relation extraction during training, and its formula is shown in Eq. (9).

$$\mathcal{L} = - \log P \left(r = r_{(e_i, e_j)} \mid e_i, e_j \right) \qquad (9)$$

4   Experiments

4.1   Experimental Setup

Datasets and Evaluation Metrics. MATRES, TCR, and HiEve are three datasets that we employ in our experiment. MATRES and TCR are used to test the performance of TRE and HiEve is used for SRE. MATRES is the latest benchmark dataset for TRE [21]. It is further improved based on TimeBank [5] and TempEval [28]. TCR [18] is another popular dataset that adopts a similar annotation scheme as MATRES. Due to its relatively small size, we use the same setup as Ning et al. (2019) [19] and only test our model with it. HiEve is a news dataset, and each article in it contains annotations of events and their subevent relationships. Table 1 provides information about these three datasets in detail. Following previous work, we assess performance based on Precision (P), Recall (R), and F1-Score (F1).

Table 1. Data statistics for dataset MATRES, TCR, and HiEve.

Dataset		Train	Dev	Test
MATRES	Document	260	21	20
	Event Pairs	10,888	1,852	840
TCR	Document	-	-	25
	Event Pairs	-	-	2,646
HiEve	Document	80	-	20
	Event Pairs	35,001	-	7,093

Parameter Settings. We use Roberta-large as initial sequence encoder. The best model is obtained using AdamW to optimize the pre-trained model and other parameters with a learning rate of 10 epochs between {1e-5, 4e-5}. We set the weight decay for all parameters to 0.01. We set the dropout rate to 0.1. The optimal number of layers of TGAGCN is 2. We obtain syntactic dependencies using the Standard CoreNLP Toolkit. For syntactic dependency parsing of cross-sentence events, we add a special relation *cross_sen* to connect two separated sentences.

Baselines. We compare TGAGCN on the MATRES dataset with four external knowledge or constraint-based methods, two multitask-based methods, and a syntax-based method. Four external knowledge or constraint-based methods include 1) **LSTM +TEMPROB+ILP** [19] which combines integer linear programming and common sense knowledge to extract temporal relationships; 2) **Joint Constrained Learning** [30] which adopts logic constraint rule and integrates event background knowledge; 3) **HGRU+ knowledge** [25] which maps events into a hyperbolic space while fusing event background knowledge; 4) **Probabilistic Box** [11] which uses probability boxes to keep relations consistent. Two multitask-based methods are 1) **Self-Training** [2] which uses ACE 2005 and TimeBank to obtain more training data by multi-task self-training; 2) **Relative Time** [32] which uses the "relative time prediction" to assist TRE. A syntax-based method is **Syntactic-Transformer** [34], which proposes a time-oriented syntactic graph transformer. In addition to the above baselines, we also compare with **Vanilla Classifier**, a vanilla RoBERTa-based classifier.

We further compare TGAGCN with external knowledge-based and context-based methods on TCR dataset. External knowledge-based approaches: **LSTM+knowledge*** [19]: a variant of LSTM+TEMPROB+ILP, incorporating time common sense knowledge in LSTM. In addition, we also compare with **LSTM+TEMPROB+ILP** [19] and **HGRU+ knowledge** [25]. Context-based approaches: 1) **LSTM*** [19]: a variant of LSTM+TEMPROB+ILP to enrich representation using LSTM. 2) **CogCompTime** [22]: a pipeline model combining semantic features and structural reasoning. 3) **Poincaré Event Embeddings** [25]: A method for learning richer event representations in hyperbolic spaces. 4) **Vanilla Classifier**.

For the Hieve dataset, we compare three types of baselines: 1) Multitask-based methods: **TACOLM** [35] (A method using duration prediction task to assist SRE) 2) Syntax-based methods: **StructLR** [9] (A supervised classifier based on syntactic features) 3) External knowledge or constraint-based methods: we use **Probabilistic Box** [11] and **Joint Constrained Learning** [30] as baselines. In addition, we also compared with **Vanilla Classifier**.

4.2   Overall Results

Tables 2, 3, and 4 show the results on the MATRES, TCR, and HiEve datasets, respectively.

MATRES & TCR. Tables 2 and 3 show the results on the task of TRE. From the tables, we can observe that our method outperforms all baselines. Compared with the state-of-the-art model, TGAGCN improves 0.3% and 1.2% on the MATRES and TCR, respectively. It demonstrates the effectiveness of our model for TRE. For TCR dataset, Poincaré Event Embeddings, which only utilizes raw contextual information, is significantly better than HGRU+knowledge, which introduces external knowledge. This shows a wealth of knowledge in the context that needs to be deeply excavated. Notably, compared to the method that introduces external knowledge [19], TGAGCN shows a near 10% improvement in P, a 5.7% improvement in F1 in MATRES, and an 8.1% improvement on the TCR. It can be seen that our proposed method can more fully exploit the rich semantic knowledge within the input text and effectively capture key task-aware information. Additionally, TGAGCN outperforms existing syntax-based approaches in F1, which is mainly attributed to our effective modeling of dependency type information.

Table 2. Comparison of various temporal relation extraction models on the MATRES. Here * denotes the results are statistically significant ($p < 0.05$).

model	P	R	F1
LSTM+TEMPROB+ILP (Ning et al., 2019)	71.3	82.1	76.3
Probabilistic Box (Hwang et al., 2022)	-	-	77.1
Joint Constrained Learning (Wang et al., 2020)	73.4	85.0	78.8
Vanilla Classifier (Wen and Ji, 2021)	78.1	82.5	80.2
Syntactic-Transformer (Zhang et al., 2022)	-	-	80.3
HGRU+knowledge (Tan et al., 2021)	79.2	81.7	80.5
Self-Training (Ballesteros et al., 2021)	-	-	81.6
Relative Time (Wen and Ji, 2021)	78.4	**85.2**	81.7
TGAGCN (Our model)	**81.1**	83.0	**82.0***

HiEve. Table 4 presents the results on the task of SRE. The statistical model StructLR based on syntactic features improves by 8.2% compared with TACOLM, which proves that syntactic features are significant for SRE. In addition, TGAGCN achieves a significant advantage over all methods, as our method can benefit from dependency types and can significantly discriminate the contribution of different dependencies.

We found an interesting phenomenon on these datasets: the Recall (R) of TGAGCN is inferior to that of the state-of-the-art method. This is probably

Table 3. Comparison of various temporal relation extraction models on the TCR. Here * denotes the results are statistically significant (p < 0.05).

model	P	R	F1
CogCompTime (Ning et al., 2018)	-	-	70.7
LSTM* (Ning et al., 2019)	79.6	75.7	77.6
LSTM+knowledge* (Ning et al., 2019)	79.3	76.9	78.1
LSTM+TEMPROB+ILP (Ning et al., 2019)	-	-	78.6
Vanilla Classifier (Wen and Ji, 2021)	89.2	76.7	82.5
HGRU+ knowledge (Tan et al., 2021)	88.3	79.0	83.5
Poincaré Event Embeddings (Tan et al., 2021)	85.0	**86.0**	85.5
TGAGCN (Our model)	**89.2**	84.3	**86.7***

because TGAGCN suffers from boundary blurring for negative categories (e.g., *Vague* and *Norel*), which is a problem faced by most methods. Although our method still has some limitations in this regard, TGAGCN still has a significant advantage in overall performance.

Table 4. Comparison of various subevent relation extraction models on the HiEve. * denotes the results are statistically significant (p < 0.05).

model	P	R	F1
TACOLM (Zhou et al., 2020)	-	-	48.9
StructLR (Glava et al., 2014)	-	-	57.7
Vanilla Classifier (Wen and Ji, 2021)	63.5	51.8	57.0
Joint Constrained Learning (Wang et al., 2020)	57.4	**61.7**	59.5
Probabilistic Box (Hwang et al., 2022)	-	-	60.6
TGAGCN (Our model)	**65.2**	58.8	**61.8***

4.3 Ablation Study

To understand how each component of TGAGCN contributes to the effectiveness, we conduct an ablation study in which several variants are built as follows:

Vanilla Classifier: Use the basic pre-trained model and standard classification layer without any extra information.

+ Standard GCN: Add a standard GCN to verify the role of the syntax-based graph neural network.

+ Event Relation Predictor: Add a special classification layer similar to Wen and Ji(2021) [32] on the standard GCN for event relation prediction.

– **Local Dependencies:** To investigate the effect of global dependencies, we remove local dependencies.

– **Global Dependencies:** To evaluate the effect of local dependencies, we remove the global dependencies.

+ **All Dependencies:** To verify the necessity of the event-specific syntactic dependency graph, we add all syntactic dependencies.

Table 5. Ablation study on MATRES and HiEve. The upper part is to gradually add a component. The lower part is the result of subtracting or adding a certain part.

Model	TEMPREL			SUBEVENT		
	P	R	F1	P	R	F1
Vanilla Classifier	78.1	82.5	80.2	63.5	51.8	57.0
+ Standard GCN	**82.2**	78.8	80.5	63.3	52.5	57.4
+ Event Relation Predictor	81.7	79.8	80.8	59.9	56.5	58.2
+ **Type-guided (Our model)**	81.1	**83.0**	**82.0**	65.2	**58.8**	**61.8**
our model − Local Dependencies	80.2	81.6	80.9	65.5	52.7	58.4
our model − Global Dependencies	81.3	81.3	81.3	**69.9**	51.9	59.5
our model + Full Dependencies	80.4	82.3	81.3	68.9	51.0	58.6

Table 5 shows the ablation study's experimental results. From the experimental results, We find the following phenomena from the table:

(1) The results of Standard GCN outperform Vanilla Classifier, which indicates syntactic dependencies are useful for TRE and SRE. However, using syntactic information indiscriminately can only improve performance to a limited extent. Introducing a special classification layer in Event Relation Predictor further improves the model, which can avoid the loss of global semantic information of the whole text.
(2) After adding the dependency type augmentation module, the model has the most significant improvement on both datasets. Compared with "+ Event Relation Predictor," the F1 value is improved by 1.2% and 3.6%, respectively. This indicates that using an effective approach to consider dependency types can help the model converge task-related dependency information to event nodes, enriching the event representation while reducing the interference of weakly relevant information.
(3) From the results of the experiments with three different range dependencies, the model with local dependencies removed decreases most severely. This suggests that local dependencies can provide more information. Adding all dependencies, as expected, introducing too many redundant dependencies will impair the model's performance. Therefore, constructing an event-specific syntactic dependency graph is necessary and effective.

4.4   The Effect of Dependency Type Guidance

To further investigate the role of our type-guided optimization of weights, we add or subtract dependency types for different ranges of dependencies. The experimental results are shown in Fig. 5. From the figure, we can find that when the dependency type guidance is removed, the model performance will drop significantly regardless of the range of dependencies The explanation is that the dependency type provides richer supplementary information about the relationship of events and can also better guide the model to optimize the weights of different dependencies. In conclusion, the use of methods to mine relational clues between events from a dependency-type perspective is significant.

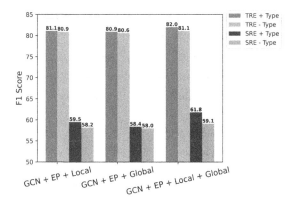

Fig. 5. The effect of dependency type guidance. where EP means event relation predictor; Local and Global represent local dependencies and global dependencies, respectively.

4.5   The Effect of Remote Event Relation Extraction

We further validate the superiority of our model for long-distance event relation extraction. We divide the two datasets (MATRES and HiEve) into three categories according to the distance between events: the number of tokens between events is less than 10, the number of tokens between events is between 10 and 20, and the number of tokens between events is greater than 20, respectively. Figure 6 shows distribution of data for each category and the variation of F1 scores for Vanilla Classifier and TGAGCN. From the figure, we can visualize that the percentage of long-distance (>20) events exceeds 50% of the entire dataset for both datasets. This proves the necessity of solving long-distance event relation extraction. Moreover, the performance of the Vanilla Classifier has a drastic decrease when the context between events exceeds 20 tokens. However, the performance of TGAGCN is less affected as the distance between events increases, indicating that the type-guided attentive GCN can effectively capture critical contextual information when the distance between events is large or the context is complex.

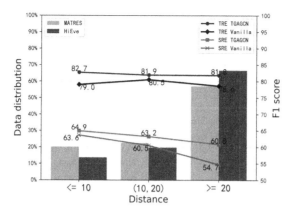

Fig. 6. Data distribution and extraction performance of relationships between distant events in MATRES and HiEve datasets.

5   Conclusion

In this paper, we propose a novel attentive graph convolutional network for event relation extraction, which relies on dependency types as drivers to address relation reasoning between long-distance events. We first mine relational clues from a dependency-type perspective. Furthermore, to fully mine the latent task-aware information in the dependency type, we optimize the model's weights for different dependencies with the dependency types as guidance. This enables the model to capture key information more accurately and avoid the interference of weakly relevant information when dealing with complex contexts. Extensive experimental results on three datasets show that our approach outperforms the recent strong baselines. In the future, we intend to study further the problem of negative class boundary blurring. We will also continue to explore the role of type-guided attentive graph convolutional network in other types of event relation extraction tasks.

Acknowledgement. This work was supported by the National Social Science Fund of China under Grant No. 20BTQ068.

References

1. Aldawsari, M., Finlayson, M.A.: Detecting subevents using discourse and narrative features. In: Proceedings of the 57th Annual Meeting of the Association for Computational Linguistics (2019)
2. Ballesteros, M., et al.: Severing the edge between before and after: neural architectures for temporal ordering of events. In: Proceedings of the 2020 Conference on Empirical Methods in Natural Language Processing (EMNLP), pp. 5412–5417 (2020)
3. Bethard, S., Martin, J.H., Klingenstein, S.: Timelines from text: identification of syntactic temporal relations. In: International Conference on Semantic Computing (ICSC 2007), pp. 11–18. IEEE (2007)

4. Cao, P., Zuo, X., Chen, Y., Liu, K., Zhao, J., Bi, W.: Uncertainty-aware self-training for semi-supervised event temporal relation extraction. In: Proceedings of the 30th ACM International Conference on Information & Knowledge Management, pp. 2900–2904 (2021)
5. Cassidy, T., McDowell, B., Chambers, N., Bethard, S.: An annotation framework for dense event ordering. In: Proceedings of the 52nd Annual Meeting of the Association for Computational Linguistics (Volume 2: Short Papers), pp. 501–506 (2014)
6. Chambers, N., Cassidy, T., McDowell, B., Bethard, S.: Dense event ordering with a multi-pass architecture. Trans. Assoc. Comput. Linguist. **2**, 273–284 (2014)
7. Do, Q., Lu, W., Roth, D.: Joint inference for event timeline construction. In: Proceedings of the 2012 Joint Conference on Empirical Methods in Natural Language Processing and Computational Natural Language Learning, pp. 677–687 (2012)
8. Flocco, D., et al.: An analysis of COVID-19 knowledge graph construction and applications. In: 2021 IEEE International Conference on Big Data (Big Data), pp. 2631–2640. IEEE (2021)
9. Glavaš, G., Šnajder, J., Kordjamshidi, P., Moens, M.F.: Hieve: a corpus for extracting event hierarchies from news stories. In: Proceedings of 9th Language Resources and Evaluation Conference, pp. 3678–3683. ELRA (2014)
10. Guo, Z., Zhang, Y., Lu, W.: Attention guided graph convolutional networks for relation extraction. In: Proceedings of the 57th Annual Meeting of the Association for Computational Linguistics, pp. 241–251 (2019)
11. Hwang, E., Lee, J.Y., Yang, T., Patel, D., Zhang, D., McCallum, A.: Event-event relation extraction using probabilistic box embedding. In: Proceedings of the 60th Annual Meeting of the Association for Computational Linguistics (Volume 2: Short Papers), pp. 235–244 (2022)
12. Jin, W., et al.: Forecastqa: a question answering challenge for event forecasting with temporal text data. In: Proceedings of the 59th Annual Meeting of the Association for Computational Linguistics and the 11th International Joint Conference on Natural Language Processing (Volume 1: Long Papers), pp. 4636–4650 (2021)
13. Li, M., et al.: Connecting the dots: event graph schema induction with path language modeling. In: Proceedings of the 2020 Conference on Empirical Methods in Natural Language Processing (EMNLP), pp. 684–695 (2020)
14. Ma, Y., et al.: MMEKG: multi-modal event knowledge graph towards universal representation across modalities. In: Proceedings of the 60th Annual Meeting of the Association for Computational Linguistics: System Demonstrations, pp. 231–239 (2022)
15. Man, H., Ngo, N.T., Van, L.N., Nguyen, T.H.: Selecting optimal context sentences for event-event relation extraction. In: Proceedings of the AAAI Conference on Artificial Intelligence, vol. 36, pp. 11058–11066 (2022)
16. Mathur, P., Jain, R., Dernoncourt, F., Morariu, V., Tran, Q.H., Manocha, D.: Timers: document-level temporal relation extraction. In: Proceedings of the 59th Annual Meeting of the Association for Computational Linguistics and the 11th International Joint Conference on Natural Language Processing (Volume 2: Short Papers), pp. 524–533 (2021)
17. Meng, Y., Rumshisky, A., Romanov, A.: Temporal information extraction for question answering using syntactic dependencies in an LSTM-based architecture. In: Proceedings of the 2017 Conference on Empirical Methods in Natural Language Processing, pp. 887–896 (2017)
18. Ning, Q., Feng, Z., Wu, H., Roth, D.: Joint reasoning for temporal and causal relations. In: Proceedings of the 56th Annual Meeting of the Association for Computational Linguistics (Volume 1: Long Papers), pp. 2278–2288 (2018)

19. Ning, Q., Subramanian, S., Roth, D.: An improved neural baseline for temporal relation extraction. In: Proceedings of the 2019 Conference on Empirical Methods in Natural Language Processing and the 9th International Joint Conference on Natural Language Processing (EMNLP-IJCNLP), pp. 6203–6209 (2019)

20. Ning, Q., Wu, H., Han, R., Peng, N., Gardner, M., Roth, D.: Torque: a reading comprehension dataset of temporal ordering questions. In: Proceedings of the 2020 Conference on Empirical Methods in Natural Language Processing (EMNLP), pp. 1158–1172 (2020)

21. Ning, Q., Wu, H., Roth, D.: A multi-axis annotation scheme for event temporal relations. In: Proceedings of the 56th Annual Meeting of the Association for Computational Linguistics (Volume 1: Long Papers), pp. 1318–1328 (2018)

22. Ning, Q., Zhou, B., Feng, Z., Peng, H., Roth, D.: Cogcomptime: a tool for understanding time in natural language. In: Proceedings of the 2018 Conference on Empirical Methods in Natural Language Processing: System Demonstrations, pp. 72–77 (2018)

23. Phung, D., Nguyen, T.N., Nguyen, T.H.: Hierarchical graph convolutional networks for jointly resolving cross-document coreference of entity and event mentions. In: Proceedings of the Fifteenth Workshop on Graph-Based Methods for Natural Language Processing (TextGraphs-15), pp. 32–41 (2021)

24. Sun, Y., Cheng, G., Qu, Y.: Reading comprehension with graph-based temporal-casual reasoning. In: Proceedings of the 27th International Conference on Computational Linguistics, pp. 806–817 (2018)

25. Tan, X., Pergola, G., He, Y.: Extracting event temporal relations via hyperbolic geometry. In: Proceedings of the 2021 Conference on Empirical Methods in Natural Language Processing, pp. 8065–8077 (2021)

26. Tian, Y., Song, Y., Xia, F.: Supertagging combinatory categorial grammar with attentive graph convolutional networks. In: Proceedings of the 2020 Conference on Empirical Methods in Natural Language Processing (EMNLP), pp. 6037–6044 (2020)

27. Tran, H.M., Phung, D., Nguyen, T.H.: Exploiting document structures and cluster consistencies for event coreference resolution. In: Proceedings of the 59th Annual Meeting of the Association for Computational Linguistics and the 11th International Joint Conference on Natural Language Processing (Volume 1: Long Papers), pp. 4840–4850 (2021)

28. UzZaman, N., Llorens, H., Derczynski, L., et al.: Semeval-2013 task 1: Tempeval-3: evaluating time expressions, events, and temporal relations. In: Second Joint Conference on Lexical and Computational Semantics (* SEM), Volume 2: Proceedings of the Seventh International Workshop on Semantic Evaluation (SemEval 2013), pp. 1–9 (2013)

29. Vaswani, A., et al.: Attention is all you need. In: Advances in Neural Information Processing Systems, vol. 30 (2017)

30. Wang, H., Chen, M., Zhang, H., Roth, D.: Joint constrained learning for event-event relation extraction. In: Proceedings of the 2020 Conference on Empirical Methods in Natural Language Processing (EMNLP), pp. 696–706 (2020)

31. Welling, M., Kipf, T.N.: Semi-supervised classification with graph convolutional networks. In: International Conference on Learning Representations (ICLR 2017)

32. Wen, H., Ji, H.: Utilizing relative event time to enhance event-event temporal relation extraction. In: Proceedings of the 2021 Conference on Empirical Methods in Natural Language Processing, pp. 10431–10437 (2021)

33. Yoshikawa, K., Riedel, S., Asahara, M., Matsumoto, Y.: Jointly identifying temporal relations with Markov logic. In: Proceedings of the Joint Conference of the 47th Annual Meeting of the ACL and the 4th International Joint Conference on Natural Language Processing of the AFNLP, pp. 405–413 (2009)
34. Zhang, S., Ning, Q., Huang, L.: Extracting temporal event relation with syntax-guided graph transformer. In: Findings of the Association for Computational Linguistics: NAACL 2022, pp. 379–390 (2022)
35. Zhou, B., Ning, Q., Khashabi, D., Roth, D.: Temporal common sense acquisition with minimal supervision. In: Proceedings of the 58th Annual Meeting of the Association for Computational Linguistics, pp. 7579–7589 (2020)

PMJEE: A Prototype Matching Framework for Joint Event Extraction

Haochen Li, Tong Mo, Di Geng, and Weiping Li[✉]

Peking University, Beijing, China
haochenli@pku.edu.cn, {motong,wpli}@ss.pku.edu.cn, gengdi@stu.pku.edu.cn

Abstract. Events are vital parts of natural language, reflecting the state changes of entities. The Event Extraction (EE) task aims to extract event triggers (the most representative words or phrases) and their arguments (participants in the event) from the given text. Most current works use sequence tagging models to solve the EE task. However, those methods only treat event and argument types as different class numbers, ignoring the semantics of those labels. However, label semantics are critical in the EE task. For example, the trigger word "fight" is semantically closer to the event type "Conflict:Attack" rather than "Life:Marriage". To emphasize the label semantics in events, we formulate EE as a prototype matching task and propose a **P**rototype **M**atching framework for **J**oint **E**vent **E**xtraction (**PMJEE**). Specifically, prototypical embeddings for both event trigger and argument types are introduced to encode their label semantics and correlations. Then a dual-channel attention layer and extraction modules are used to jointly extract event triggers and arguments. Prototypical embeddings will be optimized during training to improve the event extraction performance. Extensive experiments indicate our method achieves better performance than strong baselines, especially in data-scarce scenarios. In the detailed analysis, we verify the effectiveness of each part of the model and explore the impact of different label semantic materials.

Keywords: Event Extraction · Text Mining · Label Semantic

1 Introduction

An event is a specific occurrence involving participants, including an event trigger and several event arguments. The Event Extraction (EE) task can be divided into two subtasks, Event Detection (ED) and Argument Extraction (AE). The ED task is equivalent to identifying and classifying the event trigger, the most representative word or phrase in an event. The AE task is to identify the event's participants and classify them into the corresponding roles. The EE task is essential in information extraction, providing structured information for downstream tasks such as knowledge graph construction, auto abstracting, machine Q & A, etc. According to the synchronization of solving the two subtasks, EE models can be divided into pipeline models and joint models. The joint models usually

X. Wang et al. (Eds.): DASFAA 2023, LNCS 13946, pp. 21–36, 2023.
https://doi.org/10.1007/978-3-031-30678-5_2

perform better because they can take advantage of the information interaction between ED and AE tasks.

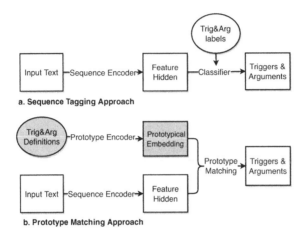

Fig. 1. Frameworks of the sequence tagging approach and Prototype Matching approach, and "Trig & Arg" means "Trigger and Argument".

Most current works use sequence tagging models to extract events [1,15,22]. The framework of the sequence tagging approach is shown in Fig. 1a: sequence encoders are used to encode the input text into hidden features, then classifiers are used to get the output trigger or argument labels. Such models primarily focus on the sequence encoders, using complex neural networks (such as LSTM, GCN, transformer, etc.) to extract features. Meanwhile, they use relatively simple classifiers, for example a *Softmax* layer, to get the label indexes for event triggers and arguments.

However, sequence tagging models treat different types of event triggers and arguments as different class indexes, ignoring their semantics and correlations which are essential to the EE task. For example, the trigger word "fight" is semantically closer to the event type "Conflict:Attack" rather than "Life:Marriage"; A "Conflict:Attack" event may probably have arguments of type "ARG-Attacker" or "ARG-Victim" rather than "ARG-Destination". Losing such semantics, sequence tagging models rely heavily on the training samples. Unfortunately, because of the datasets scarcity and imbalance, those models tend to extract event and argument types with sufficient instances but ignoring the types with few samples. Thus, those methods perform poorly in data-scarce scenarios.

Based on the same motivation, Deng et al. [2] introduced the semantical representations of event types to improve the Event Detection task and applied a memory-based mechanism to produce event prototypes, and exploited a dynamic memory network to produce more robust sentence encodings for event mentions. Furthermore, Deng et al. [3] computed the prototype for each event type by averaging its instance embeddings and used the correlation of different types of events to learn the refined prototypes (called event ontology in the paper).

However, those prototype-based methods only consider event trigger types, ignoring argument types. Thus, they can only solve the Event Detection task but cannot be generalized to the Argument Extraction task. Besides, they also lost the critical correlation between an event and its arguments. Therefore, it remains unsolved to introduce label semantics and correlations for all event components and leverage them to improve the EE task.

This paper aims to leverage label semantics to achieve Event Extraction. We formulate the EE task as a Prototype Matching task (Fig. 1b) and propose a prototype matching framework for joint event extraction called PMJEE. Specifically, we use prototype encoders to introduce prototypical embeddings for the event trigger, argument, and entity types with their definitions and correlations. Then, a dual-channel attention layer is used to encode both prototypical embeddings and input sentence word embeddings. Further, the extraction modules are used to extract event triggers and arguments jointly. The proposed prototypical embeddings will be optimized through the training process to improve the event extraction performance.

To sum up, the main contributions of the paper are as follows:

– This paper introduces a novel prototype matching framework PMJEE to solve Event Detection and Argument Extraction task jointly.
– Prototypical embeddings for the event trigger, argument, and entity types are introduced by encoding their label semantics and correlations.
– A dual-channel attention layer is proposed, which encodes both prototypical embeddings and input sentence word embeddings and provides information interaction for the ED and AE tasks.
– Extensive experimental results indicate that our method achieves better performance than strong baselines, especially in data-scarce scenarios.

2 Related Works

Studies related to this paper mainly about Event Extraction approaches and Prototypical Learning approaches.

2.1 Event Extraction

Most works on Event Extraction treat it as a Sequence Tagging task. They can be further divided into pipeline models and joint models. The pipeline models regard AE as a downstream task of ED. Chen et al. [1] used a CNN model based on dynamic pooling. Yang et al. [22] introduced BERT [4] as the sequence encoder and automatically generated labeled data to enhance the performance. In contrast, the joint models solve ED and AE tasks jointly. Nguyen et al. [15] used an RNN (Recurrent Neural Network) model for joint extraction of triggers and arguments. Yan et al. [21] leveraged the results of dependency analysis and constructed a GCN-based model. Nguyen and Nguyen [16] jointly extracted entities, triggers, and arguments based on the shared hidden representations; Wadden et al. [20] provided a graph propagation method to capture context relevant for the entity, relation, and event; Lin et al. [11] built an end-to-end information extraction system which employed global feature and beam search to extract globally optimal event structures.

Several recent research efforts have transformed the EE task into a conditional generation task. Conditional generation methods encode sentences using generative pre-trained language models such as BART [8] and T5 [6]. Paolini et al. [17] regarded event extraction as a translation task between augmented natural languages; Lu et al. [14] constructed events as event trees and used a Seq2Structure model. Li et al. [9] used a prompt-based model to generate event records with event-related knowledge injection.

2.2 Prototypical Learning

Most of the previous works utilized Prototypical Learning methods to solve named entity recognition and relation extraction tasks: Snell et al. [18] formulated Prototypical Networks for both the few-shot and zero-shot settings and analyzed the underlying distance function used in the model. Lima et al. [10] proposed a novel system that attempted to extract entities and their relations using ontology and inductive logic programming. Gao et al. [7] offered hybrid attention-based prototypical networks for noisy few-shot relation classification. Ding et al. [5] proposed a prototypical learning approach, where the prototype's representation of each relation was learned from the semantics of each statement.

Deng et al. [2] attempted to solve few-shot Event Detection using prototypical learning methods. The paper applied a memory-based mechanism to produce event prototypes and exploited a dynamic memory network to optimize embeddings for both event prototypes and event mentions. Further, Deng et al. [3] introduced the event ontology, which is initialized by averaging event instance embeddings and using the correlations of different types of events to learn the refined prototype representations.

However, the above approaches cannot be generalized to the Argument Extraction task. Thus, this paper proposes a novel prototypical learning framework to extract event triggers and arguments jointly.

3 Methodology

In this section, we first introduce the EE task and its subtasks. Then we present the prototype encoding module to introduce the prototypical embeddings for entities, arguments, and events. Finally, we construct the Prototype Matching framework PMJEE (Fig. 3) to achieve joint extraction of triggers and arguments.

The Apache troop **opened** *its tank guns*, **decimated** *that unit*, **put it**

out of existence and probably killed *some 20 Iraqis* **in the process.**

Fig. 2. An example sentence for Event Extraction. The red words represent event triggers, and the blue words represent event arguments. (Color figure online)

3.1 Task Formulation

This paper focuses on sentence-level Event Extraction task, which has two sub-tasks: Event Detection (ED) and Argument Extraction (AE).

Given the input sequence $X = \{x_1, x_2, ..., x_n\}$, where n stands for the length of the sequence, the ED task aims to extract the event trigger x_i^{trig} (e.g., "open" in Fig. 2) and its type $event_i$ (e.g., "Conflict:Attack" in Fig. 2). Given the event type $event_i$, the AE task aims to extract its event arguments x_j^{arg} (e.g., "The Apache troop" in Fig. 2) and the corresponding role types $role_j$ (e.g., "ARG-Attacker" in Fig. 2).

We formulate Event Detection and Argument Extraction as Prototype Matching tasks. First, we introduce the Prototypical Embeddings (PE) for event types and argument role types:

$$
\begin{aligned}
PE^{evt} &= \{pe_1^{evt}, pe_2^{evt}, ..., pe_{|evt|}^{evt}\} \\
PE^{arg} &= \{pe_1^{arg}, pe_2^{arg}, ..., pe_{|arg|}^{arg}\}
\end{aligned}
\tag{1}
$$

where $|evt|$ and $|arg|$ stand for the count of event types and argument role types, respectively. For the ED task, we calculate the distance between the word embedding e_i of each input token x_i and PE^{evt} to extract the event triggers and their corresponding event types. For the AE task, we calculate the distance between e_i and PE^{arg} to get the event arguments and their role types.

3.2 Prototype Encoding

The traditional approach to computing the prototypes is to average all the instance embeddings from a support set. However, a word may be an instance of multiple types of events or arguments. For example, the word "open" can trigger both a "Conflict:Attack" event (open fire) and a "Business:Start-Org" event (open business). Thus, using the averaged embedding of instances may introduce massive noise and relies heavily on the support set's quality. Further, entities, arguments, and events have various correlations, which the current approaches to getting prototypes could not capture.

To address those problems, we produce prototypical embeddings for the entity, argument, and event types by using their definitions. The definitions can be easily found from annotation guidance and are more precise than the instances. We also mix different prototypes by a transformation matrix to emphasize their correlations.

Entity Prototypical Embeddings. We use the definition of each entity type to produce entity prototypical embeddings PE^{entity}. Given an entity's definition such as *"ORG organization entities are limited to companies, corporations, agencies, institutions and other groups of people."*, we use a pre-trained language model RoBERTa [13] to get a contextual representation. The embedding of the first token is used as the prototypical embedding pe_i^{entity} for that type of entity $entity_i$.

Argument Prototypical Embeddings. Event arguments are entities that participate in a specific event. So we produce argument prototypical embeddings PE^{arg} using both the definitions of argument types and the entity prototypical embeddings PE^{entity}.

First, as the same as before, we encode an event argument's definition like *"Attacker-Arg: the attacking agent"* to get a contextual representation. The first token's embedding is used as the initial prototypical embedding pe_j^{arg*} for that type of argument arg_j. Second, an event argument can only arise from certain kinds of entities, e.g., *"Attacker-Arg"* can only arise from *"ORG, PER, and GPE"* (According to the guideline of ACE[1]). Therefore, we mix the original pe_j^{arg*} with its corresponding pe_i^{entity} to produce the final pe_j^{arg}:

$$pe_j^{arg} = \alpha pe_j^{arg*} + \frac{(1-\alpha) \sum_{i=1}^{K_j^{arg}} pe_i^{entity} M_{ij}}{K_j^{arg}} \tag{2}$$

where K_j^{arg} stands for the count of the corresponding entity types of arg_j, and $M_{ij} \in \mathbb{R}^{d \times d}$ is a transformation matrix between $entity_i$ and arg_j, in which d stands for the hidden size of RoBERTa. M_{ij} is randomly initialized and α is a mixing hyper-parameter.

Event Prototypical Embeddings. Each event type contains specific event arguments. So we produce event prototypical embeddings PE^{evt} using both the definitions of event types and the argument prototypical embeddings PE^{arg}.

First, we encode an event type's definition like *"Marry Events are official Events, where two people are married under the legal definition."* to get a contextual representation. The token embedding of "Marry" is used as the initial prototypical embedding pe_t^{evt*} for that type of event evt_t. Second, an event contains certain kinds of arguments, e.g., the event *"Life: Marry"* contains arguments *"Person-Arg, Time-Arg, and Place-Arg"*. Thus, we can get the final pe_t^{evt} by:

$$pe_t^{evt} = \beta pe_k^{evt*} + \frac{(1-\beta) \sum_{j=1}^{K_t^{evt}} pe_j^{arg} M_{jt}}{K_t^{evt}} \tag{3}$$

where K_t^{evt} stands for the count of the corresponding arguments types of evt_t, and $M_{jt} \in \mathbb{R}^{d \times d}$ is a transformation matrix between arg_j and evt_t. β is the mixing hyper-parameter.

---

[1] https://www.ldc.upenn.edu/sites/www.ldc.upenn.edu/files/english-events-guidelines-v5.4.3.pdf.

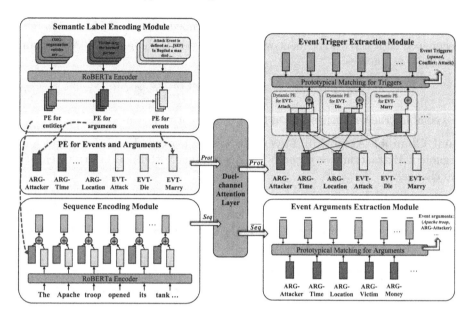

Fig. 3. The overall structure of PMJEE. On the left side, the blue boxes and orange boxes are PE^{arg} and PE^{evt} (denoted as $Prot$), while the purple boxes are sequence features (denoted as Seq). On the right side, the blue boxes and orange boxes with overlines $'-'$ stands for the sequence-aware prototype features ($\overline{Prot}$), and the purple boxes with overlines $'-'$ means prototype-aware sequence features ($\overline{Seq}$). (Color figure online)

3.3   Joint Prototype Matching Framework (PMJEE)

We provide a novel joint event extraction framework PMJEE. The overall structure is shown in Fig. 3.

First we calculate the prototypical embeddings using the method in Sect. 3.2 to get PE^{entity}, PE^{arg}, and PE^{evt}. Then, the input sentences are encoded by a Sequence Encoding Module to get sequence features. Prototypical embeddings for arguments and events ($Prot$) and the sequence features (Seq) are the two inputs of the dual-channel attention layer. Then we can get the sequence-aware prototype features ($\overline{Prot}$) and prototype-aware sequence features ($\overline{Seq}$).

$\overline{Prot}$ and $\overline{Seq}$ will be passed to the Event Trigger Extraction Module and the Argument Extraction Module, respectively. Finally, we perform Prototype Matching to extract the triggers and arguments. More details are as follows.

Sequence Encoding Module. Given an input sequence $X = \{x_1, x_2, ..., x_n\}$, we use the same RoBERTa encoder with the prototypical embeddings. After encoding, the sequence embeddings $E = \{e_1, e_2, ..., e_n\}$ can be obtained.

$$E = \{e_i\}_{i=1}^{n} = RoBERTa(X) \in \mathbb{R}^{n \times d} \tag{4}$$

where n stands for the sequence length and d stands for the size of word embeddings. At the same time, each word x_i corresponds to a certain entity type $entity_j$

(including "other" type), and the prototypical embedding pe_j^{entity} is used to represent such entity. We do element-wise addition between e_i and p_j^{entity} to obtain the sequence features:

$$Seq = \{seq_i\}_{i=1}^n = \{e_i + pe_j^{entity}\}_{i=1}^n \in \mathbb{R}^{n \times d} \tag{5}$$

Seq combine the sentence's lexical features and entity features.

Dual Channel Attention Layer. To jointly perform Event Detection and Argument Extraction, we propose the dual-channel attention layer, which has two channels:

Seq2Prot: The first channel is designed to get sequence-aware prototype features, and such features are used to do Event Detection. *Prot* (including PE^{arg} and PE^{evt}) is used as the queries Q, and *Seq* is used as the keys K and values V.

We use a similar multi-head attention mechanism with Vaswani et al. [19] to get sequence-aware prototype features $\overline{Prot}$:

$$\begin{aligned} \overline{Prot} &= Multihead(Q, K, V) = Multihead(Prot, Seq, Seq) \\ &= Concat(head_1, head_2, ..., head_h)W^O \end{aligned} \tag{6}$$

We use h independent attention heads, and for each head:

$$head_i = Attention(ProtW_i^Q, SeqW_i^K, SeqW_i^V) \tag{7}$$

where the parameter matrices $W_i^Q \in \mathbb{R}^{d \times d_m}$, $W_i^K \in \mathbb{R}^{d \times d_m}$, $W_i^V \in \mathbb{R}^{d \times d_m}$, and $W^O \in \mathbb{R}^{hd_m \times d}$ are introduced. And d_m indicates the intermediate layer dimension. Then we can get the sequence-aware prototype features $\overline{Prot} \in \mathbb{R}^{(|arg|+|evt|) \times d}$.

Prot2Seq: In the second channel, prototypical embeddings are used to produce prototype-aware sequence features, which is used in the Argument Extraction task. As a dual task, *Seq* is used to generate the queries Q, and *Prot* is used to generate the key K and value V. Using the same multi-head attention layer, we can get the prototype-aware sequence features $\overline{Seq} \in \mathbb{R}^{n \times d}$.

$$\begin{aligned} \overline{Seq} &= \{\overline{seq}_i\}_{i=1}^n = Multihead(Seq, Prot, Prot) \\ head_i &= Attention(SeqW_i^Q, ProtW_i^K, ProtW_i^V) \end{aligned} \tag{8}$$

The two channels' parameters are updated simultaneously through back propagation during training. With the joint training process, the dual-channel attention layer can simulate the interaction between ED and AE tasks.

Event Trigger Extraction Module. We extract event triggers by calculating the distance between input sequence representations and event prototypical embeddings. To further leverage the structural information, we use sequence-aware prototype features $\overline{Prot}$ to generate Dynamic Prototypical Embedding for event types. Each event's correspond arguments are specified in the annotation documents. So we use a mask matrix $M^{mask} \in \mathbb{R}^{(|arg|+|evt|)\times|evt|}$ to gather each type of event and its corresponding arguments, then we can get Dynamic PE for events:

$$\overline{PE^{evt}} = \{\overline{pe_k^{evt}}\}_{k=1}^{|evt|} = \overline{Prot}M^{mask} \tag{9}$$

Following previous works [3,7], we use Euclidean distance to conduct Prototype Matching. Specifically, given the sequence $X = \{x_1, ..., x_i, ..., x_n\}$, we treat each token x_i as a trigger candidate. We use the representation seq_i for x_i and compute its probability to trigger an event of type evt_k:

$$P(y = evt_k) = \frac{\exp(-||seq_i - \overline{pe_k^{evt}}||)}{\sum_{j=1}^{|evt|} \exp(-||seq_i - \overline{pe_j^{evt}}||)} \tag{10}$$

where $||\cdot||$ denotes the Euclidean distance and y indicates the ground truth event label for x_i. Then we use cross entropy as the loss function of Event Detection:

$$\mathcal{L}^{evt} = \sum_{k=1}^{|evt|} y \log P(y = evt_k) \tag{11}$$

Event Argument Extraction Module. Using the same Prototype Matching approach, we treat each token x_t as an argument candidate. We use the prototype-aware sequence features $\overline{seq}_t$ and compute its probability to play a role of arg_k in an event:

$$P(y = arg_k) = \frac{\exp(-||\overline{seq}_t - pe_k^{arg}||)}{\sum_{j=1}^{|arg|} \exp(-||\overline{seq}_t - pe_j^{arg}||)} \tag{12}$$

where y indicates the ground truth argument label for x_i. And the loss function of Argument Extraction can also be produced:

$$\mathcal{L}^{arg} = \sum_{k=1}^{|arg|} y \log P(y = arg_k) \tag{13}$$

Finally, the joint loss of ED and AE is calculated by:

$$\mathcal{L}^{joint} = \gamma \mathcal{L}^{evt} + (1 - \gamma)L^{arg} \tag{14}$$

where γ is the bending hyper-parameter.

4   Experiments

This section describes the dataset and experimental settings we used. Then the detailed experimental results are analyzed.

First, we conduct overall experiments with complete training data in datasets to verify the performance of PMJEE in data-sufficient scenarios. Second, we conduct few-shot experiments to ascertain the performance of our method in data-scarce scenarios. Third, detailed analyses are performed to explore the importance of each part of our method. For each experimental setup for each model, we performed ten sets of independent experiments and used the average results to eliminate the effect of randomness.

4.1   Datasets and Settings

Datasets: Our work is evaluated on the most widely used event extraction datasets ACE 2005[2] and TAC-KPB 2017[3]. As for ACE 2005, following the previous works [9,16], the 599 documents are divided into 529 training documents, 30 development documents, and 40 test documents. And for TAC-KPB 2017, following Lu et al. [14], the 458 documents are divided into 396 training documents, 31 development documents, and 31 test documents. The detailed descriptions of the datasets are presented in Table 1:

Table 1. Statistics of ACE 2005 and TAC-KPB 2017

Datasets	ACE 2005	TAC-KPB 2017
Event Type	33	38
Total Documents	599	458
Total Events	5,055	7,530
Train/Dev/Test Documents	529/30/40	396/31/31
Train/Dev/Test Sentences	9,667/644/612	8,026/683/572

Metrics. The Event Extraction task has two subtasks, Event Detection and Argument Extraction, and their standard evaluation criteria are as follows:

- **Event Detection (ED)**: The event trigger is identified and classified correctly if both its span and event type match the gold trigger;
- **Argument Extraction (AE)**: The event argument is identified and classified correctly if both its span and role type match the gold argument.

We use micro-averaged Precision (**P**), Recall (**R**), and F1 score (**F1**) in all the following evaluations.

Settings: Experiments are performed on the Nvidia GPU RTX3090, and "roberta-base" is used as the pretrained language model. We use 8-head attention in the dual-channel attention layer. The mixing ratio α (Eq. 2) and β (Eq. 3)

---

[2] https://catalog.ldc.upenn.edu/LDC2006T06.
[3] https://catalog.ldc.upenn.edu/LDC2020T13.

in the prototype encoding are set to 0.5, and the bending hyper-parameter γ in the loss function (Eq. 14) is set to 0.4. The batch size is 16, and the max sequence length is 256. The learning rate is set to 2e-5, and the dropout rate is 0.5. The training epoch is set to 50, and AdamW is used as the optimizer. Each epoch of training takes about 20 min.

Baselines. This paper choose nine strong event extraction models as the baselines: (1) Four Sequence Tagging Models: **DMCNN** [1] extracts sentence-level features by adopting a dynamic multi-pooling CNN model; **JMEE** [12] leverages the results of dependency analysis and constructs a GCN-based model; **PLMEE** [22] adopts BERT-based trigger and argument extractors; **OneIE** [11] builds an end-to-end system that employs global feature to extract globally optimal event structures. (2) Three Conditional Generation Models: **BART-Gen** [8] utilizes a conditional generation model with BART. Given the description of events, the corresponding triggers in the sentence are generated; **Text2Event** [14] constructs event structure and uses a Seq2Structure model; **KiPT** [9] uses a prompt-based generation model with event-related knowledge injection. (3) Two Prototype-based Models: **DMB-PN** [2] proposes a dynamic-memory-based prototypical network to slove few-shot event detection; **OntoED** [3] proposes a novel event detection framework with ontology embedding.

4.2 Overall Results

Table 2. The overall performances on ACE 2005 and TAC-KPB 2017 datasets with all training instances. The average score of 10 experiments is used for each model.

Models	ACE 2005						TAC-KPB 2017					
	ED			AE			ED			AE		
	P	R	F1	P	R	F1	P	R	F1	P	R	F1
Sequence Tagging Models												
DMCNN	75.6	63.6	69.1	62.2	46.9	53.5	63.2	52.0	57.3	51.1	40.6	45.2
JMEE	71.3	76.3	73.7	54.9	**66.8**	60.3	60.5	66.1	63.2	47.6	**59.1**	52.7
PLMEE	76.0	**75.4**	75.7	62.3	54.2	58.0	66.7	66.2	66.5	56.0	49.1	52.3
OneIE	-	-	74.7	-	-	56.8	-	-	67.4	-	-	54.0
Conditional Generation Models												
BART-Gen	-	-	71.1	-	-	-	-	-	67.6	-	-	-
Text2Event	67.5	71.2	69.2	46.7	53.4	49.8	64.8	67.9	66.4	46.5	50.6	48.0
KiPT	72.9	76.9	74.9	-	-	-	69.1	65.0	67.1	-	-	-
Prototype-based Models												
DMB-PN	-	-	71.8	-	-	-	-	-	65.9	-	-	-
OntoED	75.5	70.6	73.2	-	-	-	70.3	63.9	67.1	-	-	-
PMJEE (ours)	**77.3**	74.6	**75.9**	**63.6**	57.9	**60.7**	70.6	**68.2**	**69.4**	**57.2**	53.1	**55.1**

To verify the performance of our model in data-sufficient scenarios, we use all the training data and perform event detection and argument extraction tasks. The overall results are shown in Table 2, which indicate:

(1) Our method outperforms all baselines in the F1 score on both the ACE 2005 and TAC-KPB 2017 datasets. PMJEE surpasses the best baseline by 0.2% on the ED task and 0.4% on the AE task (ACE 2005), and is 2.0% ahead of the best baseline on the ED task and 1.1% on the AE task (TAC-KPB 2017).

(2) The most significant improvement is the Precision metric. Compared to the best Sequence Tagging Model, PMJEE improves on average by 2.6% on ED task and 1.3% on AE task. This validates the effectiveness of the label semantics. Further, the TAC-KPB dataset has more types of events, which is more challenging for the Sequence Tagging models. Fortunately, we can use more label semantics with our proposed prototype matching method, leading to a more significant improvement in the F1 scores.

(3) Compared to the existing Prototype-based models, PMJEE achieves the AE tasks for the first time and provides significant performance improvements on ED tasks, with an average improvement of 2.5% on the F1 score. This validates that our approach can improve the performance on both ED and AE tasks simultaneously by capturing the information interaction between the two tasks. In addition, taking advantage of the correlations between multiple types of labels, PMJEE uses a higher quality of prototype embeddings and performs better than existing prototype-based models in Precision.

4.3   Few-Shot Results

Table 3. Results for the few-shot experiments on the ACE 2005 dataset. N-way-K-shot means N different classes and K instances are used for training.

Task	Model	5way 5shot	5way 10shot	5way 15shot	10way 5shot	10way 10shot	10way 15shot
ED	PLMEE	54.3	59.5	64.3	40.2	45.4	48.3
	JMEE	58.2	62.3	65.1	44.1	49.6	54.8
	KiPT	63.2	67.0	70.1	49.1	54.6	59.8
	DMB-PN	65.1	**68.9**	71.7	52.6	58.4	61.4
	PMJEE	**67.4**	66.1	**72.2**	**59.8**	**60.3**	**62.6**
AE	PLMEE	51.5	52.2	54.2	39.3	40.9	45.8
	JMEE	43.7	45.4	50.3	33.4	35.5	40.8
	PMJEE	**53.9**	**54.2**	**56.1**	**45.7**	**44.9**	**47.8**

We conduct few-shot experiments to verify PMJEE's effectiveness in data-sparse scenarios. Following previous settings [2], we set the training data into **N-way-K-shot** settings, which means there are N different types and K instances. Because the KiPT and DMB-PN models only extract event triggers without arguments, we compare our method with four baselines in the ED task, while only compare with PLMEE and JMEE models in the AE task (Table 3).

The experimental results are shown in Table 4. PMJEE surpasses all the baselines in almost all few-shot settings in both ED and AE tasks, proving the robustness of our method. Further analyses are as follows:

K-shot Analysis: We fix the count of different types N to do K-shot analysis. When the training instances increase with N grows from 5 to 15, all models show an increasing trend in performance. Notably, when the setting is *5way5shot*, PMJEE significantly outperforms all the baselines in both ED and AE tasks, indicating the robustness of our method when data is extremely sparse. However, the performance of PMJEE is not constantly growing, and it drops slightly when the settings change from *5way5shot* to *5way10shot* in the ED task. Then PMJEE overtakes DMB-PN again with the *5way15shot* setting. The possible reason is that when training data is insufficient, its noise will pollute the prototypical embeddings through backpropagation, degrading the performance.

N-way Analysis: We compare experimental results of the same K shot setting to conduct an N-way analysis. We can observe that when N increases from 5 to 10, the performance of all models will significantly drop. This indicates that when the event and argument types increase, it becomes more difficult for the model to classify them correctly. However, PLMEE suffers less performance degradation. For example, in the AE task, when N changes from 5 to 10, our model has an average drop of -8.6%, while the average decrease of PLMEE and JMEE is -10.6% and -10.2%, respectively. We believe that as N increases, our model additionally introduces correlation between these new types through prototypical encoding, resulting in more stable performance.

4.4 Detailed Analysis

Table 4. Results for the Ablation Study. F1 scores on the ACE 2005 dataset are recorded. (.) stands for the performance drop when removing the corresponding part.

Models	ED	AE
PMJEE	**75.9**	**60.7**
$-PE^{entity}$	74.2(-1.7)	58.3(-2.4)
$-PE^{arg}$	74.6(-1.3)	59.7(-1.0)
$-PE^{evt}$	74.5(-1.4)	59.9(-0.8)
$-Argument$	75.3 (-0.6)	-
$-Event$	-	57.6 (-3.1)

Ablation Study: To clarify the effect of the prototypical embeddings and the joint matching framework we proposed, some ablation studies are conducted, and the results are listed in Table 4. In the first 3 lines, we replace our prototype embeddings for entities, argument roles, and events with random initialized vectors. And in the last two lines, we split the joint training process and do Event Detection and Argument Extraction tasks separately.

After replacing the prototypical embeddings with random initialized vectors, the F1 scores drop on both tasks. Most obviously, when "PE^{entity}" is replaced, F1 scores on the ED and AE tasks drop sharply by 1.7% and 2.4%. The results indicate that entity features are one of the most useful features. At the same time, the results prove that although the "PE^{evt}" do not directly participate in the Argument Extraction Module in PMJEE, the interaction introduced by the dual-channel attention layer also helps to improve the AE task's performance.

When the joint training process is split, the model degenerates into a sentence classification model for ED and a sequence labeling model for AE. The two independent models cannot obtain each other's interactions, causing significant drops in the experimental performances.

Table 5. Results for the Semantic Material Selection.

Semantic Material	P	R	F1
Guidelines	**77.3**	74.6	**75.9**
Event Names	72.4	**77.3**	74.8
Top Triggers	74.4	73.2	73.8
Guidelines	**59.6**	61.9	**60.7**
Argument Names	56.2	**64.7**	60.1
Top Argument Words	51.3	51.5	51.4

Label Semantic Material Selection: We explore the influence of different semantic materials used to generate the prototypical embeddings, and the results are shown in Table 5. In the first three lines, we replace the event type' **Guidelines Definitions** with **Event Names** and **Top Triggers** (Ten most frequent trigger words for each event). The results are Precision, Recall, and F1 scores on the ED task. While in the last three lines, we replace the argument type's **Guidelines Definitions** with **Argument Names** and **Top Argument Words** (Ten most frequent word spans for each role).

From the results we can see: 1) The Guideline definition material exhibits the best performance. Because it describes the event and arguments from all aspects, providing sufficient semantic features; 2) Event Names and Argument Names material achieve better performances on the Recall scores. This means that the model can generate event-related semantic information by just tag names due to pre-trained language model's large-scale training corpus; 3) Top triggers and top argument words may have different meanings depending on the context and may cause noise in event extraction, which leads to the worst results.

5   Conclusion

In this paper, we explore the potential of label semantics to solve the EE task. We formulate the EE task as a Prototype Matching task and propose a novel prototypical matching framework called PMJEE. We introduce prototypical embeddings for event triggers, argument, and entity types. A dual-channel attention layer performs information interaction, and extraction modules are used to solve the ED and AE tasks jointly. Extensive experiments prove the superiority of our model in both data-sufficient and data-scarce scenarios. The ablation study shows the effectiveness of our prototypical embeddings and the joint training framework. In the future, we will continue to unlock the potential of label semantics and apply the idea to other tasks, such as open-domain event extraction and event-event relation extraction.

Acknowledgment. This work was supported in part by National Key R&D Program of China under Grants No. 2022YFF0902703.

References

1. Chen, Y., Xu, L., Liu, K., Zeng, D., Zhao, J.: Event extraction via dynamic multi-pooling convolutional neural networks. In: Proceedings of the 53rd Annual Meeting of the Association for Computational Linguistics and the 7th International Joint Conference on Natural Language Processing (Volume 1: Long Papers), pp. 167–176 (2015)
2. Deng, S., Zhang, N., Kang, J., Zhang, Y., Zhang, W., Chen, H.: Meta-learning with dynamic-memory-based prototypical network for few-shot event detection. In: Proceedings of the 13th International Conference on Web Search and Data Mining, pp. 151–159 (2020)
3. Deng, S., et al.: OntoED: low-resource event detection with ontology embedding. In: Proceedings of the 59th Annual Meeting of the Association for Computational Linguistics and the 11th International Joint Conference on Natural Language Processing (Volume 1: Long Papers), pp. 2828–2839. Association for Computational Linguistics, Online (2021)
4. Devlin, J., Chang, M.W., Lee, K., Toutanova, K.: BERT: pre-training of deep bidirectional transformers for language understanding. In: Proceedings of the 2019 Conference of the North American Chapter of the Association for Computational Linguistics: Human Language Technologies, Volume 1 (Long and Short Papers), Minneapolis, Minnesota, pp. 4171–4186. Association for Computational Linguistics (2019)
5. Ding, N., et al.: Prototypical Representation Learning for Relation Extraction. arXiv:2103.11647 (2021)
6. Dong, L., et al.: Unified language model pre-training for natural language understanding and generation. In: Advances in Neural Information Processing Systems, vol. 32. Curran Associates, Inc. (2019)
7. Gao, T., Han, X., Liu, Z., Sun, M.: Hybrid attention-based prototypical networks for noisy few-shot relation classification. In: AAAI, vol. 33, pp. 6407–6414 (2019)
8. Lewis, M., et al.: Bart: denoising sequence-to-sequence pre-training for natural language generation, translation, and comprehension. arXiv preprint arXiv:1910.13461 (2019)

9. Li, H., Mo, T., Fan, H., Wang, J., Wang, J., Zhang, F., Li, W.: KiPT: knowledge-injected prompt tuning for event detection. In: Proceedings of the 29th International Conference on Computational Linguistics, pp. 1943–1952 (2022)

10. Lima, R., Espinasse, B., Freitas, F.: OntoILPER: an ontology- and inductive logic programming-based system to extract entities and relations from text. Knowl. Inf. Syst. **56**(1), 223–255 (2018)

11. Lin, Y., Ji, H., Huang, F., Wu, L.: A joint neural model for information extraction with global features. In: Proceedings of the 58th Annual Meeting of the Association for Computational Linguistics, pp. 7999–8009 (2020)

12. Liu, X., Luo, Z., Huang, H.: Jointly multiple events extraction via attention-based graph information aggregation. arXiv preprint arXiv:1809.09078 (2018)

13. Liu, Y., et al.: Roberta: a robustly optimized bert pretraining approach. arXiv preprint arXiv:1907.11692 (2019)

14. Lu, Y., et al.: Text2Event: controllable sequence-to-structure generation for end-to-end event extraction. In: Proceedings of the 59th Annual Meeting of the Association for Computational Linguistics and the 11th International Joint Conference on Natural Language Processing (Volume 1: Long Papers), pp. 2795–2806. Association for Computational Linguistics, Online (2021)

15. Nguyen, T.H., Cho, K., Grishman, R.: Joint event extraction via recurrent neural networks. In: Proceedings of the 2016 Conference of the North American Chapter of the Association for Computational Linguistics: Human Language Technologies, San Diego, California, pp. 300–309. Association for Computational Linguistics (2016)

16. Nguyen, T.M., Nguyen, T.H.: One for all: neural joint modeling of entities and events. In: Proceedings of the AAAI Conference on Artificial Intelligence, vol. 33, no. 01, pp. 6851–6858 (2019)

17. Paolini, G., et al.: Structured prediction as translation between augmented natural languages. arXiv preprint arXiv:2101.05779 (2021)

18. Snell, J., Swersky, K., Zemel, R.: Prototypical networks for few-shot learning. In: Advances in Neural Information Processing Systems, vol. 30 (2017)

19. Vaswani, A., et al.: Attention is all you need. In: Advances in Neural Information Processing Systems, vol. 30 (2017)

20. Wadden, D., Wennberg, U., Luan, Y., Hajishirzi, H.: Entity, relation, and event extraction with contextualized span representations. arXiv preprint arXiv:1909.03546 (2019)

21. Yan, H., Jin, X., Meng, X., Guo, J., Cheng, X.: Event detection with multi-order graph convolution and aggregated attention. In: Proceedings of the 2019 Conference on Empirical Methods in Natural Language Processing and the 9th International Joint Conference on Natural Language Processing (EMNLP-IJCNLP), Hong Kong, China, pp. 5766–5770. Association for Computational Linguistics (2019)

22. Yang, S., Feng, D., Qiao, L., Kan, Z., Li, D.: Exploring pre-trained language models for event extraction and generation. In: Proceedings of the 57th Annual Meeting of the Association for Computational Linguistics, Florence, Italy, pp. 5284–5294. Association for Computational Linguistics (2019)

Storyline Generation from News Articles Based on Approximate Personalized Propagation of Neural Predictions

Junli Wang[1], Xujian Zhao[1(✉)], Peiquan Jin[2], Chunming Yang[1], Bo Li[1], and Hui Zhang[1]

[1] Southwest University of Science and Technology, Mianyang 621010, Sichuan, China
`jasonzhaoxj@swust.edu.cn`
[2] University of Science and Technology of China, Hefei 230026, Anhui, China

Abstract. Generating a storyline is aiming to discover the evolution of events from news websites. Some existing approaches aim to automatically cluster news articles into events and connect related events in growing trees to generate storylines. Unfortunately, these methods did not perform well in learning the implicit associations of events. More recently, Graph Convolutional Network (GCN) based methods are proposed to learn the implicit associations between events. However, since the event representation in GCN model tends to be consistent after multi-layer propagation, the events cannot be correctly distinguished, which is not conducive to comprehensively learning the implicit associations between different events. In this paper, we propose an effective storyline generation method for news articles. Firstly, a novel model is presented based on Approximate Personalized Propagation of Neural Predictions for **S**tory **B**ranch **C**onstruction **M**odel, called **SBCM**, which preserves local features and can better learn the implicit association between different events. Then, we utilize a statistical method to identify transition events in news articles, and connect story branches with transition events through temporal relationships to finally generate storylines. The experimental results on two real-world Chinese news datasets show that our proposal outperforms several state-of-the-art methods.

Keywords: Text mining · Storyline generation · Graph convolutional network

1 Introduction

Nowadays, a variety of news events are generated and propagated on news websites. The public is hard to read and understand news events from such large volumes of information effectively. Thus, it has been an essential issue for web users to deeply mine and analyze the fragmented data to form a storyline reporting an event's development. Storyline generation aims to detect topic-relevant sentence-level events from news articles, then construct story branches with event

X. Wang et al. (Eds.): DASFAA 2023, LNCS 13946, pp. 37–52, 2023.
https://doi.org/10.1007/978-3-031-30678-5_3

classification, and finally assemble them into an easy-to-understand and time-based event evolution structure. Basically, a storyline can be considered as a set of events on the same topic, while a story branch can be considered as a set of events with sub-topics. Therefore, there are two main challenges for storyline generation: (1) how to construct the story branch with correct event classification? and (2) how to generate the storyline with story branches?

For the first issue (story branch construction), it is important to fully consider the associations between events and to classify them correctly into story branches. Some previous works [17,19] proposed semi-supervised models for story branch construction, they considered the similarity of explicit event content to determine story branches but not learned implicit associations between events. Recently, Zhao et al. [25] proposed an improved GCN model to generate storylines from microblog events by learning the implicit associations between events. However, the GCN-based models did not preserve local event representation, which leads to event representation tending to be consistent after multi-layer propagation, so not prefer well in dividing events to the correct story branch.

For the second issue (storyline generation), we need an effective way to smoothly and logically connect the different story branches into a storyline. Some previous works [9,14] simply connected story branches through temporal relationships, they can easily lead to the logical incoherence of the storyline generation. In addition, Lin et al. [17] proposed the minimum spanning tree algorithm to generate the storyline, but they did not smoothly express the progress of events.

In this paper, we propose an effective storyline generation method for news articles. Our basic idea for story branch construction is to devise a unified mechanism for implicit association learning without the locality weakness in propagation and exploit the supervised learning to classify events into story branches. In addition, we present the algorithm to identify the transition event connecting two story branches and link all story branches as well as transition events together smoothly and logically. Briefly, we make the following contributions in this paper.

- We present a novel model named SBCM to construct story branches by categorizing events, which effectively solves the problem that the learning model weakens the locality of nodes after multi-layer propagation. Firstly, we construct an event graph to represent global event features and then use Multi-layer Perceptron (MLP) to extract local event features. Further, events are divided into story branches using Approximate Personalized Propagation of Neural Predictions model that fully considers the implicit associations between events.
- We present an effective method to generate a logical and understandable storyline by assembling story branches and transition events chronologically. Firstly, the transition events are identified by the statistical model, and then they are sequentially connected to each story branch in the order of their timestamps to produce a reasonable storyline.
- We evaluated the proposed approaches on two real datasets and compared them with several state-of-the-art methods. The results suggest that our proposal is superior in story branch construction and storyline generation.

2   Related Work

In this section, we summarize existing works. There are mainly two research areas that are closely related to our work, i.e., Timeline Summarization and Storyline Generation.

2.1   Timeline Summarization

Timeline Summarization (TLS) aims to extract news events and cluster the related events by time to generate a timeline while [2,4,7] are highly related to the study of storyline generation. Jiang et al. [12] proposed a news corpus clustering system based on context similarity to describe the same event. Li et al. [15] identified major events from a news collection and detected them following temporal order. Using metrics based on user networks, temporal proximity, and the semantic context of events, Aansah et al. [1] constructed coherent paths and generated structural summaries of timelines to show the evolution of events over time. Li et al. [14] thought traditional methods ignore the intra-structures (arguments) and inter-structures (connections between events) of events, so they proposed to represent news reports as event graphs and consequently the summary is to compress the whole graph to the main subgraph. In their work, the main hypothesis is that events connected by common arguments and temporal sequence form the skeleton of a timeline containing events that are semantically related, temporally coherent, and structurally salient in the global event graph. Then, a time-aware optimal transport distance is introduced to learn the compression model in an unsupervised manner. Liao et al. [16] proposed a simple, fast and effective news timeline summarization method called WILSON to address the current challenges of quality and speed of news timeline generation. Nguyen et al. [22] proposed a pipeline for creating timelines consisting of date selection, sentence clustering, and sentence ranking. Quatra et al. [13] presented a paradigm shift in date selection for timeline summaries, aiming to address the original method's lack of reference to the news when selecting dates.

2.2   Storyline Generation

Different from timeline summarization, the study of storyline generation considers the evolution of events from the whole and part, and it can better meet the needs of users to understand a news event [11]. The process of Storyline Generation is consist of four steps: event extraction, event correlation analysis, story branches construction, and story branches connection. According to the different story branch construction models, the research on Storyline Generation can be divided into the following three categories.

- **Method based on the propagation model.** The method aims to model the relevance of events in a graph and construct a clear storyline structure by generating a tree structure from the graph. Researchers often use similarity to model events as graph structures and convert the construction of storylines into a graph-generated tree problem. Tree generation algorithms such as maximum spanning tree and Steiner tree [21] are commonly used to build storylines. The shortage is that it is an NP-hard problem that requires consideration of the complexity of how to generate an optimal tree in polynomial time. Lin et al. [17] early leveraged the approximation algorithm of a minimum-weight dominating set with a directed Steiner tree to figure out the issue of storyline generation.

– **Method based on similarity clustering.** The characteristic of this method is to process the data stream with incremental strategy through association analysis. For example, Liu et al. [19] proposed Story Forest using the Jaccard similarity coefficient of event keywords to link events and generate storylines. Hawwash et al. [9] processed event sentences into vectors as input to the clustering algorithm to construct storylines. Wen et al. [23] defined the granularity of events based on the cost function read by users, and computed TF-IDF weights after classifying news stories by K-means clustering algorithm with some corrections as well as model fusion by putting the cost function and event vector.

– **Method based on feature modeling.** Zhou et al. [27] proposed an unsupervised Bayesian model named DSDM to build story branches. Here the story branches are modeled as the joint distribution of named entities and topics. Additionally, the model parameters are inferred through Gibbs sampling to obtain the story branches where different events are located. Finally, events of the story branches are organized according to time to obtain the storyline. Zhou et al. [26] proposed a DSEM for the improvement of DSDM, with the document named entities are fine-grained, and the storyline is modeled as a joint distribution of events, topics, and keywords, which improves the storyline generation. Guo et al. [8] continue to refine this work by proposing a Dynamic Storyline Extraction Model (DSEM) and assuming that the prior distribution in the current period is the sum of the weights of these distributions in previous periods while using time-varying parameters to capture changes in the distribution. In addition, Zhao et al. [25] used an event graph convolutional network model to detect story branches for storyline generation from microblogs. Although the model can learn the implicit association between events in the story branch and take less time. This approach is limited in preserving local feature information to generate high-quality storylines.

3   Methodology

In this section, we detail the novel model SBCM for story branch construction and the statistics-based approach to selecting transition events for storyline generation. Figure 1 shows the framework of our proposal. It consists of three main parts: news event extraction, story branch construction and storyline generation. Firstly, news articles are cleaned through text preprocessing task. After keyword extraction, then we use top-k keywords to filter events in news, and finally obtain the event with time expression regularization. Secondly, an event

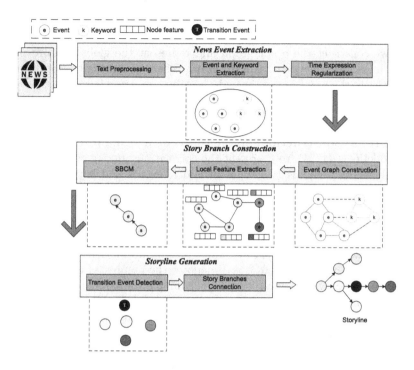

Fig. 1. An overview architecture of storyline generation.

graph is designed to depict the correlations among different events. Further, we pass the node feature matrix as input to a fully connected layer to extract each event's local representation. After that, feature propagation is carried out by Personalized PageRank, and the prediction results of story branches are output. Finally, transition nodes are detected and connected with story branches to generate a storyline.

3.1 News Event Extraction

We collect the news on the corresponding topics from Pengpai News[1] and Tencent News[2] and remove duplicate articles. Firstly, we utilize topic model based keyword extraction method to extract the keyword set. Then, we select top-5 keywords to filter irrelevant events and calculate the weighted sum of the maximum common sub-string and cosine similarity of events to remove duplicates. Finally, we assign events to time expression regularization. Time expressions are divided into regular time expressions (e.g., October 27, 2021) and implicit time expressions (e.g., Tuesday). For regular time expressions, we retain them for the event. For implicit time expressions, such as "Tuesday", we refer to the publish time "March 26, 2021" and obtain the regular time expression "March 23, 2021".

---

[1] https://www.thepaper.cn/.
[2] https://news.qq.com/.

3.2 Story Branch Construction

Firstly, we construct the event graph and represent the global features of the event with the adjacency matrix of the graph. Then, we extract the local features of the nodes in the event graph through the fully connected layer. Lastly, we feed the global features and local features into SBCM to construct story branches with correct event classification.

Event Graph Construction. In the paper, we design a heterogeneous event graph model $G = (V, \Delta)$, to represent the correlations among events. Generally, the graph model contains two types of nodes: events and keywords. We consider nouns, verbs, adjectives, adverbs, and quantifiers as keywords. Each node is denoted as $v \in V$. Here we set feature matrix X as the initial input to the SBCM model. Besides, each edge $\delta = (v, v') \in \Delta$ represents the relation between two nodes. Specifically, the edge δ has three categories. First, the weighted edge between two keywords is formed by Point Mutual Information as represented by Eq. 1, Eq. 2 and Eq. 3. Here $|\Delta(k_i, k_j)|$ denotes the number of co-occurrences of keyword k_i and k_j while $|\Delta(k_i)|$ denotes the number of k_i included in all events. Meanwhile, the total number of keywords in all events is defined as $|\Delta|$. Second, the weighted edge between the keyword and the event is calculated by the Term Frequency-Inverse Document Frequency (TD-IDF) value of the keyword in the event. Third, we calculate the weighted edge between two events using the Event Time Aging Relation model [3] defined as Eq. 4, where ε is a natural constant, and θ refers to the maximum time span factor. Here we set the parameter θ to 30 days, and when it is greater than 30 days, a weighted edge is added between events e_i and e_j. In addition, t_1 and t_2 denote the occurrence times of events e_i and e_j respectively. Consequently, the heterogeneous graph structure is constructed based on the above relation definition. When the similarity weight between two nodes is greater than a certain threshold, an undirected edge is constructed between the two nodes. Finally, we use an adjacency matrix to represent the graph structure.

$$\delta_{k_i,\, k_j} = \log \frac{P(k_i, k_j)}{P(k_i) P(k_j)} \tag{1}$$

$$P(k_i,\, k_j) = \frac{|\Delta(k_i, k_j)|}{|\Delta|} \tag{2}$$

$$P(k_i) = \frac{|\Delta(k_i)|}{|\Delta|} \tag{3}$$

$$\delta_{e_i,e_j} = \begin{cases} \varepsilon^{-\frac{|t_1 - t_2|}{\theta}} & |t_1 - t_2| > \theta \\ 1 & \text{otherwise} \end{cases} \tag{4}$$

Local Feature Extraction. PageRank algorithm, which utilizes the connection structure of network to compute global important scores, these scores can influence the ranking of search results. Personalized PageRank is the extension of the PageRank algorithm, it increased the probability of restarting to the root node and preserved the local features to improve performance on classification tasks. Since **P**ersonalized **P**ropagation of **N**eural **P**redictions (PPNP) [6] mainly solved the problem of over-smoothing in the GCN model by using the Personalized PageRank method instead of doing the graph convolution directly in the graph adjacency matrix. So we apply Personalized PageRank to story branch construction. In Eq. 5, we utilize X to represent the node features in the constructed graph. And we feed the X into a fully connected layer θ to extract the local feature L of the event is denoted as f_θ.

$$H^{(0)} = L = f_\theta(X) \tag{5}$$

$$H^{(k+1)} = (1 - \alpha)\tilde{D}^{-\frac{1}{2}}\tilde{A}\tilde{D}^{-\frac{1}{2}}H^{(k)} + \alpha H^{(0)} \tag{6}$$

Story Branch Construction Model (SBCM). PPNP first extracts the local features through neural networks and then spreads the feature information of nodes on the Personalized PageRank model to get the final prediction results. Original Personalized PageRank is calculated via $P_r = (1 - \alpha)A_r P_r + \alpha E$. E is identity matrix, $A_r = AD^{-1}$, P_r is the Personalized PageRank. PPNP follows the approach of Personalized PageRank and uses it for graph feature propagation, $P_r = \alpha(E - (1 - \alpha)\tilde{D}^{-\frac{1}{2}}\tilde{A}\tilde{D}^{-\frac{1}{2}})^{-1}$. However, the propagation process of PPNP has a matrix inversion operation, which is computationally expensive. Thus, we utilize the **A**pproximate **P**ersonalized **P**ropagation of **N**eural **P**redictions (APPNP) for feature propagation modeling on event graphs. The Personalized Propagate method is given by Eq. 6, Where k refers to the number of power iteration steps while the restart probability is represented as $\alpha \in (0, 1]$. $\tilde{D}^{-\frac{1}{2}}\tilde{A}\tilde{D}^{-\frac{1}{2}}$ is calculating by the adjacency matrix $\tilde{A}$ and the diagonal matrix $\tilde{D}$. APPNP captures higher-order information in the graph by applying the k-th power of the Personalized PageRank matrix in a single neural network, due to the addition of the restart parameter, the local features of the nodes are preserved during the propagation process, and finally, a softmax function is needed to complete the construction task.

Figure 2 shows the core idea of SBCM. Event 2, Event 5 and Event 6, first obtain the local features of the nodes through MLP, using blue, orange and flesh colors respectively, and then after the first APPNP transfer, Event 4 obtains the features of Event 2, Event 5 and Event 6, and then passes through one more time. After the second transfer, Event 4 obtains the red node feature from Event 7. After N rounds of transfer, Event 2, Event 5 and Event 6 still preserve the local features of their respective nodes. Finally, each event is classified into the corresponding story branch after multi-layer propagation.

APPNP can be defined as a more powerful graph convolutional network model for story branch construction. Consequently, considering the story branch construction based on APPNP, we need to initialize the story branch to make several nodes with labels. In the paper, first an event e_i is randomly added to the initial story branch set defined as $InitBranchSet_i$, and then the event with the largest difference from the events in $InitBranchSet$ is obtained by iterative

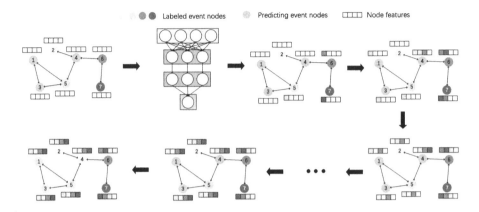

Fig. 2. The core idea of SBCM.

calculation and added to $InitBranchSet$ if the difference is greater than the threshold λ. Finally, the initial story branch set is obtained after going through all the events. Now we model the feature matrix of each event by calculating the word vector $W = \{w_1, \ldots, w_i\}$. In the paper, W is a one-hot vector of the vocabulary excluding stopwords. Words in the event sentence are represented by 1 in the vocabulary and not by 0. After constructing the set of word vectors of all events, we can obtain the input feature matrix. The details of story branch construction are shown in Algorithm 1. First, we initialize the story branches with the difference-degree comparison method and then get the predicted results from APPNP. Further, the maximization function is used to find the story branch to which an event belongs, that is, the story branch with the highest probability of classifying the event. Then, we add this event to the story branch. Finally, the set of predicted story branches is output after all iterations.

The loss function is defined as the cross-entropy cost of labeled events at Eq. 7, where Ω refers to the sets with labeled events. Meanwhile, the ground truth result and the predicted result are denoted as $Y_{e\phi}$ and $Z_{e\phi}$ respectively.

$$\min L = -\sum_{e \in \Omega} \sum_{\phi=1}^{\varphi} Y_{e\phi} \ln Z_{e\phi} \tag{7}$$

3.3 Storyline Generation

The module of storyline generation consists of two consecutive tasks: transition events detection and story branch connection. Specifically, first we deal with the issue of transition events construction based on the statistical method, and then the story branches are linked with transition events chronologically to generate a storyline.

Transition Event Detection. Although the story branch to which each event belongs is predicted in the previous module, an understandable storyline should

---

Algorithm 1. Story Branch Construction

Input: $EventSet = \{e_0, \ldots, e_m\}$: the news event set; $\hat{\tilde{A}}$: the normalized adjacency matrix; H: the feature matrix; $hlayer$: the number of hidden layers of SBCM
Output: $BranchSet = \{b_0, \ldots, b_n\}$: the predicted story branch set
Preliminary: $InitBranchSet$: the initial story branch set; $Results_{m \times n}$ (m: the number of events; n: the number of story branches): the predicted results matrix; λ: a predefined value for event variance; α: the restart parameter.

1: **for** e_i in $InitBranchSet_i$ **do**
2:    $e \leftarrow$ The event with the largest difference from e_i in $EventSet$;
3:    **if** $similarity(e, InitBranchSet_i) < \lambda$ **then**
4:       $InitBranchSet.add(e)$
5:       $EventSet.remove(e)$
6:    **end if**
7: **end for**
8: $Results \leftarrow \text{APPNP}(\hat{\tilde{A}}, H, layer, \alpha)$
9: $BranchSet \leftarrow InitBranchSet$
10: **for** i in $|EventSet|$ **do**
11:    **for** j in $|BranchSet|$ **do**
12:       $result \leftarrow Results[i, j]$
13:       **if** $result < max$ **then**
14:          $max \leftarrow result$
15:          $pos \leftarrow i$
16:       **end if**
17:       $b_j.add(e_{pos})$
18:    **end for**
19: **end for**
20: **return** $BranchSet = \{b_0, b_1, \ldots, b_n\}$

---

not only include story branches but also consider transition events that refer to the events connecting two contextual story branches. For this issue, a statistical approach represented by Eq. 8 is used to detect the transition events.

$$\tau_i = \sum_{i=1}^{n} \sum_{j=1}^{c} \min\left(P\left(a_i \mid b_j\right), 0.5\right)$$
$$T = \sum_{i=1}^{n} \tau_i \tag{8}$$

Here $P(a_i \mid b_j)$ is the probability that the event a_i is classified into the story branch b_j. In addition, the number of story branches and transition events is denoted as the symbols c and T respectively. We first execute the tenfold validated story branch prediction algorithm and store all the story branch prediction results. Then, the number of times each event is assigned to the corresponding story branch is counted. If the number of times an event is less than 0.5, the event is considered not to belong to any story branch and should be classified as a pending event. Finally, the total number of pending events is calculated,

---

Algorithm 2. Storyline Generation

Input: $BranchSet = \{b_1, \ldots, b_n\}$: the predicted story branches with events set
Output: S: the storyline with story branches and transition events
Preliminary: $T = \{t_1, \ldots, t_n\}$: the transition events set.

```
 1: for b_i in BranchSet do
 2:    for e_j in b_i do
 3:       if T_DETECTION(e_j) = true then
 4:          T.add(e_j)
 5:       end if
 6:    end for
 7: end for
 8: ORDER(T)
 9: S ← BranchSet
10: for  k in T.length do
11:    for b_k in BranchSet do
12:       if b_k.timestamp > t_k  &  b_{k+1}.timestamp < t_k  then
13:          S.insert(t_k)
14:       end if
15:    end for
16: end for
17: return  S
```

---

and events that do not belong to any story branch are designated as transition events.

Story Branches Connection. According to Sect. 3.2, the events can refer to related story branches. To ensure the consistency of the storyline, we propose the following Algorithm 2 to connect story branches and transition events to generate a high-quality storyline.

In Algorithm 2 first the predicted result by SBCM, namely the set of predicted story branches with events $BranchSet$, is used as input. Then, for each event in b_i, we detect whether it is a transition event t, and use the set of transition events T to store transition events. After the transition event detection is complete, all the events are sorted in the time order. Further, we regard the transition event t_i satisfying the timestamp greater than the story branch b_i and less than the story branch b_{i+1} as the transition event connecting two story branches. Finally, the transition event t_i is inserted into the storyline S.

4   Performance Evaluation

In the following sections, we report the experimental results of our method. In Sect. 4.1, we introduce the statistic of two real event datasets. In Sect. 4.2, we evaluate the performance of the story branch construction. And we discuss the experiments on storyline generation in Sect. 4.3.

Table 1. Overview of the datasets

Dataset	Topic	Starting and ending time	Number of news
1	Tesla owner rights incident	2021.02.21–2021.06.08	1614
2	Xinjiang Cotton Incident	2021.03.28–2021.07.01	572

4.1 Dataset

To the best of our knowledge, there are no public standard datasets available for evaluating both story branch construction and storyline generation. Therefore, we construct tailor-made datasets following Hua et al.'s work [10] for evaluating the performance of our proposal. Specifically, we collected two real-world datasets from mainstream news media such as Tecent News and Pengpai News, and the details of the two datasets are shown in Table 1.

4.2 Performance of Story Branch Construction

Baseline. To evaluate the performance of our proposal in constructing story branches, we compare our model with six baseline models. In the paper, for all the baselines, we use the same implementations or open-source codes released by authors and report the best performance of the results.

(1) **Story Forest (SF)** [19] calculates the correlation between events with the Jaccard coefficient after extracting event keywords and then clusters an event to a story branch.

(2) **Minimal Steiner Tree (MST)** [17] generates story branches via the Steiner Tree algorithm considering the minimum-weight dominating set in approximation.

(3) **Latent Dirichlet Allocation (LDA)** [20] based on the LDA topic model, it is automatically clustered into multiple clusters, each cluster is a story branch, and the number of story branches is also automatically determined..

(4) **Graph Convolutional Networks (GCN)** [25] uses GCN to learn the implicit association between events, and then organize events to generate story branches.

(5) **Graph Isomorphism Networks (GIN)** [24] is a theoretical framework for analyzing the expressive power of GNN in capturing graph structures and is defined as a Weisfeiler-Lehman graph isomorphism test. We use it to capture related events under different story branches and detect story branches.

(6) **Topology Adaptive Graph Convolutional Networks (TAGCN)** [5] is a graph convolutional network defined in the vertex domain, which provides a systematic method to design a set of fixed-size learnable filters to perform convolutions on graphs. It can learn local representations of event nodes, but tends to converge after multiple propagations in agreement.

Evaluation Metrics. To evaluate the performance of the story branch construction model, 20 researchers engaged in text mining are invited to manually construct ground-truth story branches. Specifically, we recommend that researchers refer to news reports on each topic on Wikipedia. Each story branch is composed of highly similar events and sorted in chronological order. Here, the constructed story branch that is accepted by more than half of the people is considered the ground-truth story branch. Meanwhile, we calculate the similarity between the predicted story branch and the ground-truth story branch with the Story branch mapping relationship represented by Eq. 9, where I_p and I_s refer to the set of predicted story branches and the set of standard story branches respectively.

$$\text{sim}(p, s) = \frac{2 \times |I_p \cap I_s|}{|I_p| + |I_s|} \tag{9}$$

On this basis, the performance of story branch construction is evaluated by Precision, Recall and F1 value, defined by Eq. 10, Eq. 11, and Eq. 12 respectively.

$$P = \frac{|I_p \cap I_{p \to s}|}{|I_{p \to s}|} \tag{10}$$

$$R = \frac{|I_p \cap I_{p \to s}|}{|I_p|} \tag{11}$$

$$F1 = \frac{2 \times P \times R}{P + R} \tag{12}$$

Here $I_p \to_s$ refers to the mapping of the predicted story branch set I_p to the standard story branch set I_s.

Parameter Settings. To determine the best number of story branches, the difference threshold λ is optimized by setting the step size to 0.002. We find that for dataset 1 and dataset 2, the experimental performance is relatively stable when the initial branch numbers are 12 and 9, respectively. In addition, we compare the effect of setting the maximum time span θ from 0 to 60 (in increments of 10) on the accuracy of story branch construction. Figure 3 shows that when the maximum time span θ is set to 30, the algorithm gets the best performance. Meanwhile note that if the maximum time span is set to zero, the approach does not perform well in story branch construction.

Experimental Results. Table 2 and 3 show the results of story branch construction on two datasets. Basically, our proposal outperforms the other six models in terms of all the metrics. Specifically, we consider SBCM for semi-supervised classification to solve the problem of poor learning of implicit relations between events in traditional approaches. In addition, compared with GCN-based models, our proposal significantly improves the capability of preserving node local features in graph structures and consequently improves the performance of story branch construction. Meanwhile, note that in Table 3 the Recall value of the TAGCN model is slightly higher than our model due to a small deviation in

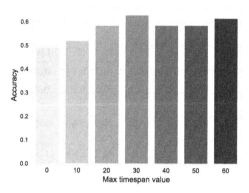

Fig. 3. The effect of θ on accuracy of story branch construction on dataset1

Table 2. Precision, Recall and F1-value of story branch construction models on dataset1

Model	P	R	F1
LDA	0.406	0.508	0.451
MST	0.496	0.467	0.481
SF	0.578	0.629	0.602
GIN	0.462	0.632	0.534
TAGCN	0.551	0.655	0.599
GCN	0.556	0.635	0.593
Ours	**0.681**	**0.668**	**0.674**

Table 3. Precision, Recall and F1-value of story branch construction models on dataset2

Model	P	R	F1
LDA	0.667	0.500	0.572
MST	0.444	0.424	0.433
SF	0.676	0.770	0.720
GIN	0.556	0.861	0.676
TAGCN	0.694	**0.963**	0.807
GCN	0.509	0.918	0.655
Ours	**0.715**	0.937	**0.811**

the initial branch numbers. Specifically, the number of news in dataset 2 is too small, making the initial story branch difficult to set.

4.3   Performance of Storyline Generation
Baseline. To evaluate the performance of our proposal in storyline generation, we compare our method with four baseline methods.

Table 4. Performance of different storyline generation methods on correct event connections

Dataset	ST	MST	SF	GCN	Ours
1	0.274	0.210	0.431	0.314	**0.451**
2	0.152	0.242	0.367	0.272	**0.424**

Table 5. Performance of different storyline generation methods on grading scores

	Metrics	ST	MST	SF	GCN	Ours
Dataset1	Reasonability	3.42	2.58	2.75	3.83	**2.42**
	Consistency	3.67	3.00	3.25	2.75	**2.33**
Dataset2	Reasonability	3.25	2.83	3.08	3.25	**2.58**
	Consistency	3.42	3.17	3.33	2.58	**2.50**

(1) **Story Timeline (ST)** links all events only according to the timestamp in events.
(2) **Story Forest (SF)** [18] compares whether a story branch is duplicated with an existing storyline, if so, merges this two story branches, otherwise selects the new parent of the storyline.
(3) **Minimal Steiner Tree (MST)** [17] aims to find the node path with the minimum weight in the constructed graph as the storyline.
(4) **Graph Convolutional Networks (GCN)** [25] obtains the storyline by assembling story branches through event-time correlation.

Evaluation Metrics. In the paper, we evaluate the performance of the storyline generation by different methods from two aspects: the correct connections between events, the consistency and reasonability of storylines. Firstly, we use the ratio of the number of correct event connections between the results generated by the algorithm and the standard results as the first indicator. Then, we invited 12 volunteers for giving the grading results of generated storylines. Specifically, the grading result by volunteers ranked the storyline generated by all methods from 1 to 5, where 1 represents the best known. The volunteers answered the following two questions: (1) Whether the entire storyline is complete and closely related and (2) Whether the storyline is easy for readers' comprehension. Finally, we use the average of the results as the second indirect metric to evaluate storyline generation methods.

Experimental Results. Table 4 shows the experiment results of the correct event connections. Obviously, our method outperforms other storyline generation methods, due to considering the transition events when connecting story branches. In addition, as Table 5 shown, our method has better performance than other methods, which means transition events can boost the quality of storyline generation in consistency and reasonability aspects.

5 Conclusions and Future Work

In the paper, we propose an effective storyline generation method for news articles. Firstly, a novel model SBCM, is presented to solve the lack of preserving local event representation and poorly learning implicit event relation problem of story branch construction. Then, we identify the transition events in articles through the statistical method, to produce a more logical and understandable storyline. Finally, experimental results demonstrate the superiority of our proposal with respect to various metrics. In the future, we will consider using various datasets to further evaluate the performance of our method. In addition, we will incorporate multimodal learning to storyline generation for better performance.

Acknowledgements. This paper is supported by the Humanities and Social Sciences Foundation of the Ministry of Education (17YJCZH260), the National Science Foundation of China (62072419), the Sichuan Science and Technology Program (2020YFS0057).

References

1. Ansah, J., Liu, L., Kang, W., Kwashie, S., Li, J., Li, J.: A graph is worth a thousand words: telling event stories using timeline summarization graphs. In: The World Wide Web Conference, pp. 2565–2571 (2019)
2. Born, L., Bacher, M., Markert, K.: Dataset reproducibility and IR methods in timeline summarization. In: Proceedings of the 12th Language Resources and Evaluation Conference, pp. 1763–1771 (2020)
3. Chen, C.C., Chen, Y.-T., Sun, Y., Chen, M.C.: Life cycle modeling of news events using aging theory. In: Lavrač, N., Gamberger, D., Blockeel, H., Todorovski, L. (eds.) ECML 2003. LNCS (LNAI), vol. 2837, pp. 47–59. Springer, Heidelberg (2003). https://doi.org/10.1007/978-3-540-39857-8_7
4. Chen, X., Chan, Z., Gao, S., Yu, M.H., Zhao, D., Yan, R.: Learning towards abstractive timeline summarization. In: IJCAI, pp. 4939–4945 (2019)
5. Du, J., Zhang, S., Wu, G., Moura, J.M., Kar, S.: Topology adaptive graph convolutional networks. arXiv preprint arXiv:1710.10370 (2017)
6. Gasteiger, J., Bojchevski, A., Günnemann, S.: Predict then propagate: graph neural networks meet personalized pagerank. In: International Conference on Learning Representations (ICLR) (2019)
7. Ghalandari, D.G., Ifrim, G.: Examining the state-of-the-art in news timeline summarization. arXiv preprint arXiv:2005.10107 (2020)
8. Guo, L., Zhou, D., He, Y., Xu, H.: Storyline extraction from news articles with dynamic dependency. Intell. Data Anal. **24**(1), 183–197 (2020)
9. Hawwash, B., Nasraoui, O.: From tweets to stories: using stream-dashboard to weave the Twitter data stream into dynamic cluster models. In: Proceedings of the 3rd International Workshop on Big Data, Streams and Heterogeneous Source Mining: Algorithms, Systems, Programming Models and Applications, pp. 182–197. PMLR (2014)
10. Hua, T., Zhang, X., et al.: Automatic storyline generation with help from Twitter. In: CIKM, pp. 2383–2388 (2016)

11. Huang, L., et al.: Optimized event storyline generation based on mixture-event-aspect model. In: Proceedings of the 2013 Conference on Empirical Methods in Natural Language Processing, pp. 726–735 (2013)
12. Jiang, H., Beeferman, D., Mao, W., Roy, D.K.: Topic detection and tracking with time-aware document embeddings. ArXiv: abs/2112.06166 (2021)
13. La Quatra, M., Cagliero, L., Baralis, E., Messina, A., Montagnuolo, M.: Summarize dates first: a paradigm shift in timeline summarization. In: Proceedings of the 44th International ACM SIGIR Conference on Research and Development in Information Retrieval, pp. 418–427 (2021)
14. Li, M., et al.: Timeline summarization based on event graph compression via time-aware optimal transport. In: Proceedings of the 2021 Conference on Empirical Methods in Natural Language Processing, pp. 6443–6456 (2021)
15. Li, M., et al.: Timeline summarization based on event graph compression via time-aware optimal transport. In: EMNLP (2021)
16. Liao, Y., Wang, S., Lee, D.: Wilson: divide and conquer approach for fast and effective news timeline summarization. In: 24th International Conference on Extending Database Technology (2021)
17. Lin, C., Lin, C., et al.: Generating event storylines from microblogs. In: CIKM, pp. 175–184 (2012)
18. Liu, B., Han, F.X., Niu, D., Kong, L., Lai, K., Xu, Y.: Story forest. ACM Trans. Knowl. Disc. Data (TKDD) **14**, 1–28 (2020)
19. Liu, B., Niu, D., Lai, K., Kong, L., Xu, Y.: Growing story forest online from massive breaking news. In: Proceedings of the 2017 ACM on Conference on Information and Knowledge Management(CIKM), pp. 777–785 (2017)
20. Mehrotra, R., Sanner, S., Buntine, W., Xie, L.: Improving LDA topic models for microblogs via tweet pooling and automatic labeling. In: Proceedings of the 36th International ACM SIGIR Conference on Research and Development in Information Retrieval, pp. 889–892 (2013)
21. Nallapati, R., Feng, A., Peng, F., Allan, J.: Event threading within news topics. In: Proceedings of the Thirteenth ACM International Conference on Information and Knowledge Management, pp. 446–453 (2004)
22. Nguyen, K.H., Tannier, X., Moriceau, V.: Ranking multidocument event descriptions for building thematic timelines. In: COLING 2014, the 25th International Conference on Computational Linguistic, pp. 1208–1217 (2014)
23. Wen, A., Lin, W., Ma, Y., Xie, H., Zhang, G.: News event evolution model based on the reading willingness and modified TF-IDF formula. J. High Speed Netw. **23**(1), 33–47 (2017)
24. Xu, K., Hu, W., Leskovec, J., Jegelka, S.: How powerful are graph neural networks? arXiv preprint arXiv:1810.00826 (2018)
25. Zhao, X., Wang, C., Jin, P., Zhang, H., Yang, C., Li, B.: Post2story: automatically generating storylines from microblogging platforms. In: Proceedings of the 29th ACM International Conference on Multimedia, pp. 2786–2788 (2021)
26. Zhou, D., Xu, H., Dai, X.Y., He, Y.: Unsupervised storyline extraction from news articles. In: IJCAI, pp. 3014–3021 (2016)
27. Zhou, D., Xu, H., He, Y.: An unsupervised Bayesian modelling approach for storyline detection on news articles. In: Proceedings of the 2015 Conference on Empirical Methods in Natural Language Processing, pp. 1943–1948 (2015)

Broaden Your Horizons: Inter-news Relation Mining for Fake News Detection

Lei Zhang[1,2], Hongbo Xu[1(✉)], Fengzhao Shi[1,2], Chaoyang Yan[1,2], and Yongxiu Xu[1]

[1] Institute of Information Engineering, Chinese Academy of Sciences, Beijing, China
{zhanglei0510,hbxu,shifengzhao,yanchaoyang,xuyongxiu}@iie.ac.cn
[2] School of Cyber Security, University of Chinese Academy of Sciences, Beijing, China

Abstract. Detecting fake news on social media is an urgent task. Some early studies focus on capturing authenticity information from news content, while the single information source results in limited clues. Recent studies have great concern about more clues derived from auxiliary knowledge, yet it is usually rare at the early stage of news propagation. Furthermore, such studies, with more attention on the intra-news properties of individual news, would impair the performance on newly emerged events. To alleviate these issues, we propose a novel **Inter-News Relation Mining (INRM)** framework to mine inter-news relations. Whether for scenarios with little auxiliary knowledge or newly emerged events, INRM can provide more effective credibility clues for verifying th e truth of news. Experiments illustrate that INRM outperforms state-of-the-art methods on both conventional tasks and newly emerged event tasks.

Keywords: Social media · Fake news · Inter-news relation

1 Introduction

The increasing convenience of social media makes fake news easier to spread. As fake news distorts and fabricates facts maliciously, its extensive dissemination can easily mislead public opinion and disrupt social order. Therefore, it is crucial to automatically identify fake news and curb its spread on social media.

Some previous studies capture credibility features from news content. However, it is non-trivial to spot valuable clues from elaborate news content itself. Recent studies tend to introduce auxiliary information for more clues, but such auxiliary information is usually limited at the early stage of news propagation. In addition, since these methods essentially focus on mining the intra-news properties of individual news (e.g., linguistic features, propagation patterns, and factual veracity), their performance inevitably degrades when dealing with newly emerged events. Therefore, in the case of little auxiliary knowledge, how to mine more credibility clues for verifying the truth of news and alleviate the challenges brought by newly emerged events, is the major issue we need to solve.

© The Author(s), under exclusive license to Springer Nature Switzerland AG 2023
X. Wang et al. (Eds.): DASFAA 2023, LNCS 13946, pp. 53–62, 2023.
https://doi.org/10.1007/978-3-031-30678-5_4

Only limited clues can be captured when there is little auxiliary knowledge, which is insufficient to verify the truth of news. Fortunately, we realize that the inter-news relations, which can be readily obtained without or with little auxiliary knowledge, can provide more clues. Furthermore, unlike models of learning intra-news properties that fail to apply to newly emerged events, inter-news relations can serve as a bridge between news on newly emerged events and previous news to promote inter-news information interaction, thereby improving model performance. We can obtain two explicit inter-news relations guided by prior knowledge: **1) Publication relations.** News published by the same user may have tendencies (*Prior Knowledge I*). For example, users who have published fake news are more likely to publish fake news again than ordinary users. That is, a news post is more likely to be fake if it is the same user as the publisher of another fake news post, and vice versa. Therefore, we regard the relation between news belonging to the same publisher as a publication relation, which is conducive to authenticity identification. **2) Engagement relations.** News shared by the same kind of users may have commonalities (*Prior Knowledge II*). For example, in order to spread the fabricated fake news widely, a quick way for counterfeiters is to hire a group of online "water army" to retweet/comment to increase exposure. If these online "water army" are also involved in another news post, it is more likely to be fake. Accordingly, we consider the relation between news with the same group of engagement users (i.e., users who retweet/comment on news) as an engagement relation, which could facilitate news verification. In addition, the explicit relations guided by prior knowledge are not absolutely reliable, so how to extract excellent relations is a challenge. Moreover, since it is not enough to rely on the above explicit relations alone, how to mine more latent inter-news relations to provide more clues for better exploration in fake news detection is another challenge.

In light of the above, we propose a novel Inter-News Relation Mining (**INRM**) framework to mine inter-news relations, which can provide more effective credibility clues for fake news detection, whether for scenarios with little auxiliary knowledge or newly emerged events. Specifically, INRM obtains two explicit inter-news relations under the guidance of prior knowledge. Then, for the explicit relations, INRM preliminarily complements relations to encourage the inter-news information transmission, and elaborates a denoise loss to remove the noise caused by unreliable relations. Next, the optimized explicit relations are used to generate multi-view information that provides local-to-global information, so as to mine the latent inter-news relations. Finally, the resulting inter-news relations, which indicate inter-news consistency and supply practical clues, are exploited to aggregate news feature representations to predict news authenticity. Our contributions are summarized as follows:

- To the best of our knowledge, this is the first work that introduces inter-news relations to provide credibility clues, which is applicable to scenarios with little auxiliary knowledge and newly emerged events.
- The proposed INRM mines latent inter-news relations by exploiting explicit relations guided by prior knowledge for fake news detection.
- Experiments demonstrate the effectiveness of INRM for fake news detection on both conventional tasks and newly emerged event tasks.

2   Related Work

Most existing studies generally focus on using news content and auxiliary knowledge. In this section, we briefly review these two categories of methods.

News content-based methods mainly extract features from the textual or visual content of news to learn the shared patterns of fake news. Text-based studies capture linguistic features [6,10] to construct more discriminative features. Visual-based studies focus on visual semantics [2]. Multimodal-based studies exploit entity consistency [8] between text modality and visual modality to identify the authenticity of news. Compared with news content-based methods that can merely capture limited credibility clues from news content, our INRM provides more effective clues for fake news detection by mining inter-news relations.

Auxiliary knowledge-based methods are learned from auxiliary relevant sources for capturing valuable clues. One line is to leverage social knowledge. Some studies explore information about user stance [15] and news credibility [12] from users' social responses. More studies analyze network patterns based on specific networks [1,14]. Another line is to introduce external knowledge [5], which usually retrieves evidence from trustworthy knowledge sources. Compared with auxiliary knowledge-based methods, our INRM does not utilize any auxiliary features, but aims to utilize prior knowledge and limited auxiliary knowledge as guiding signals to mine inter-news relations.

3   Methodology

3.1   Problem Definition

Input. 1) A fake news detection dataset $C = \{c_1, c_2, \ldots, c_m\}$ with labels $\mathcal{Y} = \{y_1, y_2, \ldots, y_m\}$, where c_i is the i-th news post to be verified and m is the number of news posts. 2) A sequence of publication users $U^p = \{u_1, u_2, \ldots, u_m\}$, where u_i refers to the user who publishes the news post c_i. 3) A sequence of engagement users $U^e = \{U_1^e, U_2^e, \ldots, U_m^e\}$, where $U_i^e = \{u_1^i, u_2^i, \ldots, u_{n_i}^i\}$ is the propagation path of the news post c_i, u_j^i represents the j-th user engaged (retweet/comment) in c_i, and n_i is the number of users engaged in c_i.

Output. The predicted label (true ($y = 1$) or fake ($y = 0$)) of each news post.

3.2   Prior Graph Constructor

We aim at making full use of prior knowledge to model the inter-news relations when the limited auxiliary knowledge makes it impossible to mine enough credibility clues from intra-news. Therefore, we initially construct prior graphs based on the explicit inter-news relations guided by prior knowledge.

Publication Graph Construction. Based on the inter-news publication relations, we construct the publication graph $\mathcal{G}^p = (\mathcal{V}, \mathcal{E}^p)$, where $\mathcal{V} = \{c_1, c_2, \ldots, c_m\}$ refers to the set of nodes and $\mathcal{E}^p$ is the set of edges. The edges describe the publication relations between nodes and can be represented by an adjacency matrix $\mathbf{A}^p \in \mathbb{R}^{m \times m}$ with $A_{ij}^p = 1$ if $u_i = u_j$, otherwise $A_{ij}^p = 0$ (Fig. 1).

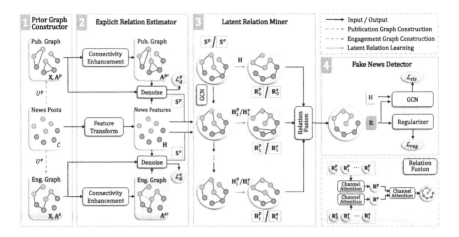

Fig. 1. An overview of our Inter-News Relation Mining (INRM) framework.

Engagement Graph Construction. We construct the engagement graph $\mathcal{G}^e = (\mathcal{V}, \mathcal{E}^e)$ based on the inter-news engagement relations, and the edges $\mathcal{E}^e$ describe the engagement relations between nodes. In the adjacency matrix $\mathbf{A}^e \in \mathbb{R}^{m \times m}$, $A_{ij}^e = 1$ if $|U_i^e \cap U_j^e| \geq \xi$, otherwise $A_{ij}^e = 0$, where ξ is the threshold. The initial node feature matrices in the publication graph and the engagement graph are the same, which are calculated in terms of TF-IDF, and denoted as $\mathbf{X} = [\mathbf{x}_1, \mathbf{x}_2, \ldots, \mathbf{x}_m] \in \mathbb{R}^{m \times z}$, where z is the number of terms in the corpus.

3.3 Explicit Relation Estimator

In order to encourage the inter-news information transmission, and considering that the explicit relations guided by prior knowledge are not absolutely reliable, the explicit relation estimator further polishes the prior graphs.

Connectivity Enhancement. Due to the poor connectivity and more isolation points in the prior graphs, it will hinder the inter-news information transmission. To enhance the connectivity, we utilize the initial node features to calculate the cosine similarity between each node pair. Then, we keep top-k similar nodes for each node, and add connected edges between them if these edges do not exist in the original graphs. Ultimately, the connectivity-enhanced publication graph with $\mathbf{A}^{p\prime} \in \mathbb{R}^{m \times m}$ and engagement graph with $\mathbf{A}^{e\prime} \in \mathbb{R}^{m \times m}$ are generated.

Denoise. It is well known that connected nodes in a graph may share similar features or attributes. However, some misconnected edges result from unreliable relations in the prior graphs. To remove these misconnected edges as much as possible, our idea is to reduce the weights of edges whose connected nodes have different features. To this end, for the publication graph, we design a denoise loss $\mathcal{L}_d^p$ (the denoise loss $\mathcal{L}_d^e$ for the engagement graph is similar):

$$\mathcal{L}_d^p = \frac{1}{2} \sum_{i,j=1}^{m} S_{ij}^p \left(\mathbf{h}_i - \mathbf{h}_j \right)^2 + \sum_{i,j=1}^{m} \left(S_{ij}^p - A_{ij}^p \right)^2 , \tag{1}$$

where $\mathbf{S}^p \in \mathbb{R}^{m \times m}$ denotes the optimized adjacency matrix, S_{ij}^p is the weight parameter of the edge between node i and j, and $\mathbf{S}^p = \mathbf{A}^{p\prime}$ is initialized. Besides, $\mathbf{h}_i$ and $\mathbf{h}_j \in \mathbb{R}^d$ are the hidden features of nodes i and j generated by the initial node features through feature transform ($\mathbf{H} = FC(\mathbf{X}) \in \mathbb{R}^{m \times d}$, where d is hidden dimension). It deserves to be noted that $\left(\mathbf{h}_i - \mathbf{h}_j \right)^2$ assesses the feature difference between nodes i and j, and $\left(S_{ij}^p - A_{ij}^p \right)^2$ measures the distance between the optimized graph and the prior graph. To get the optimized graph, If nodes i and j are connected (i.e., $S_{ij}^p \neq 0$), we expect that 1) when the feature difference is large, the edge weight S_{ij}^p should decrease; 2) when the feature difference is small, the edge weight S_{ij}^p should be preserved since it satisfies the property that the features of connected nodes should be similar. Thus, to fulfill the above expectations in the optimized graph, we should minimize $\mathcal{L}_d^p$. According to $\mathcal{L}_d^p$ and $\mathcal{L}_d^e$, the optimized explicit relations can be learned, that is, the adjacency matrix $\mathbf{S}^p$ for the publication graph and $\mathbf{S}^e$ for the engagement graph.

3.4 Latent Relation Miner

The explicit relations only implies limited clues derived from prior knowledge. To mine latent inter-news relations for more clues, our idea is to leverage explicit relations to generate rich multi-view information for capturing latent relations. Specifically, we feed the publication graph with $\mathbf{S}^p$ into the l-layer GCN. Formally, the aggregation rule of the l-th layer node feature $\mathbf{H}_l^p \in \mathbb{R}^{m \times d}$ is:

$$\mathbf{H}_l^p = \sigma \left(\hat{\mathbf{S}}^p \mathbf{H}_{l-1}^p \mathbf{W}_l^p \right) , \tag{2}$$

where $\hat{\mathbf{S}}^p \in \mathbb{R}^{m \times m}$ represents the normalization of the adjacency matrix $\mathbf{S}^p$, $\mathbf{W}_l^p \in \mathbb{R}^{d \times d}$ is a layer-wise learnable parameter matrix, and $\mathbf{H}_0^p = \mathbf{H}$. σ is the ReLU function. Similar to Eq. (2), we calculate the engagement graph with $\mathbf{S}^e$ in the same way. The node feature of l-th layer is $\mathbf{H}_l^e \in \mathbb{R}^{m \times d}$, and $\mathbf{H}_0^e = \mathbf{H}$.

After aggregation iterations, the node features of different layers provids local-to-global multi-view information. Accordingly, the node features of each layer are considered to mine latent relations. More concretely, based on the node features $\{\mathbf{H}_0^p, \mathbf{H}_1^p, \ldots, \mathbf{H}_l^p\}$ and $\{\mathbf{H}_0^e, \mathbf{H}_1^e, \ldots, \mathbf{H}_l^e\}$, we respectively calculate the node feature similarity for finding the latent relations. For the node feature $\mathbf{H}_l^p$, the latent relations between node i and node j can be calculated as follows:

$$R_l^p[i,j] = \begin{cases} \psi \left(\mathbf{H}_l^p[i], \mathbf{H}_l^p[j] \right) & \psi \left(\mathbf{H}_l^p[i], \mathbf{H}_l^p[j] \right) \geq \epsilon \\ 0 & \text{otherwise,} \end{cases} \tag{3}$$

where $\epsilon \in [0,1]$ is the threshold, and ψ is the similarity calculation function:

$$\psi \left(\mathbf{H}_l^p[i], \mathbf{H}_l^p[j] \right) = \frac{1}{T} \sum_{t=1}^{T} \cos \left(\mathbf{w}_{t,l}^p \odot \mathbf{H}_l^p[i], \mathbf{w}_{t,l}^p \odot \mathbf{H}_l^p[j] \right) , \tag{4}$$

where T refers to the number of latent relation types, and $\odot$ denotes the Hadamard product. $\mathbf{w}_{t,l}^p \in \mathbb{R}^d$ is a learnable weight vector, which is used to highlight important dimensions of node features for t-type latent relations.

With the above method, the latent relations corresponding to all views can be captured, namely $\{\mathbf{R}_0^p, \mathbf{R}_1^p, \ldots, \mathbf{R}_l^p\}$ and $\{\mathbf{R}_0^e, \mathbf{R}_1^e, \ldots, \mathbf{R}_l^e\}$. The final inter-news relations, denoted as $\mathbf{R} \in \mathbb{R}^{m \times m}$, can be generated by fusing these candidate relations through multiple channel attention layers:

$$\mathbf{R}^p = \phi\left([\mathbf{R}_0^p, \mathbf{R}_1^p, \ldots, \mathbf{R}_l^p]; \mathrm{softmax}\left(\mathbf{W}_\phi^p\right)\right), \tag{5}$$

$$\mathbf{R}^e = \phi\left([\mathbf{R}_0^e, \mathbf{R}_1^e, \ldots, \mathbf{R}_l^e]; \mathrm{softmax}\left(\mathbf{W}_\phi^e\right)\right), \tag{6}$$

$$\mathbf{R} = \phi\left([\mathbf{R}^p, \mathbf{R}^e]; \mathrm{softmax}\left(\mathbf{W}_\phi\right)\right), \tag{7}$$

where ϕ denotes a channel attention layer that performs convolution on the stacked matrix of the candidate relations. $\mathbf{W}_\phi^p \in \mathbb{R}^{1 \times 1 \times (l+1)}$, $\mathbf{W}_\phi^e \in \mathbb{R}^{1 \times 1 \times (l+1)}$ and $\mathbf{W}_\phi \in \mathbb{R}^{1 \times 1 \times 2}$ are learnable weight matrices of each channel attention layer respectively. In this way, INRM evaluates the importance of each candidate relation by learning different weights separately. After that, we get the final inter-news relations $\mathbf{R}$.

3.5   Fake News Detector

The fake news detector is a two-layer GCN, utilizing the learned inter-news relations $\mathbf{R}$ to aggregate news feature representations:

$$f(\mathbf{H}, \mathbf{R}) = \mathrm{softmax}\left(\widehat{\mathbf{R}}\sigma\left(\widehat{\mathbf{R}}\mathbf{H}\mathbf{W}_1^f\right)\mathbf{W}_2^f\right), \tag{8}$$

where $\widehat{\mathbf{R}} \in \mathbb{R}^{m \times m}$ is the normalization of $\mathbf{R}$. $\mathbf{W}_1^f \in \mathbb{R}^{d \times d}$ and $\mathbf{W}_2^f \in \mathbb{R}^{d \times 2}$ are learnable parameter matrices. Thus, the classification loss can be obtained by:

$$\mathcal{L}_{cls} = \sum_{v_i \in \mathcal{V}} \ell\left(f(\mathbf{H}, \mathbf{R})_i, y_i\right), \tag{9}$$

where ℓ is the cross-entropy loss, measuring the difference between the predicted label $f(\mathbf{H}, \mathbf{R})_i$ and the true label y_i. In summary, we can get the overall loss $\mathcal{L}$:

$$\mathcal{L} = \mathcal{L}_{cls} + \alpha\mathcal{L}_d^p + \beta\mathcal{L}_d^e + \gamma\mathcal{L}_{\mathrm{reg}}, \mathcal{L}_{\mathrm{reg}} = \|\mathbf{R}\|_1, \tag{10}$$

where $\mathcal{L}_{\mathrm{reg}}$ is a regularization term. α, β and γ are trade-off parameters that control the contributions of each loss.

3.6   Iterative Optimization

We denote the parameters of feature transform, latent relation miner and fake news detector in INRM as Θ, Ω and Γ, respectively. The constraints on $\mathbf{S}^p$ and $\mathbf{S}^e$, and the dependence between $\mathbf{S}^p$ or $\mathbf{S}^e$ and other parameters, make joint optimization of all parameters in Eq. (10) extremely difficult. Thus, we use an alternating optimization strategy to iteratively update Θ, Ω, Γ, $\mathbf{S}^p$ and $\mathbf{S}^e$.

Table 1. Performance comparison of INRM and baselines (Twitter15 and Twitter16 for *Task I*, PHEME18 for *Task II*).

	Method	Twitter15				Twitter16				PHEME18			
		Acc.	Prec.	Rec.	F1	Acc.	Prec.	Rec.	F1	Acc.	Prec.	Rec.	F1
G1	DTC	0.6024	0.6031	0.6025	0.6018	0.6533	0.6555	0.6533	0.6521	0.5746	0.5412	0.5405	0.5246
	CNN-ML	0.8558	0.8619	0.8557	0.8550	0.8762	0.8826	0.8764	0.8755	0.6082	0.5038	0.4923	0.4288
	mGRU	0.8612	0.8622	0.8611	0.8610	0.8760	0.8784	0.8760	0.8758	0.6386	0.5241	0.5213	0.4892
	EANN	0.8840	0.8868	0.8840	0.8838	0.9005	0.9039	0.9006	0.9003	0.6613	0.6955	0.7523	0.7079
	BERT	0.9125	0.8985	0.9246	0.9099	0.9271	0.9484	0.8958	0.9158	0.6606	0.7092	0.7453	0.6972
G2	dEFEND	0.8759	0.8791	0.8760	0.8757	0.9070	0.9103	0.9072	0.9069	0.6401	0.5022	0.5215	0.4833
	GCAN	0.8767	0.8257	0.8295	0.8250	0.9084	0.7594	0.7632	0.7593	–	–	–	–
	KAN	0.9250	0.9019	0.9397	0.9206	0.9363	0.9306	0.9408	0.9322	0.6513	0.6724	0.7364	0.7069
	BiGCN	0.9383	0.9308	0.9404	0.9349	0.9422	0.9339	0.9451	0.9388	0.6688	0.7079	0.7461	0.7161
	EBGCN	0.9410	0.9621	0.9266	0.9429	0.9505	0.9512	0.9506	0.9505	0.6789	0.7115	0.7975	0.7330
Ours	INRM	**0.9596**	**0.9776**	**0.9410**	**0.9587**	**0.9758**	**0.9902**	**0.9614**	**0.9753**	**0.7376**	**0.7611**	**0.8689**	**0.7976**

Update Θ, Ω, Γ. To update Θ, Ω and Γ, we fix $\mathbf{S}^p$ and $\mathbf{S}^e$ for the function:

$$\min_{\Theta, \Omega, \Gamma} \mathcal{L}_{cls}\left(\mathbf{X}, \mathbf{A}^p, \mathbf{A}^e, \mathcal{Y}; \Theta, \Omega, \Gamma\right) + \gamma \mathcal{L}_{\text{reg}}\left(\mathbf{X}, \mathbf{A}^p, \mathbf{A}^e; \Theta, \Omega\right). \tag{11}$$

Update $\mathbf{S}^p, \mathbf{S}^e$. To update $\mathbf{S}^p$ and $\mathbf{S}^e$, we fix Θ, Ω and Γ for the function:

$$\min_{\mathbf{S}^p, \mathbf{S}^e} \mathcal{L}_{cls}\left(\mathbf{X}, \mathbf{A}^p, \mathbf{A}^e, \mathcal{Y}; \mathbf{S}^p, \mathbf{S}^e\right) + \alpha \mathcal{L}_d^p\left(\mathbf{X}, \mathbf{A}^p; \mathbf{S}^p\right)$$
$$+ \beta \mathcal{L}_d^e\left(\mathbf{X}, \mathbf{A}^e; \mathbf{S}^e\right) + \gamma \mathcal{L}_{reg}\left(\mathbf{X}, \mathbf{A}^p, \mathbf{A}^e; \mathbf{S}^p, \mathbf{S}^e\right). \tag{12}$$

With these updating rules, we first initialize the model parameters. Then, we update all parameters alternatively and iteratively. Specifically, we update $\{\Theta, \Omega, \Gamma\}$ with learning rate η_1 every τ iterations, and update $\{\mathbf{S}^p, \mathbf{S}^e\}$ with learning rate η_2 every υ iterations.

4 Experiment

4.1 Experimental Setup

Datasets and Baselines. We evaluate the proposed INRM on three real-world datasets: **Twitter15**, **Twitter16** [11] and **PHEME18** [7]. In this paper, we only pick out "true" and "fake" labels as the ground truth. We compare the proposed INRM with ten baselines, which can be divided into two groups: 1) News content-based methods (**G1**): **DTC** [3], **CNN-ML** [6], **EANN** [13], **mGRU** [10], and **BERT** [4]. 2) Auxiliary knowledge-based methods (**G2**): **dEFEND** [12], **GCAN** [9], **KAN** [5], **BiGCN** [1], and **EBGCN** [14].

Implementation. We utilize two-layer GCN as the backbone. For the Twitter15, Twitter16 and PHEME18 datasets: the hidden dimension is $\{32, 64, 128\}$; the learning rate η_1 is initialized to $\{0.01, 0.001, 0.01\}$, η_2 is initialized to $\{0.001, 0.001, 0.01\}$ and the weight decay is $5e-2$; the dropout rate of 0.4 is applied; k of connectivity enhancement is set to 50 and the number of latent relation types T is set to

Table 2. Ablation study on the effect of each prior graph and each module in INRM, measured by accuracy.

| | Method | Twitter15 | Twitter16 | PHEME18 |
		Acc.	Acc.	Acc.
	INRM	**0.9596**	**0.9758**	**0.7376**
Graph	-Pub	0.9233	0.9563	0.7049
	-Eng	0.9273	0.9490	0.6926
	-Pub&Eng	0.9179	0.9466	0.6828
Module	-CE	0.9354	0.9443	0.6979
	-De	0.9327	0.9515	0.6860
	-LRM	0.9192	0.9394	0.6748
	-CE&De&LRM	0.7400	0.8422	0.6314

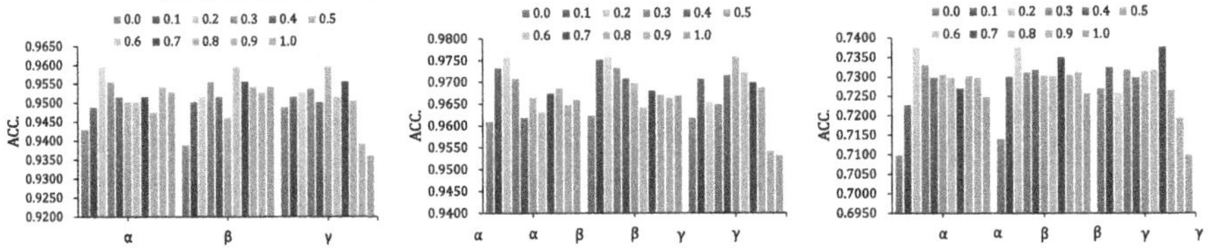

(a) Twitter15   (b) Twitter16   (c) PHEME18

Fig. 2. Sensitivity analysis with respect to trade-off parameters α, β, γ.

2; the parameters are updated with early stopping based on the Adam optimizer, where the iteration steps τ and υ are set to 5 and 10 respectively. For the grid search space of key hyper-parameters, trade-off parameters α, β, γ are searched in $[0.0, 1.0]$, engagement graph threshold ξ is chosen from 1 to 5, and similarity threshold ϵ is tuned amongst $[0.0, 0.5]$.

Evaluation Settings and Metrics. We utilize Twitter15 and Twitter16 datasets for *conventional fake news detection* tasks (*Task I*), which are randomly split into five parts for 5-fold cross-validation. Besides, we use PHEME18 dataset involving nine newly emerged events for the task of *fake news detection on newly emerged events* (*Task II*), while constructing leave-one-event-out cross-validation. The evaluation metrics for the three datasets are the same, including Accuracy, Precision, Recall and F1-measure.

4.2 Performance Comparison

Table 1 shows the results of the proposed INRM and baselines on Twitter15, Twitter16 and PHEME18 datasets. It can be observed that INRM consistently outperforms the baselines on *task I* and *task II*. In general, the INRM exploitsprior

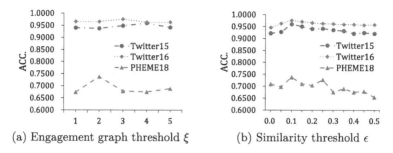

Fig. 3. The effect of the engagement graph threshold ξ and the similarity threshold ϵ on the model performance.

knowledge and limited auxiliary knowledge as guiding signals to mine inter-news relations, which are able to supply key clues for conventional fake news detection tasks and the task of fake news detection on newly emerged events.

4.3   Model Analysis

Ablation Study. We construct two ablative groups to report how each prior graph and each component contribute to the INRM. By removing the publication graph, engagement graph and both of them, "-Pub", "-Eng" and "-Pub&Eng" are obtained. "-CE", "-De", "-LRM" and "-CE&De&LRM" respectively denote the elimination components of connectivity enhancement, denoise, latent relation miner and all of them. The results presented in Table 2 indicate that each prior graph and each component indeed play an important role.

Sensitivity Analysis. To explore the sensitivity of INRM, we alter the values of hyper-parameters: 1) Trade-off parameters α, β, γ. As shown in Fig. 2, when the values of α and β approach to 0, undesirable noise is introduced due to the unreliable relations. In addition, excessive γ greatly hurts the performance, because it results in a sparse graph, making the lack of sufficient inter-news relations for validating news authenticity. 2) Threshold ξ, ϵ. The optimal engagement graph threshold ξ varies with different datasets as shown in Fig. 3(a). It can be observed from Fig. 3(b) that the model performance is likely to decrease as ϵ increases, indicating the importance of effectively learning latent relations.

5   Conclusion

In this paper, we introduce inter-news relations to solve scenarios with little auxiliary knowledge and newly emerged events. Specifically, we propose a Inter-News Relation Mining (INRM) framework to mine inter-news relations, which can provide more clues to verify the truth of news. Experiments on real-world datasets demonstrate the effectiveness of INRM for fake news detection on both conventional tasks and newly emerged event tasks.

Acknowledgements. This work is supported by the Youth Innovation Promotion Association, Chinese Academy of Sciences (No. 2020163).

References

1. Bian, T., et al.: Rumor detection on social media with bi-directional graph convolutional networks. In: Proceedings of the AAAI Conference on Artificial Intelligence, pp. 549–556 (2020)
2. Cao, J., Qi, P., Sheng, Q., Yang, T., Guo, J., Li, J.: Exploring the role of visual content in fake news detection. In: Shu, K., Wang, S., Lee, D., Liu, H. (eds.) Disinformation, Misinformation, and Fake News in Social Media. LNSN, pp. 141–161. Springer, Cham (2020). https://doi.org/10.1007/978-3-030-42699-6_8
3. Castillo, C., Mendoza, M., Poblete, B.: Information credibility on Twitter. In: Proceedings of the 20th International Conference on World Wide Web, pp. 675–684 (2011)
4. Devlin, J., Chang, M.W., Lee, K., Toutanova, K.: BERT: pre-training of deep bidirectional transformers for language understanding. arXiv preprint arXiv:1810.04805 (2018)
5. Dun, Y., Tu, K., Chen, C., Hou, C., Yuan, X.: KAN: knowledge-aware attention network for fake news detection. In: Proceedings of the AAAI Conference on Artificial Intelligence, pp. 81–89 (2021)
6. Goldani, M.H., Safabakhsh, R., Momtazi, S.: Convolutional neural network with margin loss for fake news detection. Inf. Process. Manag. **58**(1), 102418 (2021)
7. Kochkina, E., Liakata, M., Zubiaga, A.: All-in-one: multi-task learning for rumour verification. arXiv preprint arXiv:1806.03713 (2018)
8. Li, P., Sun, X., Yu, H., Tian, Y., Yao, F., Xu, G.: Entity-oriented multi-modal alignment and fusion network for fake news detection. IEEE Trans. Multimedia **24**, 3455–3468 (2021)
9. Lu, Y.J., Li, C.T.: GCAN: graph-aware co-attention networks for explainable fake news detection on social media. In: Proceedings of the 58th Annual Meeting of the Association for Computational Linguistics, pp. 505–514 (2020)
10. Ma, J., et al.: Detecting rumors from microblogs with recurrent neural networks (2016)
11. Ma, J., Gao, W., Wong, K.F.: Detect rumors in microblog posts using propagation structure via kernel learning. Association for Computational Linguistics (2017)
12. Shu, K., Cui, L., Wang, S., Lee, D., Liu, H.: dEFEND: explainable fake news detection. In: Proceedings of the 25th ACM SIGKDD International Conference on Knowledge Discovery & Data Mining, pp. 395–405 (2019)
13. Wang, Y., et al.: EANN: event adversarial neural networks for multi-modal fake news detection. In: Proceedings of the 24th ACM SIGKDD International Conference on Knowledge Discovery & Data Mining, pp. 849–857 (2018)
14. Wei, L., Hu, D., Zhou, W., Yue, Z., Hu, S.: Towards propagation uncertainty: edge-enhanced Bayesian graph convolutional networks for rumor detection. In: Proceedings of the 59th Annual Meeting of the Association for Computational Linguistics and the 11th International Joint Conference on Natural Language Processing (Volume 1: Long Papers), pp. 3845–3854 (2021)
15. Yang, R., Ma, J., Lin, H., Gao, W.: A weakly supervised propagation model for rumor verification and stance detection with multiple instance learning. arXiv preprint arXiv:2204.02626 (2022)

Cost-Effective Clustering by Aggregating Local Density Peaks

Wen-Bo Xie[1(✉)], Bin Chen[1], Jun-Hao Shi[1], Yan-Li Lee[2], Xin Wang[1], and Xun Fu[1]

[1] School of Computer Science, Southwest Petroleum University,
Chengdu 610500, China
wenboxie@swpu.edu.cn
[2] School of Computer and Software Engineering, Xihua University,
Chengdu 610039, China

Abstract. Hierarchical clustering algorithms that provide tree-shaped results can be regarded as data summarization and thus play an important role in the application of knowledge discovery and data mining. However, such structured result also brings a challenge, i.e., a difficult trade-off between complexity (time and space) and quality. To tackle of this issue, we propose a newly designed agglomerative algorithm for hierarchical clustering in this paper, which merges data points into tree-shaped sub-clusters via the operations of nearest-neighbor chain searching and determines the proxy of each sub-cluster by the process of local density peak detection. Extensive experimental studies on real-world and synthetic datasets show that our method performs well by outperforming other baselines in accuracy, response time, and memory footprint. Meanwhile, our method can scale to half a million data points on a personal computer, further verifying its cost-effectiveness.

Keywords: Hierarchical Clustering · Cost-effective Clustering · Nearest-Neighbor Graph · Density Peak

1 Introduction

Hierarchical clustering algorithms can provide tree-shaped results, a.k.a. cluster trees, which are usually regarded as the generative models of data or the summaries of data. In recent years, innovations in new technologies such as 5G and Industry 4.0 have dramatically increased the scale of data, posing new challenges to hierarchical clustering algorithms. On one hand, efforts to improve the scalability have been made in many researches, but clustering accuracy and computational costs

Corresponding author at: School of Computer Science, Southwest Petroleum University, Chengdu 610500, China. E-mail: wenboxie@swpu.edu.cn (Wen-Bo Xie). This work is supported by the Young Scholars Development Fund of SWPU under Grant No. 202199010142.

X. Wang et al. (Eds.): DASFAA 2023, LNCS 13946, pp. 63–73, 2023.
https://doi.org/10.1007/978-3-031-30678-5_5

always go against each other. On the other hand, most of the clustering results provided by hierarchical algorithms are presented by binary trees such as dendrogram [1] and CF-Tree [2]. The increase in data size will sharply rise the number of layers in the cluster tree, resulting in a significant decrease in the readability of the cluster tree in large-scale data. To tackle the above issues, we propose a cost-effective, agglomerative framework for hierarchical clustering that does not sacrifice quality and scales to a million data points on a personal computer, called ALDP (short for Clustering by **A**ggregating **L**ocal **D**ensity **P**eaks).

Our contributions are summarized as follows: (1) We propose a newly designed agglomerative hierarchical clustering algorithm to significantly reduce the number of layers in the cluster tree with a low time cost. (2) We devise an optimization method to refactor the isolated sub-clusters trees, thus can improve the clustering accuracy of ALDP. (3) Using real-world and synthetic datasets, we experimentally study the performances of our algorithm and find the followings. (a) Our algorithm shows better clustering accuracy in most cases. (b) Our algorithm scales much better than its counterparts, w.r.t. response time and memory footprint on incremental synthetic datasets.

2   Preliminary

We first review several important concepts that are going to be frequently used in the later parts:

Definition 1 Cluster Tree [3]. A cluster tree $\mathcal{T}$, of a dataset $X = \{x_i\}_1^n$, is a collection of subsets (a.k.a. sub-cluster trees, **SCTs** for short) $\tau_0 \triangleq \{x_i\}_1^n \in \mathcal{T}$ and for any $\tau_i, \tau_j \in \mathcal{T}$, either $\tau_i \subset \tau_j$, $\tau_j \subset \tau_i$ or $\tau_i \cap \tau_j = \varnothing$. In particular, there must exist a set of disjoint SCTs $\{\tau_i | i = 1, 2, \ldots, k\}$ such that $\bigcup_{i=1}^k \tau_i = \mathcal{T}$, meanwhile, $\forall \tau^* \in \mathcal{T}$, there also exist a set of disjoint STCs $\{\tau_i | i = 1, 2, \ldots, \ell\}$ satisfying $\bigcup_{i=1}^{\ell} \tau_i = \tau^*$.

Definition 2 Nearest-Neighbor-Chain (NN-Chain) [4]. Treating each data point as a node, a NN-Chain of length h, starting from node i, is denoted as $\delta_i^{(0)} \to \delta_i^{(1)} \to \cdots \to \delta_i^{(h)}$, in which $i = \delta_i^{(0)}$, $\delta_i^{(1)}$ is the nearest neighbor of $\delta_i^{(0)}$, $\delta_i^{(2)}$ is the nearest neighbor of $\delta_i^{(1)}$, and so on.

Definition 3 Reciprocal Nearest Neighbors (RNNs). Given two nodes i and j, and their nearest neighbors $\delta_i^{(1)}$ and $\delta_j^{(1)}$, respectively. If the condition $\delta_i^{(1)} = j \wedge \delta_j^{(1)} = i$ holds, we call (i, j) a pair of RNNs.

Definition 4 Isolated Sub-Cluster Tree (iSCT). If a SCT τ consists of only one pair of reciprocal nearest neighbors, we call τ an iSCT.

3   Proposed Approach

3.1   Framework

ALDP is an agglomerative algorithm that consists of three main tasks in one round of iteration: *SCTs Construction* (SCTsCons), *iSCTs Refactoring* (ISCTs REF), and *Roots Detection* (ROOTSDET).

As shown in Algorithm 1, taking the data D, a parameter α, and the iteration times t as input, the labels of data as output, ALDP uses all the nodes to initialize the roots set (line 1), then starts the iteration (lines 2–9). Firstly, ALDP obtains a collection of SCTs $\mathcal{T}$ by SCTsCons (line 3). Next, ALDP refactors iSCTs and thus updates the $\mathcal{T}$ by iSCTREF (line 4). Then, the root set $\mathfrak{r}$ will be updated by textscRootsDet (line 5). The iteration will stop according to the given t, or break when only one root is left (6–7), indicating that the full cluster tree is obtained. Finally, ALDP tags the nodes according to the roots and returns the labels (lines 9–10).

Algorithm 1. ALDP

    Input Data: D, Cutoff Threshold: α, Iteration Times: t
    Output Labels: L
1: $\mathfrak{r} \leftarrow D$
2: **for** $i{=}0$ **to** t **do**
3:     $\mathcal{T} \leftarrow$ SCTsCons($\mathfrak{r}$)
4:     update $\mathcal{T}$ with iSCTREF($\mathcal{T}$)
5:     update $\mathfrak{r}$ with RootsDet($\mathcal{T}, \alpha$)
6:     **if** $|\mathfrak{r}| = 1$ **then break**
7:     **end if**
8: **end for**
9: $L \leftarrow$ LABELING($\mathfrak{r}, \mathcal{T}$)   ▷ Labeling the data according to the roots
10: **return** L

Example 1. As shown in Fig. 1 (a), applying the three main steps, ALDP links nodes and aggregates them into seven sub-clusters. Then, treating detected roots as the proxies of sub-clusters, ALDP performs the next round of aggregation, and obtaining two clusters, as shown in Fig. 1 (b). Finally, ALDP constructs the cluster tree with tree layers, as shown in Fig. 1 (c).

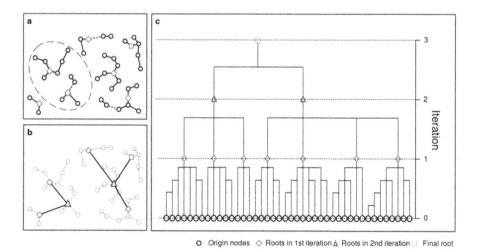

○ Origin nodes   ◇ Roots in 1st iteration △ Roots in 2nd iteration ▢ Final root

Fig. 1. Example of the iteration of ALDP.

Next, we introduce the detailed information of the three main tasks.

3.2   SCTs Construction

This step is the key to generate tree-shaped sub-clusters, in which, we generate SCTs via searching NN-Chains. Starting from an arbitrary point i that has not traversed, the searching process is completed when the following conditions meet: (a) a pair of RNNs is found, i.e., we have $\delta^{(h)} = \delta^{(h-2)}$; or (b) $\delta^{(h)}$ has been traversed. If the condition (a) holds, it indicates that a chain-shaped SCT is generated, while, if the condition (b) holds, it indicates that this NN-Chain is linked to an existing SCTs as a new branch.

3.3   iSCTs Refactoring

Based on the SCTs Construction step, a collection of SCTs are generated, most of which consist of many nodes. However, some iSCTs that consist of only a pair of RNNs (see Definition 4) are also generated. The iSCTs interfere with the Root Detection process, because, for a pair of RNNs nodes, their local densities are equal. In order to reduce the impact of iSCTs on clustering accuracy, we design ISCTsREF method to refactor iSCTs before Roots Detecting, as shown in Algorithm 2.

Algorithm 2. ISCTsREF

Input a collection of SCTs: $\mathcal{T}$
Output $\mathcal{T}$
1: extract iSCTs $\mathcal{T}^*$ from $\mathcal{T}$
2: **for each** $\tau = \{(x, y)\}$ in $\mathcal{T}^*$ **do**
3:      obtain $\delta_x^{(2)}$ and $\delta_y^{(2)}$
4:      determine $\tau_x \in \mathcal{T}$ and $\tau_y \in \mathcal{T}$, satisfying $\delta_x^{(2)} \in \tau_x$ and $\delta_y^{(2)} \in \tau_y$
5:      **if** $|\tau_x| \le |\tau_y|$ **then**
6:          update τ_x with $\tau_x \cup dege(\delta_x^{(2)}, x) \cup dege(x, y)$
7:      **else**
8:          update τ_y with $\tau_y \cup dege(\delta_y^{(2)}, y) \cup dege(y, x)$
9:      **end if**
10: **end for**
11: **return** updated $\mathcal{T}$

Taking a collection of SCTs $\mathcal{T}$ as input, updated $\mathcal{T}$ as output, ISCTsREF extracts iSCTs $\mathcal{T}^*$ from $\mathcal{T}$ (line 1) and treats each iSCTs $\tau \in \mathcal{T}^*$ with RNNs (x, y) as follows (lines 2–12): ISCTsREF first finds the 2nd-nearest-neighbors $\delta_x^{(2)}$ (resp. $\delta_y^{(2)}$) of x (resp. y) and the corresponding SCT τ_x (resp. τ_y) (lines 3–4). Then, ISCTsREF compares the size of two STCs τ_x and τ_y. If the condition $|\tau_x| \le |\tau_y|$ holds, x and y will be added to SCT τ_x; otherwise, x and y will be added to SCT τ_y (lines 5–10). Finally, ISCTsREF returns the updated collection $\mathcal{T}$. We illustrate this process in Example 2.

Example 2. As shown in Fig. 2, there are three SCTs, denoted by τ_A, τ_B, and τ_C, in which $\tau_C = (x, y)$ is an iSCT. We denote the 2nd-nearest-neighbor of x and y as $\delta_x^{(2)}$ and $\delta_y^{(2)}$, respectively. Because $|\tau_B| < |\tau_A|$, x and y are added to $|\tau_B|$ by linking y to $\delta_y^{(2)}$. Thus, we complete the refactoring of iSCTs in this example.

In this way, sub-clusters become more balance, and also, *Roots Detection* will work better. This is because using more distance samples can better distinguish the density differences between nodes, especially for those with high densities.

3.4 Roots Detection

The final task is to detect representative nodes, a.k.a. roots, which aims to determine the proxy for each sub-cluster tree. We achieve this by local density-based scoring. The node with the highest score will be regarded as the most representative node in the corresponding SCT and will be used as a proxy of the SCT for coarser hierarchical aggregation. Given a SCT τ and a node i therein, the local density-based score of i is calculated as,

$$\rho_i = \sum_{j \in \tau} \chi \left(d_{ij} - d_c \right), \tag{1}$$

in which, $\chi()$ is a polarization function,

$$\chi(x) = \begin{cases} 1, x < 0 \\ 0, x \geq 0 \end{cases} ;$$

d_{ij} is the distance between nodes i and j; and d_c is a dynamic cutoff distance that is related to the size of SCT τ, the pairwise distances in the SCT τ, and the global threshold α. Denoting the number of node pairs in SCT τ as $\beta = \frac{|\tau|(|\tau|-1)}{2}$, d_c in SCT τ equls to the $\lceil \alpha\beta \rceil^{th}$ smallest pairwise distance.

Based on the Eq. (1) we apply ROOTSDET to detect root in each SCT, as shown in Algorithm 3. Taking a collection of SCTs $\mathcal{T}$ and a cutoff threshold α as input, a set of roots $\mathfrak{r}$ as output, ROOTSDET initializes a set $\mathfrak{r}$ to store the detected roots (line 1), then starts the iteration (line 2–13). For each SCT $\tau \in \mathcal{T}$, ROOTSDET first calculates cutoff distance d_c in τ (line 3), then initializes

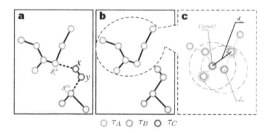

Fig. 2. Example of iSCTs Refactoring and Roots Detection

the max density score (line 4). Next, ROOTSDET calculates the density score of each node i in τ, and records the node with highest score (lines 5–11). Thereby, ROOTSDET addes the root, the local density peak in τ, in the set $\mathfrak{r}$ (line 12). Finally, ROOTSDET returns a set of roots (line 14). We illustrate this process in Exmaple 3.

Example 3. As shown in Fig. 2 (c), taking SCT τ_A as an example, we explain how ROOTSDET works. First, ROOTSDET sets the 6th smallest pairwise distance (denoted by the black line) as the cutoff distance, since there are 7 nodes and α is set to 0.25. Next, ROOTSDET calculates the local density of each node in τ_A. Finally, ROOTSDET selects node i as the root. This is because there are 4 nodes whose distance from node i is less than the cutoff distance d_c.

4   Experimental Study

Using real-world and synthetic datasets, we conduct comprehensive experimental studies to evaluate: efficiency, memory cost, effectiveness (RI, ARI and NMI) and scalability of our algorithm ALDP, compared with baseline methods.

Algorithm 3. ROOTSDET

Input a collection of SCTs: $\mathcal{T}$, Cutoff Threshold: α
Output A set of Roots: $\mathfrak{r}$
1: $\mathfrak{r} \leftarrow \varnothing$
2: **for each** τ in $\mathcal{T}$ **do**
3:      $d_c =\leftarrow \lceil \alpha\beta \rceil^{th}$ smallest pairwise distances in τ $\triangleright \beta = \frac{|\tau|(|\tau|-1)}{2}$
4:      $max_d \leftarrow 0; r \leftarrow \varnothing$
5:      **for each** i in τ **do**
6:          $\rho_i = \sum_{j \in \tau} \chi(d_{ij} - d_c)$ $\triangleright$ Eq. (1)
7:          **if** $\rho > max_d$ **then**
8:              $max_d \leftarrow \rho_i$
9:              $r \leftarrow i$
10:         **end if**
11:     **end for**
12:     update $\mathfrak{r}$ with $\mathfrak{r} \cup r$
13: **end for**
14: **return** $\mathfrak{r}$

4.1   Experimental Setting

Real-World Datasets. Ten real-world datasets selected from the University of California Irvine (UCI) [5] are used in the experiments, which represent a wide range of domains and data characteristics. Two large-scale datasets from different sources are also used: *CovType* - forest cover types [3] and *ALOI* - 3D rendering of objects, ground truth clusters refer to each object type [6]. The basic information of these datasets are shown in Table 1.

Synthetic Datasets. All synthetic datasets used in the test are generated by the well-known toolkit "sklearn" in Python, each of which has a dimension of 10 and a size of 2^n, where $n = \{n | n = 9, 10, \cdots, 17, n \in N^*\}$.

Table 1. The basic information of datasets

No.	Dataset	#Samples	#Attributes	#Classes	No.	Dataset	#Samples	#Attributes	#Classes
1	glass	214	9	6	2	vehicle	846	18	4
3	mfeat-fourier	2000	76	10	4	mfeat-karhunen	2000	64	10
5	mfeat-zernike	2000	47	10	6	segment	2310	18	7
7	optdigits	5620	62	10	8	waveform-5000	5000	40	3
9	letter	20000	16	26	10	avila	20867	10	12
11	ALOI	108 K	128	1000	12	CovType	500 K	54	7

Implementations. We implement algorithm ALDP and the following counterparts all in Python.

- Our proposed algorithm, ALDP. For UCI datasets, we get the optimized results by varying α from 0 to 0.75 in 0.05 increments. Meanwhile, for large-scale datasets, we set $\alpha = 0.5$.
- Well-known counterparts, i.e., PERCH [3], AHC [7], SCC [8], DD [9]. We set the parameters according to the recommended ones in the original papers.

In our test, the testbed includes a machine with 4.2 GHz CPU and 32 GB RAM, running python v3.9 on Windows 11. Each test runs 50 times and the average is reported.

Evaluation Metrics. We employ the rand index (RI) [10] (ranges from 0 to 1), normalized mutual information (NMI) [11] (ranges from 0 to 1) and adjusted rand index (ARI) [12] (ranges from -1 to 1) as evaluation metrics.

4.2 Experimental Results

Exp. 1: Sensitivity of Parameter α. Aiming at observing the sensitivity of parameter α, we adjust the parameter α from 0 to 0.75 in 0.05 increments to show the impact of parameter α on the clustering accuracy of our proposed algorithm. As shown in Fig. 3, the fluctuations in Rand Index are small on datasets *mefeat-ze.*, *avlia* and *letter*, but sharp on the other datasets.

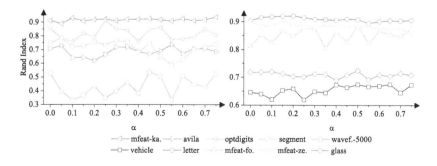

Fig. 3. Sensitivity of parameter α

Exp. 2: Clustering Accuracy. Table 2 shows the clustering accuracy (measured by RI, ARI, and NMI) of all benchmarks on the UCI datasets, in which ALDP(opt) represents the best performance of ALDP, ALDP(0.1) and ALDP(0.2)

Table 2. Clustering results on ten UCI datasets and two large-scale datasets.
The best results are highlighted in bold and the second best results are underlined.

DataSet	Algorithm	RI	ARI	NMI	DataSet	Algorithm	RI	ARI	NMI
glass	ALDP(opt)	**0.7195**	**0.2022**	**0.3203**	vehicle	ALDP(opt)	**0.6689**	**0.1343**	**0.1828**
	ALDP(0.1)	**0.7195**	0.1642	<u>0.3054</u>		ALDP(0.1)	0.6205	0.0679	0.1082
	ALDP(0.2)	<u>0.7030</u>	0.1390	0.2951		ALDP(0.2)	<u>0.6591</u>	0.0836	0.1308
	RSC	0.6307	0.0741	0.2122		RSC	0.6336	0.0557	0.1100
	AHC	0.5455	<u>0.1951</u>	0.2940		AHC	0.5346	<u>0.0916</u>	<u>0.1505</u>
	SCC	0.4411	< 0.0001	0.0453		SCC	0.5089	< 0.0001	0.0263
	DD	0.4330	< 0.0001	0.0448		DD	0.5556	0.0086	0.0030
	PERCH	0.6494	0.0023	0.0370		PERCH	0.6256	0.0004	0.0101
mfeat-fo.	ALDP(opt)	**0.8868**	**0.5533**	**0.5622**	mfeat-ka.	ALDP(opt)	**0.9513**	**0.7370**	**0.8018**
	ALDP(0.1)	0.8251	<u>0.4540</u>	0.4689		ALDP(0.1)	<u>0.9311</u>	<u>0.6265</u>	<u>0.7465</u>
	ALDP(0.2)	<u>0.8476</u>	0.4455	<u>0.5101</u>		ALDP(0.2)	0.9191	0.5932	0.7365
	RSC	0.8367	0.3195	0.5046		RSC	0.8830	0.3684	0.7110
	AHC	0.1818	0.1368	0.0816		AHC	0.5106	0.1030	0.3423
	SCC	0.1027	< 0.0001	0.1802		SCC	0.4478	< 0.0001	0.2847
	DD	0.4685	0.0109	0.1814		DD	0.6028	0.1018	0.2918
	PERCH	0.8181	0.0125	0.0878		PERCH	0.8177	0.0116	0.0745
mfeat-ze.	ALDP(opt)	**0.9126**	**0.4179**	**0.6685**	segment	ALDP(opt)	**0.8899**	**0.5538**	**0.6768**
	ALDP(0.1)	0.8855	0.2810	0.5898		ALDP(0.1)	0.7232	0.2984	0.4536
	ALDP(0.2)	0.8862	<u>0.3280</u>	0.5869		ALDP(0.2)	0.7580	<u>0.3466</u>	0.5155
	RSC	<u>0.8893</u>	0.2178	<u>0.5903</u>		RSC	0.7232	0.1722	<u>0.5329</u>
	AHC	0.5733	0.0040	0.4016		AHC	<u>0.7976</u>	0.2394	0.3535
	SCC	0.5779	< 0.0001	0.2856		SCC	0.5107	< 0.0001	0.1072
	DD	0.3633	0.0580	0.2918		DD	0.5177	0.0527	0.0039
	PERCH	0.8167	0.0084	0.0684		PERCH	0.7117	< 0.0001	0.0062
optdigits	ALDP(opt)	**0.9328**	**0.6853**	**0.8191**	wavef.	ALDP(opt)	**0.6048**	<u>0.1771</u>	**0.2126**
	ALDP(0.1)	0.7589	0.3604	0.5828		ALDP(0.1)	0.3332	< 0.0001	0.0001
	ALDP(0.2)	0.7929	0.3680	0.6255		ALDP(0.2)	0.4349	0.0582	0.0745
	RSC	<u>0.8623</u>	0.3341	<u>0.7013</u>		RSC	0.5523	0.1065	<u>0.1853</u>
	AHC	0.8559	<u>0.5517</u>	0.6821		AHC	0.3336	**0.2465**	0.0012
	SCC	0.6617	< 0.0001	0.0022		SCC	0.4902	< 0.0001	0.0012
	DD	0.1024	< 0.0001	0.0029		DD	0.3452	0.0082	0.0001
	PERCH	0.8171	< 0.0001	0.0041		PERCH	<u>0.5542</u>	< 0.0001	0.0007
letter	ALDP(opt)	**0.9344**	**0.1954**	**0.4389**	avila	ALDP(opt)	**0.7436**	**0.0670**	**0.1833**
	ALDP(0.1)	0.9183	0.1496	0.3820		ALDP(0.1)	0.6407	<u>0.0478</u>	0.1310
	ALDP(0.2)	0.9207	<u>0.1522</u>	<u>0.3971</u>		ALDP(0.2)	0.6199	0.0146	0.0720
	RSC	0.6307	0.0775	0.3836		RSC	0.5271	0.0108	<u>0.1715</u>
	AHC	0.8121	0.0575	0.3054		AHC	0.2853	< 0.0001	0.0001
	SCC	0.4076	< 0.0001	0.0025		SCC	0.2334	0.0002	0.0007
	DD	0.0425	< 0.0001	0.0030		DD	0.2639	< 0.0001	0.0001
	PERCH	<u>0.9240</u>	< 0.0001	0.0051		PERCH	<u>0.7085</u>	0.0002	0.0015

represent the basic performance of ALDP by setting $\alpha = 0.1$ and $\alpha = 0.2$, respectively. Comparing with other baselines, we find that our proposed methods outperform other baselines in most cases, even with naive parameter settings ($\alpha = 0.1$ and $\alpha = 0.2$).

Table 3 shows the clustering results on two large-scale datasets, in which ALDP ($\alpha = 0.5$) is significantly superior to other baselines in terms of clustering accuracy (measured by RI, ARI and NMI). It is noted that the results for AHC and DD are absence because they took more than 24 h to run one time in our testbed.

Table 3. Clustering results on two large-scale datasets. The best results are highlighted in bold and the second best results are underlined.

Datasets	Algorithm	RI	ARI	NMI	Datasets	Algorithm	RI	ARI	NMI
ALOI	ALDP(0.5)	**0.9990**	0.3762	**0.8265**	CovType	ALDP(0.5)	**0.6237**	**0.0487**	**0.1668**
	RSC	0.9962	0.2565	0.8004		RSC	0.5645	0.0083	0.0461
	SCC	0.9976	< 0.0001	0.6915		SCC	0.4532	< 0.0001	0.0001
	PERCH	0.9988	**0.4055**	0.8156		PERCH	0.5997	0.0328	0.1089

Exp. 3: Efficiency. We verify the efficiency of our proposed algorithm by testing the response time on incremental size of synthetic datasets. Figure 4(a) shows the response time vs. data size of our ALDP as well as other baselines, in which the relationship between response time and data size for each algorithm is fitted by a power-law. Because all the algorithms used for comparison are of polynomial time-complexity, the fitted curves in Fig. 4(a) appear to be straight lines in double logarithmic coordinates, and the slopes of the lines (shown in the legend) equal to the exponential part of the corresponding power-low function. Observing Fig. 4(a), we can find the followings: (a) the time growth rate of our ALDP (1.69) is lower than that of DD (2.17) and AHC (2.4), indicating that our algorithm is more efficient than DD and AHC; (b) although the efficiency of our algorithm is slightly inferior to that of RSC(1.7), SCC (1.41) and PERCH (1.59), the accuracy of our algorithm is much higher than that of these baselines, especially on large-scale data (recall Tables 2 and 3).

Exp. 4: Scalability. Figure 4(b) shows the response time vs. memory footprint on different size of synthetic datasets, in which the numbers beside the dots represent the size of the dataset. Observing it, we can find the followings. (a) The required memory footprint of ALDP is the smallest among the comparison algorithms for the same size dataset, indecating the best scalability. (b) The response time of ALDP is close to RSC and significantly lower than the other baselines for the same size dataset. (c) In combination with the previous experiments (recalling Table 2 and Table 3), ALDP is of cost-effectiveness because it requires less memory footprint and response time to perform clustering, and it can yield better clustering accuracy.

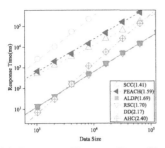
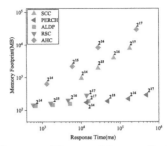

(a) Response Time vs. Data Size   (b) Response Time vs. Memory Footprint

Fig. 4. Computational costs on synthetic datasets

5   Related Works

According to the internal mechanisms, hierarchical clustering can be divided into top-down clustering algorithms (divisive algorithms) [13,14] and bottom-up clustering algorithms (agglomerative algorithms). In this paper, we focus on the agglomerative ones.

The core of agglomerative algorithms is pairwise distance calculation, which requires the majority of the total time cost. To address this issue, some algorithms regard representatives as proxies for sub-clusters to reduce computational cost, such as CURE [15] and K-Centroid Link method [16]. Recently, some NN-Chain-based algorithms are proposed to reduce the number of iterations of the aggregation process, such as the Sub-Cluster Component (SCC) [8] and the Reciprocal-nearest-neighbors Supported Clustering (RSC) [4]. SCC introduces a range of hyper-parameters to control the aggregation. RSC regards the midpoint of RNNs as the representative of a sub-cluster, thus further speeding up the iterations. Another line to handle large-scale data is incremental clustering, such as PERCH [3] and GRINCH [2]. Based on CF-tree construction, these two algorithms organize the data into a binary tree and improve the clustering accuracy by optimizing the tree structure.

6   Conclusion

In this paper, we propose a newly designed agglomerative hierarchical clustering method, in which sub-cluster trees are constructed by nearest-neighbor-chain searching, and the representative of each sub-cluster tree is selected via the local density peak detection. Extensive experiments verify that the clustering accuracy of our algorithm is superior to other baselines in most cases. Besides, our algorithm requires less time and less memory in the experiments, indicating that our algorithm is of great cost-effectiveness.

References

1. Gagolewski, M., Bartoszuk, M., Cena, A.: Genie: a new, fast, and outlier-resistant hierarchical clustering algorithm. Inf. Sci. **363**, 8–23 (2016)
2. Monath, N., Kobren, A., Krishnamurthy, A., Glass, M.R., McCallum, A.: Scalable hierarchical clustering with tree grafting. In: Proceedings of the 25th ACM SIGKDD, pp. 143–1448, New York, NY, USA (2019)
3. Kobren, A., Monath, N., Krishnamurthy, A., McCallum, A.: A hierarchical algorithm for extreme clustering. In: Proceedings of the 23rd ACM SIGKDD, pp. 255–264 (2017)
4. Xie, W.-B., Lee, Y.-L., Wang, C., Chen, D.-B., Zhou, T.: Hierarchical clustering supported by reciprocal nearest neighbors. Inf. Sci. **527**, 279–292 (2020)
5. Dua, D., Graff, C.: UCI Machine Learning Repository (2019)
6. Geusebroek, J.M., Burghouts, G.J., Smeulders, A.W.: The Amsterdam library of object images. Int. J. Comput. Vis. **61**(1), 103–112 (2005). https://doi.org/10.1023/B:VISI.0000042993.50813.60
7. Bouguettaya, A., Qi, Yu., Liu, X., Zhou, X., Song, A.: Efficient agglomerative hierarchical clustering. Expert Syst. Appl. **42**(5), 2785–2797 (2015)
8. Monath, N., et al.: Scalable hierarchical agglomerative clustering. In: Proceedings of the 27th ACM SIGKDD, pp. 1245–1255 (2021)
9. Rodriguez, A., Laio, A.: Clustering by fast search and find of density peaks. Science **344**(6191), 1492–1496 (2014)
10. Rand, W.M.: Objective criteria for the evaluation of clustering methods. J. Am. Stat. Assoc. **66**(336), 846–850 (1971)
11. Yang, Y., Shen, F., Huang, Z., Shen, H.T., Li, X.: Discrete nonnegative spectral clustering. IEEE Trans. Knowl. Data Eng. **29**(9), 1834–1845 (2017)
12. Hubert, L., Arabie, P.: Comparing partitions. J. Classif. **2**(1), 193–218 (1985)
13. Feng, L., Qiu, M.-H., Wang, Y.-X., Xiang, Q.-L., Yang, Y.-F., Liu, K.: A fast divisive clustering algorithm using an improved discrete particle swarm optimizer. Pattern Recognit. Lett. **31**(11), 1216–1225 (2010)
14. Han, X., Zhu, Y., Ting, K.M., Zhan, D.C., Li, G.: Streaming hierarchical clustering based on point-set kernel. In: Proceedings of the 28th ACM SIGKDD, pp. 525–533. Association for Computing Machinery (2022)
15. Guha, S., Rastogi, R., Shim, K.: CURE: an efficient clustering algorithm for large databases. ACM SIGMOD Rec. **27**(2), 73–84 (1998)
16. Dogan, A., Birant, D.: K-centroid link: a novel hierarchical clustering linkage method. Appl. Intell. **52**, 5537–5560 (2022)

Enlarge the Hidden Distance: A More Distinctive Embedding to Tell Apart Unknowns for Few-Shot Learning

Zhaochen Li[✉] and Kedian Mu

School of Mathematical Sciences, Peking University, Beijing, China
zhaochenli@pku.edu.cn, mukedian@math.pku.edu.cn

Abstract. Most few-shot classifiers assume consistency of the training and testing distributions. However, in many practical applications, the two distributions are often different. In this paper, we focus on the few-shot open-set recognition problem which allows that the testing categories are different from the training categories. To alleviate this problem, we take the semantic adhesion scenario as an example to analyze the influence of sample embedding vectors on the identification indicator value. Then, we propose an Extra Embedding Classification Model with an adjustment module that is trained with optional contrastive loss functions to learn distinctive features of samples in the same category. This model can enlarge the hidden distances among samples while keeping the category information. We comprehensively verified the effectiveness of our model on both the normal and the semantic adhesion scenario of the few-shot open-set recognition problem.

Keywords: Few-shot Learning · Open-Set Recognition · Data Embedding

1 Introduction

Few-shot learning has achieved appreciable results on many datasets [7]. However, most methods are devoted to improving the classification accuracy of closed-set problems where training samples and testing samples share the same data distribution. In practical applications, models often encounter the few-shot open-set recognition (FSOSR) problem where there are various categories of data in the test stage, which might be different from the distribution of the training set. This requires the classifier to have the ability to identify samples out of the known distribution.

In recent years, some methods have been proposed to solve the FSOSR problem [4–6]. However, once the relative positions of samples in the feature space are indistinguishable, they are difficult to distinguish out-of-distribution (OOD) samples and in-distribution (ID) samples by the indicator value. In this paper, we specify the difficult situation as the semantic adhesion scenario (SAS) where

X. Wang et al. (Eds.): DASFAA 2023, LNCS 13946, pp. 74–83, 2023.
https://doi.org/10.1007/978-3-031-30678-5_6

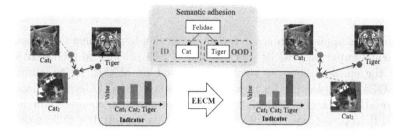

Fig. 1. EECM enlarges the relative distance between ID samples (Cat$_1$ and Cat$_2$) and OOD samples (Tiger) to increase their numerical differences in indicator values. This can further increase their numerical differences in indicator values, making it easier to distinguish the OOD samples from the known categories.

OOD samples and ID samples share the same superclass. As the left part of Fig. 1 shows, the difficulty of identifying the OOD samples in this scenario is attributed to the closeness of sample embeddings in the representation space. This leads to similar identification indicator values of OOD samples and ID samples, which makes it hard to distinguish. We have verified the problem through experiments. The results show that the feature embeddings of OOD samples provided by current classifiers are difficult to be distinguished from those of ID samples, especially in the SAS. The concrete experimental result is shown in Fig. 4(a) and more corresponding analysis will be introduced in Sect. 4.2.

Based on the above analysis, we propose an Extra-Embedding Classification Model (EECM) that computes two embedding vectors for each sample. The base embedding vector is trained to extract the semantic information of the category for classification. The extra embedding vector is used to learn the individual differences among images to enlarge the relative distance of different distribution samples in the representation space. In terms of model structure, EECM adds an adjustment module to extract the individual specific features of each sample. In terms of training strategy, we use two optional contrastive loss functions that make the model learn the differences among samples of the same categories to train the extra embedding vector for each image. (1) *Angular Normalized COntrastive Regularization (ANCOR)* [2] expands the angle among the representation vectors of the samples in the same category. (2) *Intra Category Contrastive Loss function (ICCL)* separates the relative distances among the samples in the same category in the representation space. To prevent the semantic representation ability of the network from being destroyed by the contrast loss function, we also add a cross-entropy loss function to learn the semantic information of ID categories.

We comprehensively evaluated the model on public datasets with different settings. The results show that our method makes improvements in different settings of experiments, including applying models with simple feature extractors and the SAS.

2   Related Work

Many methods applied technics in few-shot learning to overcome the difficulty of insufficient samples in FSOSR. For example, PEELER [6] and OOD-MAML [5] applied the episodic training strategy proposed by MAML [3] to sample the pseudo-OOD samples in the meta-training phase, SnaTCHer [4] adapts the transformation function trained by a popular few-shot classification method FEAT [14] to recognizes the OOD samples by comparing the differences after the transformation function. However, these methods ignore the impact of the relationship of the sample embeddings on FSOSR problem. Due to the powerful representation ability of contrastive learning, we put forward our model inspired by ANCOR [2] which is an angular normalized contrastive loss function that was proposed for the few-shot problem.

3   Proposed Method for FSOSR

We first introduce the setup of the FSOSR problem. As defined in most few-shot learning related problems, for K-way N-shot problem, the training task $\mathcal{T}^{tr}$ consists of support set $\mathcal{D}^s = \{x_i^s, y_i^s\}_{i=1}^{NK}$ and query set $\mathcal{D}^q = \{x_i^q, y_i^q\}_{i=1}^{MK}$ where M is the number of samples for each category in query set, and satisfies $\mathcal{D}^s \cap \mathcal{D}^q = \Phi$. The testing task $\mathcal{T}^{ts}$ is same as the training task except that the query set $\mathcal{D}^q = \mathcal{D}^{oq} \cup \mathcal{D}^{iq}$ where categories of $\mathcal{D}^{iq} = \{x_j^{iq}, y_j^{iq}\}_{j=1}^{MK}$ are the same as that of the support set but $\mathcal{D}^{oq} = \{x_j^{oq}, y_j^{oq}\}_{j=1}^{M'K'}$ is sampled from different categories which is regarded as OOD samples to test models' detection ability.

3.1   Extra-Embedding Classification Model

Our model consists of two branches to compute base embedding vectors and extra embedding vectors separately.

Base Embedding Vectors. As Fig. 2 shows, images are passed into backbone networks g_ϕ to obtain feature representations $\mathbf{z}$, i.e. $\mathbf{z} = g_\phi(x)$ where ϕ is the trainable parameters of the backbone. The base embedding vectors $\{\mathbf{z}^{(1)}\}$ (upper branch in Fig. 2) are computed by the feature representations after a pooling layer, i.e. $\mathbf{z}^{(1)} = \text{pooling}(\mathbf{z})$. The prototype for the c-th category is the average of the embedding vectors of the same category in the support set, shown as follows:

$$\mathbf{p}_c' = \frac{1}{N} \sum_{i=1}^{NK} \mathbf{z}_i^{s,(1)} \mathrm{I}(y_i^s = c), \tag{1}$$

where $\mathrm{I}(\cdot)$ is the indicative function, $\mathbf{z}_i^{s,(1)}$ represents the base embedding vector for the i-th image x_i^s with the label y_i^s in support set.

In order to adjust the prototypes to more discriminative positions in the embedding space, a set-to-set transform function $\mathcal{F}_\psi$ is conducted on the

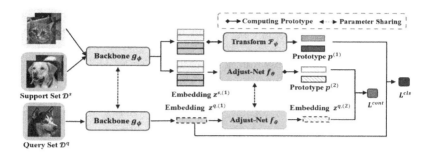

Fig. 2. The framework of the proposed method.

set of prototypes. Then we get the updated prototypes $\{\mathbf{p}_1^{(1)}, ..., \mathbf{p}_K^{(1)}\} = \mathcal{F}_\psi(\{\mathbf{p}_1', ..., \mathbf{p}_K'\})$.

In this paper, $\mathcal{F}_\psi$ is the same as the transformer module in FEAT [14]. Then the probability of (x^q, y^q) in the query set can be written as

$$P(y^q|x^q; \psi, \phi, \mathcal{D}^s)^{(1)} = \frac{\exp(-d(\mathbf{z}^{q,(1)}, \mathbf{p}_{y^q}^{(1)})/\tau)}{\sum_{c=1}^{K} \exp(-d(\mathbf{z}^{q,(1)}, \mathbf{p}_c^{(1)})/\tau)}, \tag{2}$$

where $d(\cdot, \cdot)$ is Euclidean distance, τ is the temperature parameter and $\mathbf{z}_i^{q,(1)} = \text{pooling}(g_\phi(x_i^q))$. The loss function for the base embeddings is the cross-entropy based on Eq. (2). The details can be referred to SnaTCHer [4].

Extra Embedding Vectors. The extra embedding vectors are also calculated from $\mathbf{z}$. In order to extract features with individual differences, an adjustment network module f_θ is proposed following the backbone (shown in Fig. 2 as 'Adjust-Net'), where θ is trainable parameters. In this paper, the structure of f_θ consists of a squeeze-and-excitation block and a self-attention layer. We summarize the specific calculation process as follows:

$$\mathbf{z}' = \text{pooling}(\text{Dense}(\text{pooling}(\mathbf{z}))) \odot \mathbf{z}, \tag{3}$$

where Dense is a 2-layer fully-connected network and $\odot$ is Hadamard product.

$$\mathbf{z}'' = \gamma \text{Cov}_3(\mathbf{z}')^T \sigma(\text{Cov}_1(\mathbf{z}')^T \text{Cov}_2(\mathbf{z}')) + \mathbf{z}', \tag{4}$$

where γ is a trainable parameter, Cov is a 2-d convolution layer followed by a flatten-layer and σ is a softmax function. Then, the extra embedding vector is computed as follows:

$$\mathbf{z}^{(2)} = \text{pooling}(\mathbf{z}''). \tag{5}$$

Similarly, the prototypes $\mathbf{p}_c^{(2)}$ of the extra embedding vectors and their probabilities $P(y^q|x^q; \psi, \phi, \mathcal{D}^s)^{(2)}$ are computed the same as Eq. (1) and Eq. (2).

In order to train the extra feature representations to disperse samples from each other in the embedding space, we can use two types of contrastive loss functions.

The angular normalized loss function (ANCOR) is:

$$\mathcal{L}^{ANCOR} = \mathbb{E}_{\mathcal{T}^{tr}} \mathbb{E}_{x \sim \mathcal{D}^q} - \log \frac{E(\angle x, \angle x^+)}{\sum_{j=1}^{M-1} E(\angle x, \angle x_j^-) + \sum_{j=1}^{Num} E(\angle x, \angle \hat{x}_j^-)}, \quad (6)$$

where $\angle x$ is the angular normalization of x (See ANCOR [2] for more details), x^+ and x^- are both sampled from the set of samples sharing the same label y with x. x and x^+ are two different augmentations of the same image, x^- is augmented from a different image, and $E(\angle x, \angle x^+) = \exp(-d(\angle x, \angle x^+)/\tau)$. In order to alleviate the effect of small batch size, we apply the decoupled contrastive learning [15], that is, the set $\{x_j^-\}$ doesn't contain x^+. An extra set of augmented samples $\hat{x}^-$ with scale Num is added to the few-shot contrastive function as shown in Eq. (6). The augmented samples are generated in a hidden layer where samples are embedded preliminarily by the backbone. The augmented feature vector $\hat{z}$ of $\hat{x}^-$ is formulated as:

$$\hat{z} = rg_\phi(x) + (1-r)g_\phi(x^r), \quad (7)$$

where $g_\phi(x^r)$ is sampled randomly from embedding vectors of both support and query samples the task, r is randomly sampled from the interval $[0.9, 1]$, and the label of $\hat{x}^-$ is the same as x.

Intra-Category Contrastive Loss (ICCL) is defined based on the conventional contrastive loss function with an adjustment on sampling negative data pairs. Inspired by ANCOR, in order to separate the samples in hidden space, we set the negative data pairs as samples of the same category. The contrastive loss function is formulated as follows:

$$\mathcal{L}^{ICCL} = \mathbb{E}_{\mathcal{T}^{tr}} \mathbb{E}_{x \sim \mathcal{D}^q} - \log \frac{E(x, x^+)}{\sum_{j=1}^{M-1} E(x, x_j^-) + \sum_{j=1}^{Num} E(x, \hat{x}_j^-)}, \quad (8)$$

where the augmented samples $\hat{x}$ is generated the same as Eq. (7). We also apply decoupled loss version here.

Finally, we obtain the full loss function as $\mathcal{L}^{cont} = \mathcal{L}^{cls} + \alpha \mathcal{L}^{ANCOR/ICCL}$, where $\mathcal{L}^{cls}$ is the cross-entropy classfication loss function computed by the extra embedding vectors, and α is a hyperparameter.

3.2   Detection Methods

At testing time, the value of the indicator is computed by both the two embedding vectors as follows:

$$\text{Ind}(\mathbf{z}) = \beta \text{Ind}(\mathbf{z}^{(1)}) + (1-\beta)\text{Ind}(\mathbf{z}^{(2)}), \quad (9)$$

where $\text{Ind}(\cdot)$ is the indicator, β is a hyperparameter. We can also calculate the value of the indicator by the concatenation of the two embedding vectors as $\text{Ind}(\mathbf{z}) = \text{Ind}(\text{CAT}(\mathbf{z}^{(1)}; \mathbf{z}^{(2)}))$, where $\text{CAT}(\cdot; \cdot)$ is the concatenation operation.

Table 1. Results of 1-shot 5-way open-set recognition problem. **Bold**: the reaults that exceed baselines, <u>underline</u>: the results that are in the confidence interval of SnaCTHer*

Methods	CUB*		MiniImageNet		TieredImageNet	
	Acc	AUROC	Acc	AUROC	Acc	AUROC
ProtoNet [11]	59.36 ± 0.96	54.84 ± 0.73	64.01 ± 0.88	51.81 ± 0.93	68.26 ± 0.96	60.73 ± 0.80
NN [8]	/	/	63.82 ± 0.85	56.96 ± 0.75	67.73 ± 0.96	62.70 ± 0.72
OpenMax [1]	/	/	63.69 ± 0.84	62.64 ± 0.80	68.28 ± 0.95	60.13 ± 0.74
FEAT [14]	62.84 ± 0.94	57.72 ± 0.79	67.02 ± 0.85	57.01 ± 0.84	70.52 ± 0.96	63.54 ± 0.76
PEELER [6]	59.60 ± 0.96	56.49 ± 0.77	65.86 ± 0.85	60.57 ± 0.83	69.51 ± 0.92	65.20 ± 0.76
SnaCTHer [4]	/	/	67.02 ± 0.85	68.27 ± 0.96	70.52 ± 0.96	74.28 ± 0.80
SnaCTHer*	62.84 ± 0.94	70.80 ± 0.83	65.65 ± 0.85	69.11 ± 0.86	70.84 ± 0.95	75.53 ± 0.77
ANCOR	**64.35 ± 0.88**	**72.41 ± 0.79**	<u>65.18 ± 0.82</u>	**69.35 ± 0.93**	69.71 ± 0.93	<u>74.77 ± 0.81</u>
ANCOR+	**65.09 ± 0.88**	**72.39 ± 0.78**	<u>65.32 ± 0.83</u>	**69.65 ± 0.86**	69.42 ± 0.95	74.69 ± 0.79
ICCL	**64.48 ± 0.90**	**71.60 ± 0.75**	<u>65.01 ± 0.88</u>	**70.19 ± 0.87**	69.76 ± 0.96	74.73 ± 0.82

Table 2. Results of 5-shot 5-way open-set recognition problem

Methods	CUB*		MiniImageNet		TieredImageNet	
	Acc	AUROC	Acc	AUROC	Acc	AUROC
ProtoNet	76.58 ± 0.69	63.00 ± 0.74	80.09 ± 0.58	60.39 ± 0.92	83.40 ± 0.65	64.96 ± 0.83
NN	/	/	80.12 ± 0.57	63.43 ± 0.76	83.43 ± 0.66	69.77 ± 0.75
OpenMax	/	/	80.56 ± 0.58	62.27 ± 0.71	83.48 ± 0.66	65.51 ± 0.83
FEAT	77.82 ± 0.69	64.31 ± 0.70	82.02 ± 0.53	63.18 ± 0.78	84.74 ± 0.69	70.74 ± 0.75
PEELER	76.05 ± 0.69	66.79 ± 0.68	80.61 ± 0.59	67.35 ± 0.80	84.10 ± 0.66	73.27 ± 0.71
SnaCTHer	/	/	82.02 ± 0.53	77.42 ± 0.73	84.74 ± 0.69	82.02 ± 0.64
SnaCTHer*	77.82 ± 0.69	79.13 ± 0.67	81.18 ± 0.57	77.43 ± 0.76	84.85 ± 0.63	82.02 ± 0.65
ANCOR	**79.26 ± 0.68**	**81.50 ± 0.58**	80.24 ± 0.57	**77.82 ± 0.71**	<u>84.64 ± 0.66</u>	**82.55 ± 0.64**
ANCOR+	**79.68 ± 0.62**	**81.24 ± 0.57**	<u>80.67 ± 0.55</u>	**78.64 ± 0.69**	<u>84.52 ± 0.65</u>	**82.34 ± 0.65**
ICCL	**78.78 ± 0.69**	**80.88 ± 0.63**	80.27 ± 0.57	77.13 ± 0.75	<u>84.46 ± 0.67</u>	**82.09 ± 0.66**

4   Experiments

We introduce the datasets, implementation details, and experimental results in this section.

4.1   Experimental Settings

We use MiniImageNet [12], TieredImageNet [9] and CUB [13] which are widely used in few-shot learning to evaluate our model. MiniImageNet (100 categories) and TieredImageNet (608 categories) are both subsets of ImageNet [10]. CUB is a bird species dataset that contains 11788 images of 200 bird species.

We used a 4-layer convolution network (ConvNet) as the backbone of CUB and both the ResNet-12, and ConvNet as the backbones of Mini/TieredImageNet. For the calculation of the extra embedding vectors, we set the prototype transformation as an optional factor when computing the logits, and compare the results of whether to transform or not. We used Euclidean distance to calculate all the logits in our experiments. We set α as 0.01, τ as

64, λ as 0.1, $M = M'$ as 15, $K = K'$ as 5 and Num as 20. β was selected from the set $\{0.1, 0.2, ..., 0.9, 1.0\}$ according to the corresponding performance. The optimization parameters and testing conditions are the same as SnaTCHer [4].

4.2   Results

The effectiveness of our model will be evaluated from various perspectives.

Few-Shot Open-Recognition. We use accuracy and AUROC as measurements respectively. The results of baselines are directly referred to the results in SnaTCHer [4]. The dataset and method with '*' mean that they are implemented by ourselves. ANCOR+ means that we use the prototype transformation $\mathcal{F}_\psi$ on the extra embedding vector at testing time to detect OOD samples. For the convenience of calculation, we set the anchor samples in contrastive loss function as the original samples without any augmentations on Mini/TieredImageNet.

According to the Table 1 and Table 2, we can conclude that our method improves the ability to detect the OOD samples, especially for CUB. The results of TieredImageNet may attribute to the hierarchical categories in TieredImageNet leading to that the classes of ID data and OOD data sampled in testing tasks are too different to conform with the SAS where EECM works prominently. We will further design experiments on TieredImageNet to verify the effectiveness of EECM in Sect. 4.2.

To compare the effect of the structure of the feature extractor, we also evaluated our method with ConvNet as the backbone. The results in Table 3 shows that our method leads to a more significant improvement under the simpler backbone networks.

Table 3. Results of Mini/TieredImageNet with the backbone ConvNet

Methods	MiniImageNet				TieredImageNet			
	1-shot 5-way		5-shot 5-way		1-shot 5-way		5-shot 5-way	
	Acc	AUROC	Acc	AUROC	Acc	AUROC	Acc	AUROC
SnaCTHer	53.30	62.82	69.91	69.24	52.48	63.41	69.44	71.43
EECM-ANCOR	51.70	**63.57**	68.31	**70.88**	<u>52.44</u>	**65.30**	66.19	<u>71.02</u>
EECM-ANCOR+	52.19	**64.50**	68.22	**71.12**	<u>52.19</u>	64.39	**70.53**	**73.99**
EECM-ICCL	52.02	**63.85**	68.07	**71.40**	**52.57**	**65.12**	69.90	**73.96**

Table 4. Improvements on various indicators with backbone ConvNet on 1-shot tasks

Models	MiniImageNet			TieredImageNet			CUB		
	Prob	Dist	STCH	Prob	Dist	STCH	Prob	Dist	STCH
SnaTCHer	53.43	56.44	62.82	55.18	56.47	63.41	57.72	55.98	70.80
EECM-ANCOR	**53.67**	**62.40**	**63.57**	**55.38**	**63.30**	**65.30**	**57.94**	**67.74**	**72.41**
EECM-ANCOR+	**53.43**	**60.60**	**64.50**	<u>55.07</u>	**62.29**	**64.39**	**59.04**	**60.07**	**72.39**
EECM-ICCL	**53.80**	**63.36**	**63.85**	<u>55.16</u>	**64.06**	**65.12**	**58.29**	**64.05**	**71.60**

Results on Various Indicators. Taking Probability (Prob), Distance (Dist) and SnaTCHer (STCH) as examples, we show the improvements of our method based on the framework of SnaTCHer. The indicator of Probability is to judge a sample as OOD if the probability computed by the softmax layer is low, and the indicator of Distance is to make the decision according to the distance between the sample and prototypes of categories. Taking the 1-shot task as an example, the results of Table 4 shows that our model boosts the performance of SnaCTHer on all the indicators.

Results on Resampled Tasks. In order to further verify the effectiveness of EECM in the SAS, we provide resampled tasks where the OOD images are sampled from the set of fine-grained categories that are subordinate to the same coarse-grained categories of ID images. The results in Table 5 show that EECM performs better than the baseline.

Table 5. Testing on resampled TieredImageNet tasks with backbone ResNet-12

Methods	1-shot 5-way			5-shot 5-way		
	Prob	Dist	STCH	Prob	Dist	STCH
SnaTCHer	59.64 ± 0.56	61.12 ± 0.62	64.22 ± 0.64	67.49 ± 0.54	67.35 ± 0.59	72.22 ± 0.59
ANCOR	**59.91 ± 0.59**	**62.83 ± 0.63**	**64.72 ± 0.65**	**68.09 ± 0.57**	67.50 ± 0.60	73.11 ± 0.63
ANCOR+	**59.73 ± 0.58**	**62.25 ± 0.61**	**64.48 ± 0.63**	67.66 ± 0.55	62.86 ± 0.56	**73.14 ± 0.60**
ICCL	**59.76 ± 0.60**	**62.66 ± 0.63**	**64.60 ± 0.64**	67.87 ± 0.58	67.74 ± 0.60	**73.13 ± 0.63**

Relative Distances in Representation Space. Taking ICCL for a 5-shot case as an example, we verify the separating ability of our model in embedding space on MiniImageNet with ConvNet as the backbone. In Fig. 3, we can see that EECM can capture the basic category information. When we remove the CE loss in ICCL, the representation vectors become a random distribution. In order to further verify the conclusion, we define the average of the neareast distance between OOD samples and ID samples (o2i_dist) as $\frac{1}{|\text{Ind}(\mathcal{D}^{oq})|} \sum_{j \in \text{Ind}(\mathcal{D}^{oq})} \min_{i \in \text{Ind}(\mathcal{D}^{iq})} dist^2(z_i, z_j)/\text{Dim}$ and the average of the neareast distance among ID samples (i2i_dist) is similar, replacing $\mathcal{D}^{oq}$ with $\mathcal{D}^{iq}$, where Ind($\mathcal{D}$) is the index set of samples in $\mathcal{D}$ and Dim is the dimension of each embedding vector.

The results are shown in Fig. 4, where the o2i_dist and i2i_dist of EECM were computed by the extra embeddings and the base embeddings respectively. We can see that ProtoNet and FEAT/SnaTCHer lack the ability to separate samples in hidden space, and our model can enlarge the distances between ID and OOD samples in both the SAS (Resampled TieredImageNet) and normal scenario (MiniImageNet).

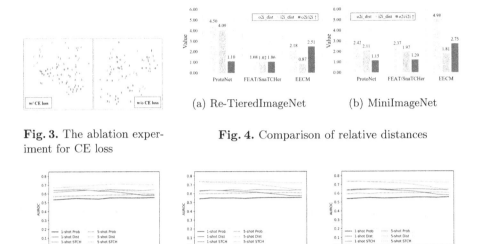

Fig. 3. The ablation experiment for CE loss

Fig. 4. Comparison of relative distances

(a) Re-TieredImageNet   (b) MiniImageNet

(a) ANCOR   (b) ANCOR+   (c) ICCL

Fig. 5. Sensitivity of β on TieredImageNet with backbone ConvNet

Sensitivity. To observe the influence of selecting parameter β in Eq. (9) on the performance of our model, we show the sensitivity of β with different loss functions in Fig. 5. With different values of parameters, we can see that the performance of our model changes slightly, which indicates the stability of our model. For the distance-based indicator, the performance decreases with the increase of the value of β, which indicates that the larger proportion of the extra embedding vector in Eq. (9) leads to better performance of the model.

5   Conclusion

This paper considered the semantic adhesion scenorio in FSOSR problem that OOD samples are entangled with ID samples in the embedding space due to distribution similarity. To alleviate the problem, we put forward the Extra Embedding Classification Model that trains extra embedding vectors for every sample to capture the distinctive features, thereby better seperate the OOD samples from ID samples in feature space. The experimental results show that our model can improve the performance of detecting OOD samples on various indicators. There remain some potential directions for future research. For example, our model focuses more on improving the OOD recognition, allowing a slight decline in classification accuracy. This problem can be solved by using more complex representation networks in the future.

Acknowledgements. This work was partly supported by the National Natural Science Foundation of China under Grant No. 61572002, No. 61690201, and No. 61732001.

References

1. Bendale, A., Boult, T.E.: Towards open set deep networks. In: Proceedings of the IEEE Conference on Computer Vision and Pattern Recognition (CVPR) (2016)
2. Bukchin, G., et al.: Fine-grained angular contrastive learning with coarse labels. In: Proceedings of the IEEE/CVF Conference on Computer Vision and Pattern Recognition (CVPR), pp. 8730–8740 (2021)
3. Finn, C., Abbeel, P., Levine, S.: Model-agnostic meta-learning for fast adaptation of deep networks. In: Precup, D., Teh, Y.W. (eds.) Proceedings of the 34th International Conference on Machine Learning. Proceedings of Machine Learning Research, vol. 70, pp. 1126–1135. PMLR (2017)
4. Jeong, M., Choi, S., Kim, C.: Few-shot open-set recognition by transformation consistency. In: 2021 IEEE/CVF Conference on Computer Vision and Pattern Recognition (CVPR), Nashville, TN, USA, pp. 12561–12570. IEEE (2021)
5. Jeong, T., Kim, H.: OOD-MAML: meta-learning for few-shot out-of-distribution detection and classification. In: Larochelle, H., Ranzato, M., Hadsell, R., Balcan, M.F., Lin, H. (eds.) Advances in Neural Information Processing Systems, vol. 33, pp. 3907–3916. Curran Associates, Inc. (2020)
6. Liu, B., Kang, H., Li, H., Hua, G., Vasconcelos, N.: Few-shot open-set recognition using meta-learning. In: Proceedings of the IEEE/CVF Conference on Computer Vision and Pattern Recognition, pp. 8798–8807 (2020)
7. Lu, J., Gong, P., Ye, J., Zhang, C.: Learning from very few samples: a survey. arXiv:2009.02653 [cs, stat] (2020)
8. Mendes Júnior, P.R., et al.: Nearest neighbors distance ratio open-set classifier. Mach. Learn. $106(3)$, 359–386 (2017). https://doi.org/10.1007/s10994-016-5610-8
9. Ren, M., et al.: Meta-learning for semi-supervised few-shot classification. In: International Conference on Learning Representations (2018)
10. Russakovsky, O., et al.: ImageNet large scale visual recognition challenge. Int. J. Comput. Vis. $115(3)$, 211–252 (2015). https://doi.org/10.1007/s11263-015-0816-y
11. Snell, J., Swersky, K., Zemel, R.: Prototypical networks for few-shot learning. In: Advances in Neural Information Processing Systems, vol. 30 (2017)
12. Vinyals, O., Blundell, C., Lillicrap, T., Wierstra, D., et al.: Matching networks for one shot learning. In: Advances in Neural Information Processing Systems, vol. 29 (2016)
13. Wah, C., Branson, S., Welinder, P., Perona, P., Belongie, S.: The Caltech-UCSD Birds-200-2011 dataset. Technical report CNS-TR-2011-001, California Institute of Technology (2011)
14. Ye, H.J., Hu, H., Zhan, D.C., Sha, F.: Few-shot learning via embedding adaptation with set-to-set functions. In: Proceedings of the IEEE/CVF Conference on Computer Vision and Pattern Recognition, pp. 8808–8817 (2020)
15. Yeh, C.H., Hong, C.Y., Hsu, Y.C., Liu, T.L., Chen, Y., LeCun, Y.: Decoupled contrastive learning (2021). arXiv:2110.06848 [cs]

Self-Training
with Label-Feature-Consistency
for Domain Adaptation

Yi Xin, Siqi Luo, Pengsheng Jin, Yuntao Du, and Chongjun Wang[✉]

National Key Laboratory for Novel Software Technology, Department of Computer
Science and Technology, Nanjing University, Nanjing, China
{mf21330098,mf21330060,mf21330042,dz1833005}@smail.nju.edu.cn,
chjwang@nju.edu.cn

Abstract. Mainstream approaches for unsupervised domain adaptation
(UDA) learn domain-invariant representations to address the domain
shift. Recently, self-training has been used in UDA, which exploits
pseudo-labels for unlabeled target domains. However, the pseudo-labels
can be unreliable due to distribution shifts between domains, severely
impairing the model performance. To address this problem, we propose a novel self-training framework-Self-Training with Label-Feature-
Consistency (ST-LFC), which selects reliable target pseudo-labels via
label-level and feature-level voting consistency principle. The former
means target pseudo-labels generated by a source-trained classifier and
the latter means the nearest source-class to the target in feature space. In
addition, ST-LFC reduces the negative effects of unreliable predictions
through entropy minimization. Empirical results indicate that ST-LFC
significantly improves over the state-of-the-arts on a variety of benchmark datasets.

Keywords: Transfer learning · Domain Adaptation · Self-Training ·
Label-Feature-Consistency

1 Introduction

Supervised deep learning methods have achieved excellent performance for tasks
such as computer vision [14], natural language processing [15]. However, such
models usually generalize poorly due to the distribution shifts between the training samples and testing samples. For example, in object recognition tasks, the
model is generally trained on images collected in fine days, but the model's
performance degrades when applying the model on rainy days and foggy days.

Unsupervised domain adaptation (UDA) is able to overcome this challenge
by transferring knowledge from a labeled source domain to the unlabeled target domain [13,35,36]. Existing UDA could be divided into moment matching,
adversarial domain adaptation, and self-training based methods. The first two

---

Y. Xin and S. Luo—Equal contribution.

types reduce domain discrepancy by reducing the statistical distribution discrepancy across domains [25,30] or perform domain adversarial training [13,27]. Different from these methods, self-training based methods [8,9,20] are inspired by semi-supervised learning (SSL) [10,19] and do not need to explicitly calculate domain discrepancy or perform complex adversarial training.

Typical self-training method firstly trains a model with labeled source samples, and then it generates the pseudo-labels for the unlabeled target samples using the source model. Considering that some pseudo-labels are incorrect, only pseudo-labels with high confidence are selected and then they are added to the training samples together with source samples to train the model, and repeat this process until the model converges. Recently, a few methods follow self-training framework and optimize this process by designing new selection strategy for target samples [21], introducing new regularization [10], and optimizing training strategy [16,17].

Although achieving remarkable progress, there are still some issues to be addressed. Firstly, self-training is a method used in semi-supervised learning, which means domain shift will not be prominent or even absent in the original application scenario. In domain adaptation, because of the existence of domain shift such as covariate shift and label shift, the distribution of pseudo-labels may be remarkably different from the target ground-truths, while purely tailoring self-training with UDA. As a result, the accuracy of pseudo-labels cannot be guaranteed without any regulars because of the accumulated error and even trivial solution, which eventually leads to misalignment of the distribution and misclassification of several classes [20]. Secondly, previous works tailor self-training for UDA by selecting pseudo-labels with confidence threshold or reweighting through information entropy or other confidence measurement methods, but it is challenging to determine a criterion for confidence judgment when it is closely related to specific tasks, causing models based on previous works are not robust. Additionally, existing self-training paradigms tend to ignore the rich information in unreliable samples. Since the ignoration of global distribution including either reliable or unreliable samples of the target domain, leading to a poor-fit between self-training and UDA, there is still a great room for improvement.

To address the above limitations, in this paper, we propose a simple yet effective method called ST-LFC (Self-Training with Label-Feature-Consistency). Instead of using confidence score to determine whether a sample is reliable to avoid generating pseudo-labels with hard-to-tweak protocols, we use label-level and feature-level consistency as a criterion. Intuitively, a successful feature extractor should generate features with greater inter-class and smaller intra-class distances. For example, during self-training process, one target sample is assigned class k by pseudo-labels, then it should be closer to source samples belong to class k and father from the source samples which belong to other classes except for k in the feature space. Motivated by [24], we use source class prototype feature to represent the class, which means the average feature embedding of the source samples with the same ground-truth labels. Moreover, ST-LFC also exploits unreliable target samples. The reason why model has a performance upper bound is the existence of these unreliable samples. The samples are unreliable because the classifier predictions are inconsistent with the nearest prototype class in the feature space, and we can't judge which is right. Our method weights the class prediction probabilities of both, constrained by entropy minimization. Finally, our method is able to incorporate any domain adaptation methods that takes into account both domain alignment and sample label utilization during training.

The effectiveness of ST-LFC is reflected by improved adaptation accuracy on popular benchmarks like Digits and Office-31 datasets. We also achieve state-of-the-art results on a challenging adaptation dataset Birds-31 [41], which indicates the usefulness of our ST-LFC in handling wide variety of scenarios.

In summary, the key highlights of the paper are:

- We propose a novel self-training framework ST-LFC for unsupervised domain adaptation. We propose a new sample strategy to effectively recognize reliable samples and we adequately utilize unrelibale samples to improve the upper bound on model performance.
- ST-LFC is designed to be general for existing domain adaptation approaches. It is able to incorporate any domain adaptation methods. In order to verify, we combine SF-LFC with two popular methods, DANN and CDAN, and observe consistent improvement over both the baselines.
- We validate the effectiveness of the proposed approach numerically by applying it on multiple tasks from various challenging benchmark datasets used for domain adaptation like Digits, Office-31 and Birds-31 and observe improved accuracies in all the cases, sometimes outperforming the state-of-the-art by a large margin.

2   Related Work

2.1   Unsupervised Domain Adaptation

Unsupervised Domain Adaptation is proposed to address the domain shift between source domains and target domains, so that networks trained on source domain can be used directly on completely unlabeled target domains [34–36]. Motivated by theoretical bound proposed in [34], Discrepancy-based approaches [30–33] measure and minimize the dissimilarity between the feature embedding of the source and target domains. DAN [33] leverages the transferability of deep neural networks [22] and introduces MMD-based regularizer to minimize the cross-domain distribution discrepancy in multiple layer of neural networks. Adversary-based approaches [25–28] utilize adversarial learning [29] to align domain distribution on feature level and pixel level [25] by obtaining domain invariant features. Motivated by image translation techniques, pixel-level alignment methods [5–7] utilize image-to-image translation network to translate images from source domain to target domain. Feature-level alignment methods [1,2,4] tend to adapt distributions of source and target images explicitly in feature space. In addition to global domain adaptation, several studies take advantage of contrastive learning to align domain distribution on class-level [23,24].

2.2   Self-Training

Self-Training [9,10,19] increases the amount of training data by iteratively select-ing pseudo-labeled target data based on model which is trained with existing labeled data. Then retrain the model with the enlarged training data. Specifi-cally, the quality of pseudo-labels of the target data and training strategy both have great impact on the performance of the model.

Existing methods can be generally catogorized into three groups. The first group tend to design selection strategies for target samples [1,3,11,21]. CAN [21] utilize domain discriminator to assign weights to target data and jointly trains the model with source and target data. The second group is introducing new regularization to assist domain adaptation. [3]uses asymmetric tritraining method to improve the accuracy of pseudo labels, which means that it utilizes three asymmetric classifiers, two networks are used to label unlabeled target samples, and one network is trained by the pseudo-labeled samples to obtain target-discriminative representations [12] first generates coarse pseudo labels by a conventional UDA method, and iteratively exploits the intra-class similarity of the target samples for improving the generated coarse pseudo labels. [10] employ confidence regularization technique to help discriminative feature representations of the source and target domains by introducing soft pseudo labels. The third group is optimizing training strategy. [17] proposed a class-balanced self-training framework to address the problem of imbalanced pseudo-labels. Motivated by co-training, CODA [16] tries to slowly adapt training set from the source domain to the target domain. [20] proposed Cycle Self-training which mainly focus on pseudo-labeling and learns to generalize the pseudo-labels across domains.

3   Method

In this section, we first give a brief overview of adversarial adaptation methods, and then introduce how to find nearest source-class in feature space. Finally, we explain our self-training with label-level and feature-level consistency method in detail.

3.1   Overview of Adversarial Domain Adaptation

In unsupervised domain adaptation, we are given a labeled source domain $\mathcal{D}^s = \{x_i^s, y_i\}_{i=1}^{|\mathcal{D}^s|}$, and an unlabeled target domain $\mathcal{D}^t = \{x_i^t\}_{i=1}^{|\mathcal{D}^t|}$. The source domain and target domain are characterized by probability distributions P_s and P_t, respectively. The goal of UDA is to train a model using $\mathcal{D}^s$ and $\mathcal{D}^t$ to make predictions on D_t. Figure 1 shows the overall architecture of our proposed method. Feature extractor $\mathcal{G}$ is shared by the source and target domains, extract-ing the low-dimensional feature representations corresponding to the inputs, given by $f = \mathcal{G}(x)$. The classifier $\mathcal{C}$ then outputs the softmax prediction dis-tribution over the classes, and is trained using the cross-entropy(CE) loss on the labeled data given by

$$\mathcal{L}_{cls} = \mathbb{E}_{(x,y) \sim \mathcal{D}^s \cup \mathcal{D}^t_{select}} [-\log[\mathcal{C}(\mathcal{G}(x))]_y] \tag{1}$$

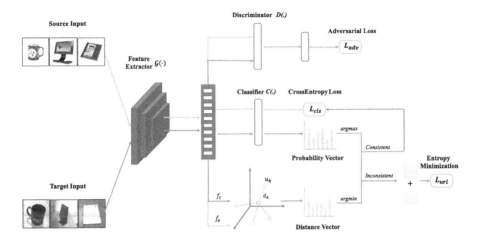

Fig. 1. Illustration of proposed ST-LFC approach. Our architecture consists of a feature extractor $\mathcal{G}$ which is shared by source and target domains. The classifier $\mathcal{C}$ is trained to classify the source images and generate target pseudo-labels using cross entropy loss $\mathcal{L}_{cls}$. The domain discriminator $\mathcal{D}$ aims to achieve domain alignment using adversarial loss $\mathcal{L}_{adv}$. Additionally, we use source features to obtain the class feature prototype μ, and compute the distance between target features and class feature prototype to get its nearest source-class. Then, we judge whether the generated pseudo-labels are consistent with the nearest source-class results. The consistent target samples are trained together with the source domain(Self-Training Process), and the inconsistent use $\mathcal{L}_{url}$ to participate in training.

where y is the ground truth labels for the source data in D_s or pseudo-labels for carefully selected target data in D_t. This is a self-training process, the model generates pseudo-labels of unlabeled data, and jointly trains the model with source labels and selected target pseudo-labels. The target domain samples selection method will be introduced in the next section. However, since $P_s \neq P_t$ and the classifier are biased to the source domain, the classifier does not generalize well to target samples. Thus, the adversarial learning strategy [28] is used to alleviate this issue. The domain discriminator D is trained by $\mathcal{L}_{DC}$ to distinguish source samples from target samples, while the feature extractor $\mathcal{G}$ is trained to generate domain-invariant features that can confuse the discriminator:

$$\min_{D} \max_{\mathcal{G}} \mathcal{L}_{DC} \tag{2}$$

$$\mathcal{L}_{DC} = -\mathbb{E}_{x \sim \mathcal{D}^s}[\log \mathcal{D}(\mathcal{G}(x))] - \mathbb{E}_{x \sim \mathcal{D}^t}[\log(1 - \mathcal{D}(\mathcal{G}(x)))] \tag{3}$$

The min-max training between feature extractor and discriminator can learn help domain-invariant features, but this is not enough for good adaptation. On the basis of learning domain-invariant features, our method utilizes the pseudo-labels of the target samples by self-training, which leads to a qualitative improvement in the performance on the target domain.

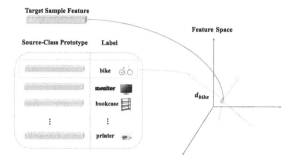

Fig. 2. Illustration of source-class prototypes. It means the average feature embedding of the source samples with the same ground-truth labels. We can find the nearest source-class for target sample in feature space.

3.2   Source-Class Prototype in Feature Space

Our source-class prototype definition is intentionally tailored to serve UDA task. Recall that during each forward pass, we have obtained source features $f_i^s = \mathcal{G}(x_i^s)$, which is generated by the feature extractor $\mathcal{G}$. We enqueue each (f_i^s, y_i^s) pair sequentially into each of the specific source-class prototypes, which means the average feature embedding of the source samples with the same ground-truth labels. The prototype of source-class k is denoted as:

$$\mu_k = \frac{1}{n_k} \sum_{y_i^s = k} f_i^s \tag{4}$$

where n_k represents the number of corresponding samples. Assuming that the source contains a total of $\mathcal{K}$ categories, $\mathcal{K}$ source-class prototypes can be obtained. Figure 2 shows the source-class prototype part in the overall framework in detail. Figure 2 takes target sample x_j^t as an example, $\mathcal{K}$-dimensional distance vectors d_j^t can be obtained by measuring the distance between target sample x_j^t feature and $\mathcal{K}$ source-class prototypes, we adopt the calculation method of squared euclidean distance, which is given by:

$$d_{j,k}^t = \left\| \mathcal{G}(x_j^t) - \mu_k \right\|^2, k = \{1, 2, 3, ..., \mathcal{K}\} \tag{5}$$

where $d_{j,k}^t$ is the kth element of $d_j^t \in \mathbb{R}^{\mathcal{K}}$. From this, we can obtain the feature distance vector between target samples and source-class prototypes. The nearest source-class has the smallest distance.

Algorithm 1. Self-training with Label-Feature-Consistency

Require: Batches for source $B_s \in D_s$; Batches for target $B_t \in D_t$; Feature extractor $\mathcal{G}$; Classifier $\mathcal{C}$; Domain discriminator $\mathcal{D}$.

Ensure: Trained feature extractor and trained classifier by ST-LFC.

1: **for** epoch=0 to MaxEpoch **do**
2:    **for** t=0 to MaxIter **do**
3:        Feature extractor $\mathcal{G}$ generates features f_s for B_s and f_t for B_t.
4:        Classifier $\mathcal{C}$ outputs probability vectors p_s for B_s and p_t for B_t.
5:        Source features f_s generate source-class prototypes by Equ 4.
6:        Calculate distance vectors from the source-class prototypes by Equ 5.
7:        Find reliable samples by Equ 6.
8:        **if** reliable sample **then**
9:            Train source feature extractor $\mathcal{G}$ and source classifier $\mathcal{C}$ by classification loss $\mathcal{L}_{rl}$.
10:       **else**
11:           Train source feature extractor $\mathcal{G}$ and source classifier $\mathcal{C}$ by classification loss $\mathcal{L}_{url}$ by Equ 8.
12:       **end if**
13:       Train source feature extractor $\mathcal{G}$ and source classifier $\mathcal{C}$ by source classification loss $\mathcal{L}_{cls}$ by Equ 1.
14:       Train source feature extractor $\mathcal{G}$ by domain adversarial loss $\mathcal{L}_{adv}$ by Equ 2.
15:    **end for**
16: **end for**
17: **return** $\mathcal{G}, \mathcal{C}$

3.3   Self-training with Label-Feature-Consistency

We first develop a voting consistency strategy to select certain pseudo-labeled target samples for self-training to adapt the model to the target domain. The certainty is decided by the consistency of two different predictions: the classifier prediction and the nearest source-class prediction. The classifier $\mathcal{C}$ outputs probability vectors $p_j^t \in \mathbb{R}^{\mathcal{K}}$ for x_j^t and distance vector $d_j^t \in \mathbb{R}^{\mathcal{K}}$ can be obtained which is introduced in last section. Then the consistency score for target sample is defined as:

$$\text{con}\left(x_j^t\right) = \begin{cases} 1 & \text{argmax}\left(p_j^t\right) == \text{argmin}\left(d_j^t\right) \\ 0 & \text{otherwise} \end{cases} \tag{6}$$

where *argmax* obtains the label with the largest predicted probability by the classifier, *argmin* obtains the nearest source-class. For target samples with consistency score of one, we consider them to be reliable samples. Reliable samples with pseudo-labels train the model together with the source by cross-entropy loss $\mathcal{L}_{rl}$. This part plays an integral role in our ST-LFC. Directly training the model with reliable target samples with pseudo-labels will make the model perform better on the target.

Existing domain adaptation methods have an upper limit, in other words, the upper limit is caused by these unreliable samples. ST-LFC also makes use of the remaining unreliable samples. In our ST-LFC, the samples are unreliable because the classifier predictions are inconsistent with the nearest prototype class in the feature space, and we can't judge which is right. The solution of our ST-LFC is to consider comprehensively, combining the probability vector $p_j^t \in \mathbb{R}^\mathcal{K}$ and distance vector $d_j^t \in \mathbb{R}^\mathcal{K}$. However, the distance vector needs to be processed and converted into a class probability value, which is given by:

$$q_{j,k}^t = \frac{e^{-d_{j,k}^t}}{\sum_{i=1}^\mathcal{K} e^{-d_{j,i}^t}}, k = \{1, 2, 3, ..., \mathcal{K}\} \tag{7}$$

where $q_{j,k}^t$ is the k_{th} element of q_j^t, which means the probability of class k considering feature distance. Then we use entropy minimization to update model, which is defined as:

$$\mathcal{L}_{url} = -\sum_{k=1}^\mathcal{K} (p_{j,k}^t + q_{j,k}^t) log(p_{j,k}^t + q_{j,k}^t) \tag{8}$$

3.4 Optimization

To sum up, the optimization of ST-LFC is mainly divided into four parts: source classification loss $\mathcal{L}_{cls}$, target reliable samples classification loss $\mathcal{L}_{rl}$; entropy minimization two level probability of the target unreliable samples $\mathcal{L}_{url}$ and domain adaptation loss $\mathcal{L}_{adv}$. Overall optimization of the model is give by:

$$\mathcal{L}_{total} = \mathcal{L}_{cls} + \mathcal{L}_{rl} + \mathcal{L}_{url} + \alpha \mathcal{L}_{adv} \tag{9}$$

where α is the trade-off parameter. Algorithm 1 depicts the complete training procedure of ST-LFC.

4 Experiments

In this section, we conduct extensive experiments on multiple domain adaptation benchmarks to verify the effectiveness of ST-LFC. We present the datasets used to evaluate our results, baselines methods we compared against, followed by results and discussion. In the experiment, we choose well-known UDA methods DANN [49] and CDAN [13] as our infrastructure respectively.

4.1 Datasets

To test the effectiveness of our method, we experiment on three different kinds of benchmark datasets used for domain adaptation, which are Digits, Office-31 and Birds-31.

Digits. We investigate three digits datasets: USPS (U), MNIST (M), and SVHN (S). We show results on three transfer tasks: M → U, U → M, S → M. USPS contains 7,438 images. MNIST is composed of 55,000 images and SVHN is composed of 73,257 images.

Table 1. Accuracy (%) on Digits for unsupervised domain adaptation.

METHOD	M→U	U → M	S → M	AVG.
Source Only	76.7	63.4	67.1	69.1
DANN	90.8	94.0	83.1	89.3
ADDA	89.4	90.1	76.0	85.2
DSN	91.3	-	82.7	-
ATT	-	-	85.0	-
CDAN	93.9	96.9	88.5	93.1
ST-LFC(with DANN)	92.9	97.3	91.5	93.9
ST-LFC(with CDAN)	**94.9**	**97.9**	**92.0**	**94.9**

Office-31 is the most widely used dataset for visual domain adaptation, with 4,652 images and 31 categories collected from three distinct domains: Amazon (A), DSLR (D) and Webcam (W). We show results for all the 6 task pairs $A \rightarrow W, D \rightarrow W, W \rightarrow D, A \rightarrow D, D \rightarrow A$ and $W \rightarrow A$. Following prior works, we report results on the complete unlabeled examples of the target domain.

Birds-31 is recently proposed by [41] for fine grained adaptation consisting of different types of birds. There are three domains in Birds-31: CUB-200-2011 (C), NABirds (N) and iNaturalist2017 (I). The numbers of images selected are 1,848, 2,988 and 2,857 respectively. We show the adaptation results on six transfer tasks formed from three domains: $C \rightarrow I, I \rightarrow C, I \rightarrow N, N \rightarrow I, C \rightarrow N$ and $N \rightarrow C$.

4.2   Setup

Baselines. We compare our method ST-LFC with state-of-art domain adaptation methods: DAN [42], DAA [45], CDAN [13], CAT [37], ALDA [38] and SRDA [39] as well as works which perform class aware alignment such as MCD [46], SimNet [47], MADA [40]. For Birds-31, we additionally verify our result with prior fine grained adaptation work, PAN [41]. Finally, we have ST-LFC with DANN, which is using ST-LFC approach on top of DANN and ST-LFC with CDAN which uses ST-LFC in combination with CDAN. We compare the task-wise accuracy and report the average accurancy across all the transfer tasks.

Implementation. We implement our method on Pytorch, we use DTN [13] architecture for digits and ResNet-50 [48] pretrained on ImageNet as the feature extractor for Office-31 and Birds-31. The classifier is made up of fully connected layers. For achieving training stability, we observe that it is essential to pretrain the model on the labeled source dataset for a few iterations before self-training process. We use mini-batch SGD with a learning rate of 0.03 for Office-31 and Birds-31. For the classifier we multiply the learning rate by 10. We use a similar annealing strategy as used in [49].

To illustrate the benefits of the proposed ST-LFC, we employ it on top of two competing adaptation benchmarks in DANN [49] and CDAN [13], while noting

Table 2. Accuracy (%) of different unsupervised domain adaptation methods on Office-31 using ResNet-50 as the backbone for 6 transfer tasks among three domains: Amazon (A), Webcam (W) and Dslr (D). Our method shows consistent improvements.

METHOD	A→W	D→W	W→D	A→D	D→A	W→A	AVG.
ResNet50	68.4	96.7	99.3	68.9	62.5	60.7	76.1
DAN	80.5	97.1	99.6	78.6	63.6	62.8	80.4
DANN	82.0	96.9	99.1	79.7	68.2	67.4	82.2
ADDA	86.2	96.2	98.4	77.8	69.5	68.9	82.9
MCD	88.6	98.5	100.0	92.2	69.5	69.7	86.5
SimNet	88.6	98.2	99.7	85.3	73.4	71.8	86.2
CDAN	93.1	98.2	100.0	89.8	70.1	68.0	86.6
CDAN+E	94.1	98.6	100.0	92.9	71.0	69.3	87.7
DAA	86.8	99.3	100.0	88.8	**74.3**	73.9	87.2
SAFN	88.8	98.4	99.8	87.7	69.8	69.7	85.7
MADA	90.0	97.4	99.6	87.8	70.3	66.4	85.2
CAT	94.4	98.0	100.0	90.8	72.2	70.2	87.6
ALDA	**95.6**	97.7	100.0	**94.0**	72.2	72.5	88.7
SRDA	95.2	98.6	100.0	91.7	74.5	73.7	89.0
ST-LFC(with DANN)	91.3	98.6	100.0	88.6	72.1	69.9	86.8
ST-LFC(with CDAN)	95.2	**99.3**	**100.0**	93.4	73.6	**75.7**	**89.5**

that our ST-LFC is general and applicable in combination with any adversarial adaptation approach. For experiments with DANN, we replace the adversarial loss with a gradient reversal layer.

4.3 Results

Digits Dataset. In Table 1, we show the results for adaptation using ST-LFC. We observe that we outperform prior methods when we use CDAN or DANN in combination with ST-LFC. ST-LFC with DANN improves average accuracy from 89.3% to 93.9% and ST-LFC with CDAN improves average accuracy from 93.1% to 94.9%, indicating the usefulness of ST-LFC for improving existing methods for domain adaptation. To better illustrate performance of ST-LFC, the results for office-31 and birds-31 are describe below.

Office-31 Dataset. We present results on the 6 transfer tasks on Office-31, including their average, in Table 2. We observe that we achieve an accuracy of 89.5% on the average (ST-LFC with CDAN), outperforming all the competing baselines. However, the A → W and A → D are slightly less effective than ALDA, and D → A is less effective than DAA. This is because we chose to build on DANN and CDAN, they are widely known UDA methods. Finally, our ST-LFC is generally applicable, it improves accuracy over both the approaches

Table 3. Results for domain adaptation on fine-frained adaptation setting, shown for 3 challenging datasets: CUB-200-2011 (C), iNaturalist2017 (I) and NABirds (N). Our ST-LFC performs consistently better than all other methods by explicitly modeling the finegrained nature of the adaptation process. All the baseline numbers taken from [41].

Method	C→I	I→C	I→N	N→I	C→N	N→C	Avg.
ResNet50	64.25	87.19	82.46	71.08	79.92	89.96	79.14
DAN	63.90	85.86	82.91	70.67	80.64	89.40	78.90
DANN	64.59	85.64	80.53	71.00	79.37	89.53	78.44
JAN	63.69	86.29	83.34	71.09	81.06	89.55	79.17
ADDA	63.03	87.26	84.36	72.39	79.69	89.28	79.33
MADA	62.03	89.99	87.05	70.99	81.36	92.09	80.50
MCD	66.43	88.02	85.57	73.06	82.37	90.99	81.07
CDAN	68.67	89.74	86.17	73.80	83.18	91.56	82.18
SAFN	65.23	90.18	84.71	73.00	81.65	91.47	81.08
PAN	69.79	90.46	88.10	75.03	84.19	92.51	83.34
ST-LFC(with DANN)	68.67	92.32	87.55	74.94	83.00	92.53	83.17
ST-LFC(with CDAN)	**71.47**	**92.80**	**90.36**	**76.82**	**85.37**	**94.53**	**85.23**

DANN [49] and CDAN [13], consistently over all the tasks (by 4.6% and 2.9% on average, respectively). It is worth noting that ST-LFC improves the accuracy of the DANN method on A → W task by 9.3% and A → D task by 8.9%.

Birds-31 Dataset. The difficulty in this setting lies in the fact that birds from same class but different domains look quite distinct, sometimes more different than images from an other class. We verify the results on all 6 transfer tasks on Birds-31 dataset in Table 3, and show that ST-LFC outperforms prior works across all the tasks. From the result, prior works that rely on global alignment objectives [36, 42] do not perform any better than a source-only model (ResNet-50 baseline), possibly because they suffer from negative alignment. However, our ST-LFC directly considers pseudo-labels, which is a more fine-grained solution for Birds-31. As a result, we improve the accuracy over DANN on all the tasks, and average accuracy from 78.44% to 83.17%. In fact, with an average accuracy of 85.23% we achieve the new state-of-the-art result using ST-SLC in combination with CDAN. More remarkably, ST-LFC even outperform PAN [41], that is specifically designed for fine-grained adaptation. This result underlinesthat ST-LFC is able to perform well on fine-grained vision categorization despite the domain shift.

4.4   Insight Analysis

Ablation Study. A highlight of ST-LFC compared with other methods based on Self-Training is that it utilizes both carefully selected reliable target and con-

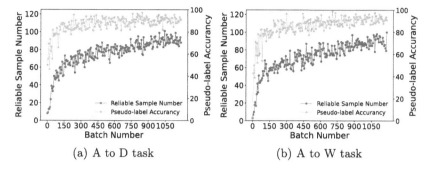

(a) A to D task   (b) A to W task

Fig. 3. (a) shows the number of selected samples and the accuracy of pseudo-labels during training in A to D task; (b) shows the number of selected samples and the accuracy of pseudo-labels during training in A to W task.

Table 4. Results for ablation study. To verify the effectiveness of the ST-LFC method, we experiment with DANN, ST-LFC(without exploiting unreliable samples) and our ST-LFC(with both) on Office-31 and Birds-31.

Reliable	Unreliable	Office-31	Birds-31
✗	✗	82.2	78.4
✓	✗	85.5	82.5
✓	✓	**86.8**	**83.2**

straints on unreliable target. We design experiments on the Office31 and Birds-31 datasets to verify the importance of these two parts, and the results are shown in Table 4. The experiment is divided into three parts: DANN method, ST-LFC(with DANN) without using unreliable target, our ST-LFC(with DANN). ST-LFC without using unreliable target is improved on the basis of DANN methods by 3.3% and 4.1% respectively, which shows that ST-LFC is effective for the utilization of reliable target. In addition, the performance of ST-LFC is improved after adding constraints on the unreliable target by 1.3% and 0.7% respectively, which demonstrates the effectiveness of unreliable target exploitation.

Target Pseudo-Labels Analysis. For all methods based on self-training, the accuracy of pseudo-labels and the number of selected samples are the key factors affecting the performance of the method, and our method is no exception. Figure 3 shows the accuracy rate of pseudo-labels and the variation in the number of samples selected per batch during ST-LFC training. We can see that as the number of ST-LFC training rounds continues to increase, the number of target samples selected in each batch also increases. Each batch of this experiment contains 124 samples, from which a maximum of about 110 reliable samples can be selected in both A → D and A → W task, and the pseudo-label accuracy of reliable samples can reach about 90% with increasing training. In this way, our ST-LFC is very successful in the selection of reliable target samples, and not only performs well in the selection accuracy, but also selects a sufficient number of samples.

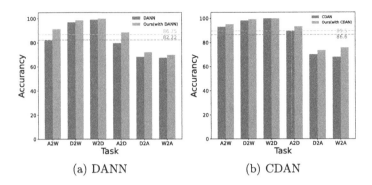

Fig. 4. (a) shows the results of ST-LFC with DANN and DANN comparison; (b) shows the results of ST-LFC with CDAN and CDAN comparison.

Generality Analysis. In order to analysis ST-LFC can be applied to enhance any existing domain adaptation approach, we combine two well-known domain adaptation methods DANN and CDAN as examples. We plot the improvement of Office-31 by our ST-LFC, as shown in Fig. 4. The previous results section mainly describes the experimental data, while this part mainly analyzes the results. There are six tasks in Office-31, and it can be seen from the Fig. 4 that our ST-LFC has a certain magnitude of performance improvement for each task. The performance improvement for DANN is better, because the upper limit of DANN is lower than CDAN, the improvement space of CDAN is smaller. In addition, among the six tasks, A to W and A to D have the best improvement effect, which also shows that ST-LFC is friendly to tasks with large improvement space. In fact, ST-LFC can combine any state-of-the-art methods and can further improve performance.

5   Conclusion

In this work, we propose Self-Training with Label-Feature-Consistency(ST-LFC), which is designed to be general and can be applied to enhance any existing adaptation approach. Firstly, we design a new selection strategy for reliable target samples, which uses label-level and feature-level voting consistency principle. This selection strategy lays the foundation for ST-LFC to perform well in UDA problems. Secondly, ST-LFC does not adopt the abandonment strategy for unreliable samples, but using entropy minimization to constrain the class probability of two levels. Finally, the combination of ST-LFC and adaptation methods can both enable domain alignment and exhibit the powerful advantages of self-training. We show numerical results on various challenging benchmark datasets and perform favorably against many existing adaptation methods.

Limitations and Future Work. Although our ST-LFC performs well on UDA problems, it is not sufficiently applicable. Recently, Test Time Adaptation [50] has been proposed, which uses source to train the model and uses unlabeled target to adjust model only when testing. The scenario solution of this problem is similar to self-training, and we hope that ST-LFC can go a step further in this problem.

Acknowledgements. This paper is supported by the National Key Research and Development Program of China (Grant No. 2018YFB1403400), the National Natural Science Foundation of China (Grant No. 62192783, 61876080), the Key Research and Development Program of Jiangsu(Grant No. BE2019105), the Collaborative Innovation Center of Novel Software Technology and Industrialization at Nanjing University.

References

1. Wang, Z., et al.: Differential treatment for stuff and things: a simple unsupervised domain adaptation method for semantic segmentation. In: Proceedings of the IEEE/CVF Conference on Computer Vision and Pattern Recognition (2020)
2. Du, L., et al.: SSF-DAN: separated semantic feature based domain adaptation network for semantic segmentation. In: Proceedings of the IEEE/CVF International Conference on Computer Vision (2019)
3. Saito, K., Ushiku, Y., Harada, T.: Asymmetric tri-training for unsupervised domain adaptation. In: International Conference on Machine Learning. PMLR (2017)
4. Zhang, Yi., et al. Fully convolutional adaptation networks for semantic segmentation. In: Proceedings of the IEEE Conference on Computer Vision and Pattern Recognition (2018)
5. Murez, Z., et al.: Image to image translation for domain adaptation. In: Proceedings of the IEEE Conference on Computer Vision and Pattern Recognition (2018)
6. Li, Y., Yuan, L., Vasconcelos, N.: Bidirectional learning for domain adaptation of semantic segmentation. In: Proceedings of the IEEE/CVF Conference on Computer Vision and Pattern Recognition (2019)
7. Chen, Y.C., et al.: Crdoco: pixel-level domain transfer with cross-domain consistency. In: Proceedings of the IEEE/CVF Conference on Computer Vision and Pattern Recognition (2019)
8. Li, B., et al.: Rethinking distributional matching based domain adaptation. arXiv preprint arXiv:2006.13352 (2020)
9. Kumar, A., Ma, T., Liang, P.: Understanding self-training for gradual domain adaptation. In: International Conference on Machine Learning. PMLR (2020)
10. Zou, Y., et al.: Confidence regularized self-training. In: Proceedings of the IEEE/CVF International Conference on Computer Vision (2019)
11. Chen, C., et al.: Progressive feature alignment for unsupervised domain adaptation. In: Proceedings of the IEEE/CVF Conference on Computer Vision and Pattern Recognition (2019)
12. Wang, J., Zhang, X.L.: Improving pseudo labels with intra-class similarity for unsupervised domain adaptation. arXiv preprint arXiv:2207.12139 (2022)
13. Long, M., et al.: Conditional adversarial domain adaptation. In: Advances in Neural Information Processing Systems, vol. 31 (2018)
14. Voulodimos, A., et al.: Deep learning for computer vision: a brief review. Comput. Intell. Neurosci. 2018 (2018)
15. Torfi, A., et al.: Natural language processing advancements by deep learning: A survey. arXiv preprint arXiv:2003.01200 (2020)

16. Chen, M., Weinberger, K.Q., Blitzer, J.: Co-training for domain adaptation. In: Advances in Neural Information Processing Systems, vol. 24 (2011)
17. Zou, Y., et al.: Unsupervised domain adaptation for semantic segmentation via class-balanced self-training. In: Proceedings of the European conference on computer vision (ECCV) (2018)
18. Riloff, E., Wiebe, J., Wilson, T.: Learning subjective nouns using extraction pattern bootstrapping. In: Proceedings of the Seventh Conference on Natural Language Learning at HLT-NAACL 2003 (2003)
19. Prabhu, V., et al.: Sentry: selective entropy optimization via committee consistency for unsupervised domain adaptation. In: Proceedings of the IEEE/CVF International Conference on Computer Vision (2021)
20. Liu, H., Wang, J., Long, M.: Cycle self-training for domain adaptation. Adv. Neural. Inf. Process. Syst. **34**, 22968–22981 (2021)
21. Zhang, W., et al.: Collaborative and adversarial network for unsupervised domain adaptation. In: Proceedings of the IEEE Conference on Computer Vision and Pattern Recognition (2018)
22. Yosinski, J., et al.: How transferable are features in deep neural networks?. In: Advances in Neural Information Processing Systems, vol. 27 (2014)
23. Sharma, A., Kalluri, T., Chandraker, M.: Instance level affinity-based transfer for unsupervised domain adaptation. In: Proceedings of the IEEE/CVF Conference on Computer Vision and Pattern Recognition (2021)
24. Chen, Y., et al.: Transferrable contrastive learning for visual domain adaptation. In: Proceedings of the 29th ACM International Conference on Multimedia (2021)
25. Bousmalis, K., et al.: Unsupervised pixel-level domain adaptation with generative adversarial networks. In: Proceedings of the IEEE Conference on Computer Vision and Pattern Recognition (2017)
26. Bousmalis, K., et al.: Domain separation networks. In: Advances in Neural Information Processing Systems, vol. 29 (2016)
27. Cui, S., et al.: Gradually vanishing bridge for adversarial domain adaptation. In: Proceedings of the IEEE/CVF Conference on Computer Vision and Pattern Recognition (2020)
28. Ganin, Y., et al.: Domain-adversarial training of neural networks. J. Mach. Learn. Res. **17**(1), 2030–2096 (2016)
29. Goodfellow, I., et al.: Generative adversarial networks. Commun. ACM **63**(11), 139–144 (2020)
30. Kang, G., et al.: Contrastive adaptation network for unsupervised domain adaptation. In: Proceedings of the IEEE/CVF Conference on Computer Vision and Pattern Recognition (2019)
31. Gretton, A., et al.: A kernel method for the two-sample-problem. In: Advances in Neural Information Processing Systems, vol. 19 (2006)
32. Zhang, X., et al.: Deep transfer network: unsupervised domain adaptation. arXiv preprint arXiv:1503.00591 (2015)
33. Long, M., et al.: Learning transferable features with deep adaptation networks. In: International Conference on Machine Learning. PMLR (2015)
34. Ben-David, S., et al.: Analysis of representations for domain adaptation. In: Advances in Neural Information Processing Systems, vol. 19 (2006)
35. Ben-David, S., et al.: A theory of learning from different domains. Mach. Learn. **79**(1), 151–175 (2010)
36. Long, M., et al.: Deep transfer learning with joint adaptation networks. In: International Conference on Machine Learning. PMLR (2017)

37. Deng, Z., Luo, Y., Zhu, J.: Cluster alignment with a teacher for unsupervised domain adaptation. In: Proceedings of the IEEE/CVF International Conference on Computer Vision (2019)
38. Chen, M., et al.: Adversarial-learned loss for domain adaptation. In: Proceedings of the AAAI Conference on Artificial Intelligence, vol. 34. no. 04 (2020)
39. Wang, S., Zhang, L.: Self-adaptive re-weighted adversarial domain adaptation. arXiv preprint arXiv:2006.00223 (2020)
40. Pei, Z., et al.: Multi-adversarial domain adaptation. In: Thirty-Second AAAI Conference on Artificial Intelligence (2018)
41. Wang, S., et al.: Progressive adversarial networks for fine-grained domain adaptation. In: Proceedings of the IEEE/CVF Conference on Computer Vision and Pattern Recognition (2020)
42. Jia, Y., et al.: Caffe: convolutional architecture for fast feature embedding. In: Proceedings of the 22nd ACM International Conference on Multimedia (2014)
43. Long, M., et al.; Unsupervised domain adaptation with residual transfer networks. In: Advances in Neural Information Processing Systems, vol. 29 (2016)
44. Sankaranarayanan, S., et al.: Generate to adapt: aligning domains using generative adversarial networks. In: Proceedings of the IEEE Conference on Computer Vision and Pattern Recognition (2018)
45. Kang, G., et al.: Deep adversarial attention alignment for unsupervised domain adaptation: the benefit of target expectation maximization. In: Proceedings of the European Conference on Computer Vision (ECCV) (2018)
46. Saito, K., et al.: Maximum classifier discrepancy for unsupervised domain adaptation. In: Proceedings of the IEEE Conference on Computer Vision and Pattern Recognition (2018)
47. Pinheiro, P.O.: Unsupervised domain adaptation with similarity learning. In: Proceedings of the IEEE Conference on Computer Vision and Pattern Recognition (2018)
48. He, K., et al.: Deep residual learning for image recognition. In: Proceedings of the IEEE Conference on Computer Vision and Pattern Recognition (2016)
49. Ganin, Y., Lempitsky, V.: Unsupervised domain adaptation by backpropagation. In: International Conference on Machine Learning. PMLR (2015)
50. Sun, Y., et al.: Test-time training with self-supervision for generalization under distribution shifts. In: International Conference on Machine Learning. PMLR (2020)

Meta Pseudo Labels for Anomaly Detection via Partially Observed Anomalies

Sinong Zhao[1], Zhaoyang Yu[1], Xiaofei Wang[1], Trent G. Marbach[2], Gang Wang[1], and Xiaoguang Liu[1(✉)]

[1] College of Computer Science, NanKai-Orange D.T. Joint Lab, Nankai University, Tianjin, China
{zhaosn,yuzz,wangxf,wgzwp,liuxg}@nbjl.nankai.edu.cn
[2] Department of Mathematics, Toronto Metropolitan University, Toronto, Canada

Abstract. General anomaly detection based on weakly supervised or partially observed anomalies has been an important research. However, most such algorithms treat the unlabeled set as a substitute for normal samples and ignore the potential anomalies in it, which fails make full use of the abnormal supervision information. To address this issue, we propose a meta-pseudo-label based framework for anomaly detection (MPAD). The framework strives to obtain effective pseudo anomalies from the unlabeled samples to supplement the observed anomaly set. Specifically, a teacher network is improved based on the feedback of a student network on a validation set, thereby generating more conducive pseudo anomalies to assist the student network while incurring less confirmation bias. Extensive experiments show that the proposed MPAD algorithm outperforms current popular algorithms on five real datasets.

Keywords: Anomaly Detection · Semi-Supervised Learning · Meta Pseudo-Label

1 Introduction

Anomalies are generally defined as behaviors or events that are different from most normal situations which are rare but extremely harmful. Therefore accurate detection of anomalies are essential within many environments. Such environments may include fraud detection in finance [1], disease detection in clinical medicine [2], web intrusion detection [3] in network security, etc.

There have been many traditional anomaly detection algorithms based on unsupervised learning [4–6] or only normal class observed [7–9]. They usually assume that there are no observed anomalies during training and lose the chance to take advantage of the abnormal information. Consequently, a series of anomaly detection algorithms recently emerged that train models via a large number of unlabeled samples along with a few observed anomalies [10–12]. This setting is more in line with actual application scenarios, which can not only make up for the

X. Wang et al. (Eds.): DASFAA 2023, LNCS 13946, pp. 100–109, 2023.
https://doi.org/10.1007/978-3-031-30678-5_8

lack of supervision information in the unsupervised algorithms, but also reduce the burden of abnormal label collection in supervised schemes. Nevertheless, most of them directly regard unlabeled samples as a normal set, which may be unreasonable for some datasets containing a non-negligible amount of anomalies in unlabeled set. The core problem is finding out how to leverage the unlabeled samples to enhance the anomaly detection models.

Semi-supervised learning [13,14] is an appropriate choice to apply to anomaly detection due to its adequate mining of unlabeled samples. Some previous anomaly detection works [15,16] which are in semi-supervised frame simply applied unsupervised algorithms on unlabeled samples. In this paper, we employ a pseudo-label algorithm on the unlabeled set to find a set of *pseudo anomalies*.

Meta pseudo label (MPL) [17] takes the idea of meta-learning, which the teacher network continuously adjusts to reduce the confirmation bias using the feedback of the student network on the labeled samples. Inspired by MPL, we introduce a Meta-Pseudo-Label Anomaly Detection (MPAD) method in this paper. MPAD exploits the feedback of the student network on pseudo anomalies to influence the update of the teacher network. Meta pseudo anomalies (MPAs) then generated by the teacher network not only have less confirmation bias but also assist the student network to be more generalized on test set. In our implementation, we withhold a fixed validation set to judge the detection performance of the student network, and in turn the difference in performance is treated as a reward or punishment during the training of the teacher network.

The major contributions of this paper are summarized as follows:

– We propose an anomaly detection framework with partially observed anomalies which employs the pseudo-label algorithm to increase the content and quality of observed anomalies, thereby improving the accuracy of the anomaly detection model;
– The feedback of student model is used to correct the update direction of the teacher network, so that the teacher network can generate more beneficial pseudo anomalies;
– Extensive experimental results on datasets in five different fields show that the proposed MPAD framework exceeds five most currently popular algorithms in effectiveness.

2   Related Work

2.1   Anomaly Detection Methods

Traditional anomaly detection algorithms mainly follow the unsupervised setting. They cannot take advantage of existing anomaly information. Similar settings to this paper are semi-supervised or weakly-supervised based anomaly detection. One class of semi-supervised anomaly detection methods assumes that only normal samples are available when building a model. The classic algorithms are OCSVM [7] and deep support vector data description (SVDD) [8]. As they only learn patterns of the normal category, any pattern that differs from the normal ones is considered as an anomaly. The advantage of this approach is that it can reduce the overfitting problem of abnormal learning. They generally assume that the data are similar within a class and they are mostly applicable to situations with a large number of positive samples. Another class of semi-supervised anomaly detection methods presumes that a small amount of labeled normal and anomalies are available in addition to unlabeled ones, e.g., DeepSAD [15] and the method in [16]. They are both based on SVDD. Generally speaking, these models outperform unsupervised algorithms due to the presence of supervised information. Some work [10–12,19,20] have the same detection settings as our MPAD and focus on a small number of observable anomalies and unlabeled samples. Yet, most of these works assume that unlabeled samples are normal, and our model extracts reliable pseudo anomalies from unlabeled samples to enhance the utilization of supervised information.

2.2 Semi-supervised Methods

At present, semi-supervised algorithms [13,14] are mainly based on consistency, pseudo-labels, and a class of hybrid algorithms. Consistency algorithms are mainly based on the assumption that different representations of the same sample can yield the same results on downstream tasks. Many of them rely on rich data augmentation. But pseudo-labels methods have no such problem. The meta pseudo label [17] method used the results of a student network on the labeled samples as the feedback to a teacher network, reducing the pseudo labels' confirmation bias. To the best of our knowledge, there are currently no anomaly detection algorithms based on partially observed anomalies that use pseudo-label algorithms. We propose a general framework of MPAD based on MPL, which can employ any network structure as the teacher and the student network, and is compatible with various types of data.

3 Methods

3.1 Preliminaries

We follow the setting that partially anomalies are observed in anomaly detection. Notationally, the dataset is represented by $\mathcal{D} = \{\mathcal{D}_{\mathrm{L}}, \mathcal{D}_{\mathrm{U}}\}$. $\mathcal{D}_{\mathrm{L}}$ notes the partially known anomalies set and $\mathcal{D}_{\mathrm{U}}$ is the unlabeled set in which normal samples are much more than anomalies. $\mathcal{D}_{\mathrm{L}} = \{(x_1, y_1), \cdots, (x_K, y_K)\}$, $\mathcal{D}_{\mathrm{U}} = \{x_{K+1}, \cdots, x_{K+N}\}$, where $x_i \in \mathcal{X}$, $\mathcal{X} = \mathbb{R}^d$, $y_i = 1$, $y_i \in \mathcal{Y}$, $\mathcal{Y} = \{0,1\}$. Additionally, we follow the description of the models in pseudo-label algorithms. We define the teacher models that provide pseudo labels T, and their parameters θ_T. Student models that take pseudo labels, which in this paper are the anomaly detection models, are called S and the corresponding parameters are θ_S. We expect to train an anomaly detection model leveraging the dataset $\mathcal{D}$ and implement the model on the test set to examine its performance.

3.2   Meta Pseudo Anomaly Detection Scheme

We first introduce the basic pseudo-label algorithm, which obtains the distribution probability of the sample from a neural network, and gets the hard pseudo-label y^{PL} by a threshold λ:

$$y^{\mathrm{PL}} = \mathbb{1}\left[T(x_u; \theta_T) \geqslant \lambda\right], \tag{1}$$

in which $x_u \in \mathcal{D}_{\mathrm{U}}$ and $T(x_u; \theta_T)$ is the probability that x_u belongs to a particular class output by the teacher network. This formula is also used to generate pseudo anomalies when applied in anomaly detection.

This simplest pseudo-label method works well in anomaly detection, but the performance of student detection models are limited by the accuracy of the pseudo-labels produced by the teacher network. To improve this accuracy, we borrow the idea of Meta Pseudo Labels (MPL) to the pseudo anomalies generation which we called Meta Pseudo Anomalies (MPA).

We first utilize the teacher network to generate pseudo anomalies following Eq. (1). Here the teacher network refers to the self-training schedule, i.e., executing two steps in a loop: (1) Train a classifier using an already labeled dataset. Here we treat the unlabeled set as normal; (2) Use the trained classifier to label the unlabeled data, and add those with high prediction confidence to the labeled set. Based on these pseudo anomalies, the optimization objective θ_S^{MPA} of the student network is:

$$\theta_S^{\mathrm{MPA}} = \operatorname*{argmin}_{\theta_S} \mathcal{L}_{\mathrm{CN}}\Big(S\big([x_u, x_l, \mathrm{MPA}]; \theta_S\big), y\Big), \tag{2}$$

where $x_l \in \mathcal{D}_{\mathrm{L}}$ are labeled anomalies and MPA is added to this set when training. $\mathcal{L}_{\mathrm{CN}}$ is the loss of student network presented in Eq. (10) and y is the true labels with the pseudo labels. So far this is a standard pseudo-label algorithm using self-training.

Seeing that the ultimate purpose of the student network is to improve the generalization effect on the test data. We expect the teacher network to generate pseudo anomalies that meet this goal. We manage to separate part of data called $\mathcal{D}_{\mathrm{V}}$ from $\mathcal{D}$ to do this. Since MPA is generated according to θ_T as in Eq. (1), the optimization result for student network can be seen as a function of θ_T which we write it as $\theta_S^{\mathrm{MPA}}(\theta_T)$. The overall goal is to minimize the loss of the student network on $\mathcal{D}_{\mathrm{V}}$:

$$\min_{\theta_S, \theta_T} \mathcal{L}_{\mathrm{CN}}\Big(S\big(\mathcal{D}_{\mathrm{V}}; \theta_S^{\mathrm{MPA}}(\theta_T)\big), y_v\Big). \tag{3}$$

We expect that this objective will correct the update direction of the teacher network and further improve the performance of the detection network.

There are two variables in the target at the same time, so the parameters cannot be updated directly by calculating the derivative. Here we update the two parameters step-by-step depending on meta-learning. In order to achieve the approximate optimization, we let θ_T and θ_S update alternately. And only one step is updated each time along the gradient direction rather than directly updating to the current optimal. This is because the current optimum is only a local optimum of the objective function according to the meta-learning theory. θ_S update one step to θ'_S first:

$$\theta'_S = \theta_S - \eta_S \nabla_{\theta_S} \mathcal{L}_{\mathrm{CN}}(\theta_T, \theta_S). \tag{4}$$

The contrastive above is applied with MPA generated by θ_T. θ_T is updated leveraging the updated student network θ'_S:

$$\theta'_T = \theta_T - \eta_T \nabla_{\theta_T} \mathcal{L}_T(\theta'_S). \tag{5}$$

We denote the objective function as H and split the derivative into a product of two derivatives:

$$\frac{\partial H}{\partial \theta_T} = \frac{\partial H}{\partial \theta'_s} \cdot \frac{\partial \theta'_s}{\partial \theta_T} = h \cdot \frac{\partial \mathcal{L}_{\mathrm{CE}}(\hat{y}_u, T(x_u; \theta_T))}{\partial \theta_T}, \tag{6}$$

where $\hat{y}_u$ is the pseudo-labels and $h = \mathcal{L}_{\mathrm{CN}}(\theta_S) - \mathcal{L}_{\mathrm{CN}}(\theta'_S)$ following the Taylor's Formula. Both two contrastive losses are computed on the validation set. The second term in the last equation is the cross-entropy loss between the teacher network output and the pseudo-labels. In addition, we also trained the teacher network with the loss on the labeled samples. The total loss is as follows:

$$\mathcal{L}_T = \mathcal{L}_{\mathrm{CE}}(T(x_l; \theta_T), y_l) + \big(\mathcal{L}_{\mathrm{CN}}(\theta_S) - \mathcal{L}_{\mathrm{CN}}(\theta'_S)\big) \times \mathcal{L}_{\mathrm{CE}}(T(x_u; \theta_T), \hat{y}_u). \tag{7}$$

Here we assume that the unlabeled set is normal, and calculate the standard cross-entropy loss together with the labeled anomalies as the first term above.

3.3   Student Anomaly Learner

We chose DevNet [11] as the student model, which itself is a model based on a small number of observed anomalies. DevNet makes efficient use of observed anomalies. Its performance tends to increase on most datasets with the increase of observed anomalies. The main principles are: First, L abnormal scores of normal samples r_i are sampled from a standard Gaussian distribution, and the mean value is used as the reference score μ_r of the normal points:

$$\mu_r = \frac{1}{L} \sum_{i=1}^{L} r_i, r_i \sim \mathcal{N}(\mu = 0, \sigma = 1). \tag{8}$$

Then the z-score is applied to calculate the gap between the training data z_i and the reference score,

$$\mathrm{dev}(z_i) = \frac{z_i - \mu_r}{\sigma_r}. \tag{9}$$

Finally, the distance is increased between the abnormal points and the reference score while reducing the gap between the normal points and the reference score through the contrastive loss $\mathcal{L}_{\mathrm{CN}}$:

$$\mathcal{L}_{\mathrm{CN}} = (1 - y_i) \cdot |\mathrm{dev}(z_i)| + y_i \cdot \max\big(0, \delta - \mathrm{dev}(z_i)\big). \tag{10}$$

3.4 Total Flow of MPAD

We first initialize the teacher network and the student network with the observed anomalies and unlabeled samples, respectively. During training, batches are obtained from the initial dataset in a one-to-one ratio of normal and abnormal. The supervised loss of the teacher network is calculated within a batch. At the same time, a one-step update is made to the student network, and loss $(\mathcal{L}_{\mathrm{CN}}(\theta_S))$ on the validation set are recorded. The teacher network generates pseudo anomalies (MPA) according to the given probability threshold $\mathcal{P}^{\mathrm{MPA}}$. We add these MPAs to the observed anomalies set to secondly update the student network, and also record the loss $(\mathcal{L}_{\mathrm{CN}}(\theta_S'))$ on the validation set. Finally, the teacher network is updated using the deviation of the loss on the student network and its own loss on labeled set.

4 Experiments

4.1 Experimental Settings

Datasets. We evaluate the proposed MPAD on five public datasets covering different fields.

Census is a dataset of US Census from 1994 and 1995 which includes 500 variables related to demographics and employment. Among them, very few people with an income more than 50,000 are regarded as anomalies for detection.

Campaign comes from a telemarketing campaign of a Portuguese bank. It contains 62 attributes such as customer information and economic activities. A small number of users who chose to subscribe to the banking product are identified as abnormal.

Thyroid is established to study whether patients had hypothyroidism. There are three categories which are normal, hyperfunctioning and dysfunctional. Here we merge the latter two categories as anomalies.

Arrhythmia is a dataset for studying arrhythmia and contains information about the patients' physical conditions and heart rates. Patients are classified into one normal class ECGs and 15 different types of arrhythmias. Here we combine the arrhythmia classes as anomalies.

Pima is a research dataset of diabetes in Pima Indian women, which comes from the UCI repository. Here we label those with diabetes as anomalies.

Baselines

- **DevNet** [11]: focuses on learning the anomaly scores directly rather than improving the representations. It designs a reference score of the normal samples according to the data distribution, and combines the contrastive loss to isolate the anomaly scores of normal samples and abnormal samples. It is an end-to-end anomaly detection algorithm based on partially observed anomalies and is also the student model of our MPAD.

- **DeepSAD** [15]: is a semi-supervised version of SVDD. It builds models with both unlabeled and labeled data. It places the normal samples close to the center of the hypersphere, while the abnormal samples are far from the surface of the hypersphere according to the label information, which improves the performance of SVDD.
- **SS-DGM** [21]: is a semi-supervised deep generative model. It combines a discriminative model of latent features with a generative semi-supervised model. This paper follows the setting of SS-DGM in [15] and applies it to anomaly detection.
- **OCSVM** [7]: is a classic single-class anomaly detection model which only use normal samples for training. It builds a hyperplane to segment samples, which maximize the separation between positive and negative samples.
- **iForest** [18]: is an efficient unsupervised model in anomaly detection. It achieves the isolation of anomalies by recursive segmentation of eigenvalues.

Implementation Details. We apply an MLP with a hidden layer as the teacher network in the implementation of MPAD. The architecture of the student model (DevNet) is the same as the teacher network, except that the teacher network outputs a two-dimensional vector, and the DevNet outputs a single-dimensional vector to calculate different losses. The number of neurons in the hidden layer is 64. In addition, the teacher network and the student network are optimized using the SGD and Adam optimizers, with learning rate of 0.03 and 0.001, respectively. The training and test sets of all algorithms are in a ratio of 8:2 with the random state of 42. DeepSAD, SS-DGM and the proposed MPAD are implemented with pytorch, while iForest and OCSVM are achieved with sklearn.

Metrics. AUC-ROC: is the area under the curve with the false positive rate as the abscissa and the true positive rate as the ordinate. It is a comprehensive evaluation criterion which represents the expected generalization performance of the model in different situations. Generally, if one curve can completely surround the other, it means that the former performs better than the latter, so the area under the curve is a good representation of the pros and cons of a model. In anomaly detection, it tends to show the ability to recognize normal classes due to extreme class imbalance.

AUC-PR: is the area under the curve drawn with the recall of the positive samples as the abscissa and the precision as the ordinate. It only pays attention to positive samples (anomalies). Similar to AUC-ROC, one curve is wrapped by another, indicating that the latter is more capable of achieving high recall and precision at the same time. In anomaly detection, we focus more on the detection ability of the anomaly category, so we care more about AUC-PR values than AUC-ROC values.

4.2   Effectiveness Results

The number of available anomalies in this comparative experiment is 30, and the noise of the training set is 0.02. Our MPAD and all baselines pick the best per-

Table 1. The performance w.r.t. AUC-ROC and AUC-PR among the proposed MPAD and the baselines on five tabular datasets with 30 labeled anomalies and 2% noise injection for training. The best performance for each dataset is boldfaced.

Datasets	AUC-ROC						AUC-PR					
	MPAD	DevNet	DeepSAD	SS-GDM	OCSVM	iForest	MPAD	DevNet	DeepSAD	SS-GDM	OCSVM	iForest
Census	**0.8906**	0.8284	0.7354	0.7683	0.5352	0.6165	**0.4118**	0.2895	0.0682	0.0420	0.0746	0.0744
Campaign	**0.8399**	0.8073	0.6337	0.7047	0.6942	0.7160	**0.4458**	0.3694	0.2334	0.2411	0.2530	0.2892
Thyroid	**0.9136**	0.8790	0.8957	0.7355	0.6153	0.7105	**0.6720**	0.3361	0.5921	0.4268	0.1225	0.2173
Arrhythmia	0.7557	0.7300	0.7751	**0.8107**	0.6844	0.7684	**0.5655**	0.4275	0.3912	0.4617	0.3774	0.5122
Pima	0.6845	0.7182	0.6698	**0.7598**	0.5487	0.6423	**0.6787**	0.6315	0.5788	0.5847	0.4378	0.4912

forming hyperparameters and demonstrate the optimal performance in Table 1. It can be seen that the proposed MPAD has achieved the best AUC-PR on all five datasets and the best AUC-ROC on three datasets. Among them, AUC-PR of MPAD exceeds the optimal result on each dataset by 12.2%, 7.6%, 8%, 5.3%, and 4.7%, respectively. This illustrates the advancement achieved by our algorithm. In general, the unsupervised algorithm iForest and the single-class model OCSVM do not perform as well as the first three baselines. Since Deep-SAD and SS-GDM algorithms are also semi-supervised methods, they show the performance only second to MPAD on the *Thyroid* dataset. DevNet obtains the second position on the rest of the datasets. Among them, SS-GDM shows higher AUC-ROC on *Arrhythmia* and *Pima*, proving that it has better recognition of normal samples.

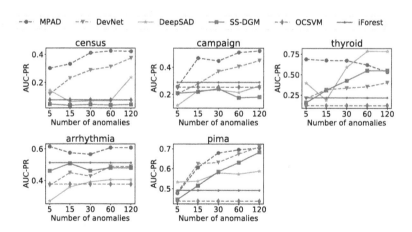

Fig. 1. AUC-PR w.r.t. No.labeled anomalies on five datasets.

4.3 Data Efficiency Study

This experiment aims to test how the performance of algorithms change as the observed anomalies increase. The noise ratio is fixed at 0.02 during this experiment, and the observed anomalies are changed from 5 to 15, 30, 60, and 120 for modeling. It can be seen from Fig. 1 that our MPAD always maintains a high AUC-PR on *Census*, *Campaign*, and *Arrhythmia* datasets, and the results on *Pima* have a significant upward trend with the increase of observed anomalies. As for *Thyroid*, MPAD maintains a high level when there are fewer visible anomalies, and the effect decreases when the number of anomalies increases. The number of visible anomalies will act on the initialization effect of the teacher network. This result indicates that there are visible anomalies overlapping with the normal ones, making the teacher's performance drop further affecting the result of student. The two algorithms, OCSVM and iForest, are not influenced by the number of visible anomalies. They have certain advantages when there are few visible anomalies, yet they cannot make effective use of this information and lose their odds as the number of anomalies increases.

5   Conclusion

We introduce pseudo-label algorithms to partially-observed-anomalies anomaly detection. Thus, unlabeled data is used reasonably, and valuable pseudo anomalies can be extracted to assist the establishment of anomaly detection models. The most important part is that the proposed meta pseudo anomalies generation procedure makes the teacher network and the student network update alternately, and the teacher network is subject to both supervised information and the student network's feedback. Comprehensive experiments show that the pseudo anomalies generated in this way are better than the general pseudo labels, and our framework outperforms the other state-of-the-art anomaly detection methods on five public datasets.

Acknowledgements. This work is supported by the National Natural Science Foundation of China (62272253, 62272252, 62141412) and Fundamental Research Funds for the Central Universities.

References

1. West, J., Bhattacharya, M.: Intelligent financial fraud detection: a comprehensive review. Comput. Secur. **57**, 47–66 (2016)
2. Fernando, T., Gammulle, H., Denman, S., Sridharan, S., Fookes, C.: Deep learning for medical anomaly detection-a survey. ACM Comput. Surv. (CSUR) **54**(7), 1–37 (2021)
3. Khraisat, A., Gondal, I., Vamplew, P., Kamruzzaman, J.: Survey of intrusion detection systems: techniques, datasets and challenges. Cybersecurity **2**(1), 1–22 (2019). https://doi.org/10.1186/s42400-019-0038-7
4. Hoffmann, H.: Kernel PCA for novelty detection. Pattern Recogn. **40**(3), 863–874 (2007)
5. Chen, J., Sathe, S., Aggarwal, C., Turaga, D.: Outlier detection with autoencoder ensembles. In: Proceedings of the 2017 SIAM International Conference on Data Mining, pp. 90–98. SIAM (2017)

6. Zhou, C., Paffenroth, R.C.: Anomaly detection with robust deep autoencoders. In: Proceedings of the 23rd ACM SIGKDD International Conference on Knowledge Discovery and Data Mining, pp. 665–674 (2017)
7. Li, K.L., Huang, H.K., Tian, S.F., Xu, W.: Improving one-class SVM for anomaly detection. In: Proceedings of the 2003 International Conference on Machine Learning and Cybernetics (IEEE Cat. No. 03EX693), vol. 5, pp. 3077–3081. IEEE (2003)
8. Ruff, L., et al.: Deep one-class classification. In: International Conference on Machine Learning, pp. 4393–4402. PMLR (2018)
9. Bergman, L., Hoshen, Y.: Classification-based anomaly detection for general data. arXiv preprint arXiv:2005.02359 (2020)
10. Zhang, Y.L., Li, L., Zhou, J., Li, X., Zhou, Z.H.: Anomaly detection with partially observed anomalies. In: Companion Proceedings of the The Web Conference 2018, pp. 639–646 (2018)
11. Pang, G., Shen, C., van den Hengel, A.: Deep anomaly detection with deviation networks. In: Proceedings of the 25th ACM SIGKDD International Conference on Knowledge Discovery & Data Mining, pp. 353–362 (2019)
12. Pang, G., van den Hengel, A., Shen, C., Cao, L.: Toward deep supervised anomaly detection: reinforcement learning from partially labeled anomaly data. In: Proceedings of the 27th ACM SIGKDD Conference on Knowledge Discovery & Data Mining, pp. 1298–1308 (2021)
13. Yang, X., Song, Z., King, I., Xu, Z.: A survey on deep semi-supervised learning. arXiv preprint arXiv:2103.00550 (2021)
14. Chapelle, O., Scholkopf, B., Zien, A.: Semi-supervised learning (Chapelle, O. et al., EDS.; 2006)[book reviews]. IEEE Trans. Neural Netw. **20**(3), 542–542 (2009)
15. Ruff, L., et al.: Deep semi-supervised anomaly detection. arXiv preprint arXiv:1906.02694 (2019)
16. Görnitz, N., Kloft, M., Rieck, K., Brefeld, U.: Toward supervised anomaly detection. J. Artif. Intell. Res. **46**, 235–262 (2013)
17. Pham, H., Dai, Z., Xie, Q., Le, Q.V.: Meta pseudo labels. In: Proceedings of the IEEE/CVF Conference on Computer Vision and Pattern Recognition, pp. 11557–11568 (2021)
18. Liu, F.T., Ting, K.M., Zhou, Z.H.: Isolation-based anomaly detection. ACM Trans. Knowl. Discov. Data (TKDD) **6**(1), 1–39 (2012)
19. Pang, G., Cao, L., Chen, L., Liu, H.: Learning representations of ultrahigh-dimensional data for random distance-based outlier detection. In: Proceedings of the 24th ACM SIGKDD International Conference on Knowledge Discovery & Data Mining, pp. 2041–2050 (2018)
20. Pang, G., Ding, C., Shen, C., van den Hengel, A.: Explainable deep few-shot anomaly detection with deviation networks. arXiv preprint arXiv:2108.00462 (2021)
21. Kingma, D.P., Mohamed, S., Jimenez Rezende, D., Welling, M.: Semi-supervised learning with deep generative models. In: Advances in Neural Information Processing Systems, vol. 27 (2014)

Learning Enhanced Representations via Contrasting for Multi-view Outlier Detection

Xiaocong Chen[1], Xinye Wang[1(✉)], Yang Wang[1], Chao Han[2,3], and Lei Duan[1(✉)]

[1] School of Computer Science, Sichuan University, Chengdu, China
{chenxiaocong,wangxinye,wangyang}@stu.scu.edu.cn, leiduan@scu.edu.cn
[2] School of Computer Science and Engineering, University of Electronic Science and Technology of China, Chengdu, China
scuhanchao@163.com
[3] Nuclear Power Institute of China, Chengdu, China

Abstract. Multi-view outlier detection has attracted rapidly growing attention to researchers due to its wide applications. However, most existing methods fail to detect outliers in more than two views. Moreover, they only employ the clustering technique to detect outliers in a multi-view scenario. Besides, the relationships among different views are not fully utilized. To address the above issues, we propose ECMOD for learning enhanced representations via contrasting for multi-view outlier detection. Technically, ECMOD leverages two channels, the reconstruction and the constraint view channels, to learn the multi-view data, respectively. The two channels enable ECMOD to capture the rich information better associated with outliers in a latent space due to fully considering the relationships among different views. Then, ECMOD integrates a contrastive technique between two groups of embeddings learned via the two channels, serving as an auxiliary task to enhance multi-view representations. Furthermore, we utilize neighborhood consistency to uniform the neighborhood structures among different views. It means that ECMOD has the ability to detect outliers in two or more views. Meanwhile, we develop an outlier score function based on different outlier types without clustering assumptions. Extensive experiments on real-world datasets show that ECMOD significantly outperforms most baselines.

Keywords: Outlier detection · Multi-view data · Contrastive learning

1 Introduction

Outlier detection has been extensively studied and applied in various applications due to its ability to detect outliers based on a single-view data source, such as intrusion detection systems [9], credit-card fraud [1], and medical diagnosis [14]. However, in the real world, the same instance described from different views can produce rich feature information. Therefore, outlier detection based on multi-view has emerged as a crucial research task [7]. For multi-view outlier detection, the abnormal condition is usually divided into: (i) class outliers, which are the instances that have different neighborhood relationships in different views, *i.e.*, they show inconsistent behaviors; (ii) Attribute outliers, which are often very different from normal instances in each view; (iii) Class-attribute outliers, which exhibit the class outliers behaviors in some views and exhibit the attribute outliers behaviors in the others. The three abnormal conditions can reflect the dataset's topological structure in the original space, which is the key to outlier detection. Therefore, we focus on the three types of outliers simultaneously.

Existing multi-view outlier detection methods [4, 8, 10–12, 15, 16] follow the same way in the previous work [6] to generate multi-view data in given outlier ratios. Then, they explore the consensus properties of multiple views with the information among them, achieving a better performance than single-view methods [3, 5]. They have made some success partly, but still leave some challenges. *How to perform better with the arbitrary number of views?* These approaches [6, 15, 16] usually consider the two-view scenario and capture the consensus nature of multiple views by considering the information between pairs of views. This means that the performance of these methods is invalidated when faced with three or more views due to the exponential computational complexity of using the pairwise approach. *How to detect outliers without clustering assumptions?* Some methods [10, 11] have shown the performance on datasets under the clustering scenario. However, when the distribution of the dataset in the original space is more dispersed, the clustering algorithm will obtain a poor accuracy. Therefore, detecting outliers without clusters needs to be addressed urgently in multi-view data. *How to learn a data representation with rich feature information from different views?* The key of multi-view outlier detection is to fully integrate the rich feature information contained in different views. However, some existing methods typically employ autoencoders to extract intrinsic information under the assumption that each view is independent, so that the relations among different views are not considered.

To address the challenges above, we propose a novel method, named ECMOD, for learning enhanced representations via contrasting for multi-view outlier detection. The key characteristics of ECMOD include: (i) it can detect outliers on more than two views due to the ability of neighborhood consistency which tries to uniform the neighborhood structures among different views; (ii) It can measure the difference between the representations via contrastive learning to ensure that the distribution of instances in the original space is similar to that in the latent space with the constraint between views; (iii) It designs the reconstruction and constrained view channels respectively to learn two informative representations of multi-view data for capturing the structure of each view and the relationships between views. Extensive experiments show that ECMOD achieves significant improvements on real-world datasets.

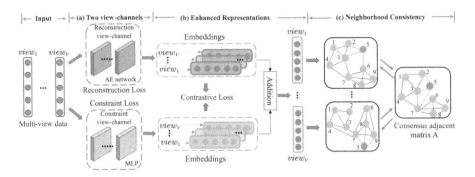

Fig. 1. Illustration of the framework of ECMOD. (a) Two channels to learn embeddings of multi-view data; (b)contrastive learning to enhance representations; (c) trying to uniform the neighborhood structure of different views (the orange arrows). The outlier score jointly considering with reconstruction error and neighborhood distance in the latent space can characterize the multi-view outliers explicitly.

2   The Design of ECMOD

First, suppose there are V views and N objects, we can get a multi-view dataset $\mathcal{X} = \{x_1^1, \cdots, x_1^V, x_2^1, \cdots, x_2^V, x_N^1, \cdots, x_N^V\}$ containing $N \times V$ d-dimensional instances, where d is the dimension of each view which differs in each view. x_i^v denotes the i-th instance in the v-th view. Let $\mathrm{X} = \{x^1, \cdots, x^V\}$ denote a set of an object's feature vectors in all views. Then $\widetilde{x} = [x^1; \cdots; x^V]$ denotes a feature vector of an object in all views. The framework of ECMOD is shown in Fig. 1.

2.1   Channel-Specific Representations

To preserve the intrinsic information within the view and capture the constraint relationships between different views, we adopt the reconstruction and the constrained view-channels to generate the data representations purposefully.

Reconstruction View-Channel. In order to handle high-noisy features in the dataset, we first design a reconstruction view-channel. In this channel, we utilize an autoencoder (AE) network to learn a low-dimensional feature vector with little noise, which is denoted as $(f(\cdot); g(\cdot))$, where $f(\cdot), g(\cdot)$ are the encoder and decoder for each view, respectively. Let $f(x)$ be the latent reperesentation of each object in each view, and $\hat{x} = g(f(x))$ be the recoverd reperesentations of AE.

This channel train multi-view data with normal and abnormal instances to ensure that the intrinsic information is preserved in the learned latent space. To preserve the intrinsic information, the reconstruction loss should be minimized:

$$\mathcal{L}_1 = \sum_{x \in \mathcal{X}} \|x - \hat{x}\|^2, \tag{1}$$

Generally, the reconstruction loss of a normal instance should be smaller than that of an abnormal one.

Constrained View-Channel. We develop another constrained view-channel in order to obtain view-specific representations. This channel is composed of multi-layer perceptron (MLP) networks including a common one and V view-specific ones. For an instance, the common network generates a common representation and each view-specific network generates an offset across the constrained view-channel.

To generate the view-specific representation, we add the corresponding offset to the common representation. View-specific representations are essential in data reconstructions, as they can satisfy the differences between different views. Thus, the representation of x in the latent space can be captured as $h = \mathrm{MLP}(\tilde{x}) + \mathrm{MLP}(x)$, $x \in \mathrm{X}$. Note that the hyper-parameters of these two MLP are different. Then, the constraint representation set is denoted as follows:

$$\mathcal{H} = \left\{ h_1^1, \cdots, h_1^V, h_2^1, \cdots, h_2^V, h_N^1, \cdots, h_N^V \right\}, \tag{2}$$

where $\mathcal{H}$ contains $N \times V$ l-dimensional feature vectors, where l is the dimension of the output layer of the MLP network.

The constraint relations between instances are reflected in the view-specific representations which can preserve the structure of instances in the original view space. Specifically, to make the structures in the latent space closer to the original space, the constraint in all views is denoted as:

$$\mathcal{L}_2 = \sum_{v=1}^{V} \sum_{i,j=1}^{N} \left| \frac{\|x_i^v - x_j^v\|^2}{d} - \frac{\|h_i^v - h_j^v\|^2}{l} \right|, \tag{3}$$

where $x_i^v, x_j^v \in \mathcal{X}$ and $h_i^v, h_j^v \in \mathcal{H}$.

2.2 The Enhancement of Representations

The primary challenge of this part is about how to complement the two groups of view embeddings to enhance the latent representations. Thus, we devise a contrastive technique between the two groups of view embeddings.

Contrastive Learning. Contrastive learning is a kind of self-supervised learning [13]. We regard the two channels in ECMOD as two aspects characterizing different aspects of multi-view data with three types of outliers. We then contrast the two groups of embeddings learned via two channels. A standard binary cross-entropy loss is adopted in all views as our learning objective:

$$\mathcal{L}_3 = \sum_{v=1}^{V} \sum_{i=1}^{N} - \log\left(\sigma(\varphi\langle f(x_i^v), h_i^v \rangle)\right) - \log\left(1 - \sigma(\varphi\langle \tilde{f}(x_i^v), h_i^v \rangle)\right), \tag{4}$$

where σ is the sigmoid function and $\tilde{f}(x_i^v)$ (or $\tilde{h}_i^v$) is the negative instance obtained by corrupting $f(x^v)$ (or h^v) with row-wise and column-wise shuffling, and $\varphi\langle\cdot\rangle : \mathbb{R}^l \times \mathbb{R}^l \longmapsto \mathbb{R}$ is the discriminator function that takes two vectors as the input and then scores the agreement between them. We simply implement the discriminator as the inner product between the two vectors. Then we add up the two representations to get the enhanced representation:

$$z = f(x) + h, \tag{5}$$

where $h \in \mathcal{H}$ and $x \in \mathcal{X}$. We denote the enhanced representation set as:

$$\mathcal{Z} = \left\{ z_1^1, \cdots, z_1^V, z_2^1, \cdots, z_2^V, z_N^1, \cdots, z_N^V \right\}. \tag{6}$$

2.3   Neighborhood Consistency

After obtaining the enhanced representations, following the work [4], the neighborhood consensus matrix in the latent space can be computed.

Specifically, we use k nearest neighbors to refer to the neighborhood structure, which indicates that an instance has similar k nearest neighbors across different views while an outlier does not. We adopt the Gaussian kernel-based affinity matrix $\mathbf{A}$ to be the consensus adjacent matrix for each view, where $\mathbf{A}_{ij} = \exp\left(-\|z_i - z_j\|^2/2\delta^2\right)$ if z_j is the k nearest neighbors of z_i in this view or $\mathbf{A}_{ij} = 0$ otherwise.

Because $\mathbf{A}$ is approximate, the distances of the neighbors for each view should be minimized:

$$\min_{\mathbf{A}} \sum_{v=1}^{V} \sum_{i,j=1}^{N} \mathbf{A}_{ij} \|z_i^v - z_j^v\|^2. \tag{7}$$

Since the initial value of $\mathbf{A}$ is unknown, we first pre-train the deep AE and MLP network without the neighborhood consensus part on all multi-view data. Then we utilize the enhanced embeddings and k to initialize the consensus adjacent matrix $\mathbf{A}$. We optimize our objective function by using an alternating direction minimization (ADM) strategy. The final objective function of the proposed ECMOD is:

$$\min_{\mathbf{A}, f, g, \text{MLP}} \mathcal{L}_1 + \mathcal{L}_2 + \alpha \mathcal{L}_3 + \beta \sum_{v=1}^{V} \sum_{i,j=1}^{N} \mathbf{A}_{ij} \|z_i^v - z_j^v\|^2, \tag{8}$$

where $\alpha = 0.05$ is the hyper-parameter and β is the trade-off parameter. δ appeared in $\mathbf{A}$ is the bandwidth parameter. We set $\delta = 1.0$ by default.

2.4   Outlier Score Inference

Suppose that we can remove the abundant information by mapping the original view into a latent space, the multiple views of a normal instance have a similar neighborhood structure. Inspired by [4], after learning enhanced latent representations, for an instance, the outlier score with the reconstruction error $S_r(\widetilde{x})$ and the sum distance to the k nearest neighbors $S_k(\widetilde{x})$ can be defined as:

$$S(\widetilde{x}) = S_r(\widetilde{x}) + \beta S_k(\widetilde{x}),$$

$$s.t. \quad S_r(\widetilde{x}) = \sum_{v=1}^{V} \|x^v - \hat{x}^v\|^2, S_k(\widetilde{x}) = \sum_{v=1}^{V} \sum_{z' \in knn(z^v)} \|z^v - z'\|^2 \tag{9}$$

Table 1. Data characteristics of experimental datasets.

Dataset	#Classes	#Attributes	#Outliers	#Instances
Zoo	7	16	–	101
Wine	3	12	–	178
Wdbc	2	30	–	569
Pima	2	8	–	768
Arrhythmia	16	274	66	452
Yeast	11	8	64	1364
Letter	26	32	100	1600
Musk	2	166	97	3062
Thyroid	3	6	93	3772
Optdigits	10	64	150	5216
Mammography	2	6	260	11183
Shuttle	5	9	3511	49097
ForestCover	7	10	2747	286048

where β is the same trade-off parameter as Eq. (8). This strategy can effectively identify three outliers simultaneously.

3 Experiments

3.1 Datasets and Preprocessing

First, 4 widely used small-scale datasets from UCI[1] are utilized to demonstrate the effectiveness of ECMOD. Furthermore, we select 9 large-scale datasets from ODDS[2] for exhibiting the performance of our model on large datasets. The detailed descriptions of the datasets are shown in Table 1. The UCI datasets are originally multi-category datasets without outliers, however, the ODDS datasets have different number of outliers. We first clear outliers from the ODDS datasets where the label value is 1. Then, for "ForestCover" and "Shuttle" datasets, we randomly select 20000 and 10000 instances from the original dataset due to the programs cannot finish within expected time, respectively. And for other ODDS datasets except for "Arrhythmia", we randomly select 1000 instances from different classes.

For experimental purposes, due to the datasets are all single-view ones, we follow the method in [6] to generate multi-view data to simulate data from multiple sources. The next step is to generate outliers. We follow the method in [8,10] to pre-process the data with three types of multi-view outliers. Specifically, we released the source codes of ECMOD at Github[3].

---

[1] https://archive.ics.uci.edu/ml/datasets.php.
[2] http://odds.cs.stonybrook.edu/.
[3] https://github.com/scu-kdde/OAM-ECMOD-2023.

Table 2. Experimental results in AUC (%) on 4 UCI datasets in 2 or 3 views with different outlier ratios for attribute outliers, class-attribute outliers and class outliers.

Dataset			$HOAD_L$	$HOAD_G$	AP_L	AP_G	MLRA	LDSR	MODDIS	NCMOD	ECMOD
5%-5%-5%	Zoo	v = 2	55.0±8.0	54.0 ± 8.0	77.0 ± 6.0	79.0 ± 5.0	70.0 ± 6.0	85.0 ± 3.0	87.0 ± 4.0	90.4 ± 3.7	**95.5 ± 2.7**
		v = 3	48.0 ± 8.0	73.0 ± 6.0	87.0 ± 4.0	87.0 ± 3.0	71.0 ± 7.0	83.0 ± 4.0	87.0 ± 5.0	89.6 ± 2.9	**95.0 ± 2.7**
	Wine	v = 2	61.0 ± 5.0	59.0 ± 6.0	63.0 ± 9.0	63.0 ± 9.0	71.0 ± 5.0	78.0 ± 3.0	89.0 ± 3.0	86.2 ± 5.8	**94.3 ± 3.2**
		v = 3	61.0 ± 6.0	61.0 ± 6.0	65.0 ± 8.0	66.0 ± 8.0	77.0 ± 4.0	78.0 ± 3.0	88.0 ± 4.0	85.3 ± 9.2	**93.9 ± 3.6**
	Wdbc	v = 2	55.0 ± 6.0	63.0 ± 5.0	72.0 ± 4.0	72.0 ± 4.0	71.0 ± 3.0	95.0 ± 2.0	94.0 ± 2.0	95.2 ± 2.2	**98.4 ± 4.5**
		v = 3	59.0 ± 9.0	54.0 ± 13.0	63.0 ± 4.0	62.0 ± 4.0	69.0 ± 5.0	92.0 ± 2.0	94.0 ± 1.0	96.6 ± 2.4	**98.5 ± 3.8**
	Pima	v = 2	59.0 ± 7.0	58.0 ± 4.0	59.0 ± 3.0	59.0 ± 3.0	62.0 ± 4.0	71.0 ± 2.0	88.0 ± 2.0	88.2 ± 1.5	**89.4 ± 1.7**
		v = 3	64.0 ± 5.0	61.0 ± 3.0	50.0 ± 4.0	50.0 ± 4.0	64.0 ± 2.0	69.0 ± 2.0	87.0 ± 1.0	87.5 ± 2.2	**88.5 ± 1.7**
2%-5%-8%	Zoo	v = 2	49.0 ± 9.0	56.0 ± 7.0	77.0 ± 6.0	80.0 ± 5.0	67.0 ± 6.0	85.0 ± 4.0	86.0 ± 5.0	91.4 ± 3.3	**97.2 ± 3.4**
		v = 3	48.0 ± 8.0	66.0 ± 6.0	86.0 ± 4.0	87.0 ± 4.0	67.0 ± 6.0	84.0 ± 3.0	84.0 ± 5.0	93.1 ± 3.2	**95.4 ± 3.2**
	Wine	v = 2	57.0 ± 6.0	53.0 ± 6.0	69.0 ± 6.0	69.0 ± 6.0	64.0 ± 5.0	73.0 ± 5.0	84.0 ± 4.0	82.1 ± 3.9	**84.4 ± 4.3**
		v = 3	57.0 ± 6.0	55.0 ± 6.0	69.0 ± 5.0	71.0 ± 5.0	70.0 ± 5.0	71.0 ± 4.0	80.0 ± 4.0	76.7 ± 14.7	**86.8 ± 4.6**
	Wdbc	v = 2	56.0 ± 9.0	62.0 ± 4.0	81.0 ± 3.0	81.0 ± 3.0	66.0 ± 3.0	91.0 ± 3.0	90.0 ± 2.0	89.3 ± 2.7	**97.3 ± 6.0**
		v = 3	52.0 ± 5.0	53.0 ± 10.0	71.0 ± 4.0	70.0 ± 4.0	72.0 ± 5.0	91.0 ± 3.0	89.0 ± 2.0	90.1 ± 2.7	**96.6 ± 5.3**
	Pima	v = 2	48.0 ± 7.0	56.0 ± 4.0	63.0 ± 3.0	63.0 ± 3.0	59.0 ± 6.0	67.0 ± 2.0	**82.0 ± 2.0**	80.4 ± 2.6	81.0 ± 2.3
		v = 3	62.0 ± 5.0	59.0 ± 3.0	55.0 ± 3.0	55.0 ± 3.0	61.0 ± 2.0	64.0 ± 3.0	80.0 ± 2.0	79.4 ± 2.2	**81.0 ± 2.3**
8%-5%-2%	Zoo	v = 2	50.0 ± 10.0	49.0 ± 11.0	77.0 ± 4.0	79.0 ± 4.0	76.0 ± 5.0	86.0 ± 4.0	88.0 ± 4.0	90.5 ± 4.0	**95.1 ± 3.0**
		v = 3	47.0 ± 11.0	79.0 ± 6.0	86.0 ± 4.0	86.0 ± 4.0	75.0 ± 5.0	83.0 ± 4.0	88.0 ± 4.0	86.0 ± 3.0	**93.5 ± 2.4**
	Wine	v = 2	63.0 ± 6.0	62.0 ± 5.0	60.0 ± 7.0	60.0 ± 7.0	76.0 ± 5.0	81.0 ± 3.0	93.0 ± 3.0	93.0 ± 2.1	**98.4 ± 2.1**
		v = 3	64.0 ± 6.0	65.0 ± 5.0	65.0 ± 9.0	66.0 ± 9.0	81.0 ± 5.0	81.0 ± 4.0	93.0 ± 2.0	92.8 ± 2.3	**98.5 ± 2.4**
	Wdbc	v = 2	55.0 ± 7.0	63.0 ± 4.0	66.0 ± 4.0	66.0 ± 4.0	63.0 ± 7.0	96.0 ± 2.0	98.0 ± 1.0	98.7 ± 1.3	**99.9 ± 2.2**
		v = 3	54.0 ± 5.0	57.0 ± 16.0	57.0 ± 4.0	56.0 ± 4.0	72.0 ± 6.0	95.0 ± 1.0	97.0 ± 1.0	98.5 ± 1.3	**99.6 ± 1.7**
	Pima	v = 2	53.0 ± 10.0	60.0 ± 4.0	54.0 ± 3.0	54.0 ± 3.0	59.0 ± 4.0	75.0 ± 2.0	94.0 ± 1.0	**96.2 ± 1.1**	95.6 ± 1.1
		v = 3	66.0 ± 9.0	63.0 ± 3.0	48.0 ± 4.0	48.0 ± 4.0	64.0 ± 4.0	73.0 ± 2.0	94.0 ± 1.0	94.1 ± 3.8	**96.6 ± 1.0**
Avg.			55.9 ± 7.2	60.0 ± 6.3	67.5 ± 4.8	67.9 ± 4.7	68.6 ± 4.8	81.1 ± 2.9	88.9 ± 2.7	89.6 ± 3.5	**93.8 ± 2.9**

3.2   Experimental Settings

Baselines and Implementation Details. We compare our model with baselines as following which were carried out according to the parameter configuration in the original work: HOAD [6], AP [2] (we use two similarity matrices including the negative L2-norm and the Gauss kernel for HOAD and AP), MLRA [11], LDSR [10], MODDIS [8], NCMOD [4]. In ECMOD, the embedding dimensions were all set to 32. The model was optimized with the Adam optimizer during training. The batch size was set to 128 and the learning rate was set to 0.0001 for all datasets. The value of k was set to 8 and the parameter α and β were set to 0.05 and 1, respectively. To avoid randomness, following [6,8], we repeat the process of randomly generating data with outliers 50 times for each dataset and calculate the mean values and standard deviations of the running results. We use AUC [4,8,15,16] as evaluation metric in our experiments. For more comprehensive experimental verifications, we consider some sets of outlier ratios for three types of outliers to help us discover ECMOD adapts to datasets with different anomaly rates.

3.3   Main Results

Comparison experiments are conducted from different perspectives. To observe the impact of the outlier rates on the algorithm, we investigate the cases when the three types of outliers have different outlier ratios. For the fairness of the experiment, we set ratios of attribute outliers, class-attribute outliers, and class outliers to 5%-5%-5%, 2%-5%-8% and 8%-5%-2%, respectively. Then we compare the AUC of all the competitors on UCI datasets. From Table 2, we have the following observations:

- In the cases of three different outlier rates, ECMOD performs better than other competitors on the majority of dataset settings in Table 2. As we can see, ECMOD can detect three types of outliers simultaneously. Overall, ECMOD improve over the best baseline by 4.2%. This verifies the significance of neural network in learning representations and the contrastive technique in enhancing representations.
- In addition, MODDIS and NCMOD are even lower than the clustering method LDSR in some settings. This illustrates that existing neural network-based methods perform poorly in multi-view outlier detection. On the contrary, ECMOD still shows competitive results. We attribute this to the constraints between views and the neighborhood consistency, which better preserve the information between different views.
- Furthermore, we compute the average AUC in the 2-view and 3-view cases and get the mean AUC as 0.938 and 0.935, respectively. This illustrates that the more views, the more chaotic the data group structure is. And ECMOD can adapt to scenarios with more than two views.

3.4   Ablation Study

In order to verify the importance of variables in ECMOD, due to limited space, we discuss the ablation studies only on UCI datasets in two additions, which are denoted as SVL (the Specific View Loss) of the constrained view-channel in Sect. 2.1 and CL (the Contrastive Learning) in Sect. 2.2. Firstly, we examine the performance of ECMOD's variants by removing the two parts from the full approach. To be specific, we consider the following variants of ECMOD for comparison:

- w/o SVL: This variant removes the loss Eq. (3) in the constrained view-channel of channel-specific representations module.
- w/o CL: This variant removes the binary cross-entropy loss Eq. (4) between two channels in the data enhancement module. It means the representations in the latent space are not enhanced.

As listed in Table 3, the performance of the variant w/o SVL gets worse on 14 of 24 datasets, so that the information from the constrained view-specific channel is essential for the representation in the latent space. The results of the variant w/o CL show that the contrasted loss plays an important role in performance improvement. Because the contrasted loss enhances the latent representations for neighborhood consistency. These ablation studies confirm that the two variants of ECMOD are useful to significantly improve the outlier detection.

Table 3. AUC (%) of ablation study on different ratios of outliers in 2 or 3 views.

Ratio (%)	Method	v = 2				v = 3			
		zoo	wine	wdbc	pima	zoo	wine	wdbc	pima
5-5-5	w/o SVL	0.842	0.832	0.754	0.620	0.939	0.895	0.846	0.833
	w/o CL	0.842	0.852	0.703	0.543	0.905	0.897	0.848	0.841
	ECMOD	**0.941**	**0.924**	**0.903**	**0.924**	**0.977**	**0.985**	**0.888**	**0.879**
2-5-8	w/o SVL	0.846	0.856	0.739	0.687	0.837	0.840	0.751	0.761
	w/o CL	0.861	0.877	0.730	0.610	0.843	0.842	0.747	0.756
	ECMOD	**0.951**	**0.943**	**0.833**	**0.796**	**0.903**	**0.929**	**0.783**	**0.789**
8-5-2	w/o SVL	0.826	0.792	0.461	0.427	0.955	0.935	0.934	0.869
	w/o CL	0.817	0.804	0.416	0.383	0.961	0.951	0.935	0.894
	ECMOD	**0.943**	**0.899**	**0.930**	**0.931**	**0.993**	**0.994**	**0.956**	**0.966**

Table 4. Experimental results in AUC (%) on 9 datasets from ODDS.

Dataset	NCMOD	**ECMOD**
Arrhythmia2	81.6 ± 2.3	**87.7 ± 2.1**
Arrhythmia3	74.3 ± 4.6	**84.0 ± 2.2**
Yeast2	84.1 ± 1.5	**86.6 ± 1.1**
Yeast3	84.0 ± 1.4	**87.5 ± 1.4**
Letter2	85.4 ± 2.1	**86.2 ± 1.5**
Letter3	86.7 ± 1.7	**87.3 ± 1.5**
Musk2	63.4 ± 9.4	**63.5 ± 8.2**
Musk3	**58.6 ± 10.2**	58.5 ± 9.3
Thyroid2	84.3 ± 1.9	**84.4 ± 1.3**
Thyroid3	87.0 ± 1.5	**87.8 ± 1.1**
Optdigits2	60.5 ± 3.2	**82.1 ± 2.5**
Optdigits3	61.6 ± 10.6	**85.9 ± 1.1**
Mammography2	67.2 ± 12.1	**67.9 ± 9.3**
Mammography3	72.7 ± 9.8	**73.2 ± 8.7**
Shuttle2	86.2 ± 1.7	**91.7 ± 1.4**
Shuttle3	84.4 ± 1.5	**85.9 ± 1.4**
ForestCover2	73.5 ± 11.4	**84.4 ± 5.3**
ForestCover3	77.1 ± 17.3	**84.1 ± 8.7**

3.5   Analysis on Large-scale Datasets

Based on the above analysis, MODDIS, NCMOD, and the proposed ECMOD are not based on the clustering hypothesis. In order to further compare the performance of the neural network-based methods, we conduct experiments on some large-scale ODDS datasets in 5%-5%-5%. The results are listed in Table 4. Since the average performance of NCMOD is better than other baselines, we only compare ECMOD with NCMOD. From Table 4, ECMOD achieves the best.

4 Conclusion

We propose a novel multi-view outlier detection method named ECMOD, which utilizes the autoencoder network and the MLP networks as two channels to represent the multi-view data in different ways. Then we adopt a contrastive technique to complement learned representations via two channels. Therefore, the intrinsic information with the constraint is embedded into a comprehensive latent space for each view. In addition, we uniform neighborhood structures among different views to detect three types of outliers in multi-view data. Extensive experimental results verify the superiority of the proposed ECMOD.

Acknowledgements. This work was supported in part by the National Natural Science Foundation of China (61972268), and the Joint Innovation Foundation of Sichuan University and Nuclear Power Institute of China.

References

1. Agarwal, A., Ratha, N.K.: Black-box adversarial entry in finance through credit card fraud detection. In: CIKM, p. 3052 (2021)
2. Alvarez, A.M., Yamada, M., Kimura, A., Iwata, T.: Clustering-based anomaly detection in multi-view data. In: CIKM, pp. 1545–1548 (2013)
3. Breunig, M.M., Kriegel, H., Ng, R.T.: LOF: identifying density-based local outliers. In: SIGMOD, pp. 93–104 (2000)
4. Cheng, L., Wang, Y., Liu, X.: Neighborhood consensus networks for unsupervised multi-view outlier detection. In: AAAI, pp. 7099–7106 (2021)
5. Duan, L., Tang, G., Pei, J., Bailey, J., Campbell, A., Tang, C.: Mining outlying aspects on numeric data. Data Min. Knowl. Discov. **29**(5), 1116–1151 (2015)
6. Gao, J., Fan, W., Turaga, D.S., Parthasarathy, S., Han, J.: A spectral framework for detecting inconsistency across multi-source object relationships. In: ICDM, pp. 1050–1055 (2011)
7. Huang, L., Wang, C., Chao, H.: oComm: overlapping community detection in multi-view brain network. IEEE ACM Trans. Comput. Biol. **18**(4), 1582–1595 (2021)
8. Ji, Y., et al.: Multi-view outlier detection in deep intact space. In: ICDM, pp. 1132–1137 (2019)
9. Kocher, G., Kumar, G.: Machine learning and deep learning methods for intrusion detection systems: recent developments and challenges. Soft. Comput. **25**(15), 9731–9763 (2021)
10. Li, K., Li, S., Ding, Z., Zhang, W., Fu, Y.: Latent discriminant subspace representations for multi-view outlier detection. In: AAAI, pp. 3522–3529 (2018)
11. Li, S., Shao, M., Fu, Y.: Multi-view low-rank analysis for outlier detection. In: SDM, pp. 748–756 (2015)

12. Sheng, X., Zhan, D., Lu, S., Jiang, Y.: Multi-view anomaly detection: neighborhood in locality matters. In: AAAI, pp. 4894–4901 (2019)
13. Xia, X., Yin, H., Yu, J., Wang, Q., Cui, L., Zhang, X.: Self-supervised hypergraph convolutional networks for session-based recommendation. In: AAAI, pp. 4503–4511 (2021)
14. Xing, J., Gao, C., Zhou, J.: Weighted fuzzy rough sets-based tri-training and its application to medical diagnosis. Appl. Soft Comput. **124**, 109025 (2022)
15. Zhao, H., Fu, Y.: Dual-regularized multi-view outlier detection. In: IJCAI, pp. 4077–4083 (2015)
16. Zhao, H., Liu, H., Ding, Z., Fu, Y.: Consensus regularized multi-view outlier detection. IEEE Trans. Image Process. **27**(1), 236–248 (2018)

Powering Fine-Tuning: Learning Compatible and Class-Sensitive Representations for Domain Adaption Few-shot Relation Extraction

Yijun Liu[1,2], Feifei Dai[2(✉)], Xiaoyan Gu[2], Haihui Fan[2], Dong Liu[2], Bo Li[2], and Weiping Wang[2]

[1] School of Cyber Security, University of Chinese Academy of Sciences, Beijing, China
`liuyijun@iie.ac.cn`
[2] Institute of Information Engineering, Chinese Academy of Sciences, Beijing, China
`{daifeifei,guxiaoyan,fanhaihui,liudong0039,libo,wangweiping}@iie.ac.cn`

Abstract. Relation extraction (RE) is an important task in information extraction that has drawn much attention. Although many RE models have achieved impressive performance, their performance drops dramatically when adapting to the new domain and under few-shot scenarios. One reason is that the huge gap in semantic space between different domains makes the model obtain suboptimal representations in the new domain. The other is the inability to learn class-sensitive information with only a few samples, which makes the instances with confusing factors hard to be distinguished. To address these issues, we propose a **C**ontrastive learning-based **F**ine-**T**uning approach with **K**nowledge **E**nhancement (CFTKE) for the Domain Adaptation Few-Shot RE task (DAFSRE). Specifically, we fine-tune the model in a contrastive-learning way to refine the semantic space in the new domain, which can bridge the gap between different domains and obtain better representations. To enhance the stability and learning ability of contrastive learning-based fine-tuning, we design the data augmentation mechanism and type-aware networks to enrich the instances and stand out the class-sensitive features. Extensive experiments on the DAFSRE benchmark dataset demonstrate that our approach significantly outperforms the state-of-the-art models (by 2.73% on average).

Keywords: Few-shot learning · Information extraction · Transfer learning

1 Introduction

Relation extraction (RE) aims to identify the relation between two entities in a given sentence, which is an important task in information extraction. However, RE usually suffers from data scarcity and domain adaption due to the massive cost of labour and the complex realistic scenes. To solve the above problems, many scholars [4,6,19] have turned their attention to the task of Domain Adaption Few-Shot RE (DAFSRE), which extracting relation under low resources and domain shift. According to the information utilized by the models, existing approaches for DAFSRE can be roughly divided into meta-learning-based approaches [3,12,13] and knowledge enhancement-based approaches [17–19]. As for meta-learning [15], it encourages the model to learn the ability of rapid learning from previous experience and generalize to new concepts, so that the model can adapt to novel classes quickly. Recently, many works [5,11,16] are in line with this idea and have achieved remarkable performance. However, it is often hard to do classification only depending on the information in the sentence since the information contained in a single sentence is limited. As there is rich semantic information in the knowledge base (KB), knowledge enhancement-based approaches are proposed, which focus on introducing external knowledge to enhance the representations of entities or relations. For instance, Concept-FERE [18] and TD-proto [17] introduce the entity concepts and entity description information to enhance the entity representation. KEFDA [19] introduces domain-specific knowledge to enhance the representation of entities and relations, achieving state-of-the-art performance.

Although the DAFSRE task has made great progress, there are still two problems that hinder its development and have not been studied adequately.

First, semantic spaces in common classes and novel classes are incompatible, so the common semantic space cannot be applied to novel classes directly. In DAFSRE, most works [2,10,19] pre-train the feature extractor in common classes and straightway apply it in novel classes. However, the feature distribution between different domains has a huge gap. Directly applying the feature extractor in novel classes will obtain suboptimal representations, resulting in a significant decline in classification performance. Fine-tuning can learn from the new samples and refine the semantic space which may alleviate this problem. Traditional fine-tuning-based approaches usually fine-tune the feature extractor and re-train a new classification layer. However, in DAFSRE, there are only a few samples in novel class, making it impossible to train the new classification layer. For this reason, fine-tuning for refining the semantic space is a difficult challenge.

Second, the inability to learn class-sensitive information in novel classes may lead to relation confusion, making the instances hard to distinguish. Since some previous works [4,12,13] only consider the context information without learning any class-sensitive information in novel classes, they may be unable to distinguish confusing relations. To learn the class-sensitive information, we may need to establish associations between samples of the same class or design some new layers. As there are only a few labeled samples in novel classes, it is hard to control what to learn, and the model is easy to fall into the risk of overfitting. Thus, learning class-sensitive information in few-shot scenarios remains a challenge.

In this paper, we propose a **C**ontrastive learning-based **F**ine-**T**uning approach with **K**nowledge **E**nhancement (CFTKE), which focuses on fine-tuning the model with only a few samples to bridge the gap in semantic space between different domains and learn new class-sensitive information from novel classes for DAFSRE task. Figure 1 shows the overall framework of CFTKE. Specifically, to obtain better representations, we pre-train the feature extraction module in a large-scale common domain dataset, which can capture the context information and combine the background knowledge for each instance. Next, we propose a contrastive learning-based fine-tuning (CLFT) module to bridge the gap in semantic space between common classes and novel classes. We model the fine-tuning as a contrastive learning task, which calculates the supervised contrastive loss by instances pair and refines the whole model. In this way, the model can refine the semantic space and obtain better representations in novel classes without introducing any additional classification layers. Meanwhile, the type-aware networks and data augmentation mechanism are designed to enhance the learning ability and stability of CLFT. Finally, we use a classification module to obtain the relation prototypes and enhanced queries that pay more attention to class-sensitive information for improving classification performance.

In summary, the main contributions of this paper are as follows:

- To the best of our knowledge, this is the first work to consider the incompatibility of semantic spaces between different domains in DAFSRE. We fine-tune the model by contrastive learning to refine the feature distribution, which can bridge the gap in semantic space with only a few samples.
- To learn the class-sensitive information and distinguish the confusing relations, we design the type-aware networks and data augmentation mechanism which can enhance the learning ability and stability of CLFT.
- Experimental results on the FewRel 2.0 benchmark show that our approach significantly outperforms the baseline models and achieves the state-of-the-art DAFSRE performance.

2   Methodology

2.1   Problem Definition

DAFSRE aims to predict the novel classes with only a few labeled instances where the domain of the test set is different from the training set. N-way-K-shot setting has been widely used to simulate low resource RE scenarios, which contains a support set $\mathcal{S}$ and a query set $\mathcal{Q}$. $\mathcal{S}$ includes N novel classes, each with quite small K labeled instances. The instances in $\mathcal{Q}$ need to be classified through the given $N \times K$ support instances. However, there are so few instances in $\mathcal{S}$ that it may not even be able to train a feature extractor. An auxiliary dataset $\mathcal{D}_{train}$ is given, containing abundant common classes, each class has a large number of labeled instances. It should be noted that common classes and novel classes are disjoint with each other to ensure the FSL scenarios. The goal of DAFSRE is to optimize the following objective function:

$$\mathcal{L} = -\frac{1}{R} \sum_{q \in \mathcal{Q}} P(y_q \mid \mathcal{S}, q). \tag{1}$$

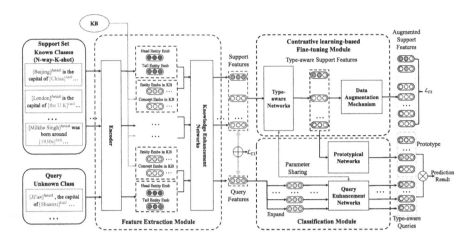

Fig. 1. The overall architecture of the proposed model CFTKE. The $\mathcal{L}_{c1}$ and $\mathcal{L}_{c2}$ are supervised contrastive loss for pretraining and finetuning.

2.2   Model Overview

As shown in Fig. 1, the proposed model mainly consists of three modules: 1)The feature extraction module is used to obtain contextual information and combine background knowledge for better entity representation. 2)The CLFT module is designed to bridge the gap in semantic space between different domains and power fine-tuning. The type-aware networks and the augmentation mechanism in this module can be beneficial to distinguish the subtle difference between confusing relations and enhance the stability of CLFT. 3)The classification module can pay more attention to class-sensitive features when comparing the query with relation prototype and achieve better classification performance. Next we will elaborate these modules.

2.3   Feature Extraction Module

In the pre-training phase, we use $\mathcal{D}_{train}$ to train the feature extraction module for better entity representations. This module consists of two components: the encoder and the knowledge enhancement networks.

Encoder: The encoder can convert each word in a sentence into a low-dimensional embedding. Following the previous works [4,13], we employ BERT [7] as the encoder to obtain contextualized embeddings for sentence x. After inputting the tokens of the sentence into BERT, the embedding of each token can be obtained:

$$\mathbf{h}_1, \mathbf{h}_2, ..., \mathbf{h}_m = Encoder(x), \mathbf{h}_i \in \mathbb{R}^{d_e}, \tag{2}$$

where d_e is the dimension of the entity embedding.

Knowledge Enhancement Networks: As there is rich semantic information in the KB, this component is designed to introduce external knowledge, which can enhance the representation of entity. Given the pre-trained embeddings of head and tail entity $\mathbf{h}_{head}^{KB}$, $\mathbf{h}_{tail}^{KB}$ in KB and the concept set of entity e_i of various hierarchical $C_i = \{h_{c_1}, h_{c_2}, ...\}$.[1] We fuse these features by concatenating operation and FFNN function to obtain the final representation $\hat{\mathbf{h}}$ for each instance:

$$\mathbf{h}_{ep} = \text{FFNN}(\mathbf{h}_{head} \oplus \mathbf{h}_{tail} \oplus \mathbf{h}_{head}^{KB} \oplus \mathbf{h}_{tail}^{KB}), \tag{3}$$

$$\mathbf{h}_{C_i} = \frac{1}{|C_i|} \sum_{\mathbf{h_c} \in C_i} \mathbf{h_c}. \tag{4}$$

$$\hat{\mathbf{h}} = \text{Norm}\left(\text{FFNN}(\mathbf{h}_{ep} \oplus \mathbf{h}_{C_{head}} \oplus \mathbf{h}_{C_{tail}})\right), \tag{5}$$

where $\text{Norm}\,(\cdot)$ indicates the normalization function.

Loss Function: Essentially, DAFSRE is a comparison-based task. As supervised contrastive loss [8] is calculated by comparison, we take it as the loss function of our approach during the pre-training phase. The supervised contrastive loss needs to pair all instances in $\mathcal{S}$ and $\mathcal{Q}$, then calculate the loss by these instances pairs. In this way, we can model the interaction between all samples in $\mathcal{S}$ and $\mathcal{Q}$ so that providing more supervised information than the traditional cross-entropy loss. Given the final instance representation $\hat{\mathbf{h}}$ with corresponding label y_i in $\mathcal{S}$ and $\mathcal{Q}$. The supervised contrastive loss is calculated by the following formula:

$$\mathcal{L}_{C1} = \sum_{i \in \{\mathcal{S} \cup \mathcal{Q}\}} \frac{-1}{|J(i)|} \sum_{j \in J(i)} \log \frac{\exp\left(\hat{\mathbf{h}}_i \cdot \hat{\mathbf{h}}_j / \tau\right)}{\sum_{a \in A(i)} \exp\left(\hat{\mathbf{h}}_i \cdot \hat{\mathbf{h}}_a / \tau\right)}, \tag{6}$$

where $J(i)$ is the set of indices of all instances that the label are the same as instance i, and $A(i)$ is the set of indices of all instances exclude i. τ is the temperature in contrastive learning.

2.4   Contrastive Learning-Based Fine-Tuning Module

To bridge the gap of feature distribution between different domains and learn new class-sensitive information, we propose a CLFT module to power fine-tuning and refine the model. The type-aware networks and data augmentation mechanism in this module can enhance the learning ability and stability of CLFT.

Type-Aware Networks: These networks aim to learn class-sensitive information at the feature level in novel classes, which could be beneficial to distinguish the subtle difference of confusing relations. It is based on a proposition: for a feature vector, there are some positions encoding class-sensitive information and some encoding domain information. For different classes in the same domain, the

---

[1] The pre-trained embeddings of entities and concepts are provided by Zhang et al. [19]

feature of the class-sensitive positions could contribute significantly to the classification. However, the features of the location encoding the domain information may have a lower contribution.

Based on this, we design the type-aware networks to stand out the class-sensitive features of instances in each novel class. The type-aware networks consist of multiple class-specific trainable vectors. Each class-specific vector corresponds to a relation, which can assign higher weight to class-sensitive feature positions and lower weight to useless information positions. The way we stand out the class-sensitive features of each instance can be expressed as follows:

$$\tilde{\mathbf{h}} = \hat{\mathbf{h}} \otimes \mathbf{v}_i, i = 1, ..., N \tag{7}$$

where $\mathbf{v}_i \in \mathbb{R}^{1 \times d_h}$ is a trainable weight vector initialized to 1 and the dimension is the same as $\hat{\mathbf{h}}$. $\otimes$ indicates the dot product operation.

Data Augmentation Mechanism: This mechanism is designed to enrich the instances of novel classes and ensure the existence of positive and negative instance pairs in contrastive learning. Normally, each class needs at least two instances to form a mini-batch for calculating the loss in contrastive learning. However, in N-way-K-shot setting, when $K = 1$, each novel class contains only one instance, which means that the $\mathcal{S}$ contains no positive instances pair. Directly calculating the loss and optimizing the model will lead the model to collapse.

To tackle this, we augment the features of instances with a simple and efficient method of dropout and perturbation. The way we augment the features can be expressed as follows:

$$\tilde{\mathbf{h}}' = \text{dropout}\left(\tilde{\mathbf{h}} + U\left(-\frac{\lambda}{2}, \frac{\lambda}{2}\right) * \text{std}\left(\tilde{\mathbf{h}}\right)\right), \tag{8}$$

where $\text{std}(\cdot)$ stands for standard deviation. $U(a, b)$ represents uniform distribution noise ranged from a to b. [2] dropout $(\cdot)$ indicates the dropout function [14] and λ is a hyperparameter that controls the relative noise intensity.

Here, we establish the enhanced support set $\mathcal{S}' = \{s'; s \in \mathcal{S}\}$ and combine it with $\mathcal{S}$ to form new support set $\mathcal{S}_{new} = \mathcal{S} \cup \mathcal{S}'$. In this way, there are at least two instances of each class in $\mathcal{S}_{new}$.

Loss Function: We use the supervised contrastive loss as our loss function as we did in the pre-training phase. Because the labels in the query set are unavailable, we only calculate the loss in $\mathcal{S}_{new}$ and fine-tune the model. The loss $\mathcal{L}_{C2}$ for fine-tuning is calculated by the following formula:

$$\mathcal{L}_{C2} = \sum_{i \in \mathcal{S}_{new}} \frac{-1}{|J(i)|} \sum_{j \in J(i)} \log \frac{\exp\left(\tilde{\mathbf{h}}_i \cdot \tilde{\mathbf{h}}_j / \tau\right)}{\sum_{a \in A(i)} \exp\left(\tilde{\mathbf{h}}_i \cdot \tilde{\mathbf{h}}_a / \tau\right)}. \tag{9}$$

---

[2] $U(a, b)$ is a matrix with the same shape with $\tilde{\mathbf{h}}$ rather than a scalar.

2.5   Classification Module

In the inference phase, we design the query enhancement networks and proto-typical networks to enhance the query and obtain prototypes. By comparing prototypes with type-aware queries, we can pay more attention to the class-sensitive features and achieve better classification performance.

Prototypical Networks: To obtain each relation prototype efficiently, we adopt prototypical networks for aggregating all instances in a relation. The pro-totype of relation r can be formulated as:

$$\tilde{\mathbf{h}}^r = \frac{1}{|\mathcal{S}_{new}^r|} \sum_{i \in \mathcal{S}_{new}^r} \tilde{\mathbf{h}}_i, \tag{10}$$

where $\hat{\mathbf{h}}^r$ is the prototype representation of relation r, $\mathcal{S}_{new}^r = \{s_k^r | k = 1, 2, ..., 2N\}$ is a subset of $\mathcal{S}_{new}$ with all instances have the same label r.

Query Enhancement Networks: When comparing with relation prototype, paying more attention to the class-sensitive feature locations is beneficial to distinguish the query instance. To do this, for each query instance, we first make N copies of it and then stand out the class-sensitive features for each copy by query enhancement networks. The way we stand out class-specific features and obtain N type-aware queries $\tilde{\mathbf{h}}_q^r$ can be expressed as follow:

$$\tilde{\mathbf{h}}_q^r = \hat{\mathbf{h}}_q \otimes \mathbf{v}_i, i = 1, 2, ..., N. \tag{11}$$

Moreover, to leverage knowledge from fine-tuning phase, these networks share parameters with the type-aware networks.

Finally, to measure the distance between the query and each prototype, we adopt dot product $d(\cdot)$ to calculate the similarity scores. The softmax function is performed to get the label probability of each query:

$$p(y = r \mid \mathbf{S}, q) = \frac{\exp\left(d\left(\tilde{\mathbf{h}}_q^r, \tilde{\mathbf{h}}^r\right)\right)}{\sum_{i=1}^{N} \exp\left(d\left(\tilde{\mathbf{h}}_q^i, \tilde{\mathbf{h}}^i\right)\right)}. \tag{12}$$

3   Experiments

3.1   Dataset and Evaluation

Dataset: We conduct the experiments on FewRel 2.0 dataset [4] as it is the only qualified large-scale dataset for evaluating the DAFSRE task. FewRel 2.0 contains 64/10/15 relations, a total of 44800/1000/1500 labeled instances for training/validation/test. The training set is from the common domain, but the validation and test set are from medical domain.

Evaluation: The most common metric for evaluating the DAFSRE model is the average accuracy of the N-way-K-shot task. According to the previous work

[4,6], we set N to 5 and 10, K to 1 and 5 for forming four FSL scenarios. Since the label of the test set in the FewRel 2.0 is not publicly available, we report the final test accuracy by submitting the model predictions to FewRel Leaderboard[3].

3.2   Implementation Details

We use BERT-base-uncased as the encoder. The iteration number in our pre-training phase is set to 5000, and validation is conducted every 500 iterations. The learning rate is 2e-5 when pre-training in $\mathcal{D}_{train}$. In fine-tuning phase, the learning rate is 0.1 for the type-aware networks and 5e-5 for others. The iteration number in the fine-tuning phase is set to 1 when K = 1, 5 when K = 5. For the data augmentation mechanism, the dropout rate and noisy factor λ are set to 0.1. The temperature τ in contrastive learning is 0.1. The batch size is set to 1 to prevent label confusion. It is worth noting that in the test phase, every time we perform classification, we will initialize the model to prevent the model from remembering the test instances by fine-tuning.

Table 1. Accuracy (%) of models on FewRel 2.0 test set under N-way-K-shot settings, the best results are in bold.

Model	5-Way-1-Shot	5-Way-5-Shot	10-Way-1-Shot	10-Way-5-Shot	Average
Proto	40.12	51.50	26.45	36.93	38.75
Proto-ADV	41.90	54.74	27.36	37.40	40.35
BERT-PAIR	67.41	78.57	54.89	66.85	66.93
DaFeC	61.20	76.99	47.63	64.79	62.65
MTB	74.7	87.9	62.5	81.1	76.6
CP	79.7	84.9	68.1	79.8	78.1
HCPR	76.34	83.03	63.77	72.94	74.02
GTPN	80.0	92.6	69.3	86.9	82.2
KEFDA	87.81	95.00	81.84	90.63	88.82
FAFE	73.58	90.10	62.98	80.51	76.79
CFTKE	**91.24**	**96.50**	**85.09**	**93.50**	**91.58**

3.3   Comparison with Previous Work

We consider two categories of DAFSRE models as baselines. Models without introducing any external information: BERT-PAIR [4], DaFeC [1], Proto and Proto-ADV [12]. External information enhanced models: MTB [13], CP [10], GTPN [9], HCPR [6], FAEA [2], KEFDA [19].

Table 1 shows the accuracy of all models on official testing tasks of DAFSRE under the four few-shot settings. From the table, it can be seen that our approach

---

[3]   https://thunlp.github.io/fewrel.html.

dramatically improves the accuracy of the DAFSRE task and achieve state-of-the-art performance. Specifically, CFTKE improves the accuracy by an average of 2.76% compared to the second-best model KEFDA, which demonstrates the effectiveness of our approach.

Furthermore, our approach significantly outperforms the traditional models that do not introduce external information by 24.65% on average accuracy. As for the methods such as MTB and CP that used lots of external data for pre-training, our approach also improves the average accuracy by more than 10%. It indicates that fine-tuning with contrastive learning can indeed help us obtain better representations and improve classification performance. Compared to the KEFDA model, which also utilizes domain-specific KB, CFTKE achieves 3.34% and 2.19% gain in 1-shot and 5-shot settings, respectively. Theoretically, the 1-shot RE task is more difficult than the 5-shot RE task. It means that our approach is suitable for more difficult few-shot scenarios with the help of class-sensitive information and refinement in semantic space.

Table 2. Experimental results of the ablation studies on FewRel 2.0 test set.

Model	5-way-1-shot	5-way-5-shot	10-way-1-shot	10-way-5-shot	Average
CFTKE	**91.24**	**96.50**	**85.09**	**93.50**	**91.58**
-CLFT	90.29($\downarrow$ 0.95)	94.13($\downarrow$ 2.37)	84.19($\downarrow$ 0.90)	89.56($\downarrow$ 3.94)	89.54
-Type-aware	90.75($\downarrow$ 0.49)	95.60($\downarrow$ 0.90)	84.64($\downarrow$ 0.45)	92.65($\downarrow$ 0.85)	90.91
-Data augmentation	20.14($\downarrow$ 71.10)	96.23($\downarrow$ 0.27)	10.09($\downarrow$ 75.00)	93.17($\downarrow$ 0.33)	54.91
-Knowledge enhancement	75.37($\downarrow$ 15.87)	90.47($\downarrow$ 6.03)	66.31($\downarrow$ 18.87)	85.82($\downarrow$ 7.68)	79.49

3.4   Ablation Study

Table 2 shows the results of ablation studies on FewRel 2.0 test set. From these ablations, we find that: (1)CLFT can significantly improve the DAFSRE performance with only a few samples, especially in 5-shot setting. When removing the CLFT, the accuracy drops by 2.37% and 3.94% in 5-way-5-shot and 10-way-5-shot settings. (2) Data augmentation mechanism and Type-aware networks are essential for powering CLFT. Without the help of the data augmentation mechanism, fine-tuning on one instance will lead the model to collapse and dramatically decrease the model performance. In the 5-shot setting, removing the type-aware networks will cause the accuracy drops by 0.88%. It indicates that learning class-sensitive information is beneficial to distinguish the subtle difference of confusing relations that improve the model performance. (3) Knowledge enhancement networks play an important role in DAFSRE, especially in 1-shot setting. This certifies that external knowledge is essential when data is scarce.

4   Conclusion

In this paper, we propose a novel CFTKE approach that refines the semantic space and learns class-sensitive information in novel classes for the DAFSRE task. By modeling fine-tuning as a contrastive learning task, our approach can leverage a few labeled instances to bridge the gap in semantic space between different domains. Meanwhile, the data augmentation mechanism and type-aware networks can power fine-tuning, which enhances the stability and learning ability of CLFT. Experimental results demonstrate the effectiveness of our approach and achieve state-of-the-art performance on the DAFSRE benchmark.

Acknowledgement. This work was supported by No.XDC02050200.

References

1. Cong, X., Yu, B., Liu, T., Cui, S., Tang, H., Wang, B.: Inductive unsupervised domain adaptation for few-shot classification via clustering. In: Hutter, F., Kersting, K., Lijffijt, J., Valera, I. (eds.) ECML PKDD 2020. LNCS (LNAI), vol. 12458, pp. 624–639. Springer, Cham (2021). https://doi.org/10.1007/978-3-030-67661-2_37
2. Dou, C., et al.: Function-words enhanced attention networks for few-shot inverse relation classification. IJCAI (2022)
3. Finn, C., Abbeel, P., Levine, S.: Model-agnostic meta-learning for fast adaptation of deep networks. In: ICML, pp. 1126–1135. PMLR (2017)
4. Gao, T., et al.: FewRel 2.0: towards more challenging few-shot relation classification. In: EMNLP, pp. 6250–6255 (2019)
5. Gao, T., Han, X., Liu, Z., Sun, M.: Hybrid attention-based prototypical networks for noisy few-shot relation classification. In: AAAI, vol. 33, pp. 6407–6414 (2019)
6. Han, J., Cheng, B., Lu, W.: Exploring task difficulty for few-shot relation extraction. In: EMNLP, pp. 2605–2616 (2021)
7. Kenton, J.D.M.W.C., Toutanova, L.K.: Bert: pre-training of deep bidirectional transformers for language understanding. In: NAACL, pp. 4171–4186 (2019)
8. Khosla, P., et al.: Supervised contrastive learning. In: NeurIPS, pp. 18661–18673 (2020)
9. Liu, F., et al.: From learning-to-match to learning-to-discriminate: global prototype learning for few-shot relation classification. In: Li, S., et al. (eds.) CCL 2021. LNCS (LNAI), vol. 12869, pp. 193–208. Springer, Cham (2021). https://doi.org/10.1007/978-3-030-84186-7_13
10. Peng, H., et al.: Learning from context or names? An empirical study on neural relation extraction. In: EMNLP, pp. 3661–3672 (2020)
11. Qu, M., Gao, T., Xhonneux, L.P., Tang, J.: Few-shot relation extraction via Bayesian meta-learning on relation graphs. In: ICML, pp. 7867–7876. PMLR (2020)
12. Snell, J., Swersky, K., Zemel, R.: Prototypical networks for few-shot learning. In: Advances in Neural Information Processing Systems, vol. 30 (2017)
13. Soares, L.B., Fitzgerald, N., Ling, J., Kwiatkowski, T.: Matching the blanks: distributional similarity for relation learning. In: ACL, pp. 2895–2905 (2019)
14. Srivastava, N., et al.: Dropout: a simple way to prevent neural networks from overfitting. J. Mach. Learn. Res. **15**(1), 1929–1958 (2014)
15. Vinyals, O., Blundell, C., Lillicrap, T., Wierstra, D., et al.: Matching networks for one shot learning. In: Advances in Neural Information Processing Systems, vol. 29 (2016)

16. Wang, Y., et al.: Learning to decouple relations: few-shot relation classification with entity-guided attention and confusion-aware training. In: COLING, pp. 5799–5809 (2020)
17. Yang, K., et al.: Enhance prototypical network with text descriptions for few-shot relation classification. In: CIKM, pp. 2273–2276 (2020)
18. Yang, S., Zhang, Y., Niu, G., Zhao, Q., Pu, S.: Entity concept-enhanced few-shot relation extraction. In: IJCNLP, pp. 987–991 (2021)
19. Zhang, J., et al.: Knowledge-enhanced domain adaptation in few-shot relation classification. In: SIGKDD, pp. 2183–2191 (2021)

Accelerating Exact K-Means++ Seeding Using Lower Bound Based Framework

Haowen Zhang[1(✉)] and Jing Li[2]

[1] College of Computer Science and Technology,
Zhejiang Sci-Tech University, Hangzhou, China
`zhanghw@zstu.edu.cn`
[2] College of Computer Science and Technology,
Zhejiang University, Hangzhou, China
`zjujing@zju.edu.cn`

Abstract. The k-means++ seeding is a widely used approach to obtain reasonable initial centers of k-means clustering, and it performs empirical well. Nevertheless, the time complexity of k-means++ seeding makes it suffer from being slow on large datasets. Therefore, it is necessary to improve the efficiency of k-means++ seeding to make it scale well for large datasets. Most of the previous work focused on approximating k-means++ seeding. They can achieve speedup effectively; however, they are inherently approximation methods and cannot get the same results as k-means++ seeding. In this paper, a lower bound based framework (LBF) is proposed to accelerate k-means++ seeding. LBF is an *exact* approach which can perform k-means++ seeding in less time while guaranteeing the same results. Two specific lower bound schemes based on progressive partial distance and piecewise aggregate approximation are embedded into the LBF, and we provide the theoretical analysis to show their lower bound property and computational efficiency. Experimental results on various synthetic and real-world datasets demonstrate the effectiveness of the proposals compared with the baseline k-means++ seeding and approximate method.

Keywords: K-means++ seeding · Lower bound · Progressive partial distance · Piecewise aggregate approximation

1 Introduction

Lloyd's algorithm [11] (also known as the k-means clustering) is a fundamental unsupervised approach for partitioning unlabeled data into k clusters. Its performance crucially depends on the k chosen initial centers [12]. The simple initialization scheme (e.g., randomly selecting initial centers) can yield arbitrarily bad results. Therefore, many studies concentrated on improving its initialization step (also called the *seeding* stage) [6,14]. Among these optimizations, the k-means++ [14] is the most widely used and is still considered to have state-of-the-art performance until now.

Compared with Lloyd's algorithm, k-means++ has an additional step, *Seeding*, in which the initial chosen centers are selected via the D^2-*sampling* [14].

X. Wang et al. (Eds.): DASFAA 2023, LNCS 13946, pp. 132–141, 2023.
https://doi.org/10.1007/978-3-031-30678-5_11

The k-means++ is easy to implement and performs empirical well. However, given n data points in d-dimensional Euclidean space, the time complexity of k-means++ seeding is $O(nkd)$, which becomes impractical for large datasets. One iteration of Lloyd's algorithm also has the running time of $O(nkd)$, but Lloyd's algorithm can be parallelized easily in the distributed platform(e.g., MapReduce) or replaced by efficient methods such as mini-batch k-means [1]. Nevertheless, the k-means++ seeding's sequential nature hinders its parallelization[1]. Besides, many applications require efficient seeding without running subsequent Lloyd's algorithm [2]. Thus, it is necessary to improve the efficiency of k-means++ seeding so that it can work well on large datasets.

The performance bottleneck of k-means++ seeding is caused by the D^2-*sampling* step as all pairwise Euclidean Distances between data samples and selected centers need to be computed. Therefore, instead of using costly D^2-*sampling*, previous works [1,2,7] mainly focus on adopting efficient sampling strategies to speed up k-means++ seeding. For example, the *Markov Chain Monte Carlo sampling* (MCMC) is utilized in literature [1] to obtain fast seeding. These optimizations can accelerate k-means++ seeding effectively; however, they are inherently approximation algorithms and cannot get the same results as k-means++ seeding because the D^2-*sampling* is replaced.

Contributions. This paper proposes a *lower bound based framework (LBF)* for speeding up k-means++ seeding. LBF is an *exact* approach which can perform k-means++ seeding in less time while guaranteeing the same results as D^2-*sampling*. In original D^2-*sampling*, a large number of expensive pairwise Euclidean Distances need to be examined. Like classical k-means++ seeding, LBF also uses D^2-*sampling* to generate a new cluster center. Nevertheless, in LBF, we can safely decrease many Euclidean Distance calculations using the lower bound technique. Two lower bound schemes based on progressive partial distance [10] and piecewise aggregate approximation [9], respectively, are presented in this paper and embedded into LBF. The experimental results empirically validate the effectiveness of the proposals.

2   Preliminary and Related Work

Given an integer $k \geq 1$ and n data points $X = \{\mathbf{x}_1, \mathbf{x}_2, ..., \mathbf{x}_n\}$ in $\mathbb{R}^d$, the k-means problem is to discovery k cluster centers $C = \{\mathbf{c}_1, \mathbf{c}_2, ...\mathbf{c}_k\}$ in $\mathbb{R}^d$ which can minimize the quantization error $Q_C(X) = \sum_{\mathbf{x} \in X} d(\mathbf{x}, C)^2$, in which $d(\mathbf{x}, C)^2 = \min_{\mathbf{c} \in C} \|\mathbf{x} - \mathbf{c}\|^2$ and $\|\mathbf{x} - \mathbf{c}\|^2 = \sum_{i=1}^{d}(x_i - c_i)^2$. The $\mathbf{c}^*$ is called the cluster center of $\mathbf{x}$ if $\mathbf{c}^* = \arg\min_{\mathbf{c} \in C} \|\mathbf{x} - \mathbf{c}\|^2$. Finding the optimal solution for k-means problem is time-consuming. For this reason, a heuristic approach named Lloyd's algorithm [11] is widely used to approximate the solution. The initial selection of k cluster centers significantly influences the performance of Lloyd's algorithm. Therefore, many methods have been proposed to improve its seeding

---

[1] The k-means|| [3] is a parallel implementation of k-means++ seeding; however, it cannot guarantee the same results as k-means++ seeding.

stage. Among these methods, k-means++ seeding is the most widely used and has the most far-reaching impact.

The main idea of k-means++ seeding is selecting cluster centers one-by-one via D^2-sampling. Algorithm 1 shows its pseudo-code. From Algorithm 1, we can find that the ith selected center depends on previous centers, and when using D^2-sampling, k-means++ seeding has to scan full dataset k times; thus, its time complexity is $O(nkd)$, which becomes impractical for large datasets.

In the past few years, many optimizations have been proposed to improve k-means++ seeding's efficiency [1,2,7]. Instead of using costly D^2-sampling, these methods adopt efficient sampling schemes to generate cluster centers. For example, the *MCMC sampling* [1,2] and *reject sampling* [7] are used to replace D^2-sampling to accelerate seeding procedure. The running time of computing the Euclidean Distance $\|\mathbf{x} - \mathbf{c}\|^2$ is $O(d)$, which is time-consuming when data points are in high-dimensional space. Therefore, embedding dimensionality reduction techniques into D^2-sampling is also an effective way to improve the efficiency of k-means++ seeding [5]. These methods perform faster than original k-means++ seeding; however, they are inherently approximation approaches and cannot get the same results as the original algorithm. Although k-means++ seeding is not apparently parallelizable due to its sequential nature, there exist some parallel implementations of k-means++ seeding [3]. Nevertheless, they also cannot guarantee the exact results as k-means++ seeding because the original D^2-sampling is modified.

As for the exact algorithm, the literature [13] adopts triangle inequality to accelerate k-means++ seeding. Unlike it, this paper uses the lower bound based framework (LBF) for speeding up k-means++ seeding. The LBF has already been successfully used in fast clustering [4] and efficient exact nearest neighbor search [17]; however, to our best knowledge, we are unaware of any work speeding up the exact k-means++ seeding with the framework introduced in this paper.

3 The Proposed Method

3.1 Intuition Behind Our Method

Let $C_{i-1} = \{\mathbf{c}_1, \mathbf{c}_2, ..., \mathbf{c}_{i-1}\}$ denote the cluster centers selected in the first $i-1$ iterative steps, $\mathbf{c}_i$ represent the currently added cluster center, and D_j^i denote the Euclidean Distance between $\mathbf{x}_j$ and its cluster center in ith iteration. Obviously, we have $C_i = C_{i-1} \cup \mathbf{c}_i$ and $D_j^{i-1} = d(\mathbf{x}_j, C_{i-1})^2$. For any data point $\mathbf{x}_j$ in the dataset, by caching $d(\mathbf{x}_j, C_{i-1})^2$ into a variable D_j^{i-1}, we can compute $d(\mathbf{x}_j, C_i)^2$ efficiently. Specifically, $D_j^i = d(\mathbf{x}_j, C_i)^2 = \min(\|\mathbf{x}_j - \mathbf{c}_i\|^2, D_j^{i-1})$.

When a new cluster center $\mathbf{c}_i$ is added into C, the data point $\mathbf{x}_j$'s cluster center may become $\mathbf{c}_i$. To determine whether the cluster center corresponding to $\mathbf{x}_j$ has changed, the classical k-means++ seeding must calculate $\|\mathbf{x}_j - \mathbf{c}_i\|^2$. If $\|\mathbf{x}_j - \mathbf{c}_i\|^2 < D_j^{i-1}$, then $\mathbf{x}_j$'s cluster center is $\mathbf{c}_i$ and $D_j^i = \|\mathbf{x}_j - \mathbf{c}_i\|^2$; otherwise, the cluster center of $\mathbf{x}_j$ remains unchanged and $D_j^i = D_j^{i-1}$.

Algorithm 1. The pseudo-code of k-means++ seeding

Input: X: *Dataset which includes n data points;* k: *Cluster number*
Output: C_k: *Selected cluster centers;*
1: Choose the first center $\mathbf{c}_1$ randomly from X and set $C_1 = \{\mathbf{c}_1\}$;
2: **for** $i = 2$ to k **do**
3:     $\forall \mathbf{x} \in X$, Compute $d(\mathbf{x}, C_{i-1})^2$;
4:     Sample $\mathbf{x} \in X$ with probability $p(\mathbf{x}) = \frac{d(\mathbf{x}, C_{i-1})^2}{\sum_{\hat{\mathbf{x}} \in X} d(\hat{\mathbf{x}}, C_{i-1})^2}$; $//D^2$-sampling
5:     $C_i = C_{i-1} \cup \mathbf{x}$;
6: **end for**
7: **return** C_k;

Computing the Euclidean Distance between two data points in the high-dimensional space is costly, and in k-means++ seeding, every data point in the dataset must calculate the Euclidean Distance from $\mathbf{c}_i$. Therefore, reducing the number of Euclidean Distance calculations is an effective method to improve the efficiency of k-means++ seeding.

The basic idea of our proposed framework is to determine whether the cluster center corresponding to $\mathbf{x}$ changes without calculating the Euclidean Distance between $\mathbf{x}$ and currently selected cluster center. To achieve this, we first define the lower bound of Euclidean Distance.

Definition 1. $\forall \; \boldsymbol{x}, \boldsymbol{c} \in \mathbb{R}^d$, *if* $0 \leq LB(\boldsymbol{x}, \boldsymbol{c}) \leq \|\boldsymbol{x} - \boldsymbol{c}\|^2$ *holds, then we call the function $LB(\cdot)$ is the lower bound of Euclidean Distance in $\mathbb{R}^d$.*

Given a lower bound $LB(\cdot)$, if $LB(\mathbf{x}_j, \mathbf{c}_i) > D_j^{i-1}$ holds, then we have $\|\mathbf{x}_j - \mathbf{c}_i\|^2 \geq LB(\mathbf{x}_j, \mathbf{c}_i) > D_j^{i-1}$. Thus, we know that the cluster center of $\mathbf{x}_j$ has not changed without computing the Euclidean Distance $\|\mathbf{x}_j - \mathbf{c}_i\|^2$. However, if $LB(\mathbf{x}_j, \mathbf{c}_i) \leq D_j^{i-1}$, the Euclidean Distance $\|\mathbf{x}_j - \mathbf{c}_i\|^2$ has to be computed in this case to determine whether $\|\mathbf{x}_j - \mathbf{c}_i\|^2$ is greater than D_j^{i-1}.

It is worth noting that before introducing the specific calculation method of $LB(\cdot)$, we assume that we now have an oracle $LB(\cdot)$ function. Based on lower bound $LB(\cdot)$, we formally introduce our framework for speeding up k-means++ seeding in the following part.

3.2   The LBF for Accelerating K-Means++ Seeding

This subsection presents LBF, our accelerated version for k-Means++ seeding. Algorithm 2 presents the details of LBF. The value of D_j represents the Euclidean Distance between data point $\mathbf{x}_j$ and its corresponding cluster center. For the newly added cluster center $\mathbf{c}_i$, Algorithm 2 first calculates the lower bound between $\mathbf{c}_i$ and data point $\mathbf{x}_j$ (line 6). If $LB(\mathbf{x}_j, \mathbf{c}_i) > D_j$, we can conclude that $\mathbf{c}_i$ can not be the cluster center of $\mathbf{x}_j$ because the inequality $\|\mathbf{x}_j - \mathbf{c}_i\|^2 \geq LB(\mathbf{x}_j, \mathbf{c}_i) > D_j$ holds. Nevertheless, if $LB(\mathbf{x}_j, \mathbf{c}_i) \leq D_j$, we can not conclude that $\|\mathbf{x}_j - \mathbf{c}_i\|^2$ is greater than D_j; thus, the Euclidean Distance $\|\mathbf{x}_j - \mathbf{c}_i\|^2$ has to be computed. If $\|\mathbf{x}_j - \mathbf{c}_i\|^2 < D_j$, $\mathbf{c}_i$ becomes $\mathbf{x}_j$'s cluster center and D_j is updated (line 7–11). Like classical k-means++ seeding, Algorithm 2 finally selects cluster center by D^2-*sampling* (line 14–15).

Algorithm 2. The pseudo-code of LBF

Input: X: *Dataset which includes n data points;* k: *Cluster number*
Output: C: *Selected cluster centers;*
1: Pre-processing step for $LB(\cdot)$;
2: Choose the first center $\mathbf{c}_1$ randomly from X and set $C = \{\mathbf{c}_1\}$;
3: Initialize D_i with the value of $+\infty$, $i = 1$ *to* n;
4: **for** $i = 1$ to $k - 1$ **do**
5:    **for** $j = 1$ to n **do**
6:      $lower\_bound = LB(\mathbf{x}_j, \mathbf{c}_i)$;
7:      **if** $lower\_bound \leq D_j$ **then**
8:        $temp\_distance = \|\mathbf{x}_j - \mathbf{c}_i\|^2$;
9:        **if** $temp\_distance < D_j$ **then**
10:          $D_j = \|\mathbf{x}_j - \mathbf{c}_i\|^2$;
11:        **end if**
12:      **end if**
13:    **end for**
14:    $\mathbf{c}_{i+1} = \mathbf{x}_j$, where j is sampled with probability $p(j) = \frac{D_j}{\sum_{j=1}^{n} D_j}$;   //$D^2$-*sampling*

15:    $C = C \cup \mathbf{c}_{i+1}$;
16: **end for**
17: **return** C;

The difference between LBF and conventional k-means++ seeding is that LBF adopts the lower bound technique to reduce the number of Euclidean Distance computations. Suppose that the cost for calculating lower bound and Euclidean Distance is $COST_{LB}$ and $COST_{ED}$, respectively. Let N and M denote the number of times the Euclidean Distance needs to be calculated in k-means++ seeding and LBF, respectively. $F = \frac{N-M}{N} \in [0,1]$ represents the filtering ratio. Then, theoretically, classical k-means++ seeding takes $TIME_{classical} = N \times COST_{ED}$ time for Euclidean Distance computation and LBF takes $TIME_{LBF} = N \times COST_{LB} + N \times (1 - F) \times COST_{ED}$ time for Euclidean Distance computation.

An ideal lower bound $LB(\cdot)$ must be efficient and tight, simultaneously. An efficient lower bound indicates its computational cost is much less than Euclidean Distance and a tight lower bound means its value is close to real Euclidean Distance. The lower bound function $LB(\cdot)$ is a critical building block for LBF. The following subsection will introduce two specific implementations of $LB(\cdot)$.

3.3   Two Lower Bound Strategies

This subsection introduces two lower bound schemes used in LBF, a simple one is based on Progressive Partial Distance (PPD) [10], and a sophisticated one is based on Piecewise Aggregate Approximation (PAA) [9] (also known as the Segment Mean [16]).

Lower Bound via Progressive Partial Distance. Given two data points $\mathbf{x} = (x_1, x_2, ..., x_d)$ and $\mathbf{c} = (c_1, c_2, ..., c_d)$, and an integer $S < d$, the PPD between $\mathbf{x}$ and $\mathbf{c}$ is computed as follows: $PPD(\mathbf{x}, \mathbf{c}, S) = \sum_{i=1}^{S}(x_i - c_i)^2$. Obviously, $\|\mathbf{x} - \mathbf{c}\|^2 = PPD(\mathbf{x}, \mathbf{c}, S) + \sum_{i=S+1}^{d}(x_i - c_i)^2 > PPD(\mathbf{x}, \mathbf{c}, S)$; thus $PPD(\mathbf{x}, \mathbf{c}, S)$ is a lower bound of $\|\mathbf{x} - \mathbf{c}\|^2$ and can be computed more efficient than $\|\mathbf{x} - \mathbf{c}\|^2$. Besides, given an integer $N > 1$ and suppose $NS \leq d$, we have

$$PPD(\mathbf{x}, \mathbf{c}, NS) = PPD(\mathbf{x}, \mathbf{c}, (N-1)S) + \sum_{i=(N-1)S+1}^{NS}(x_i - c_i)^2. \qquad (1)$$

Therefore, the PPD between $\mathbf{x}$ and $\mathbf{c}$ can be obtained step-by-step.

When embedding PPD into LBF, we first check whether $PPD(\mathbf{x}_j, \mathbf{c}_i, S) > D_j$. Once it is greater than D_j, the process of computing lower bound PPD will be terminated, and Euclidean Distance $\|\mathbf{x}_j - \mathbf{c}_i\|^2$ can be skipped. Otherwise, we have to compute $PPD(\mathbf{x}_j, \mathbf{c}_i, 2S)$ until the whole distance (i.e., $PPD(\mathbf{x}_j, \mathbf{c}_i, d)$) is computed.

Lower Bound via Piecewise Aggregate Approximation. We first give a simple but loose lower bound and then introduce how to tighten this lower bound via PAA. Given two data points $\mathbf{x} = (x_1, x_2, ..., x_d)$, $\mathbf{c} = (c_1, c_2, ..., c_d)$, and their arithmetic mean $\bar{\mathbf{x}}$, $\bar{\mathbf{c}}$, we have

$$LB\_MEAN(\mathbf{x}, \mathbf{c}) \equiv d \times \|\bar{\mathbf{x}} - \bar{\mathbf{c}}\|^2 \leq \|\mathbf{x} - \mathbf{c}\|^2. \qquad (2)$$

Proof. $f(\cdot) = \|\cdot\|^2$ is a convex function and according to convex function's property, we have $f(\sum_{i=1}^{d}\lambda_i x_i) \leq \sum_{i=1}^{d}\lambda_i f(x_i)$, in which $\lambda_i \geq 0$ and $\sum_{i=1}^{d}\lambda_i = 1$. Therefore, we have $\|\bar{\mathbf{x}} - \bar{\mathbf{c}}\|^2 = \|\sum_{i=1}^{d}\frac{1}{d}(x_i - c_i)\|^2 \leq \frac{1}{d}\sum_{i=1}^{d}\|x_i - c_i\|^2$. Thus, $d \times \|\bar{\mathbf{x}} - \bar{\mathbf{c}}\|^2 \leq \sum_{i=1}^{d}\|x_i - c_i\|^2 = \|\mathbf{x} - \mathbf{c}\|^2$, which concludes the proof.

$LB\_MEAN(\mathbf{x}, \mathbf{c})$ is a lower bound of $\|\mathbf{x} - \mathbf{c}\|^2$ and can be computed more efficient than $\|\mathbf{x} - \mathbf{c}\|^2$. However, it is too loose to be effective in general. We next introduce how to tighten $LB\_MEAN(\cdot)$ via PAA.

Given a data point $\mathbf{x} = (x_1, x_2, ..., x_d)$, suppose $\mathbf{x}$ can be divided into B disjoint segment and the length of each segment is l, i.e., $l = \frac{d}{B}$. In this way, $\mathbf{x}$ can be written as $\mathbf{x} = (\mathbf{x}_{(1)}, \mathbf{x}_{(2)}, ..., \mathbf{x}_{(B)})$, where $\mathbf{x}_{(i)} = (x_{l*(i-1)+1}, x_{l*(i-1)+2}, ..., x_{l*i})$. Note that if d can not be dividable by l, 0 can be appended to $\mathbf{x}$. The PAA representation of $\mathbf{x}$ is $PAA(\mathbf{x}) = (\bar{\mathbf{x}}_1, \bar{\mathbf{x}}_1, ..., \bar{\mathbf{x}}_B)$, in which $\bar{\mathbf{x}}_i$ is the arithmetic mean value of $\mathbf{x}_{(i)}$, i.e., $\bar{\mathbf{x}}_i = \frac{1}{l}\sum_{j=1}^{l}x_{l*(i-1)+j}$.

Given two data points $\mathbf{x}$, $\mathbf{c}$ and their PAA representations $PAA(\mathbf{x})$, $PAA(\mathbf{c})$, we have *PAA inequality* shown as follows.

$$LB\_PAA(\mathbf{x}, \mathbf{c}) \equiv l \times \|PAA(\mathbf{x}) - PAA(\mathbf{c})\|^2 \leq \|\mathbf{x} - \mathbf{c}\|^2. \qquad (3)$$

The proof of *PAA inequality* can be found in [9], and due to limited space, we refer the reader to [9] for further details. The time complexity of computing $LB\_PAA(\mathbf{x}, \mathbf{c})$ is $O(B)$, which performs faster than than $\|\mathbf{x} - \mathbf{c}\|^2$. We can find that $LB\_MEAN(\mathbf{x}, \mathbf{c})$ is a special case of $LB\_PAA(\mathbf{x}, \mathbf{c})$ when $l = d$. Besides, we can also proof that $LB\_PAA(\mathbf{x}, \mathbf{c}) \geq LB\_MEAN(\mathbf{x}, \mathbf{c})$ [9].

It is worth noting that there exists additional pre-processing cost when embedding *LB_PAA* into LBF. Specifically, given a datasets $X = \{\mathbf{x}_1, \mathbf{x}_2, ..., \mathbf{x}_n\}$, LBF has to transform X into $PAA(X) = \{PAA(\mathbf{x}_1), PAA(\mathbf{x}_2), ..., PAA(\mathbf{x}_n)\}$. The time complexity of this transformation is $O(nd)$, which is much less than classical k-means++ seeding ($O(nkd)$).

4 Experiments

4.1 Experimental Setup

Algorithms. Four algorithms are used in our experimental evaluation. **CKM** is the classical k-means++ seeding. **LBF_PPD** and **LBF_PAA** are two algorithms proposed in this paper which embed progressive partial distance and piecewise aggregate approximation into LBF, respectively, for accelerating exact classical k-means++ seeding. **K-MC²** [1,2] is a representative approximation approach which adopts *Markov Chain Monte Carlo Sampling* to replace exact D^2-*Sampling* used in **CKM**. **K-MC²** can speed up **CKM** effectively; however, it is inherently an approximation algorithm and cannot get the same results as **CKM**. All methods are implemented in Python on a laptop PC (Windows 10 enterprise, i7-9750H CPU, and 16 GB memory).

Datasets. Two synthetic and six real-world datasets are used to evaluate the performance of the algorithms mentioned above. The *Gaussian* and *MixedSinx* are two synthetic datasets. *Gaussian* contains 100000 data points, and each data point has a dimension of 128. To generate data point $\mathbf{x}_i$ in *Gaussian*, we select two integers μ_i from $[-10, 10]$ and σ_i from $[1, 5]$ randomly. Each element in $\mathbf{x}_i$ is sampled from Gaussian distribution $N(\mu_i, \sigma_i^2)$. *MixedSinx* includes 100000 data points with dimension 256. The ith data point in *MixedSinx* is generated using function $\mathbf{x}_i = A \times sin(2\pi \times f \times t)$, in which A and f are selected from $[1, 10]$ randomly, and $t = [0, \frac{1}{256}, 2 \times \frac{1}{256}, ..., 1)$. Six real-world datasets come from various domains and vary in sizes and dimensions. Datasets *Crop*, *Wafer*, *StarLightCurves* and *U.W.G.LibraryAll* are collected from UCR Archive [8]. Datasets *usps* and *coil100* can be download from website[2,3], respectively. Table 1 provides detailed information about the used datasets.

Parameter Setup. In **LBF_PPD** and **LBF_PAA**, we should determine the parameters S and l before experimental evaluation. It is worth noting that the optimal parameters depend on the dataset structure. Nevertheless, the same parameters are utilized in our experiments for all datasets to avoid manual tuning. Specifically, we set $S = 16$ and $l = 16$ on all datasets. In **K-MC²**, the MCMC chain length m should also be given before evaluation. From literature [2], we know that the performance of **K-MC²** can be improved and converged to that of **CKM** as m increases. In [2], we can find that when $m = 200$, **K-MC²** achieves minor relative error compared with **CKM**. Therefore, we set $m = 200$ in our evaluation. A wide variety of values for the number of clusters are considered for each dataset. Specifically, we set cluster number $k = \{8, 16, 32, 64, 128\}$.

[2] https://hastie.su.domains/StatLearnSparsity_files/DATA/zipcode.html.
[3] https://www1.cs.columbia.edu/CAVE/software/softlib/coil-100.php.

Table 1. Datasets descriptions.

Name	Database size, n	Dimension, d	Type
Crop	24000	46	Image
Wafer	7164	152	Sensor
StarLightCurves	9236	1024	Sensor
U.W.G.LibraryAll	4478	945	Motion
coil100	7200	1024	Image
usps	9298	256	Image
Gaussian	100000	128	Synthetic
MixedSinx	100000	256	Synthetic

4.2 Running Time

This part evaluates the running time of the proposals and related competitors by conducting the k-means seeding on synthetic and real-world datasets. The *speedup rate*, which evaluates the running time of an algorithm relative to the running time of **CKM**, is reported. The results of the speedup rate can be found in Fig. 1. To illustrate experimental results more clearly, the *Wilcoxon signed-rank test* [15], which returns a p-value between two methods in terms of running time, is provided. If $p < 0.05$, there is a significant difference between the two compared methods. Besides we report the average rank of each method. The results can be found in Table 2.

Table 2. Wilcoxon signed-rank test and average rank.

	CKM	LBF_PAA	LBF_PPD	K-MC^2
CKM		$p < 0.05$	$p < 0.05$	$p < 0.05$
LBF_PAA			$p < 0.05$	$p = 0.6868$
LBF_PPD				$p = 0.2883$
K-MC^2				
Average rank	3.65	1.65	2.725	1.975

From Table 2, we can find that all algorithms significantly outperform baseline **CKM** method with the p-value smaller than 0.05, which demonstrates the effectiveness of these optimization methods. Our proposed **LBF_PAA** method performs best among all algorithms with a best average rank of 1.65, followed by **K-MC^2**, whose average rank is 1.975. Actually, Table 2 shows that **LBF_PAA** is not significantly better than **K-MC^2** ($p = 0.6868$). **K-MC^2** performs better than our **LBF_PPD**; however, no significant difference exists between **K-MC^2** and **LBF_PPD** ($p = 0.2883$).

Figure 1 provides more details on the performance comparison of algorithms on different datasets. We can find from Fig. 1 that **K-MC^2** performs best when k is small ($k = 8, 16$). However, its performance degrades significantly with the increase of k. When $k = 128$, on some datasets (e.g., Wafer, StarLightCurves), the performance of **K-MC**^{2} is even worse than baseline method **CKM** (*speedup* < 1). This is because the time complexity of **K-MC**^{2} is linear with $O(k^2)$. Therefore, **K-MC**^{2} does not scale well for large values of k.

Figure 1 shows that the proposed method **LBF_PAA** performs faster than **CKM** on all datasets for all values of k (except when $k = 8$ on dataset usps[4]). Our inferior method **LBF_PPD** performs worse than **CKM** in most datasets when $k = 8$. However, it outperforms **CKM** as k increases. Note that **LBF_PAA** and **LBF_PPD** behave differently with **K-MC**^{2}; their efficiency improve as k increases. We explain this phenomenon in detail as follows.   Suppose c_i is the

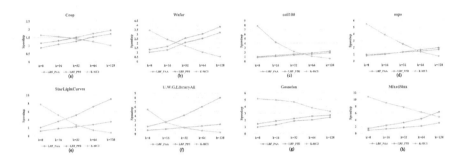

Fig. 1. The speedup rate of **LBF_PPD**, **LBF_PAA** and **K-MC**^{2} on all used datasets.

newly added cluster center. Given a data point x_j and the currently cluster center set C, LBF does not need to compute the Euclidean Distance between c_i and x_j if $LB(c_i, x_j) > D_j$, in which $D_j = d(x_j, C)^2 = \min_{c \in C} \|x - c\|^2$. Let $C^\star = C \cup c^\star$, then we have $D_j^\star = \min_{c \in C^\star} \|x - c\|^2 \leq D_j$. Therefore, $P(LB(c_i, x_j) > D_j) \leq P(LB(c_i, x_j) > D_j^\star)$, in which $P(A)$ represents the probability of event A occurring. Thus, the larger the value of k, the greater the probability that we do not need to calculate the Euclidean Distance between c_i and x_j, so the efficiency is improved.

5   Conclusion and Future Work

In this paper, a framework called LBF is proposed to accelerate classical k-means++ seeding. LBF adopts the lower bound technique to filter out many Euclidean Distance computations that are redundant. LBF adopts the D^2-*sampling* to generate cluster centers; thus it is an exact approach that can obtain the same results as the original k-means++ seeding. Two lower bounds built on progressive partial distance and piecewise aggregate approximation are embedded into the LBF. Experimental results validate the effectiveness and superiority of the proposals. Many Euclidean Distance lower bounds have been proposed

---

[4] In this case, the performance of algorithm **LBF_PAA** is slightly weaker than that of **CKM** and the speedup rate of **LBF_PAA** is 0.96.

during the past decades. Therefore, embedding these lower bounds into our proposed LBF is a future work worth investigating.

Acknowledgement. This work is supported by Science Foundation of Zhejiang Sci-Tech University (ZSTU) under Grant No. 22232264-Y

References

1. Bachem, O., Lucic, M., Hassani, H., Krause, A.: Fast and provably good seedings for k-means. In: Advances in Neural Information Processing Systems, vol. 29 (2016)
2. Bachem, O., Lucic, M., Hassani, S.H., Krause, A.: Approximate k-means++ in sublinear time. In: Thirtieth AAAI Conference on Artificial Intelligence (2016)
3. Bahmani, B., Moseley, B., Vattani, A., Kumar, R., Vassilvitskii, S.: Scalable k-means+. In: Proceedings of the VLDB Endowment, vol. 5, no. 7 (2012)
4. Bottesch, T., Bühler, T., Kächele, M.: Speeding up k-means by approximating Euclidean distances via block vectors. In: International Conference on Machine Learning, pp. 2578–2586. PMLR (2016)
5. Chan, J.Y., Leung, A.P.: Efficient k-means++ with random projection. In: 2017 International Joint Conference on Neural Networks (IJCNN), pp. 94–100. IEEE (2017)
6. Choo, D., Grunau, C., Portmann, J., Rozhon, V.: k-means++: few more steps yield constant approximation. In: International Conference on Machine Learning, pp. 1909–1917. PMLR (2020)
7. Cohen-Addad, V., Lattanzi, S., Norouzi-Fard, A., Sohler, C., Svensson, O.: Fast and accurate k-means++ via rejection sampling. In: Advances in Neural Information Processing Systems, vol. 33, pp. 16235–16245 (2020)
8. Dau, H.A., et al.: Hexagon-ML: The UCR time series classification archive (2018)
9. Keogh, E., Chakrabarti, K., Pazzani, M., Mehrotra, S.: Dimensionality reduction for fast similarity search in large time series databases. Knowl. Inf. Syst. **3**(3), 263–286 (2001)
10. Liu, Y., Wei, H., Cheng, H.: Exploiting lower bounds to accelerate approximate nearest neighbor search on high-dimensional data. Inf. Sci. **465**, 484–504 (2018)
11. Lloyd, S.: Least squares quantization in PCM. IEEE Trans. Inf. Theor **28**(2), 129–137 (1982)
12. Olukanmi, P., Nelwamondo, F., Marwala, T.: k-means-mind: comparing seeds without repeated k-means runs. Neural Comput. Appl., 1–15 (2022)
13. Raff, E.: Exact acceleration of k-means++ and k-means ||. arXiv preprint arXiv:2105.02936 (2021)
14. Vassilvitskii, S., Arthur, D.: k-means++: The advantages of careful seeding. In: Proceedings of the Eighteenth Annual ACM-SIAM Symposium on Discrete Algorithms, pp. 1027–1035 (2006)
15. Wilcoxon, F.: Individual Comparisons by Ranking Methods. In: Kotz, S., Johnson, N.L. (eds.) Breakthroughs in Statistics. Springer Series in Statistics, pp. 196–202. Springer, New York (1992). https://doi.org/10.1007/978-1-4612-4380-9_16
16. Yi, B.K., Faloutsos, C.: Fast time sequence indexing for arbitrary Lp norms (2000)
17. Zhang, H., Dong, Y., Xu, D.: Accelerating exact nearest neighbor search in high dimensional Euclidean space via block vectors. Int. J. Intell. Syst. **37**(2), 1697–1722 (2022)

Modeling Intra-class and Inter-class Constraints for Out-of-Domain Detection

Shun Zhang[1], Jiaqi Bai[1], Tongliang Li[1], Zhao Yan[2(✉)], and Zhoujun Li[1(✉)]

[1] State Key Lab of Software Development Environment, Beihang University, Beijing, China
{shunzhang,bjq,tonyliangli,lizj}@buaa.edu.cn
[2] Tencent Cloud Xiaowei, Beijing, China
zhaoyan@tencent.com

Abstract. Out-of-Domain (OOD) detection aims to identify whether a query falls outside the predefined intent set, which is crucial to maintaining high reliability and improving user experience in a task-oriented dialogue system. The key challenge is how to learn discriminative intent representations that are beneficial for distinguishing in-domain (IND) and OOD intents. However, previous methods ignore the compactness between instances and dispersion among categories, which limits the OOD detection performance. In this paper, we propose a novel Hybrid Contrastive Learning (HybridCL) framework to model both intra-class and inter-class constraints in OOD detection. Specifically, we first propose an intra-class constraint contrastive learning (Intra-CCL) objective, which encourages instances with the same class to be close to their prototypes, forming compact clusters. Then, we present an inter-class constraint contrastive learning (Inter-CCL) objective to effectively enlarge the discrepancy among different classes as much as possible, enforcing the strong separability for different classes in the intent embedding space. Besides, to further enhance the discriminative representation capability of the encoder, we employ an intent-wise attention mechanism to capture the relationships between the intents and the corresponding labels. Experiments and analysis on two public benchmark datasets show the effectiveness of our approach.

Keywords: Out-of-domain Detection · Contrastive Learning

1 Introduction

Out-of-Domain (OOD) detection is a key step in developing a Natural Language Understanding (NLU) module at the core of any modern Conversational AI system [15,23]. Effectively identifying the OOD intent can improve customer satisfaction and discover potential user needs. Different from traditional intent detection tasks, the number of OOD instances is always unknown in practical scenarios and extensive OOD instances can barely annotate, which makes it

X. Wang et al. (Eds.): DASFAA 2023, LNCS 13946, pp. 142–158, 2023.
https://doi.org/10.1007/978-3-031-30678-5_12

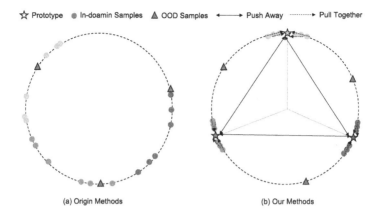

Fig. 1. We address the task of unsupervised OOD detection. Left: Existing contrastive learning-based methods show the distributions of in-domain and OOD samples, but fail to extract discriminative features due to a lack of mandatory intra- and inter-class constraints. Right: Our method constrains high intra-class compactness and inter-class separation to learn discriminative intent representations that benefit OOD detection.

challenging to identify OOD intent in the practical dialogue system. Recently, there has been growing research interest in the OOD detection task (Fig. 1).

In general, the existing OOD detection methods can be roughly divided into two categories, i.e., supervised methods and unsupervised methods. Most of the supervised methods try to construct pseudo-OOD instances for $(C+1)$-way training, where C is the number of IND classes and the additional class represents the OOD intents, such as SELFSUP [28], POG [30] and ODIST [18]. However, the supervised method in the OOD detection task requires large human laborers to annotate OOD instances, which is time- and resource-consuming. Moreover, as stated in [26], manually constructed OOD samples endowed with artificial inductive bias cannot cover all unknown classes in the actual environment, so this kind of method has its limitations. In contrast, the unsupervised methods do not need OOD instances. It instead leverages IND data to conduct a two-stage training procedure, such as DOC [19], OpenMAX [1], DeepUNK [14], SEG [25] and SCL [26]. Typically, the unsupervised methods first learn intent representations via the IND data and then adopt detecting algorithms to distinguish IND and OOD intents. The key challenge of unsupervised methods is to learn discriminative intent representations that benefit OOD detection. In this paper, we mainly focus on the unsupervised OOD detection method.

Although previous unsupervised OOD detection methods have achieved remarkable progress, these methods always suffer from the following drawbacks: (1) **Low Intra-class Compactness:** To distinguish IND and OOD intents, existing works [14,26] mainly estimate a decision boundary from the general instance-aware perspective. But none of them consider the following question: what kind of intent representation is more compact to construct a tighter decision boundary? Although they show effectiveness for classifying known IND classes, such learned representations with low intra-class compactness are limited for separating IND and OOD intents. (2) **Low Inter-class Dispersion:** To effectively distinguish semantically similar OOD and IND intents, one of the key challenges is to enlarge the discrepancy among different classes as much as possible. To learn the separation between classes, [14] introduce margin cosine loss, [26] further proposes SCL, which jointly optimizes contrastive learning [5,10] and margin cosine loss to learn discriminative intent representations. However, these methods fail to impose mandatory constraints on the distances between different classes in the intent semantic space, resulting in the potential overlap in distributions of some classes with similar intent categories, which is not conducive to OOD detection. All problems mentioned above limit the performance of OOD detection.

To address the aforementioned problems, in this paper, we propose a novel Hybrid Contrastive Learning (HybridCL) framework for OOD detection, which aims to learn discriminative intent representations with high intra-class compactness and high inter-class separation. Figure 2 shows the process of our approach. Specifically, to address the first issue, we propose the intra-class constraint contrastive learning (Intra-CCL) objective, which makes it prone to generating a more compact cluster for instances of each class, via *explicitly* pulling instances together with the prototypes of the same classes and pushing them away from prototypes of different classes. This essentially constraints on the compactness of the distribution of intent representations around the prototype from a prototype-aware perspective, assisting in constructing clear decision boundaries, thereby improving the performance of OOD detection. We further design an Inter-class constraint contrastive learning (Inter-CCL) objective to solve the second issue, which aims to enlarge the angular distances and discrepancy between different classes as much as possible. It forces the intention instances from different classes to be larger than a margin. According to our experiments, Intra-CCL and Inter-CCL are complementary in a theoretically sound way, and such two contrastive learning objectives are helpful for constraining clear decision boundaries to generate more discriminative intent representations for OOD detection. Additionally, to better enhance the discriminative representation capability of the encoder, we employ an intent-wise attention mechanism to capture the relationship between intents and labels by highlighting the alignment between intent and its corresponding label information.

To summarize, the contributions of this work are:

- We propose a novel Hybrid Contrastive Learning (HybridCL) framework for unsupervised OOD detection, which aims to model both intra-class and inter-class constraints.
- Compared to baseline models, our method achieves high intra-class compactness and high inter-class dispersion to learn discriminative intent representations.
- Extensive experiments show that our framework outperforms all the baseline models by a large margin. Detailed analyses further study the impact of two kinds of constraints on the OOD detection task.

2   Related Work

2.1   OOD Detection

OOD detection is a new task attracted wide attention in recent years. Methodologies in OOD detection can be divided into supervised and unsupervised settings according to the presence of training data from OOD. Since the scope of OOD covers nigh infinite space, gathering the data in the whole OOD space is infeasible. For this realistic reason, the most recent OOD detection studies generally distinguish OOD intent in an unsupervised manner. For instance, [8] simply uses the maximum value of the neural Softmax output as the detection score and set a threshold to identify OOD samples. Further developments [24] use Gaussian Discriminant Analysis (GDA) in latent space to detect OOD intents. To learn discriminative semantic representation, [14] first adopts a margin loss on the intent classifier, then the local outlier factor (LOF) [2] was used to detect the low-density instances as the OOD intent. [25] tried to incorporate the class label information via the Gaussian mixture model to obtain more suitable intent representations. However, all of these methods only capture shallow semantic features, which are insufficient to extract discriminative intent representations and make it hard to transfer prior IND knowledge to OOD detection. Recently, [27] combined cross-entropy loss on labeled IND data and instance-wise contrastive learning loss on unlabeled data. To maximize inter-class variance and minimize intra-class variance from the instance-aware perspective, [26] tried to use a supervised contrastive learning objective to learn discriminative intent representations. However, the key drawback of the above method is that they do not consider both intra-class compactness and inter-class dispersion.

2.2   Contrastive Learning

Contrastive learning has proven effective to learn unsupervised representations for downstream tasks. The key idea is to construct multiple positive and negative pairs, which are then used to learn to optimize a distance metric that brings the positive pairs closer while pushing the negative pairs away. Self-supervised contrastive learning regards two augmented views of a sample as positive pairs and views of different samples as negative pairs, which mainly learn instance-level features [3,7]. However, such methods heavily rely on data augmentation of instances and fail to leverage supervised signals, moreover, suffering from the false-negative problem. As an extension, supervised contrastive learning insists the representations of examples in the same class are similar and those for different classes be distinct [5,10]. These methods aim to minimize intra-class variance and maximize inter-class variance from the instance-aware perspective. To further enlarge the discrepancy among classes, [31] propose a margin-based contrastive loss. Recently, another series of contrastive approaches [6,13] seek to explicitly exploit semantic structures by pulling query samples closer to the prototype of the same class and further away from those of different classes. In this paper, inspired by [13,31], we propose a HybridCL framework to generate more discriminative intent representations for both intra-class and inter-class constraints for OOD detection.

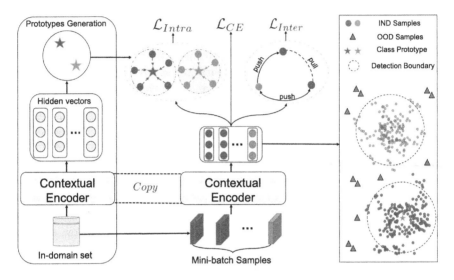

Fig. 2. The overall architecture of our proposed unsupervised OOD detection approach. During the training stage, we train an intent representation extractor using three objectives. Once trained, we adopt it to extract the intent representations and apply a detection algorithm to detect OOD instances during the test stage.

3   Methodology

In this section, we describe the proposed HybridCL framework for OOD detection in detail. As demonstrated in Fig. 2, the architecture of the HybridCL framework contains four main components: 1) *prototypes generation*, which calculates the prototype of each type of intent through IND training data; 2) *intra-class constraint contrastive learning*, which performs prototypical contrastive learning based on the supervised signal of intent labels for high intra-class compactness; 3) *inter-class constraint contrastive learning*, which performs the margin-based contrastive learning strategy based on the label information of intent for high inter-class dispersion; 4) *classifier*, which detects the OOD intent based on the generated discriminative intent representations.

3.1   Problem Definition

Formally, an unsupervised OOD detection task can be formulated as follows: in the training stage, the model is trained only on the IND training set $\mathcal{D}_m =$

$\{\boldsymbol{x}_i, y_i\}_{i=1}^{N}$ contains N training samples, where $\boldsymbol{x}_i \in \mathbb{R}^L$ is the input sample consisting of L words and $y_i \in Y_{IND} = \{y_1^s, y_2^s, \ldots, y_I^s\}$ is the intent labels. During testing, the model is expected to detect IND intents and decide whether a sample belongs to the OOD intents. The prediction space is $\{y_1^s, y_2^s, \ldots, y_I^s, y_{OOD}\}$, where all the OOD intents are grouped into a single class y_{OOD}.

Figure 2 shows the overall architecture of our method. During the training stage, we first pre-train the intent classifier using intra-CCL and inter-CCL objectives, then finetune the model using CE, both on the IND data. Once trained, we adopt it to extract the intent representations and apply OOD detection algorithms to detect OOD intents.

3.2 Encoder Module

Following the setting suggested in [14], we use the pre-trained BERT [4] as the main encoder structure for the OOD detection. As shown in Fig. 2, we feed the IND instances into the BERT and obtain the hidden semantic representation of each input. Given the i-th instance $x_i = \{t_1, t_2, \cdots, t_L\}$, the input of the BERT model is formed as:

$$\boldsymbol{x}_i = ([\text{CLS}], x_1^t, x_2^t, \ldots, x_L^t, [\text{SEP}]) \tag{1}$$

where $\{x_1^t, x_2^t, \ldots, x_L^t\}$ is the word piece embeddings of the intent instance. [CLS] denotes a special token and [SEP] denotes a separator. Encoded by the multi-layer self-attention structure, BERT outputs the contextual representation for each context token as $\boldsymbol{h} = [[\text{CLS}], h_1, h_2, \ldots, h_L] \in \mathbb{R}^{(L+1) \times H}$, where H is the hidden layer size. For the IND training set $\mathcal{D}_m$, the hidden representations of the training instances can be represented as $\mathcal{H} = \{\boldsymbol{h}_i\}_{i=1}^{N}$.

Intent-Wise Representation Extraction. To further enhance the discriminative representation capability of the encoder, inspired by [16,21], we argue that label information of pre-defined IND data includes important clues for learning discriminative intent representations. Therefore, we design an intent-wise attention mechanism (IW-AM) to enhance the intent representations by utilizing the learned discriminative label embeddings. Specifically, we randomly initialize a trainable vector $\mathbf{u}_c$ for each IND intent c and apply the vector to calculate attention over hidden states. After that, we aggregate them to get the intent-wise intent representation $\mathbf{r}_c$ for the intent c:

$$\boldsymbol{\alpha}_i = \frac{\exp\left(\mathbf{u}_c^T \mathbf{h}_i\right)}{\sum_{j=1}^{n} \exp\left(\mathbf{u}_c^T \mathbf{h}_j\right)} \tag{2}$$

$$\boldsymbol{r}_c = \sum_{i=1}^{N} \alpha_i \boldsymbol{h}_i \tag{3}$$

Finally, as suggested in [29], we operate mean-pooling on this representation of instances to get the averaged representation z_i:

$$z_i = \text{Mean-Pooling}\left([[\text{CLS}], r_1, r_2, \ldots, r_L]\right), \tag{4}$$

3.3   Intra-class Constraint Contrastive Learning

To learn the decision boundaries with the high intra-class compactness, from a prototype-aware perspective, we propose an intra-class constraint contrastive learning (Intra-CCL) objective to explicitly encourage instance-to-prototype aggregation. It constrains instances from the same class to be close to the corresponding prototype, forming compact clusters centered on the prototype.

High Intra-class Compactness. Specifically, as shown in Fig. 2, given the hidden vectors $\{z_i\}_{i=1}^N$ in a mini-batch $\mathcal{B}$ (here, N is the size of mini-batch) and an *anchor* of prototype $p_{c(i)}$, $z_i, p_{c(i)}$ is considered as a *positive* pair (z_i is an instance vector belonging to the corresponding prototype $p_{c(i)}$), while the instances z_i and the prototype p_j are treated as *negative* pair with respect to the anchor (z_i is not an instance vector belonging to the corresponding prototype p_j). Then, the intra-CCL loss is defined as:

$$\mathcal{L}_{\text{intra}}(z_i, y_i) = \sum_{i \in N} -\log \frac{\exp\left(z_i \cdot p_{c(i)}/\tau\right)}{\sum_{c=1}^C \exp\left(z_i \cdot p_j/\tau\right)} \tag{5}$$

where $p_{c(i)}$ is the representation of intent prototype with label c, τ is temperature factor and C is the number of intent categories. By optimizing the above objective function $\mathcal{L}_{\text{intra}}$, intent instances can be close to the corresponding prototype in semantic space.

Prototype Estimation and Update. During training, an important step is to estimate the prototype $\mathbf{p}_c$ for each class. Inspired by previous work, [20] derived the prototype of each class by computing the mean vector of the embedded support points belonging to the class and acquire better results. Therefore, we estimate the prototypes to use the mean vector of all the representations of the IND training instances $\mathcal{H} = \{z_i\}_{i=1}^N$ to generate k prototypes for each class $\mathcal{C} = \{p_i\}_{i=1}^k$:

$$\mathbf{p}_c = \frac{1}{M} \sum_{i \in \mathcal{D}(c)} \mathbf{z}_i, c = \{1, 2, \dots, k\} \tag{6}$$

where $\mathbf{p}_i$ is the prototype for class c, M represents the number of instances contained in each class. Here, a prototype is defined as a *representative embedding for a group of semantically similar instances* [13]. Note that we dynamically update the prototypes at the beginning of each training epoch.

3.4   Inter-class Constraint Contrastive Learning

To construct the decision boundaries with high dispersion among classes, inspired by [5], we further design an inter-class constraint contrastive learning (Inter-CCL) objective, which aims to enlarge the angular distances and discrepancy between different classes as much as possible. Intuitively, we desire embeddings where different intent classes are relatively far apart in the semantic space.

High Inter-class Dispersion. Specifically, we use the label information of intent to construct positive and negative pairs. In order to improve the high separation among classes, we introduce a dynamic inter-class margin constraint, which ensures that instances from the same class be smaller than instances from different classes. In this way, we make instances more discriminative and produce farther margins to separate semantically similar OOD and IND intents. Then the inter-CCL loss is calculated as follows:

$$\ell_i^p = \frac{1}{|\mathcal{P}(i)|} \sum_{p \in P(i)} \|z_i - z_p\|^2 \tag{7}$$

$$\ell_i^n = \frac{1}{|\mathcal{N}(i)|} \sum_{n \in N(i)} (\delta - \|z_i - z_n\|^2)_+ \tag{8}$$

$$\mathcal{L}_{\text{inter}} = \frac{1}{dN} \left(\sum_{i \in \mathcal{B}} \ell_i^p + \sum_{i \in \mathcal{B}} \ell_i^n \right) \tag{9}$$

where $\mathcal{P}(i) = \{p \in A(i) : y_i = y_p\}$ is the set of anchor samples from the same class as i, $\mathcal{N}(i) = \{n \in A(i) : y_i \neq y_n\}$ is the set of anchor instances from other classes than y_i, $A(i) = \{1, \ldots, N\} \backslash \{i\}$ is the set of all intent anchor samples. δ is a margin, d is the number of dimensions of h.

Margin Estimation and Update. Note that we only perform HybridCL on the IND data, so it is hard to accurately estimate the margin. Therefore, we define the adaptive margin, which is the maximum distance between pairs of instances from the same class in the batch:

$$\delta = \max_{i=1}^{M} \max_{p \in P(i)} \|z_i - z_p\|^2 \tag{10}$$

3.5   IND Pre-training

To facilitate both intra-class compactness and inter-class dispersion for improving OOD detection, the learning objective of our proposed model is to train the model by jointly minimizing the two losses generated by intra-CCL and inter-CCL. Thus, the overall loss $\mathcal{L}$ is formulated by summing up two losses:

$$\mathcal{L} = \alpha \mathcal{L}_{\text{intra}} + (1 - \alpha) \mathcal{L}_{\text{inter}} \tag{11}$$

where α is a hyper-parameter to balance two objectives.

3.6   IND Finetuning

Above contrastive loss mainly focuses on learning discriminative intent features, whereas there can not classify directly. Therefore, we adopt the standard cross-entropy loss to finetune the model on the IND data:

$$\mathcal{L}_{\text{ce}}(\theta) = -\frac{1}{N} \sum_{i=1}^{N} \log \frac{\exp(\mathcal{F}_{y_i}(z_i^*))}{\sum_{j=1}^{[C]} \exp(\mathcal{F}_j(z_i^*))} \tag{12}$$

where $\mathcal{F}(\cdot)$ denotes linear classifier, z_i^* is the result of z_i after pre-training and $\mathcal{F}_j(z_i^*)$ denotes the logit of the jth class and θ denotes the parameters.

Table 1. Statistics of experimental datasets.

Type	Class	Training	Validation	Test	Vocabulary Size	Length (Avg)
CLINC-FULL	150	15100	3100	5500	8288	8.32
CLINC-SMALL	150	7600	3100	5500	7017	8.31

4 Experiments

4.1 Experimental Settings

Datasets. We conduct experiments on two different and challenging real-world datasets to evaluate the proposed HybridCL framework. 1) *CLINC-FULL* [12] is a dataset that has been annotated and refined manually for evaluating the ability of OOD detection. It has 150 different intents covering 10 various domains and contains 22500 IND samples, and 1200 OOD samples respectively. 2) *CLINC-SMALL* [12] is a variant version of CLINC-FULL and is to measure the ability of OOD detection of the model in the case of insufficient samples. The data also has 150 intents, but each type contains only 50 samples. The detailed statistics are shown in Table 1.

Evaluation Metrics. For all datasets, following [26,28], we group all OOD classes as one rejected class. We calculate accuracy, Recall, and F1-score in the same way as [26]. We use four common metrics for OOD detection to measure the performance, including IND metrics: Accuracy and macro F1, and OOD metrics: Recall and macro F1, represented by ACC-IND, F1-IND, R-OOD, and F1-OOD respectively. R-OOD and F1-OOD are the main evaluation metrics. To comprehensively evaluate the performance of our model, we also compare the macro F1-score (F1-ALL) overall classes (IND and OOD classes).

Implementation Details. The pre-trained uncased BERT-base [4] is used as the encoder module in which each word token is mapped to a 768-dimensional embedding. To conduct a fair comparison, we follow a similar implementation setting as [26], and we use a single-layer BiLSTM [9] as a feature extractor. We use Adam [11] as the optimizer, the learning rate is set to 3e-5 and the dropout rate is set to 0.5. For contrastive losses, the temperature parameter τ is set to 0.08. We use the best F1 scores on the development set to calculate the MSP, GDA, and LOF thresholds adaptively. Each result of the experiments is tested 10 times under the same setting and gets the average value. We use the HuggingFace [22] PyTorch [17] implementation on Tesla V100 GPU.

4.2   Baselines

In the training stage, we compare HybridCL with CE and SCL [26]. In the detection stage, to verify the generalization of our proposed models, we use three OOD detection algorithms: Maximum Softmax Probability (MSP) [8], GDA [24], and LOF [2]:

- **MSP.** [8] applies a threshold on the maximum softmax probability. We use the best F1 scores on the validation set to calculate the threshold adaptively.
- **GDA.** [24] is a generative distance-based classifier for OOD detection with Euclidean space. It estimates the class-conditional distribution on feature spaces of DNNs via Gaussian discriminant analysis to avoid over-confidence problems and uses Mahalanobis distance to measure the confidence score of whether a test sample belongs to OOD.
- **LOF.** [2] uses the local outlier factor to detect unknown intents. The motivation is that if an example's local density is significantly lower than its k-nearest neighbor's, it is more likely to be considered as the unknown intent.

Besides, we extensively compare the proposed HybridCL with the following state-of-the-art (SOTA) methods:

- **DOC.** [19] rejects the OOD class by calculating different probability thresholds of each known class with Gaussian fitting.

Table 2. Performance comparison on CLINC-FULL and CLINC-SMALL datasets for the BERT-based model. The baseline with † means the model is re-implemented by ourselves. The best scores are highlighted in boldface.

Database	Detection	Training	ACC-IND	F1-IND	R-OOD	F1-OOD	F1-ALL
CLINC-FULL	MSP	CE†	91.64	86.48	31.00	44.93	78.75
		SCL†	89.16	88.10	40.30	52.75	83.47
		HybridCL	**92.07**	**89.13**	**51.32**	**61.77**	**87.70**
	GDA	CE†	89.44	89.08	60.00	63.29	87.42
		SCL†	91.33	88.68	48.80	58.30	82.60
		HybridCL	**92.87**	**90.35**	**60.02**	**70.14**	**89.92**
	LOF	CE†	85.67	87.30	74.70	69.10	87.18
		SCL†	85.49	88.11	79.40	69.77	88.47
		HybridCL	**86.33**	**90.12**	**87.40**	**72.92**	**90.01**
CLINC-SMALL	MSP	CE†	90.22	86.19	37.40	49.54	78.60
		SCL†	90.47	87.27	43.60	54.47	81.93
		HybridCL	**92.48**	**87.88**	**45.96**	**56.14**	**83.66**
	GDA	CE†	86.68	86.20	61.64	62.58	86.07
		SCL†	87.87	88.30	64.64	65.01	87.15
		HybridCL	**90.77**	**90.13**	**66.77**	**68.93**	**88.13**
	LOF	CE†	83.49	85.94	76.50	67.64	86.19
		SCL†	84.31	87.17	78.30	67.97	87.77
		HybridCL	**85.93**	**88.42**	**79.90**	**70.77**	**88.30**

Table 3. Performance compared with previous state-of-the-art baselines for the BERT-based model.

Model	25%		50%		75%		CLINC-FULL	
	F1-OOD	F1-IND	F1-OOD	F1-IND	F1-OOD	F1-IND	F1-OOD	F1-IND
MSP	50.88	47.53	57.62	70.58	59.08	82.59	–	–
DOC	81.98	65.96	79.00	78.25	72.87	83.69	–	–
SEG	47.23	60.59	62.52	61.13	42.50	41.60	–	–
OpenMAX	75.76	61.62	81.89	80.54	76.35	73.13	–	–
DeepUNK	87.33	70.73	85.85	82.11	81.15	86.27	66.80	89.12
SCL	93.40	77.16	86.42	84.55	73.98	86.89	69.77	88.11
HybridCL	**93.97**	**77.31**	**87.63**	**86.77**	**75.88**	**89.43**	**72.92**	**90.12**

- **SEG.** It incorporates the semantic information of each class into the large-margin Gaussian mixture loss [49] for feature representation, followed by a LOF detector.
- **OpenMAX.** [1] is an open set detection method in computer vision, and we adapt it for OOD detection. We first use the CE loss to train a classifier on IND instances, then fit a distribution to the classifier's output logits.
- **DeepUNK.** [14] learns the deep intent features with the margin loss and detects the unknown intent with the LOF.
- **SCL.** [26] first uses a supervised contrastive learning objective to learn discriminative intent features and then applies a detection algorithm to detect OOD instances.

4.3   Main Results

HybridCL Achieves the Best Results Under All Detection Algorithms. The main comparison results of HybridCL on two benchmark datasets with three different OOD detection algorithms are reported in Table 2. It can be observed from the experimental results, our proposed HybridCL framework performs consistently the best results under all detection algorithms. Specifically, our method achieves 9.02%, 11.84%, and 3.15% improvements over the SCL threshold on F1-OOD under three OOD detection algorithms on the CLINC-FULL dataset and 1.67%, 3.92%, and 2.80% on the CLINC-SMALL dataset. Similar improvements were observed in the CE method. This verifies the effectiveness of our HybridCL framework in the OOD detection task. We highlight the improvement benefits from two scientific contributions: the intra-CCL and the inter-CCL objectives, which show that constraining both high intra-class compactness and inter-class separation helps improve OOD detection.

HybridCL Significantly Outperforms Previous Baselines. To compare with previous baselines, we use the BERT as the backbone of the model, and sample 25%, 50%, 75% or Full (100%) classes of the CLINC-FULL dataset as the IND classes, the other as OOD. As exhibited in Table 3, the significance tests of

HybridCL over the SOTA baseline models show that our HybridCL significantly outperforms the baseline models. Specifically, comparing the latest work [26], we achieve 0.57%, 1.21%,1.90%, and 3.15% performance gains on F1-OOD for the 25%, 50%, 75%, and Full settings.

Table 4. Experimental results of ablation study.

Model	CLINC-FULL			CLINC-SMALL		
	F1-IND	F1-OOD	F1-ALL	F1-IND	F1-OOD	F1-ALL
HybridCL	**90.12**	**72.92**	**90.01**	**88.42**	**70.77**	**88.30**
w/o Intra-CCL	89.18	71.65	89.03	88.22	69.38	87.42
w/o Inter-CCL	88.72	70.59	88.60	87.88	69.63	87.97
w/o IW-AM	89.52	72.38	89.77	87.72	70.21	88.17

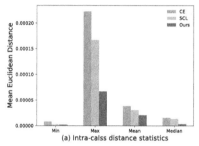 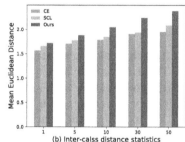

Fig. 3. Statistics of intra-class and inter-class distances.

Similar improvements are observed on other baselines. This also indicates that exploring high intra-class compactness and inter-class separation to model discriminative intent representation can better distinguish IND and OOD intents.

4.4 Analysis

In this section, we further perform a series of detailed analyses on the proposed HybridCL to confirm its effectiveness.

Ablation Study of HybridCL. To analyze the impact of different components in our proposed HybridCL on the performance, we conduct an ablation study and report the results in Table 4. We can observe that the removal of Intra-CCL or Inter-CCL objective sharply reduces the performance in all evaluation metrics and across two datasets. This indicates that intra-class compactness and

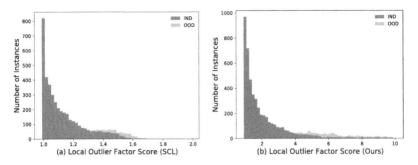

Fig. 4. The LOF score distribution in CLINC-FULL. Left: SCL training. Right: HybridCL training. The X-axis represents the LOF score, and the Y-axis represents the number of samples falling into the corresponding interval.

inter-class separation are indispensable for improving the performance of OOD detection. Jointly optimizing the two objectives helps to learn a discriminative intent representation. Meanwhile, the intent-wise attention mechanism (IW-AM) contributes obviously by 0.54% and 0.56% to the performance. This implies that highlighting the alignment between samples and labels improves the performance of OOD detection.

HybridCL Achieves High Intra-class Compactness and Inter-class Dispersion. To quantitatively measure how HybridCL affects the quality of discriminative intent representations, inspired by [26], we demonstrate the effectiveness of HybridCL for OOD detection by measuring intra-class and inter-class variance. (1) We introduce the compactness score, which calculates the variance of the distances of all intent samples to the corresponding class prototypes as an evaluation metric for intra-class compactness. Obviously, we can see from Fig. 3(a) that all the minimum and median, mean and max values of our method are lower than CE and SCL, suggesting that HybridCL makes the intent representation within a single class tighter. (2) We employ the dispersion score, where we compute the inter-class distance by averaging the euclidean distances from the central class to its K nearest classes. We can see from Fig. 3(b) that HybridCL consistently achieves larger inter-class distances compared to previous methods. The above results are consistent with our conjecture that HybridCL explicitly optimizes the intra-class compactness and inter-class dispersion, which especially benefits OOD detection.

Visualizing the Ability to Detect OOD. A key contribution of the proposed HybridCL is to integrate two kinds of constraint in contrastive tasks. To better verify the effectiveness of HybridCL to characterize the difference between IND and OOD intents, we adopt the Local Outlier Factor(LOF) as the score function and compare the LOF score distribution of IND and OOD data under different training objectives in Fig. 4. Compared with the SCL (left) (Fig. 4(a)), first, we can observe that HybridCL (right) (Fig. 4(b)) has less overlap between IND and OOD in the LOF score distribution, which verifies the effect of our proposed high intra-class compactness constraint objective. Then, the HybridCL is uniform in LOF score distribution and distributed in a wide area, which verifies the feasibility of our proposed high inter-class separation constraint objective. The above results prove that enforcing intra-class and inter-class constraints helps distinguish IND and OOD intents.

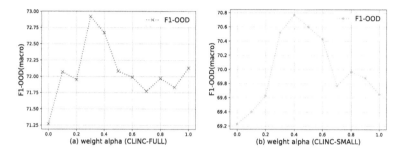

Fig. 5. Effect of α. The X-axis represents the value of α, and Y-axis represents the F1-OOD score.

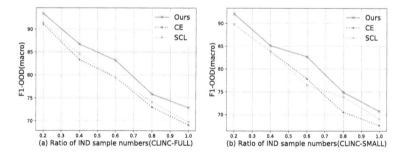

Fig. 6. Impact of in-domain classes. The X-axis represents the value of IND sample numbers, and Y-axis represents the F1-OOD score.

Impact of the Coefficient α. While we want to improve the intra-class compactness (helpful to construct a tighter decision boundary that is beneficial to separate IND and OOD), we also need to balance it to improve the inter-class dispersion (ensure the detection quality of hard OOD) over the training data. In the overall loss defined in Eq. 11, the coefficient α can balance the two losses for discriminative intent representations modeling. To analyze the effect of α, we vary it from 0 to 1 and report the results of CLINC-FULL and CLINC-SMALL in Fig. 5(a) and Fig. 5(b), respectively. We find with the increment of α, the model gradually reaches the best compactness-dispersion trade-off, which means the model can ensure the classification quality of IND and OOD detection effectively. It is worth noting that only intra-CCL or inter-CCL can not make the model achieve the best results.

Impact of the In-Domain Classes. To analyze the impact of using different sizes of IND intent classes on the performance of OOD detection, we conduct experiments on the CLINC-FULL and CLINC-SMALL datasets and show the results in Fig. 6. specifically, we sample the IND classes ratio in the range of 0.2, 0.4, 0.6, 0.8, and 1.0 on two datasets. In the training phase, we compare HybridCL with CE and SCL, and in the detection phase, we use LOF for a fair comparison. Figure 6(a) and Fig. 6(b) show the performance of F1-OOD and the number of different IND classes on CLINC-FULL and CLINC-SMALL, respectively. We can see that although all the methods achieve poorer performance along with the increase of IND intent classes, the performance of HybridCL is stable and achieves the best performance in all settings. This verifies the effectiveness and generalizability of HybridCL in dealing with few-shot OOD detection.

5   Conclusion

In this work, we propose a novel contrastive learning framework, named Hybrid Contrastive Learning (HybridCL), to learn discriminative intent representations for OOD detection. We consider the constraints of intent representation from the two aspects of intra-class and inter-class, respectively. First, to achieve high compactness between instances, we develop an intra-class contrastive learning constraint objective that encourages instances to be close to their corresponding prototypes. Then, to achieve high dispersion among classes, we derive an inter-class contrastive learning constraint objective that enlarges the discrepancy among different classes as much as possible. Finally, we employ an intent-wise attention mechanism to further enhance the discriminative representation capability of the encoder. Extensive experiments and detailed analysis on two benchmark datasets verify the effectiveness of the proposed HybridCL.

Acknowledgements. This work was supported in part by the National Natural Science Foundation of China (Grant Nos. 62276017, U1636211, 61672081), the 2022 Tencent Big Travel Rhino-Bird Special Research Program, and the Fund of the State Key Laboratory of Software Development Environment (Grant No. SKLSDE-2021ZX-18).

References

1. Bendale, A., Boult, T.E.: Towards open world recognition. In: Proceedings of the CVPR (2015)
2. Breunig, M.M., Kriegel, H., Ng, R.T., Sander, J.: LOF: identifying density-based local outliers. In: Proceedings of the SIGMOD (2000)
3. Chen, T., Kornblith, S., Norouzi, M., Hinton, G.E.: A simple framework for contrastive learning of visual representations. In: Proceedings of the ICML (2020)

4. Devlin, J., Chang, M., Lee, K., Toutanova, K.: BERT: pre-training of deep bidirectional transformers for language understanding. In: Proceedings of the NAACL (2019)
5. Gunel, B., Du, J., Conneau, A., Stoyanov, V.: Supervised contrastive learning for pre-trained language model fine-tuning. In: Proceedings of the ICLR (2021)
6. Guo, Y., et al.: HCSC: hierarchical contrastive selective coding. In: Proceedings of the CVPR (2022)
7. He, K., Fan, H., Wu, Y., Xie, S., Girshick, R.B.: Momentum contrast for unsupervised visual representation learning. In: Proceedings of the CVPR (2020)
8. Hendrycks, D., Gimpel, K.: A baseline for detecting misclassified and out-of-distribution examples in neural networks. arXiv preprint: arXiv:1610.02136 (2016)
9. Hochreiter, S., Schmidhuber, J.: Long short-term memory. Neural Comput. **9**, 1735–1780 (1997)
10. Khosla, P., et al.: Supervised contrastive learning. In: Proceedings of the NeurIPS (2020)
11. Kingma, D.P., Ba, J.: Adam: a method for stochastic optimization. In: Proceedings of the ICLR (2015)
12. Larson, S., et al.: An evaluation dataset for intent classification and out-of-scope prediction. In: Proceedings of the EMNLP (2019)
13. Li, J., Zhou, P., Xiong, C., Hoi, S.C.H.: Prototypical contrastive learning of unsupervised representations. In: Proceedings of the ICLR (2021)
14. Lin, T., Xu, H.: Deep unknown intent detection with margin loss. In: Proceedings of the ACL (2019)
15. Liu, B., Mazumder, S.: Lifelong and continual learning dialogue systems: learning during conversation. In: Proceedings of the AAAI (2021)
16. Mullenbach, J., Wiegreffe, S., Duke, J., Sun, J., Eisenstein, J.: Explainable prediction of medical codes from clinical text. In: Proceedings of the NAACL (2018)
17. Paszke, A., et al.: Pytorch: an imperative style, high-performance deep learning library. In: Proceedings of the NeurIPS (2019)
18. Shu, L., Benajiba, Y., Mansour, S., Zhang, Y.: ODIST: open world classification via distributionally shifted instances. In: Proceedings of the EMNLP Findings (2021)
19. Shu, L., Xu, H., Liu, B.: DOC: deep open classification of text documents. In: Proceedings of the EMNLP (2017)
20. Snell, J., Swersky, K., Zemel, R.: Prototypical networks for few-shot learning. In: Proceedings of the NeurIPS (2017)
21. Wang, G., et al.: Joint embedding of words and labels for text classification. In: Proceedings of the ACL (2018)
22. Wolf, T., et al.: Transformers: state-of-the-art natural language processing. In: Proceedings of the EMNLP (2020)
23. Wu, C.S., Madotto, A., Hosseini-Asl, E., Xiong, C., Socher, R., Fung, P.: Transferable multi-domain state generator for task-oriented dialogue systems. In: Proceedings of the ACL (2019)
24. Xu, H., He, K., Yan, Y., Liu, S., Liu, Z., Xu, W.: A deep generative distance-based classifier for out-of-domain detection with Mahalanobis space. In: Proceedings of the COLING (2020)
25. Yan, G., et al.: Unknown intent detection using gaussian mixture model with an application to zero-shot intent classification. In: Proceedings of the ACL (2020)
26. Zeng, Z., et al.: Modeling discriminative representations for out-of-domain detection with supervised contrastive learning. In: Proceedings of the ACL (2021)
27. Zeng, Z., He, K., Yan, Y., Xu, H., Xu, W.: Adversarial self-supervised learning for out-of-domain detection. In: Proceedings of the NAACL (2021)

28. Zhan, L., Liang, H., Liu, B., Fan, L., Wu, X., Lam, A.Y.S.: Out-of-scope intent detection with self-supervision and discriminative training. In: Proceedings of the ACL (2021)
29. Zhang, H., Xu, H., Lin, T.: Deep open intent classification with adaptive decision boundary. In: Proceedings of the AAAI (2021)
30. Zheng, Y., Chen, G., Huang, M.: Out-of-domain detection for natural language understanding in dialog systems. IEEE ACM Trans. Audio Speech Lang. Process. **28**, 1198–1209 (2020)
31. Zhou, W., Liu, F., Chen, M.: Contrastive out-of-distribution detection for pre-trained transformers. In: Proceedings of the EMNLP (2021)

CB-GAN: Generate Sensitive Data with a Convolutional Bidirectional Generative Adversarial Networks

Richa Hu[1], Dan Li[1(✉)], See-Kiong Ng[2,3], and Zibin Zheng[1]

[1] School of Software Engineering, Sun Yat-Sen University, Zhuhai 528478, China
hurch3@mail2.sysu.edu.cn, {lidan263,zhzibin}@mail.sysu.edu.cn
[2] Institute of Data Science, National University of Singapore,
Singapore 117602, Singapore
seekiong@nus.edu.sg
[3] School of Computing, National University of Singapore,
Singapore 117417, Singapore

Abstract. In the era of big data, numerous data measurements collected from all walks of life are playing important roles in various data mining applications. Not all data owners (or keepers) could develop feasible learning models for knowledge discovery's sake. Oftentimes, the original data need to be passed to or shared with researchers or data scientists for better mining insights, especially in the medical, financial, and industrial fields. However, concerns about sensitivity and privacy limit the availability and completeness of shared (or passed) data and the quality of mining results. In this paper, we propose a novel Convolutional Bidirectional Generative Adversarial Networks (CB-GAN) framework to generate sensitive synthetic data. The Convolutional Neural Networks are utilized to capture the feature correlations of the original data, and the Generative Adversarial Networks with Autoencoders are combined to synthesize realistically distributed data. To demonstrate the feasibility of the model, we evaluated it from three aspects: how similar are the distributions of the synthetic data to the original data, how well can the synthetic data accomplish future data mining tasks, and how much sensitive information has been hidden. Various experimental results showed the superiority of the proposed method compared with the state-of-the-art methods.

Keywords: Synthetic Data Generation · Data Sensitivity · Privacy Protection · Generative Adversarial Networks

1 Introduction

With the development of sensor networks and smart devices, numerous data measurements are being collected from all walks of life, which drove practitioners to coin the term big data and develop tools to analyze it over the past years

This research was supported by the Guangdong Natural Science Foundation General Program (Grant No. 2022A1515011713).

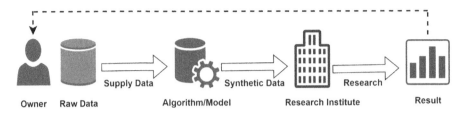

Fig. 1. A typical data sharing process with protection of sensitive information.

[26]. Data-driven tasks, such as statistical analysis, data mining, pattern recognition, knowledge discovery, etc., are becoming popular trends for algorithm-aided decision-making, industrial predictive maintenance, automated process monitoring, and more recently the opportunity for novel commercial Artificial Intelligence (AI) models. However, due to privacy or security considerations, a significant amount of data is siloed and exploited solely by the institutions or individuals that host it [23], who usually cannot harness and dig the potential and innovations adequately. Taking the Electronic Health Records (EHR) for medical mining as an example, access to publicly available EHR databases has motivated the medical research [6,19]. However, adopting the EHR database does not allow researchers to access the original complete EHR data easily and fully mine the underline knowledge due to its privacy and sensitive information concerns [8,17]. As a result, sharing adequate high-quality data with the premise of protected sensitivity and privacy from institutions or individuals that collect and hold data to institutions or researchers that can materialize their professional advantages in data mining and analysis has significant benefits in increasing the efficiency of business processes, facilitating growth by unlocking new business models, and profiting from data sharing trading. A typical data sharing process is shown in Fig. 1. First, the raw data is processed by or supplied to the algorithm or model to hide sensitive information (in our case, it will be supplied to the proposed CB-GAN model to generate synthetic data), and then the processed data (or the synthetic data) will be sent to the relevant professional institutions or personnel for research to obtain useful knowledge and fed back to the data owner.

At the early stage, the most common and intuitive way to protect the privacy and sensitive information of shared data is direct de-identification [15], including randomization, generalization, and deletion of sensitive attributes. [11] reduce the precision of data fields (feature) through generalisation. Specifically, in the Ontario birth registry dataset, the date of birth's precision was reduced to a year, the age was categorized as $\leq$19, 20–30, 30–40, or >40 years, and the postal code was represented by the first three characters. [9] deleted 18 private and sensitive variables, including the date of birth and the postal code in perspective of protecting sensitive information. However, these approaches usually result in significant information loss [9] and still are not impregnable to attacks since attackers could easily recover some of the prepossessed personal identifiers based on other unattended attributes. For example, [27] proposed the notion of "corruption", where the adversary might have already obtained individuals' sensitive values before digging into the released sharing data. Under this circumstance, the conventional generalization method may lead to severe privacy leakage.

Next, another branch of approaches has arisen with the development of mathematical modeling algorithms and deep learning techniques, namely, the Synthetic Data Generation (SDG) [7,22], which aims at generating high-quality synthetic data in a privacy-preserving manner. In [22], the Content Modelling for Synthetic E-Health Records (CoMSER) method was proposed to generate synthetic EHRs data based on the Health Information Statistics (HIS) dataset and corresponding expert knowledge, which showed promising results in generating the RS-EHR, However, CoMSER is domain-specific and cannot be transplanted to other fields. Later, the notion of Differential Privacy (DP) was proposed to protect privacy by bounding the possible impact of any variable might have on the output of an algorithm [10]. [21] focused on marginal-based approaches and proposed a DP-based synthetic data generation model called AIM. However, its input data was assumed to be discrete, and the raw data with numerical attributes (namely continuous variables) must be appropriately discretized before running the proposed model.

Recently, GAN-based methods were proposed to generate synthetic data due to the empirical advantages of GANs as generative models for arbitrary data distributions [8,28,29,32]. Unlike most of the previous works that are highly disease-specific [7,22,25,31], MedGAN [8] and CorGAN [28] could capture general informative features about the EHR data without focusing on specific diseases, which enables them be suitable for more diverse applications. To protect the privacy of the original data during the model training process, differential privacy is implemented in [29,32], which can better prevent exposing the sensitive information of the original data from the synthetic data. These aforementioned methods often generate domain-specific field data, and in contrast, the quality of the synthetic data was not satisfying in assessing it by accomplishing machine learning tasks.

Thus, a feasible generative model should meet the following criterion: it should be a general framework that could be applied to generate synthetic data for various fields, it should produce high-quality synthetic data (without hindering too much original information), and it should hide sensitive information as much as possible at the same time. In this work, we introduced a Convolutional Bidirectional Generative Adversarial Networks (CB-GAN) framework, which consists of a convolutional Autoencoder and a convolutional bidirectional Generative Adversarial Network. Using a 1-D Convolutional Neural Network (1-D CNN) as its core model, the Autoencoder could map whatever data types to a continuous feature space, thus capturing both the local spatial correlations and temporal feature information of the input data. The contributions of our work are as follows:

1. The proposed CB-GAN utilizes 1-D CNN as the core neural network to capture the local correlations and temporal information of the original data.
2. A Bidirectional Generative Adversarial Network is adopted to generate high-quality synthetic data compared with existing GAN-based methods.

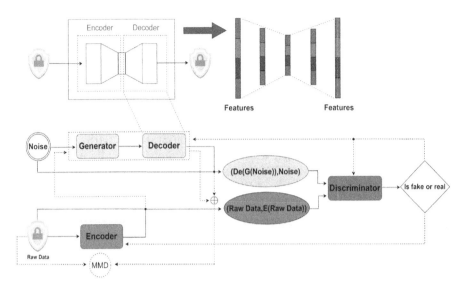

Fig. 2. Framework of the proposed Convolutional Bidirectional Generative Adversarial Networks (CB-GAN). The three lock symbol represent the raw data.

3. An Autoencoder was pre-trained to enable the Bidirectional Generative Adversarial Network to generate discrete data.
4. A desensitization loss was added to the framework to ensure the generated data to be too value-similar to the original data.
5. The proposed CB-GAN is a general framework that was evaluated with various datasets from different fields.

With a slight abuse of notation, the terms privacy, private information, personal identifiers, and sensitive information are noted interchangeably - they all stand for what the individuals or institutions want to hide or protect before they pass the data to others. The remaining part of this work is organized as follows: The proposed CB-GAN model is introduced with related methodology in Sect. 2. Experiments on various datasets to evaluate the proposed model are presented in Sect. 3. Section 4 summarizes the work and proposes possible future work.

2   CB-GAN Framework

The proposed framework shown in Fig. 2 consists of two parts, the Autoencoder Pre-training part (shown as the upper part of Fig. 2) for feature mapping and the Bidirectional Generative Adversarial Networks for Synthetic Data Generation part (shown as the lower part of Fig. 2). To deal with discrete data, 1-D CNN is adopted as the core model to jointly capture the local spatial pattern and temporal information of the correlated features. More specifically, since this work aims to build a general model which can generate high-fidelity synthetic data for various fields (where attribute types might be various, such as discrete, continuous, or a mixture of both types) in a privacy-preserving manner, the pre-training part is designed to transform the input space into a continuous space regardless of the input data types. Thus, the generated fake data is decoded by the pre-trained Autoencoder before being fed to the discriminator. As for the bidirectional generative adversarial networks, the random noise from the latent space is fed to the generator as input to generate synthetic data, and inversely the raw data is encoded to the latent space. Then the discriminator tries to distinguish fake (synthetic data and random noise) from real (original raw data and its counterpart in the latent space). In the following parts, we will drive the Autoencoder pre-training and bidirectional synthetic data generation with details.

2.1   Autoencoder Pre-training

Algorithm 1. 1-D CNN Autoencoder Pre-Training

Inputs: Real data X, Real data's size N, n_a is the number of epochs for autoencoder pre-training, batch size of real data r.

1: **for** $i = 1 \rightarrow n_a$ **do**
2:     Partition X into $X_1,...,X_k$ where $k = \frac{N}{r}$
3:     **for** $l = 1 \rightarrow k$ **do**
4:         $\hat{X}_l = Dec(Enc(X_l))$
5:         $Aloss = MSE(\hat{X}_l, X_l)$
6:         Update autoencoder parameters by minimizing $Aloss$:
7:         min $Aloss$
8:     **end for**
9: **end for**

As shown in Algorithm 1, during the pre-training process, the raw data is first fed into the 1-D convolutional Autoencoder, which presses the original data features to a sub-space and then reconstructs it with the least error. The loss function of the Autoencoder is defined as Mean Square Error (MSE):

$$MSE(\hat{x}, s) = \frac{1}{n} \sum_{i=1}^{n} |\hat{x}_i - x_i|^2 \tag{1}$$

where x represents the raw data and $\hat{x}$ represents the reconstructed data. With adequate pre-training, the convolutional layers of the decoder part will be directly utilized in the following data generation processing.

2.2   Bidirectional Generation

As shown in Algorithm 2, the random noise $\mathbf{z}$ sampled from the latent space $\mathcal{N}(0,1)$ is firstly fed to the generator G to obtain the primary generated samples $G(z)$, which is then mapped by pre-trained decoder De and output the generated

Algorithm 2. 1-D CNN BiGAN Training

Inputs: Random noise $\mathbf{z} \sim \mathcal{N}(0, 1)$, Real data X, Real data's size N, number of epochs
for BiGAN training n_{bigan}, batch size of real data r.
1: **for** $i = 1 \rightarrow n_{bigan}$ **do**
2:     Partition X into $X_1,...,X_k$ where $k = \frac{N}{r}$
3:     **for** $l = 1 \rightarrow k$ **do**
4:         $D_{real} = D(< X_l, E(X_l) >)$
5:         $D_{fake} = D(< De(G(\mathbf{z})), \mathbf{z} >))$
6:         $Dloss = BCE(D_{real}, 1.0) + BCE(D_{fake}, 0.0)$
7:         Update discriminator parameters by minimizing $Dloss$:
8:         $\min Dloss$
9:         $Gloss = BCE(D(< De(G(\mathbf{z})), \mathbf{z} >), 1.0)$
10:        $Eloss = BCE(D(< X_l, E(X_l) >), 0.0)$
11:        $Mloss = MMD(X_l, De(G(\mathbf{z}))) + MMD(X_l, De(G(E(X_l))))$
12:        $Rloss = RMSE(X_l, De(G(\mathbf{z})))$
13:        $loss = Gloss + Eloss + Mloss - Rloss$
14:        Update generator,decoder and encoder parameters by minimizing $loss$:
15:        $\min loss$
16:     **end for**
17: **end for**

---

synthetic data $De(G(z))$. Simultaneously, as mentioned in Sect. 1, to achieve the
generator's better understanding of the real data distribution (namely, to aid the
generator to produce synthetic data with higher quality), the original raw data
is $\mathbf{x} \in \mathbf{R}^d$ is encoded to the latent space by the encoder E(Different from the
Fig. 2 upper part encoder). Then, the discriminator D is applied to distinguish
between $< \mathbf{x}, E(\mathbf{x}) >$ and $< De(G(\mathbf{z})), \mathbf{z} >$. Following a standard min-max
training objective of GAN, here the D will be trained to recognize $< \mathbf{x}, E(\mathbf{x}) >$
as real and $< De(G(\mathbf{z})), \mathbf{z} >$ as fake, while G will be trained to "fool" the
discriminator by making it believe that $< De(G(\mathbf{z})), \mathbf{z} >$ is sampled from the
original data space. Thus, the min-max function V (E, G, D) of CB-GAN can
be derived as follows,

$$
\min_{G} \max_{D} V(E, G, D) =
$$
$$
\mathbf{E}_{\mathbf{x} \sim p_x}[\log D(< \mathbf{x}, E(\mathbf{x}) >)] + \mathbf{E}_{\mathbf{z} \sim p_z}[\log(1 - D(< De(G(\mathbf{z})), \mathbf{z} >))]
\tag{2}
$$

where p_x is the distribution of the real samples and p_z is the distribution of the
random latent space. With adequate iterative training, the generation ability
of the generator and the encoder will be improved gradually until the model
outputs satisfy synthetic samples.

Note that we adopt the Binary Cross Entropy (BCE) as the discrimination
error during the training processing,

$$
BCE(\mathcal{X}, y) = -\frac{1}{n} \sum_{i}^{n} (x_i \log(y_i) + (1 - x_i) \log(1 - y_i))
\tag{3}
$$

where $\mathcal{X}$ represents the data need to be discriminated and y represents the assumed labels.

Besides, the Maximum Mean Discrepancy (MMD) is employed as an auxiliary condition to limit the similarity between the synthetic data and the original data.

$$MMD = \left\| \frac{1}{n} \sum_{i=1}^{n} \phi\left(x_i\right) - \frac{1}{m} \sum_{j=1}^{m} \phi\left(y_j\right) \right\|_{\mathcal{H}} \tag{4}$$

where x and y are the original data and the synthetic data, respectively. Here, our goal is to minimize MMD, thus making the distribution of the synthetic data more similar to the original data.

What's more, in the perspective of privacy-preserving, we utilize the Root Mean Square Error (RMSE) to evaluate the data desensitization,

$$RMSE = \sqrt{\frac{1}{n} \sum_{i=1}^{n} \left(x_i - y_i\right)^2} \tag{5}$$

where larger RMSE means less privacy is contained in the synthetic data.

3   Experiments

In this section, we will first introduce the details of the experimental setup and introduce the datasets involved. Then we will answer the following Research Questions (RQs) via explaining various experimental results:

– **RQ1-** Data quality measure: compared with baseline methods, how similar are the distributions of the synthetic data to the original data? Here, larger similarity means higher synthetic data quality.
– **RQ2-** Data mining evaluation: compared with baseline methods, how well can the synthetic data accomplish future data mining tasks?
– **RQ3-** Data desensitization assessment: compared with baseline methods, can CB-GAN hide more sensitive information?

3.1   Experimental Setup

For every involved dataset, We randomly split it into train and test sets, where the train set is used to train the synthetic data generation models, and then the random noise is fed to the trained models to generate synthetic samples. The Adam optimizer [20] is employed during the iterations of model training to find better hyper-parameters. To make the comparison fair, all the learning rates of the involved models (including CB-GAN and the baseline models) are set as 0.005. The Batch Normalization [18] is applied to improve the training process for all the involved models. The experiments are conducted on a GeForce GTX 1650 NVIDIA GPU.

3.2 Datasets

To comprehensively evaluate the aim of our model – to generate high-quality synthetic data in a privacy-preserving manner for various fields, datasets from the medical, financial and industrial fields were selected (statistics of those datasets are summarized in Table 1). As described in the following part, datasets (1)-(3) are collected from the medical field, (6) and (7) are datasets collected from Ethereum blockchain (financial field), and the rest are datasets collected from cyber-physical systems (industrial field). Among the aforementioned datasets, data collected from the financial and industrial fields are time-series. To be specific, they are:

(1) **Kaggle Cardiovascular Disease(KCD)** [30]: The Kaggle dataset determines if a patient has cardiovascular disease or not by a variety of features such as weight, age, diastolic blood pressure, and smoke or not. This dataset has a mixture of continuous and discrete attributes.

(2) **UCI Epileptic Seizure Recognition(UCI)** [4]: In the dataset, the task is to classify the seizure activity. The features are the Electroencephalogram (EEG) records at various time points and values are continuous.

(3) **Kaggle Cervical Cancer(KCC)** [13]: The Kaggle Cervical Cancer dataset covers patients' records and uses the Biopsy feature to classify. There are continuous and discrete features in the dataset.

(4) **SWaT** [14]: The aim of the SWaT dataset is to support research in the design of secure Cyber-Physical Systems. The data collection process was implemented on a six-stage Secure Water Treatment (SWaT) testbed. The features are values recorded by various sensors and actuators. Labels of the dataset indicate whether the system is attacked or not.

(5) **WADI** [2]: WADI is an extension of the SWAT. The water distribution testbed (WADI) is used to conduct security analysis for water distribution networks and evaluate detection mechanisms for potential cyber-attacks or physical defections. Labels of the dataset indicate whether the system is attacked or not.

(6) **A Labeled Transactions-Based Dataset(LTBD)** [3]: The dataset is a benchmark dataset of the transactions-based dataset of the Ethereum network for the tuning and assessing any proposed intrusion detection system used in Blockchain networks. The features include gas, block number, and so on. The classification task is to distinguish whether the receiver address is normal or abnormal.

(7) **BlockTransaction(Block)** [33]: Ethereum is one of the most popular permission-less blockchains. The dataset is built by collecting and processing the on-chain data from Ethereum. The label "is Error" indicates whether the transaction is normal.

(8) **EPIC** [1]: Electrical Power and Intelligent Control (EPIC) Testbeds which realistically mimic the operation of critical infrastructure are of significant value to researchers. The dataset has no label for classification.

(9) **IoT** [5]: Internet of Things (IoT) dataset collected by high-interaction IoT honeypots deployed in the wild. The IoT dataset has no label as well.

Table 1. The statistics of the datasets.

Dataset	KCD	UCI	KCC	Block	LTBD	SWaT	WADI	EPIC	IoT
Size	70,000	11,500	858	1,048,576	71,251	449,920	172,801	863	4432
Features	14	178	36	18	18	53	106	266	11
Positive Labels	35,000	2300	52	3272	11,649	53,900	10,347	None	None

3.3 Baselines

The baselines considered in this work are some recent GAN-based synthetic data generation methods, such as MedGAN [8], CorGAN [28], DPGAN [32] and the Convolutional Generative Adversarial Networks(RDP-CGAN) [29].

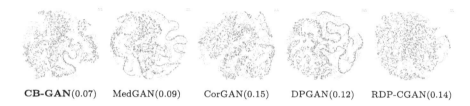

CB-GAN(0.07)   MedGAN(0.09)   CorGAN(0.15)   DPGAN(0.12)   RDP-CGAN(0.14)

Fig. 3. The distributions of real (light red points) and synthetic (grey points) **Labeled Transactions-Based data** by t-SNE. The values are Peacock test of distribution difference. The smaller value indicates that the distributions of the tested datasets are more similar, namely implies higher quality of the synthetic data (Color figure online).

3.4 Distributions of the Synthetic Data (RQ1)

We adopted the t-distributed Stochastic Neighbor Embedding(t-SNE) and Peacock test of distribution difference to assess the generated data's quality. The t-distributed Stochastic Neighbor Embedding(t-SNE) is a nonlinear dimension reduction algorithm that is suitable for high-dimensional data reduction to 2 dimensions and for visualization. Such as input data with N samples and M features (N_samples, M_features), after t-SNE calculation, the result is (N_samples, 2). In Fig. 3, we visualized the t-SNE results of the Transaction-Based data for CB-GAN and baselines, where The red points represent the raw data and the grey points represent the synthetic data. It is shown that the CB-GAN-generated data more attentively repeated the distribution of the raw data compared with that of baselines. Since visualization by t-SNE does not quantitatively show their similarity well, we additionally adopted the Peacock test to complement this drawback. A smaller value of the Peacock test, which is a multi-dimensional variation of the well-known Kolmogorov-Smirnov test [12], indicates a smaller distribution difference (indicating the high quality of the synthetic data). The Peacock test results are marked in Fig. 3, where we can see that CB-GAN is the smallest. Due to the page limitation, we did not show the t-SNE visualization of other datasets. Results of the Peacock test on all the considered datasets are summarised in Table 2, where the quality of the synthetic data generated by CB-GAN is shown to be the highest in all the cases.

Besides, we also adopted the Maximum Mean Discrepancy (MMD), which is usually applied to statistically measure the difference of sequence data, to indicates to which extent our model captures the statistical distribution of the real data. Since a lower MMD score indicates higher distribution similarity and better generation, MMD values of the CB-GAN-generated data and the raw data are the smallest as shown in Table 2, which presents the best generation performance. For simplicity, the best and second best results are presented in bold and underline for Tables in the rest of the paper.

Table 2. Comparison of different methods using MMD metric (Peacock test).

Dataset	MedGAN	CorGAN	DPGAN	RDP-CGAN	CB-GAN
KCD	0.112 (0.142)	0.212 (0.132)	0.41 (0.157)	0.335 (0.130)	**0.072 (0.090)**
UCI	0.517 (0.191)	0.607 (0.380)	0.775 (0.316)	1.107 (0.40)	**0.25 (0.181)**
KCC	0.184 (0.284)	0.179 (0.482)	0.515 (0.366)	0.777 (0.279)	**0.175 (0.231)**
Block	0.097 (0.299)	0.053 (0.309)	0.179 (0.342)	0.881 (0.375)	**0.046 (0.288)**
LTBD	0.129 (0.090)	0.179 (0.147)	0.333 (0.123)	1.168 (0.139)	**0.084 (0.069)**
SWaT	0.141 (0.184)	0.19 (0.299)	0.097 (0.241)	0.673 (0.187)	**0.034 (0.133)**
WADI	0.129 (0.076)	0.086 (0.087)	0.158 (0.098)	0.299 (0.102)	**0.064 (0.069)**
EPIC	0.249 (0.294)	1.011 (0.423)	0.245 (0.389)	0.851 (0.326)	**0.222 (0.233)**
IoT	1.106 (0.326)	0.358 (0.222)	0.73 (0.320)	0.998 (0.305)	**0.345 (0.190)**

3.5   Performance of the Synthetic Data on Machine Learning Tasks (RQ2)

In this part, we considered three machine-learning tasks to assess the quality of the synthetic data, namely, (1) the classification task, (2) the feature correlation analysis, and (3) the linear regression task.

Classification. Here, we utilized the Gradient Boosting (GB) [24] to evaluate the classification task. Firstly, the classifier was trained with the real training data and then applied to classify the test data to obtain an upper bound for classification accuracy, namely, high-quality synthetic data is assumed to achieve similar (or at least not too worse) classification accuracy with that obtained by the original data. Then the classifier is trained on the synthetic data (generated by CB-GAN and baseline models) and classified on the test data to show how well can the synthetic data accomplish the classification task. We report the Area Under the ROC curve (AUROC) and Area Under the Precision-Recall Curve (AUPRC) in Tables 3 and 4. As is shown in Table 3, CB-GAN over-performed all the baseline methods. As for the AUPRC metric, CB-GAN over-performed others except on the Labeled Transactions-Based Dataset and WADI, which are two imbalanced datasets.

Table 3. Performance comparison of different methods under the AUROC metric.

Dataset	real data	MedGAN	CorGAN	DPGAN	RDP-CGAN	CB-GAN
KCD	79.81%	72.94%	72.85%	67.97%	60.71%	**76.16%**
UCI	98.43%	86.55%	83.72%	78.81%	62.8%	**89.78%**
KCC	92.97%	88.76%	88.87%	89.07%	88.84%	**92.88%**
Block	99.94%	98.56%	98.44%	96.33%	76.45%	**99.3%**
LTBD	94.69%	83.15%	84.05%	84.31%	82.88%	**85.67%**
SWaT	89.01%	84.34%	81.05%	85.32%	83.48%	**85.93%**
WADI	98.79%	90.42%	87.55%	91.69%	77.32%	**91.94%**

Table 4. Performance comparison of different methods under the AUPRC metric.

Dataset	real data	MedGAN	CorGAN	DPGAN	RDP-CGAN	CB-GAN
KCD	79.93%	72.27%	73.58%	68.47%	61.07%	**76.63%**
UCI	95.17%	81.84%	74.82%	80.04%	67.02%	**84%**
KCC	65.47%	60.03%	60.57%	61.3%	64.36%	**64.78%**
Block	99.78%	95.12%	95.19%	93.28%	64.02%	**99.2%**
LTBD	88.16%	**79.19%**	71.38%	73.74%	65.88%	77.35%
SWaT	90.13%	83.35%	82.31%	77.89%	76.22%	**83.94%**
WADI	98.17%	**74.65%**	70.86%	69.21%	52.48%	73.28%

Feature Correlation Analysis. The feature correlation analysis is to rank data features according to their importance. Here, we assume that the larger similarity between the feature ranking lists obtained based on the synthetic data and the original data, the better the synthetic data generation model can preserve the features' correlation of the original data. For page limitation, here we only show the testing results on the BlockTransaction data are shown in Fig. 4, and we can see that CB-GAN outperformed others.

Regression. Unlike the classification model's purpose of predicting discrete labels, the Linear regression model tries to predict the ground truth values of a feature on the basis of other correlated features. In our experiments, we select meaningful features (namely, features that are significantly based on domain knowledge) as the prediction objectives. For instance, in the Kaggle Cardiovascular Disease dataset, we set the regression task as: to predict the values of diastolic blood pressure on the basis of other variables. The sum of the squared difference between the real and predicted values is employed to demonstrate the regression performance. As shown in Table 5, CB-GAN outperformed others

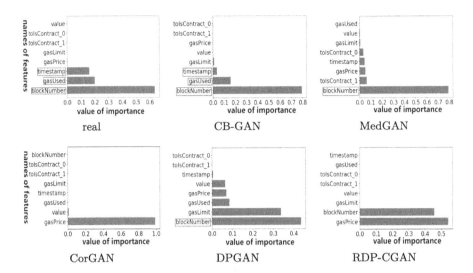

Fig. 4. The feature importance of real and synthetic **BlockTransaction data**. The horizontal and vertical axes represent the value of importance and the names of features, respectively. Features are arranged from bottom to top by the magnitude of importance and mark the same features as the feature importance ranking of the real data.

except on the Transactions-Based Dataset and WADI (where we just slightly lost the competition). This is the Transactions-Based Dataset is continuous, and the fully connected layer used by MedGAN is more suitable for processing purely continuous data compared to the 1D CNN adopted by CB-GAN, which is better at dealing with a mixture of continuous and discontinuous data. As for WADI, most of its attributes are recorded by actuators as ON/OFF status (discontinuous), and CB-GAN's generation performance is not outstanding enough for almost purely discontinuous data.

Table 5. Prediction error of real and synthetic data in linear regression tasks.

Dataset	real data	MedGAN	CorGAN	DPGAN	RDP-CGAN	CB-GAN
KCD	1.862	9.904	3.039	_2.303_	6.833	**2.095**
UCI	0.124	4.677	_3.308_	6	30.279	**3.205**
KCC	1.301	8.541	11.435	_3.774_	14.29	**3.706**
Block	7.599	7.689	7.647	7.657	_7.644_	**7.64**
LTBD	0.057	**0.058**	0.113	2.049	0.171	_0.087_
SWaT	165.343	703.409	_698.386_	715.89	10322.08	**298.177**
WADI	37.57	83.954	81.915	77.179	**75.347**	_76.549_
EPIC	0.173	_0.47_	1.707	2.49	1.508	**0.371**
IoT	0.275	0.763	_0.67_	1.256	2.118	**0.441**

3.6   Data Desensitization Assessment (RQ3)

The core motivation of this paper is to utilize synthetic data generation to meet the privacy requirements for common data-sharing issues. Here, we assume that the synthetic data should be "far away" from the original data in terms of value-based evaluation metrics. Unlike [8] which utilized the hamming distance to calculate the similarity between the raw and synthetic data based on its binary premise (discrete), we chose the RMSE as a desensitization assessment metric since involved datasets for this work are continuous or a mixture of continuous and discrete data. For each record of the original data set, we calculated its RMSE to its corresponding sample in the synthetic dataset and sum them together. Table 6 presents that the desensitization of our method is the second best on most datasets, while RDP-CGAN performs well on many datasets since it protects privacy by clipping gradients and adding noise. However, RDP-CGAN cuts the gradient and adds noise during the training process, the model can not capture the distribution of the original data well, so the quality of the generated synthetic data is relatively poor, different from the distribution of the original data as shown in Sect. 3.4, and certainly can not accomplish the machine learning tasks well as discussed in Sect. 3.5.

In summary, qualified synthetic data (that can achieve better machine learning performance) generated by the baseline methods are poorly desensitized, and vice versa. As a comparison, the data quality and the desensitization degree achieved by the proposed CB-GAN come to a promising compromise.

Table 6. Comparison of different methods' data desensitization.

Dataset	MedGAN	CorGAN	DPGAN	RDP-CGAN	CB-GAN
KCD	98.195	110.309	**117.237**	110.347	<u>110.91</u>
UCI	11.391	11.092	<u>19.313</u>	13.264	**21.87**
KCC	5.567	6.458	8.363	<u>8.514</u>	**9.208**
Block	24.476	24.765	23.551	**32.927**	<u>24.85</u>
LTBD	57.142	57.312	53.984	**118.816**	<u>60.553</u>
SWaT	211.727	233.024	234.401	**331.433**	<u>238.124</u>
WADI	79.571	83.137	76.164	**89.462**	<u>84.198</u>
EPIC	6.193	**11.294**	6.58	7.518	<u>9.76</u>
IoT	7.302	3.055	11.365	**34.151**	<u>20.062</u>

3.7   Compare LSTM and 1-D CNN for Time Series Data

Since the datasets involved in this work are mostly time series data, and the LSTM [16] is famous for its ability of capturing temporal correlations, here, we replaced the 1-D CNN in CB-GAN with LSTM as abortion experiment on time series data. As shown in Table 7, 8 and 9, for multivariate time series datasets involved in this work, the 1-D CNN-based synthetic data generation framework generally generates synthetic data with higher quality than that generated by the LSTM-based framework. This is because of 1-D CNN's ability to catch temporal and spatial correlation, demonstrating that the proposed CB-GAN with 1-D CNN as its core model could a generalized framework.

Table 7. We replaced the 1-D CNN in CB-GAN with LSTM, called LSTM-based and comparison of two methods' data MMD.

Dataset	LSTM-based	CB-GAN
Block	0.06	**0.046**
LTBD	**0.027**	0.084
SWaT	0.053	**0.034**
WADI	**0.033**	0.064
EPIC	0.223	**0.222**
IoT	0.412	**0.345**

Table 8. We replaced the 1-D CNN in CB-GAN with LSTM, called LSTM-based and comparison of two methods' data Linear Regression.

Dataset	real data	LSTM-based	CB-GAN
Block	7.599	8.403	**7.64**
LTBD	0.057	0.088	**0.087**
SWaT	165.343	**229.604**	298.177
WADI	37.57	77.411	**76.549**
EPIC	0.173	0.863	**0.371**
IoT	0.275	0.557	**0.441**

Table 9. We replaced the 1-D CNN in CB-GAN with LSTM, called LSTM-based and comparison of two methods' data AUROC and AUPRC.

Dataset	AUROC			AUPRC		
	real data	LSTM-based	CB-GAN	real data	LSTM-based	CB-GAN
Block	99.94%	98.7%	**99.3%**	99.78%	95.65%	**99.2%**
LTBD	94.69%	84.89%	**85.67%**	88.16%	74.51%	**77.35%**
SWaT	89.01%	81.43%	**85.93%**	90.13%	83.5%	**83.94%**
WADI	98.79%	**94.02%**	91.94%	98.17%	72.49%	**73.28%**

4   Conclusion

In this paper, we proposed a novel framework called CB-GAN, which adopts the 1-D convolutional neural network to capture local features and temporal information of the original data. We conducted various experiments on various datasets to evaluate the performance of the proposed CB-GAN by answering three research questions: how similar are the distributions of the synthetic data to the original data, how well can the synthetic data accomplish future data mining tasks, and how much sensitive information can CB-GAN hide. The experimental results showed that the proposed CB-GAN outperformed baseline methods by achieving a promising compromise between the synthetic data quality and the desensitization degree.

References

1. Adepu, S., Kandasamy, N.K., Mathur, A.: EPIC: an electric power testbed for research and training in cyber physical systems security. In: Katsikas, S.K., et al. (eds.) SECPRE/CyberICPS -2018. LNCS, vol. 11387, pp. 37–52. Springer, Cham (2019). https://doi.org/10.1007/978-3-030-12786-2_3

2. Ahmed, C.M., Palleti, V.R., Mathur, A.P.: WADI: a water distribution testbed for research in the design of secure cyber physical systems. In: Proceedings of the 3rd International Workshop on Cyber Physical Systems for Smart Water Networks, pp. 25–28 (2017)

3. Al-E'mari, S., Anbar, M., Sanjalawe, Y., Manickam, S.: A labeled transactions-based dataset on the Ethereum network. In: Anbar, M., Abdullah, N., Manickam, S. (eds.) ACeS 2020. CCIS, vol. 1347, pp. 61–79. Springer, Singapore (2021). https://doi.org/10.1007/978-981-33-6835-4_5

4. Andrzejak, R.G., Lehnertz, K., Mormann, F., Rieke, C., David, P., Elger, C.E.: Indications of nonlinear deterministic and finite-dimensional structures in time series of brain electrical activity: dependence on recording region and brain state. Phys. Rev. E **64**(6), 061907 (2001)

5. Aung, Y.L., Tiang, H.H., Wijaya, H., Ochoa, M., Zhou, J.: Scalable VPN-forwarded honeypots: dataset and threat intelligence insights. In: Sixth Annual Industrial Control System Security (ICSS) Workshop, pp. 21–30 (2020)

6. Botsis, T., Hartvigsen, G., Chen, F., Weng, C.: Secondary use of EHR: data quality issues and informatics opportunities. Summit Transl. Bioinform. **2010**, 1 (2010)

7. Buczak, A.L., Babin, S., Moniz, L.: Data-driven approach for creating synthetic electronic medical records. BMC Med. Inform. Decis. Mak. **10**(1), 1–28 (2010)

8. Choi, E., Biswal, S., Malin, B., Duke, J., Stewart, W.F., Sun, J.: Generating multi-label discrete patient records using generative adversarial networks. In: Machine Learning for Healthcare Conference, pp. 286–305. PMLR (2017)

9. Clause, S.L., Triller, D.M., Bornhorst, C.P., Hamilton, R.A., Cosler, L.E.: Conforming to HIPAA regulations and compilation of research data. Am. J. Health Syst. Pharm. **61**(10), 1025–1031 (2004)

10. Dwork, C., Roth, A.: The algorithmic foundations of differential privacy. Theoret. Comput. Sci. **9**(3–4), 211–407 (2013)

11. El Emam, K., Rodgers, S., Malin, B.: Anonymising and sharing individual patient data. BMJ **350** (2015)

12. Fasano, G., Franceschini, A.: A multidimensional version of the Kolmogorov-Smirnov test. Mon. Not. R. Astron. Soc. **225**(1), 155–170 (1987)

13. Fernandes, K., Cardoso, J.S., Fernandes, J.: Transfer learning with partial observability applied to cervical cancer screening. In: Alexandre, L.A., Salvador Sánchez, J., Rodrigues, J.M.F. (eds.) IbPRIA 2017. LNCS, vol. 10255, pp. 243–250. Springer, Cham (2017). https://doi.org/10.1007/978-3-319-58838-4_27

14. Goh, J., Adepu, S., Junejo, K.N., Mathur, A.: A dataset to support research in the design of secure water treatment systems. In: Havarneanu, G., Setola, R., Nassopoulos, H., Wolthusen, S. (eds.) CRITIS 2016. LNCS, vol. 10242, pp. 88–99. Springer, Cham (2017). https://doi.org/10.1007/978-3-319-71368-7_8

15. U.S. Dept. of Health and Human Services: Guidance regarding methods for de-identification of protected health information in accordance with the health insurance portability and accountability act (HIPAA) privacy rule. HIPAA) Privacy Rule (2012)

16. Hochreiter, S., Schmidhuber, J.: Long short-term memory. Neural Comput. **9**(8), 1735–1780 (1997)
17. Hodge, J.G., Jr., Gostin, L.O., Jacobson, P.D.: Legal issues concerning electronic health information: privacy, quality, and liability. JAMA **282**(15), 1466–1471 (1999)
18. Ioffe, S., Szegedy, C.: Batch normalization: accelerating deep network training by reducing internal covariate shift. In: International Conference on Machine Learning, pp. 448–456. PMLR (2015)
19. Jensen, P.B., Jensen, L.J., Brunak, S.: Mining electronic health records: towards better research applications and clinical care. Nat. Rev. Genet. **13**(6), 395–405 (2012)
20. Kingma, D.P., Ba, J.: Adam: a method for stochastic optimization. arXiv preprint: arXiv:1412.6980 (2014)
21. McKenna, R., Mullins, B., Sheldon, D., Miklau, G.: Aim: an adaptive and iterative mechanism for differentially private synthetic data. arXiv preprint: arXiv:2201.12677 (2022)
22. McLachlan, S., Dube, K., Gallagher, T.: Using the CareMap with health incidents statistics for generating the realistic synthetic electronic healthcare record. In: 2016 IEEE International Conference on Healthcare Informatics (ICHI), pp. 439–448. IEEE (2016)
23. Miller, A.R., Tucker, C.: Health information exchange, system size and information silos. J. Health Econ. **33**, 28–42 (2014)
24. Natekin, A., Knoll, A.: Gradient boosting machines, a tutorial. Front. Neurorobot. **7**, 21 (2013)
25. Park, Y., Ghosh, J., Shankar, M.: Perturbed Gibbs samplers for generating large-scale privacy-safe synthetic health data. In: 2013 IEEE International Conference on Healthcare Informatics, pp. 493–498. IEEE (2013)
26. S. Oliveira, M.I., Barros Lima, G.D.F., Farias Lóscio, B.: Investigations into data ecosystems: a systematic mapping study (2019)
27. Tao, Y., Xiao, X., Li, J., Zhang, D.: On anti-corruption privacy preserving publication. In: 2008 IEEE 24th International Conference on Data Engineering, pp. 725–734. IEEE (2008)
28. Torfi, A., Fox, E.A.: CorGAN: correlation-capturing convolutional generative adversarial networks for generating synthetic healthcare records. In: The Thirty-Third International Flairs Conference (2020)
29. Torfi, A., Fox, E.A., Reddy, C.K.: Differentially private synthetic medical data generation using convolutional GANs. Inf. Sci. **586**, 485–500 (2022)
30. Ulianova, S.: Cardiovascular disease dataset. Data retrieved from the Kaggle dataset (2018)
31. Walonoski, J., et al.: Synthea: an approach, method, and software mechanism for generating synthetic patients and the synthetic electronic health care record. J. Am. Med. Inform. Assoc. **25**(3), 230–238 (2018)
32. Xie, L., Lin, K., Wang, S., Wang, F., Zhou, J.: Differentially private generative adversarial network. arXiv preprint: arXiv:1802.06739 (2018)
33. Zheng, P., Zheng, Z., Wu, J., Dai, H.N.: XBlock-eth: Extracting and exploring blockchain data from Ethereum. IEEE Open J. Comput. Soc. **1**, 95–106 (2020)

Rainfall Spatial Interpolation with Graph Neural Networks

Jia Li[1], Yanyan Shen[2(✉)], Lei Chen[1], and Charles Wang Wai Ng[3]

[1] Department of Computer Science and Engineering, HKUST, Hong Kong, China
{jlidw,leichen}@cse.ust.hk
[2] Department of Computer Science and Engineering, SJTU, Shanghai, China
shenyy@sjtu.edu.cn
[3] Department of Civil and Environmental Engineering, HKUST, Hong Kong, China
cecwwng@ust.hk

Abstract. Rainfall spatial interpolation is a crucial task to infer rainfall distribution in space for hydrological studies and natural disaster prevention. However, obtaining accurate interpolation results is a nontrivial task due to the complex and dynamic changing spatial correlations of rainfall. Besides, the practical interpolation will be more intractable when there is a lack of auxiliary variables that can help characterize spatial correlations. The performance of traditional interpolation methods is limited by deterministic formulations and statistical assumptions on modeling spatial correlations. Given the huge success of Graph Neural Networks (GNNs), researchers have exploited GNNs for spatial interpolation tasks. However, existing works usually assume the existence of node attributes and rely on a fixed adjacency matrix to guide the message passing among nodes, thus failing to handle practical rainfall interpolation well. To address these limitations, we propose a novel GSI (**G**raph for **S**patial **I**nterpolation) model, which focuses on learning the spatial message-passing mechanism. By constraining the message passing flow and adaptive graph structure learning, GSI can perform effective interpolation by modeling spatial correlations of rainfall adaptively. Extensive experiments show that our approach outperforms the state-of-the-art methods on two real-world raingauge datasets.

Keywords: Rainfall · Spatial interpolation · Graph neural network

1 Introduction

As one of the most fundamental meteorological and hydrological elements, precipitation plays a key role in triggering natural disasters (e.g., floods, landslides) [28]. Acquiring reliable rainfall spatial information on high temporal resolution (e.g., hourly) is essential to improve the ability of disaster pre-warning [25]. Nowadays, many countries and regions have built monitoring stations to record rainfall amounts. As the instrument to directly measure liquid precipitation, rain gauges are the most used and reliable information source of ground rainfall [19]. However, considering monetary cost and the constraints on land use,

(a) July 14, 2021 14:00   (b) July 14, 2021 15:00

Fig. 1. Demonstration of spatial rainfall in Hong Kong. The white area represents the land, and the colors above indicate the rainfall accumulation in the past one hour: the darker the color, the higher the value.

it is impractical to install rain gauges in every corner. Hence, it is a vital issue to accurately estimate rainfall information at locations without rain gauges, i.e., rainfall spatial estimation.

Spatial interpolation is the traditional method for rainfall spatial estimation, which is a process of using values at observed locations to estimate values at unobserved ones in geographic space [7,13]. The key to rainfall spatial interpolation is to estimate the potential spatial correlations (also known as spatial dependence [5]) between locations, which directly affects the accuracy of the interpolation results. However, it is a non-trivial task to model the spatial dependence for rainfall due to the following three challenging factors. First, rainfall is influenced by various complex factors and shows the irregular distribution in space. Figure 1 shows the rainfall spatial distributions at two time slots in Hong Kong. We have observed that spatial rainfall is irregularly distributed, hence the geographic distance cannot fully reflect the similarity between locations. Second, rainfall is a constantly changing phenomenon [20], which means space patterns of rainfall events may change greatly through time. As shown in Fig. 1, rainfall distribution changes significantly even in the adjacent two hours, such dynamic nature means that a fixed spatial representation cannot handle different rainfall fields. Besides, in practical interpolation, it is non-trivial to obtain other observed variables that are helpful for characterizing spatial correlations of rainfall (like wind, cloud cover), since it means more types of monitoring stations are required. The dilemma of insufficient auxiliary attributes makes it more challenging to model spatial relationships accurately.

Traditional spatial interpolation approaches can be generally classified into two categories [15,22]: (i) deterministic approaches; (ii) geostatistical approaches. Deterministic approaches [4,8,21] rely on pre-defined formulations to directly define the spatial correlation, For example, Inverse Distance Weighting (IDW) [13], utilizes the functions of inverse distances to directly define the spatial correlation. Geostatistical approaches [26,27] are based on statistical assumptions to estimate the spatial correlation. In geostatistical techniques, the fundamental assumption is that observation data are from an underlying

Gaussian process. Obviously, these pre-set formulations and statistical assumptions do not hold in many real-world situations. Due to these unrealistic settings, their performance is still limited [13,16]. In recent years, Graph Neural Networks (GNNs) have been getting more and more attention due to their great expressive power on graph-based problems [11,31,32]. While GNNs were initially developed for explicit graph data, they have been applied to many other applications where the data can be transformed into a graph. However, to the best of our knowledge, there has been no GNN-based work specialized in handling the rainfall spatial interpolation task. The two most relevant works, KCN [1] and IGNNK [29], are designed to solve general interpolation tasks. Specifically, they suffer from one or two of the following drawbacks. The first is to assume the existence of node attributes, which makes it inapplicable in rainfall interpolation scenarios lacking auxiliary attributes. Second, their message passing is only based on the fixed adjacency matrix, which lacks the flexibility to model the complex and dynamic changing spatial correlation of rainfall.

In this paper, we aim to develop an effective end-to-end framework to solve the rainfall spatial interpolation task[1]. Inspired by the superiority of GNNs on modeling spatial dependencies, we propose to extend GNN to handle rainfall spatial interpolation in a data-driven way. To overcome the limitations of existing methods, we argue that a model should be well designed to have the abilities to: (i) achieve a direct interpolation in the absence of auxiliary attributes; (ii) adaptively capture the latent spatial correlation of rainfall. To this end, we propose a new model, the **GSI** (**G**raph for **S**patial **I**nterpolation). Specifically, we train the GSI to infer the values of unlabeled nodes in a transductive style: (1) by adding constraints on message passing, GSI ensures effective implementation of direct interpolation; (2) GSI adopts an adaptive graph learning mechanism to capture the latent spatial correlation in an end-to-end learning manner. To further reduce the estimation errors, we propose a residual correction as the post-processing to rectify the rainfall field. The experiments show that our methods outperform various existing interpolation methods.

The **contributions** of this paper are summarized as follows:

- We identify the challenges of rainfall spatial interpolation and provide insights to design the GNN model, that is, the abilities of direct interpolation and adaptively capturing the spatial correlation.
- We propose the GSI model, which adopts constraining message passing flow and adaptive graph structure learning to effectively solve the practical rainfall interpolation task. Besides, we introduce a residual correction to further improve the interpolation accuracy.
- We conduct extensive experiments on two real-life hourly raingauge datasets. The experiment results show that the proposed method outperforms state-of-the-art baselines.

---

[1] Although this work is proposed for rainfall spatial interpolation, it can be applied for other similar spatial interpolation problems (like temperature, humidity) due to the methodological generality.

Table 1. Notation table.

Notations	Descriptions
V	The set of locations.
V_L	The set of locations with rain gauges.
V_U	The set of locations without rain gauges.
X_L	The observed rainfall values at locations with rain gauges.
X_U	The unobserved rainfall values at locations without rain gauges.
v_i	The i-th location.
x_i	The rainfall value at location v_i.
A_i	The adjacency matrix of the i-th layer.
$\tilde{A}_i$	The new adjacency matrix of the i-th layer, $\tilde{A}_i = A_i \odot \tilde{M}^{(i)}$.
$\hat{A}_i$	The normalized $\tilde{A}_i$.
$W^{(i)}$	The learnable transition matrix of the i-th layer.
$M^{(i)}$	The learnable mask of the i-th layer.
$H^{(i)}$	The output of the i-th layer

The rest of this paper is organized as follows. The problem analysis is described in Sect. 2. In Sect. 3, we propose our methods. We analyze the experimental results in Sect. 4. Section 5 introduces the related works. Finally, we conclude our work in Sect. 6.

2 Preliminaries

2.1 Problem Statement

Table 1 summarizes the frequently-used notations[2] used in the following sections. Let V be a location set, where V_L denotes the locations with rain gauges, and V_U is the locations without rain gauges. $X_L = \{x_l | v_l \in V_L\}$ are the observed rainfall values while $X_U = \{x_u | v_u \in V_U\}$ are unobserved. The rainfall spatial interpolation problem is to estimate X_U according to V and X_L, which is a transductive setting since all the locations are given in advance.

2.2 Challenging Issues

Lack of Auxiliary Attributes. As mentioned earlier, real-world interpolation often encounters the dilemma of insufficient auxiliary attributes. In the absence of other detailed meteorological observed variables, the variable most frequently used for enhancing interpolation is elevation [12]. In KCN [1], authors also take elevation as the node attribute and evaluate their model on the monthly average precipitation data. So should the elevation be used as a node attribute in the GNN model for hourly rainfall interpolation?

---

[2] "locations" might also be called "nodes" or "points", we use those terms interchangeably in this paper.

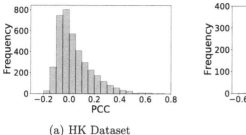
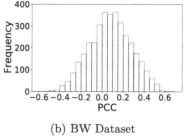

(a) HK Dataset   (b) BW Dataset

Fig. 2. Histogram of Pearson correlation coefficients between hourly rainfall and elevation.

According to the existing hydrology-related studies [6,14], when the observation time steps are less than a month, the relation between precipitation and elevation is likely to be less obvious and less useful. To explore the correlation between hourly rainfall amount and elevation, we calculate the Pearson correlation coefficient (PCC) of each rain hour on two region datasets (see dataset details Sect. 4.1) and plot the histograms. As shown in Fig. 2, more than 94% (83%) of rainy hours have a $|PCC| < 0.3$ on the HK (BW) dataset, which means the correlation between hourly rainfall and elevation is extremely weak. Besides, when calculating PCC across the whole dataset, the poor correlation is much more obvious: only 0.015 on the HK dataset and 0.047 on the BW dataset. Given the poor correlation, in this work, we focus on the direct interpolation task, in which the rainfall values of unsampled locations are inferred by other known observations.

Complex and Dynamic Spatial Correlations. As discussed above, the spatial correlation of rainfall varies over time. To capture such dynamic spatial correlations, the model should have the ability to adaptively learn the graph structure. There are three groups of graph structure learning [33]: metric-based approaches, neural approaches, and direct approaches. Metric-based and neural approaches use a parameterized network to derive edge weights based on the node representations, while direct approaches ignore node features and directly optimize the adjacency matrix. In view of unavailable node attributes, in this work, we utilize the direct approach to enjoy more flexibility in capturing the latent spatial correlations.

3   Methodology

Figure 3 depicts an overview of our proposed framework for the rainfall spatial interpolation task. It consists of four components: data preparation, initial graph construction, GNN for interpolation, and correction mechanism. The first part includes data cleaning and data standardization. Then an initial graph is constructed based on the geographic distance. To implement an effective method for

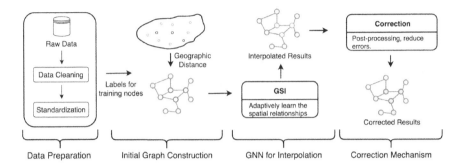

Fig. 3. Framework overview.

rainfall spatial interpolation, we propose GSI, which makes two improvements to GNNs: constraining message passing flow and adaptive graph structure learning. The GSI model can adaptively capture the spatial correlations and infer the values (in green) at unlabeled locations. Finally, we introduce residual correction as a post-processing step to learn the residual structure and rectify estimates (in red) for a more accurate result.

3.1   Data Preparation

Data preparation mainly includes two steps: data cleaning and standardization. Abnormal damage or the aging of instruments may cause some missing values or rainfall values not to be fully recorded. Simply filling with certain values may have a negative impact on performance, since these values are most likely not true. Hence, we first remove these missing and incomplete values, and the spatial interpolation is only based on reliable rainfall values. Besides, in order to accelerate the convergence of models, the data of each raingauge graph are normalized before being input into the network: $x' = \frac{x - \mu_x}{\sigma_x}$, where x is the raw data, μ_x and σ_x are the mean and standard deviation.

3.2   Initial Graph Construction

According to the first law of geography [23], the correlations between different locations decay with distance increasing. Based on such an inverse correlation, we use the following formulation as a rough measure of the spatial correlations to build the initial adjacency matrix: $A[i,j] = \frac{d_{ij}^{-2}}{\sum_{k=1}^{n} d_{ik}^{-2}}$, if $d_{ij} \leq \tau$, otherwise 0. Here, $A[i,j]$ denotes the edge weight between node v_i and node v_j, d_{ij} is the geographic distance between v_i and v_j. Considering that the spatial correlation of rainfall between locations that are too far apart can be negligibly small, a threshold τ is set to cut off the connections between these nodes. Besides, each unlabeled node $v_u \in V_U$ does not own the observation, we initialize $x_u = 0$ as its input value.

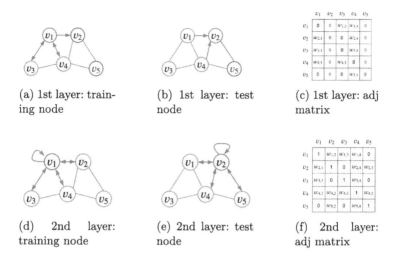

(a) 1st layer: train-
ing node

(b) 1st layer: test
node

(c) 1st layer: adj
matrix

(d)   2nd   layer:
training node

(e) 2nd layer: test
node

(f)   2nd   layer:
adj matrix

Fig. 4. Message passing flow. Training nodes: v_1, v_3, v_4; test nodes: v_2, v_5. **Blue** color means that the node values are the real observed values; **Gray** color means no observations (initialized to 0); **Green** color means the aggregated values. The currently focused node is marked by a **Red Circle**, and **Red Lines** with arrows denoting the message flows. (Color figure online)

3.3   GNN for Spatial Interpolation

As one type of GNNs, Graph Convolutional Network (GCN) [11] has achieved remarkable performance in modeling spatial data. The layer-wise GCN can be formulated as follows:

$$H^{(i)} = \sigma \left(\hat{A} H^{(i-1)} W^{(i)} \right) \tag{1}$$

where $\hat{A}$ denotes a normalized adjacency matrix; the ith layer takes $H^{(i-1)} \in \mathbb{R}^{n \times d_{i-1}}$ as the input and outputs the matrix $H^{(i)} \in \mathbb{R}^{n \times d_i}$, $H^{(0)} = X$; $W^{(i)}$ is the layer-specific trainable transition matrix; $\sigma(\cdot)$ is a non-linear activation function. However, it is not feasible to directly apply the vanilla GCN for rainfall interpolation, cause this will lead to two major problems: (1) the GCN fails to learn since training nodes take the labels as input; (2) the vanilla GCN lacks the mechanism to dynamically learn the graph structure. Hence, we develop GSI, which adopts two techniques to solve the above problems: (i) constraining message passing flow, and (ii) adaptive graph structure learning. Like most of the existing works [11,31], GSI adopts two graph convolutional layers.

Constraining Message Passing Flow. To implement direct interpolation, we aim to use labeled nodes as sources and rely on them to estimate the values of unlabeled nodes. To this end, two issues need to be considered in the message passing process of the first layer: (1) training nodes should not take their own labels as input, otherwise self-information will weaken the message passing between nodes; (2) test nodes have no real observations, so their information should not be passed to the other nodes. Hence, unlike other applications, we use two adjacency matrices (i.e., A_1 and A_2) differently in the two layers. Let A denote the initial adjacency matrix without self-loops. The adjacency matrix A_1 is obtained by cutting off the message flow from unlabeled nodes to other nodes (assign 0 to the corresponding edge weights) based on A. In the 2nd layer, $A_2 = A + I$ is used, and the added self-loops mean that the aggregated value from the 1-hop neighbors can be used again. The message passing flows and adjacency matrices on different layers are shown in Fig. 4. For a clear illustration, in each sub-figure, we only focus on one node and show its message flow. Let's take the training node v_1 and the test node v_2 as examples. In the 1st layer, the training nodes have real observations, but the test nodes do not:

- the training node v_1 provides information for its all neighbors, but only receives information from neighbors that are training nodes (Fig. 4a);
- the test node v_2 does not provide information for its neighbors, and only receives information from neighbors that are training nodes (Fig. 4b);
- self-loops are not allowed, avoiding that training nodes take their own labels as input (Fig. 4a and 4b).

In the 2nd layer, the values of all nodes become the aggregated values from the 1-hop neighbors:

- each node can provide/receive information to/from its all neighbors (Fig. 4d and 4e);
- the self-loops ensure that the information from 1-hop will not be ignored by themselves (Fig. 4d and 4e).

Adaptive Graph Structure Learning. Since the spatial structure of rainfall field is dynamically changing, a fixed adjacency matrix cannot fully denote the real spatial relationships between locations. In this work, we hope that edge weights can be adjusted adaptively in a data-driven manner.

Full Parameterization (FP) Learner. Considering the dilemma of unavailable node attributes in practical interpolation, we adopt an FP method to directly learn the latent adjacency relationships in the raingauge graph. To retain the asymmetry of the adjacency matrix (i.e., A_1) in the first layer, we propose to use the dynamic mask $M^{(i)} \in \mathbb{R}^{n \times n}$ as the adjacency learner of the i-th layer to adjust the initial adjacency matrix via the element-wise product manner: $\tilde{A}_i = A_i \odot M^{(i)}$.

Post-processor. In real-world spatial rainfall, the spatial relationships between locations are bi-directional. Hence, we perform the symmetrization for the dynamic mask $M^{(i)}$ to adjust both edges equally between two locations. Besides, the edge weights should be non-negative, an activation is performed for $M^{(i)}$. The symmetrization and activation are expressed as:

$$\tilde{M}^{(i)} = \frac{\sigma_p\left(M^{(i)}\right) + \sigma_p\left(M^{(i)}\right)^\top}{2} \qquad (2)$$

where σ_p is a non-linear activation. Here, we adopt the ELU [3] function instead of ReLU to prevent the gradient from disappearing for the FP learner. Then we perform element-wise product to generate the new adjacency matrix $\tilde{A}_i$ and apply the normalization to make the normalized one $\hat{A}_i$:

$$\tilde{A}_i = A_i \odot \tilde{M}^{(i)} \tag{3}$$

$$\hat{A}_i = \tilde{D}_i^{-1} \tilde{A}_i \tag{4}$$

where $\tilde{D}_i$ is the degree matrix of $\tilde{A}_i$.

Multi-head Mechanism. Considering the complex spatial variability and increasing the model's flexibility, we employ an approach similar to multi-head attention mechanism [24] in the first layer to generate K masks. Finally, the layers of GSI model can be formulated as follows:

$$H^{(1)} = \|_{k=1}^{K} \sigma \left(\tilde{D}_{1,k}^{-1} \tilde{A}_{1,k} H^{(0)} W^{(1,k)} \right) \tag{5}$$

$$H^{(2)} = \sigma \left(\tilde{D}_2^{-1} \tilde{A}_2 H^{(1)} W^{(2)} \right) \tag{6}$$

where $\tilde{A}_{1,k} = A_1 \odot \tilde{M}^{(1,k)} = A_1 \odot \frac{\sigma_p\left(M^{(1,k)}\right) + \sigma_p\left(M^{(1,k)}\right)^{\top}}{2}$, $M^{(1,k)}$ represents k-th learnable mask in the first layer; $W^{(1,k)}$ is the trainable transition matrix of the k-th branch in the first layer; $\|$ denotes the *concatenation* operation; σ is a non-linear activation, here we use ELU. $H^{(0)} = X$ is the input; the output of the first layer, $H^{(1)}$, will consist of K aggregation branches for each node; $\hat{Y} = H^{(2)}$ is the final output.

The graph structure is jointly determined by the initial adjacency matrix and the adaptive masks, which will be optimized in an end-to-end manner.

3.4   Correction Mechanism

As a post-processing step, residual correction is applied to further improve the accuracy of rainfall spatial interpolation. As discussed in Sect. 3.3, in GSI, the interpolated value of each node is inferred based on its neighbors. For training nodes, we can also obtain their estimated values and then calculate the corresponding residuals: $R_L = Y_L - \hat{Y}_L$. Residual correction aims to learn the residual structure and estimate the residuals on test nodes, then correct the test nodes by using the estimated residuals. Algorithm 1 summarizes the residual correction algorithm, where the estimated results $\hat{Y}$ from GSI is the input, the corrected results of unlabeled nodes $\hat{Y}_U'$ is the output.

Since the residual estimation is also a spatial interpolation task, it can be implemented by different methods. Therefore, our GSI model can also combine with Kriging to do a residual correction. In Sect. 4, two settings are implemented and compared, i.e., residual correction with GSI itself and residual correction with Kriging.

Algorithm 1: Residual Correction

Input: Rainfall observations X where X_U is null for all unseen nodes;
        Estimated values for all nodes $\hat{Y}$.

Output: Corrected results for unlabled nodes $\hat{Y}'_U$.

1 $Y_L \leftarrow X_L$;   ▷ set observations as labels of training nodes

2 $R \leftarrow 0$ ▷ create the residual vector

3 $R_L \leftarrow Y_L - \hat{Y}_L$ ▷ calculate residuals of training nodes

4 $\hat{R}_U \leftarrow Interpolate(R)$ ▷ get residuals for test nodes

5 $\hat{Y}'_U \leftarrow \hat{Y}_U + \hat{R}_U$ ▷ obtain corrected results

6 **return** $\hat{Y}'_U$

3.5 Model Training and Complexity

Training. We summarize all learnable parameters in GSI model as $\Theta = \{W^{(i,k)}, M^{(i,k)}\}$, $i = 1, 2; k = 1, .., K$. The model is trained in the transductive setting: for the raingauge graph of each time, Θ are optimized by minimizing the training loss via backpropagation:

$$\mathcal{L} = \frac{1}{|V_L|} \sum_{v_i \in V_L} (y_i - \hat{y}_i)^2 + \alpha \|\Theta\|_2^2 \tag{7}$$

where the first term is the mean squared error of training set V_L, y_i and $\hat{y}_i$ are the observed value and estimated value of training node v_i; the second term denotes L2 regularization, α is the weight decay factor. During the training of our GSI model, no validation set is used since the reduction of the information source (training nodes) will decrease the interpolation effect. Besides, the early stopping mechanism is adopted to avoid overfitting. For residual correction with GSI, the second GSI model is trained from scratch to capture the spatial correlations of residual errors.

Computation Complexity. When training the GSI model on one instance, the computation of the adjacency matrix takes time $O(n^2)$ where n is the node number. The computation of each layer with one mask takes time $O(n^2 + n^2d + nd)$, where d is the hidden dimension; $O(n^2)$ is for the element-wise product of the adjacency matrix and the mask, then we can get a $n \times n$ graph structure matrix; $O(nd)$ is for $H^{(i)}W^{(i)}$ to get a $n \times d$ support matrix, $O(n^2d)$ is for the multiplication of graph structure matrix and support matrix. Hence, the computation complexity for each layer with one mask is $O(n^2d)$. In GSI, the first layer has K-head masks, and the second layer only includes one mask. Hence, the computation complexity of forward computation and backpropagation for one instance is $O(Kn^2d)$.

Table 2. Detailed information of the datasets.

Dataset	Time Span	#Rain hours	#Raingauges	#Training nodes	#Test nodes
HK	2008–2012	3855	123	98	25
BW	2012–2014	3640	132	106	26

4   Experiments

4.1   Experiment Setup

Datasets. We evaluate our model using two real-world hourly raingauge datasets from Hong Kong (HK), China, and Baden-Württemberg (BW), Germany.

- HK data are collected by the Hong Kong Observatory (HKO) and the Geotechnical Engineering Office (GEO), of which data precision is 0.1-mm. The rainfall data from 123 rain gauges are available. We pick rainy hours from 2008 to 2012 to form the final dataset, which includes a total of 3855 valid hours.
- BW data are open data provided by the Climate Data Center (CDC)[3] of the German Weather Service (DWD). The rainfall data are based on 0.01-mm precision. There is a total of 132 rain gauges available. 3640 valid rainy hours from 2012 to 2014 are selected to be the final dataset.

On both datasets, each raingauge record is associated with its ID, timestamp, longitude, latitude, elevation, and rainfall value. To assess the performance of interpolation approaches, we randomly sample 20% nodes as the test set, the rest acts as the training set. The dataset details are shown in Table 2.

Baselines. We compare our models with six interpolation methods which can be classified into two categories: **traditional interpolation approaches:** (i) Thin Plate Spline (TPS) [8], (ii) Triangular Irregular Network (TIN) [4], (iii) Inverse Distance Weighting (IDW) [21], (iv) Ordinary Kriging (OK) [26]; **GNN-based approaches:** (v) KCN [1] and (vi) IGNNK [29]. Besides, three variants of our proposed methods are evaluated: (1) **GSI**, the basic solution without any correction; (2) **GSI-RC-G**, GSI with residual correction and the residual correction is implemented by GSI network; (3) **GSI-RC-K**, GSI with residual correction and residual correction is implemented by Kriging.

Metrics. We use three metrics to evaluate the performance: (i) Root Mean Squared Error (RMSE); (ii) Mean Absolute Error (MAE); (iii) Nash-Sutcliffe's Efficiency (NSE) coefficient [17]. NSE is generally used to assess the performance of hydrological models, and its value ranges from $-\infty$ to 1. For RMSE and MAE, the smaller, the better; for NSE, the closer to 1, the better.

---

[3] https://www.dwd.de/EN/climate_environment/cdc/cdc_node_en.html.

Implementation Details. For traditional approaches, we report their best performance by running different settings: the power parameter of IDW is 2, the variogram model of OK is spherical, and the type of TIN interpolating function is linear; there is no parameters need manual tuning for TPS. For GNN-based baselines, we follow the original papers to set or tune all hyper-parameters. The time dimension of IGNNK is set as one to compare with other spatial interpolators. For the GSI model, we use Adam [10] as the optimizer. During training, there is no validation set, and training loss is used as the performance measure for early stopping. We optimize the hyper-parameters with the help of HyperOpt [2]. The tuning parameters include learning rate, weight decay, dropout rate, hidden dimension, and the number of multi-head masks. Table 3 summarizes the value ranges of the hyperparameter search. All algorithms are written in Python. All neural networks are conducted on the PyTorch, and we run all the experiments on a CentOS 7.9.2009 server equipped with a 72-core Intel(R) Xeon(R) Gold 6240 CPU and one Tesla V100 GPU.

Table 3. The search space for the hyperparameters.

Hyperparameter	Range
Learning rate	(0, 0.02)
Weight decay	(0, 1e-3)
Dropout rate	(0, 0.5)
Hidden dimension	{2, 4, 8, 16}
#Multi-head	{2, 4, 8, 16}

4.2   Performance Comparison

Table 4 shows the overall performance of all methods on two datasets.

Our approach achieves the best performance on both raingauge datasets: reducing the RMSE score by 6.24% on the HK dataset and 2.15% on the BW dataset when compared with the best-performed baselines. We can see that the GSI model itself can achieve good interpolation results, and the residual correction can further improve the performance which indicates the importance of capturing residual correlations in geospatial data. More specifically, the residual correction with Kriging (GSI-RC-K) always achieves the best performance, but the residual correction with GSI (GSI-RC-G) may lead to slight performance degradation on the BW dataset. We conjecture that one method has the limitation in correcting its own residuals, but the diversity of the interpolators can complement each other.

Among all baselines, traditional interpolation methods generally outperform GNN-based solutions. Specifically, TPS is the best-performed baseline on the HK dataset, while IDW and TIN obtain the lowest RMSE and MAE on the BW dataset. The MAE of TIN is even lower than the one of our GSI-RC-K, but

Table 4. Overall performance. The best and the second-best results are marked by **bold** and <u>underline</u>. "↑" means the score is higher the better while "↓" means the score is lower the better.

Algorithm	HK Dataset			BW Dataset		
	RMSE↓	MAE↓	NSE↑	RMSE↓	MAE↓	NSE↑
TIN	3.0088	0.9684	0.7538	1.0985	**0.3494**	0.4008
IDW	2.9171	1.1056	0.7686	1.0493	0.3917	0.4533
TPS	2.6594	<u>0.8953</u>	0.8076	1.0985	<u>0.3537</u>	0.4008
OK	2.8661	1.0001	0.7766	1.0804	0.3647	0.4203
KCN	6.4833	2.2946	-0.1432	1.0892	0.3930	0.4109
IGNNK	3.4698	1.8856	0.6726	1.1616	0.6188	0.3300
GSI	2.5283	0.9056	0.8261	<u>1.0367</u>	0.3908	<u>0.4663</u>
GSI-RC-G	<u>2.5268</u>	0.8997	<u>0.8264</u>	1.0584	0.3803	0.4438
GSI-RC-K	**2.4934**	**0.8834**	**0.8309**	**1.0267**	0.3546	**0.4766**

Table 5. Ablation study results. The percentages in "()" represent error increase for RMSE/MAE and performance degradation for NSE.

Settings	HK Dataset			BW Dataset		
	RMSE↓	MAE↓	NSE↑	RMSE↓	MAE↓	NSE↑
GSI	**2.5283**	**0.9056**	**0.8261**	**1.0367**	**0.3908**	**0.4663**
GSI-init-ones	3.6719 (+45.2%)	1.6004 (+76.7%)	0.6333 (-23.3%)	1.4559 (+40.4%)	0.7280 (+86.3%)	-0.0525 (-111.3%)
GSI-NC	2.9149 (+15.3%)	0.9984 (+10.2%)	0.7689 (-6.9%)	1.1607 (+12.0%)	0.3950 (+1.1%)	0.3311 (-29.0%)
GSI-NM	2.7740 (+9.7%)	1.0194 (+12.6%)	0.7907 (-4.3%)	1.1772 (+13.6%)	0.4801 (+22.9%)	0.3119 (-33.1%)
GSI-NC-NM	3.4604 (+36.9%)	1.1729 (+29.5%)	0.6743 (-18.4%)	1.2276 (+18.4%)	0.4826 (+23.5%)	0.2517 (-46.0%)

TIN tends to achieve a much larger RMSE. The best traditional methods are different in the two datasets, which means the pre-settings (i.e., deterministic formulations or statistical assumptions) in these methods are limited and cannot handle various rainfall fields. Two GNN-based models, KCN and IGNNK, fail to handle the hourly rainfall interpolation task well. Their poor results on two datasets show that learning contraction mapping related to node attributes is not the right direction for interpolation tasks when there are no sufficient and useful node attributes to rely on.

4.3 Ablation Study

Effects of Initial Graph Construction. In this part, we aim to show the importance of a distance-based initial graph. As a contrast, we set all non-zero edge weights of A_1 and A_2 to 1, called GSI-init-ones, and keep other settings unchanged to conduct the experiment. The comparison results are shown in Table 5. We can see that when discarding the distance-based rough measure on spatial relationships: the performance drops drastically, where RMSE increased by 45.2% and 40.4%, and MAE increased by 76.7% and 86.3% on the HK and

BW datasets, respectively. The poor results are even much worse than compared baselines (see Table 4). Hence, although the geographic distance cannot fully capture the correlations between locations, the distance factor cannot be ignored entirely in geographic space. Taking such a rough measure of spatial correlation as the initial graph structure can help the model converge to achieve more accurate interpolation results.

Effects of Two Major Techniques in GSI. Two major techniques include: (i) constraining message passing flow; (ii) dynamic graph structure learning via adaptive masks. We reconfigure GSI to create three variants: a) GSI-NC: GSI with **N**o **C**onstraints on message passing, i.e., $A_1 = A_2 = \tilde{A} = A + I$; b) GSI-NM: GSI with **N**o adaptive **M**asks; c) GSI-NC-NM: GSI with **N**o **C**onstraints on message passing, also **N**o adaptive **M**asks. Table 5 presents the comparison results of GSI and three variants. We can see that when removing one of these two techniques, the performance decrease to varying degrees. Without message passing constraints in the first layer (GSI-NC), the performance decreases significantly for all metrics. It is in agreement with our earlier argument: the graph structure is dynamically optimized through message passing between nodes; however, incorrect message flow will negatively affect information aggregation and make GCN fail to accurately model spatial correlations. When removing the adaptive masks (GSI-NM), the performance also drops a lot, showing the importance of dynamic graph structure learning. Removing both techniques (GSI-NC-NM) yields much worse results, of which the performance degradation is around a sum of that of GSI-NC and GSI-NM. This result shows the designs of the GSI model do matter for spatial interpolation, and directly applying vanilla GCN to spatial interpolation cannot achieve better results than traditional methods.

5   Related Work

Rainfall Spatial Interpolation. There are two main categories for traditional spatial interpolation algorithms : (1) deterministic approaches, like Inverse Distance Weighting (IDW) [21], Triangular Irregular Network (TIN) [4], Spline [8]; (2) geostatistical approaches, such as Ordinary Kriging (OK) [26], Universal Kriging (UK) [27]. Deterministic methods are directly pre-defined formulas, while geostatistical methods utilize statistical models to describe spatial correlations. Kriging is a generic name of geostatistical techniques [9]. OK is the basic form of Kriging, and UK incorporates the additional variables on the basis of OK. Further reviews of spatial interpolation can be found in [13, 22].

Graph Neural Networks. Various variants of GNNs have been proposed, such as Graph Convolutional Networks (GCNs) [11], Graph Attention Networks (GATs) [24], and Spatial-temporal Graph Neural Networks (STGNNs) [30]. This work is more related to GCNs. There are mainly two streams of GCNs: spectral and spatial. Spectral GCNs define the graph convolution in the form of spectral analysis [11], while spatial GCNs take a more intuitive way to perform convolution directly on graph nodes and their neighbors [18]. In this work, we follow the spatial perspective to construct the convolution operations on the raingauge graph.

GNNs for Spatial Interpolation. Recently, two GNN-based methods, KCN [1] and IGNNK [29], have been proposed for spatial/spatiotemporal interpolation tasks. KCN constructs a local subgraph and trains a GNN to predict each center node's label based on node attributes and neighboring labels. IGNNK focuses on the spatiotemporal interpolation task, in which time-series signals at each node are regarded as the node attributes and a GNN is trained to reconstruct all signals on each sample subgraph.

6 Conclusions

In this paper, we explored the rainfall spatial interpolation task and proposed a novel GNN model, GSI, to effectively solve the practical interpolation task lacking auxiliary attributes. Specifically, GSI employs two major techniques, i.e., constraining the message passing flow and adaptive graph structure learning to perform the spatial interpolation. The former ensures effective implementation of direct interpolation in the absence of available attributes, and the latter enables GSI to model complex spatial correlations of rainfall adaptively. To further reduce estimation errors, we introduce a residual correction mechanism to rectify the estimated rainfall field. The empirical study shows that our proposed method outperforms the state-of-the-art baselines in two large real-life raingauge datasets.

Acknowledgements. The authors would like to thank the anonymous reviewers for their insightful reviews. This work is supported by the National Key Research and Development Program of China (2022YFE0200500), Shanghai Municipal Science and Technology Major Project (2021SHZDZX0102) and SJTU Global Strategic Partnership Fund (2021 SJTU-HKUST). Lei Chen's work is partially supported by National Science Foundation of China (NSFC) under Grant No. U22B2060, the Hong Kong RGC GRF Project 16209519, RIF Project R6020-19, AOE Project AoE/E-603/18, Theme-based project TRS T41-603/20R, China NSFC No. 61729201, Guangdong Basic and Applied Basic Research Foundation 2019B151530001, Hong Kong ITC ITF grants MHX/078/21 and PRP/004/22FX, Microsoft Research Asia Collaborative Research Grant and HKUST-Webank joint research lab grants.

References

1. Appleby, G., Liu, L., Liu, L.P.: Kriging convolutional networks. In: AAAI (2020)
2. Bergstra, J., et al.: Making a science of model search: hyperparameter optimization in hundreds of dimensions for vision architectures. In: ICML (2013)
3. Clevert, D., Unterthiner, T., Hochreiter, S.: Fast and accurate deep network learning by exponential linear units (elus). In: ICLR (2016)

4. De Floriani, L., Magillo, P.: Triangulated Irregular Network. Springer, New York (2018)
5. Getis, A.: A history of the concept of spatial autocorrelation: a geographer's perspective. Geogr. Anal. **40**(3), 297–309 (2008)
6. Goovaerts, P.: Geostatistical approaches for incorporating elevation into the spatial interpolation of rainfall. J. Hydrol. **228**(1–2), 113–129 (2000)
7. Hu, Q., Li, Z., Wang, L., Huang, Y., Wang, Y., Li, L.: Rainfall spatial estimations: a review from spatial interpolation to multi-source data merging. Water **11**(3), 579 (2019)
8. Hutchinson, M.F.: Interpolating mean rainfall using thin plate smoothing splines. Int. J. Geogr. Inf. Syst. **9**(4), 385–403 (1995)
9. Jewell, S.A., Gaussiat, N.: An assessment of kriging-based rain-gauge-radar merging techniques. Q. J. R. Meteorol. Soc. **141**(691), 2300–2313 (2015)
10. Kingma, D.P., et al.: Adam: a method for stochastic optimization. In: ICLR (2015)
11. Kipf, T.N., Welling, M.: Semi-supervised classification with graph convolutional networks. In: ICLR (2017)
12. Kyriakidis, P.C., et al.: Geostatistical mapping of precipitation from rain gauge data using atmospheric and terrain characteristics. JAMC **40**(11), 1855–1877 (2001)
13. Li, J., Heap, A.D.: A review of spatial interpolation methods for environmental scientists. Geosci. Austral. (2008)
14. Lloyd, C.: Assessing the effect of integrating elevation data into the estimation of monthly precipitation in great Britain. J. Hydrol. **308**(1–4), 128–150 (2005)
15. Ly, S., Charles, C., Degré, A.: Different methods for spatial interpolation of rainfall data for operational hydrology and hydrological modeling at watershed scale: a review. Biotechnol. Agron. Soc. Environ. **17**(2) (2013)
16. Ma, J., Ding, Y., Cheng, J.C., Jiang, F., Wan, Z.: A temporal-spatial interpolation and extrapolation method based on geographic long short-term memory neural network for pm2. 5. J. Clean. Prod. **237**, 117729 (2019)
17. Nash, J.E., Sutcliffe, J.V.: River flow forecasting through conceptual models part i-a discussion of principles. J. Hydrol. **10**(3), 282–290 (1970)
18. Niepert, M., Ahmed, M., Kutzkov, K.: Learning convolutional neural networks for graphs. In: ICML (2016)
19. Organization, W.M.: Guide to instruments and methods of observation volume 1-measurement of meteorological variables. WMO, Geneva (2018)
20. Rodriguez-Iturbe, I., Febres De Power, B., Sharifi, M.B., Georgakakos, K.P.: Chaos in rainfall. Water Resour. Res. **25**(7), 1667–1675 (1989)
21. Shepard, D.: A two-dimensional interpolation function for irregularly-spaced data. In: Proceedings of the 1968 23rd ACM National Conference (1968)
22. Sluiter, R.: Interpolation methods for climate data: literature review. KNMI, De Bilt (2009)
23. Tobler, W.R.: A computer movie simulating urban growth in the Detroit region. Econ. Geogr. **46**, 234–240 (1970)
24. Velickovic, P., Cucurull, G., Casanova, A., Romero, A., Liò, P., Bengio, Y.: Graph attention networks. In: ICLR (2018)
25. Verworn, A., et al.: Spatial interpolation of hourly rainfall-effect of additional information, variogram inference and storm properties. Hydrol. Earth Syst. Sci. **15**(2), 569–584 (2011)
26. Wackernagel, H.: Ordinary Kriging. Springer, Berlin (1995)
27. Wackernagel, H.: Universal Kriging. Springer, Berlin (1995)

28. Wu, H., Yang, Q., Liu, J., Wang, G.: A spatiotemporal deep fusion model for merging satellite and gauge precipitation in china. J. Hydrol. **584**, 124664 (2020)
29. Wu, Y., Zhuang, D., Labbe, A., Sun, L.: Inductive graph neural networks for spatiotemporal kriging. In: AAAI (2021)
30. Yan, S., Xiong, Y., Lin, D.: Spatial temporal graph convolutional networks for skeleton-based action recognition. In: AAAI (2018)
31. Zhang, M., Cui, Z., Neumann, M., Chen, Y.: An end-to-end deep learning architecture for graph classification. In: AAAI (2018)
32. Zhang, M., et al.: Link prediction based on graph neural networks. In: NeurIPS (2018)
33. Zhu, Y., et al.: A survey on graph structure learning: Progress and opportunities. arXiv (2021)

A Knowledge-Driven Memory System for Traffic Flow Prediction

Binwu Wang[1], Yudong Zhang[1], Pengkun Wang[1], Xu Wang[1], Lei Bai[2(✉)], and Yang Wang[1(✉)]

[1] University of Science and Technology of China, Hefei 230000, China
{wbw1995,zyd2020,pengkun,wx309}@mail.ustc.edu.cn, angyan@ustc.edu.cn
[2] Shanghai AI Laboratory, Shanghai 200000, China
baisanshi@gmail.com

Abstract. Traffic flow prediction is critical for intelligent transportation systems. Recent studies indicate that performance improvement by designing new models is becoming marginal. Instead, we argue that the improvement can be achieved by using traffic-related facts or laws, which is termed exogenous knowledge. To this end, we propose a knowledge-driven memory system that can be seamlessly integrated into GCN-based traffic forecasting models. Specifically, the memory system includes three components: access interface, memory module, and feedback interface. The access interface based on the attention mechanism and the feedback interface based on the gate mechanism are used to guide the model to extract useful patterns and integrate these patterns into the model to enhance spatiotemporal representation respectively. The memory module is used to learn specific knowledge-based patterns, and this is achieved by constraining the learning process with unsupervised loss functions formulated inspired by exogenous knowledge. We construct three kinds of memory modules driven by different exogenous knowledge: the long-term trend memory to learn periodic patterns, the hierarchical effect memory to capture coarse-grained region patterns, and the representative pattern memory to extract representative patterns. Experiments combined with multiple existing models demonstrate the effectiveness of the memory system.

Keywords: Traffic forecasting · Spatiotemporal data mining · Graph convolutional network

1 Introduction

Traffic forecasting plays a fundamental role in intelligent transportation systems (ITS) which is beneficial for practical traffic applications. For instance, road traffic speed and occupancy forecasting can provide insights for urban planning, dynamic management of urban traffic, the efficiency of the logistics industry, and route planning public.

To achieve accurate traffic forecasting, most researchers are devoted to developing complex spatiotemporal learning models. Machine learning methods in this field mainly use time series analysis models, e.g, Auto-Regressive Integrated Moving Average (ARIMA) and Support Vector Regression (SVR), which fail to model complex spatiotemporal correlation among nodes of the traffic network and time points along the temporal dimension. In recent years, with the rise of deep learning, researchers [1–3] introduce various cutting-edge deep-learning models to learn spatiotemporal correlation, and then generated spatiotemporal representation is used as input to decoders (e.g fully connected layers) to predict traffic. For example, [1,4] utilize convolutional neural networks (CNN) to learn spatial dependencies and combine CNN with time series models (e.g long short-term memory (LSTM) or temporal convolutional network(TCN)) to capture temporal dependencies. Recently, impressed by the promising performance of graph convolutional neural networks (GCN), researchers [5–8] move to integrate GNN into traffic forecasting for capturing dependencies among nodes. For example, STGCN [9] constructs a graph topology based on the road network and then uses GCN for graph representation learning. STSGCN [10] which is a well-designed synchronous model expands GCN into the spatiotemporal dimension to synchronously capture local spatiotemporal correlation with a local spatiotemporal graph.

However, recent studies indicate that the gain of the forecasting performance induced by modifying neural network structures has become marginal [11], and hence it is in great need to seek alternative approaches to further boost the performance of the traffic forecasting models. To this end, we note an overlooked aspect in the field of traffic: exogenous knowledge, which refers to the facts or laws related to traffic and is the external abstraction of the internal features of traffic data. Therefore, a natural idea is to introduce exogenous knowledge to help analyze the evolution laws of traffic networks, which can provide inspiration to learn more comprehensive spatiotemporal correlation. For example, based on the fact that traffic data is periodic, some models [1,4] integrate different methods to explicitly capture periodic dependencies, which is proven to be effective for modeling more robust temporal dependencies.

In this paper, instead of designing advanced spatiotemporal learning models, we move to investigate another aspect: how to effectively leverage traffic exogenous knowledge to improve the prediction performance of the model, and finally propose a general module, knowledge-driven memory system, which uses the memory as the backbone due to its flexible capability of storing, abstracting and organizing the knowledge into a structural and addressable form. According to the exogenous knowledge, unsupervised loss functions are formulated to constrain the memory system to learn specific patterns, which are termed as knowledge-based patterns. The model can extract these patterns to enhance spatiotemporal representation.

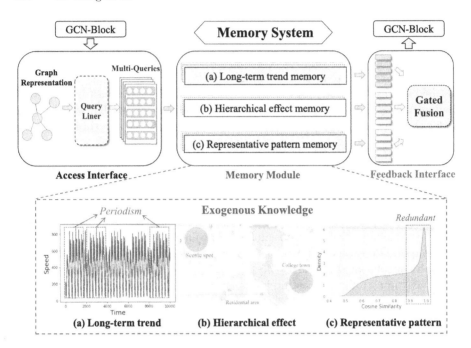

Fig. 1. The details of the memory system. Subfigure (a) illustrates that the traffic is periodic. Subfigure (b) shows the macro level of the transportation system (three hotspots). Subfigure (c) shows the cosine similarity distribution of spatiotemporal patterns, most of which have extremely high similarity.

The memory system includes three carefully-designed components: access interface, memory module, and feedback interface. The access interface provides a specific access address based on the query information from the model to guide the model to extract useful patterns, and the memory module is parameterized and end-to-end updated with models to learn knowledge-based patterns according to exogenous knowledge. The feedback interface refers to how to integrate the information from the memory system into the models. Specifically,

Access Interface. The access interface is based on the attention mechanism. The advantage is that the model can adaptively extract useful patterns by matching the query vector of the model with the memory module.

Memory Module. Three types of traffic exogenous knowledge (as shown in Fig. 1) are introduced to enrich the memory system ecology. That is, three kinds of memory modules are configured in the memory system to store the corresponding patterns.

- **Long-term trend memory**. Based on the fact that traffic data is periodic (as shown in Fig. 1(a)), long-term trend memory is used to explicitly store periodic patterns of the traffic network, which can be used to model stronger temporal dependence.

- **Hierarchical effect memory.** Urban traffic network is a hierarchical structure (as shown in Fig. 1(b)), including not only micro-structure where fine-grained roads or nodes are regarded as entities, but also macro-structure with coarse-grained regions gathered by micro-entities as entities. In a coarse-grained region, the associations among nodes may be more intimate. To this end, we propose the hierarchical effect memory to model this effect and learn coarse-grained patterns of the traffic network. Considering the availability of external data (e.g POI), we introduce a graph pooling loss function to constrain the model to adaptively learn a friendly hierarchical structure of the traffic network.
- **Representative pattern memory.** A recent study [12] reveals that traffic patterns of road networks are redundant, and the traffic status of the entire road network can be effectively represented by generalizing a set of representative patterns. Based on this discovery, we propose the representative pattern memory to extract representative patterns of the traffic network.

Feedback Interface. To efficiently integrate extracted information from the memory module into the model, we provide a feedback interface based on the gating mechanism to filter out redundant information and achieve efficient information fusion.

In conclusion, we propose a novel memory system driven by exogenous knowledge for traffic forecasting. Our contributions are summarized as follows:

- We investigate leveraging exogenous knowledge to improve the prediction performance of the model and propose a knowledge-driven memory system that can broadly boost the representational power of GCN-based traffic forecasting models.
- We carefully customize three components of the memory system. The access interface based on the attention mechanism and the feedback interface are used to extract knowledge-based patterns from memory modules and integrate these patterns into the models to enhance the spatiotemporal representation. And three kinds of memory modules driven by exogenous knowledge are introduced to learn and store periodic patterns, coarse-grained patterns, and representative patterns respectively.
- The memory system is deployed to diverse traffic forecasting models to evaluate the validity, and experiments on two real-world datasets demonstrate that traffic forecasting models can widely benefit from the memory system.

2 Preliminaries

In this section, we first define some terms that will be used in the problem statement and then formulate the traffic forecasting problem.

Def.1 (Traffic Network). We use a graph $\mathcal{G} = (\mathcal{V}, \mathcal{E}, \mathcal{A})$ to denote a traffic network, where $\mathcal{V}$ is the node (e.g, traffic sensors) set with $|\mathcal{V}| = N$ nodes. $\mathcal{E}$ is a set of edges representing the connectivity among vertices, and $\mathcal{A} \in R^{N \times N}$ is the adjacency matrix of the graph.

Def.2 (Traffic Data). Traffic data is collected from devices (e.g, traffic sensors) deployed on roads. We denote the traffic condition at time step t as a graph signal $X_t \in R^{N \times C}$, where C is the number of traffic conditions of interest (e.g traffic speed, traffic flow, etc.).

Problem Formulation.1 (Traffic Forecasting). In this paper, traffic forecasting with the memory system can be formulated as: **Input**: a GCN-based traffic forecasting model Ψ, the memory system $\mathcal{M}$, and the observed data of L time steps of graph $\mathcal{G}$, $X = (X_{t-L+1}, X_2, ..., X_t) \in R^{L \times N \times C}$. **Output**: a forecasting function Ψ with the memory system which can effectively infer the traffic data next P time-steps $\Psi(X) = (X_{t+1}, X_2, ..., X_{t+P-1}) \in R^{P \times N \times C}$:

$$\Psi^*, \mathcal{M}^* = \arg \min_{\Psi, \mathcal{M}} \|\Psi(X) - Y\|^2, \tag{1}$$

where Ψ^* and $\mathcal{M}^*$ mean the optimized model function and memory system.

3   Method

In this section, we first introduce the learning process of the GCN-based traffic forecasting model, then show the interaction between the model and the memory system (as shown in Fig. 1). Finally, three core components of the memory system are elaborated.

3.1   GCN-Based Models for Traffic Forecasting

Recently, researchers move to study GCN-based traffic forecasting models due to the powerful capability of modeling graph structure data. They modify or extend GNNs to extract spatial features and combine GCNs with time series models (e.g RNN or Transformer) to learn spatiotemporal correlation. Finally, the generated spatiotemporal representation is used as input to a decoder (e.g fully connected layers or more complicated designs) to predict future traffic states. Given the spatiotemporal representation $\mathbf{X}$ as input, GCN performs convolution on the graph topology $\mathcal{G}$ and aggregates features from the neighborhood. The calculation process of general GCN can be represented as:

$$\bar{\mathbf{X}} = \sigma \left(\mathbf{W} \cdot \mathcal{F}_g \left(\{\mathbf{X}_v\} \cup \{\mathbf{X}_u, \forall u \in \mathcal{N}(v)\} \right) \right) \tag{2}$$

where $\mathbf{W}$ are learnable parameters. $\mathbf{X}_v$ specifies the representation of node v. $\mathcal{N}(v)$ means neighborhood of node v. $\mathcal{F}_g(\cdot)$ is a aggregate function (such as the mean function) and σ is the activation function. $\bar{\mathbf{X}} \in R^{N \times D_h}$ represents the generated spatiotemporal representation, where D_h is the number of channels.

3.2   The Model with Memory System

The memory system, which includes three components: access interface, memory module, and feedback interface, can be integrated into GCN-based traffic forecasting models to boost the representational power.

Specifically, the spatiotemporal learning model use generated spatiotemporal representation $\mathbf{X}$ as a query vector to retrieve knowledge-based patterns stored in the memory module, which are used to obtain enhanced spatiotemporal representation. The process of interactions between the model and the memory system is as follows:

$$\bar{\mathbf{X}} = \mathcal{F}_m(\mathbf{C}_\star \mathcal{M}_\star; \mathbf{X}) \tag{3}$$

where $\mathcal{M}_\star$ represents one kind of memory module, where $\star \in \{L, H, R\}$ means the long-term trend memory, the hierarchical effect memory, and the representative pattern memory. $\mathbf{C}_\star$ is an access matrix provided by the access interface and records the slot location information that the model should access according to the query vector X. Based on the access matrix, the model extracts stored features from the memory module $\mathcal{M}_\star$. The feedback interface function $\mathcal{F}_m(\cdot)$ is used to integrate the extracted information into the model to obtain enhanced spatiotemporal representation $\bar{\mathbf{X}}$.

Access Interface. The access interface is based on the attention mechanism and returns an access matrix $\mathbf{C}$ according to the query vector to guide the model to access the memory module, ensuring that the model can accurately extract useful information and prevent the disturbance of other irrelevant features.

Specifically, we first linearly map the query vector $X^{(l)}$ to a high-dimensional space and the result is denoted as $\mathbf{Q}_\star \in R^{N \times d_q}$. The patterns stored in the memory are treated as key vectors, then, the similarity between the query vector and key vectors can be computed by dot product operation:

$$C_\star(k) = \frac{\exp\left(\langle \mathbf{Q}_\star, \mathcal{M}_\star(k) \rangle\right)}{\sum_{k'=1}^{K} \exp\left(\langle \mathbf{Q}_\star, \mathcal{M}_\star(k') \rangle\right)} \tag{4}$$

where $\mathcal{M}_\star(k)$ means k-th slot of the memory module. To increase the capacity, we model the query vector as a multi-head array and can obtain a similarity matrix sequence $[C_1, ..., C_h] \in R^{N_h \times N \times K}$, where N_h is the number of heads. The access matrix $\mathbf{C}$ can be obtained by aggregating the similarity sequence with convolution operation across heads:

$$\mathbf{C}_\star = \text{softmax}\left(\Gamma_\phi\left(\|_{m=0}^{N_h} C_m\right)\right) \tag{5}$$

where Γ_ϕ means the convolution operator with 1×1 size kernel. With the access matrix, the model can extract patterns stored in the slots:

$$\widetilde{\mathbf{X}} = \mathbf{C}_\star \mathcal{M}_\star \tag{6}$$

Feedback Interface. The feedback interface function $\mathcal{F}_m(;)$ is used to integrate the extracted patterns $\widetilde{\mathbf{X}}$ from the memory module into the model to enhance spatiotemporal representation $\mathbf{X}$. In order to filter out redundant information and achieve effective integration, we use the gate mechanism as the feedback interface function $\mathcal{F}_m(;)$. Specifically, we first compute a filter gate $\mathbf{O}$:

$$\mathbf{O} = \sigma\left(\mathbf{W}_g\left[\widetilde{\mathbf{X}},\mathbf{X}\right] + \mathbf{b}_g\right) \tag{7}$$

where $\mathbf{W}_g$ and $\mathbf{b_g}$ are learnable parameters. Based on the filter gate $\mathbf{O}$, we integrate two representation vectors to obtain an enhanced spatiotemporal representation:

$$\mathcal{F}_m(\mathbf{X}_1;\mathbf{X}_2) = \mathbf{O} \odot \widetilde{\mathbf{X}} + \mathbf{X} \tag{8}$$

where $\odot$ is Hadamard product.

3.3   Knowledge-Driven Memory Module

The memory module $\mathcal{M}$ is used to store knowledge-based patterns and initialized as the parameterized matrix, which can be updated end-to-end with the model. To effectively learn these patterns, we formulate the loss function based on exogenous knowledge to constrain the learning of the memory. As mentioned before, we consider three kinds of exogenous knowledge and construct different memory modules. For example, for exogenous knowledge that traffic data is periodic, the long-term trend memory $\mathcal{M}_L \in R^{K_L \times D_m}$ is used to learn periodic patterns, where K_L is the number of slots in $\mathcal{M}_L$ and D_L means the number of pattern channels. Similarly, we construct a hierarchical effect memory $\mathcal{M}_H$ to model macro-regional patterns and a representative pattern memory $\mathcal{M}_R$ to capture representative patterns.

Inspired by three types of traffic exogenous knowledge, we reconstruct the input sequence from different perspectives and design special loss functions to align the mapping matrix and the access matrix provided $\mathbf{C}$ by the access interface, ensuring the memory to learn structural and addressable patterns.

1. Long-term Trend Memory. Traffic data is considered to be periodic [13, 14], thus, long-term trends play an important role in the traffic forecasting task. Driven by this knowledge, we propose long-term trend memory $\mathcal{M}_L$ to capture periodic patterns of the traffic network.

The periodic patterns of the nodes with close spatial functionality are consistent [10], thus, to improve the learning effect, we first daily traffic patterns of each node into K_L clusters, which can reflect the functional properties of the nodes. And the clustering matrix is denoted as $\mathbf{S}_L \in R^{N \times K_L}$, where $\mathbf{S}_L[i,g] = 1$ means that the daily traffic patterns of node v_i belongs to the g-th cluster. In order to constrain each slot of the memory $\mathcal{M}_L$ to store corresponding long-term trends of a cluster, we propose a clustering loss function which is calculated by the clustering matrix $\mathbf{S}_L$ and the access matrix $\mathbf{C}_L$ of memory $\mathcal{M}_L$:

$$\mathcal{L}_L = \sum_i \sum_j \mathbf{S}_L[i,j] \log\left(\mathbf{C}_L[i,j]\right) \tag{9}$$

2. Hierarchical Effect Memory. The transportation system is a hierarchical structure that includes not only basic micro-levels (e.g nodes or road networks) but also macro-levels (e.g hot spots) [15]. In a macro region, the correlation between nodes may be closer, thus, modeling the hierarchical effect and learning coarse-grained patterns of the macro region can provide a broader perspective for capturing the spatial correlation among nodes. Some researchers use external information (e.g POIs, land attributes, or population density) to analyze the macro-structure of road networks. However, the information may be not readily available due to privacy policies. We propose a hierarchical effect memory $\mathcal{M}_H$ which can adaptively learn the coarse-grained patterns of the road network without external information.

Specifically, rethinking the access matrix $\mathbf{C}_H \in R^{N \times K_H}$ of hierarchical effect memory $\mathcal{M}_H$, if we set K_H much smaller than the number of nodes N, $\mathbf{C}_H$ can be viewed as the mapping matrix of microscopic nodes to the macroscopic regions, and features stored in the memory $\mathcal{M}_H$ can reflect hierarchical information of the road network. To promote the model to learn friendly coarse-grained patterns in the latent space, we introduce an unsupervised graph clustering loss widely used in deep clustering methods [16–18]. Specifically, we first model auxiliary target distribution $\mathbf{V}$ as an auxiliary which can be computed as:

$$\mathbf{V}[i,j] = \frac{(\mathbf{C}_H[i,j])^2 / \sum_i \mathbf{C}_H[i,j]}{\sum_{j'} (\mathbf{C}_H[i,j'])^2 / \sum_i \mathbf{C}_H[i,j']} \tag{10}$$

The auxiliary target distribution $\mathbf{V}$ can improve the cluster purity by normalizing the contributions. The Kullback-Leibler (KL) divergence between $\mathbf{V}$ and the access matrix $\mathbf{C}_H$ is used as unsupervised loss:

$$\mathcal{L}_H = \mathrm{KL}\left(\mathbf{V} \| \mathbf{C}_H\right) = \sum_i \sum_j \mathbf{V}[i,j] \log \frac{\mathbf{V}[i,j]}{\mathbf{C}_H[i,j]} \tag{11}$$

3. Representative Pattern Memory. Recently, researchers discovered that the traffic patterns of the entire road network are extremely redundant [12], so a few representative patterns shared by all nodes can effectively prompt the spatiotemporal information of the entire road network. And these representative patterns can help the model better understand the spatiotemporal state of the road network. Thus, we propose a representative pattern memory $\mathcal{M}_R$ to store the high-dimensional representation of representative patterns of the road network.

Specifically, for the road network $\mathcal{G}$, we first calculate the daily average flow vector F of each node. If the sensor records data every five minutes (e.g PeMS system), the shape of F is equal to $(N \times 288)$. Then we split it to obtain the pattern set using a time window, where the length of the time window is equal to the time step of the input sequence L. The pattern set is denoted as $\mathbf{B} \in R^{N \times L_k}$, where $L_k = \lfloor \frac{288}{L} \rfloor$, and this set is proved redundant [12]. We show this fact with PeMS dataset (as shown in Fig. 1(c)), which shows the cosine similarity distribution between the various patterns in the pattern set **B**. We can see that pattern set is biased distribution. So we perform cluster-based downsampling and the center vector of each cluster is regarded as a representative pattern. Thus, we use $\mathbf{P} \in R^{N_P \times L}$ to denote the representative pattern set, where N_P means the number of representative patterns.

To retrieve the representative pattern that best matches the input traffic features, We take a pattern $x \in R^{1 \times L}$ of the input sequence $X \in R^{N \times L}$ as an example. First, we compute the cosine similarity between it and representative patterns. Then we select the **top-k** representative patterns with the highest cosine similarity into the candidate set, and the corresponding similarity matrix is denoted as $s_r \in R^{1 \times N_P}$. If a representative pattern is in the candidate set, the corresponding position of s_r is the cosine similarity of the two patterns. Otherwise, it is equal to 0. Similarly, for the entire input sequence X, we get the matching degree matrix $\mathbf{S}_r \in R^{N \times N_P}$.

Each slot in representative pattern memory $\mathcal{M}_R$ is used to store the high-dimensional representation of each representative pattern in **P**. This is achieved through a loss function which can force the model to align access matrix $\mathbf{C}_r$ and matching degree matrix $\mathbf{S}_r$, ensuring that the model only accesses the slots which store representative patterns matching with the input sequence. Specifically, the following loss function is computed:

$$\mathcal{L}_P = \|S_r - \mathbf{C}_r\|^2 \tag{12}$$

3.4   Loss Function with Memory System

For three kinds of exogenous knowledge, we design different loss functions respectively to constrain the model to store specific features. Thus, the total loss function for deploying the memory system to a traffic forecasting model can be defined as:

$$\mathcal{L} = \left\|\hat{Y} - Y\right\|^2 + \alpha\mathcal{L}_L + \beta\mathcal{L}_H + \mu\mathcal{L}_P \tag{13}$$

where the first part represents the loss between the predicted values and the ground-true values. α, β, and μ are hyperparameters to balance each loss.

4   Experiment

4.1   Experiment Settings and Traffic Forecasting Models

Dataset. We evaluate the effectiveness of the memory system on two widely used public traffic network datasets, PeMSD3 and PeMSD7. All datasets are collected from the Caltrans Performance Measurement System (PeMS) and aggregated into 5-minutes windows, thus, there are 288 data points per day. The peMSD3 dataset records traffic data from 358 sensors from September 1, 2018, to November 30, 2018. And the PeMSD7 dataset collects monitoring data from 883 sensors from July 1, 2016, to August 31, 2016.

Data Preprocess. Linear interpolation is utilized to fill in the missing values in each dataset. Min-max normalization is adopted to normalize the data into the range of $[-1; 1]$ to make the training process more stable. And two datasets are divided into training sets, validation sets, and testing sets according to the ratio of 6:2:2 in chronological order, i.e., the earliest 60% of samples are split into the training set, the subsequent 20% of samples are used for validation, and the last 20% of samples are used for testing. And we use one-hour historical data to predict the traffic data after one hour (ie. $L = P = 12$).

Experiment Settings. We optimize all the models with the Adamw optimizer. The initial learning rates in the PeMSD3 dataset and PeMSD7 dataset are set to 0.008 and 0.005 respectively. And the learning rate decays to 1% of the initial value if the loss on the validation set does not decrease for 15 epochs. The hyperparameters of the models are chosen through a carefully parameter-tuning process on the validation set. The number of heads in the access interface N_h is set to 4. And the number of slots in each memory module (i.e. K_L, K_H, and K_R) on the PeMSD3 dataset are equal to 8, 12, and 50 respectively, and they on the PeMSD7 dataset are set to 12, 16, and 64 respectively.

Metrics. Three metrics - Mean Absolute Error (MAE), Root Mean Square Error (RMSE), and Mean Absolute Percentage Error (MAPE) are used to evaluate the prediction performance of the models.

Traffic Forecasting Models. The memory system is deployed into existing advanced GCN-based traffic forecasting models.

- **STGCN** [9] uses graph convolution and temporal convolution for learning spatial and temporal dependencies, respectively.
- **DCRNN** [9] combines diffusion graph convolution and recurrent units to capturing spatiotemporal correlation.
- **ASTGCN** [19] is a traffic predicting model based on self-attention, which learns dynamic spatiotemporal correlation in a flexible manner.
- **GraphWaveNet** [6] proposes node embedding vectors to construct graph structures, and uses GCN and dilated casual convolution to predict traffic.
- **STGNN** [20] uses GCN to learn spatial correlation with GRU and Transformer to learn global and local temporal dependencies.
- **AGCRN** [21] proposes an adaptive graph learning method for GCN to capture the dynamic features of the traffic road network and combines it with RNN for traffic forecasting.
- **HGCN** [15] designs a hierarchical GCN to learn the hierarchical features of traffic networks.
- **STFGNN** [10] constructs temporal graphs based on DTW algorithm and distance-based spatial graphs to learn spatiotemporal correlation.

Table 1. The results of the models with the memory system on two datasets. The front and second parts of each metric are the original performance of the models and the performance of the models with the memory system respectively.

Model	PeMSD3								
	MAE			RMSE			MAPE		
STGCN	17.55	16.76	+ 4.50%	30.82	29.23	+ 5.16%	17.34	16.67	+ 3.86%
DCRNN	17.98	17.18	+ 4.45%	30.31	29.40	+ 3.00%	18.34	17.73	+ 3.32%
ASTGCN	17.34	16.87	+ 2.71%	29.56	28.78	+ 2.64%	17.21	16.90	+ 1.80%
GraphWaveNet	19.12	18.45	+ 3.50%	32.77	31.21	+ 4.76%	19.37	18.59	+ 4.03%
STGNN	17.24	17.31	− 0.41%	29.62	29.23	+ 1.32%	17.38	17.02	+ 2.07%
AGCRN	15.98	15.54	+ 2.75%	28.25	27.65	+ 2.11%	15.34	15.48	− 0.91%
HGCN	17.21	16.41	+ 4.65%	29.34	27.87	+ 5.01%	17.15	16.56	+ 3.44%
STFGNN	16.77	16.33	+ 2.62%	28.34	27.81	+ 1.87%	16.30	16.18	+ 0.74%
Model	PEMSD7								
	MAE			RMSE			MAPE		
STGCN	25.33	24.32	+3.99%	39.34	37.65	+4.30%	11.21	10.66	+4.91%
DCRNN	25.21	24.23	+3.89%	38.61	37.09	+3.94%	11.82	11.48	+2.88%
ASTGCN	24.21	23.44	+3.18%	37.87	36.21	+4.38%	10.73	10.33	+3.73%
GraphWaveNet	26.39	24.96	+5.42%	41.50	39.41	+5.04%	11.97	11.54	+3.59%
STGNN	24.23	24.19	+0.17%	38.22	37.61	+1.60%	12.01	11.98	+0.25%
AGCRN	22.37	21.87	+2.24%	36.55	35.98	+1.56%	9.12	9.14	− 0.22%
HGCN	26.61	25.11	+5.63%	40.03	38.59	+3.60%	11.57	10.87	+6.05%
STFGNN	23.46	22.91	+2.34%	36.62	35.80	+2.24%	9.21	9.01	+2.17%

4.2 Experiment Result Analysis

The experiment results on the two datasets are shown in Table 1. We observe that the memory module has a positive effect on the predictive performance of the models, because various exogenous knowledge provides insights from multiple perspectives on analyzing the evolutionary patterns of traffic data, and thus knowledge feature stored in the memory system can help models learn comprehensive spatiotemporal correlation and enhance prediction performance of the models. We find that STGCN only constructs the graph structure based on geographic coordinates to model spatial dependencies, and it fails to learn complex spatiotemporal correlation. And the memory system can guide STGCN to learn comprehensive spatiotemporal correlation with exogenous knowledge.

Although DCRNN and AGCRN integrate RNN to capture long-term temporal trends, hierarchical effect memory and representative pattern memory in the memory system can provide them with hierarchical features and representative pattern perspectives of the traffic network, and these features as supplements can boost spatiotemporal representation learning of model. HGCN uses multi-level graphs to extract the hierarchical features of the traffic network, so it achieves better prediction performance than the simple GCN-based model STGCN, and it still can benefit from the other two memory modules (i.e. long-term trend memory and representative pattern memory) in the memory system.

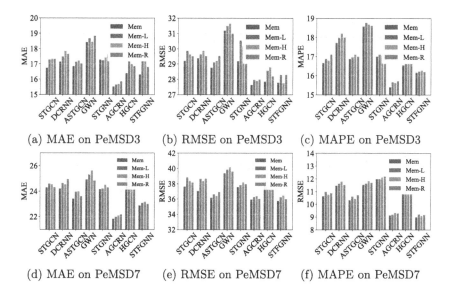

Fig. 2. Three kinds of memory modules validity analysis.

4.3   Ablation Experiment Analysis

In this section[1], we evaluate the effectiveness of three types of memory modules on two datasets (i.e. long-trend memory, hierarchical effect memory, and representative pattern memory). We remove each kind of memory module respectively, and the variants are denoted as Mem-L, Mem-H, and Mem-R. The experiment results are shown in Fig. 2, which show that each kind of memory module is beneficial to the improvement of prediction performance.

For capturing long-term trends, ASTGCN forms periodic sequences as input by sampling data points one week apart. DCRNN and AGCRN rely on the special components RGU to capture long-term trends of traffic data. However, we find that each variant Mem-L, which combines the memory system without long-term trend memory, achieves higher errors than each model with three kinds of memory modules Mem. Because the long-term trend memory module provides more accurate periodic insights of the entire road network by aggregating daily patterns of all nodes.

To model complex spatial correlation among nodes, GraphWaveNet, STGNN, AGCRN, and STFGNN design various methods to generate graph structures,

---

[1] We abbreviate **GraphWaveNet** as GWN.

but they only consider microscopic nodes as entities and fail to learn the macroscopic features of the traffic network, and the hierarchical effect memory can complement these features to these models. So the models without the hierarchical effect memory achieve higher errors. For HGCN which is a multi-level GCN-based model to capture the hierarchical structure, the hierarchical effect memory can still improve the prediction performance. It may be that HGCN performs spectral clustering on the adjacency matrix to get the multi-layer graphs, which are static and may not accurately describe the hierarchical structure of the road network. On the contrary, the memory module can adaptively perceive the road network structure.

The experiment shows representative pattern memory module is widely applicable because learned representative patterns can represent the traffic state of the road network and help models more accurately infer the future traffic.

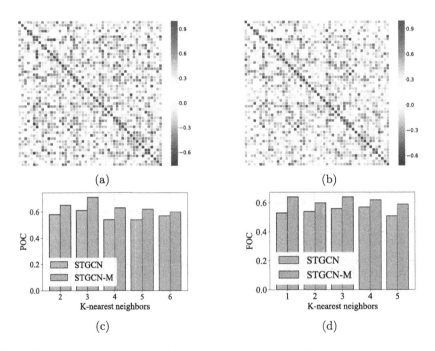

Fig. 3. Subgraph (a) shows the similarity heatmap of the spatiotemporal representation of 100 nodes by STGCN with the memory system. Subgraph (b) shows the similarity heatmap of spatiotemporal representation by only STGCN. Subgraph (c) and Subgraph (d) show the traffic similarity between each node and its k-nearest neighbors in the embedding space.

4.4   Case Study

It is crucial to accurately describe a high-dimensional spatiotemporal representation of nodes for traffic forecasting. A good spatiotemporal learning model should learn node representation which can reflect traffic pattern similarity [22]. In this section, we use STGCN as an example to investigate the effect of the memory system on the spatiotemporal representation learning of the models.

First, we show two heatmaps of node representation learned by STGCN with the memory system (as shown in Fig. 3(a)) and node representation learned by STGCN (as shown in Fig. 3(b)) on the test dataset. We find that the memory system can help the model learn a well-discriminated representation space. That is, the representation of nodes with similar traffic patterns is as close as possible, which benefits the decoder to analyze the representation for predicting traffic.

We further calculate the predicted traffic similarity between each node and its neighbors. Pearson correlation (POC) and the First-order temporal correlation (FOC) [22] are used as similarity functions. We observe that the node embedding of STGCN with the memory system (STGCN-M) shows significant improvement over embeddings of only STGCN. And it indicates that accessing the memory system can effectively learn better traffic-related representation.

4.5   Related Work

Traffic Forecasting. In recent years, with the development of deep learning, researchers are devoted to designing advanced deep learning models for traffic forecasting. For example, ST-ResNet [1] exploits convolutional neural networks to mine spatiotemporal correlation for predicting the inflow and outflow of each region. DMVST-Net [4] proposes a local CNN module to learn local spatial dependencies, while LSTM is integrated to learn temporal dependencies. ST-GSP [23] is a semantic encoder composed of ResNet to capture urban-scale spatial correlation and the influence of external factors.

However, CNNs cannot effectively process graph-structured data. Driven by recent advances in graph convolutional neural networks, GCNs are introduced to model spatial dependencies among nodes. For example, T-GCN [24] integrates GCN and GRU to learn spatiotemporal correlation for traffic forecasting. DCRNN [25] proposes a GCN-based layered coupling method for adaptively capturing multi-level spatial correlation of traffic networks. ST-GDN [26] uses diffusion graph convolution to learn local regional geographic dependencies and global spatial semantics. ST-ChebNet [27] uses Chebyshev graph neural network to learn complex topology in traffic networks.

Neural Networks with Memory. Researchers combine memory modules with neural networks for more powerful learning and reasoning capabilities to solve several challenging tasks such as one-shot learning [28,29] and question answering [30,31]. [31] designs a memory based network that designs an inference components with a readable and writable memory module to remember historical supporting information for question answering. [29] designs a memory module that can record network activations of rare events for one-shot learning.

4.6 Conclusion

In this paper, we investigate leveraging exogenous traffic knowledge to improve model prediction performance and propose a knowledge-driven memory system, which can be easily deployed to GCN-based traffic forecasting models to boost representational power. Three components of the system are carefully designed, and the access interface is based on the attention mechanism to provide precise access information for models. Three memory modules including long-term trend memory, hierarchical effect memory, and representative pattern memory are used to learn and store knowledge-based patterns according to different exogenous knowledge, and the models can enhance spatiotemporal representation by accessing these patterns. And the feedback interface based on the gate mechanism is used to integrate extracted information from the memory system into the model. To evaluate the effectiveness of the memory system, we apply the memory system to existing traffic forecasting models and conduct experiments on two datasets, which demonstrate the effectiveness of the memory system.

Acknowledgements. This paper is partially supported by the National Natural Science Foundation of China (No.12227901 and No.62072427), the Project of Stable Support for Youth Team in Basic Research Field, CAS (No.YSBR-005) Academic Leaders Cultivation Program, USTC.

References

1. Zhang, J., Zheng, Y., Qi, D.: Deep spatio-temporal residual networks for citywide crowd flows prediction. In: Proceedings of the AAAI (2017)
2. Liu, L., Qiu, Z., Li, G., Wang, Q., Ouyang, W., Lin, L.: Contextualized spatial-temporal network for taxi origin-destination demand prediction. IEEE Trans. Intell. Transp. Syst. **20**, 3875–3887 (2019)
3. Chu, K.-F., Lam, A.Y., Li, V.O.: Deep multi-scale convolutional LSTM network for travel demand and origin-destination predictions. IEEE Trans. Intell. Transp. Syst. **21**, 3219–3232 (2019)
4. Yao, H., et al.: Deep multi-view spatial-temporal network for taxi demand prediction. In: Proceedings of the AAAI (2018)
5. Li, Z., et al.: A hybrid deep learning approach with GCN and LSTM for traffic flow prediction. In: 2019 IEEE Intelligent Transportation Systems Conference (ITSC) (2019)
6. Wu, Z., Pan, S., Long, G., Jiang, J., Zhang, C.: Graph WaveNet for deep spatial-temporal graph modeling. arXiv preprint arXiv:1906.00121 (2019)
7. Zhang, X., et al.: Traffic flow forecasting with spatial-temporal graph diffusion network (2020)
8. Ali, A., Zhu, Y., Zakarya, M.: Exploiting dynamic spatio-temporal graph convolutional neural networks for citywide traffic flows prediction. Neural Netw. **145**, 233–247 (2022)
9. Yu, B., Yin, H., Zhu, Z.: Spatio-temporal graph convolutional networks: a deep learning framework for traffic forecasting. arXiv preprint arXiv:1709.04875 (2017)
10. Li, M., Zhu, Z.: Spatial-temporal fusion graph neural networks for traffic flow forecasting. In: Proceedings of the AAAI Conference on Artificial Intelligence, vol. 35(5), pp. 4189–4196 (2021)

11. Tang, Y., Qu, A., Chow, A.H., Lam, W.H., Wong, S.C., Ma, W.: Domain adversarial spatial-temporal network: a transferable framework for short-term traffic forecasting across cities. arXiv preprint arXiv:2202.03630 (2022)
12. Lee, H., Jin, S., Chu, H., Lim, H., Ko, S.: Learning to remember patterns: pattern matching memory networks for traffic forecasting. In: Proceedings of the ICLR (2021)
13. Zheng, M., Ruan, Z., Tang, M., Do, Y., Liu, Z.: Influence of periodic traffic congestion on epidemic spreading. Int. J. Mod. Phys. C **27**, 1650048 (2016)
14. Zonoozi, A., Kim, J.-J., Li, X.-L., Cong, G.: Periodic-CRN: a convolutional recurrent model for crowd density prediction with recurring periodic patterns. In: Proceedings of the IJCAI (2018)
15. Guo, K., Hu, Y., Sun, Y., Qian, S., Gao, J., Yin, B.: Hierarchical graph convolution networks for traffic forecasting. In: Proceedings of the AAAI (2021)
16. Xie, J., Girshick, R., Farhadi, A.: Unsupervised deep embedding for clustering analysis. In: Proceedings of the ICML (2016)
17. Razavi, A., Van den Oord, A., Vinyals, O.: Generating diverse high-fidelity images with VQ-VAE-2. In: Proceedings of NeurIPS (2019)
18. Aljalbout, E., Golkov, V., Siddiqui, Y., Strobel, M., Cremers, D.: Clustering with deep learning: taxonomy and new methods. arXiv preprint arXiv:1801.07648 (2018)
19. Guo, S., Lin, Y., Feng, N., Song, C., Wan, H.: Attention based spatial-temporal graph convolutional networks for traffic flow forecasting. In: Proceedings of the AAAI (2019)
20. Wang, X., et al.: Traffic flow prediction via spatial temporal graph neural network. In: Proceedings of the WWW (2020)
21. Bai, L., Yao, L., Li, C., Wang, X., Wang, C.: Adaptive graph convolutional recurrent network for traffic forecasting. In: Proceedings of the NeurIPS (2020)
22. Pan, Z., Liang, Y., Wang, W., Yu, Y., Zheng, Y., Zhang, J.: Urban traffic prediction from spatio-temporal data using deep meta learning. In: Proceedings of the KDD (2019)
23. Zhao, L., Gao, M., Wang, Z.: ST-GSP: spatial-temporal global semantic representation learning for urban flow prediction. In: Proceedings of the WSDM (2022)
24. Zhao, L., et al.: T-GCN: a temporal graph convolutional network for traffic prediction. IEEE Trans. Intell. Transp. Syst. **21**, 3848–3858 (2019)
25. Ye, J., Sun, L., Du, B., Fu, Y., Xiong, H.: Coupled layer-wise graph convolution for transportation demand prediction. In: Proceedings of the AAAI (2021)
26. Zhang, X., et al.: Traffic flow forecasting with spatial-temporal graph diffusion network. In: Proceedings of the AAAI (2021)
27. Yan, B., Wang, G., Yu, J., Jin, X., Zhang, H.: Spatial-temporal Chebyshev graph neural network for traffic flow prediction in IoT-based its. IEEE Int. Things J. **9**, 9266–9279 (2021)
28. Santoro, A., Bartunov, S., Botvinick, M., Wierstra, D., Lillicrap, T.: Meta-learning with memory-augmented neural networks. In: Proceedings of the ICML (2016)
29. Kaiser, Ł., Nachum, O., Roy, A., Bengio, S.: Learning to remember rare events. arXiv preprint arXiv:1703.03129 (2017)
30. Sukhbaatar, S., Weston, J., Fergus, R., et al.: End-to-end memory networks. In: Proceedings of the NeurIPS (2015)
31. Weston, J., Chopra, S., Bordes, A.: Memory networks. arXiv preprint arXiv:1410.3916 (2014)

Weakly-Supervised Multi-action Offline Reinforcement Learning for Intelligent Dosing of Epilepsy in Children

Zhuo Li[1]($\boxtimes$), Yifei Shen[2], Ruiqing Xu[3], Yu Yang[2], Jiannong Cao[2], Linchun Wu[4], and Qing Wu[5]

[1] School of Computer Science and Technology, Chongqing University of Posts and Telecommunications, Chongqing, China
lizhuo@cqupt.edu.cn
[2] Department of Computing, The Hong Kong Polytechnic University, Hong Kong, China
{yifei.shen,cs-yu.yang}@polyu.edu.hk, csjcao@comp.polyu.edu.hk
[3] Department of Computer Science, University College London, London, UK
[4] Donglian Information Technology Co. Ltd., Chengdu, China
wulinchun@dliotlink.com
[5] Department of Pharmacy, Children's Hospital of Chongqing Medical University, National Clinical Research Center for Child Health and Disorders, Chongqing, China

Abstract. Epilepsy in childhood is a common neurological disorder in children. Most cases are benign childhood epilepsy, which can be controlled with medication by adaptive adjustment of the dosage of antiepileptic drugs (AEDs). Recently, reinforcement learning-based intelligent dosing has attracted increasing attention. In clinical practice, patients usually take more than one drug at a time, however, conventional reinforcement learning algorithms do not sufficiently address the combination of two or more active drugs. In this paper, we propose the multi-action offline reinforcement learning (MA-ORL) model to solve this problem. Concretely, MA-ORL inherits the basic framework of the actor-critic network. For the patient's health and safety concerns, MA-ORL abandons the intrusive and high-risk "trial-and-error" interactions with the environment but directly learns from the offline clinical dataset in the initial phase until the model is sufficiently trained and ready to use. Besides, to choose multiple actions simultaneously, we replace the actor's output in standard reinforcement learning with a 2D matrix indicating the mixed feature representation of all different actions, then multiply it with separate masks to obtain separated actions. In addition to reaching an optimal return (i.e., the reduction of seizure frequency), MA-ORL also emphasizes the accuracy of the recommended dosage. Therefore, we introduce weak supervision to the learning objective to restrict the range of the learning outcome. It guarantees the recommended dosage by MA-ORL is as close as the one prescribed by experienced physicians. We conduct extensive experiments on clinical medical records containing 245 cases of epilepsy in childhood. Experimental results show MA-ORL reaches the highest cumulative return among all baseline models. Moreover, the suggested amount of medication taken a day by MA-ORL is more accurate than any other benchmark.

Keywords: Multi-Action Reinforcement Learning · Offline Reinforcement Learning · Weakly-Supervised Learning · Epilepsy in Children

© The Author(s), under exclusive license to Springer Nature Switzerland AG 2023
X. Wang et al. (Eds.): DASFAA 2023, LNCS 13946, pp. 208–223, 2023.
https://doi.org/10.1007/978-3-031-30678-5_16

1 Introduction

Epilepsy is one of the most common neurological disorders of the brain characterized by recurrent epileptic seizures [1,2]. Statistics from [3] show that around 50 million people of all ages suffer from it worldwide. The incidence of epilepsy in children ranges from 41 to 187 per 100,000 which is much higher than that in the adult group [4]. Childhood epilepsy has several typical syndromes whose root causes are complex and varied [5]. Although it requires long episodes of treatment and cannot be rooted out easily [6], fortunately, about 70% of the patients living with epilepsy, based on the estimation of WHO, could live seizure-free if diagnosed and treated properly [3].

At present, antiepileptic drugs are widely used in the treatment of epilepsy [7–9]. The accuracy of the prescribed daily dosage of some antiepileptic drugs should measure down to the milligram level. Noted that precise diagnosis and treatment rely heavily on the experience of medical experts that can only be gained through many years of hands-on practice. It does not facilitate passing down experiences from senior physicians to young doctors. However, the disadvantage of the apprenticeship-like training system for medical personnel is the advantage of machine learning techniques that can directly learn the experience of the medical experts from massive medical archives. This is called supervised learning in the realm of machine learning, which is widely applied in healthcare and medication and has achieved remarkable results [10–13].

Alongside supervised learning, reinforcement learning is another paradigm of machine learning, and also has many instantiations of applications in healthcare and medication, such as [14–17]. In this learning scheme, the algorithm learns by making decisions through "trial-and-error", where it is rewarded for correct decisions and penalized for wrong ones. On the other hand, the drug dosage recommendation for epilepsy in children is a typical sequence decision-making problem. This is because the condition of children with epilepsy is followed carefully at the hospital at 1–2 months intervals during the treatment, and physicians regularly adjust the dosage of the antiepileptic drugs based on the examination and laboratory test results of the follow-ups. Naturally, leveraging reinforcement learning technique to recommend the drug dosage for chronic diseases has been widely studied. On-policy is the most common form of reinforcement learning. The agent plays an online game that interacts with the environment frequently to collect training samples [18]. It achieves stunning results in chess games, computer games, and other simulation-based applications, but fails in most real-life scenarios because traditional on-policy strategy is intrusive when interacting with the environment. However, in healthcare and medication, the target of action is the human body, hence traditional "trial-and-error" approach to collecting samples may damage both the agent and the environment. In addition, in many practices of reinforcement learning, an agent needs to control multiple actions concurrently. For example, for a simple exploration robot, you could tell it to ride forward one meter while turning its camera 90°C clockwise. In particular, during the treatment of epilepsy, patients will take a combination of antiepileptic drugs, which means the reinforcement learning agent needs to suggest the dosage of multiple drugs. Thus, it is a challenge for the standard version of reinforcement learning algorithms to control multiple actions concurrently.

To address the aforementioned limitations, in this paper, we propose a multi-action offline deep reinforcement learning (MA-ORL) model for suggesting the daily dosage of multiple antiepileptic drugs. Other than using the intrusive "trial-and-error" way to interact with the environment, MA-ORL adopts a pure offline mode that trains the policy from offline medical records, which is more efficient and safe during the learning process. The reinforcement learning agent aims to learn the way senior doctors prescribe drugs to achieve maximum cumulative reward, that is, the seizure frequency reduces or remains stable for children with epilepsy after taking a course of treatment. In terms of the model architecture, MA-ORL inherits the framework of the actor-critic-based batch-constrained deep Q-learning (BCQ) model [19], but is distinguished from it. To be more specific, we make meticulous modifications to the structure of the actor of BCQ by replacing the output scalar of the actor with a 2D matrix indicating the output mixed feature representation of multiple actions. To separate each action from the representation mixture, MA-ORL introduces a group of 0–1 masks with 1 indicating the elements of interest and 0 otherwise. Each mask multiplies with the representation mixture to get the latent feature of each action. Seeing that conventional reinforcement learning prefers choosing actions that merely lead to higher rewards regardless of the accuracy and rationality of the picked action, MA-ORL adds weak supervision to the actor's loss to restrict the learning outcome. It plays a role to guide the learning process of intelligent dosing such that the recommended dosage of multiple drugs is as close as the prescribed dosage by doctors.

To assess the performance of MA-ORL, we conduct extensive experiments on a clinical dataset consisting of 245 cases of children with epilepsy from the Children's Hospital of Chongqing Medical University, the National Clinical Research Center for Child Health and Disorders, China, which are in close cooperation with our research group, from January 1, 2016, to April 29, 2021. We select two common antiepileptic drugs, named Levetiracetam and Oxcarbazepine, for the multi-drug recommendation experiment. Experimental results show the MA-ORL model obtains the best results in terms of the discounted cumulative return and multi-drug dosage recommendation accuracy among all baseline models.

2   Related Work

In a multi-action reinforcement learning scheme, an agent aims to control multiple actions at the same time. Some researchers turn it into a single-action problem. Specifically, instead of seeing multiple actions performed simultaneously, conceptually, it is easier to think of this as conventional single-action reinforcement learning problem, whose action space consists of all subsets of different actions, but the size of the combined action set is much larger than the original multi-action set. Some researchers treat the multi-action paradigm as a multi-agent problem that studies how multiple agents interact in a shared stochastic environment, each of which tackles an action independently [20]. The key to multi-agent reinforcement learning is investigating how different agents cooperate and compete with each other with the aim to maximize each agent's long-term reward. According to different collaborative strategies, multi-agent reinforcement learning can be divided into: (1) experience replay-based approach [21], (2) value decomposition-based approach [22,23], and (3) centralized critic-based approach [24]. Multi-agent reinforcement learning has found incredible success in popular strategy games [25,26], and has also been widely applied to healthcare and medication [16]. Besides, other people formulate it as a multi-task learning problem that uses a single shared machine learning model to handle multiple different tasks in parallel [27]. In multi-task learning, a shared feature representation layer is employed to learn latent features shared between different tasks [28]. Parameter regularization is used to discriminate the distance between the parameters of the individual models to the overall training objective [29].

3   Preliminaries

Considering a Markov Decision Process (MDP) in standard reinforcement learning setting, it can be described by a 5-element tuple $< S, A, T, R, \gamma >$, where S and A are finite sets of states and actions, $T : S \times A \times S \to [0, 1]$ is the transition function, T is the transition function (i.e., a set of conditional transition probabilities among states), $R : S \times A \times S \to R$ is the reward function, and $\gamma \in [0, 1]$ is the discount factor [30,31]. Reinforcement learning aims to learn the optimal policy π^* to maximize the expected return in the future:

$$R_t = \sum_{t=t'}^{T} \gamma^{t-t'} r_t. \tag{1}$$

It is based on the value functions to evaluate how good the agent performs an action under certain circumstances, including the state-action value function $Q^\pi(s, a)$:

$$Q^\pi(s, a) = \mathbb{E}\big[R_t | S_t = s, A_t = a, \pi\big]. \tag{2}$$

Then, the optimal $Q^*(s, a)$ function can be obtained by:

$$Q^*(s, a) = \max_\pi Q^\pi(s, a). \tag{3}$$

By taking the argmax$(\cdot)$ of the optimal expected return of $Q^*(s, a)$, we can obtain the optimal policy $\pi^*(s)$:

$$\pi^*(s) = \arg\max_{a \in A} Q^*(s, a). \tag{4}$$

4   Multi-action Offline Reinforcement Learning

4.1   Problem Formulation

A multi-action reinforcement learning problem can still be defined as an MDP transition $< S, A, T, R, \gamma >$. Distinguished from the single-action setting, whose output of the reinforcement learning actor A_{SA} is a scalar with its value indicating an action (e.g., 0 represents turning left and 1 turning right), in the multi-action setting, $A_{MA} = \big[A_1, ..., A_n, ..., A_N\big]$ $(n = 1, 2, \dots, N)$ is a set of N concurrent actions.

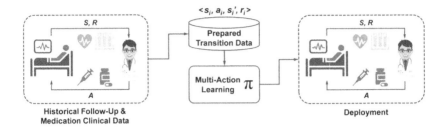

Fig. 1. Demonstration of the offline reinforcement learning.

A_n ($A_n = \{a_{n,1}, a_{n,2}, \ldots, a_{n,K_n}\}$) is the n-th action, where $a_{n,k}$ ($k = 1, 2, \ldots, K_n$) denotes the k-th possible value of A_n and K_n is the dimension of A_n, $K_n = Dim(A_n)$ (i.e., the number of all possible values of A_n).

Our goal is to learn an optimal policy $\pi^*(s)$ based on the observations of the current state s to generate concurrent multiple actions $A_1^*, A_2^*, \ldots, A_N^*$ shown in Eq. (5). For example, in the application of intelligent dosing, the agent needs to suggest multiple drugs' daily dosages as accurately as the prescription given by experienced physicians.

$$[A_1^*, A_2^*, \ldots, A_N^*] = \pi^*(s). \tag{5}$$

4.2 Offline Reinforcement Learning Framework

In this section, we will introduce the multi-action offline reinforcement learning (MA-ORL) model in detail, and Fig. 1 gives a vivid illustration of it.

MA-ORL inherits the basic framework of the actor-critic network that meticulously combines both the value-based and the policy-based reinforcement learning strategies. MA-ORL abandons the traditional on-policy strategy and adopts the offline learning mode, which is trained on a pre-collected dataset. It is because the target of the study of intelligent dosing is the human body such that the on-policy strategy that uses the intrusive "trial-and-error" approach to collecting training samples has high risks of endangering patients' lives. Training the MA-ORL model on a fixed offline dataset can avoid inefficient exploration in the early stage of training, and gain experience in diagnosing, treating disease, and prescribing drugs from medical experts quickly. The model can be deployed online for further fine-tuning after it is well trained.

Though offline reinforcement learning is safer and more efficient, as a tradeoff, it also leads to extrapolation error which results from the mismatch between the dataset and the true state-action pair of the current policy [18, 19]. To address this issue, MA-ORL draws on the successful experience of BCQ [19]. Concretely, we introduce conditional variational auto-encoder (VAE) $\mathcal{G}_{VAE}$ [32, 33], a representative generative model. Given a state s, $\mathcal{G}_{VAE}$ generates M possible candidates $A^{\mathcal{G}}$ with similar data distribution to the fixed offline dataset, i.e., $A^{\mathcal{G}} = \{a_1^{\mathcal{G}}, \ldots, a_m^{\mathcal{G}}, \ldots, a_M^{\mathcal{G}} | a_m^{\mathcal{G}} \sim \mathcal{G}_{VAE}(s)\}$, then we pick the action having the highest Q value. To increase the diversity of the generated results, a perturbation network $\xi_\phi(s, a_m^{\mathcal{G}} | \Phi)$ is introduced to add some noises to the selected action a_i, then clipped by $[-\Phi, \Phi]$ (Φ is a small constant). This secures that the generated actions can fall into a constrained region without having to sample from $\mathcal{G}_{VAE}$ excessive times to produce qualified actions. Equation (6) gives the expression below:

$$\pi^*(s) = \underset{a_m^{\mathcal{G}} + \xi_\phi(s, a_m^{\mathcal{G}}, \Phi)}{\arg\max} \quad Q_\theta^*\big(s, a_i^{\mathcal{G}} + \xi_\phi(s, a_m^{\mathcal{G}}, \Phi)\big), \tag{6}$$

where θ is the parameter of the Q network.

Illustration of Masked Multi-Action Separation

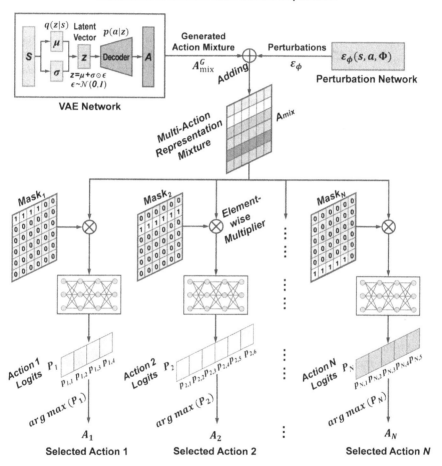

Fig. 2. An illustration of using masking for the multi-action separation.

4.3   Masking for Multi-action Separation

Figure 2 illustrates how masking is used for separating different actions from the representation mixture. Specifically, we make substantial modifications to the actor's structure. In the standard version of reinforcement learning, the actor is a fully-connected neural network in nature whose last layer outputs a scalar. In MA-ORL, we replace the last layer of the neural network with a $NL \times 1$ 1D vector, then reshape it into an $N \times L$ 2D matrix-like mixed feature representation $A_{mix}^{\mathcal{G}}$, where N indicates how many actions we want to maneuver simultaneously and L denotes the maximum value of the dimension of N actions (the maximum number of logits), i.e.,

$$
\begin{aligned}
L &= \max\big(Dim(A_1), \ldots, Dim(A_n), \ldots, Dim(A_N)\big) \\
&= \max\big(K_1, \ldots, K_n, \ldots, K_N\big),
\end{aligned}
\tag{7}
$$

where K_n ($n = 1, \ldots, N$) is the dimension of the n-th action.

Then perturbation ξ_ϕ is added to $A_{mix}^{\mathcal{G}}$ to get the perturbated representation mixture A_{mix}, i.e., $A_{mix} = A_{mix}^{\mathcal{G}} + \xi_\phi$. To separate the computed logits of each action from the representation mixture, we introduce a group of masks, $Mask_1, Mask_2, \ldots, Mask_N$, and multiply each of them with A_{mix} to obtain the separated logits. Each mask has the same size as A_{mix} that consists of 0 and 1 with 1 indicating the logits of interest of the an action that is to be separated from the mixture and 0 otherwise. Take the n-th mask $Mask_n$ as an example, we multiply it with the mixed feature representation A_{mix} and obtain the logits of the n-th action P_n ($P_n = \{p_{n,1}, \ldots, p_{n,k}, \ldots, p_{n,K_n}\}$), where $p_{n,k}$ represents the calculated probability of the k-th possible value (i.e., $a_{n,k}$) of A_n:

$$
P_n = A_{mix} \otimes Mask_n \ (n = \{1, 2, \ldots, N\}),
\tag{8}
$$

where $\otimes$ is the element-wise multiplier.

Then, the best next move of A_n^* can be selected by picking the value with the highest probability, i.e., take the $\arg\max(\cdot)$ of P_n:

$$
A_n^* = \arg\max(P_n).
\tag{9}
$$

For example, assume the maximum value of P_n is $p_{n,k}$, i.e., $\max(P_n) = p_{n,k}$, that corresponds to the k-th possible value of A_n, i.e., $a_{n,k}$, then $a_{n,k}$ will be selected as the optimal n-th action A_n^*, i.e., $A_n^* = a_{n,k}$.

Likewise, masking is used to separate the logits of other actions and obtain their best next move subsequently.

4.4 Weakly-Supervised Learning in Training Objective

MA-ORL follows the actor-critic framework that the actor network chooses actions on the basis of the state information and the critic network evaluates how good is the action in terms of the Q value, and the actor and the critic update its parameters alternately.

Similar to the DQN model [36], the critic loss $\mathcal{L}_{critic}$ minimizes the mean square error (MSE) between the target Q value y_m and the estimated Q value $Q_\theta(s_m, a_m)$:

$$
\min \mathcal{L}_{critic} = \min\Big(\frac{1}{M}\sum_{m=1}^{M}\big(y_m - Q_\theta(s_m, a_m)\big)^2\Big),
\tag{10}
$$

where $y_m = r + \gamma Q'_{\theta'}(s'_m, a'_m)$ with s'_m the next state and a'_m the next action of the m-th sample ($m = 1, 2, \ldots, M$, and M is the total number of training samples). r is the reward and γ is the discount factor.

For the actor, it is a policy network in nature, so the gradient ascend method is used to optimize the actor loss function $\mathcal{L}_{actor}$. We will not discuss the detailed derivation process of the policy gradient here.

$$\max \mathcal{L}_{actor} = \max\left(\frac{1}{M} \sum_{m=1}^{M} Q_\theta\big(s_m, a_m^{\mathcal{G}} + \xi_\phi(s_m, a_m^{\mathcal{G}}|\Phi))\big)\right), \tag{11}$$

where $a_m^{\mathcal{G}} \sim \mathcal{G}_{\text{VAE}}(s_m)$ is the action generated by the VAE network $\mathcal{G}_{\text{VAE}}$, and ξ_ϕ is the perturbation network.

In training MA-ORL, instead of using the hard-update method, a soft-update approach is used to update the parameters of the target network:

$$\begin{aligned} \theta' &\leftarrow \tau\theta + (1 - \tau)\theta', \\ \phi' &\leftarrow \tau\phi + (1 - \tau)\phi', \end{aligned} \tag{12}$$

where $\tau \in [0, 1]$ is an update coefficient, usually a small number.

Moreover, the goal of reinforcement learning is to learn an overall optimal policy. This principle guides the sequential decision-making on action selection. It stresses the ultimate goal for the whole episode while ignoring the accuracy and rationality of the selected move every round interacting with the environment. In particular, intelligent dosing emphasizes recommendation accuracy, that is, the recommended drug dosages should satisfy medical common sense.

To solve this problem, we take advantage of the successful experience of imitation learning [34,35]. Different from reinforcement learning which starts from scratch to learn a policy, imitation learning directly learns the trajectory of the sequential decisions made by experienced physicians and imitates them to decide the amount of medicine. Specifically, we introduce weak supervision to the learning objective to guarantee the difference between the dosage suggested by the algorithm and prescribed by the doctors is as small as possible. In clinical practice, drug dosage taken one time is a discrete value, like 25mg, 50mg, 100mg, etc., so it becomes a multi-class classification problem in supervised learning. We introduce the cross-entropy loss $\mathcal{L}_{CE}$ to the actor's objective function $\mathcal{L}_{actor}$ in Eq. (11) so as to restrict the selection of actions:

$$\min \mathcal{L}'_{actor} = \min\left(-\mathcal{L}_{actor} + \sum_{n=1}^{N} \alpha_n \mathcal{L}_{CE,n}\right), \tag{13}$$

where $\mathcal{L}_{CE,n}$ represents the cross-entropy loss of the n-th action and α_n is the corresponding weighting factor. Noted that maximizing $\mathcal{L}_{actor}$ in Eq. (11) is equivalent to minimizing $-\mathcal{L}_{actor}$ in Eq. (13).

Cross-entropy compares the similarity between two distributions and is widely used as the objective function in supervised multi-class classification problems. The cross-entropy loss of the n-th concurrent action $\mathcal{L}_{CE,n}$ is computed as in Eq. (14):

$$\mathcal{L}_{CE,n} = CrossEntropy\left(P_n, \overline{A_n}\right)$$

$$= -\frac{1}{M} \sum_{m=1}^{M} \sum_{k=1}^{K_n} \bar{a}_{n,k,m} \log(p_{n,k,m}), \tag{14}$$

where $P_n = \{p_{n,1}, \ldots, p_{n,k}, \ldots, p_{n,K_n}\}$ ($p_{n,k} = p_{n,k,1}, \ldots, p_{n,k,m}, \ldots, p_{n,k,M}$) is the calculated action logits of the n-th action A_n, and $\overline{A_n} = \{\bar{a}_{n,1}, \ldots, \bar{a}_{n,k}, \ldots, \bar{a}_{n,K_n}\}$ ($\bar{a}_{n,k} = \{\bar{a}_{n,k,1}, \ldots, \bar{a}_{n,k,m}, \ldots, \bar{a}_{n,k,M}\}$) is the ground truth labels in one-hot encoding.

4.5   MA-ORL in Intelligent Dosing of Epilepsy

In the application of intelligent dosing of epilepsy in children, we want to study combination therapy, particularly the daily dosage recommendation for multiple drugs, to control and reduce seizures. In intelligent dosing setting, the reinforcement learning environment refers to the patient, and the action is the recommended daily dosage of antiepileptic drugs. After patients take drugs, we can observe the patient's condition, say, the patient's subjective feelings, the objective laboratory test results and the doctor's diagnosis, etc. The feedback after receiving the treatment is regarded as the reward in reinforcement learning. The reward is set to 100 if the child has no seizure, -100 if the child has a seizure, and 0 otherwise. Our collected dataset records whether the patients have had seizures in the past three months r_1, six months r_2, and since the last follow-up r_3. In practice, we take the weighted sum of r_1, r_2, and r_3 to compute the reward r in each transition:

$$r = \alpha_1 \times r_1 + \alpha_2 \times r_2 + \alpha_3 \times r_3, \tag{15}$$

where α_1, α_2 and α_3 are the corresponding weighting coefficients, and empirically we set α_1, α_2 and α_3 to 50%, 30% and 20%, respectively.

5   Experiment

In this section, we will introduce the experimental part. First of all, we have a brief description of the dataset as well as the evaluation metric, then introduce the baseline models, and finally present the experimental results in detail.

5.1   Dataset

We collect the clinical data comprising 245 cases of children with epilepsy from the Children's Hospital of Chongqing Medical University, the National Clinical Research Center for Child Health and Disorders, China, that has close cooperation with us, from January 1, 2016, to April 29, 2021. The dataset records the patient follow-up information of benign childhood epilepsy with centrotemporal spikes, including the patient complaints, physical examination, past history, laboratory test results, doctor's diagnosis, treatment plan, medication, and feedback on treatment. Specifically, the archive stores information on the use of medicine, e.g., drug names, dosage, frequency, duration, etc. Levetiracetam (LEV) and Oxcarbazepine (OXC) are two frequently used medicines to treat epilepsy and alleviate symptoms. In the experiment, we want to precisely control the dosage of LEV and OXC taken a day to reduce the seizure frequency and keep patients in good condition.

5.2 Evaluation Metrics

To evaluate the performance of MA-ORL in terms of the accuracy of the recommended dosage taken per day, we use the following evaluation metric:

$$\begin{cases} Accuracy = (TP + TN)/(TP + TN + FP + FN), \\ ErrorRate = 1 - Accuracy, \\ Precision = TP/(TP + FP), \\ Recall = TP/(TP + FN), \\ F1 = (2 * Precision * Recall)/(Precision + Recall), \end{cases} \qquad (16)$$

where TP, FP, TN, FN represent true positive, false positive, true negative and false negative samples, respectively.

5.3 Baseline Models

In the experiment, we use the following baseline models:

- **MA-DQN:** The multi-action deep Q learning model, which introduces the multi-action selection module to the classical DQN model [36].
- **MA-DDPG:** The multi-action deep deterministic policy gradient model, which also introduces the multi-action selection module to the classical DDPG model [37].

5.4 Results

In this section, we first compare the performance among different models in drug dosage recommendations, then demonstrate the trend of the expected cumulative return of all models, show the advantages of weakly-supervised learning in intelligent dosing, and finally depict the curve of training loss.

Performance Comparison Across Different Models: Table 1 shows the performance comparison results between MA-ORL and baseline models. MA-ORL achieves the highest accuracy (87.42% and 94.87%) and the lowest error rate (12.58% and 5.13%) for the suggested dosage of LEV and OXC, respectively, among all different models. The precision by MA-DQN is the best, 63.13% and 77.07%, for LEV and OXC, respectively, while MA-ORL ranks last, only 44.31% and 42.42%, and MA-DDPG lies in between them. A reverse pattern can be observed for recall that MA-ORL performs the best which is followed by MA-DDPG and MA-DQN. It is because precision and recall are a pair of contradictory evaluation metrics, in other words, a higher value of precision usually means a smaller value of recall. If we inspect the results case by case, we can find that there are a large number of False Positives (FP) samples and a small number of False Negatives (FN) samples, which means the classification threshold of MA-ORL is set to a low value. To have a fair evaluation, we introduce the F1 score, the harmonic mean of precision and recall, which is a more balanced metric to reflect the classification results. Comparison results show MA-ORL reaches about 0.51 and 0.48 in F1 score for LEV and OXC, surpassing MA-DQN and MA-DDPG by 0.07 and 0.05, and 0.16 and 0.13, respectively. Therefore, the overall performance ranking for all models is MA-ORL > MA-DDPG > MA-DQN.

Performance by Converting Multi-action RL to Conventional Single-Action RL:
Conventionally, people solve the multi-action reinforcement learning problem by transferring it to a single-action problem. Instead of seeing recommending LEV and OXC
as two independent actions, the action space of the transformed single-action scheme
consists of all subsets of two actions performed simultaneously. Table 2 presents the
recommended results using single-action reinforcement learning. We can see that in our
experiment, the action space is the combination of all possible daily dosages of LEV
and OXC, which is much larger than the original action set. For example, if doctors prescribe neither of the two drugs, we can use the combination (LEV: 0 mg, OXC: 0 mg)
to represent the selected single action. If the prescribed dosages of LEV and OXC are
50 mg and 100 mg respectively, we use (LEV: 50 mg, OXC: 100 mg) to denote the chosen single action, and so on. As shown in Table 2, the average precision and recall for
different combinations of LEV and OXC are 23.98% and 12.11%, respectively, which
are much lower than the results achieved by MA-ORL. Noted that the precision and
recall of most drug dosage combinations are low (many of them even are 0), which
means there are few True Positive (TP) samples, i.e., a small number of recommendations are predicted correctly. In terms of a more balanced metric, the F1-score, the
single-action reinforcement learning achieves 0.1287 which falls far behind MA-ORL.
From the obtained results above, we can conclude that leveraging MA-ORL to suggest
drug dosages is more accurate than converting it back to the single-action one.

Table 1. Performance comparison between MA-ORL and baseline models.

Drugs	Models	Accuracy	Error Rate	Precision	Recall	F1 Score
LEV	**MA-DQN**	70.19%	29.81%	63.13%	32.78%	0.4315
	MA-DDPG	77.39%	22.61%	52.58%	39.78%	0.4529
	MA-ORL	87.42%	12.58%	44.31%	60.00%	0.5097
OXC	**MA-DQN**	87.53%	12.47%	77.07%	20.26%	0.3209
	MA-DDPG	89.74%	10.26%	75.20%	23.47%	0.3578
	MA-ORL	94.87%	5.13%	42.42%	55.90%	0.4824

The Impact of Weak Supervision on the Dosage Recommendation Results: The
goal of conventional reinforcement learning is to achieve the optimal return, i.e., the
max cumulative Q value, regardless of the rationality of the selected action. For example, in intelligent dosing, reinforcement learning is prone to suggest higher doses to
patients, which seems to take effect quickly, but may exceed the acceptable range and
do harm to the human body. To help the agent to imitate the behavior of experienced
physicians, we introduce weak supervision to the learning objective of the actor to minimize the difference between the suggested and the prescribed daily dosages.

To show the advantage of weakly-supervised learning, we compare the performance
of the complete MA-ORL and its reduced case by removing the weakly-supervised
($\mathcal{WS}$) learning module in Table 3. MA-ORL without using weak supervison (w/o $\mathcal{WS}$)

just passes 60% in accuracy for both LEV and OXC, which is approximately 23% and 32% lower than the complete model. MA-ORL outperforms the reduced case a lot in both precision and recall. In terms of the F1 score, MA-ORL w/o $\mathcal{WS}$ achieves 0.1 and 0.04 only for suggesting the dosage of LEV and OXC, respectively, which fall far behind the complete model.

Table 2. Multi-drug dosage recommendation results after converting using the multi-action RL to the single-action RL.

No.	Combined Single Action		Accuracy	Error Rate	Precision	Recall	F1 Score
1	**LEV: 0 mg,**	**OXC: 0 mg**	85.20%	14.80%	45.65%	32.81%	0.3818
2	**LEV: 0 mg,**	**OXC: 250 mg**	96.59%	3.41%	56.76%	40.38%	0.4719
3	**LEV: 0 mg,**	**OXC: 500 mg**	84.98%	15.02%	37.84%	23.33%	0.2887
4	**LEV: 0 mg,**	**OXC: 750 mg**	84.25%	15.75%	60.00%	8.07%	0.1423
5	**LEV: 0 mg,**	**OXC: ≥1000 mg**	45.28%	54.72%	30.57%	87.92%	0.4536
6	**LEV: 250 mg,**	**OXC: 0 mg**	95.94%	4.06%	44.44%	7.27%	0.1250
7	**LEV: 250 mg,**	**OXC: 500 mg**	99.93%	0.07%	100.00%	50.00%	0.6667
8	**LEV: 250 mg,**	**OXC: 750 mg**	99.71%	0.29%	0.00%	0.00%	0.0000
9	**LEV: 250 mg,**	**OXC: ≥1000 mg**	99.35%	0.65%	0.00%	0.00%	0.0000
10	**LEV: 500 mg,**	**OXC: 0 mg**	94.70%	5.30%	62.50%	6.67%	0.1205
11	**LEV: 500 mg,**	**OXC: 250 mg**	99.85%	0.15%	0.00%	0.00%	0.0000
12	**LEV: 500 mg,**	**OXC: 500 mg**	99.78%	0.22%	0.00%	0.00%	0.0000
13	**LEV: 500 mg,**	**OXC: 750 mg**	99.71%	0.29%	0.00%	0.00%	0.0000
14	**LEV: 500 mg,**	**OXC: ≥1000 mg**	99.85%	0.15%	0.00%	0.00%	0.0000
15	**LEV: 750 mg,**	**OXC: 0 mg**	91.65%	8.35%	50.00%	4.35%	0.0800
16	**LEV: 750 mg,**	**OXC: 500 mg**	99.78%	0.22%	0.00%	0.00%	0.0000
17	**LEV: 750 mg,**	**OXC: 750 mg**	98.91%	1.09%	0.00%	0.00%	0.0000
18	**LEV: 750 mg,**	**OXC: ≥1000 mg**	99.06%	0.94%	0.00%	0.00%	0.0000
19	**LEV: ≥1000 mg,**	**OXC: 0 mg**	94.85%	5.15%	40.00%	5.80%	0.1013
20	**LEV: ≥1000 mg,**	**OXC: 500 mg**	99.93%	0.07%	0.00%	0.00%	0.0000
21	**LEV: ≥1000 mg,**	**OXC: 750 mg**	99.85%	0.15%	0.00%	0.00%	0.0000
22	**LEV: ≥1000 mg,**	**OXC: ≥1000 mg**	99.93%	0.07%	0.00%	0.00%	0.0000

Table 3. Performance comparison between MA-ORL and its reduced case without using weakly-supervised learning.

Drugs	Models	Accuracy	Error Rate	Precision	Recall	F1 Score
LEV	**MA-ORL**	87.42%	12.58%	44.31%	60.00%	0.5097
	MA-ORL w/o $\mathcal{WS}$	64.40%	35.60%	11.54%	9.07%	0.1015
OXC	**MA-ORL**	94.87%	5.13%	42.42%	55.90%	0.4824
	MA-ORL w/o $\mathcal{WS}$	62.35%	37.64%	2.68%	19.28%	0.0471

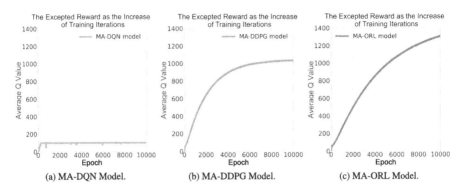

Fig. 3. Comparison of the expected discounted cumulative reward (average Q value) across different models.

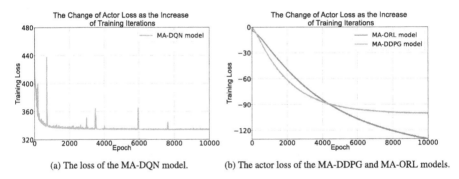

Fig. 4. The change of the actor loss as the increase of training epochs.

The Change of the Expected Return Among Different Models: Figure 3 demonstrates the change of the expected discounted cumulative return, i.e., the average Q value, as the increase of the training iterations of all models. In Fig. 3 (a), it depicts the curve of the discounted cumulative reward of the MA-DQN model, which starts from 0, then goes up quickly in the first 200 rounds, and levels off at around 100 rounds. Figure 3 (b) and (c) plot the expected returns of MA-DDPG and MA-ORL, respectively. Both curves grow fast in the beginning, but the orange curve (MA-DDPG) ascends more steeply in the first 4000 rounds of training, then the slope decreases and converges to around 1000 in the end. MA-ORL (blue curve) climbs up not as fast as MA-DDPG at first, but it keeps the momentum, exceeds MA-DDPG near the 5000-th iteration, and finally stops at around 1300 in average Q value.

The Change of the Training Loss of MA-ORL: Figure 4 illustrates the change of the actor loss as the increase of training iterations of different models. As we know that DQN is a value-based reinforcement learning algorithm, and it adopts a delayed update strategy, that is, the current network is updated every iteration, but the target network is updated every, say, 100 rounds of training. MA-DQN inherits the essence of DQN, whose loss function compares the difference between the current and target networks. As Fig. 4 (a) shows, the loss of MA-DQN, denoted by the pink curve, drops from more than 480, and declines very fast in the first 1000 rounds of training, then flattens out at around 340. Some intermittent upward spikes can be observed because MA-DQN involves the ϵ-greedy strategy in action selection. In other words, the MA-DQN algorithm has the probability of ϵ choosing a random action, and the randomly picked action may not benefit the long-term optimized goal.

Different from MA-DQN, MA-DDPG and MA-ORL are developed on the basis of the actor-critic network, where the actor takes care of choosing an action while the critic is responsible for criticizing the selected bad actions. The actor is actually a policy gradient network mapping the state information to the action directly. It is virtually a neural network whose network parameters are updated using gradient ascent. As aforementioned, we introduce weak supervision to the actor's loss in Eq. (13) such that the recommended drug dosages are within the normal ranges. As shown in Fig. 4 (b), both the actor loss of MA-DDPG (orange curve) and MA-ORL (blue curve) decreases from below 0. MA-DDPG drops rapidly at first, flattens afterward, and converges to -100 in the end. The slope of MA-ORL is not as steep as MA-DDPG, but it keeps the downward momentum as the increase of training epochs.

6   Conclusion

This paper proposes a multi-action offline reinforcement learning (MA-ORL) model to precisely control the daily doses of multiple antiepileptic drugs for epilepsy in children. For patient safety, MA-ORL learns the experience of medical experts from clinical medical records directly instead of taking the intrusive and inefficient "trial-and-error" manner to collect training samples. To separate the latent features of different actions from the shared representation, a group of 0–1 masks is introduced and multiplied with the mixed representation mixture to get the separated features. Also, we introduce weakly-supervised learning to impose a restriction on the learning process of action selection, which avoids suggesting irrational dosages and minimizes the difference between the prescribed dosage and the recommended one. Now, MA-ORL has been deployed to the Children's Hospital for further testing and assisting doctors in decision-making.

Acknowledgments. This work is supported by the Ph.D., Scientific Research Foundation, The Chongqing University of Posts and Telecommunications (No. E012A2022026), National Key Research and Development Program of China (No. 2019YFE0110800), National Natural Science Foundation of China (No. 61972060 and 62027827), Natural Science Foundation of Chongqing (No. cstc2020jcyj-zdxmX0025), Research Institute for Artificial Intelligence of Things (Project No. CD5H) and Research and Innovation Office (Project No. BD4A), The Hong Kong Polytechnic University.

References

1. Chang, B.S., Lowenstein, D.H.: Epilepsy. N. Engl. J. Med. **349**(13), 1257–1266 (2003)
2. Fisher, R., et al.: ILAE official report: a practical clinical definition of epilepsy. Epilepsia **55**(4), 475–482 (2014)

3. Epilepsy. World Health Organization (WHO) (2022). https://www.who.int/news-room/fact-sheets/detail/epilepsy

4. Camfield, P., Camfield, C.: Incidence, prevalence and Aetiology of seizures and epilepsy in children. Epileptic Disord. **17**(2), 117–123 (2015)

5. Stafstrom, C.E., Carmant, L.: Seizures and epilepsy: an overview for neuroscientists. Cold Spring Harb. Perspect. Med. **7**(5), 1–19 (2015)

6. Minardi, C., et al.: Epilepsy in children: from diagnosis to treatment with focus on emergency. J. Clin. Med. **8**(1), 39 (2019)

7. George, J., Kulkarni, C., Sarma, G.: Antiepileptic drugs and quality of life in patients with epilepsy: a tertiary care hospital-based study. Value Health Reg. Issues **6**, 1–6 (2015)

8. Bittigau, P., et al.: Antiepileptic drugs and apoptotic neurodegeneration in the developing brain. Proc. Natl. Acad. Sci. **99**(23), 15089–15094 (2002)

9. Liu, G., Nicole, S., Perkins, A.: Epilepsy: treatment options. Am. Fam. Physician **96**(2), 87–96 (2017)

10. Topol, E.J.: High-performance medicine: the convergence of human and artificial intelligence. Nat. Med. **25**(1), 44–56 (2019)

11. He, J., Baxter, S.L., Xu, J., Xu, J., Zhou, X., Zhang, K.: The practical implementation of artificial intelligence technologies in medicine. Nat. Med. **25**(1), 30–36 (2019)

12. Esteva, A., et al.: A guide to deep learning in healthcare. Nat. Med. **25**(1), 24–29 (2019)

13. Taghanaki, S.A., Abhishek, K., Cohen, J.P., Cohen-Adad, J., Hamarneh, G.: Deep semantic segmentation of natural and medical images: a review. Artif. Intell. Rev. **54**(1), 137–178 (2020)

14. Wang, L., Zhang, W., He, X., Zha, H.: Supervised reinforcement learning with recurrent neural network for dynamic treatment recommendation. In: Proceedings of the 24th ACM SIGKDD International Conference on Knowledge Discovery & Data Mining, pp. 2447–2456 (2018)

15. Zhao, Y., Kosorok, M.R., Zeng, D.: Reinforcement learning design for cancer clinical. trials. Stat. Med. **28**(26), 3294–3315 (2009)

16. Yu, C., Liu, J., Nemati, S., Yin, G.: Reinforcement learning in healthcare: a survey. ACM Comput. Surv. **55**(1), 1–36 (2021)

17. Komorowski, M., Celi, L.A., Badawi, O., Gordon, A.C., Faisal, A.A.: The artificial intelligence clinician learns optimal treatment strategies for sepsis in intensive care. Nat. Med. **24**(11), 1716–1720 (2018)

18. Levine, S., Kumar, A., Tucker, G., Fu, J.: Offline reinforcement learning: tutorial, review, and perspectives on open problems. CoRR abs/2005.01643 (2020)

19. Fujimoto, S., Meger, D., Precup, D.: Off-policy deep reinforcement learning without exploration. In: Proceedings of the 36th International Conference on Machine Learning (ICML2019), pp. 2052–2062 (2019)

20. Busoniu, L., Babuska, R., De Schutter, B.: A comprehensive survey of multiagent reinforcement learning. IEEE Trans. Syst. Man, Cybern. Part C (Appl. Rev.) **38**(2), 156–172 (2008)

21. Foerster, N.J., et al.: Stabilising experience replay for deep multi-agent reinforcement learning. In: Proceedings of the 34th International Conference on Machine Learning (ICML2017), pp. 1146–1155 (2017)

22. Sunehag, P., et al.: Value-decomposition networks for cooperative multi-agent learning. In: Proceedings of the 17th International Conference on Autonomous Agents and Multiagent Systems (AAMAS2018), pp. 2085–2087 (2017)

23. Rashid, T., Samvelyan, M., Witt, D.S.C., Farquhar, G., Foerster, N.J., Whiteson, S.: QMIX: monotonic value function Factorisation for deep multi-agent reinforcement learning. In: Proceedings of the 35th International Conference on Machine Learning (ICML2018), pp. 4292–4301 (2018)

24. Ryan, L., Yi, W., Aviv, T., Jean, H., Pieter, A., Igor, M.: Multi-agent actor-critic for mixed cooperative-competitive environments. In: Proceedings of the 31st International Conference on Neural Information Processing Systems (NIPS2017), pp. 6382–6393 (2017)
25. Vinyals, O., et al.: StarCraft II: a new challenge for reinforcement learning. CoRR abs/1708.04782 (2017)
26. Baker, B., Kanitscheider, I., Markov, M.T., Wu, Y., Powel, G.: Emergent tool use from multi-agent Autocurricula. In: Proceedings of the 8th International Conference on Learning Representations (ICLR2020) (2020)
27. Zhang, Y., Yang, Q.: A survey on multi-task learning. IEEE Trans. Knowl. Data Eng. **34**(12), 5586–5609 (2022)
28. Sebastian R.: An overview of multi-task learning in deep neural networks. CoRR abs/1706.05098 (2017)
29. Yang, Y., Hospedales, T.M.: Trace norm Regularised deep multi-task learning. In: Proceedings of the 5th International Conference on Learning Representations (ICLR2017), Workshop Track Proceedings (2017)
30. Sutton, R.S., Barto, A.G.: Reinforcement Learning: An Introduction, 2nd edn. MIT Press, Cambridge (2018)
31. François-Lavet, V., Henderson, P., Islam, R., Bellemare, M.G., Pineau, J.: An introduction to deep reinforcement learning. Found. Trends Mach. Learn. **11**(3–4), 219–354 (2018)
32. Kingma, D.P., Welling, M.: Auto-encoding Variational Bayes. In: Proceedings of the 2nd International Conference on Learning Representations (ICLR2014) (2014)
33. Sohn, K., Lee, H., Yan, X.: Learning structured output representation using deep conditional generative models. In: Proceedings of the 28th International Conference on Neural Information Processing Systems (NIPS2015), pp. 3483–3491 (2015)
34. Hussein, A., Gaber, M.M., Elyan, E., Jayne, C.: Imitation learning: a survey of learning methods. ACM Comput. Surv. **50**(2), 1–35 (2017)
35. Yue Y., Le, H.M.: Imitation learning tutorial. ICML Tutorial (2018)
36. Mnih, V., et al.: Playing Atari with deep reinforcement learning. CoRR abs/1312.5602 (2013)
37. Lillicrap, P.T., et al.: Continuous control with deep reinforcement learning. In: Proceedings of the 4th International Conference on Learning Representations (ICLR2016) (2016)

Fine-Grained Software Vulnerability Detection via Neural Architecture Search

Qianjin Du[1], Xiaohui Kuang[2], Xiang Li[2], and Gang Zhao[2(✉)]

[1] The Department of Computer Science and Technology, Tsinghua University,
Beijing 100084, China
dqj20@mails.tsinghua.edu.cn
[2] National Key Laboratory of Science and Technology on Information System
Security, Beijing 100084, China
xhkuang@bupt.edu.cn, bisezhaog@163.com

Abstract. Vulnerability detection methods based on the deep learning have achieved remarkable performance improvements compared to traditional methods. Current deep learning-based detectors mostly use a single RNN or its variants (i.e., LSTM or GRU) to detect vulnerabilities. However, vulnerability detection is a multi-domain problem. Different types of vulnerabilities have different characteristics. Using a single neural network cannot perform well in all types of vulnerability detection tasks. Manually designing a matching neural network for each type of vulnerability detection task not only requires a lot of trials and computational resources but also highly relies on the knowledge and design experience of experts. To address this issue, in this paper, we propose a novel fine-grained vulnerability detection framework named A-DARTS, which is capable of searching for well-performed neural network architectures for different vulnerability detection tasks by introducing neural network architecture search (NAS) techniques. Specifically, we design a more efficient search space to ensure superior neural network architectures can be found by the search algorithm. Besides, we propose an adaptive differentiable search algorithm to search for superior neural network architectures. Experimental results show that searched models consistently outperform all baseline models and achieve significant performance improvements.

Keywords: Software Vulnerability · Deep Learning · Neural Architecture Search

1 Introduction

Software vulnerabilities are flaws or weaknesses present in the software, which allow attackers to cause harm to the stakeholders of the software. Over the years, there are numerous attack events that utilize vulnerabilities in software to obtain sensitive user information, which caused enormous financial loss. Therefore, detecting vulnerabilities in software as early as possible to avoid being

X. Wang et al. (Eds.): DASFAA 2023, LNCS 13946, pp. 224–238, 2023.
https://doi.org/10.1007/978-3-031-30678-5_17

```
void bad_heap()
{
    int *buffer = (int *) malloc(10 * sizeof(int));
    for(int i = 0;i <= 10;i++)
    {
        buffer[i] = i;
    }
}
```

(a) CWE-119

```
String exampleFun(boolean flag, String s1, String s2) {
    String s3 = null;
    if (flag) {
        s3 = s1 + s2;
    }
    return s3.trim();
}
```

(b) CWE-476

Fig. 1. Two different types of vulnerability examples.

attacked, has become an urgent matter. Traditional vulnerability detection methods can be categorized into static approaches and dynamic approaches. Static approaches, such as rule-based analysis methods [2,8] and code similarity-based methods [5,18], discover the vulnerabilities in software by analyzing the source code of the software. Dynamic approaches, including fuzz testing [1,20,21] and taint analysis [13,16], verify or discover vulnerabilities of software by executing the programs. Traditional vulnerability detection methods commonly require manually defining vulnerability rules or generating test cases, and highly rely on the knowledge and practical experience of experts.

With the development of deep learning, some studies have attempted to introduce deep learning to improve detection efficiency. VulDeePecker [7] proposed a novel detection framework, which extracted *code gadgets* consisting of semantically related code to represent programs, converted them to vectors using the Word2Vec tool, fed them into the deep learning model to detect vulnerabilities. VulDeePecker achieved obvious performance improvements compared to traditional methods. After that, μVulDeePecker [27] was proposed to detect multiple types of vulnerabilities. VulDeeLocater [6] aimed to pin down locations of vulnerabilities. Besides, Liu *et al.* [12] proposed a system named CD-VulD, which utilized deep learning models to perform cross-domain vulnerability detection. All of these studies utilize commonly used RNN or its variants (LSTM, GRU, etc.) to detect different types of vulnerabilities. However, multi-class vulnerability detection is a multi-domain problem, that is, the data with different vulnerability types are inconsistent. Figure 1 illustrates two code examples with different types of vulnerabilities. For the buffer overflow error (CWE-119) vulnerability detection, it requires the deep learning model to track the usage of buffer. For the NULL pointer dereference error (CWE-476) vulnerability detection, it requires the model to check if a pointer is NULL. Different types of vulnerability detection tasks have different requirements for the deep learning model. A single neural network, such as RNN or LSTM, is unable to successfully handle all types of vulnerability detection tasks. For this problem, it is natural to design a matching neural network for each type of vulnerability detection tasks. Nevertheless, manually designing a superior model for each type of vulnerability detection is inefficient, which not only consumes massive computing resources to conduct trials but also highly relies on the knowledge and design experience of experts. To address this issue, we introduce NAS (Neural Architecture Search) techniques to automatically design superior deep learning models for different types of vulnerability detection tasks (see Fig. 2). NAS is a technique that aims to automatically design efficient neural network architectures for a given task. It has achieved great success in the computer vision field [11,15], e.g., searched models achieve comparable (even superior) performance compared to human-designed models. However, NAS for vulnerability detection is poorly explored. Inspired by this idea, in this paper, we introduce NAS techniques to automatically design superior models for automated vulnerability detection. Specifically, we design an efficient search space, which consists of multiple feature transformation functions to enlarge the representational space of the search space. Then, we propose an adaptive differentiable search algorithm to search for powerful model architectures.

We conduct search experiments on the CGD dataset to verify the effectiveness of our proposed method. Experimental results show that searched models outperform all baseline models and achieve significant performance improvements. In general, our contributions can be concluded as follows:

- We introduce NAS techniques to automatically design well-performed models for vulnerability detection tasks, which avoids the heavy labor of manually designing detection models. To the best of our knowledge, this is the first study that NAS techniques are introduced in vulnerability detection tasks.
- We design an efficient search space to ensure that searched models with the superior representational capacity can discover precise vulnerability patterns. In addition, we propose an adaptive differentiable search algorithm to search for powerful model architectures.
- We conduct extensive experiments to verify the effectiveness of our approach. Experimental results demonstrate that searched models achieve significant performance improvements compared to baseline models.

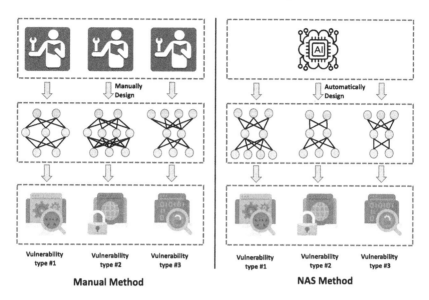

Fig. 2. The comparison between the manual design method and the NAS method. Manually designing neural networks for different vulnerability detection tasks requires a lot of trials and relies on the knowledge and experience of experts. The NAS method is capable of designing neural networks automatically, which greatly improves design efficiency.

2   Related Work

2.1   Vulnerability Detection

Traditional vulnerability detection methods can be divided into two categories: static methods and dynamic methods. A promising static approach in vulnerability detection is the code similarity method [5,18], which focuses on detecting vulnerabilities resulting from code cloning. Despite its high precision in locating vulnerabilities [6], it is limited in its ability to detect vulnerabilities that are not caused by code cloning. Another commonly utilized static approach is the rule-based method [2,8]. This method leverages the expertise of security experts to create templates, or rules, to identify potential vulnerable code segments. Although static methods have broad coverage and have been successfully applied in many vulnerability detection scenarios, they often have a high risk of false-positive errors [10]. Dynamic methods, such as fuzz testing [1,20,21] and symbolic execution [22], detect software vulnerabilities by executing the program. These methods have a low false-positive rate and can accurately uncover potential software vulnerabilities. However, the effectiveness of these methods heavily depends on the quality of the testing inputs, as they determine the execution path to areas of interest [23]. Generating high-quality testing inputs with high program coverage is a challenging task, and it often requires a significant amount of human effort and time.

Deep learning-based detection methods aim to automatically learn vulnerability patterns rather than rely on human experts to manually define rules or generate testing cases. VulDeePecker [7] is the first work leveraging deep learning techniques to perform automated fine-grained software vulnerability detection. It builds a binary classifier based on LSTM to detect whether a piece of code is vulnerable or not. CD-VulD [12] is proposed to perform cross-domain vulnerability detection, which utilizes domain adaptation techniques to learn cross-domain program representations.

2.2 Neural Architecture Search

NAS (**N**eural **A**rchitecture **S**earch) is a technique that aims to automatically search for efficient neural network architectures for a given task. The early technique route is based on reinforcement learning (RL) to learn promising architectures in a discrete search space [15, 26]. But these methods consume enormous computing resources, e.g., NAS-Net used 800 GPUs for 28 days to search for a SOTA network architecture on the CIFAR-10 dataset. Besides RL-based methods, another technique route is the evolutionary algorithm (EA). EA-based methods have demonstrated comparable performance to RL-based methods, while incurring lower search costs. To speed up the search process, a gradient-based Neural Architecture Search method, DARTS, has been introduced [9, 11]. DARTS accelerates the search process by relaxing the search space to be continuous, thus achieving higher search efficiency.

3 Proposed Approach

In this study, we propose a novel adaptive neural architecture search method named A-DARTS for vulnerability detection. The proposed method is comprised of two components: **Search Space Design** and **Search Algorithm**.

3.1 Search Space Design

Existing deep learning-based vulnerability detection methods mostly use RNN or its variants to discover vulnerabilities in the software. In this paper, our goal is to search for superior neural network architectures based on RNNs by introducing neural architecture search (NAS) techniques. Given an input x_t and hidden state h_{t-1} where t and $t-1$ denote time steps, the computation of RNN is defined as $h_t = f(x_t, h_{t-1})$ where $f(\cdot, \cdot)$ is the feature transform function of the RNN cell. Actually, the objective of NAS is to seek multiple superior feature transformation functions. The search space is represented as a directed acyclic graph (DAG) [15] (see Fig. 3(a)). The candidate child models can be obtained by taking a sub-graph of the DAG. In other words, DAG contains all possible child models. The DAG consists of N nodes and V edges. The node j represents

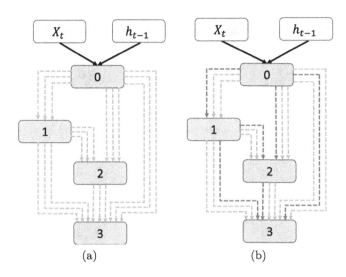

Fig. 3. The illustration of DARTS: (a) Each edge is a mixture of candidate transformation functions during the search. Each edge associates with the architecture parameter α to measure the importance of different candidate functions. The architecture parameters are optimized by gradient descent. (b) After the search, each edge selects the optimal candidate function to generate the final architecture.

the latent feature representation. The edge $o^{(i,j)}(\cdot)$ represents candidate feature transformation functions from node i to node j. For node j, it sums over vectors from all pre-decessor nodes $(j \leq i)$, followed by a linear transformation with a parameter matrix $w_{i,j}$. Formally, let s_i denote the latent representation of node i and s_j denote the latent representation of node j. And s_j is computed by:

$$s_j = \sum_{i<j} o^{(i,j)}(s_i \cdot w_{i,j}) \tag{1}$$

In the DAG, each edge $o^{(i,j)}(\cdot)$ is associated with multiple candidate feature transformation functions that transform the input $(s_i \cdot w_{i,j})$. In DARTS, candidate feature transformation functions only include *Tanh* and *Relu* two activation functions, which limits the feature representation of the search space. To improve the representational capability of the search space, in our work, we introduce multiple powerful feature transform functions, including *Tanh*, *Relu*, *Sigmoid*, *Identity*, *LeakyReLU*, *ELU*, *GELU* and *None*.

3.2 Search Algorithm

The Search Algorithm of DARTS. As elaborated in Sect. 3.1, each edge $o^{(i,j)}(\cdot)$ is associated with a set of candidate feature transformation functions $\mathcal{O}(\mathcal{O} = \{o_k^{(i,j)}\}, for \ k \in [1, K])$. A child model can be obtained by choosing a

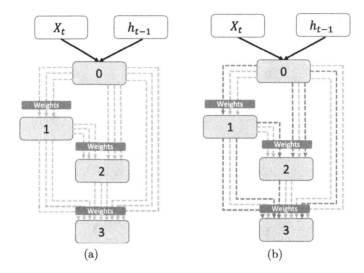

Fig. 4. The illustration of A-DARTS: (a) Each incoming edge is associated with the weight $\beta^{(i,j)}$ to measure the importance of the incoming edge. (b) After the search, each edge adaptively selects optimal feature transformation functions by clustering architecture parameters to generate the final architecture.

sub-set of edges (i.e., picking one candidate function $o_k^{(i,j)}$ for each edge $o^{(i,j)}(\cdot)$). Thus, our goal is to seek the optimal candidate feature transformation functions $o_k^{*(i,j)}$ for each edge $o^{(i,j)}(\cdot)$ and the output of the search is the searched model architecture. To improve search efficiency and make the search space continuous, DARTS relaxes the categorical choice of a particular candidate feature transformation function to a softmax over all possible candidate feature transformation functions:

$$o^{(i,j)}(s_i \cdot w_{i,j}) = \sum_{o \in \mathcal{O}} \frac{exp(\alpha_o^{(i,j)})}{\sum_{o' \in \mathcal{O}} exp(\alpha_{o'}^{(i,j)})}(s_i \cdot w_{i,j}) \tag{2}$$

where $\alpha^{(i,j)} \in \mathbb{R}^{|\mathcal{O}|}$ is referred to as the architecture parameter indicating the importance of different candidate feature transformation functions. The architecture search problem is transformed into a learning problem of continuous variables $\alpha^{(i,j)}$, which can be optimized by using gradient descent. The optimizations of the architecture parameters α and the model weights w are formulated as a bi-level optimization:

$$\min_{\alpha} \mathcal{L}_{val}(w^*(\alpha), \alpha) \tag{3}$$

$$s.t.\ w^*(\alpha) = \arg\min_{w} \mathcal{L}_{train}(w, \alpha) \tag{4}$$

where $\mathcal{L}_{train}$ and $\mathcal{L}_{val}$ represent the training loss and the validation loss respectively. The goal of the architecture search is to obtain the optimal α^* by minimizing the validation loss $\mathcal{L}_{val}(w^*, \alpha^*)$, and the model's weights w^* associated with the architecture are obtained by minimizing the training loss $\mathcal{L}_{train}$. After

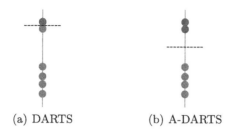

(a) DARTS   (b) A-DARTS

Fig. 5. An instance of the distribution information of architecture parameter $\alpha^{(i,j)}$.

the search, each edge selects an optimal feature transformation function $o^{*(i,j)}$ by $o^{*(i,j)} = \arg\max_{o \in \mathcal{O}} \alpha_o^{(i,j)}$ to generate the final searched model architecture (see Fig. 3(b)).

The Adaptive Search Algorithm of A-DARTS. Although the most efficient NAS method DARTS achieves superior performance compared to human-designed models, there still remain some problems to be optimized: (1) In DARTS, every node equally receives incoming edges from all of its pre-decessor nodes and does not differentiate the importance of different incoming edges. This easily generates redundant or sub-optimal searched model architectures because some incoming edges may be invalid or unnecessary compared to other incoming edges. (2) To generate the final searched model architecture, DARTS directly selects the feature transformation function with the largest architecture parameter for each edge. This ignores the distribution information of architecture parameters, which easily generate sub-optimal model architectures. Figure 5(a) illustrates the distribution information of architecture parameters for an edge, in which each dot denotes a feature transformation function. We can find that the first dot (red dot) and the second dot are very close, and both of them represent valid (important) feature transformation functions. DARTS only chooses the first dot and discards the second dot, which easily generates sub-optimal searched model architectures.

To address the above two issues, we propose an adaptive search algorithm named A-DARTS. Firstly, we introduce incoming edge weights to measure the importance of incoming edges (see Fig. 4(a)). In this way, the search algorithm can adaptively discover important incoming edges and mask off invalid incoming edges of each node. Secondly, after the search, each edge adaptively chooses one or multiple optimal feature transformation functions to generate the final model architecture by clustering architecture parameters (see Fig. 5(b)). In the following part, we will present the details of A-DARTS.

In order to adaptively seek valid (or important) incoming edges of each node, we introduce incoming edge weights to measure the importance of incoming edges for each node. We denote incoming edge weights as β. The node j is re-defined as:

$$s_j = \sum_{i<j} \frac{\beta^{(i,j)}}{\sum_{s=0}^{K-1} \beta^{(s,j)}} o^{(i,j)}(s_i \cdot w_{i,j}) \tag{5}$$

where $\beta^{(i,j)}$ denotes the weight of the incoming edge $o^{(i,j)}(\cdot)$, and K denotes the number of incoming edges of node j. The optimization of A-DARTS is formulated as:

$$\min_{\alpha,\beta} \mathcal{L}_{val}(w^*(\alpha,\beta), \alpha, \beta) \tag{6}$$

$$s.t.\ w^*(\alpha,\beta) = \arg\min_{w} \mathcal{L}_{train}(w, \alpha, \beta) \tag{7}$$

where α denotes the architecture parameters, and β denotes the incoming edge weights, and w denotes the model parameters.

After searching, the architecture parameters $\alpha^{(i,j)}$ of each edge are divided into two categories by using the K-Means clustering algorithm. The one of categories is the valid class, representing corresponding feature transformation functions are valid. The other class represents invalid feature transformation functions. To generate the final searched model architecture, for each edge, we select feature transformation functions whose corresponding architecture parameters belong to the valid class (see Fig. 4(b)).

4   Experiments

We conduct search experiments on the published fine-grained vulnerability detection dataset CGD [7]. The CGD dataset is the largest vulnerability dataset, which consists of two types of vulnerabilities: the buffer overflow error vulnerability (CWE-119) and the resource management error vulnerability (CWE-399).

4.1   Experimental Setup

Our experiments are implemented based on the open-source deep learning framework PyTorch [14]. We follow the same data pre-process as VulDeePecker [7]. Our experiments consist of two phases: model search and model evaluation. In the model search phase, we search for superior model architectures by using A-DARTS. In the model evaluation phase, we fully train searched models and then evaluate their corresponding performance. The searched RNN cell consists of N nodes, and N ranges from 3 to 7. The search procedure consists of a total of 300 epochs with a batch size of 256 and begins to optimize the architecture parameters from the 50_{th} epoch for a stable search. We optimize the model weights w using SGD with an initial learning rate of 0.01, and optimize the architecture parameters α and incoming edge weights β with Adam. The embedding dimension of searched models is set to 40, and the hidden size is set to 40.

After the search, we obtain searched model architectures according to the learned architecture parameters. Then we fully train searched models to evaluate the corresponding performance. The total training epoch is 300 with a batch size of 1024. We use SGD with an initial learning rate of 0.01 to optimize model weights. The remaining hyper-parameter settings are consistent with the hyper-parameter settings of the model search phase.

4.2 Baselines

We conduct extensive experiments to verify the effectiveness of our method. We adopt the following models as our baselines:

* Rats/Flawfinder [19,24] are widely-used open source vulnerability detection tools.
* Support Vector Machine (SVM) and Random Forest (RF) [17,25] are token-based detection methods, which use SVM/RF as classifiers to detect code vulnerabilities.
* CodeBERT [3] is a powerful pre-trained model for programming language, which is trained in six programming languages.
* GraphCodeBERT [4] is a new pre-trained programming language model, extending CodeBERT to consider the inherent structure of code data flow into the training objective.
* VulDeePecker [7] is the current top-performing slice-level fine-grained detection method.

4.3 Results

The final performance for the CWE-119 vulnerability detection and CWE-399 vulnerability detection are reported in Table 1 and Table 2, respectively. For the CWE-119 vulnerability detection (see Table 1), we search for three superior models named A-DARTS-A, A-DARTS-B and A-DARTS-C respectively. It can be observed that searched models consistently outperform all baseline methods, obtaining significant performance improvements. The searched model A-DARTS-A achieves an F1-score of 77.05%, surpassing its counterparts VulDeePecker (Bi-LSTM, 2-Layers) and VulDeePecker (Bi-GRU, 2-Layers) by 4.01% and 4.25%. It is worth mentioning that we cannot reimplement the original results of VulDeePecker reported in paper [7] for CWE-119 vulnerability detection, because VulDeePecker does not publish the source code and not provide the splitting of training and testing datasets based on the CGD dataset. But the comparison is fair. The difference between A-DARTS and VulDeePecker is the model architecture, and the remaining components are consistent. For the CWE-399 vulnerability detection (see Table 2), we search for three models named A-DARTS-D, A-DARTS-E and A-DARTS-F respectively. Our searched models still outperform all baseline models. Searched model A-DARTS-D achieves an F1-score of 94.17%, surpassing the best baseline model VulDeePecker (Bi-LSTM, 2-Layers) by 4.45% with about 50% parameter reduction (13.12K VS 26.32K) and surpassing VulDeePecker (Bi-LSTM, 1-Layer) by 4.52% with equivalent parameter size (13.12K VS 13.20K). Similarly, other searched models, such as A-DARTS-E and A-DARTS-F, also achieve superior performance. A-DARTS-E obtains F1-score gains of 5.36% and 3.99% compared to VulDeePecker (Bi-GRU, 2-Layers) and VulDeePecker (Bi-LSTM, 2-Layers). An interesting finding is that, among all searched models, A-DARTS-A and A-DARTS-D with the fewest parameters achieve the best performance in the CWE-119 vulnerability detection task and the CWE-399 vulnerability detection task, respectively. We argue that the reason might be that large searched models easily face the problem of overfitting. Generally, the architecture of RNN or its variants is fixed, which is difficult to avoid the problem of overfitting. A-DARTS directly searches for the model architecture on the specific datasets, which is more flexible and could adaptively discover more matching model architectures according to the characteristics of datasets. For example, the search algorithm could prefer models with fewer parameters to avoid the problem of overfitting.

Table 1. Results for CWE-119 vulnerability detection

Models	Params(K)	F1
Rats	–	18.20
FlawFinder	–	26.06
SVM	–	53.49
RF	–	60.09
CodeBERT	–	65.66
GraphCodeBERT	–	67.98
VulDeePecker (Bi-RNN, 1-Layer)	3.36	67.07
VulDeePecker (Bi-RNN, 2-Layers)	6.64	66.87
VulDeePecker (Bi-LSTM, 1-Layer)	13.20	72.92
VulDeePecker (Bi-LSTM, 2-Layers)	26.32	73.04
VulDeePecker (Bi-GRU, 1-Layer)	9.92	72.68
VulDeePecker (Bi-GRU, 2-Layers)	19.76	72.80
A-DARTS-A	15.26	**77.05**
A-DARTS-B	18.88	76.36
A-DARTS-C	16.76	76.04

In general, extensive experimental results demonstrate that the models searched by our proposed method achieve better performance compared to baseline methods, suggesting the effectiveness of our proposed method for fine-grained vulnerability detection.

4.4   Comparison with NAS Methods

In order to further verify the effectiveness of our method, we conduct several search experiments to compare with current NAS methods (including the random

Table 2. Results for CWE-399 vulnerability detection

Models	Params(K)	F1
Rats	–	42.60
FlawFinder	–	47.88
SVM	–	78.77
RF	–	69.62
CodeBERT	–	80.72
GraphCodeBERT	–	83.36
VulDeePecker (Bi-RNN, 1-Layer)	3.36	88.55
VulDeePecker (Bi-RNN, 2-Layers)	6.64	88.87
VulDeePecker (Bi-LSTM, 1-Layer)	13.20	89.65
VulDeePecker (Bi-LSTM, 2-Layers)	26.32	89.72
VulDeePecker (Bi-GRU, 1-Layer)	9.92	88.58
VulDeePecker (Bi-GRU, 2-Layers)	19.76	88.35
A-DARTS-D	13.12	**94.17**
A-DARTS-E	22.96	93.71
A-DARTS-F	26.25	93.65

Table 3. The comparison between different NAS methods.

Models	F1	Search Cost (GPU days)
Random RNN	90.22	–
E-NAS	91.35	0.50
DARTS	91.76	0.20
DARTS+	91.96	0.13
A-DARTS	**94.17**	**0.10**

search algorithm, ENAS [15], DARTS [11] and DARTS+ [9]) for the CWE-399 vulnerability detection. We search for multiple models by using different NAS methods and then fully train these searched models. The final performance of searched models is shown in Table 3. The model searched by our method achieves the best performance with the search cost of 0.10 GPU days, outperforming all counterparts. Our method surpasses E-NAS and DARTS by 2.82% and 2.41% with the less search cost.

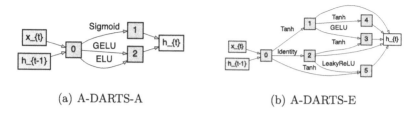

(a) A-DARTS-A   (b) A-DARTS-E

Fig. 6. Illustrations of searched models.

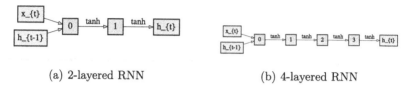

(a) 2-layered RNN   (b) 4-layered RNN

Fig. 7. Illustrations of RNNs.

4.5 Illustrations of Searched Architectures

We visualize searched models A-DARTS-A and A-DARTS-E in Fig. 6. The architecture of RNN with multiple layers is illustrated in Fig. 7. We can find that searched models tend to use wide and shallow network architecture rather than deep network architecture. A reasonable explanation is that there exist some long-term control/data dependencies that need to be handled by models in vulnerability detection tasks, while using the wide network architecture is helpful for memorizing long-term information.

5   Conclusion

In this paper, we propose a NAS-based approach for fine-grained vulnerability detection, which is capable of searching for superior network architectures for vulnerability detection. The advantage of our method is that it can adaptively search for different network architectures according to the characteristics of different types of vulnerability detection tasks. Experimental results show that searched models achieve significant and universal performance improvements compared to baseline models.

References

1. Cadar, C., Dunbar, D., Engler, D.R., et al.: KLEE: unassisted and automatic generation of high-coverage tests for complex systems programs. In: OSDI, vol. 8, pp. 209–224 (2008)
2. Engler, D., Chen, D.Y., Hallem, S., Chou, A., Chelf, B.: Bugs as deviant behavior: a general approach to inferring errors in systems code. ACM SIGOPS Oper. Syst. Rev. **35**(5), 57–72 (2001)

3. Feng, Z., et al.: CodeBERT: a pre-trained model for programming and natural languages. arXiv preprint: arXiv:2002.08155 (2020)
4. Guo, D., et al.: GraphCodeBERT: pre-training code representations with data flow. arXiv preprint: arXiv:2009.08366 (2020)
5. Kim, S., Woo, S., Lee, H., Oh, H.: VUDDY: a scalable approach for vulnerable code clone discovery. In: 2017 IEEE Symposium on Security and Privacy (SP), pp. 595–614. IEEE (2017)
6. Li, Z., Zou, D., Xu, S., Chen, Z., Zhu, Y., Jin, H.: VulDeeLocator: a deep learning-based fine-grained vulnerability detector. IEEE Trans. Dependable Secure Comput. **19**, 2821–2837 (2021)
7. Li, Z., et al.: VulDeePecker: a deep learning-based system for vulnerability detection. arXiv preprint: arXiv:1801.01681 (2018)
8. Li, Z., Zhou, Y.: PR-miner: automatically extracting implicit programming rules and detecting violations in large software code. ACM SIGSOFT Softw. Eng. Notes **30**(5), 306–315 (2005)
9. Liang, H., et al.: DARTS+: improved differentiable architecture search with early stopping. arXiv preprint: arXiv:1909.06035 (2019)
10. Lin, G., Wen, S., Han, Q.L., Zhang, J., Xiang, Y.: Software vulnerability detection using deep neural networks: a survey. Proc. IEEE **108**(10), 1825–1848 (2020)
11. Liu, H., Simonyan, K., Yang, Y.: DARTS: differentiable architecture search. arXiv preprint: arXiv:1806.09055 (2018)
12. Liu, S., et al.: CD-VuLD: cross-domain vulnerability discovery based on deep domain adaptation. IEEE Trans. Dependable Secure Comput. (2020)
13. Newsome, J., Song, D.X.: Dynamic taint analysis for automatic detection, analysis, and Signaturegeneration of exploits on commodity software. In: NDSS, vol. 5, pp. 3–4. Citeseer (2005)
14. Paszke, A., et al.: Pytorch: An imperative style, high-performance deep learning library. arXiv preprint: arXiv:1912.01703 (2019)
15. Pham, H., Guan, M., Zoph, B., Le, Q., Dean, J.: Efficient neural architecture search via parameters sharing. In: International Conference on Machine Learning, pp. 4095–4104. PMLR (2018)
16. Portokalidis, G., Slowinska, A., Bos, H.: Argos: an emulator for fingerprinting zero-day attacks for advertised honeypots with automatic signature generation. ACM SIGOPS Oper. Syst. Rev. **40**(4), 15–27 (2006)
17. Rigatti, S.J.: Random forest. J. Insur. Med. **47**(1), 31–39 (2017)
18. Roy, C.K., Cordy, J.R., Koschke, R.: Comparison and evaluation of code clone detection techniques and tools: a qualitative approach. Sci. Comput. Program. **74**(7), 470–495 (2009)
19. Secure, S.: Rough auditing tool for security (rats) (2010)
20. Sen, K., Marinov, D., Agha, G.: Cute: a concolic unit testing engine for C. ACM SIGSOFT Softw. Eng. Notes **30**(5), 263–272 (2005)
21. Serebryany, K.: libFuzzer-a library for coverage-guided fuzz testing. LLVM Project (2015)
22. Stephens, N., et al.: Driller: augmenting fuzzing through selective symbolic execution. In: NDSS, vol. 16, pp. 1–16 (2016)
23. Suneja, S., Zheng, Y., Zhuang, Y., Laredo, J., Morari, A.: Learning to map source code to software vulnerability using code-as-a-graph. arXiv preprint: arXiv:2006.08614 (2020)
24. Wheeler., D.A.: Flawfinder (2012)

25. Xue, H., Yang, Q., Chen, S.: SVM: support vector machines. In: The Top Ten Algorithms in Data Mining, pp. 51–74. Chapman and Hall/CRC, Boca Raton (2009)
26. Zoph, B., Le, Q.V.: Neural architecture search with reinforcement learning. arXiv preprint: arXiv:1611.01578 (2016)
27. Zou, D., Wang, S., Xu, S., Li, Z., Jin, H.: μvuldeepecker: a deep learning-based system for multiclass vulnerability detection. IEEE Trans. Dependable Secure Comput. **18**, 2224–2236 (2019)

Adversarial Learning-Based Stance Classifier for COVID-19-Related Health Policies

Feng Xie⬤, Zhong Zhang, Xuechen Zhao, Haiyang Wang, Jiaying Zou, Lei Tian, Bin Zhou$^{(\boxtimes)}$, and Yusong Tan

College of Computer, National University of Defense Technology, Changsha, China
{xiefeng,zhangzhong,zhaoxuechen,wanghaiyang19,zoujiaying20,leitian129, binzhou,ystan}@nudt.edu.cn

Abstract. The ongoing COVID-19 pandemic has caused immeasurable losses for people worldwide. To contain the spread of the virus and further alleviate the crisis, various health policies (e.g., stay-at-home orders) have been issued which spark heated discussions as users turn to share their attitudes on social media. In this paper, we consider a more realistic scenario on stance detection (i.e., cross-target and zero-shot settings) for the pandemic and propose an adversarial learning-based stance classifier to automatically identify the public's attitudes toward COVID-19-related health policies. Specifically, we adopt adversarial learning that allows the model to train on a large amount of labeled data and capture transferable knowledge from source topics, so as to enable generalize to the emerging health policies with sparse labeled data. To further enhance the model's deeper understanding, we incorporate policy descriptions as external knowledge into the model. Meanwhile, a GeoEncoder is designed which encourages the model to capture unobserved background factors specified by each region and then represent them as non-text information. We evaluate the performance of a broad range of baselines on the stance detection task for COVID-19-related health policies, and experimental results show that our proposed method achieves state-of-the-art performance in both cross-target and zero-shot settings.

Keywords: Natural Language Processing · Stance Detection · Public Health Informatics · COVID-19 Pandemic · Health Policy

1 Introduction

The coronavirus disease 2019 (COVID-19) pandemic has brought serious challenges to human health, society, and the economy. To curb the spread of the virus and alleviate the crisis, policymakers and public authorities have imposed corresponding health policies (e.g., stay-at-home orders, vaccination) based on the dynamics of the epidemic situation in response to complex virus challenges.

F. Xie and Z. Zhang—Equal contributions.

X. Wang et al. (Eds.): DASFAA 2023, LNCS 13946, pp. 239–249, 2023.
https://doi.org/10.1007/978-3-031-30678-5_18

Table 1. The examples of stance classification task. Given a tweet and the involved topic, the stance classifier is capable of detecting the **stance label** automatically.

Example 1: Don't be selfish. Stay home, reduce the spread, and safe lives. If you have to go out, please wear a mask and gloves.
Topic: Stay at Home Orders   **Stance label:** Favor
Example 2: There is no way in hell I would put a vaccine into my body that comes out under or is advocated by the trump regime. I trust COVID-19 more than I trust the trump regime.
Topic: Vaccination   **Stance label:** Against

The promulgation of these health policies has sparked discussion on the Internet, but the public's attitudes towards them vary. Stance detection is of great practical value as an effective tool for Internet public opinion monitoring, which detects the attitude (i.e., *in favor of*, *against*, or *neutral*) of an opinionated text toward a pre-defined topic automatically [2,16], as is shown in Table 1.

Nevertheless, for newly proposed health policies, the available labeled data is limited, and it is also infeasible to annotate a large amount of data in a short period of time. Existing studies adopt in-target[1] setting for tracking the public's stances [14,15], that is, the training and testing phase are under the same health policy. However, they usually require adequate labeled data to achieve decent performance, which limits their application range. Therefore, it is imperative to propose a more practical and robust stance classifier to be applied to emerging health policies. The alternative strategies are to conduct cross-target [16,17] or zero-shot settings [1,2]. Cross-target stance detection trains on one target and tests on a related target in a one-to-one way, while zero-shot stance detection aims to train on multiple targets with labeled data and evaluate on an unseen target. Both cross-target and zero-shot settings are more challenging principally because the language models may not be compatible between different targets [17]. Since COVID-19-related health policies are inherently correlated, adopting cross-target or zero-shot stance detection becomes viable and could provide more timely and useful guidance for administrative decision-making.

In this paper, we consider a more realistic scenario on stance detection (i.e., cross-target and zero-shot settings) for the pandemic and propose an adversarial learning-based stance classifier to detect the public's attitudes toward COVID-19-related health policies. The domain adaptation technique is one of the best solutions applied to cross-target and zero-shot conditions [2,16], where domain-invariant information is responsible for ensuring the transferability across different domains (topics). We treat each health policy as a topic and model topic transfer as domain adaptation. Specifically, we embed a text and related policy descriptions jointly via Bidirectional Encoder Representations from Transformers (BERT) [7], and then we employ a feature separation module to extract and distinguish topic-specific and topic-invariant information. Moreover, since people's attitudes across different regions are influenced by unobserved background

---

[1] In this paper, we will use the terms: target and topic interchangeably.

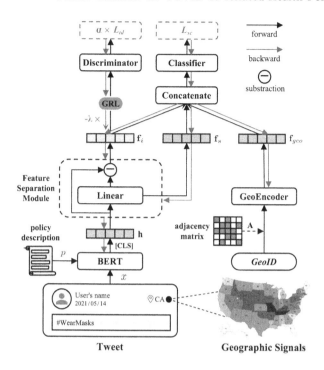

Fig. 1. Overall architecture of proposed model.

factors, such as cultural background, political ideologies, and regional epidemic situation (i.e., epidemiological contexts), a GeoEncoder is devised which encodes geographic signals as non-text representations to learn regional background information and thus improve the model's understanding. Following the success of adversarial learning for domain adaptation [6,9], we integrate a topic discriminator into the model for adversarial training to better capture topic-invariant information, hence enhancing the transferability of applying it to the emerging health policies. Experiments conducted on COVID-19 stance datasets demonstrate that our proposed method achieves state-of-the-art performance in both cross-target and zero-shot settings against a broad range of baselines.

2   Methodology

Problem Formulation. Given a set of labeled texts from the source target(s) $D_s = \{(t_s^i, x_s^i, y_s^i)\}_{i=1}^{N_s}$ and a set of unlabeled texts from a destination target (i.e., unseen target) $D_d = \{(t_d^i, x_d^i)\}_{i=1}^{N_d}$, where t is the target involved in the text x, and y is the corresponding stance label. We also have a large set of unlabeled texts from both targets $D_u = \{(t_u^i, x_u^i)\}_{i=1}^{N_u}$ (called external data) for adversarial training, where each sample belongs to the source target(s) or the destination target. Cross-target stance detection task is using one source target to predict stance labels (i.e., *"favor"*, *"against"*, or *"none"*) of texts in D_d, while zero-shot setting is using multiple source targets. Note that, each sample in D_s or D_d is associated with a Geographic Identifier (*GeoID*) indicating the user's location. In Fig. 1, we illustrate an overview architecture of our proposed model.

2.1   Encoder Module

BERT [7] has shown tremendous success in various Natural Language Processing (NLP) tasks. To take full advantage of contextual information, we could jointly condition and embed topic t and text x using BERT. However, the original topic words are too short, which results in limited targeted background knowledge available for the stance classifier. Since the proposal of health policies is accompanied by the corresponding descriptions and people's attitudes are based on the contents or even the details of the policies, instead of using the topic words directly, we leverage external knowledge about the topic (i.e., policy descriptions) to enhance the model's deeper understanding of health policies. We tokenize the text x and policy-related descriptions p as input sentence pair:

$$\mathbf{h} = \text{BERT}([CLS]p[SEP]x[SEP])_{[CLS]}, \tag{1}$$

where $[CLS]$ is a special symbol that be used as the aggregate representation for the overall semantic context, while $[SEP]$ is the special separator token. $\mathbf{h}$ is the hidden state of $[CLS]$ in the final layer and we use it for subsequent modeling.

2.2   Feature Separation Module

The contextual representation generated by BERT contains both topic-specific and topic-invariant information. To allow the model to generalize to unseen topics, it is essential and effective to learn and utilize transferable topic knowledge (i.e., topic-invariant information). Inspired by [6], we employ a simple linear transformation to separate and distinguish topic-specific and topic-invariant information, which can reduce the transfer difficulty without removing stance cues. First, we employ a linear layer to extract topic-specific features $\mathbf{f}_s$:

$$\mathbf{f}_s = \mathbf{W}_{fs}\mathbf{h} + \mathbf{b}_{fs}, \tag{2}$$

where $\mathbf{W}_{fs}$ and $\mathbf{b}_{fs}$ are weight parameters. By removing target-specific features from the $\mathbf{h}$, results in the constructed target-invariant representation $\mathbf{f}_i$:

$$\mathbf{f}_i = \mathbf{h} - \mathbf{f}_s. \tag{3}$$

The target-invariant features will enable the transferability of the model and serve as the input of the stance classifier and topic discriminator.

2.3   GeoEncoder Module

Geographic signals are spatial information, which can reflect potential characteristics and profiles of groups. Infusing geographic signals can provide region-specific features for model learning since people's attitudes across different regions are influenced by regional background factors, such as cultural background, political ideologies, and epidemiological contexts. Besides, graph structures (e.g., geographical topology) that describe the connectivity among regions grant us to explore underlying relationships between different regions and also can reflect some unobserved background factors. Inspired by this, we propose a GeoEncoder which utilizes geographic signals to capture the hidden background factors as non-text features $\mathbf{f}_{geo}$. Given the *GeoID* of a text and the geographical adjacency matrix $\mathbf{A}^2$, we leverage Graph Convolution Network (GCN) [13] to make regions aggregate information from their neighboring regions:

$$\mathbf{E}^{(l)} = \sigma(\mathbf{A}\mathbf{E}^{(l-1)}\mathbf{W}^{(l-1)}), \tag{4}$$

where $\mathbf{W}^{(l)}$ is a layer-specific weight matrix, $\sigma(\cdot)$ is ReLU activation function, and $\mathbf{E}^{(l)}$ is the embedding matrix at l-th layer, with $\mathbf{E}^{(0)} = \mathbf{E}$. $\mathbf{E} \in \mathbb{R}^{N \times F}$ are learnable embedding vectors specified by the *GeoIDs* and $\mathbf{f}_{geo} = \mathbf{E}^{(l)}_{[GeoID]}$, N means the number of total regions, F is the hidden dimension of $\mathbf{f}_{geo}$.

2.4   Stance Classifier

We apply a linear layer with softmax as the stance classifier to predict the stance labels of texts. We combine three features ($\mathbf{f}_i$, $\mathbf{f}_s$, and $\mathbf{f}_{geo}$) to obtain the final representation for joint modeling. We use Cross-Entropy loss as loss function:

$$\hat{y}_{sc} = \text{Softmax}(\mathbf{W}_{sc}(\mathbf{f}_i \oplus \mathbf{f}_s \oplus \mathbf{f}_{geo}) + \mathbf{b}_{sc}), \tag{5}$$

$$\mathcal{L}_{sc} = \sum_{x \in D_s} \text{CrossEntropy}(y_{sc}, \hat{y}_{sc}), \tag{6}$$

where y_{sc} is the ground truth stance label, $\mathbf{W}_{sc}$ and $\mathbf{b}_{sc}$ are parameters of stance classifier, and $\oplus$ represents the concatenation operation.

2.5   Topic Discriminator

To further ensure the topic-invariant representation can distill more transferable topics knowledge to facilitate model adaptation across different topics, inspired by [2,16], we use a linear network with softmax as a topic discriminator to classify the corresponding topic label based on topic-invariant features. The topic discriminator is also trained by minimizing the Cross-Entropy loss:

$$\hat{y}_{td} = \text{Softmax}(\mathbf{W}_{td}\mathbf{f}_i + \mathbf{b}_{td}), \tag{7}$$

---

^{2} By default, each region is adjacent to itself.

Table 2. Dataset summary.

Topic	#Unlabeled	#Labeled (Favor/Against/None)	Keywords°
Stay at Home (SH)	778	420 (194/113/113)	lockdown, stayhome
Wear Masks (WM)	1030	756 (173/288/295)	mask, facemasks
Vaccination (VA)	1535	526 (106/194/226)	vaccine, vaccination

° For more keywords of crawler, please refer to the source code(See footnote 4)

$$\mathcal{L}_{td} = \sum_{x \in D_s \cup D_u} \text{CrossEntropy}(y_{td}, \hat{y}_{td}), \tag{8}$$

where y_{td} is the ground truth topic label, and $\mathbf{W}_{td}$ and $\mathbf{b}_{td}$ are parameters of topic discriminator. The training process of discriminator and topic-invariant features $\mathbf{f}_i$ is adversarial. Specifically, target-invariant features aim to become generic enough to confuse topic discriminator (maximize $\mathcal{L}_{td}$) while topic discriminator makes efforts to correctly classify the topic labels (minimize $\mathcal{L}_{td}$). Following [9], we adopt the gradient reversal layer (GRL) in our model, which is a widely used technique in transfer learning-based methods. The adversarial training process is essentially a minmax game as follows:

$$\min_{\Theta_M} \max_{\mathbf{W}_{td}, \mathbf{b}_{td}} \mathcal{L}_{sc} - \alpha \mathcal{L}_{td}, \tag{9}$$

where Θ_M including fine-tunable BERT, GeoEncoder, $\mathbf{W}_{fs}$, $\mathbf{b}_{fs}$, $\mathbf{W}_{sc}$, and $\mathbf{b}_{sc}$, and α is a trade-off parameter. During the forward propagation, GRL acts as an identity transform. During the back-propagation though, GRL takes the gradient from the subsequent level, multiplies it by $-\lambda$, and passes it to the preceding layer for more stable training and update [9,16].

3   Experiments

3.1   Experimental Setup

Data. Previous studies [5,10] have built COVID-19-related stance datasets, where tweets are manually annotated as three stance labels: *favor*, *against*, and *none* (i.e., neutral). In this work, we follow their research and adopt three health policies: (I) Stay at Home Order, (II) Wear Masks and (III) Vaccination, and we only select labeled tweets posted in the USA. Meanwhile, we also collect unlabeled tweets for these three policies via Twitter API[3]. The statistics of prepared data are summarized in Table 2. Due to space limitations, please refer to our source code for detailed descriptions of the dataset (policy descriptions, keywords, etc.). Codes and other resources are publicly available.[4]

---

[3] https://developer.twitter.com/.
[4] https://github.com/Xiefeng69/stance-detection-for-covid19-related-health-policies.

Baselines and Evaluation Metrics. We select the following methods in the literature as comparison baselines: (1) neural network-based methods BiLSTM [12], BiCond [3], and TextCNN [4]; (2) attention-based methods TAN [8] and CrossNet [17]; (3) BERT-based methods BERT [7] and WS-BERT [11]. Besides, we adopt WS-BERT-S and WS-BERT-D which encode Wikipedia knowledge in the single manner and dual manner, respectively [11]. Following previous works [2,17], we utilize the average F1-score (denoted as F_{avg}) and the average of both micro-average and macro-average F1-scores (denoted as F_m) as evaluation metrics.

Table 3. Performance comparison of cross-target stance detection measured by F_{avg} and F_m. **Bold face** indicates the best result of each column and <u>underlined</u> the second-best. Relative gain is compared with the second best result.

Models	Cross-target settings (%)											
	SH→WM		SH→VA		WM→SH		WM→VA		VA→SH		VA→WM	
	F_{avg}	F_m	F_{avg}	F_m	F_{avg}	F_m	F_{avg}	F_m	F_{avg}	F_m	F_{avg}	F_m
BiLSTM	25.4	30.6	25.6	30.5	39.1	45.1	40.8	47.5	31.5	38.1	33.0	38.6
BiCond	29.0	33.1	30.1	34.5	37.3	42.1	37.5	44.4	33.8	40.2	35.5	40.9
TextCNN	34.6	37.8	31.5	36.6	39.4	43.9	37.6	42.5	30.7	33.3	35.8	38.5
TAN	44.3	46.2	34.5	39.0	45.5	47.4	45.1	48.5	37.7	38.2	42.6	44.1
CrossNet	<u>45.7</u>	<u>49.9</u>	<u>39.4</u>	43.6	43.4	47.3	47.7	50.7	37.7	38.3	46.7	48.1
BERT	44.7	49.3	34.9	41.2	44.3	49.7	52.6	55.3	<u>44.4</u>	<u>45.6</u>	<u>53.7</u>	<u>55.1</u>
WS-BERT-S	45.4	49.1	**40.3**	**44.9**	41.9	48.0	51.0	54.8	39.9	41.4	47.2	49.9
WS-BERT-D	40.1	47.1	30.5	38.9	<u>48.2</u>	<u>52.5</u>	<u>55.4</u>	<u>57.5</u>	43.5	44.9	49.5	51.1
Ours	**47.6**	**51.9**	<u>39.4</u>	<u>44.4</u>	**50.9**	**54.1**	**57.6**	**59.3**	**46.1**	**47.4**	**54.5**	**56.3**
Improve (%)	4.2%	4.0%	–	–	5.6%	3.0%	3.9%	3.1%	3.8%	3.9%	1.5%	2.1%

Implementation Details. All the experiments are performed in *Pytorch* on *NVIDIA GeForce 3090 GPU*. The reported results are the averaged score of 5 runs with different random initialization. In cross-target setting, the models are trained and validated on one topic and evaluated on another. There can be categorized into six source→destination tasks for cross-target evaluation: **SH→WM**, **SH→VA**, **WM→SH**, **WM→VA**, **VA→SH**, and **VA→WM**. In zero-shot setting, the models are trained and validated on multiple topics and tested on one unseen topic. We use the unseen topic's name as the task's name, thus, the zero-shot evaluation can be set into: **SH**, **WM**, and **VA**. For all tasks, the batch size is set to 16, the dropout rate is set to 0.1, and the input texts are truncated or padded to a maximum of 100 tokens. We train all models using AdamW optimizer with weight decay 5e-5 for a maximum of 100 epochs with patience of 10 epochs, and the learning rate is chosen in {1e-5, 2e-5}. In our model, we adopt the pre-trained uncased BERT-base as the encoder. The maximum length of policy descriptions is fixed at 50, the layer number l of GCN is set to 2, the trade-off parameter α is set to 0.01, the GRL's parameter λ is set to 0.1, and the hidden dimension of GeoEncoder is optimized in {128, 256}. Please refer to our project repository (See footnote 4) for more experimental implementation details (hyper-parameter settings, training details, data preparation, etc.).

3.2 Experimental Results

We evaluate all models both in cross-target and zero-shot settings, and the results are reported in Table 3 and Table 4, respectively. There is an overall phenomenon that the accuracy in zero-shot setting is better than cross-target setting. This is because zero-shot stance detection leverages a broader source of supervision, which allows models have better semantic understanding across different topics. Our proposed method outperforms comparison baselines on most tasks and improves the average F_{avg} and F_m by 3.5% and 2.7%, respectively.

Table 4. Performance comparison of zero-shot stance detection.

Models	Zero-shot settings (%)					
	SH		**WM**		**VA**	
	F_{avg}	F_m	F_{avg}	F_m	F_{avg}	F_m
BiLSTM	45.6	49.6	25.7	31.7	36.7	42.4
BiCond	45.7	50.1	29.8	34.6	29.6	35.2
TextCNN	41.0	41.5	35.7	37.8	34.8	39.2
TAN	45.8	47.7	50.2	51.7	46.5	49.3
CrossNet	45.9	49.3	55.6	56.8	45.1	48.2
BERT	49.6	53.8	<u>63.4</u>	<u>64.6</u>	<u>57.5</u>	<u>59.4</u>
WS-BERT-S	48.6	53.0	61.0	62.4	55.6	57.9
WS-BERT-D	<u>51.6</u>	<u>55.2</u>	61.6	63.3	55.3	57.6
Ours	**53.3**	**56.2**	**65.1**	**66.4**	**58.9**	**59.9**
Improve (%)	3.3%	1.8%	2.7%	2.8%	2.4%	0.8%

We observe that BiLSTM, BiCond, and TextCNN overall perform worst mainly because they do not explicitly use topic information or have limited perception of unknown destination topic, which affirms the importance of designing a topic-aware approach for cross-target and zero-shot settings. For TAN and CrossNet, by designing an attention mechanism in the network to notice the importance of each word and reflect users' concerns, they can provide interpretable evidence to enable semantic understanding across different topics. BERT is still a comprehensive baseline, even though it ignores transferable knowledge between topics, but has strong generalization ability because it learns from a large-scale unsupervised corpus. Therefore, BERT exploits rich semantic information to perform relatively good performance in both scenarios. WS-BERT is a recent powerful stance classifier that leverages external knowledge from Wikipedia as a bridge to enable model's deeper understanding and then precisely captures the stance towards a topic. However, since WS-BERT does not make constraints for learning transferable knowledge between different topics in the training phase, it will prone to fitting to a specific topic, which results in bringing performance drops when topic transfer. Compared with the above baselines, our proposed model takes topic information, external knowledge (i.e., policy descriptions), and non-text information (i.e., geographic signals) into account to improve model's discriminability. Meanwhile, adversarial learning is applied as a constraint to learn topic-invariance to facilitate enhancing the model's transferability.

3.3   Model Size and Efficiency

In Table 5, we record and compare the model size and efficiency of all methods on **SH** task, which uses the most samples for training. For the legibility of the comparison results, we plot the relationship between model performance, model size, and training time in Fig. 2. BERT-based models have large trainable parameters because increasing model size when pretraining natural language representations often result in improved performance on downstream tasks. Compared to WS-BERT, our proposed model has an acceptable model size. Among all models, TextCNN is the fastest in terms of training time and inference time, as it applies convolutional neural networks to capture word coherence in parallel. Our model has the longest training time per epoch, this is because, during the adversarial learning process, it is necessary to leverage a large amount of unlabeled data to train the discriminator for domain adaptation. Therefore, the increment of training samples will inevitably increase the training time per epoch. Despite this,

Table 5. Comparison of model size and efficiency of training and testing on SH.

Models	Parameters (*Million*)	Training Time (*s/epoch*)	Inference Time (*s*)
BiLSTM	39.6	2.6	0.21
BiCond	41.5	2.9	0.23
TextCNN	40.4	1.1	0.03
TAN	40.2	2.3	0.24
CrossNet	41.6	4.1	0.24
BERT	109.5	11.2	0.61
WS-BERT-S	218.9	13.7	1.01
WS-BERT-D	218.9	15.6	1.28
Ours	110.1	23.7	0.86

adversarial learning has no obvious adverse effect on inference time. Thus, the increase in training time can be seen as a trade-off against the performance gains obtained by training the discriminator to better extract transferable knowledge.

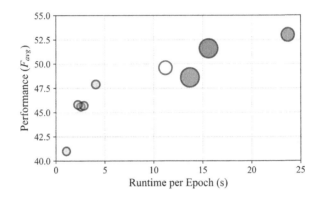

Fig. 2. Model size and efficiency comparison on SH task.

4   Conclusion and Future Work

In this paper, we consider a more realistic scenario on stance detection (i.e., cross-target and zero-shot settings) for the pandemic and propose a novel and practical stance classifier to automatically identify the public's attitudes toward COVID-19-related health policies. Specifically, we apply adversarial learning to ensure model's transferability to emerging policies. Moreover, to enhance the model's deeper understanding, we infuse policy descriptions as external knowledge and devise a GeoEncoder to capture regional background factors as non-text features. Extensive experiments demonstrate that our proposed method achieves state-of-the-art performance in both cross-target and zero-shot settings.

In the future, we will devote to building a high-accuracy stance detection application for public health policies towards COVID-19 by utilizing multi-aspect information (e.g., stance expressions, sentiments, and epidemic trends).

Acknowledgments. We thank reviewers for their helpful feedback. This work is supported by the National Natural Science Foundation of China No. 62172428.

References

1. Allaway, E., Mckeown, K.: Zero-shot stance detection: a dataset and model using generalized topic representations. In: Proceedings of EMNLP (2020)
2. Allaway, E., Srikanth, M., McKeown, K.: Adversarial learning for zero-shot stance detection on social media. In: Proceedings of NAACL (2021)
3. Augenstein, I., Rocktäschel, T., Vlachos, A., Bontcheva, K.: Stance detection with bidirectional conditional encoding. In: Proceedings of EMNLP (2016)

4. Chen, Y.: Convolutional neural network for sentence classification. Master's thesis, University of Waterloo (2015)
5. Cotfas, L.A., Delcea, C., Roxin, I., Ioanăş, C., Gherai, D.S., Tajariol, F.: The longest month: analyzing COVID-19 vaccination opinions dynamics from tweets in the month following the first vaccine announcement. IEEE Access **9**, 33203–33223 (2021)
6. Cui, S., Wang, S., Zhuo, J., Su, C., Huang, Q., Tian, Q.: Gradually vanishing bridge for adversarial domain adaptation. In: Proceedings of CVPR (2020)
7. Devlin, J., Chang, M.W., Lee, K., Toutanova, K.: Bert: pre-training of deep bidirectional transformers for language understanding. In: Proceedings of NAACL (2018)
8. Du, J., Xu, R., He, Y., Gui, L.: Stance classification with target-specific neural attention networks. In: Proceedings of IJCAI (2017)
9. Ganin, Y., Lempitsky, V.: Unsupervised domain adaptation by backpropagation. In: Proceedings of ICML (2015)
10. Glandt, K., Khanal, S., Li, Y., Caragea, D., Caragea, C.: Stance detection in COVID-19 tweets. In: Proceedings of ACL (2021)
11. He, Z., Mokhberian, N., Lerman, K.: Infusing knowledge from wikipedia to enhance stance detection. arXiv preprint arXiv:2204.03839 (2022)
12. Hochreiter, S., Schmidhuber, J.: Long short-term memory. Neural Comput. **9**(8), 1735–1780 (1997)
13. Kipf, T.N., Welling, M.: Semi-supervised classification with graph convolutional networks. arXiv preprint arXiv:1609.02907 (2016)
14. Küçük, D., Arıcı, N.: Sentiment analysis and stance detection in Turkish tweets about COVID-19 vaccination. In: Handbook of Research on Opinion Mining and Text Analytics on Literary Works and Social Media (2022)
15. Mather, B., Dorr, B.J., Rambow, O., Strzalkowski, T.: A general framework for domain-specialization of stance detection: a COVID-19 response use case. In: The International FLAIRS Conference Proceedings (2021)
16. Wei, P., Mao, W.: Modeling transferable topics for cross-target stance detection. In: Proceedings of SIGIR (2019)
17. Xu, C., Paris, C., Nepal, S., Sparks, R.: Cross-target stance classification with self-attention networks. In: Proceedings of ACL (2018)

Automatic ICD Coding Based on Segmented ClinicalBERT with Hierarchical Tree Structure Learning

Beichen Kang[1], Xiaosu Wang[1(✉)], Yun Xiong[1], Yao Zhang[1], Chaofan Zhou[1], Yangyong Zhu[1], Jiawei Zhang[2], and Chunlei Tang[3]

[1] Shanghai Key Laboratory of Data Science, School of Computer Science, Fudan University, Shanghai, China
`xswang19@fudan.edu.cn`
[2] IFM Lab, Department of Computer Science, University of California, Davis, CA, USA
[3] Harvard Medical School, Boston, MA, USA

Abstract. Automatic ICD coding aims at assigning the international classification of disease (ICD) codes to clinical notes documented by clinicians, which is crucial for saving human resources and has attracted much research attention in recent years. However, facing the challenges brought by the complex long textual narratives in clinical notes and the long-tailed data distribution in ICD codes, existing studies are ineffectual in the struggle to extract key information from the clinical notes and handle large amounts of small-data learning problems on the tail codes, which makes it hard to achieve satisfactory performance. In this paper, we present a ClinicalBERT-based model for automatic ICD coding, which can effectively cope with complex long clinical narratives via a segmentation learning mechanism and take advantage of the tree-like structure of ICD codes to transmit information among code nodes. Specifically, a novel hierarchical tree structure learning module is proposed to enable each code to utilize information both from upper and lower nodes of the tree, so that better code classifiers are learned for both head and tail codes. Experiments on MIMIC-III dataset show that our model outperforms current state-of-the-art (SOTA) ICD coding methods.

Keywords: Automatic ICD coding · Segmented ClinicalBERT · Hierarchical tree structure learning

1 Introduction

A clinical note is usually composed of multiple long and heterogeneous textual narratives (e.g., discharge summary, past medical history, history of present illness) generated by clinicians during patient encounters, and can be assigned a set of codes from the international classification of diseases (ICD), which represent diagnostic and procedural information during patient visits, as shown in

X. Wang et al. (Eds.): DASFAA 2023, LNCS 13946, pp. 250–265, 2023.
https://doi.org/10.1007/978-3-031-30678-5_19

Input: Clinical Note

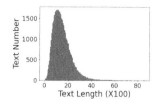

ICD Coding

Output: Predicted ICD Codes

ICD–9 Codes	Disease Name
790.01	Precipitous drop in hematocrit
428.31	Diastolic heart failure
331.4	Obstructive hydrocephalus
003.22	Salmonella pneumonia
...	...

(a) ICD coding.

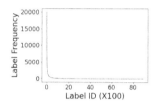

(b) The text length distribution for MIMIC-III.

(c) Long-tailed data distribution in MIMIC-III ICD codes.

Fig. 1. ICD Coding task and the statistics for MIMIC-III dataset.

Fig. 1(a). ICD codes can be exploited in a number of ways, ranging from multinational medical cooperation communication to predictive modeling of patient state. Manual ICD coding is not only time-consuming and labor-intensive, but also prone to errors, and thus brings huge economic losses [20]; which makes automatic ICD coding a desired technology [7].

Automatic ICD coding is a multi-label classification task, which aims at assigning a set of associated ICD codes to a clinical note. Automatic ICD coding task requires a model to accurately summarize the key information of clinical notes, understand the medical semantics corresponding to ICD codes, and perform precise matching based on the understanding of the two, which is extremely challenging. As shown in Fig. 1(b), the average length of clinical notes exceeds 1500 words, but only a small part of them are relevant to the ICD coding task; furthermore, the clinical notes contain a multitude of medical abbreviations and synonyms, which leads to a semantic gap between clinical note descriptions and ICD code descriptions and causes ambiguity and imprecision when matching ICD codes to clinical notes. Early researches build text feature extractors based on CNN [4,14,19,30,34] or RNN [24,28,32]. The CNN-based methods capture the context information of a word by aggregating the information of the left and right words; however, limited by the size of convolution kernel, CNN-based methods can only capture the local information of words. The RNN-based methods can capture better long-distance features, but there are still shortcomings of short-term memory. Models based on CNN or RNN cannot cope with the challenges brought by the complex long textual narratives in clinical notes due to their own shortcomings.

As the pre-trained language models (PLM) making great progress on general domain natural language understanding tasks, active works [3,9,11,21] employ transformer-based PLMs, e.g. BERT, as text feature extractors for ICD coding task. Limited by the upper limit of the text length that a PLM can handle, the paper [5] finds that truncating text to control the length of the input sequence loses a lot of important information and results in worse performance than the previous SOTA results [28,30]. Instead of directly truncating the clinical notes, in this paper, we split a clinical note into multiple fixed-length text segments, and set a certain length of overlap between each adjacent segment, so as to prevent the loss of important n-gram information. Moreover, ClinicalBERT [1], as a clinical-domain PLM, can better understand the semantics of clinical text and is exploited to better extract features from each segmentation; subsequently, full-document information is summarized by an aggregation layer consisting of RNNs.

Although extracting key information from long and semantically complex clinical notes is already difficult, another challenge is the long-tailed data distribution in ICD codes. As shown in Fig. 1(c), the code label space is very high-dimensional, where there are over 15,000 codes in the ICD-9 taxonomy and over 140,000 codes in the ICD-10 taxonomy; the number of sample instances of ICD codes presents a long-tailed distribution, which means that the majority of ICD codes are not frequently used. Therefore, an automatic ICD coding model not only needs to be able to extract more informative features from long and semantically complex clinical notes, but also needs to deal with large amounts of small-data learning problems on the tail codes. Whereas existing studies [4,25,30] treat the tree-like hierarchical structure of the ICD codes as an undirected graph without distinguishing the meaning of connecting edges between parent and child nodes, which ignores the hierarchical information transmission that helps alleviate the small-data learning problem on tail ICD codes. For example, the ICD codes 275.1 ("Disorders of copper metabolism") and 275.4 ("Disorders of calcium metabolism") are the child categories of the ICD code 275 ("Disorders of mineral metabolism"); the understanding of ICD code 275.4 can help the understanding of its sibling code 275.1 by passing information through its parent code 275. In this paper, we propose Hierarchical Tree structure learning based on a Hierarchical Information Transmission (**THIT**) module, which takes the textual descriptions of the ICD codes and their hierarchical structure as input and produces a latent representation for each code. The hierarchical approach enables our model to exploit information both from upper and lower nodes of the tree for a particular node. In this way, the representation simultaneously captures the semantics of each code and the hierarchical relationship among codes.

Our major contributions can be summarized as follows:

– We propose a novel ClinicalBERT-based model for automatic ICD Coding which employs a segmentation mechanism to solve the problem that BERT is difficult to handle long texts.

- We incorporate tree structure learning based on hierarchical information transmission to capture the hierarchical structure information and assist codes with less samples to learn their representation.
- Extensive experiments are conducted on the benchmark MIMIC-III dataset, and the results demonstrate the effectiveness and superiority of our model.

2 Related Work

2.1 Pretrained Language Models

Since the appearance of the Transformer model [27], pretrained language models (PLMs) with the transformer architecture have led to consistent improvements across almost every area of Natural Language Processing. BERT [8], a bidirectional transformer, revealed that pretraining models can produce contextual representations and transfer well to many tasks. Variants such as XLNet [31], RoBERTa [16] are pretrained on larger general corpus to enhance the ability to model textual data. Nonetheless, these PLMs can not achieve satisfactory results in the medical domain because healthcare data contains large specialized vocabulary and general domain PLMs do not possess specific medical knowledge. Therefore, prior works adapted such models in biomedical domain, biomedical-domain PLMs, such as BioBERT [13], ClinicalBERT [1] and PubMedBERT [10], have been proposed. These PLMs are pretrained on medical text and improve performance on many biomedical tasks, including question answering, document classification and evidence-based medical information extraction.

In order to extract better biomedical-specific contextual representations for ICD coding task, we use ClinicalBERT [1] as the backbone of our text feature extraction module. Due to the limit of the input sequence length, we adopt a novel segmentation mechanism and then through label-wise attention layer to learn label-specific document representations insead of directly using chunk level representations as in BERT-XML [33].

2.2 ICD Coding

Methods Based on CNN or RNN. The study of automatic ICD coding can be traced back to the late 1990s [7]. Most of these studies regard ICD coding as a multi-label text classification problem. Extensive works have been done on applying deep learning approaches to automatic ICD coding. Many of these approaches rely on variants of Convolutional Neural Networks (CNNs) and Long Short-Term Memory Networks (LSTMs). Mullenbach [19] proposed an attentional convolutional network (CAML) to generate label-dependent representation for each label. As convolution neural network with fixed window size can not handle different lengths of input texts, [14] used a multi-filter convolutional layer to capture various text patterns with different lengths. JointLAAT [28] utilized a bidirectional LSTM with an improved label-aware attention mechanism to learn label-specific vectors of clinical text for each label. Recently, ISD [34] employed an interactive shared representation network and a self-distillation mechanism to alleviate the noisy text and the long-tailed problem. Although ISD used transformer as the attention layer, it still used CNN to extract text features, so we classify it as a CNN-based method.

Except for using CNN or LSTM for document feature learning, some works tried to consider the hierarchical structure of ICD codes. Xie et al. [30] leveraged graph convolutional neural network to capture both the hierarchical relationship among medical codes to tackle the imbalanced label distribution problem. Similarly, Cao et al. [4] embedded ICD codes into hyperbolic space and constructed a graph which considered the co-occurrence correlation. JointLAAT [28] also proposed a hierarchical joint learning architecture to handle the tail codes. Different from these works, we utilize ICD codes tree hierarchy for tree structure learning, which enables a particular node to learn from both upper and lower nodes of the tree. In this way, we argue that the tail codes can capture information of other nodes and alleviate long-tailed data distribution problem.

Transformer-Based Methods. Though PLMs achieved state-of-the-art performance on many NLP tasks, applying PLMs to ICD coding is still challenging. Pascual et al. [21] used PubMedBERT for ICD coding task and found that the main limitation was clinical notes usually exceed BERT maximum input length. They devised five different strategies to split long text but still had a gap with current state-of-the-art models. Feucht et al. [9] used pretrained model Longformer [2] combined with a description-based label attention classifier (DLAC) to achieve better performance. Biswas et al. [3] traind a Transformer encoder from scratch to overcome sequence length limitation and captured contextual word representations. Huang et al. [11] focused on leveraging different PLMs for text encoder and predicted ICD codes after a label attention layer. In contrast, our model takes advantage of the hierarchical structure of ICD codes and employs ClinicalBERT to extract contextual representations.

2.3   Tree Structure Representation Learning

ICD coding follows a tree-like structure to ensure the functional integrity of classification, and the codes to be classified are leaf nodes. Therefore, the tree structure representation learning is helpful to predict ICD codes, which can capture the hierarchical association between codes and the meanings of each code at the same time. Tree-LSTM [26] was first proposed to model the dependency parse trees of sentences and transfer information upward according to the structure of the syntax tree to obtain the representation of the whole syntax tree. For ICD coding task, Tree-LSTM was applied in [6] to encode ICD coding structure. Other researchers proposed a tree-of-sequences LSTM [29] to encode code descriptions and promote the co-assignment of clinically relevant codes, but their model did not perform well on infrequent codes. To address the long-tailed data distribution issue in ICD coding, we incorporate the tree structure learning to transfer information between parent and child nodes and learn better classifiers for infrequently occurring codes.

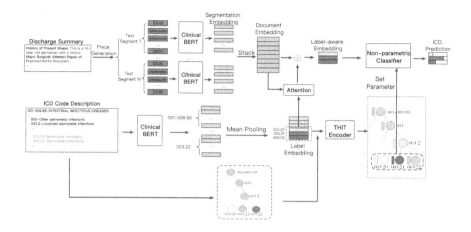

Fig. 2. An overview of our SCB-T architecture for automatic ICD coding. SCB-T uses segmented learning and label-wise attention to extract the features of the text, and then learns the label embedding through THIT module, and finally use a linear classifier to predict the probability of each code.

3   Methods

In this section, we describe our new Segmented ClinicalBERT with Hierarchical Tree Structure Learning model (SCB-T) for automated ICD coding. We first introduce an overview of our proposed method, and then a detailed description of each component.

3.1   Overview

The overall model is illustrated in Fig. 2. We regard this ICD coding task as a multi-label text classification task [18]. Let $\mathbf{L}$ represent the set of all ICD codes. The clinical document can be represented as $\{w_1, w_2, \ldots, w_m\}$. In addition, each code $l \in \mathbf{L}$ has an offical description which can be represented as $\{w_1, w_2, \ldots, w_{n_i}\}$ Our goal is to determine $y_l \in \{0, 1\}$ for all $l \in \mathbf{L}$.

3.2   Text Feature Extraction Layer

In this layer, our model needs to input both the medical record texts and ICD code description texts. On the one hand, the complexity of transformers scales quadratically with the length of their input, which restricts the maximum number of words that they can process at once [21], and clinical notes usually exceed this maximum input length. We propose a novel segmentation learning mechanism which allows ClinicalBERT to perform text feature extraction on long texts. On the other hand, healthcare data contains a specific vocabulary that is not common within a general pretraining corpus, leads to many out of vocabulary(OOV) words [33]. Therefore, we use ClinicalBERT to replace the common language pre-trained BERT, which alleviate the semantic gap between common language and medical.

Document Feature Extraction. Given a clinical note $\{w_1, w_2, \ldots, w_m\}$. We specify the segment length as s, and divide it into equal-length text segments. As directly truncate the text will lose a lot of information, we use overlap at the cut-off point to avoid losing important n-gram information in this segmentation operation. We use an overlap of k words specifically, taking $k = 3$ as an example, we split the clinical note into $\{w_1, w_2, \ldots, w_s\}; \{w_{s-3}, w_{s-2}, \ldots, w_{2s-4}\}; \cdots$ to ensure that the n-gram phrase is complete in at least one segment. We use pre-trained ClinicalBERT to get the embedding of each segment and then we use an aggregation layer to obtain document-level text feature. This process can be formally defined as:

$$
\begin{aligned}
\mathbf{X}_i^{seg} &= ClinicalBERT(w_1, w_2, \ldots, w_s) \\
\mathbf{X}_{doc} &= Stack(\mathbf{X}_1^{seg}, \mathbf{X}_2^{seg}, \ldots)
\end{aligned}
\tag{1}
$$

where $\mathbf{X}_i^{seg}$ refers to the embedding of the ith segment. Because the clinical note is divided into multiple segments and the segments can not sense each other. We add an aggregation layer to perform global feature learning for text segments, where $\mathbf{X_{doc}}$ refers to the document-level text feature. In the aggregation layer we use LSTM or GRU.

Label Feature Extraction. Our goal is to obtain the embeddings of ICD code description texts as label features. Given a code description text $\{w_1, w_2, \ldots, w_{n_i}\}$, In order to maintain the consistency between the label features and the document-level text features, we also use ClinicalBERT to get the embedding of code description $\mathbf{X}_i^{label} = [x_1, x_2, \ldots, x_{n_i}]$. Because the text length of the code description information is inconsistent, we average all word embeddings of a code description to get v_i as the label feature, the calculation method is as follows:

$$
v_i = \frac{1}{n_i} \sum_{j=1}^{n_i} x_j
\tag{2}
$$

3.3 Attention Layer

When assigning multiple labels (i.e., ICD codes) to each document, the labels can affect each other. For example, when the clinical note contains ICD codes "288.51" and "414.11", where "288.51" represents "Lymphocytopenia" and "414.11" represents "Aneurysm of coronary vessels". When classifying the label "Aneurysm of coronary vessels", the description of the symptom "Lymphocytopenia" in the clinical note is redundant. For this reason, we apply the label attention mechanism to make each ICD code attend to different parts of the document feature $\mathbf{X_{doc}}$. Firstly, we need to calculate the attention weights:

$$
\begin{aligned}
\boldsymbol{\alpha} &= softmax(\boldsymbol{V} \boldsymbol{X}^{doc}) \\
\boldsymbol{\alpha}_i &= [\alpha_i^1, \alpha_i^2, \ldots, \alpha_i^m]_{L \times m}
\end{aligned}
\tag{3}
$$

where $\boldsymbol{\alpha}$ represents the attention weights for each pair of an ICD code and a word, $\mathbf{V}$ is the embedding of all labels; $\boldsymbol{\alpha}_i$ is then used to compute vector representation for each label, L is the total number of the labels. After that, the attention weight matrix $\boldsymbol{\alpha}$ is then multiplied with the document feature $\mathbf{X_{doc}}$ to select label-specific vectors representing the input clinical note as:

$$\boldsymbol{X}^{attn} = Matmul(\boldsymbol{\alpha} \boldsymbol{X}^{doc})$$
$$\boldsymbol{X}^{attn} = [\boldsymbol{x}_1^{attn}, \boldsymbol{x}_2^{attn}, \ldots, \boldsymbol{x}_L^{attn}]_{L \times d} \tag{4}$$

each ith column $\boldsymbol{x}_i^{attn}$ of the matrix $\mathbf{X^{attn}}$ is a representation of the input clinical note mapped to the ith label feature space and thus avoiding unrelated labels interfering with each other during classification.

3.4 Hierarchical Tree Structure Learning Layer

As the hierarchical structure of ICD codes is a tree, previous works [4,25,30] treated the hierarchy of codes as an undirected graph, without distinguishing the meaning of connecting edges between parent and child nodes. Due to the dimensionality of the label space, most of the ICD codes are not frequently used leading to an extremely imbalanced data problem. We can utilize the hierarchical structure among the codes to help the model work better for infrequent codes. For example, the ICD codes 275.1 ("Disorders of copper metabolism") and 275.4 ("Disorders of calcium metabolism") are the child categories of the ICD code 275 ("Disorders of mineral metabolism"); the understanding of ICD code 275.4 can help the understanding of its sibling code 275.1 by passing information through its parent code 275.

To this end, we propose Hierarchical Tree structure learning based on Hierarchical Information Transmission(**THIT**) module, which takes the textual descriptions of the ICD codes and their hierarchical structure as input and produces a latent representation for each code. The hierarchical approach enables the tree structure to use information both from upper and lower nodes of the tree for a particular node. In this manner, the representation simultaneously captures the semantics of each code and the hierarchical relationship among codes. Our module consists of an upward stage and a downward stage, which produces two hidden states $\boldsymbol{h}_p'$ and $\boldsymbol{h}_c'$ at each node in the tree, as shown in Fig. 3.

Upward Stage. During the Upward stage, every activation takes the code hierarchy, node original feature $\boldsymbol{v}_i$ and previous hidden state $\boldsymbol{h}_i$ which are found lower on the structure in a bottom up fashion as inputs. To update the node representation of all parent nodes from leaves to root, every hidden state has been computed using the GRUCell module, then the root hidden state is used as input of the Downward stage. This process is calculated as follows:

$$
\begin{aligned}
\boldsymbol{r} &= \sigma(\boldsymbol{W}_{ir}\boldsymbol{v}_i + \boldsymbol{b}_{ir} + \boldsymbol{W}_{hr}\boldsymbol{h}_i + \boldsymbol{b}_{hr}) \\
\boldsymbol{z} &= \sigma(\boldsymbol{W}_{iz}\boldsymbol{v}_i + \boldsymbol{b}_{iz} + \boldsymbol{W}_{hz}\boldsymbol{h}_i + \boldsymbol{b}_{hz}) \\
\boldsymbol{n} &= tanh(\boldsymbol{W}_{in}\boldsymbol{v}_i + \boldsymbol{b}_{in} + \boldsymbol{r} * (\boldsymbol{W}_{hn}\boldsymbol{h}_i + \boldsymbol{b}_{hn})) \\
\boldsymbol{h}_i' &= (1 - \boldsymbol{z}) * \boldsymbol{n} + \boldsymbol{z} * \boldsymbol{h}_i
\end{aligned}
\tag{5}
$$

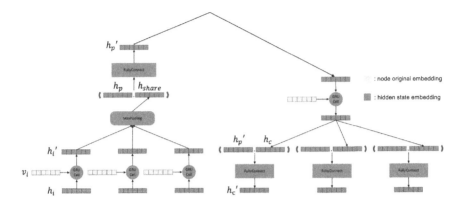

Fig. 3. An overview of the upward and downward stage of THIT module. The gray square represents the hidden state embedding, and the yellow square represents the node original embedding. (Color figure online)

where h_i' is the hidden state of the next stage, W are component-specific weight matrices and b are bias vectors. Assume that the current parent node is p, the hidden states of all its children obtained through GRUCell are used to update the features of its parent node p. The update method is as follows:

$$h^{share} = MaxPooling(h_1', h_2', \ldots, h_{\|C_p\|}')$$
$$h_p' = FC(Concat(h_{share}, h_p))$$

(6)

where h_{share} is the common feature of all child nodes using maxpooling, C_p is the set of child nodes of node p and h_p' represents the hidden state of next stage of node p, which is concatenated by the common feature and the previous hidden state.

Because the label nodes is the leaf nodes in the tree structure, the hidden state of the root node captures the semantics of all codes in this hierarchy during the above upward stage, which is then needed to be transmitted downwards to each individual code via the following downward stage.

Downward Stage. The Downward stage calculates the activations for every child of a node using content from the parent in a top down fashion. The process of computing the hidden states between the two stages is separated, so after the upward stage is completed, then the downward representations are computed. The downward stage is similar to the upward stage, through GRUCell so that the hidden state of the parent node can retain the information which can be supplemented to the child nodes. The child node is updated as follows:

$$h_c' = FC(Concat(h_c, h_p'))$$

(7)

where h_c is the previous hidden state of the child node, h_c' is the hidden state of the next stage of the child node. We concatenate the hidden states of the parent and the child to obtain the updated hidden state of each leaf code. By passing information from top to the bottom, the child nodes can capture the abstract information of the parent nodes. After obtaining the label features incorporated with the tree structure information, we extract leaf node features as code classifiers $\boldsymbol{W}_{cls}$, which is used for final multi-label classification:

$$\boldsymbol{v}_i' = Concat(\boldsymbol{v}_i, \boldsymbol{h}_i')$$
$$\boldsymbol{W}_{cls} = Stack(\boldsymbol{v}_1', \boldsymbol{v}_2', \dots, \boldsymbol{v}_L')$$

$$(8)$$

where $\boldsymbol{v}_i'$ is concatenated by the original node feature $\boldsymbol{v}_i$ and the hidden state $\boldsymbol{h}_i'$ learned by THIT.

In conclusion, the Upward Stage composes the semantics of children (representing sub-diseases) and merges them into the current node, which hence captures child-to-parent relationship. The Downward Stage makes each node inherit the semantics of its parent, which captures parent-to-child relation. As a result, the hierarchical relationship among codes are encoded in the hidden states and better code classifiers are learned for infrequently occurring codes.

3.5   Output Layer

Given the label-specific document representation from attention layer which is represented as $\boldsymbol{X}^{attn}$, the output layer computes a probability for each label using another linear layer and a sigmoid transformation:

$$\boldsymbol{X}^{cls} = \boldsymbol{X}^{attn}\boldsymbol{W}_{proj} \tag{9}$$

$$\hat{\boldsymbol{p}} = Sigmoid(\boldsymbol{X}^{cls}\boldsymbol{W}_{cls}^{\mathrm{T}}) \tag{10}$$

where $\boldsymbol{W}_{proj}$ is a mapping matrix for linear transformation, $\boldsymbol{X}^{cls}$ is the final document representation for classification, and $\hat{\boldsymbol{p}}$ is the probability of each label. Note that since the number of positive instances of labels presents a long-tailed distribution, which means the network is strongly biased towards negative predictions. Therefore, we use focal loss [15] to minimize the binary cross entropy loss between the prediction $\hat{p}$ and the target y as:

$$\boldsymbol{L} = -\alpha * (1 - \hat{p}_i)^\gamma * log(\hat{p}_i) * y_i - (1 - \alpha) * \hat{p}_i^{\ \gamma} * log(1 - \hat{p}_i) * (1 - y_i) \tag{11}$$

where α and γ are the hyper-parameters of focal loss.

Table 1. Descriptive statistics for MIMIC-III dataset.

Dataset	Split	Samples	Average Tokens	Labels	Unique Labels	Average Labels
MIMIC-III-full	train	47,723	1,434	748,306	8,692	15.68
	test	3,372	1,731	60,666	4,085	17.99
	valid	1,631	1,724	28,402	3,012	17.41
MIMIC-III-50	train	8,066	1,478	45,927	50	5.69
	test	1,729	1,763	10,428	50	5.87
	valid	1,573	1,739	9,241	50	6.03

4   Experiments

4.1   Dataset

We evaluate our model on the publicly available MIMIC-III [12] dataset for ICD-9-CM coding. We focus on discharge summaries, which condense all the information during a patient stay into a single document. In addition, we also utilize the ICD codes description and ICD hierarchy structure in our model. In MIMIC-III, each document was annotated by human coders with several ICD-9-CM codes. There were 52,722 discharge summaries and 8,929 unique codes in total. Following the previous work [19], we use the train, validation, and test splits which are split using patient ID thus no patient appears in both the training and test sets.

For comparison with prior works, we conduct experiments using **MIMIC-III-full** as well as **MIMIC-III-50**. Detailed statistics for the two settings are summarized in Table 1.

4.2   Implementation Details

We implement our model using PyTorch [22]. Our model is trained on 4 T V100-32G GPUs with a batch size of 16. The text segment length s is set to 128. The Adam optimizer [17] is employed to perform optimization, and the parameters of Adam are $\beta_1 = 0.9$, $\beta_2 = 0.999$. The learning rate update method adopts the warming-up strategy, the maximum learning rate is $5e^{-5}$, the number of warm-up steps is set to 3000. Every 1000 steps of training, the validation set will be used to test the performance of the model. In order to prevent overfitting, we employ the early stop mechanism and stop the training if there is no improvement in the validation set after three tests.

4.3   Evaluation Metrics and Baselines

To make a comprehensive comparison with the previous works on ICD coding, we employ a variety of metrics, including macro- and micro-averaged F1 and AUC (area under the ROC curve). The micro F1 score is calculated by treating each (text, code) pair as a separate prediction, while macro F1 score computes

Table 2. Results on MIMIC-III-full test set. The bold scores indicate the best results for each metric. The scores of baseline models are from the corresponding papers.

Model	AUC		F1		P@k	
	Macro	Micro	Macro	Micro	P@8	P@15
CAML	0.895	0.986	0.088	0.539	0.709	0.561
DR-CAML	0.897	0.985	0.086	0.529	0.690	0.548
MASTT-KG	0.910	**0.992**	0.090	0.553	0.728	0.581
HyperCore	0.930	0.989	0.090	0.551	0.722	0.579
LAAT	0.919	0.988	0.099	0.575	0.738	0.591
JointLAAT	0.921	0.988	0.107	0.575	0.735	0.590
ISD	0.938	0.990	0.119	0.559	0.745	–
SCB-T (ours)	**0.951**	**0.992**	**0.121**	**0.581**	**0.754**	**0.610**

a simple average over all labels. We also report precision at k (P@k $\in \{5,8,15\}$), which computes the precision of the top-k predicted labels with the highest predictive probabilities.

We select some representative works as our baselines, including CNN-based models (CAML [19], DR-CAML [19], HyperCore [4], MASTT-KG [30], ISD [34]), LSTM-based models (LAAT [28], JointLAAT [28]) and transformer-based models (BERT-ICD [21], Longformer-DLAC [9], TransICD [3]). Please note that the transformer-based models only report the results on the MIMIC-III-50 dataset.

4.4   Results and Analysis

MIMIC-III-Full. Table 2 shows the results on the MIMIC-III-full dataset. Our SCB-T model gives the strongest results in all evaluation metrics. Specifically, SCB-T outperforms the current SOTA model ISD [34] by 1.3% on Macro-AUC, 0.2% on Macro-F1, 2.2% on Micro-F1, and 0.9% on P@8. Compared to the LSTM-based or CNN-based feature extractors, the clinical-domain PLMs show superior ability in learning domain-specific representations. Due to the emphasis of macro-metrics on rare-label performance [23], this indicates that SCB-T does better for the infrequent codes by utilizing the tree structure of ICD codes to transmit hierarchical information and thus alleviates the long-tailed data distribution problem.

MIMIC-III-50. To compare with prior published works, we also evaluate on the MIMIC-III-50 dataset, as shown in Table 3. It is worth noticing that all previous transformer-based methods perform worse than other neural networks based methods, which reveals that the segmentation mechanism proposed in our paper together with the attention mechanism result in a significant improvement. Furthermore, the P@5 of our model is 69.3%, indicating that 3.465 out of 5 predictions with the top probabilities are correct. This score is relatively higher than all other baselines.

Table 3. Results on MIMIC-III-50 test set. The bold scores indicate the best results for each metric. The scores of baseline models are from the corresponding papers.

Model	AUC		F1		
	Macro	Micro	Macro	Micro	P@5
BERT-ICD	0.845	0.887	–	–	–
Longformer-DLAC	0.870	0.910	0.522	0.620	0.610
TransICD	0.894	0.923	0.562	0.644	0.617
HyperCore	0.895	0.929	0.609	0.663	0.632
MASTT-KG	0.914	0.936	0.638	0.684	0.644
LAAT	0.925	0.946	0.666	0.715	0.675
JointLAAT	0.925	0.946	0.661	0.716	0.671
ISD	**0.935**	0.949	0.679	0.717	0.682
SCB-T(ours)	**0.935**	**0.953**	**0.685**	**0.727**	**0.693**

Table 4. Ablation results on the MIMIC-III-full dataset.

Model	AUC		F1		P@k	
	Macro	Micro	Macro	Micro	P@8	P@15
SCB-T w/o ClinicalBERT	0.925	0.986	0.109	0.579	0.737	0.592
SCB-T w/o segmentation	0.897	0.984	0.089	0.537	0.702	0.553
SCB-T w/o THIT	0.920	0.982	0.104	0.562	0.733	0.589
SCB-T(ours)	**0.951**	**0.992**	**0.121**	**0.581**	**0.754**	**0.610**

4.5   Ablation Study

To verify the effectiveness of our proposed components of the method, we further perform an ablation study on the MIMIC-III-full dataset. The corresponding results are reported in Table 4.

The Effect of ClinicalBERT. In particular, for the first setting, namely SCB-T w/o ClinicalBERT, we replace ClinicalBERT with CNN-based text feature extractor proposed in [19]. We notice a substantial drop in every metric when ClinicalBERT is not adopted. In another way, ClinicalBERT improves the performance because it can extract better biomedical-specific contextual representations.

The Effect of Segmentation Mechanism. When we adopt ClinicalBERT but remove the segmentation mechanism (SCB-T w/o segmentation) and truncate the text length to 512, the results are even worse than SCB-T w/o ClinicalBERT. This clearly demonstrates that truncating the text directly will lose a lot of important information, and the proposed segmentation mechanism with overlapping words operation alleviates this problem well because it preserves important n-gram information.

The Effect of THIT. For the third setting, we discard the whole THIT part (SCB-T w/o THIT) and directly feed the extracted text features to the multi-label classifier. This further hurts the performance of the model. THIT can increase performance by 1.7% on Macro-F1 and 2.1% on Micro-F1. This verifies that THIT module can take advantage of the hierarchy structure of ICD codes to learn better code representations for classification. Specifically, THIT module can utilize the tree structure of ICD codes to transmit hierarchical information, enables a particular node to use information both from upper and lower nodes of the tree and thus helps model to achieve better results.

5   Conclusion

In this paper, we propose a segmented ClinicalBERT-based method and a tree structure learning module to extract key information from the clinical notes and handle large amounts of small-data learning problems on the tail codes for the automatic ICD coding task. The model takes advantage of ClinicalBERT based on segmentation learning, which extracts better representations for lengthy medical narratives. And the tree structure learning module based on hierarchical information transmission captures the parent-child relationship among codes and transmit the information to help the model work better for infrequent codes. Experimental results on MIMIC-III dataset indicate that our proposed model outperforms previous state-of-the-art methods. Further ablation studies demonstrate that ClinicalBERT with segmentation mechanism and the tree structure leaning in SCB-T have improved the performance.

Acknowledgements. This work is partially supported by the National Key Research and Development Plan Project 2022YFC3600901, CNKLSTISS, and NSF through grants IIS-1763365, IIS-2106972.

References

1. Alsentzer, E., et al.: Publicly available clinical BERT embeddings. arXiv:1904.03323 (2019)
2. Beltagy, I., Peters, M.E., Cohan, A.: Longformer: The long-document transformer. arXiv:2004.05150 (2020)
3. Biswas, B., Pham, T.-H., Zhang, P.: TransICD: transformer based code-wise attention model for explainable ICD coding. In: Tucker, A., Henriques Abreu, P., Cardoso, J., Pereira Rodrigues, P., Riaño, D. (eds.) AIME 2021. LNCS (LNAI), vol. 12721, pp. 469–478. Springer, Cham (2021). https://doi.org/10.1007/978-3-030-77211-6_56
4. Cao, P., Chen, Y., Liu, K., Zhao, J., Liu, S., Chong, W.: Hypercore: hyperbolic and co-graph representation for automatic ICD coding. In: Proceedings of the 58th Annual Meeting of the Association for Computational Linguistics, pp. 3105–3114 (2020)
5. Chalkidis, I., Fergadiotis, M., Kotitsas, S., Malakasiotis, P., Aletras, N., Androutsopoulos, I.: An empirical study on large-scale multi-label text classification including few and zero-shot labels. In: EMNLP (2020)

6. Chen, Y., Ren, J.: Automatic ICD code assignment utilizing textual descriptions and hierarchical structure of ICD code. In: 2019 IEEE International Conference on Bioinformatics and Biomedicine (BIBM), pp. 348–353. IEEE (2019)

7. De Lima, L.R., Laender, A.H., Ribeiro-Neto, B.A.: A hierarchical approach to the automatic categorization of medical documents. In: Proceedings of the Seventh International Conference on Information and Knowledge Management, pp. 132–139 (1998)

8. Devlin, J., Chang, M.W., Lee, K., Toutanova, K.: Bert: Pre-training of deep bidirectional transformers for language understanding. arXiv:1810.04805 (2018)

9. Feucht, M., Wu, Z., Althammer, S., Tresp, V.: Description-based label attention classifier for explainable ICD-9 classification. arXiv:2109.12026 (2021)

10. Gu, Y., et al.: Domain-specific language model pretraining for biomedical natural language processing. ACM Trans. Comput. Healthc. (HEALTH) 3(1), 1–23 (2021)

11. Huang, C.W., Tsai, S.C., Chen, Y.N.: PLM-ICD: automatic ICD coding with pretrained language models. arXiv:2207.05289 (2022)

12. Johnson, A.E., et al.: MIMIC-III, a freely accessible critical care database. Sci. Data 3(1), 1–9 (2016)

13. Lee, J., et al.: BioBERT: a pre-trained biomedical language representation model for biomedical text mining. Bioinformatics 36(4), 1234–1240 (2020)

14. Li, F., Yu, H.: ICD coding from clinical text using multi-filter residual convolutional neural network. In: Proceedings of the AAAI Conference on Artificial Intelligence, vol. 34, pp. 8180–8187 (2020)

15. Lin, T.Y., Goyal, P., Girshick, R., He, K., Dollár, P.: Focal loss for dense object detection. In: Proceedings of the IEEE International Conference on Computer Vision, pp. 2980–2988 (2017)

16. Liu, Y., et al.: Roberta: a robustly optimized BERT pretraining approach. arXiv:1907.11692 (2019)

17. Loshchilov, I., Hutter, F.: Decoupled weight decay regularization. arXiv:1711.05101 (2017)

18. McCallum, A.K.: Multi-label text classification with a mixture model trained by EM. In: AAAI 99 Workshop on Text Learning (1999)

19. Mullenbach, J., Wiegreffe, S., Duke, J., Sun, J., Eisenstein, J.: Explainable prediction of medical codes from clinical text. arXiv:1802.05695 (2018)

20. O'malley, K.J., Cook, K.F., Price, M.D., Wildes, K.R., Hurdle, J.F., Ashton, C.M.: Measuring diagnoses: ICD code accuracy. Health Serv. Res. 40(5p2), 1620–1639 (2005)

21. Pascual, D., Luck, S., Wattenhofer, R.: Towards BERT-based automatic ICD coding: limitations and opportunities. arXiv:2104.06709 (2021)

22. Paszke, A., et al.: Pytorch: an imperative style, high-performance deep learning library. In: Advances in Neural Information Processing Systems, vol. 32 (2019)

23. Schütze, H., Manning, C.D., Raghavan, P.: Introduction to Information Retrieval, vol. 39. Cambridge University Press, Cambridge (2008)

24. Shi, H., Xie, P., Hu, Z., Zhang, M., Xing, E.P.: Towards automated ICD coding using deep learning. arXiv:1711.04075 (2017)

25. Song, C., Zhang, S., Sadoughi, N., Xie, P., Xing, E.: Generalized zero-shot text classification for ICD coding. In: Proceedings of the Twenty-Ninth International Conference on International Joint Conferences on Artificial Intelligence, pp. 4018–4024 (2021)

26. Tai, K.S., Socher, R., Manning, C.D.: Improved semantic representations from tree-structured long short-term memory networks. arXiv:1503.00075 (2015)

27. Vaswani, A., et al.: Attention is all you need. In: Advances in Neural Information Processing Systems, vol. 30 (2017)

28. Vu, T., Nguyen, D.Q., Nguyen, A.: A label attention model for ICD coding from clinical text. arXiv:2007.06351 (2020)

29. Xie, P., Xing, E.: A neural architecture for automated ICD coding. In: Proceedings of the 56th Annual Meeting of the Association for Computational Linguistics (Volume 1: Long Papers), pp. 1066–1076 (2018)

30. Xie, X., Xiong, Y., Yu, P.S., Zhu, Y.: EHR coding with multi-scale feature attention and structured knowledge graph propagation. In: Proceedings of the 28th ACM International Conference on Information and Knowledge Management, pp. 649–658 (2019)

31. Yang, Z., Dai, Z., Yang, Y., Carbonell, J., Salakhutdinov, R.R., Le, Q.V.: XLNet: generalized autoregressive pretraining for language understanding. In: Advances in Neural Information Processing Systems, vol. 32 (2019)

32. Yuan, Z., Tan, C., Huang, S.: Code synonyms do matter: Multiple synonyms matching network for automatic ICD coding. arXiv:2203.01515 (2022)

33. Zhang, Z., Liu, J., Razavian, N.: Bert-xml: Large scale automated ICD coding using BERT pretraining. arXiv:2006.03685 (2020)

34. Zhou, T., et al.: Automatic ICD coding via interactive shared representation networks with self-distillation mechanism. In: Proceedings of the 59th Annual Meeting of the Association for Computational Linguistics and the 11th International Joint Conference on Natural Language Processing (Volume 1: Long Papers), pp. 5948–5957 (2021)

Hierarchical Encoder-Decoder with Addressable Memory Network for Diagnosis Prediction

Mingxia Wang[1], Yun Xiong[1(✉)], Yao Zhang[1], Philip S. Yu[2], and Yangyong Zhu[1]

[1] Shanghai Key Laboratory of Data Science, School of Computer Science, Fudan University, Shanghai, China
{wangmx20,yunx,yaozhang,yyzhu}@fudan.edu.cn
[2] Department of Computer Science, University of Illinois Chicago, Chicago, USA
psyu@uic.edu

Abstract. Deep learning methods have demonstrated success in diagnosis prediction on Electronic Health Records (EHRs). Early attempts utilize sequential models to encode patient historical records, but they lack the ability to identify critical diseases for patient health conditions. Besides, some works focus on the hierarchical structure of EHR data during learning patient representations, but neglect it in diagnosis prediction process, which leads to insufficient utilization of hierarchical information. To tackle these challenges, we propose a new **H**ierarchical Encoder-Decoder with **A**ddressable **M**emory **Net**work for Diagnosis Prediction named HAMNet. Specifically, we employ a hierarchical encoder-decoder framework to utilize the hierarchical structure of historical visits both during representation learning and diagnosis prediction. Furthermore, we propose the addressable memory network to distinguish core diseases towards patient health status at next visit through the well-designed addressing mechanism. It employs feed-forward layers to build the memory network, which can automatically update the memory module through gradient backpropagation without explicit read/write operations. Finally, we evaluate the performance of HAMNet on a large real-world EHR dataset MIMIC-III, and it achieves the state-of-the-art performance compared with baselines.

Keywords: Diagnosis Prediction · Bioinformatics · Data Mining

1 Introduction

Benefiting from the widespread use of healthcare systems, Electronic Health Records are widely collected including diagnoses, procedures and medications. EHR data has a two-level hierarchy as shown in Fig. 1. A large amount of EHR data contains valuable information, and it provides great opportunity for diagnosis prediction task [1,2,4,11–13], which refers to predicting the future diagnosis according to historical medical records.

X. Wang et al. (Eds.): DASFAA 2023, LNCS 13946, pp. 266–275, 2023.
https://doi.org/10.1007/978-3-031-30678-5_20

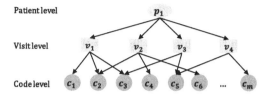

Fig. 1. An example shows the hierarchy structure of EHR data. Patient p_1 has four historical visit records, and each visit contains several medical codes.

Early attempts use sequential models [1–3,12,13] to encode sequences of visits, but lack consideration of complex relationships between diseases. Thus some works [5,8,14] introduce external medical knowledge to capture complex relationships between diseases, but they ignore the hierarchical structure of EHR data. Therefore, some methods [6,7,17] design hierarchical networks to identify important diagnosis codes and visits. Although these methods have achieved relatively good results, there are still the following challenges.

Absence of Hierarchy in Prediction. Some methods [6,7,17] employ hierarchical information to learn current patient conditions. However, they neglect to learn the next visit-level representation of the patient, which is also important for diagnosis prediction. Therefore, the hierarchy structure of EHR data should also be considered during prediction.

Lack of Identification of Critical Diseases for Patient Health Status. Existing methods [1–3,8,12,13] treat diseases in each visit equally. Some works [6,7,17] employ attention mechanism to identify important diseases within each visit locally. However, the patient health status is often complex and changeable, and it is still a challenge to distinguish key diseases from a global perspective.

To address the above challenges, we propose a novel **H**ierarchical Encoder-Decoder with **A**ddressable **M**emory **N**etwork (HAMNet for short). The following contributions are made in this paper:

- We design a hierarchical encoder-decoder structure to make full use of hierarchical information both in representation learning and diagnosis prediction, which obtains the representations of corresponding level and then enhances the performance of diagnosis prediction.
- We propose the addressable memory network to identify crucial diseases for patient health conditions at next visit. It employs the well-designed addressing mechanism and updates memory automatically with simpler design.
- We demonstrate the validity of our proposed approach on a real-world dataset MIMIC-III, and it surpasses the state-of-the-art methods.

2   Problem Definition

Diagnosis Codes. Let $C = \{c_1, c_2, ..., c_m\}$ represents all diagnosis codes. $c_j = [0, ..., 1, ..., 0]^{\mathrm{T}}$ is a one-hot vector, where 1 appears in the j-th row.

Historical Medical Records. The sequence $V = [v_1, v_2, ..., v_n]$ denotes the historical medical records of patient p, where $v_i \in \{0,1\}^m$. If the j-th diagnosis code $c_j \in C$ appears in the i-th visit, then $v_{ij} = 1$, otherwise $v_{ij} = 0$. Each visit v_i has a timestamp $t_i \in \mathbb{R}$.

Diagnosis Prediction. Given the historical records V and corresponding timestamps, the goal is to predict diagnosis of the next visit $\hat{v}_{n+1} = [\hat{c}_1, ..., \hat{c}_m]$, where $\hat{c}_i \in \mathbb{R}$ represents the probability that code c_i appears in next visit.

3 Methodology

3.1 Architecture Overview

As shown in Fig. 2, HAMNet adopts the framework of hierarchical encoder and decoder. Firstly, hierarchical encoder exploits attention mechanism during code2visit stage and visit2patient stage, respectively identifying the core diseases and important visits towards patient health conditions. Secondly, hierarchical decoder utilizes hierarchy structure during prediction including patient2visit stage and visit2code stage. Specially, in patient2visit stage we design the addressable memory network, which has the ability to distinguish important diseases towards patient health conditions.

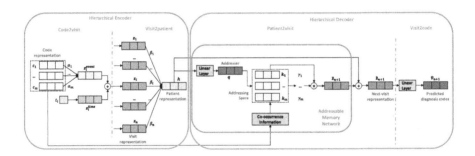

Fig. 2. The proposed HAMNet model.

3.2 Hierarchical Encoder

Hierarchical encoder designs hierarchical attention mechanism to select important codes and visits and learn patient representation.

Code2visit Stage. We employ the code-level attention mechanism to obtain important codes from each visit, and then integrate the representations of diagnosis codes to obtain the representation of each visit. For each diagnosis code c_j, we first get its dense embedding $e_j \in \mathbb{R}^d$ as follows:

$$e_j = W_1 c_j + b_1, \tag{1}$$

where $W_1 \in \mathbb{R}^{d \times m}$ is weight matrix, and $b_1 \in \mathbb{R}^d$ is bias vector. The embeddings of diagnosis code set c can be represented as $E = [e_1, e_2, .., e_m]$. For visit $v_i \in \{0, 1\}^m$, we get its dense embedding set $S = [s_1, s_2, .., s_m] \in \mathbb{R}^{m \times d}$, where $s_j = e_j$ if $v_{ij} = 1$, otherwise $s_j = 0$. After that we can get the attention score $f_i \in \mathbb{R}$ for each code embedding s_i through

$$f_i = PReLU(W_2 s_i + b_2), \tag{2}$$

where $W_2 \in \mathbb{R}^d$, and $b_2 \in \mathbb{R}$ are both parameters. If $s_i = 0$, f_i is set to $-\infty$. Then we get the normalized weight through a softmax layer:

$$\alpha_i = Softmax(f_i) = \frac{exp(f_i)}{\sum_{i=1}^{m} exp(f_i)}. \tag{3}$$

Then we obtain the representation $z_i^{event} \in \mathbb{R}^d$ of i-th visit by conducting a weighted sum operation.

$$z_i^{event} = \sum_{i=1}^{m} \alpha_i \cdot s_i. \tag{4}$$

Considering the temporal nature of visit sequence, we encode the timestamp of each visit through the following time encoding function [9] to obtain time embedding vector:

$$\mathbf{t2v}(t)[j] = \begin{cases} \omega_j t + \varphi_j & if\, j = 0 \\ sin(\omega_j t + \varphi_j) & if\, 1 \leq j \leq d \end{cases}$$

where $\mathbf{t2v}(t)[j]$ is the j-th element of $\mathbf{t2v}(t)$, ω_j and φ_j are frequency and phase-shift of sine function, and d is the dimension of $\mathbf{t2v}(t)$. Therefore we obtain $z_i^{time} = \mathbf{t2v}(t_i) \in \mathbb{R}^d$. Finally we can get final visit representation $z_i = z_i^{event} + z_i^{time}$.

Visit2patient Stage. We aim to learn patient representation based on representations of visits obtained above. First we calculate the attention score $g_i \in \mathbb{R}$ for each visit,

$$g_i = PReLU(W_3 z_i + b_3), \tag{5}$$

where $W_3 \in \mathbb{R}^d$, and $b_3 \in \mathbb{R}$ are both parameters. Next we utilize a softmax layer to normalize the attention weight,

$$\beta_i = Softmax(g_i) = \frac{exp(g_i)}{\sum_{i=1}^{n} exp(g_i)}, \tag{6}$$

where $\beta_i \in \mathbb{R}$. Finally, the representation $h \in \mathbb{R}^d$ of patient p learned from hierarchy structure of EHR data can be generated through

$$h = \sum_{i=1}^{n} \beta_i \cdot z_i. \tag{7}$$

3.3   Hierarchical Decoder

Hierarchical decoder contains patient2visit stage and visit2code stage during prediction. We first predict the representation of next visit through the well-designed addressable memory network, and then predict the diseases that may occur in the next visit through a linear layer.

Patient2visit Stage. To learn the next-visit representation of patient, we propose the **addressable memory network**. It employs the memory network to record information about diseases of patient, and utilizes the addressing mechanism, which aims to identify diseases that are critical to the patient health. Compared with the key-value memory network, the addressable memory network can update the memory module without read and write operations, instead it takes advantage of feed-forward layers to automatically update the memory module by gradient backpropagation during training. The proposed addressable memory network is composed of three components: addressing space, addresser and addressing mechanism.

- **Addressing Space.** The embeddings of diseases form the addressing space in addressable memory network, each of which represents a specific disease. We use co-occurrence information referring to GRAM [5] to initialize the embeddings of diagnosis codes, and the basic embeddings are $K = [k_1, k_2, ..., k_m]$, where $k_i \in \mathbb{R}^l$.
- **Addresser.** The representation of patient learned from hierarchy structure is treated as an addresser in addressable memory network. A linear layer is employed to align dimensions of initialized diagnosis codes embeddings and patient representation:

$$q = W_4 h + b_4, \tag{8}$$

where $W_4 \in \mathbb{R}^{l \times d}$ and $b_4 \in \mathbb{R}^l$. $q \in \mathbb{R}^l$ is seen as the addresser.
- **Addressing Mechanism.** In order to identify diseases that are critical to patient health status, we design the addressing mechanism. The addresser queries each memory slot in addressing space to obtain weights, which represent the importance of each disease. After that, memory slots and weights are summed together to obtain the patient representation at next visit. Specifically, the weight of each memory slot can be obtained by

$$\gamma_i = Softmax(q^{\mathrm{T}} k_i) = \frac{exp(q^{\mathrm{T}} k_i)}{\sum_{i=1}^{m} exp(q^{\mathrm{T}} k_i)}. \tag{9}$$

Finally, the patient health condition of next visit $\bar{z}_{n+1} \in \mathbb{R}^l$ can be obtained as follows:

$$\bar{z}_{n+1} = \sum_{i=1}^{m} \gamma_i \cdot k_i, \tag{10}$$

In order to obtain the health status of patient at next visit more comprehensively, we concatenate the patient representation learned from (7) and the next visit representation learned from (10) to get $z'_{n+1} = [h, \bar{z}_{n+1}] \in \mathbb{R}^{d+l}$, and then we obtain the representation of the next visit $\hat{z}_{n+1}$ through

$$\hat{z}_{n+1} = W_5 z'_{n+1} + b_5, \tag{11}$$

where $W_5 \in \mathbb{R}^{d \times (d+l)}$ and $b_5 \in \mathbb{R}^d$, $\hat{z}_{n+1} \in \mathbb{R}^d$.

Visit2code Stage. We employ the representation of patient at next visit to predict diagnosis codes as follows,

$$\hat{v}_{n+1} = Sigmoid(W_6 \hat{z}_{n+1} + b_6), \tag{12}$$

where $Sigmoid(x) = \frac{1}{1+exp(-x)}$, $W_6 \in \mathbb{R}^{m \times d}$, $b_6 \in \mathbb{R}^m$, and $\hat{v}_{n+1} \in \mathbb{R}^m$.

To train the model, we use cross-entropy between the ground truth v_{n+1} and the predicted $\hat{v}_{n+1}$ to calculate the loss,

$$\mathcal{L}(\theta) = -\frac{1}{|P|} \sum_{i=1}^{|P|} ((v_{n+1}^i)^{\mathrm{T}} \log \hat{v}_{n+1}^i + (1 - v_{n+1}^i)^{\mathrm{T}} \log(1 - \hat{v}_{n+1}^i)), \tag{13}$$

where θ represents all parameters in HAMNet, $|P|$ represents the number of patients.

4 Experiments

4.1 Experimental Setup

We conduct experiments on MIMIC-III[1] (Medical Information Mart for Intensive Care) dataset, which is a public large-scale EHR dataset. The diagnosis codes are represented following the ICD-9[2] high-dimensional coding system. We group the ICD-9 codes by Clinical Classification Software (CCS) single-level diagnosis grouper[3] and replace the original ICD codes with their group codes following [8]. We remove patients with fewer than two medical records and diagnosis codes that occur less than five times followed by [16]. Table 1 illustrates the detailed statistics of dataset.

We divide benchmarks into three categories: methods without hierarchy structure and relationship between diseases (Retain [3], Dipole [13], T-LSTM [2], RetainEX [11], Timeline [1], HiTANet [12]), method with relationship between diseases (CAMP [8]), and method with hierarchy structure (LSAN [17]). Besides, We use **Recall@k** and **MAP@k** as evaluation metrics referring to [8]. We randomly divide the dataset into training set, validation set and test set according to the ratio of 80%:10%:10%. Furthermore, we only predict CCS single-level diagnosis grouper in diagnosis prediction task, not specific ICD-9 codes instead

---

[1] https://mimic.physionet.org/.
[2] http://www.icd9data.com/.
[3] https://www.hcup-us.ahrq.gov/toolssoftware/ccs/AppendixASingleDX.txt.

Table 1. Statistics of MIMIC-III dataset.

Dataset	MIMIC-III
patients	7,486
visits	19,884
average visits per patient	2.66
maximum visits per patient	42
minimum visits per patient	2
unique ICD-9 codes	2,437
average ICD-9 codes per visit	11.05
maximum ICD-9 codes per visit	34
minimum ICD-9 codes per visit	1
unique CCS group codes	243
average CCS group codes per visit	11.07
maximum CCS group codes per visit	34
minimum CCS group codes per visit	1

followed by [8]. For all methods, we implement them in PyTorch [15] framework and conduct experiments on a GeForce GTX 1080Ti GPU. We use Adam [10] optimizer to train parameters. Batch size is set to 64 and the dimension of patient hidden state is set to 128. We run all methods five times and report the mean and standard deviation of model performance on test set. Since our method does not introduce external knowledge, we remove ICD-9 tree structure of CAMP and set the demographic information of all patients to be the same for fairness.

4.2   Performance Comparison

Table 2. Diagnosis prediction performance on MIMIC-III dataset.

Methods	Recall@k			MAP@k		
	Recall@5	Recall@20	Recall@25	MAP@5	MAP@20	MAP@25
Retain	0.257±0.003	0.590±0.004	0.652±0.002	0.224±0.003	0.409±0.005	0.433±0.004
Dipole	0.267±0.003	0.601±0.003	0.659±0.003	0.232±0.002	0.421±0.003	0.444±0.003
T-LSTM	0.230±0.005	0.542±0.007	0.606±0.006	0.191±0.004	0.347±0.007	0.370±0.007
RetainEX	0.278±0.003	0.604±0.002	0.661±0.002	0.242±0.004	0.432±0.004	0.455±0.004
Timeline	0.239±0.003	0.541±0.005	0.600±0.004	0.205±0.004	0.358±0.005	0.379±0.004
HiTANet	0.220±0.022	0.512±0.041	0.571±0.044	0.181±0.023	0.324±0.043	0.344±0.045
CAMP	0.260±0.003	0.583±0.004	0.644±0.004	0.225±0.003	0.401±0.004	0.425±0.004
LSAN	0.275±0.005	0.606±0.004	0.666±0.003	0.241±0.005	0.432±0.005	0.456±0.005
HAMNet	**0.280±0.002**	**0.616±0.001**	**0.675±0.001**	**0.247±0.002**	**0.442±0.002**	**0.466±0.002**

The diagnosis prediction results of HAMNet and other baselines are illustrated in Table 2. HAMNet outperforms all benchmarks on all metrics with stable performance, which demonstrates the effectiveness of HAMNet. It adopts a hierarchical encoder and decoder framework, which takes full advantage of hierarchical structure information of EHR data during learning patient representations and making prediction. Furthermore, it utilizes the addressable memory network, which has the ability to identify core diseases for patient health status. Both of the above parts benefits the performance of diagnosis prediction task.

4.3 Ablation Study

In this section, we compare HAMNet and its variants to examine the contribution of each individual component to the whole framework.

HAMNet/A means removing addressable memory network from HAMNet, which only employs hierarchical encoder to obtain patient representation. Specifically, HAMNet/A replaces (11) into $\hat{z}_{n+1} = W_5 h + b_5$.

HAMNet/AD represents removing hierarchy decoder and addressable memory network from HAMNet, which applies the learned patient representation from hierarchical encoder to make one-step diagnosis prediction. Therefore, (12) is changed into $\hat{v}_{n+1} = Sigmoid(W_6 h + b_6)$.

Table 3. Ablation study on diagnosis prediction.

Methods	Recall@k			MAP@k		
	Recall@5	Recall@20	Recall@25	MAP@5	MAP@20	MAP@25
HAMNet	**0.280±0.002**	**0.616±0.001**	**0.675±0.001**	**0.247±0.002**	**0.442±0.002**	**0.466±0.002**
HAMNet/A	0.272±0.004	0.603±0.007	0.663±0.006	0.237±0.004	0.427±0.007	0.451±0.007
HAMNet/AD	0.272±0.005	0.597±0.010	0.656±0.009	0.238±0.004	0.423±0.010	0.446±0.009

Table 3 presents the results of HAMNet and its variants. First, HAMNet outperforms all other variants on all metrics, which suggests that hierarchical encoder-decoder structure and the addressable memory network can promote each other. Second, compared with HAMNet, the performance of HAMNet/A drops, which validates that our designed addressable memory network is necessary for HAMNet. Through identifying diseases that are critical to patient health conditions, addressable memory network can improve the predictive power of the model. Third, compared with HAMNet/A, HAMNet/AD performs worse on most metrics, which demonstrates the indispensability of hierarchical decoder module in HAMNet. Compared with one-step prediction, hierarchical decoder can improve the accuracy of prediction results.

4.4 Memory Network Analysis

In order to further verify the superiority of our designed memory network based on addressing mechanism, we design the following variant: replace the memory network in HAMNet with the key-value memory network in CAMP, denoted as HAMNet-KVM. The results are recorded in Table 4, which shows that HAMNet performs better than HAMNet-KVM in all metrics. It proves that the memory network based on the addressing mechanism is better than the traditional key-value memory network. Through addressing mechanism, critical diseases towards the patient health status can be identified with simpler design, facilitating patient representation learning at next visit.

Table 4. Memory network analysis.

Methods	Recall@k			MAP@k		
	Recall@5	Recall@20	Recall@25	MAP@5	MAP@20	MAP@25
HAMNet-KVM	0.271±0.002	0.600±0.002	0.660±0.002	0.237±0.002	0.421±0.003	0.445±0.002
HAMNet	**0.280±0.002**	**0.616±0.001**	**0.675±0.001**	**0.247±0.002**	**0.442±0.002**	**0.466±0.002**

5 Conclusion

Diagnosis prediction from EHR data is one of the most important tasks in health condition analysis. Existing methods only consider the hierarchical information of EHR data in modeling patient health status, and do not take it into consideration during prediction. Furthermore, these methods are not able to distinguish crucial diseases towards patient health conditions. In this paper, we propose a novel **H**ierarchical Encoder-Decoder with **A**ddressable **M**emory **N**etwork called HAMNet to handle existing challenges. HAMNet adopts a hierarchical encoder-decoder framework to take full advantage of hierarchical structure of EHR data not only in learning patient representation but also in making prediction. Furthermore, it designs the addressing memory network during patient2visit stage, which can capture important diseases for patient health status at next visit. We evaluate HAMNet on a real world EHR data MIMIC-III. The results demonstrate that the proposed HAMNet outperforms state-of-the-art models and achieves stable improvements in diagnosis prediction task.

Acknowledgements. This work is supported in part by he National Key Research and Development Plan Project 2022YFC3600901, and NSF under grants III1763325, III-1909323, III-2106758, and SaTC-1930941.

References

1. Bai, T., Zhang, S., Egleston, B.L., Vucetic, S.: Interpretable representation learning for healthcare via capturing disease progression through time. In: Proceedings of the 24th ACM SIGKDD International Conference on Knowledge Discovery & Data Mining, pp. 43–51 (2018)
2. Baytas, I.M., Xiao, C., Zhang, X., Wang, F., Jain, A.K., Zhou, J.: Patient subtyping via time-aware LSTM networks. In: Proceedings of the 23rd ACM SIGKDD International Conference on Knowledge Discovery and Data Mining, pp. 65–74 (2017)

3. Choi, E., Bahadori, M.T., Kulas, J.A., Schuetz, A., Stewart, W.F., Sun, J.: Retain: an interpretable predictive model for healthcare using reverse time attention mechanism. In: Proceedings of the 30th International Conference on Neural Information Processing Systems, pp. 3512–3520 (2016)
4. Choi, E., Bahadori, M.T., Schuetz, A., Stewart, W.F., Sun, J.: Doctor AI: predicting clinical events via recurrent neural networks. In: Machine Learning for Healthcare Conference, pp. 301–318. PMLR (2016)
5. Choi, E., Bahadori, M.T., Song, L., Stewart, W.F., Sun, J.: Gram: graph-based attention model for healthcare representation learning. In: Proceedings of the 23rd ACM SIGKDD International Conference on Knowledge Discovery and Data Mining, pp. 787–795 (2017)
6. Choi, E., Xiao, C., Stewart, W., Sun, J.: Mime: multilevel medical embedding of electronic health records for predictive healthcare. In: Advances in Neural Information Processing Systems, vol. 31 (2018)
7. Du, J., et al.: An interpretable outcome prediction model based on electronic health records and hierarchical attention. Int. J. Intell. Syst. **37**(6), 3460–3479 (2021)
8. Gao, J., et al.: Camp: co-attention memory networks for diagnosis prediction in healthcare. In: 2019 IEEE International Conference on Data Mining (ICDM), pp. 1036–1041. IEEE (2019)
9. Kazemi, S.M., et al.: Time2vec: learning a vector representation of time. arXiv preprint arXiv:1907.05321 (2019)
10. Kingma, D.P., Ba, J.: Adam: a method for stochastic optimization. arXiv preprint arXiv:1412.6980 (2014)
11. Kwon, B.C., et al.: RetainVis: visual analytics with interpretable and interactive recurrent neural networks on electronic medical records. IEEE Trans. Vis. Comput. Graph. **25**(1), 299–309 (2018)
12. Luo, J., Ye, M., Xiao, C., Ma, F.: HiTANet: hierarchical time-aware attention networks for risk prediction on electronic health records. In: Proceedings of the 26th ACM SIGKDD International Conference on Knowledge Discovery & Data Mining, pp. 647–656 (2020)
13. Ma, F., Chitta, R., Zhou, J., You, Q., Sun, T., Gao, J.: Dipole: diagnosis prediction in healthcare via attention-based bidirectional recurrent neural networks. In: Proceedings of the 23rd ACM SIGKDD International Conference on Knowledge Discovery and Data Mining, pp. 1903–1911 (2017)
14. Ma, F., You, Q., Xiao, H., Chitta, R., Zhou, J., Gao, J.: Kame: knowledge-based attention model for diagnosis prediction in healthcare. In: Proceedings of the 27th ACM International Conference on Information and Knowledge Management, pp. 743–752 (2018)
15. Paszke, A., et al.: Pytorch: an imperative style, high-performance deep learning library. In: Advances in Neural Information Processing Systems, vol. 32 (2019)
16. Peng, X., Long, G., Shen, T., Wang, S., Jiang, J.: Self-attention enhanced patient journey understanding in healthcare system. In: Hutter, F., Kersting, K., Lijffijt, J., Valera, I. (eds.) ECML PKDD 2020. LNCS (LNAI), vol. 12459, pp. 719–735. Springer, Cham (2021). https://doi.org/10.1007/978-3-030-67664-3_43
17. Ye, M., Luo, J., Xiao, C., Ma, F.: LSAN: modeling long-term dependencies and short-term correlations with hierarchical attention for risk prediction. In: Proceedings of the 29th ACM International Conference on Information & Knowledge Management, pp. 1753–1762 (2020)

DP-MHAN: A Disease Prediction Method Based on Metapath Aggregated Heterogeneous Graph Attention Networks

Zhe Qu[1], Lizhen Cui[1,2(✉)], and Yonghui Xu[2]

[1] School of Software, Shandong University, Jinan, China
quzhe@mail.sdu.edu.cn, clz@sdu.edu.cn
[2] Joint SDU-NTU Centre for Artificial Intelligence Research (C-FAIR),
Shandong University, Jinan, China

Abstract. Disease prediction as an important component of medical assistant diagnostic systems has received much attention from researchers. Many studies attempt to extract disease-related features from electronic health records (EHRs). However, the heterogeneity of entities (e.g., diseases, symptoms, medications, and other treatment items) in EHRs is usually neglected in these studies. To tackle this issue, we propose a Disease Prediction method based on Metapath aggregated Heterogeneous graph Attention Networks (DP-MHAN) to improve the final prediction performance. First, we extract entities and the relationships between entities from the EHRs and construct a heterogeneous medical graph. Second, we define three special metapaths and aggregate the embedding of entities on the metapaths, in which the embedding of each entity can be calculated via its directly connected neighbors. Moreover, the attention mechanism is adopted to locate the most effective embedding, which is utilized for disease prediction. Experimental results show that our proposed method outperforms state-of-the-art methods.

Keywords: Metapath · Heterogeneous information network · Graph neural network · Attention mechanism · Disease prediction

1 Introduction

Disease prediction aims to perform health event prediction using patient-related valuable clinical data such as electronic health records (EHRs). With the development of deep learning and data mining technology, various models have been proposed to address temporal events prediction tasks, such as mortality rate prediction [9], readmission prediction [10], length of stay [5], disease risk prediction [17,20]. An excellent disease prediction model can assist physicians to identify disease risks in a timely manner and avoid misdiagnosis.

In the past few years, many deep learning-based disease prediction models have been proposed. Early deep learning-based disease prediction models adopt Convolutional Neural Networks (CNN) to measure the similarity of patients

ⓒ The Author(s), under exclusive license to Springer Nature Switzerland AG 2023
X. Wang et al. (Eds.): DASFAA 2023, LNCS 13946, pp. 276–285, 2023.
https://doi.org/10.1007/978-3-031-30678-5_21

[17], but this kind of method make it difficult to extract temporal features from EHRs. In order to deal with this issue, some researchers adopt Recurrent Neural Networks (RNN) and Long-Short Term Memory (LSTM), or combine the CNN with RNN in their research works [12,14]. Meanwhile, Graph Neural Network (GNN) has attracted the focus of researchers as graphs are a good portrayal to model the entities' relationships in EHRs, and various Graph-based models [3,16,18] have achieved good prediction results.

However, these deep learning models mentioned above still have several challenges for disease prediction: (1) Most of the existing CNN/RNN-based models treat diseases independently, but the disease interactions are neglected. (2) Graph-based models learn node embeddings using the aggregation of its neighboring nodes' features, but most of these models are designed for homogeneous data structure. Heterogeneity is an intrinsic characteristic of EHR data, and heterogeneous information networks (HIN) consists of multi-typed nodes and edges, which is more suitable to represent the EHRs than homogeneous networks for accurate disease risk prediction [6]. In this paper we propose a Disease Prediction method based on Metapath aggregated Heterogeneous graph Attention Networks (DP-MHAN). The main contributions of this study are summarized as follows: (1) We construct a heterogeneous medical graph, and a three-metapath-based graph neural network is designed for disease prediction. (2) We use an attention mechanism to learn the weights between various entities, which is beneficial for aggregating the representation of the patient's neighbors to form entity embeddings. (3) We evaluate the performance of DP-MHAN on a real-world EHRs dataset, and the results outperform those of the baseline models.

2   Related Work

CNN/RNN-Based Disease Prediction. Suo et al. [17] build a time fusion CNN framework to measure patient similarity via EHRs and utilize the similarity scores to perform disease risk prediction. Reference [12] propose a risk prediction method using RNN and CNN as the basic predictive models. Rajput et al. [14] adopt LSTM to deal with the error of exponential decay of backpropagation in RNNs, which is used to identify risk factors of heart disease. Niu et al. [13] propose a disease risk prediction model using both time-series health status indicators and medical notes from EHRs. Although these CNN/RNN-based methods have achieved good results, they are still limited by their ability to extract the longitudinal interactions in EHRs.

Graph-Based Disease Prediction. Choi et al. [3] propose a graph-based attention model (GRAM) to predict diseases, and this approach represents a medical concept as a combination of its ancestors in the ontology via an attention mechanism. Valenchon et al. [18] apply a multiple-graph recurrent graph convolutional neural network (GCN) architecture to predict disease outcomes for datasets with missing values. In [16], GNN is employed to learn the latent node embeddings (i.e., symptoms, users, and diseases) to predict disease for new patients using EHRs. Zhang et al. [19] propose a meta-learning framework for

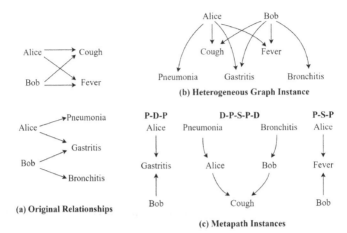

Fig. 1. Depiction of (a) the original relationships, (b) the heterogeneous graph instance, and (c) the metapath instances that we adopted. P, S, D denote patient, symptom, and disease, respectively.

clinical risk prediction from longitudinal patient EHRs. As mentioned above, most of these methods are proposed for homogeneous data structure

3   Problem Formulation

Patients can suffer from multiple diseases and be diagnosed with multiple symptoms. We model this interrelationship of patients with diseases and symptoms as a heterogeneous medical graph, which can be denoted as $G = (V, E)$, V and E are the node set and edge set. V consists of three subsets V_p, V_d, and V_s, which are the node sets of patients, diseases, and symptoms, respectively. Figure 1 depicts an overview of the construction method of the heterogeneous medical graph. Metapath is proposed to capture the structural and semantic relationships between entities [7]. A metapath P is denoted as $A_1 \rightarrow A_2 \rightarrow ... \rightarrow A_n$ (abbreviated as $A_1 A_2...A_n$). Our study aims to learn a HIN-based model via the given heterogeneous medical graph G = (V, E). Then, for a new patient, the learned model is adopted to predict the possible disease.

4   The Proposed Method

The framework of the proposed DP-MHAN is illustrated in Fig. 2, and the detailed descriptions are provided in the following subsections.

4.1   Unifying Entity Dimension

We use one-hot coding to initially encode different entities in a heterogeneous information network and represent them with embeddings. As the different number of patients, diseases, and symptoms in the EHR dataset, the dimensionality

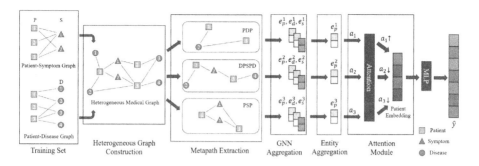

Fig. 2. The framework of our proposed disease prediction method.

of the initially generated embedding is different, which causes inconvenience to the fusion algorithm later in our model. To solve this problem, we need to project the embedding of different dimensions into the same low-dimensional feature space. For an entity embedding e, the new equal-dimension embedding e' is given by the following formula:

$$e'_{\{pds\}} = W_{\{pds\}} e_{\{pds\}} \tag{1}$$

where p, d, and s represent patient, disease and symptom respectively, and W is the matrix of weights to be learned. After the projection operation, we can extract the key features from the initial embedding e.

4.2 Heterogeneous Medical Graph Construction

EHR dataset contains the time of each patient visit, the disease, and the patient's symptoms. However, the symptoms and the disease are recorded separately, and it is necessary to consider the information of the patient, symptoms, and disease simultaneously. We create patient-symptom graph G_{ps} and patient-disease graph G_{pd} separately based on the EHR dataset. $G_{ps} \in \mathbb{R}^{|P| \times |S|}$, $G_{pd} \in \mathbb{R}^{|P| \times |D|}$. If patient i suffers from disease m in a visit, then $G_{pd}[i][m] = 1$, this is also true for patient j and symptom n. Next, we fuse G_{ps} and G_{pd} to construct the heterogeneous medical graph G. We use the adjacency matrix A to denote G, $A \in \mathbb{R}^{(|D|+|S|+|P|) \times (|D|+|S|+|P|)}$, and if any two entities between patients, diseases, and symptoms are related in patient-symptom graph or patient-disease graph, the corresponding value of these two entities in matrix A is set as 1.

4.3 Node Aggregation via Metapath

In the heterogeneous medical graph, we use the metapath defined in Sect. 3 for entity information extraction. We define the three metapaths as follows:

- **Patient-Disease-Patient (PDP):**
  Patients suffering from the same disease are correlated.

- **Patient-Symptom-Patient (PSP):**
  Similarly, patients with the same symptoms are also correlated.
- **Disease-Patient-Symptom-Patient-Disease (DPSPD):**
  Diseases with similar symptoms are also correlated, but we do not have direct disease-symptom data, so we rely on the Patient as an intermediate node.

Then, we traverse the heterogeneous medical graph according to the metapaths and record all the entities that are related to the metapaths. After that, based on the entity embedding constructed in Sect. 4.1, we aggregate the embeddings on the metapath.

For a specific node m, if n is the end point of a metapath P starting from m, we use $P(m,n)$ to denote all the nodes (including m and n) that the metapath P passes through from m to n. We aggregate the embedding of the entities on this metapath into a single embedding using the following formula:

$$e_{P(m,n)} = mean(\{e'_i | i \in P(m,n)\}) \tag{2}$$

After encoding the entire metapath, we need to continue to propagate the information of the metapath. Since entities of different types and distances on the metapath have different effects on other entities, a natural idea is to use attention mechanisms to distinguish these relationships. And hence, graph attention is adopted in our model, as shown in Eq. (3)-Eq. (5).

$$e_{mn}^P = LeakyReLU\left(w_P^\top \cdot [e'_m || e_{P(m,n)}]\right) \tag{3}$$

$$\alpha_{mn}^P = \frac{exp\left(e_{mn}^P\right)}{\sum_{s \in N_m^P} exp\left(e_{ms}^P\right)} \tag{4}$$

$$e_m^P = \sigma\left(\sum_{n \in N_m^P} \alpha_{mn}^P \cdot e_{mn}^P\right) \tag{5}$$

where $w_P^\top$ is a weighted attention vector to be learned, and $||$ is the vector concatenation operator. We splice the embedding of the target entity with the embedding of the metapath, and then reduce it to its original size by multiplying it by $w_P^\top$. And hence, in the subsequent attention score calculation, the information of the whole metapath is taken into account. Next, we calculate the attention score α_{mn}^P of all entities in the other metapath of the same type according to Eq. (4). Finally, the embedding of the target entity m about the metapath P is obtained using Eq. (5), which is denoted as e_m^P.

4.4   Aggregation of Different Metapaths

Sect. 4.3 is the information extraction process for a single metapath. To calculate the final entity embedding, we use the three metapaths simultaneously, and after the process in Sect. 4.3 we obtain three embeddings for each entity separately. Naturally, we need to measure the importance of each metapath for the final embedding calculation before aggregation, and the attention mechanism is utilized to highlight which metapath that are more important to the patient. For a patient denoted as p, we have three different metapath embeddings generated after the step of Sect. 4.3, i.e., e_p^1, e_p^2, e_p^3. Then, the fused metapath embedding q is calculated as follows:

$$q = mean(e_p^1, \ e_p^2, \ e_p^3) \tag{6}$$

We calculate the score of each metapath and generate the final patient embedding e_p with the following formula:

$$score(e_p^i, \ q) = v^\top tanh(W_1 e_p^i + W_2 q) \tag{7}$$

$$a_i = softmax(score(e_p^i, \ q)) \tag{8}$$

$$e_p = \sum_{i=1}^{|P|} a_i \cdot e_p^i \tag{9}$$

where W_1, W_2, $v^\top$ are learnable weight. a_i is the weight coefficient corresponding to a certain metapath P_i calculated by softmax. We weighted the sum of all the metapath-specific embeddings of patient p.

4.5   Loss Function

Disease prediction is a multi-label classification task. We use the MLP module to process the final patient embedding e_p, and MLP projects e_p into a vector space with the same dimensions as the number of diseases. The output of MLP is regarded as the predicted probability $\hat{y}$:

$$\hat{y} = MLP(e_p) \tag{10}$$

The loss function L of classification used in our model is cross-entropy loss:

$$L = CrossEntropy(\hat{y}, \ y) \tag{11}$$

where y is the ground-truth label of diseases. We use the patient's latest disease as the ground-truth label. In the training phase, we train and get the patient, symptom, and disease embeddings. After that, the trained embedding is used to compute new patient embedding via metapath in the inference phase. Finally, the prediction result is obtained.

5   Experiments

5.1   Datasets

The performance of the proposed model is evaluated on MIMIC-III [8], which contains over 58,000 hospital admission records. We select patients which suffer from at least three diseases, and randomly split the dataset into training/validation/test sets with patient numbers as 6000/125/1000. In our experiments, we use the latest disease that the patient suffers from as the patient's

Table 1. Prediction performance of compared methods in terms of w-F1 and R@k.

Models	w-F1	R@10	R@20
RETAIN	19.71	24.04	33.97
Dipole	18.44	22.98	32.93
Timeline	19.49	23.69	33.74
GRAM	20.31	24.39	34.63
G-BERT	19.54	23.79	34.19
DP-MHAN	**20.65**	**25.87**	**35.28**

label, and all other diseases before the latest disease are treated as patient-related diseases. Please note that if the number of the prior diseases is more than three, only the most recent three ones are utilized. The patient's symptoms are extracted from the diagnostic records, which are described in separate words. We adopt word vectors from the NLP domain to model these symptom words.

5.2   Baseline Methods

To validate the effectiveness of the proposed disease prediction model, we compare our method with five state-of-the-art methods. In the comparison methods, **RETAIN** [4], **Dipole** [11], and **Timeline** [1] are CNN/RNN-based methods. For Graph-based methods, we select **GRAM** [3] and **G-BERT** [15].

5.3   Evaluation Metrics

The evaluation metrics for disease prediction are weighted F1 score (w-F1) and top k recall (R@k). w-F1 is a weighted sum of F1 scores for all diseases. R@k is an average ratio of desired diseases in top k predictions by the total number of desired medical codes in each visit. We set $k = [10, 20]$ for R@k.

5.4   Experimental Setting

In our experiment, the model parameters are initialized randomly. We set the embedding dimension of all the entities as 64, and the dimension of the attention vector is chosen as 128. We conduct experiments based on Python 3.8 and PyTorch 1.8, using an Nvidia Geforce RTX 3090 with 256GB memory. Our model is trained with 100 epochs and uses ADAM as the optimizer. We set the dropout rate to 0.5. The learning rate is initially set to 0.01 and then decays to 0.001.

5.5   Result Analysis

Table 1 shows the disease prediction results, and it can be seen that the performance of our model exceeds all baseline models. This is because DP-MHAN can better capture the relationship between the patient, disease, and symptoms.

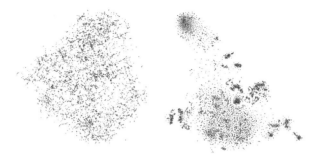

Fig. 3. Embedding visualization before (left) and after (right) metapath aggregation.

To demonstrate the effectiveness of the proposed model distinctly, the patient embedding visualization is shown in Fig. 3. The high dimensional embeddings are reduced by t-SNE [2] to 2D space for visualization. A clear tendency is that embedding space becomes more and more squeezed after metapath aggregation. Patient embeddings after graph convolution aggregate together obviously indicate that the learned embedding can adequately represent the entities' interrelationships. Therefore, we can infer that learning proper embedding that reflects entity correlations is beneficial to disease predictions.

5.6   Ablation Study

To validate the effectiveness of each component of our model, we further conduct experiments on different DP-MHAN variants. Let DP-MHAN be the reference model, we consider the model's components from the following perspectives:

Aggregation Methods. The aggregation function we use in the proposed model is *mean*. We also consider other aggregation functions:

- Linear aggregation (DP-MHAN$_{lin}$):

$$e_{P(m,n)} = \frac{1}{|P(m,n)|} \sum_{i=1}^{|P(m,n)|} w_l^{(i)} e' + b_l^{(i)} \tag{12}$$

where $w_l^{(i)}$ and $b_l^{(i)}$ are learnable parameters.
- Concatenation aggregation (DP-MHAN$_{cat}$):

$$e_{P(m,n)} = w^c(e_p^1 || e_p^2 || ... || e_p^{|P(m,n)|}) \tag{13}$$

where w^c is learnable parameter and is responsible for mapping the concatenated embedding back to the initial feature space.
- Max aggregation (DP-MHAN$_{max}$):

$$e_{P(m,n)} = max(e_p^1, e_p^2, ..., e_p^{|P(m,n)|}) \tag{14}$$

where the *max* function is element-wise, which is inspired by CNN.

Table 2. Experimental result of ablation study.

Models	w-F1	R@10	R@20
DP-MHAN$_{lin}$	20.23	25.16	34.27
DP-MHAN$_{cat}$	19.90	24.69	33.74
DP-MHAN$_{max}$	19.35	23.53	33.63
DP-MHAN	**20.65**	**25.87**	**35.28**

Table 2 shows the prediction results of different aggregation methods, and it can be seen that the other three aggregation methods are inferior to the *mean* method. This is because that linear aggregation makes the model more complicated, concatenation aggregation brings useless information, and max aggregation ignores valid information.

Different Embedding Dimension. In our experiments, we vary the embedding dimension as 16, 32, 64, 128, and 256 respectively, and we find that the model performs best when the value of the initial patient embedding dimension is 64. It is speculated that an oversized embedding will result in the model learning redundant information and affect the model's performance negatively.

6   Conclusion

This paper proposes a metapath-based heterogeneous graph attention network to learn the representations of entities in EHR data. We define three metapaths and aggregate the embeddings of entities on the metapaths. Attention mechanism is applied to locate the most effective embedding, which is used to perform disease prediction. Experimental results prove the effectiveness of DPMHAN against state-of-the-art models.

Acknowledgements. This work is supported by the National Key R & D Program of China (No. 2021YFF0900800), the Shandong Provincial Key Research and Development Program (Major Scientific and Technological Innovation Project) (No. 2021CXGC010108), the Shandong Provincial Natural Science Foundation (No. ZR202111180007), and the Fundamental Research Funds of Shandong University.

References

1. Bai, T., Zhang, S., Egleston, B.L., Vucetic, S.: Interpretable representation learning for healthcare via capturing disease progression through time. In: ACM SIGKDD, pp. 43–51 (2018)
2. Cho, K., et al.: Learning phrase representations using RNN encoder-decoder for statistical machine translation. In: EMNLP, pp. 1724–1734 (2014)

3. Choi, E., Bahadori, M.T., Song, L., Stewart, W.F., Sun, J.: Gram: graph-based attention model for healthcare representation learning. In: ACM SIGKDD, pp. 787–795 (2017)
4. Choi, E., Bahadori, M.T., Sun, J., Kulas, J., Schuetz, A., Stewart, W.: RETAIN: an interpretable predictive model for healthcare using reverse time attention mechanism. In: NIPS, pp. 3512–3520 (2016)
5. Darabi, S., Kachuee, M., Fazeli, S., Sarrafzadeh, M.: Taper: time-aware patient EHR representation. IEEE J. Biomed. Health **24**(11), 3268–3275 (2020)
6. Gao, J., Tian, L., Wang, J., Chen, Y., Song, B., Hu, X.: Similar disease prediction with heterogeneous disease information networks. IEEE Trans. Nanobiosci. **19**(3), 571–578 (2020)
7. Huang, C., Fang, Y., Lin, X., Cao, X., Zhang, W.: Able: meta-path prediction in heterogeneous information networks. ACM Trans Knowl. Discov. D **16**(4), 1–21 (2022)
8. Johnson, A.E., et al.: MIMIC-III, a freely accessible critical care database. Sci. Data **3**(1), 1–9 (2016)
9. Jun, E., Mulyadi, A.W., Choi, J., Suk, H.I.: Uncertainty-gated stochastic sequential model for EHR mortality prediction. IEEE Trans. Neural Net. Learn.Syst. **32**(9), 4052–4062 (2021)
10. Lee, H.K., et al.: An analytical framework for TJR readmission prediction and cost-effective intervention. IEEE J. Biomed. Health **23**(4), 1760–1772 (2019)
11. Ma, F., Chitta, R., Zhou, J., You, Q., Sun, T., Gao, J.: Dipole: diagnosis prediction in healthcare via attention-based bidirectional recurrent neural networks. In: ACM SIGKDD, pp. 1903–1911 (2017)
12. Ma, F., Gao, J., Suo, Q., You, Q., Zhou, J., Zhang, A.: Risk prediction on electronic health records with prior medical knowledge. In: ACM SIGKDD, pp. 1910–1919 (2018)
13. Niu, S., Yin, Q., Song, Y., Guo, Y., Yang, X.: Label dependent attention model for disease risk prediction using multimodal electronic health records. In: IEEE ICDM, pp. 449–458 (2021)
14. Rajput, K., Chetty, G., Davey, R.: Risk factors identification for heart disease in unstructured dataset using deep learning approach. In: IEEE ICDMW, pp. 1056–1059 (2019)
15. Shang, J., Ma, T., Xiao, C., Sun, J.: Pre-training of graph augmented transformers for medication recommendation. In: IJCAI, pp. 5953–5959 (2019)
16. Sun, Z., Yin, H., Chen, H., Chen, T., Cui, L., Yang, F.: Disease prediction via graph neural networks. IEEE J. Biomed. Health **25**(3), 818–826 (2021)
17. Suo, Q., et al.: Personalized disease prediction using a CNN-based similarity learning method. In: IEEE BIBM, pp. 811–816 (2017)
18. Valenchon, J., Coates, M.: Multiple-graph recurrent graph convolutional neural network architectures for predicting disease outcomes. In: IEEE ICASSP, pp. 3157–3161 (2019)
19. Zhang, X.S., Tang, F., Dodge, H.H., Zhou, J., Wang, F.: MetaPred: meta-learning for clinical risk prediction with limited patient electronic health records. In: ACM SIGKDD, pp. 2487–2495 (2019)
20. Zheng, S., et al.: Multi-modal graph learning for disease prediction. IEEE Trans. Med. Imaging **41**(9), 2207–2216 (2022)

A Topic-Aware Data Generation Framework for Math Word Problems

Tianyu Zhao[1], Chengliang Chai[2], Jiabin Liu[1], Guoliang Li[1(✉)], Jianhua Feng[1], and Zitao Liu[3]

[1] Tsinghua University, Beijing, China
{zhaoty17,liujb19}@mails.tsinghua.edu.cn,
{liguoliang,fengjh}@tsinghua.edu.cn
[2] Beijing Institute of Technology, Beijing, China
ccl@bit.edu.cn
[3] TAL Education Group, Beijing, China
liuzitao@100tal.com

Abstract. In educational practice, math word problems (MWPs) are important material to train pupils' to solve real-world math problems. In this paper, our focus is on the learning-based topic-aware MWP generation problem. To summarize, there are several challenges with respect to the problem. The first one is how to design the generation model considering the structural characteristics of math equations as well as the topic information. Second, how to generate multiple MWPs covering diverse topics is another challenge. To address the above challenges, we propose a novel framework, **GenMWP**, to automatically generate high-quality MWPs. First, we use the Tree-LSTM model to capture the structure of the equation. Second, we design a topic embedding model to consider the topic information, allowing the user to input any words or sentences to describe the topic. Third, given only an equation, we also support generating multiple MWPs covering different topics. The automatic evaluation result shows that **GenMWP** outperforms other baseline methods, and the human evaluation result shows that the MWPs generated by **GenMWP** have a higher rating in fluency, consistency, equation relevance, and topic relevance.

Keywords: Math Word Problem · Natural Language Generation · Topic model

1 Introduction

Recently, intelligent education has drawn huge attention [15,16], in which math word problems are important research materials. A math word problem (MWP) is a coherent narrative text that describes a story and raises a question about an unknown quantity. It also provides known quantities and logic inside, which makes it possible for students to solve the problem by assembling an equation. Table 1 shows an illustrative example of an MWP and the corresponding solution. MWPs are essential material used in primary education, which develops

X. Wang et al. (Eds.): DASFAA 2023, LNCS 13946, pp. 286–302, 2023.
https://doi.org/10.1007/978-3-031-30678-5_22

the pupils' skills to understand and solve real-world problems. Therefore, the automatic generation of MWPs is an important task in intelligent education, which will both release the heavy burden of teachers for designing many MWPs and provide students with the opportunity to practice with themselves.

In addition, an MWP is related to a certain topic (*e.g.*, daily life, sports, sales, etc.), and thus easy to understand for pupils. Hence, in this paper, we study the problem of topic-aware MWP generation. Intuitively, given an equation associated with a topic, a high-quality generated MWP typically satisfies the following properties. (1) The generated MWP should be coherent and easy to understand. (2) The described problem can be solved by the equation. (3) The MWP is highly related to the input topic. For example, given the equation $x = 24.01 \div (1 - 30\%)$ and the topic word "Shopping", the MWP in Table 1 is a good generation result.

Table 1. An example of MWP

Math Word Problem: Jenny bought ski gloves that were marked down 30% to \$24.01. What was the price of the gloves before the markdown?
Equation: $x = 24.01 \div (1 - 30\%)$
Solution: $x = 34.3$

Given an equation, there are two challenges when generating MWPs. On the one hand, how to take the structural information of the equation and the topic information into consideration so as to improve the quality of MWPs (**C1**). On the other hand, how to generate multiple MWPs with diverse topics is also a challenge because the teachers may need to generate many math problems for one equation (**C2**).

To address these challenges, we propose a novel MWP generation framework **GenMWP**, which takes as input an equation and a piece of topic text (can be several words or a sentence) as user input to generate an MWP. First, we design an encoder-decoder model to conduct MWP generation, where the encoder is a tree-structured Long Short-Term Memory (Tree-LSTM) model to capture the structural information of an equation (for **C1**). We design a topic embedding model to map the input topic into a vector, which is fed into the encoder to guide the topic of generated MWP. Afterwards, we use a sequential decoder to generate the MWP. Moreover, even the user does not provide any input for the topic, our model can produce multiple MWPs covering diverse topics for a single equation (for **C2**).

To summarize, this paper makes the following contributions:

- We propose a learning-based MWP generation framework **GenMWP** that utilizes the Tree-LSTM model to well capture the structural characteristics of math equations.
- We introduce a topic embedding model to encode the user input w.r.t the topic information into a topic embedding in an unsupervised manner. The input can be either words or sentences flexibly.

– We also design a topic inference model to infer topic embeddings based on the equation, where these embeddings can be provided to the MWP generation model to generate diverse MWPs covering different topics.

2   Related Work

The math word problem generation task can be treated as a controllable text generation task. In general, there are two main categories for MWP generation: (1) template-based methods and (2) learned-based methods.

Template-Based Methods. Early studies generate math word problems using template-based methods (*e.g.,* [9, 11, 14]). In these methods, a lot of pre-defined MWP templates are usually involved. In this case, the generator just needs to select proper words to fill out the blanks in the MWP template. The main disadvantage of template-based methods is that the generated MWPs usually share highly similar sentence structures, which lacks semantic richness and flexibility. Also, to deal with different types of problems, it is often necessary to pre-define a large number of MWP templates manually, which is a high-cost task.

Learned-Based Methods. Recent studies use recurrent neural network based approaches to treat this task as a controllable natural language generation problem, where usually a deep learning model is used to learn how to generate an MWP by an equation given a large training dataset. Zhou *et al.* [19] propose a seq2seq based model to generate MWP, where the equation encoder learns the representation of the given equation and the topic encoder encodes several topic words provided by the user. Then the equation encoder and the topic encoder are fused together and pass the output to the decoder to generate the MWP. However, this method has the following disadvantages: (a) It uses a sequential model to encode the equation input, which is unnatural since it does not consider the tree structure of the equation; and (b) The topic information of MWP can only be encoded by topic words, which is user-unfriendly and lack of flexibility since it needs the user to manually select topic words to generate an MWP. Liu *et al.* [5] propose a conditional VAE model for MWP generation, which applies an external commonsense-based knowledge graph to generate commonsense-related information in the MWP. However, the commonsense knowledge has to be created manually and it only support generating MWPs from one equation template $N_1x + N_2y = N_3; N_4x + N_5y = N_6$.

3   Preliminaries

3.1   Problem Definition

The MWP generation task can be formalized as a conditional natural language generation task. That is, given a math equation and a topic as input, it generates a math word problem that can be solved by the equation and the description of the problem is highly related to the input topic. In this section, we first describe the above factors in detail, and then formally give the definition of our task.

Equation. An equation is a mathematical statement, where two expressions are connected by the equal sign "=", *e.g.*, $9x + 5 = 32$. In this paper, following previous works, we mainly consider the unary equation *i.e.*, an equation with only one variable x. To make the problem easier to solve, we can rewrite the equation into a uniform form like "$x = (an \ expression)$" through simple mathematical derivation. For example, the above equation can be rewritten as $x = (32 - 5) \div 9$.

The Math Word Problem (MWP). An MWP is a piece of natural language text describing a scenario that can be solved by an equation.

Example 1. Locations A and B are 32 km apart. Starting from A, Tom ran at a speed of 9 km per hour and stopped at a place C, the middle of A and B, where C is 5 km away from B. How many hours did Tom run?

Clearly, the above MWP can be solved by the equation $x = (32 - 5) \div 9$, where the variable x denotes how many hours Tom ran.

The topic of an MWP. Apparently, an equation can be described by many different MWPs with respect to different topics. Let us see another example.

Example 2. Alice has 32 candies. She gave the candies evenly among her 9 friends, and finally, there were 5 candies left. How many candies did each of her friends get?

The above two problems are related to completely different topics, but share the same equation $x = (32 - 5) \div 9$ as their solution. Therefore, it is better to allow the user to specify the topic so that we can generate MWP flexibly based on the user's requirement. To generate an MWP with a specific topic, the user needs to input a piece of text (words or sentences) T, indicating the topic. In this paper, we further represent the topic as a dense embedding vector. The details of how to embed the topic will be described in Sect. 6.1.

Problem Definition. As follows, we formally define the topic-aware MWP generation problem. Given an equation E and a piece of topic text T as input, we aim to generate a math word problem text P, such that (1) P is a piece of smooth natural language; (2) The problem P can be solved by equation E; (3) the topic of P is highly related to that of T.

Example 3. Given the equation $x = 100 - 13 \times 3$ and $T = $ "He bought movie tickets", we generate the following text: $P = $ "Alex has \$100 saved up. He uses his money to buy 3 movie tickets. Each ticket costs \$13. How much money does he have left?"

Besides, we also support the scenario that the user does not specify the topic T. For this case, we can generate multiple MWPs covering different topics, so as to improve the generation diversity (See Sect. 6.2).

3.2 Preprocessing Steps

At a high level, we need to train a model to achieve automatic MWP generation by taking the equation and topic information as inputs. To this end, we should

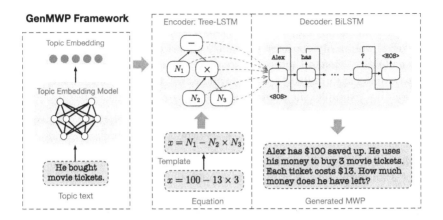

Fig. 1. The framework of MWP generative model

first prepare a dataset $\mathcal{D} = \{(E_i, T_i, P_i) : 1 \le i \le N\}$, where E_i is an equation, T_i is the topic text, and P_i is the corresponding MWP text. However, we need to perform some preprocessing steps on $\mathcal{D}$ to better adapt to our model.

Equation Template. Considering numeric values in equations barely contain no semantic information, we replace them with placeholders for ease of training, producing an equation template. For instance, in Example 3, we convert the equation into a template $x = N_1 - N_2 \times N_3$. Correspondingly, numbers in the MWP text should also be replaced with the same placeholders. In the rest of the paper, the word "equation" actually means an equation template for convenience.

Topic Text Extraction. In most MWP datasets, only the equations and MWP text are given, lacking the topic information. That is, for each training instance, initially, only (E_i, P_i) is available. To address this issue, given a tuple of (E_i, P_i), we propose to extract some sentences and words, add to the tuple respectively, and construct multiple triple training instances. More concretely, to collect sentences, we can simply randomly select some sentences in P_i. For words, we use tf-idf to select top-k important words in P_i.

4   Overall Framework

At a high level, given a training instance $(E, P, T) \in \mathcal{D}$, **GenMWP** learns a model that takes as input the equation E, topic T, and then outputs the MWP P. As shown in Fig. 1, in our framework, T is fed into a topic embedding model, which generates an embedding that is used to guide the topic of the MWP to be generated. Besides, the main part of **GenMWP** is an encoder-decoder architecture, where the encoder is to model the input equation E, meanwhile, considering the topic embedding, so as to learn a representation for an equation. Then the representation is sent to the decoder to generate a sequence of tokens, *i.e.,* MWP.

There are several challenges here. First, how to design a suitable equation encoder model for MWP generation is a challenge. To address this, considering that the equation can be well expressed by a tree, we use a Tree-LSTM as the encoder. Second, how to accurately learn the topic embedding and combine it with the Tree-LSTM model for generation is challenging. Third, how to leverage the topic embedding model to generate multiple MWPs with diverse topics.

Tree-LSTM Encoder. Different from the natural language, adjacent tokens in equations may not be directly related, which means adopting a sequential model (*e.g.,* LSTM or GRU) to encode the equation is not suitable. However, by adopting Tree-LSTM, only directly related tokens (*i.e.,* operator and its operands) in the equation are connected. Therefore, it is reasonable to use Tree-LSTM to generate MWPs and we leave the details in Sect. 5.

Topic Embedding Model. As shown in Fig. 1, given a topic text T_i as the input, the topic embedding model embeds it to an embedding vector t_i. The topic embedding vector acts as a summarization of T_i, and thereby text with similar topics should have similar embeddings. Details on the topic embedding model are introduced in Sect. 6.1.

Diverse MWPs Generation. In real-world scenarios, the user may not be willing to provide topic text when generating MWPs. In this case, we have to tackle the problem that given only an equation, how to generate MWPs that matches the equation. In order to consider the topic information without the user input, we need to capture the equation-to-topic relationship by *learning a mapping from an equation to a topic embedding*. Then the embedding will be served as the input of the MWP generation model to generate an MWP. More generally, the user does not provide the topic input, and thus it is necessary to generate multiple MWPs (say k MWPs) with different topics given one equation for her to pick some of them. We tackle this scenario by adopting a topic inference model, which will be described in detail in Sect. 6.2.

5   MWP Generation Model

In this section, we formally introduce the MWP generation model, which takes as input the equation template $E = (x_1, x_2, \cdots, x_m)$ and a topic embedding vector t (generated by T_i using the topic model in Sect. 6) to generate the MWP text $P = (y_1, y_2, \cdots, y_m)$.

As we introduced, the MWP generation is a conditional natural language generation task. Therefore, we can easily fit this task into a seq2seq model, where the encoder and decoder are both sequence models (such as LSTM and GRU). In standard LSTM, given a sequence $(x_1, x_2, \cdots, x_m)$, at each time-step t, the LSTM unit updates its hidden state h_t by taking the token x_t and the hidden state at the previous time h_{t-1} as input. Specifically, each LSTM unit uses an input gate i_t, an output gate o_t, and a forget gate f_t to update its hidden state. However, it only accepts sequence data as input and does not fit the tree structure. Relatively, a tree-structured model is more suitable to understand the equation. Therefore, we use Tree-LSTM [12] as our equation encoder.

Tree-structured Encoder. Different from the standard LSTM, the update of the gating vectors and cell state in the Tree-LSTM unit depends on the state of all its children units. There are two different variants of Tree-LSTM called Child-Sum Tree-LSTM and N-ary Tree-LSTM. The children of the Child-Sum Tree-LSTM model are unordered while those of N-ary Tree-LSTM are ordered. Because the order of the operands is important for some operations (*e.g.*, $-$ and $\div$) and the maximum number of operands in an operation is 2, we use the N-ary Tree-LSTM. Given an expression tree of an equation template, we build a Tree-LSTM model with the same structure. Tree-LSTM units correspond to nodes in the expression tree from the same position and are indexed from 1 to m in the postorder traversal. For a node, the transition equations are as follows:

$$i_t = \sigma \left(\boldsymbol{W}^i \boldsymbol{x}_t + \boldsymbol{U}_L^i \boldsymbol{h}_{Lt} + \boldsymbol{U}_R^i \boldsymbol{h}_{Rt} + \boldsymbol{b}^i \right) \tag{1}$$

$$o_t = \sigma \left(\boldsymbol{W}^o \boldsymbol{x}_t + \boldsymbol{U}_L^o \boldsymbol{h}_{Lt} + \boldsymbol{U}_R^o \boldsymbol{h}_{Rt} + \boldsymbol{b}^o \right) \tag{2}$$

$$\boldsymbol{f}_{Lt} = \sigma \left(\boldsymbol{W}^f \boldsymbol{x}_t + \boldsymbol{U}_{LL}^f \boldsymbol{h}_{Lt} + \boldsymbol{U}_{LR}^f \boldsymbol{h}_{Rt} + \boldsymbol{b}^f \right) \tag{3}$$

$$\boldsymbol{f}_{Rt} = \sigma \left(\boldsymbol{W}^f \boldsymbol{x}_t + \boldsymbol{U}_{RL}^f \boldsymbol{h}_{Lt} + \boldsymbol{U}_{RR}^f \boldsymbol{h}_{Rt} + \boldsymbol{b}^f \right) \tag{4}$$

$$\boldsymbol{u}_t = \tanh \left(\boldsymbol{W}^u \boldsymbol{x}_t + \boldsymbol{U}_L^u \boldsymbol{h}_{Lt} + \boldsymbol{U}_R^u \boldsymbol{h}_{Rt} + \boldsymbol{b}^u \right) \tag{5}$$

$$\boldsymbol{c}_t = \boldsymbol{i}_t \odot \boldsymbol{u}_t + \boldsymbol{f}_{Lt} \odot \boldsymbol{c}_{Lt} + \boldsymbol{f}_{Rt} \odot \boldsymbol{c}_{Rt} \tag{6}$$

$$\boldsymbol{h}_t = \boldsymbol{o}_t \odot \tanh \left(\boldsymbol{c}_t \right) \tag{7}$$

where $\boldsymbol{h}_{Lt}$ and $\boldsymbol{c}_{Lt}$ denote the hidden state and memory cell of the left child of node t, while $\boldsymbol{h}_{Rt}$ and $\boldsymbol{c}_{Rt}$ are used respectively for its right child; the matrices $\boldsymbol{W}, \boldsymbol{U}$ and $\boldsymbol{b}$ with different subscripts and superscripts are all trainable parameters; $\sigma(\cdot)$ denotes the Sigmoid function. If t is a leaf node, the hidden state and memory cell passed from its children are all set to zero vectors, *i.e.*, $\boldsymbol{h}_{Lt} = \boldsymbol{c}_{Lt} = \boldsymbol{h}_{Rt} = \boldsymbol{c}_{Rt} = \boldsymbol{0}$. If x_t is a unary operator (*e.g.*, square root or unary minus), we specify the only operand as the right child of the node t and set $\boldsymbol{h}_{Lt} = \boldsymbol{c}_{Lt} = \boldsymbol{0}$.

Fusion with Topic Embedding t. Since we can encode the topic text T into a topic embedding vector $\boldsymbol{t}$ using our topic embedding model (Sect. 6.1). To integrate the topic information into the encoder, we can just concatenate the input token embedding and the topic embedding vector, and take it as the input of each Tree-LSTM cell, *i.e.*, $\boldsymbol{x}_t = [e(x_t); \boldsymbol{t}]$, where $e(\cdot)$ denotes the embedding vector of an input token.

Decoder. After encoding the equation and topic in the encoder, we employ LSTM as the decoder to generate MWP text $P = (y_1, y_2, \cdots, y_n)$. Specifically, at each time-step t, the decoder updates its hidden state using the previous hidden state $\boldsymbol{s}_{t-1}$, the previous output y_{t-1}, and the context vector $\boldsymbol{c}_t$ (introduced in the next paragraph Attention Mechanism):

$$\boldsymbol{s}_t = \mathrm{LSTM}(\boldsymbol{s}_{t-1}, [e(y_{t-1}); \boldsymbol{c}_t]) \tag{8}$$

where the initial hidden state is determined by the hidden state of the root node in the encoder, *i.e.*, $s_0 = h_m$. Then the decoder predicts the next word using a softmax layer over the whole MWP vocabulary:

$$Pr(y_t|P_{<t}, E, t) = \text{softmax}(\tanh(W_c[s_t; c_t]))|_{y_t} \qquad (9)$$

where W_c is a learnable matrix to apply a linear transformation over the LSTM output.

Attention mechanism. We leverage the attention mechanism from the encoder to the decoder to compute the context vector c_t. By using the attention mechanism, we can make the decoder dynamically focus on some subtree of the tree-structured encoder when predicting a word. the context vector c_t is updated as:

$$c_t = \sum_{i=1}^{m} \alpha_{it} h_i \qquad (10)$$

$$\alpha_{it} = \frac{\exp(e_{it})}{\sum_{k=1}^{m} \exp(e_{kt})} \qquad (11)$$

$$e_{it} = v_a^{\mathsf{T}} \tanh(W_a[s_{t-1}; h_i]) \qquad (12)$$

where v_a^{T} and W_a are learnable weights.

5.1 Model Training and Inference

Model Training. Given a training dataset $\mathcal{D}' = \{(E_i, t_i, P_i) : 1 \leq i \leq N\}$, the MWP generation model aims to maximize the negative log likelihood of $\mathcal{D}'$:

$$\mathcal{J} = - \sum_{(E,t,P) \in \mathcal{D}'} \log Pr(P|E, t) \qquad (13)$$

In the training stage, given each training sample $(E, t, P) \in \mathcal{D}'$, the Tree-LSTM encodes the equation E and a topic vector t, and the decoder generates a token of the MWP at each step correspondingly. Therefore the probability $Pr(P|E, t)$ can be further expressed as:

$$Pr(P|E, t) = \prod_{y_t \in P} Pr(y_t|P_{<t}, E, t; \theta) \qquad (14)$$

where θ is the set of model parameters. The conditional probability $Pr(y_t|P_{<t}, E, t; \theta)$ is estimated by the encoder-decoder model in Eq. (9). We then use the cross-entropy loss function and stochastic gradient descent to update the model parameters.

Model Inference. In the inference stage, given an equation and a topic embedding vector, the model uses the beam search to continuously predict the top-k candidate words with highest probabilities at each step and finally generate the MWP text with the highest probability from the candidates.

6   Topic Model for MWP Generation

In this section, we introduce the topic embedding model to generate a topic embedding t as one of the inputs of the encoder (Sect. 6.1). Then we introduce the topic inference model to generate diverse queries with different topics in Sect. 6.2.

6.1   Topic Embedding Model

In this part, we study how to turn the topic text T_i into a dense embedding vector t_i, so as to guide the decoder to generate an MWP satisfying the topic. To train our topic embedding model, we use all the topic texts as the training set $\mathcal{T}$, i.e., $\mathcal{T} = \{T_i | (E_i, T_i, P_i) \in \mathcal{D}\}$.

Topic Embedding Model. To compute the topic embedding, we leverage the method proposed in Distributed Memory Model of Paragraph Vector (PV-DM) [3,18] because it is a typical unsupervised approach for modeling the topic from a collection of documents, without any training data. In this model, every sentence or word (*i.e.*, topic text T_i) is embedded into a vector of topic embedding. The general idea is to predict a word using the surrounding word embeddings within a sliding window, as well as the topic embedding. The topic embedding acts as a summarization of the entire text, and thereby text with similar topics have similar embeddings.

Specifically, each topic text T_i corresponds to a topic embedding vector t_i and each word w_i also corresponds to a word embedding vector v_{w_i}, which are randomly initialized. To train this model, the word vectors of surrounding words within a window size along with the topic vector are concatenated to predict the target word. Formally, given an MWP text $T_i = (w_1, w_2, w_3, \cdots, w_N)$, the model aims to maximize the log probability:

$$\sum_{j=k+1}^{N-k} \log Pr(w_j | P_i, \{w_s\}_{s=j-k}^{j-1}, \{w_s\}_{s=j+1}^{j+k}) \qquad (15)$$

where k denotes the maximum distance between the target word and context words used for prediction (*i.e.*, the window size is $2k+1$). We can easily use a classifier to implement this model:

$$y = \text{softmax}(U[t_i; v_{w_{j-k}} : v_{w_{j-1}}; v_{w_{j+1}} : v_{w_{j+k}}] + b)|_{w_j} \qquad (16)$$

where U and b are the learnable parameters.

Note that this model is prepared for sentence-formed topic text. For the topic text composed of unordered topic words, we make them into a "fake sentence" by randomly arranging them many times and assembling them together. Also, all numbers in the text are treated as a special token $\langle \text{NUM} \rangle$ to avoid overfitting since different numbers should not affect the overall topic semantics.

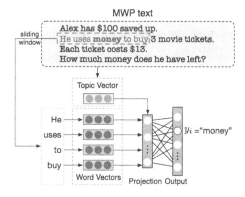

Fig. 2. the topic embedding model

Topic Embedding Model Training. In the training stage, given an MWP text, we keep using a sliding window with a length of $2k + 1$ over the text and continuously pick the predicted word and the context words, and feed them into the topic embedding model, as shown in Fig. 2. We then apply the cross-entropy loss function on the model and use stochastic gradient descent to continuously update the model parameters (including word embedding vectors, topic vectors, and classifier parameters U and b). Considering that the dimension of the output vector (*i.e.*, the vocabulary size) is quite large, so hierarchical softmax or negative sampling is used to accelerate the training in practice [8].

Topic Embedding Model Inference. In the inference stage, given a new topic text, it should generate a topic embedding vector by fixing the word embedding vectors and classifier parameters and uses stochastic gradient descent to continuously update the topic embedding vector. Now, we can apply the topic embedding model on the entire training dataset $\mathcal{D} = \{(E_i, T_i, P_i) : 1 \leq i \leq N\}$ to get $\mathcal{D}' = \{(E_i, t_i, P_i) : 1 \leq i \leq N\}$, where t_i is the topic embedding vector of topic text T_i.

6.2 Topic Inference Model

In this part, we study how to generate diverse MWPs by applying a topic inference model. That is, given only an equation E, the model outputs k topic embeddings that are both in diverse topics and related to the equation E.

Topic Clustering. We use a clustering algorithm to divide the topic embeddings in the training set into several topic clusters $\{\text{TC}_1, \text{TC}_2, \cdots, \text{TC}_L\}$. In our practice, we first use the UMAP algorithm [7] to perform dimension reduction on topic embedding vectors and apply HDBSCAN [6] to cluster them, where one of the advantages is that it does not need to manually specify the number of clusters.

Model Implementation and Training. After the clustering, to ensure the diversity of the output, we design the topic inference model as a classification task, that is, given an equation template $E = (x_1, x_2, \cdots, x_m)$, the model maximizes the probabilities that the topic of the MWP generated by this equation belongs to each topic cluster:

$$\mathcal{J} = - \sum_{(E,t,P) \in \mathcal{D}'} \log Pr(\text{cluster}(\boldsymbol{t})|E) \tag{17}$$

We implement this model by simply using a multilayer perceptron (MLP) classifier with a softmax layer after a Tree-LSTM equation encoder from the MWP generation model to estimate the conditional probability. In addition, in order to pay different attention to each sub-expression of the equation, we adopt the *self-attention* mechanism to the equation encoder. Formally, the topic inference model is implemented as:

$$Pr(\text{cluster}(\text{TC}_i)|E) = \text{softmax}(\text{MLP}([\boldsymbol{h}_m, \boldsymbol{a}_m]))|_{\text{TC}_i} \tag{18}$$

$$\boldsymbol{a}_m = \sum_{i=1}^{m} a_{im} \boldsymbol{h}_i \tag{19}$$

$$a_{im} = \frac{\exp(g_{im})}{\sum\limits_{k=0}^{m} \exp(g_{km})} \tag{20}$$

$$g_{im} = \boldsymbol{v}_c^{\mathsf{T}} \tanh(\boldsymbol{W}_c[\boldsymbol{h}_i; \boldsymbol{h}_m]) \tag{21}$$

where $\boldsymbol{h}_i$ is the hidden state of the i-th cell of the Tree-LSTM encoder, and $\boldsymbol{W}_c$ and $\boldsymbol{v}_c$ are learnable parameters of the self-attention mechanism. Before training, we initialize the encoder parameters (*i.e.*, the token embedding and the gate matrices) from the well-trained MWP generation model to accelerate the training process (Fig. 3).

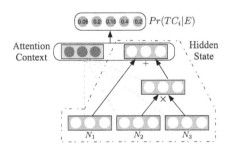

Fig. 3. the topic inference model

Model Inference. In the inference stage, given an equation template E, the model output the probabilities $Pr(\text{TC}_i|E), i \in [1, L]$ and picks k clusters with the highest probabilities. Then, to generate a topic embedding vector from each selected topic cluster, we randomly select several topic embeddings from the

Table 2. Summary of Datasets

Dataset	language	# MWP
Dolphin18K	English	18,460
Math23K	Chinese	23,162
Ape210K	Chinese	210,488

training set in this cluster and compute the average as the output topic embedding in this cluster. We then take the equation and the inferred topic embeddings and input them into the MWP generation model to generate diverse MWPs.

To summarize, the topic inference model generates several topic embeddings with different topics, which considers both the correlation between the equation and the topic, and the topic diversity. This model can provide the topic embedding vectors when the user is not willing to input the topic text, which will be further fed into the MWP generation model.

7   Experiments

7.1   Datasets and Experiment Setting

Datasets. We conduct the experiments on three datasets: Dolphin18K [2], Math23K [13] and Ape210K [17]. The details of these datasets are listed in Table 2.

Experiment Setting. The models proposed in this paper are implemented using Python 3.7 and PyTorch 1.8.1. All the experiments are conducted in a machine with 3.10 GHz Intel Xeon CPU 6242R, 256 GB RAM, and an NVIDIA RTX 3090 GPU, running Ubuntu 14.04.

Implementation details. The dimension of the all embedding vectors are set to 256. The window size of the topic embedding model is set to 13. For the MWP generation model, the hidden state size is set to 512, dropout rate is set to 0.5, and the beam size in the decoding stage is set to 5.

7.2   Experiment on MWP Generation

In this part, we evaluate the scenario that given an equation and a piece of topic text, aims to generate an MWP using the topic embedding model and the MWP generation model.

Table 3. Automatic evaluation on MWP generation

	Dolphin18K			Math23K			Ape210K		
	BLEU	ROUGE-L	METEOR	BLEU	ROUGE-L	METEOR	BLEU	ROUGE-L	METEOR
seq2seq	5.21	29.28	14.17	10.43	32.39	28.82	13.34	43.06	40.56
tree2seq	5.48	30.01	14.89	11.21	33.47	29.88	13.97	44.02	42.23
seq2seq-te	18.22	48.20	28.83	28.21	53.37	39.02	49.46	75.24	72.58
MAGNET	15.03	42.89	24.71	27.51	53.47	38.75	48.99	74.96	71.29
GenMWP	**19.28**	**48.66**	**29.73**	**29.99**	**57.22**	**42.60**	**54.96**	**77.52**	**75.88**

Table 4. Human evaluation on MWP generation

		Fluency		Equation Rel.		Topic Rel.		Consistency	
		rating	κ	rating	κ	rating	κ	rating	κ
Dolphin18K	seq2seq	3.88	0.75	2.38	0.69	2.66	0.88	3.27	0.66
	tree2seq	3.87	0.68	2.87	0.66	2.63	0.80	3.31	0.62
	seq2seq-te	**4.03**	0.75	3.13	0.58	**4.88**	0.86	3.81	0.52
	MAGNET	3.97	0.75	3.16	0.48	4.82	0.90	3.88	0.56
	GenMWP	**4.03**	0.79	**3.27**	0.55	**4.88**	0.86	**4.05**	0.52
Math23K	seq2seq	4.37	0.69	2.86	0.63	2.41	0.84	3.88	0.71
	tree2seq	4.32	0.71	2.92	0.66	2.02	0.78	3.89	0.77
	seq2seq-te	4.48	0.62	3.22	0.58	4.77	0.83	4.07	0.58
	MAGNET	4.51	0.66	3.16	0.52	4.72	0.86	4.11	0.62
	GenMWP	**4.58**	0.71	**3.34**	0.56	**4.89**	0.82	**4.12**	0.77
Ape210K	seq2seq	4.34	0.78	3.09	0.66	2.23	0.90	3.78	0.70
	tree2seq	4.22	0.66	3.17	0.68	2.37	0.86	3.66	0.66
	seq2seq-te	4.56	0.77	3.44	0.62	4.86	0.82	3.89	0.77
	MAGNET	4.56	0.74	3.45	0.56	4.86	0.82	**4.11**	0.72
	GenMWP	**4.62**	0.83	**3.62**	0.48	**4.90**	0.90	**4.11**	0.68

Comparison Methods. We compare our method with the following methods:

- **seq2seq**: a baseline method using a vanilla seq2seq model where standard LSTM is used as both the encoder and decoder;
- **seq2seq-te**: uses seq2seq model, where the topic embedding is passed to the standard LSTM encoder;
- **tree2seq**: uses Tree-LSTM as encoder but does not consider the topic embedding into the model;
- **MAGNET**: a state-of-the-art method proposed in [19], which also takes the equation and topic text as input to generate MWPs.

Automatic Evaluation Metrics. We use three automatic evaluation metrics to evaluate the performance of our method following recent works [5,19], *i.e.*, **BLEU** [10], **ROUGE-L** [4], and **METEOR** [1], all of which are widely used metrics in natural language generation. We compare the generated MWP with the reference MWP in the test dataset by the above metrics. All these metrics have output values between 0 and 100 (in percentage), with a higher value denoting a text that is more similar to the reference.

Human Evaluation. Like other natural language generation tasks, we also conduct human evaluation studies to evaluate the generated MWPs in different ways: (1) **Fluency**: whether the generated MWP conforms to the grammar of natural language and fluent; (2) **Equation Relevance**: whether the generated MWP is relevant to the given equation input; (3) **Topic Relevance**: whether the

Table 5. Evaluation result of diverse MWPs generation

		Fluency		Equation Rel.		Consistency		Topic Diversity	
		rating	κ	rating	κ	rating	κ	rating	κ
Dolphin18K	rand-emb	3.62	0.70	2.57	0.62	3.75	0.56	3.23	0.65
	rand-clus	3.58	0.62	2.25	0.58	3.62	0.52	**3.62**	0.61
	GenMWP	**3.99**	0.81	**3.08**	0.66	**3.85**	0.56	**3.62**	0.58
Math23K	rand-emb	3.88	0.66	2.30	0.58	3.66	0.61	3.39	0.68
	rand-clus	3.92	0.74	2.12	0.51	3.58	0.59	**3.99**	0.61
	GenMWP	**4.21**	0.71	**3.12**	0.62	**4.01**	0.67	3.98	0.59
Ape210K	rand-emb	4.41	0.67	2.12	0.55	3.72	0.53	3.41	0.55
	rand-clus	4.38	0.72	2.08	0.58	3.67	0.55	3.88	0.52
	GenMWP	**4.56**	0.72	**3.55**	0.61	**4.05**	0.67	**4.01**	0.49

generated MWP is relevant to the topic text input; (4) **Consistency**: whether the generated MWP is coherent and whether it contains semantic conflict. We recruit 10 volunteers to rate the generated MWPs from these aspects with scores ranging from 1 to 5, and compute the average score as the human evaluation result. In addition, to measure the agreement among different annotators, we also report the Fleiss's κ coefficient.

Automatic Evaluation Result. The experiment results of automatic evaluation on the 3 datasets are listed in Table 3. We make the following observations. (1) Both **seq2seq** and **tree2seq** have small evaluation results on 3 datasets. This is because, without topic embedding as model input, the model cannot directly learn topic information and thus the words in generated MWP have little matching with the reference MWP. (2) **tree2seq** also has a small improvement than **seq2seq** on these datasets. The reason is that by replacing the standard LSTM into Tree-LSTM as the encoder of the MWP generation model, the model can have a more clear view of the equation structure and thus can have slightly better generation results. (3) **seq2seq-te** has a great improvement in these metrics compared with **seq2seq** and **tree2seq**. The reason is that by passing the topic embedding into the model, it can learn which words are more likely to appear under this topic context, and therefore it can generate an MWP that has more tokens matching the reference MWP, which proves the effectiveness of our proposed topic embedding model. (4) The performance of **MAGNET** is slightly worse on these metrics compared with **seq2seq-te**. Although **MAGNET** also takes the topic information into consideration by adopting a topic encoder that encodes several topic words into word embeddings and uses an attention mechanism to pass the topic context into the decoder, it does not establish a semantic representation of the topic in a vector space, where similar topics are expressed as similar vectors. (5) **GenMWP** has an improvement in all the metrics compared with **MAGNET** and **seq2seq-te** on all 3 datasets. This also proves the effectiveness of the proposed tree-LSTM encoder and the topic embedding model.

Human Evaluation Result. The experiment results of human evaluation on the 3 datasets are listed in Table 4. We make the following observations. (1) On all the datasets, **GenMWP** has higher ratings compared with other comparison methods in fluency, equation relevance, topic relevance and, consistency, which indicates the effectiveness of **GenMWP** from different aspects. (2) In terms of fluency and consistency, all methods have rather good ratings (greater than 3.8), because, under the support of a large-scale training dataset, an LSTM decoder can easily generate sentences which is fluent and coherent with the context. (3) In terms of equation relevance, all methods have relatively low ratings (compared with the full rating 5.0), which is because learning mathematical logic in the equation is rather challenging. But compared with other methods, **GenMWP** still has a relatively better evaluation. (4) In terms of topic relevance, the ratings of **seq2seq** and **tree2seq** are significantly lower (about 2.0-) than the other 3 methods because they do not accept any topic information as input, and the highest rating of **GenMWP** indicates that the topic information is well captured and represented by our proposed topic embedding model. (5) The Fleiss's κ coefficients on all models vary from 0.45 to 0.9, revealing moderate to substantial agreements among the raters.

In summary, automatic evaluation results and human evaluation results both indicate that **GenMWP** outperforms the other methods when generating an MWP given an equation and a piece of topic text.

7.3   Experiment on Diverse MWPs Generation

In this part, we evaluate the diverse MWPs generation scenario, that is, given only an equation, generating k MWPs with diverse topics. In this experiment, k is set to 3.

Comparison Methods. Because **MAGNET** does not support diverse MWPs generation, we compare our methods with the following baselines:

– **rand-emb**: randomly select k topic embeddings from the training set and pass it to the MWP generation model.
– **rand-clus**: randomly select k topic clusters first, and then randomly select one topic embedding from each cluster.

Evaluation Metrics. Since the topic is not specified under this setting, using the corresponding MWP in the test dataset as the conference is not appropriate considering that the generated MWPs may have different topics from it. Therefore, we only use human evaluation to measure the quality of the generated MWPs and add a metric (*i.e.,* **Topic Diversity**; rating from 1 to 5) to measure the diversity of the generated MWPs.

Evaluation Result. The experiment results of generating diverse MWPs are shown in Table 5. We make the following observations. (1) In terms of fluency and consistency, **GenMWP** has higher ratings (about 0.2+ to 0.4+) than the

2 baseline methods. That is because the baseline methods pick the topic embeddings by a totally random policy, therefore these topic embeddings may not be suitable to this equation, thus causing the generated MWP to be less fluent and coherent. (2) In terms of equation relevance, **GenMWP** also has higher ratings (about 0.5+ to 1.0+) than the baseline methods. That is also because the randomly selected embeddings may not be related to the given equation. (3) In terms of topic diversity, **GenMWP** has higher ratings (about 0.4+ to 0.6+) than **rand-emb**, but has similar ratings with **rand-clus**. The reason is that **GenMWP** and **rand-clus** both select topic embeddings from different topic clusters. However, considering **rand-clus** may select from an unrelated topic and cause less fluency, consistency, and equation relevance, it is not a practical method to use. (4) The Fleiss's κ coefficients on all evaluations vary from 0.5 to 0.8, revealing moderate to substantial agreements among the raters.

In conclusion, evaluation results show that **GenMWP** is a better method for diverse MWPs generation than the other two baseline methods, proving the effectiveness of the proposed topic inference model.

8   Conclusion

In this paper, we study the problem of topic-aware automatic MWP generation using a typical encoder-decoder framework. First, we use the Tree-LSTM as the encoder to capture the structural information of math equations. Then we design a topic embedding model to make the generated MWP satisfy the given topic. We also support the scenario that the user does not provide any input for the topic, and we can output multiple MWPs with diverse topics. Experiments on three large-scale math word problem datasets illustrate that our proposed framework can produce fluent and coherent MWPs that are also related to the given topic and equation.

Acknowledgements. This paper was supported by National Key R&D Program of China (2020AAA0104500), National Natural Science Foundation of China (61925205, 62232009), Huawei, TAL Education Group, and Beijing National Research Center for Information Science and Technology.

References

1. Banerjee, S., Lavie, A.: METEOR: an automatic metric for MT evaluation with improved correlation with human judgments. In: Proceedings of the ACL Workshop on Intrinsic and Extrinsic Evaluation Measures for Machine Translation and/or Summarization, pp. 65–72 (2005)
2. Huang, D., Shi, S., Lin, C.Y., Yin, J., Ma, W.Y.: How well do computers solve math word problems? Large-scale dataset construction and evaluation. In: Proceedings of the 54th Annual Meeting of the Association for Computational Linguistics (Volume 1: Long Papers), pp. 887–896 (2016)
3. Le, Q., Mikolov, T.: Distributed representations of sentences and documents. In: International Conference on Machine Learning, pp. 1188–1196. PMLR (2014)

4. Lin, C.Y.: Rouge: A package for automatic evaluation of summaries. In: Text Summarization Branches Out, pp. 74–81 (2004)
5. Liu, T., Fang, Q., Ding, W., Liu, Z.: Mathematical word problem generation from commonsense knowledge graph and equations. arXiv preprint arXiv:2010.06196 (2020)
6. McInnes, L., Healy, J., Astels, S.: HDBScan: hierarchical density based clustering. J. Open Source Softw. **2**(11), 205 (2017)
7. McInnes, L., Healy, J., Melville, J.: UMAP: Uniform manifold approximation and projection for dimension reduction. arXiv preprint arXiv:1802.03426 (2018)
8. Mikolov, T., Chen, K., Corrado, G., Dean, J.: Efficient estimation of word representations in vector space. arXiv preprint arXiv:1301.3781 (2013)
9. Nandhini, K., Balasundaram, S.R.: Math word question generation for training the students with learning difficulties. In: Proceedings of the International Conference & Workshop on Emerging Trends in Technology, pp. 206–211 (2011)
10. Papineni, K., Roukos, S., Ward, T., Zhu, W.J.: BLEU: a method for automatic evaluation of machine translation. In: Proceedings of the 40th annual meeting of the Association for Computational Linguistics, pp. 311–318 (2002)
11. Polozov, O., O'Rourke, E., Smith, A.M., Zettlemoyer, L., Gulwani, S., Popović, Z.: Personalized mathematical word problem generation. In: Twenty-Fourth International Joint Conference on Artificial Intelligence (2015)
12. Tai, K.S., Socher, R., Manning, C.D.: Improved semantic representations from tree-structured long short-term memory networks. arXiv preprint arXiv:1503.00075 (2015)
13. Wang, Y., Liu, X., Shi, S.: Deep neural solver for math word problems. In: Proceedings of the 2017 Conference on Empirical Methods in Natural Language Processing, pp. 845–854 (2017)
14. Williams, S.: Generating mathematical word problems. In: 2011 AAAI Fall Symposium Series (2011)
15. Zhao, T., et al.: Towards automatic mathematical exercise solving. Data Sci. Eng. **4**(3), 179–192 (2019)
16. Zhao, T., et al.: MathGraph: a knowledge graph for automatically solving mathematical exercises. In: Li, G., Yang, J., Gama, J., Natwichai, J., Tong, Y. (eds.) DASFAA 2019. LNCS, vol. 11446, pp. 760–776. Springer, Cham (2019). https://doi.org/10.1007/978-3-030-18576-3_45
17. Zhao, W., Shang, M., Liu, Y., Wang, L., Liu, J.: Ape210k: A large-scale and template-rich dataset of math word problems. arXiv preprint arXiv:2009.11506 (2020)
18. Zhao, Z., Liu, T., Li, S., Li, B., Du, X.: Guiding the training of distributed text representation with supervised weighting scheme for sentiment analysis. Data Sci. Eng. **2**(2), 178–186 (2017)
19. Zhou, Q., Huang, D.: Towards generating math word problems from equations and topics. In: Proceedings of the 12th International Conference on Natural Language Generation, pp. 494–503 (2019)

SACA: An End-to-End Method for Dispatching, Routing, and Pricing of Online Bus-Booking

Yucen Gao, Yulong Song, Xikai Wei, Xiaofeng Gao[✉], and Guihai Chen

Shanghai Jiao Tong University, Shanghai 200240, China
{guo_ke,sylacd,weixikai}@sjtu.edu.cn, {gao-xf,gchen}@cs.sjtu.edu.cn

Abstract. Bus-booking is a novel service proposed to handle the massive traffic demand. Passengers can reserve bus seats through a bus-booking platform and it will place multiple passengers on one bus with a customized route. Order dispatching, route planning, and trajectory pricing are three key problems for this service, which interact intimately and are inappropriate to be solved separately in multi-stage. Consequently, we propose a Soft Actor-Critic (SAC) and attention based framework called SACA to solve these three problems jointly in an end-to-end manner. With the solutions of order dispatching and route planning obtained by an attention mechanism, SACA determines the order price for higher future revenue by a SAC network. SACA can deal with online scenarios and through various experiments based on real-world data, we prove its effectiveness compared with other multi-stage methods.

Keywords: Bus-booking platform · Order dispatching · Route planning · Trajectory Pricing · Deep reinforcement learning · Attention

1 Introduction

With the rapid expansion of urban scale and population growth, the demand for the timeliness and convenience of diversified urban transportation has been growing fast in recent years. However, existing transportation system is insufficient to meet the demand of passengers, especially for some transportation hubs at specific periods, such as airports at midnight, when the subways and buses are out of service [1]. To cope with such scenarios, a novel service called bus-booking was introduced to provide an efficient and economical way of public transportation [2]. A bus-booking platform can dispatch multiple passenger orders to a single bus with a customized route plan, then escort passengers to their destinations in sequence. Combing the benefits of bus service and car-hailing together,

This work was supported by National Key R&D Program of China [2020 YFB1707900]; National Natural Science Foundation of China [62272302], Shanghai Municipal Science and Technology Major Project [2021 SHZDZX0102] and DiDi GAIA Research Collaboration Plan [202204].

this new service provides flexible bus route and is cheaper than taxi, which contributes to a better user experience [3].

A bus-booking platform faces three key challenges: order dispatching, route planning and trajectory pricing. As shown in Fig. 1, the platform collects the bus supply and user demand information at each time slot, and utilizes some algorithms to match orders with buses, plan routes for different buses, and determine trajectory prices for different orders. The bus supply information at the next time slot is affected by the solution of the current time slot.

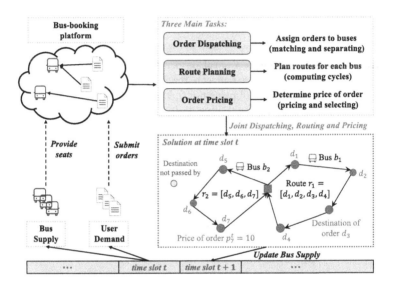

Fig. 1. Flowchart of a bus-booking platform.

Some airports have already explored such service, such as the customized bus and chartered bus service in the PEK Airport[1]. Although those specialized services address some traffic demand, pricing fairness and how to serve individual users still need to be studied. Because order dispatching, route planning and order pricing affect each other, a multi-stage approach may miss some interaction information, leading to performance degradation. Currently, some researchers have tried to solve two problems simultaneously [2], while no previous work considered all three problems jointly.

In the paper, we propose a Soft Actor-Critic (SAC) network and attention based end-to-end method, called SACA, to solve the three issues simultaneously. We first formulate the Joint order Dispatching, Route planning, and trajectory Pricing problem (JDRP), which is proven to be NP-hard. SACA determines the order price and the accepted order set according to the demand and supply relationship through a SAC framework, and then give the order dispatching

---

[1] https://en.bcia.com.cn/wljc.html.

and route planning solution with an attention mechanism, thus getting the new state of SAC. The experimental results based on a real-world dataset from DiDi demonstrate that SACA can achieve better performance in terms of overall revenue compared to the baseline methods.

2   Related Work

Order Dispatching: Order dispatching problem is usually formulated as a constrained optimization problem. In the early days, dynamic matching is the common way [4]. However, these methods are limited by reliance on supply and demand forecasts. When the forecast error is large, it is difficult to achieve good long-term revenue. Recently, Multi Agent Reinforcement Learning (MARL) methods were proposed. Jin et al. proposed a hierarchical MARL framework to make joint order dispatching and driver repositioning decisions [5]. Different from the problem of dispatching vehicles between different regions, we need to control the number of departure buses in each time slot to obtain long-term revenue.

Route Planning: In the early days, exact algorithms are proposed to solve route planning related problems. As the scale increased, heuristic algorithms were proposed to speed up the solving process. In recent years, machine learning techniques have been used to give high quality route planning solutions quickly [7]. For the bus-booking service, we need to modify the optimization objective and mask mechanism of ML methods.

Trajectory Pricing: Dynamic pricing was the main strategy for the trajectory pricing problem in the early days [9]. However, it usually does not take into account long-term revenue because it is highly susceptible to short-term fluctuations in supply and demand. Therefore, reinforcement learning was proposed to achieve a higher long-term revenue. However, these RL-based methods are not suitable for bus-booking pricing since Q-learning framework is designed for discrete action space. We therefore apply the SAC network to the continuous action (pricing) space and design novel demand and supply functions.

In recent years, order dispatching is usually considered jointly when solving route planning or trajectory pricing problems. Machine learning methods were proposed to solve the joint problems of the two issues. However, there exists no learning-based approach to solve the joint problem of the three issues in an end-to-end manner. Compared with the multi-stage methods for the joint problem, our proposed end-to-end framework is able to explore more feasible spaces to discover globally better solutions.

3   Problem Statement

We firstly introduce the entities in Sect. 3.1, and then give the formal definition of the Joint order Dispatching, Route planning, and trajectory Pricing problem (JDRP) problem and prove that it is NP-hard in Sect. 3.2.

3.1   Entity Models

To cope with online scenarios of the bus-booking service, we adopt the round-based model. In each round, the platform operates on pended orders and available buses to optimize some objectives. Some basic features are introduced here.

Definition 1 (Time slot). *A time slot is the time span for the platform to dispatch existing orders to different buses, plan routes, and determine prices.*

Referring to [6], we focus on assigning orders at a time slot, after which the platform batches new orders until the next time slot. In each time slot, the input consists of an order set $O = \{o_j\}$ and a bus set $B = \{b_i\}$.

Definition 2 (Order). *An order o_j is represented by $\langle d_j, c_j^o \rangle$, where d_j is the geographical location of its destination and c_j^o is the passenger number of o_j.*

Definition 3 (Bus). *A bus b_i is represented by $\langle r_i, L_i, c_i^b, v_i \rangle$, where r_i is b_i's route, consisting of destinations of orders dispatched to b_i; L_i is b_i's capacity limit; c_i^b is the number of passengers assigned to b_i; v_i is the average bus speed.*

Definition 4 (Platform's revenue). *The revenue of the bus-booking platform in each slot is equal to*

$$R^t = \sum_{o_j \in \hat{O}} p_j - \sum_{b_i} cost_i \tag{1}$$

where $\hat{O}$ represents the set of dispatched orders, p_j represents the price of o_j, and $cost_i = \beta_d \cdot dis_i$ represents the cost for bus b_i. β_d is the cost to bus per unit travel distance and dis_i is the travel distance of b_i during journey.

3.2   Problem Formulation

We first define the long-term revenue for the platform, which is the objective of the JDRP problem. Four constraints are then introduced.

Definition 5 (Long-term revenue). *The long-term revenue of the bus-booking platform is defined as the sum of revenue in each time slot, i.e.,*

$$R^{all} = \sum_{t=1}^{T} R^t \tag{2}$$

Then, four constraints are required to be satisfied in the JDRP problem:

Uniqueness constraint. $\forall o_j$, it can only be assigned to one bus.
Capacity constraint. $\forall b_i$, the number of assigned orders cannot exceed L_i.
Precedence constraint. $\forall o_j$ assigned to b_i, d_j should be included in r_i.
Loop constraint. All buses depart from the same departure station and are required to eventually return to the departure station.

Definition 6 (Joint order Dispatching, Route planning, and trajectory Pricing problem). *Given inputs of departure station DS, order set $\mathcal{O}$ and bus set $\mathcal{B}$, the JDRP problem is to determine prices, dispatch orders to buses and plan bus routes such that the long-term revenue R^{all} is maximized, while satisfying four constraints mentioned above. The formulation is as follows.*

$$\max \sum_{t=1}^{T}\left(\sum_{j}(p_j \cdot \sum_{i} x_{ij}) - \sum_{i}\beta_d \cdot dis_i\right)$$
$$s.t. \qquad \sum_{i} x_{ij} \leq 1, \forall o_j$$
$$\sum_{j} x_{ij} \leq L_i, \forall b_i$$
$$d_j \in r_i, \forall x_{ij} == 1$$
$$DS = r_i[0] \quad \& \quad DS = r_i[-1], \forall r_i$$

where $x_{ij} = 1$ means that order o_j is assigned to bus b_i.

Proof of NP-Hardness. We prove that the JDRP problem is NP-hard with a reduction from the Travelling Salesman Problem (TSP). Assuming that there exists one time slot and one bus, we can construct a complete graph $G = (V, E)$, where $v \in V$ represents the departure station and destinations of passengers, $cost(e)$ represents the distance. When we set the bus capacity to be ∞, the price for each passenger to be 1, β_d to be 0.5 and $cost(e)$ for each e to be 1, we need to accept all passengers to obtain the maximum revenue. In the case, to solve the JDRP problem means solving the TSP problem, a famous NPC problem. Hence the JDRP problem is no easier than TSP, which means it is NP-hard.

4   Model Framework

The framework of our proposed method is shown in Fig. 2. The method can be divided into three parts, a candidate order set generation module, a trajectory pricing module, and an order dispatching and route planning module.

4.1   Candidate Order Set Generation

At each time slot, there are some potential passengers with different destinations. We denote the number of potential passengers with the destination k by N_{pk}^t. Assuming that a function $F_{pk}(p)$ represents the relationship between demand and price for passengers with the destination k, the number of passengers willing to pay for the bus-booking platform at time t can be represented as D_k^t.

$$D_k^t(p_k^t) = N_{pk}^t F_{pk}(p_k^t) \tag{3}$$

where $F_{pk}(p) = 1 - \frac{dist_{max} + dist_{min} - dist_k}{dist_{max}} \cdot p^2$. $dist_k$ indicates the distance between station k and DS, $dist_{max} = \max(dist_k), \forall k$ and $dist_{min} = \min(dist_k), \forall k$. The multiplier is in the range of $[\frac{dist_{min}}{dist_{max}}, 1]$. It indicates the influence of distance to destinations on the expected price of passengers.

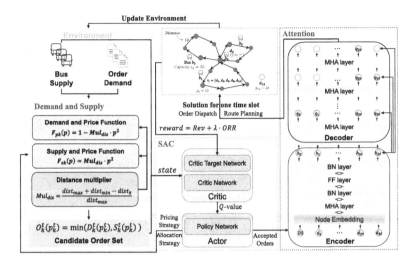

Fig. 2. Overview of the SACA method.

We use $N_s{}^t$ and F_{sk} to represent the number of available seats and the relationship between supply and price. Similarly, $F_{sk} = \frac{dist_{max} + dist_{min} - dist_k}{dist_{max}} \cdot p^2$. S_k^t is used to denote the supply to d_k.

$$S_k^t(p_k^t) = N_s{}^t a_k^t F_{sk}(p_k^t) \tag{4}$$

where a_k^t represents the fraction of assigned seats to d_k and $\sum_k a_k^t = 1$.

Hence, the total number of successfully dispatched orders and revenue of time slot t can be computed:

$$O_k^t(p_k^t) = \min(D_k^t(p_k^t), S_k^t(p_k^t)) \tag{5}$$

$$R^t = \sum_k p_k^t O_k^t(p_k^t) - \sum_{b_i} cost_i \tag{6}$$

4.2 Trajectory Pricing with SAC

We use SAC to determine prices. The reasons are as follows. Firstly, SAC can deal with the continuous action space and the pricing task has a continuous adjustment range. Secondly, SAC can automatically tune its hyper-parameters, saving the time of manual parameter tuning. Thirdly, SAC maximizes the reward and entropy simultaneously, which has good performance in both stability and exploration. The specific components of SAC are described as follows:

State: The state reflects the demand and supply at time slot t. Specifically, $x_s^t = \langle N_p{}^t, B^t \rangle$, where $N_p{}^t = [N_{p_k}^t]_{k \in N}$ represents the potential demand vector. $B^t = \{\tau_i^t\}_{b_i \in \mathcal{B}}$ represents the status of buses, where τ_b is the remaining time for a bus to return to the departure station.

Action: The action reflects the prices and scheduled seats for different destinations. Specifically, $x_a^t = \langle P^t, A^t \rangle$, where $P^t = [p_k^t]_{k \in N}$ represents the price vector, and $A^t = [a_k^t]_{k \in N}$ represents the distribution of seats for different destinations, subjecting to the constraint of $\sum_k a_k^t = 1$.

Transition: Assuming that the duration of each time slot is l and the dispatched buses corresponding to the accepted orders $\sum_k O_k^t(p_k^t)$ are B_d^t, we can get the available seats at the next time slot.

$$N_{s_k}^{t+1} = N_{s_k}^t - \sum_{b_i \in B_d^t} L_i + \sum_i (I(\tau_i^t \leq l) \cdot L_i) \tag{7}$$

Reward: The reward consists of the *Revenue Rev* and the *Order Response Rate (ORR)*. Since the demand varies with the increase of the time slot, the revenue could be obtained in each time slot differs. To make the SAC model converge, we add *ORR* to the reward since it can promote the model to accept more orders to get a higher reward in each time slot. The reward function is:

$$Reward^t = R^t + \lambda ORR^t \tag{8}$$

where λ represents the importance level of *ORR* and $ORR^t = \frac{\sum_k O_k^t}{\sum_k D_k^t}$.

4.3   Order Dispatching and Route Planning with Attention

Given the pricing solution and the accepted order set, we are required to dispatch orders and plan routes for buses that result in a shorter route length and a lower cost. Referring to the reinforcement learning model [7], we realize an attention mechanism, which can handle varying input sequence lengths (bus destination sets) and get good order dispatching and route planning solutions quickly.

Encoder: Assuming that the coordinate of destinations is denoted by x_k, the hidden state $h_k = W x_k + b$ through a node embedding layer.

Attention Layer: Referring to the Transformer [7], each attention layer consists of two sublayers: a multi-head attention layer(MHA) and a fully connected feedforward layer (FF). The batch normalization (BN) is used to replace the common layer normalization.

$$\hat{h}_k = BN^{l+1}(h_k^l + MHA_k^{l+1}(h_1^l, \cdots, h_N^l)) \tag{9}$$

$$h_k^{l+1} = BN^{l+1}(\hat{h}_k + FF^{l+1}(\hat{h}_k)) \tag{10}$$

where l indicates the l_{th} attention layer.

Decoder: With the following formulas, the decoder selects the element with the largest weight as the current output element.

$$u_k^i = v^T \tanh(W_1 h_k + W_2 h_i^d), j \in (1, 2, \cdots, n) \tag{11}$$

$$p(C_i | C_1, \cdots, C_{i-1}, P) = \text{softmax}(u^i) \tag{12}$$

where h_i^d is the i_{th} decoder hidden state. v, W_1, and W_2 are learnable parameters of the output model. u_k^i indicates the degree of relevance between the k_{th} input element and the i_{th} output element. (P, C^P) represents a training pair.

We use the mask mechanism to mask destinations other than the departure station in order to satisfy the capacity constraint. Simultaneously, the optimization goal is modified to minimize the route length for buses. The trained attention mechanism can be used directly to dispatch orders and plan routes for buses with a pricing solution and the accepted order set.

5   Experiments

We first describe the real dataset and illustrate some preprocessing measures. Then, we introduce the station selection process. Afterwards, we conduct experiments with different numbers of buses and compare SACA with other baselines.

5.1   Dataset and Evaluation Criteria

Dataset: We conduct experiments on a real-world order data from Didi Chuxing GAIA Initiative[2]. It is a car-hailing order dataset in Chengdu city. It contains various fields, including record ID, billing time, longitude and latitude. It covers 30 days with 7,065,937 pieces of records.

Data Preprocessing: Data preprocessing includes data filtering and data expansion. After eliminating missing values and outliers, we only consider orders during nighttime to adapt the data to our scenario. Since there are not many orders that meet our requirements in one day, we perform reasonable interpolation to generate more orders, thus expanding the effective order data set.

5.2   Baselines and Experimental Setting

Compared Methods: We combine the pricing method and routing method to realize the baseline methods.

- **DPEA+SubBus:** DPEA optimizes the order price and bus scheduling by adopting TRPO [8]. We use the SubBus method to plan routes for buses [3].
- **PODP+L2i:** PODP is a dynamic pricing method that uses the predicted future demand [9]. L2i is a learning-based iterative method to plan routes [10].

Configurations: Our experiments are conducted on a Dell G15-5515 laptop with 8 processors (8 × AMD Core R7-5800H, 3.20 GHz) and one graphics card (NVIDIA GeForce RTX 3060 Ti). The code is mainly written with Python 3.8. We set the bus capacity to be 30 and the price to be in the range of $[0, 1]$.

---

[2] Didi Chuxing GAIA Initiative: https://outreach.didichuxing.com/.

5.3   Stations Selection

Here, we use a DBSCAN-PAM hybrid clustering algorithm to explore hot-spots from real car-hailing order records, thus obtaining the station network with 30 destination stations in Fig. 3. The algorithm consists of two steps:

1. The DBSCAN algorithm is used to eliminate the noise data.
2. The PAM algorithm is used to cluster the data after noise elimination and uses the medoid of a cluster as a candidate station.

Fig. 3. 30 selected destination stations.

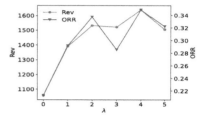

Fig. 4. Sensitivity analysis of λ.

5.4   Results

Solution Quality: As shown in Table 1, SACA performs better in terms of the revenue and total distance. Compared with DEPA+SubBus, SACA has better performance in both *Rev* and *Length* metrics because of the better implementation of order dispatching and route planning. Compared with PODP+L2i, though the baseline achieves a higher *ORR*, SACA can more reasonably accept the proper passengers to achieve higher revenue and shorter total distance.

Table 1. The Overall Revenue Comparison

Number of buses	10			20		
Model/Metric	Rev	ORR	Length	Rev	ORR	Length
DPEA+SubBus	342.21	0.13	57.09	948.98	0.33	150.65
PODP+L2i	606.11	0.26	113.13	1410.91	0.53	242.04
SACA	675.78	0.13	66.01	1639.95	0.34	173.22

Sensitivity Analysis: In the design of reward, λ is an adjustable hyperparameter, which reflects the relative importance level of *Rev* and *ORR*. Therefore, we study the sensitivity of λ. Figure 4 illustrates the performances of SACA

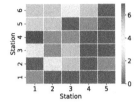

Fig. 5. Potential demand.

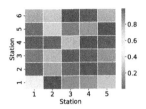

Fig. 6. Pricing of SACA.

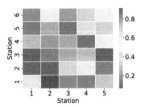

Fig. 7. Pricing of PODP.

under different values of λ. Therefore, finding a suitable λ in combination with specific data could lead to better performance of SACA in practice.

Policy Analysis: We analyze the differences between pricing policies of SACA and PODP. Figure 5 reflects the potential demand of 30 stations at 1 : 00 AM. Figure 6, 7 illustrate the pricing policy of SACA and PODP. For the station $(1, 4)$ and $(1, 5)$, SACA determines a higher price, which means that SACA can explore higher prices for stations with high demand. For the stations in column 4 and 5, SACA determines a high price to avoid accept improper orders which may cause the bus to take a detour, while PODP determines a low price.

6 Conclusion

In this paper, we focus on a novel transportation service named Bus-Booking. We formulate the joint order dispatching, route planning and trajectory pricing problem for the bus-booking service as the JDRP problem and propose a SAC and attention based end-to-end framework named SACA to solve the JDRP problem. Specifically, with the solutions of order dispatching and route planning obtained by the attention mechanism, SAC determines the price and the accepted order set for higher long-term revenue. Experiments conducted on the real-world dataset show that our proposed method has better performance than the baseline methods in terms of the long-term revenue and the total distance.

References

1. Zhou, H., Gao, Y., Gao, X., Chen, G.: Real-time route planning and online order dispatch for bus-booking platforms. In: Li, G., Yang, J., Gama, J., Natwichai, J., Tong, Y. (eds.) DASFAA 2019. LNCS, vol. 11447, pp. 748–763. Springer, Cham (2019). https://doi.org/10.1007/978-3-030-18579-4_44

2. Gao, Y., Gao, Y., Li, Y., Gao, X., Li, X., Chen, G.: An attention-based bi-gru for route planning and order dispatch of bus-booking platform. In: Jensen, C.S., et al. (eds.) DASFAA 2021. LNCS, vol. 12681, pp. 609–624. Springer, Cham (2021). https://doi.org/10.1007/978-3-030-73194-6_40

3. Kong, X., Li, M., Tang, T., Tian, K., Moreira-Matias, L., Xia, F.: Shared subway shuttle bus route planning based on transport data analytics. IEEE Trans Autom. Sci. Eng. **15**(4), 1507–1520 (2018)

4. Yan, C., Zhu, H., Korolko, N., Woodard, D.: Dynamic pricing and matching in ride-hailing platforms. Naval Res. Logist. **67**, 705–724 (2020)

5. Jin, J., et al.: CoRide: joint order dispatching and fleet management for multi-scale ride-hailing platforms. In: International Conference on Information and Knowledge Management (CIKM), pp. 1983–1992 (2019)

6. Zheng, L., Chen, L., Ye, J.: Order dispatch in price-aware ridesharing. In: International Conference on Very Large Data Bases (VLDB), vol. 11 no. 8, pp. 853–865 (2018)

7. Kool, W., van Hoof, H., Welling, M.: Attention, learn to solve routing problems! In: International Conference on Learning Representations (ICLR) (2019)

8. Turan, B., Pedarsani, R., Alizadeh, M.: Dynamic pricing and management for electric autonomous mobility on demand systems using reinforcement learning. arXiv:1909.06962 (2019)

9. Asghari, M., Shahabi, C.: ADAPT-pricing: a dynamic and predictive technique for pricing to maximize revenue in ridesharing platforms. In: ACM International Conference on Advances in Geographic Information Systems (SIGSPATIAL), pp. 189–198 (2018)

10. Lu, H., Zhang, X., Yang, S.: A learning-based iterative method for solving vehicle routing problems. In: International Conference on Learning Representations (ICLR) (2020)

Chinese Medical Nested Named Entity Recognition Model Based on Feature Fusion and Bidirectional Lattice Embedding Graph

Qing Cong[1][iD], Zhiyong Feng[1,3][iD], Guozheng Rao[1,3(✉)][iD], and Li Zhang[2][iD]

[1] College of Intelligence and Computing, Tianjin University, Tianjin 300350, China
{chf,zyfeng,rgz}@tju.edu.cn
[2] School of Economics and Management, Tianjin University of Science and Technology, Tianjin 300457, China
zhangli2006@tust.edu.cn
[3] Tianjin Key Laboratory of Cognitive Computing and Applications, Tianjin 300350, China

Abstract. Medical named entity recognition can assist doctors to quickly identifying key content and improving clinical work efficiency. Chinese named entity recognition methods based on pre-trained language models have achieved remarkable performance. However, most of these models have the following problems for medical named entity recognition: these models are designed for flat named entity recognition tasks but not for nested entities. Furthermore, the medical entities are hard to be recognized due to the lack of medical domain knowledge. To tackle these problems, we propose a Chinese medical nested named entity recognition model based on feature fusion and a bidirectional lattice embedding graph. The problem of poor recognition of medical entities due to the lack of medical domain knowledge is solved by introducing a medical lexicon. The problem of Chinese polyphonic characters with different meanings in the same form is solved by introducing pinyin information. The model considers the similarity between different entity types to improve the model's effectiveness. The results on a Chinese medical nested named entity dataset CBLUE-CMeEE demonstrate the outperform performance and effectiveness of the model.

Keywords: nested named entity recognition · Chinese medical entity · bidirectional lattice embedding graph · biaffine network

1 Introduction

Currently, Chinese named entity recognition methods based on pre-trained language models [2,7,12,14] have better performance than those based on BiL-STM [20] or CNN [5]. However, they still experience some shortcomings: First,

X. Wang et al. (Eds.): DASFAA 2023, LNCS 13946, pp. 314–324, 2023.
https://doi.org/10.1007/978-3-031-30678-5_24

the named entity recognition task needs to determine the start and end positions of entities in the text. i.e., the entity boundary information. However, in Chinese, word segmentation is usually using word segmentation tools, and the resulting problem of segmentation error propagation affects model performance [11]. Meanwhile, these methods do not adequately capture the interaction information between characters and words. Second, the phenomenon of compounding also exists in Chinese. In other words, the same word has multiple pronunciations but expresses different meanings. This phenomenon also affects the model's recognition ability. Third, with the application of named entity recognition technology in different fields, methods that can only handle flat named entities can no longer meet some of the complex requirements [17]. The number of parameters in the existing biaffine network-based named entity recognition model [9,18] increases with an increase in entity recognition types, leading to an increase in training cost.

To address these problems, we propose a feature fusion and bidirectional lattice embedding graph (FFBLEG) for Chinese named entity recognition. In this paper, our contributions are as follows:

1. We propose a feature fusion and bidirectional lattice embedding graph (FFB-LEG), a model for Chinese flat and nested named entity recognition.
2. We apply BiLSTM to encode pinyin sequences, obtain pinyin embeddings and fuse them with character embeddings to obtain character node features to solve the Chinese homophone problem.
3. We conducted experiments on a Chinese medical nested dataset CBLUE-CMeEE. Compared with other comparative models, our model exhibits better performance.

2   Related Work

NER played significant roles in many fields, such as information extraction, knowledge graph construction, event extraction, and precision medicine. Summarily, the primary methods of Chinese Medical NER are rule-based, statistical, and deep learning.

The rule-based methods are present early. It designed the writing rules to extract language features for the Chinese Medical NER tasks. For example, Zhang et al. [21] analyzed the characteristics of medical institution names, summarized their corresponding rules, combined them with the conditional random field model to identify Chinese medical institution names, and achieved better results.

Many statistical machine learning methods have been applied to Chinese Medical NER to tackle the problem above. For example, Tang et al. [15] proposed a medical NER system based on Support Vector Machines(SVM) and Conditional Random Fields (CRFs) to extract clinical entities in hospital discharge summaries.

Deep learning-based methods have been widely used in natural language processing in recent years, and better results have been achieved on most tasks.

Researchers have proposed various sequence annotation models for the named entity recognition task based on convolutional neural networks, recurrent neural networks, and pre-training methods, achieving better performance than traditional methods. For example, Wang et al. [16] combined dictionaries into deep neural networks for the Chinese clinical NER.

More and more pre-training based methods were applied to Chinese medical NER tasks. Bidirectional Encoder Representations from Transformers (BERT) [6] has shown marvelous improvements across various NLP tasks, and consecutive variants such as RoBERTa [12] ALBERT-tiny/xxlarge [8]. ZEN [4], MacBERT [2], BERT-Biaffine [18], and ChineseBERT [14] have been proposed to further improve the performance of the pre-trained language models. For example, Cui et al. [2] revisited Chinese pre-trained language models to propose a simple but effective model called MacBERT. Luo et al. [13] achieved some performance gains by recognizing named entities in Chinese electronic medical records based on stroke ELMo and multi-task learning methods with additional features such as word vectors and radicals. Li et al. [10] proposed a unified framework capable of handling both flat and nested NER tasks that transforms NER tasks from sequence labeling tasks to machine reading comprehension tasks.

Generally, the Chinese medical NER has attracted much research attention in many fields and has become a hot topic in the current medical text mining. However, the word-based NER model still has many challenges in Chinese medical NER. Consequently, Chinese word segmentation is required, and segmentation errors do occur. Moreover, there is a problem of insufficient domain knowledge. The Chinese medical nested NER is still a vast challenge [19].

3   Model Design

The overall architecture of the feature fusion and bidirectional lattice embedding graph(FFBLEG) model is shown in Fig. 1. It consists of four modules: The first module is the lattice graph construction which is applied to save all words matching the lexicon. The second module is the feature fusion, which obtains the embedding of characters by fusing the pinyin features and character features. The third module is used for bidirectional lattice embedding graph for graph representation learning and the fourth module is the biaffine prediction layer, which scores entities and decodes them to obtain the final results. To speed up the convergence of BERT [6], we add the output of feature fusion and the output of the bidirectional lattice embedding graph into the biaffine network through the residual structure.

3.1   Lattice Graph Construction

We use medical lexicon matching to introduce word information. This method aims to speed up the preprocessing procedure of the model. Let us take the text "giant mediastinal mass" as an example. The specific steps used in word case

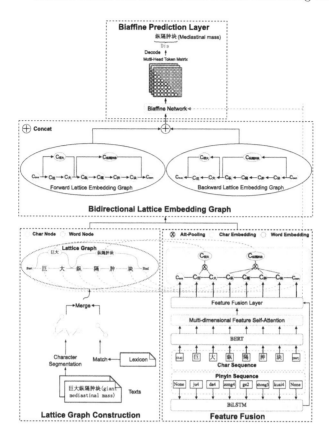

Fig. 1. BiLECB model structure.

construction are as follows: First, the model segments the text to obtain the character sequence X =['giant', 'large', 'mediastinal', 'septal', 'swelling', 'mass']. In the word segmentation step, the word sequence Y =['huge', 'mediastinal mass'] is formed by matching the text against a dictionary and finding all of the words that can be matched. Then the character sequence X and the word sequence Y are merged. To ensure that each word lattice graph has only one head node and one tail node, two additional characters [START] and [END] are added as the head and tail nodes. Finally, a lattice graph is shown in the Fig. 1.

3.2 Feature Fusion

To obtain more features and thereby enhance the feature representation of the named entities, the character embeddings obtained by BERT and the multidimensional feature self-attention mechanism are spliced with the pinyin embeddings. The specific flow is shown in the feature fusion section in Fig. 1.

(1) Character features: First, the input text is segmented according to the input to the BERT model to construct the required embedding. After inputting

this embedding into the BERT model, the first hidden layer state H_i^1 and the last hidden layer state H_i^{12} are used for attention aggregation, and the calculation formula is shown in Eq. 1. Then the character embedding C is obtained using multidimensional feature self-attention, and the calculation formula is shown in Eq. 2:

$$H = \text{Att-Pooling}(H^1 + H^{12}) \tag{1}$$

$$C = \sum_{n=i}^{l} a_{c_i} \odot H_i \tag{2}$$

In Eq. 2, l denotes the length of the input character sequence, H_i denotes the embedding of the ith character in the input character sequence, and a_{c_i} denotes the normalized multidimensional feature attention weight of the i-th character, which is calculated as shown in Eq. 3:

$$a_c' = \text{Softmax}(W_1{}^T \sigma(W_0 H_i + b_0) + b_1) \tag{3}$$

In Eq. 3, $a_c' \in \mathbb{R}^{l \times d_n}$ denotes the unnormalized multidimensional feature attention weights, $W_0 \in \mathbb{R}^{d_n \times l}$ and $W_1 \in \mathbb{R}^{l \times d_n}$ denote trainable parameters, b_0 and b_1 denote bias, and $\sigma(\cdot)$ denotes the activation function.

(2) Pinyin features: For each character in the character sequence, the pinyin of the Chinese character is extracted using the PyPinyin library. When extracting the pinyin of a character, the library will determines whether a word can be formed based on the information before and after the individual character and adjusts the final pinyin output of the polyphonic character based on the information from word.

After obtaining the pinyin sequence of the character sequence, the pinyin embedding is obtained through the BiLSTM network. Before being input into the BiLSTM, the pinyin sequences also need to be encoded. In this chapter, we choose to segment each pinyin sequence by its characters and then encode it using numerical mapping. We define the pinyin sequence of a single Chinese character as $X_p = [x_1, x_2, ..., x_L]$, where L is the length of the pinyin sequence.

Firstly, the pinyin sequence is mapped into a vector sequence and then the feature vector is obtained by BiLSTM. The feature vector is calculated as shown in Eq. 4:

$$h_{pi} = [\overrightarrow{h}_i \oplus \overleftarrow{h}_i] \tag{4}$$

In Eq. 4, $\overrightarrow{h}_i$ denotes the forward feature vector obtained by BiLSTM, $\overleftarrow{h}_i$ denotes the backward feature vector obtained by BiLSTM, and h_{pi} denotes the phonetic feature vector obtained by splicing the forward and backward feature vectors.

The final formula for obtaining the phonetic embedding P using maximum pooling is shown in Eq. 5:

$$P = \text{Max-Pooling}(h_{p1}, h_{p2}, ..., h_{pL}) \tag{5}$$

(3) Feature fusion: The initial embedding of the word nodes in the word lattice graph $E = [e_1, e_2, ..., e_l]$ is obtained from the character embedding C

and the phonetic embedding P after stitching by a multilayer perceptron. The computational formula is shown in Eq. 6.

$$e_i = \text{MLP}(c_i \oplus p_i) \tag{6}$$

In Eq. 6, $c_i \in \mathbb{R}^{d_n}$ denotes the character embedding of the ith character, $p_i \in \mathbb{R}^{d_p}$ denotes the phonetic embedding of the ith character, and $e_i \in \mathbb{R}^{d_n}$ denotes the word embedding of the i-th character.

(4) Word embedding: The word embedding of a word node in the lattice graph is obtained by aggregating the word embedding of all of the words that make up the word in the text using an attention mechanism.

3.3   Bidirectional Lattice Embedding Graph

After obtaining the initial embedding of the lattice nodes, we can compute the final representation of each node using graph representation learning. We design a bidirectional lattice embedding graph attention network based on the graph attention network [1] for learning the forward and backward graph features of texts.

The bidirectional lattice embedding graph attention network is divided into two separate graph attention networks: the forward lattice embedding graph attention network and the backward lattice embedding graph attention network. In the bidirectional lattice embedding graph attention network, the initial contextual embedding of each lattice node v_i is H_0^i is obtained from the multi-granularity embedding component. Now, we introduce the specific message aggregation update mechanism for the forward lattice embedding graph attention network as an example. The backward lattice embedding graph attention network uses the same mechanism, but the edges are in opposite directions.

4   Experiments

4.1   Datasets

CBLUE-CMeEE[1]: The CBLUE-CMeEE dataset was derived from the Chinese medical information processing evaluation benchmark CBLUE [19]. The dataset contains 23,000 texts from 938 medical documents in which the named entities in medical texts are classified into 9 categories. Among them, the naming entity "clinical manifestation" may be nested with the other 8 types of named entities.

4.2   Baseline

To demonstrate the effectiveness and advancement of the FFBLEG model proposed in this paper, we selected several different models for comparison:

---

[1] https://github.com/CBLUEbenchmark/CBLUE.

- BERT [6]: We use the base model with 12 layers, 768 hidden layers, 12 heads, and 110 million parameters.
- BERT-wwm-ext-base [3]: A Chinese pre-trained BERT model with whole word masking.
- RoBERTa-large [12] : Compared with BERT, RoBERTa removes the next sentence prediction objective and dynamically changes the masking pattern applied to the training data.
- RoBERTa-wwm-ext-base/large. RoBERTawwm-ext is an efficient pre-trained model which integrates the advantages of RoBERTa and BERT-wwm.
- ALBERT-tiny/xxlarge [8]. ALBERT is a pre-trained model with two objectives: Masked Language Modeling (MLM) and Sentence Ordering Prediction (SOP).
- ZEN [4]: A BERT-based Chinese text encoder enhanced by N-gram representations, where different combinations of characters are considered during training.
- PCL-MedBERT(https://code.ihub.org. cn/projects/1775): A pre-trained medical language model proposed by the Peng Cheng Laboratory.
- MacBERT [2]: MacBERT is an improved BERT with novel MLM as a correction pre-training task.
- BERT-Biaffine [18]: It used contextual embeddings as input to a multilayer BiLSTM. It emploied a biaffine model to assign scores for all spans in a sentence.
- ChineseBERT [14]: It incorporated both the glyph and pinyin information of Chinese characters into language model pre-training.

4.3 Experimental Results and Analysis

First, to verify the effectiveness and advancement of the FFBLEG model proposed in this paper on the nested named entity recognition task, experiments were conducted on the CBLUE-CMeEE dataset. Table 1 shows the experimental results of the FFBLEG model and the comparison model on the CBLUE-CMeEE dataset. For the CBLUE-CMeEE dataset, we used a medical lexicon for matching.

As shown in Table 1, the FFBLEG model proposed in this chapter outperforms other comparative models on both recall and F1 metrics. Compared with the RoBERTa-wwm-ext-base and BERT-Biaffine model, there is a relative improvement of 3.86% and 4.05% in the F1 value. It indicates that the FFBLEG model explicitly introduces medical word information through the bidirectional word lattice embedding graph, which enables the model to capture the features of medical entities better and thus improves the recognition ability of the model.

4.4 Ablation Study

To verify the validity of each module in the FFBLEG model, we conducted a series of ablation experiments, and the results are shown in Table 2. The specific description of each ablation model is as follows:

Table 1. Experimental results of the FFBLEG models and comparison models on the CBLUE-CMeEE nested datasets (Note: all comparison model results are based on open-source code)

Model	P(%)	R(%)	F1(%)	△F1(%)
BERT-base	53.41	63.32	57.95	−11.84
BERT-wwm-ext-base	–	–	61.7	−5.04
RoBERTa-large	–	–	62.1	−4.36
RoBERTa-wwm-ext-base	–	–	62.4	−3.86
RoBERTa-wwm-ext-large	–	–	61.8	−4.87
ALBERT-tiny	–	–	50.5	−28.34
ALBERT-xxlarge	–	–	61.8	−4.87
ZEN	–	–	61.0	−6.25
PCL-MedBERT	–	–	60.6	−6.95
MacBERT	60.45	63.53	61.95	−4.62
ChineseBERT	55.12	49.80	52.33	−23.84
BERT-Biaffine	64.17	61.29	62.29	−4.05
FFBLEG	**64.70**	**64.92**	**64.81**	–

- FFBLEG W/O Pinyin is based on the FFBLEG model in this paper, and the pinyin embedding module is removed.
- FFBLEG W/O Word is based on the FFBLEG model in this paper. The word nodes in the word lattice graph are removed, and the word information is not introduced.
- FFBLEG W/O Bw is based on the FFBLEG model in this paper. the backward word lattice embedding graph is removed, and only the forward context information captured by the forward word lattice graph is considered.
- FFBLEG W/O Fw is based on the FFBLEG model in this paper. The forward lattice embedding graph is removed, and only the forward context information captured by the backward lattice graph is considered.

The numbers in parentheses in the experimental results in Table 2 indicate the absolute change in F1 values for the ablation model compared to the full model. From the table, we can learn that removing any module from the FFBLEG model introduced in this paper leads to a decrease in the model-type tendency. The F1 value of the FFBLEG W/O Pinyin model decreases to some extent on all three datasets compared with the FFBLEG model, which indicates that the introduction of pinyin information can solve the problem of Chinese polyphonic characters with different meanings. The FFBLEG W/O Word model on the CBLUE-CMeEE dataset shows a large degree of performance degradation, which indicates that the introduction of medical domain knowledge through medical dictionaries can enhance the model's ability to recognize medical entities. From the results of the two ablation models, FFBLEG w/o Fw and FFB-

LEG w/o Bw, it is clear that the bidirectional contextual information captured through the bidirectional word lattice embedding graph enhances the features useful for entity classification.

Table 2. Results of ablation study (F1(%))

Model	CBLUE-CMEEE
FFBLEG	64.81
- W/O Pinyin	64.15(−1.03)
- W/O Word	63.61(−1.88)
- W/O Bw	63.72(−1.71)
- W/O Fw	63.53(−2.02)

4.5   Conclusion

In this paper, we propose a new Chinese named entity recognition model based on feature fusion and bidirectional lattice embedding graph. The model introduces word information and character information through a lattice graph to enhance the features that the model can capture. To better model the information interaction between words and characters and avoid the propagation of incorrect word segmentation information. We address the Chinese homophone problem by introducing pinyin information. The input to this network is the addition of the output of the graph attention network and the previous feature fusion layer, which are used to accelerate the convergence speed of BERT. Experimental results on the CBLUE-CMeEE dataset demonstrate our effectiveness and that our model is state-of-the-art.

However, our model still has several limitations. for example, word embeddings are obtained by aggregating character embeddings, and the features have a certain similarity. One solution is to use pre-trained word vectors to obtain word embeddings to enhance word features.

References

1. Brody, S., Alon, U., Yahav, E.: How attentive are graph attention networks? In: International Conference on Learning Representations (2022). https://openreview.net/forum?id=F72ximsx7C1
2. Cui, Y., Che, W., Liu, T., Qin, B., Wang, S., Hu, G.: Revisiting pre-trained models for Chinese natural language processing. In: Findings of the Association for Computational Linguistics, EMNLP 2020, pp. 657–668, November 2020
3. Cui, Y., Che, W., Liu, T., Qin, B., Yang, Z.: Pre-training with whole word masking for Chinese bert. IEEE/ACM Trans. Audio Speech Lang. Process. **29**, 3504–3514 (2021)

4. Diao, S., Bai, J., Song, Y., Zhang, T., Wang, Y.: Zen: Pre-training chinese text encoder enhanced by n-gram representations. In: Findings of the Association for Computational Linguistics: EMNLP 2020. pp. 4729–4740 (2020)
5. Gui, T., Ma, R., Zhang, Q., Zhao, L., Jiang, Y.G., Huang, X.: Cnn-based chinese ner with lexicon rethinking. In: the 28th International Joint Conference on Artificial Intelligence, pp. 4982–4988 (2019)
6. Kenton, J.D.M.W.C., Toutanova, L.K.: Bert: pre-training of deep bidirectional transformers for language understanding. In: Proceedings of NAACL-HLT, pp. 4171–4186 (2019)
7. Lai, Y., Liu, Y., Feng, Y., Huang, S., Zhao, D.: Lattice-bert: leveraging multi-granularity representations in chinese pre-trained language models. In: Proceedings of the 2021 Conference of the North American Chapter of the Association for Computational Linguistics: Human Language Technologies, pp. 1716–1731 (2021)
8. Lan, Z., Chen, M., Goodman, S., Gimpel, K., Sharma, P., Soricut, R.: Albert: a lite bert for self-supervised learning of language representations (2020)
9. Li, J., et al.: Unified named entity recognition as word-word relation classification. arXiv preprint arXiv:2112.10070 (2021)
10. Li, X., Feng, J., Meng, Y., Han, Q., Wu, F., Li, J.: A unified MRC framework for named entity recognition. In: Proceedings of the 58th Annual Meeting of the Association for Computational Linguistics, pp. 5849–5859 (Jul 2020)
11. Li, Y., Yu, B., Mengge, X., Liu, T.: Enhancing pre-trained Chinese character representation with word-aligned attention. In: Proceedings of the 58th Annual Meeting of the Association for Computational Linguistics, pp. 3442–3448, July 2020
12. Liu, Y., et al.: Roberta: a robustly optimized bert pretraining approach. arXiv preprint arXiv:1907.11692 (2019)
13. Luo, L., Yang, Z., Song, Y., Li, N., Lin, H.: Chinese clinical named entity recongnition based on stroke elmo and multi-task learning. Chin. J. Comput. **43**(10), 1943–1957 (2020)
14. Sun, Z., et al.: Chinesebert: chinese pretraining enhanced by glyph and pinyin information. In: Proceedings of the 59th Annual Meeting of the Association for Computational Linguistics and the 11th International Joint Conference on Natural Language Processing (Volume 1: Long Papers), pp. 2065–2075 (2021)
15. Tang, B., Cao, H., Wu, Y., Jiang, M., Xu, H.: Recognizing clinical entities in hospital discharge summaries using structural support vector machines with word representation features. In: BMC medical informatics and decision making, vol. 13, pp. 1–10. BioMed Central (2013)
16. Wang, Q., Zhou, Y., Ruan, T., Gao, D., Xia, Y., He, P.: Incorporating dictionaries into deep neural networks for the chinese clinical named entity recognition. J. Biomed. Inf. **92**, 103133 (2019)
17. Wang, Y., Tong, H., Zhu, Z., Li, Y.: Nested named entity recognition: a survey. ACM Trans. Knowl. Discov, Data (feb (2022)
18. Yu, J., Bohnet, B., Poesio, M.: Named entity recognition as depe, ndency parsing. In: Proceedings of the 58th Annual Meeting of the Association for Computational Linguistics. pp. 6470–6476 July 2020
19. Zhang, N., et al.: Cblue: a chinese biomedical language understanding evaluation benchmark. In: Proceedings of the 60th Annual Meeting of the Association for Computational Linguistics (Volume 1: Long Papers), pp. 7888–7915 (2022)

20. Zhang, Y., Yang, J.: Chinese ner using lattice lstm. In: Proceedings of the 56th Annual Meeting of the Association for Computational Linguistics (Volume 1: Long Papers), pp. 1554–1564 (2018)
21. Jinlong, Z., Shi, W.Q.C.: CRF and rules-based recognition of medical institutions name in chinese. Comput. Appl. Softw. **31**(3), 159–162 (2014)

Two-stage Interest Calibration Network for Reranking Hotels

Denghao Ma[1]([envelope]), Jiajia Sun[2], Yueguo Chen[2], Liang Shen[1], and Genliang Yi[1]

[1] Meituan, Beijing, China
{madenghao,shenliang03,yigenliang}@meituan.com
[2] DEKE Lab, Renmin University of China, Beijing, China
{jiajiasun,chenyueguo}@ruc.edu.cn

Abstract. As one key task of hotel search, interest modelling is to capture users' interests over their historical clicks. To this end, it is important yet challenging to understand the searching intention behind each click (interest focus). Besides, the query input by a user represents the user's interests to some extent. The differences between historical queries and the current query result in the gap between historical interests and the current interests (interest gap). Because historical clicks stem from historical interests, using them to model the current interests inevitably suffers from the interest gap challenge. To capture the interest focus and address the interest gap, we propose the two-stage interest calibration network (TCN), i.e., search-internal and search-external. In the search-internal calibration, we propose new insights of using the divergences among clicks and unclicks to model interest focus, and then develop a divergence-based calibration network. In the search-external calibration, inspired by the smoothing techniques for language models, we propose the interest smoothing principle to bridge interest gap: *the interests learnt from historical clicks + smoothing factor ≈ current interests*. To implement this principle, we develop an interest smoothing network by reusing the query data. In the network, the interest domination unit is developed to learn user's interests from historical clicks, and the interest smoothing unit is developed to construct the smoothing factor. Extensive offline experiments and online A/B testing are performed and show that TCN significantly outperforms the state-of-the-art baselines. Besides, our model has been deployed in a hotel e-commerce platform and brought 2.90% CTCVR and 1.53% CTR lifts.

1 Introduction

Users (guests) have become accustomed to searching accommodations through hotel e-commerce platforms, such as *Airbnb*, *Ctrip* and *Meituan-Hotel*. Due to the importance of hotel e-commerce platforms, a growing number of researchers [1,2] pay attention to the hotel search task. A typical pipeline of hotel search systems contains three main stages: 1) Recall, retrieving relevant hotels from a

---

D. Ma and J. Sun—Equal Contribution.

© The Author(s), under exclusive license to Springer Nature Switzerland AG 2023
X. Wang et al. (Eds.): DASFAA 2023, LNCS 13946, pp. 325–335, 2023.
https://doi.org/10.1007/978-3-031-30678-5_25

large database of hotels; 2) Rank, selecting the top thousand from the relevant hotels as candidates; 3) Rerank, sorting the candidates by fine-grained features (e.g., behavior-based interest features) and returning the top hundred as output.

Interest modelling aims to construct users' interest representations over their behavior sequences, and suffers from two key challenges:

Challenge One: Interest Focus. In hotel e-commerce platforms, a hotel has many attributes and each attribute has multiple values. When a user clicked the hotel, she might only like some attributes and values (interest focus). Which attributes and values did the user like? Modelling the interest focus on a click is critical for modelling the user's interests on all clicks, but is very challenging. For a hotel, different users who clicked it may have different interest focuses, because they care about different attributes and values. How to use the same clicked hotel to model different interest focuses for different users?

Challenge Two: Interest Gap. A user searches hotels by entering a query, e.g., the check-in location, number of guests and check-in/checkout dates. The query represents user's interests to some extent. For instance, if the number of guests is two and the check-in date is Valentine's day, the user's interests in the room type may be double rooms, while that may be single rooms if the number of guests is one. For a user, the current query is different from her historical queries, and thus the current interests may be different from historical interests (interest gap). Because historical clicks stem from historical interests, using them to model the current interest inevitably suffers from the interest gap challenge.

In hotel search, the interest gap has not been explored. In product recommendation, one approach is modelling the interest evolving process [3]. However, this approach can not work well in the hotel domain because 1) the click sequences in the hotel domain are much shorter than those in the product domain. The short sequences may not provide sufficient statistics for the interest evolution learning; 2) Time intervals between clicks in the hotel domain are much longer than those in the product domain. The long time intervals severely weaken the relationship between clicks, and thus hinder the modelling of interest evolution.

Solution. We focus on addressing the above two challenges and propose a two-stage interest calibration network to address them.

Search-internal Calibration. In a search, user's interests often focus on some attribute values of hotels. Therefore, there is high similarity among the clicked hotels in the search, while there is high divergence between clicked hotels and unclicked hotels. To model the interest focus on a click, we propose three insights by using the search data. In a search, for an attribute (e.g., price), 1) if its attribute values on clicked hotels are more similar, the user cares more about this attribute, and thus it is more important for constructing the interest focus; 2) if its values on the clicked hotels are more different from those on the unclicked hotels, the attribute is more important; 3) if one attribute value is more similar to the values of clicked hotels and more different from the values of unclicked hotels, the value is more important. Accordingly, we develop a divergence-based calibration network to model the interest focus on a click.

Search-external Calibration. Inspired by the smoothing techniques for language models [4], we propose an interest smoothing principle to bridge the interest gap: *the interests learnt from historical clicks + smoothing factor ≈ current interests*. To implement this principle, we develop the interest smoothing network by reusing the query data. Specifically, the interest domination unit is developed to learn the interest representation from historical clicks. It uses a click-extension approach and query-extension approach to estimate the relevance between the current query and historical clicks. The interest smoothing unit is developed to construct the smoothing factor by reusing the query data. Like the query-extension approach, it derives the smoothing factor from both the current query itself and the relevant hotels of the current query.

Our contributions are concluded as follows:

- We define the interest modelling task by emphasizing the search and query data which have been neglected by most interest modelling methods, and identify two new challenges for the task, i.e., interest focus and interest gap.

- We are the first to explore the interest focus in the task of interest modelling, propose three insights based on search data, and develop the divergence-based calibration network to capture the interest focus.

- To bridge the interest gap challenge, we propose the interest smoothing principle, and then develop the interest smoothing network by reusing the query data.

- Both offline experiments and online A/B testing have been conducted to validate the effectiveness of our model. Our model has been deployed in Meituan-Hotel and brought 2.90% CTCVR and 1.53% CTR lifts, which are significant to the hotel business.

2 Preliminaries

Reranking Stage. Because hotel business cares about both click through rate (CTR) and click through&conversion rate (CTCVR), we adopt a multi-task learning approach to optimize CTR and CTCVR prediction tasks, i.e., MMoE [10]. CTCVR prediction is to estimate the probability $p_{cr}(d|u,q)$ of a user u clicking and reserving a hotel d under query q. CTR prediction is to estimate the probability $p_c(d|u,q)$ of u clicking d under q. We estimate them as follows:

$$p_{cr}(d|u,q) = p_c(d|u,q) \cdot p_r(d|u,q), \tag{1}$$

$$p_c(d|u,q) = \sigma_c(\overline{d}, \overline{u}, q, \Psi_c); \quad p_r(d|u,q) = \sigma_r(\overline{d}, \overline{u}, q, \Psi_r), \tag{2}$$

where $p_r(d|u,q)$ is the probability that a user u reserves a hotel d, after clicked it. The $\overline{d}$ and $\overline{u}$ are the features of d and u. The $\sigma_*()$ with parameters Ψ_* is a $m \times n \times 1$ MLP with ReLU, ReLU and Sigmoid activation functions. The feature $\overline{u} = (\hat{u}, R(u))$ consists of the profile features $\hat{u}$ and the sequential feature $R(u)$.

The $\hat{u}$ may include gender, age level, purchase power, preferred categories and brands. The item features $\bar{d}$ contain category, brand, cumulative order number, the lowest price, and so on.

In a search q, given a candidate hotel $d \in D$, a user $u \in U$ may click d, reserve or not. We can use a tuple to record the response of u on d, i.e., (u, d, q, y_c, y_{cr}) where $y_c = 1$ if u clicks d, otherwise $y_c = 0$; $y_{cr} = 1$ if u clicks and reserves d, otherwise $y_{cr} = 0$. Based on examples C, we apply the cross-entropy loss function [9] to optimize two tasks, and the final loss is sum of loss on two tasks.

Problem Definition. Because the reranking stage receives candidates T_q from the rank stage, interest modelling in the reranking stage can take T_q as input. Since the selection of T_q in rank stage is based on the current query q, T_q are relevant to q and represent user's interests to some extent. So we define the task: *The input consists of 1) a user $u \in U$; 2) the sequence S, Q, X and Y where each $S_k \in S$ contains $q_k \in Q$, $X_k \in X$ and $Y_k \in Y$; 3) the current query q and the relevant candidate hotels T_q of q. Both historical query q_k and the current query q consist of the check-in location, check-in and checkout dates, price, star and the number of guests. The output is an embedded representation $R(u)$ that represents the current interest of the user u.*

3   Solution: Two-stage Interest Calibration Network

We propose a two-stage interest calibration network to learn $R(u)$, i.e., search-internal calibration for modelling the interest focus and search-external calibration for bridging the interest gap.

3.1   Search-internal Calibration

To accurately model interest focus, we propose three insights. In a search, for an attribute, if its values on the clicked hotels are more similar, the user cares more about this attribute and thus the attribute is more important for modelling the interest focus (**Insight one**). For an attribute, if its values on clicked hotels are more different from those on unclicked hotels, the attribute is more important for modelling the interest focus. This tells us that the differences of attribute values between clicked hotels and unclicked hotels can be used for identifying the liking attributes and estimating the liking degrees on the attributes (**Insight two**). If an attribute value is more similar to the attribute values of clicked hotels, but more different from the attribute values of unclicked hotels, the value is more likely to be the interest focus (**Insight three**).

According to the above three insights, we propose the divergence-based calibration network, and formulate the network as follows:

$$v(x_k^i) = W_{x_k^i} \circ E_{x_k^i} = f_1(q_k \in Q, X_k \in S_k, Y_k \in S_k, x_k^i \in X_k, E) \circ E_{x_k^i}, \quad (3)$$

where x_k^i is a click in search k, $\circ$ is Hadamard dot and E is a trainable embedding dictionary. In E, every feature is initialized as an embedding. The $E_{x_k^i}$ is the embedding of click x_k^i and is consistent for all users. To get $E_{x_k^i}$, we concatenate the embeddings of all features on x_k^i. The W is a learnable weight vector and has the same dimensions of $E_{x_k^i}$. It ensures that different users who clicked the same hotel may have different interest focuses.

Let $X_k^j = [E_{x_k^1}^j; \cdots; E_{x_k^n}^j]$ denote the elements of j-th dimension of the clicked hotels X_k. To capture the insight one, the relative divergence of X_k^j is used for estimating the importance of the j-th dimension:

$$W_{X \to X}(X_k^j) = \frac{1}{RD(X_k^j) + \varepsilon}, \quad RD(X_k^j) = \frac{1}{n}\sum_{i=1}^{n} \frac{|E_{x_k^i}^j - \overline{X_k^j}|}{|\overline{X_k^j}| + \varepsilon}, \tag{4}$$

where ε is a very small numeric constant for preventing the denominator from being zero, e.g., $1e-6$. $\overline{X_k^j}$ is the average value of the elements in X_k^j, and $E_{x_k^i}^j$ is the j-th element in $E_{x_k^i}$. The $|\cdot|$ is a sign of computing the absolute value.

According to the insight two, we estimate the importance of a dimension by using the element distribution divergence between the clicked hotels and the unclicked hotels. The element distribution divergence is estimated as follows:

$$W_{Y \to X}(X_k^j) = \sum_{i=1}^{n}\sum_{z=1}^{m} |E_{x_k^i}^j - E_{y_k^z}^j|/n * m, \tag{5}$$

where $E_{y_k^z}^j$ is the j-th dimension of $E_{y_k^z}$; n and m are the sizes of X_k and Y_k.

According to the insight three, we estimate the importance of each embedding element to the interest focus. To estimate the similarity between $E_{x_k^i}^j$ and X_k^j, we firstly compute the difference between $E_{x_k^i}^j$ and X_k^j, and then use the reciprocal of the difference as the similarity. We combine the similarity and difference to estimate the importance of $E_{x_k^i}^j$, and formulate the importance as follows:

$$W_{XY \to X}(E_{x_k^i}^j) = \frac{\sum_{z=1}^{m} |E_{x_k^i}^j - E_{y_k^z}^j|/m}{|E_{x_k^i}^j - \overline{X_k^j}| + \varepsilon}, \tag{6}$$

According to the importance $W_{X \to X}(X_k^j)$ and $W_{Y \to X}(X_k^j)$ of the dimension j, and element importance $W_{XY \to X}(E_{x_k^i}^j)$, we rewrite f_1 in Eq. 3 as follows:

$$v(x_k^i) = f_1(q_k, X_k, Y_k, x_k^i, E) \circ E_{x_k^i} = [w(E_{x_k^i}^1), \cdots, w(E_{x_k^i}^n)]^T \circ [E_{x_k^i}^1, \cdots, E_{x_k^i}^n]^T \tag{7}$$

$$w(E_{x_k^i}^j) = \sigma_1(w_1[W_{Y \to X}; W_{X \to X}; W_{XY \to X}; E_{x_k^i}^j; \overline{X_k^j}; E_{q(x_k^i)}] + b_1), \tag{8}$$

where $[]^T$ denotes a transpose. The σ_1 with parameters w_1 and b_1 is a $64 \times 32 \times 1$ MLP with activation function ReLU. The $q(x_k^i)$ is the corresponding query of x_k^i, and $E_{q(x_k^i)}$ is the embedding of $q(x_k^i)$. To construct $E_{q(x_k^i)}$, we concatenate the embeddings of all components in $q(x_k^i)$.

3.2 Search-external Calibration

To bridge interest gaps, we propose an interest smoothing principle: *the interests learnt from historical clicks + smoothing factor ≈ current interests*. To implement it, we develop the interest smoothing network and formulate it as follows:

$$R(u) = R_X + \lambda R_q, \tag{9}$$

where R_X is the interests learnt from historical clicks and R_q is the smoothing factor. Let $X = \{x_1^1, \cdots, x_1^n, \cdots x_m^1, \cdots, x_m^n\}$ denote the historical clicks. We design the interest domination unit to construct R_X and interest smoothing unit to construct R_q. The λ is a learnable weight.

Interest Domination Unit. To estimate the importance of a click, we argue 1) if a click is more relevant to the current query, it is more important for representing the current interests; 2) if a click is more relevant to the whole click sequence, it is more important. Accordingly, we derive R_X from the click sequence X and current query q as follows:

$$R_X(u, q, X) = \sum_{x_i \in X} r(q, x_i) r(X, x_i) v(x_i), \tag{10}$$

where $r(q, x_i)$ is the relevance of x_i to q and $r(X, x_i)$ is the relevance of x_i to X. The $v(x_i)$ is the calibrated embedding of x_i, and is learned by Eq. 7.

To estimate $r(X, x_i)$, there are three intuitions: 1) if x_i appears in X more frequently, x_i is more relevant to X; 2) if x_i is more relevant to the other clicks in X, x_i is more relevant to X; 3) if the relevant clicks of x_i appear in X more frequently, x_i is more relevant to X. Accordingly, we estimate $r(X, x_i)$ as follows:

$$r(X, x_i) = \sum_{x_j \in X} n(x_i, X) n(x_j, X) r(x_j, x_i), \tag{11}$$

where $n(x_i, X)$ is the number of x_i appearing in X. The $r(x_j, x_i)$ is the relevance of x_j to x_i, and we estimate it as follows:

$$r(x_j, x_i) = \sigma_2(w_2[v(x_j); v(x_i); |v(x_j) - v(x_i)|] + b_2), \tag{12}$$

where σ_2 is a three-layer MLP with same activation functions of σ_1 in Equ. 8.

To estimate $r(q, x_i)$, we propose two approaches—click extension and query extension. The first approach uses the corresponding historical query $q(x_i)$ of x_i to extend the semantics of x_i, and then uses the relevance between q and $q(x_i)$ to estimate $r(q, x_i)$. The second approach extends the query q by using top-k relevant hotels T_q of q, and uses the relevance between T_q and x_i to estimate $r(q, x_i)$. Because our model is in the rerank stage and thus can take the candidates sorted by rank stage as input, we can use the top-k candidates as the relevant hotels of q (T_q) (default value of k is 20). We estimate $r(q, x_i)$ as: $r(q, x_i) = r(T_q, x_i) r(q, q(x_i))$, where $r(q, q(x_i))$ is the relevance between the current query q and the historical query $q(x_i)$. We estimate it as $\sigma_3(w_3[E_q; E_{q(x_i)}; |E_q - E_{q(x_i)}|] + b_3)$, where σ_3 is a three-layer MLP with the same activation functions of σ_1 in Eq. 8. Similar to Eq. 11, we estimate $r(T_q, x_i)$ as follows:

$$r(T_q, x_i) = \sum_{t_j \in T_q} n(t_j, T_q) r(t_j, x_i), \tag{13}$$

where $r(t_j, x_i)$ is estimated by using Eq. 12.

Interest Smoothing Unit. We construct the smoothing factor R_q from two perspectives. First, the current query q contains multiple important information, such as number of guests and check-in/checkout dates. These information can represent user's current interests to same extent. For example, if the number of guests is two and the check-in date is Valentine's day, the user's interests in room type may be a double room. Second, the candidate hotels T_q are selected by the rank stage based on q, and thus T_q represent user's current interests to same extent. Accordingly, we formulate R_q as follows:

$$R_q(u, q, X) = [E_q; \sum_{t_i \in T_q} r(T_q, t_i) r(X, t_i) E_{t_i}], \tag{14}$$

where $r(T_q, t_i)$ is the relevance of t_i to T_q and estimated by Eq. 13. The $r(X, t_i)$ is the relevance of t_i to X and estimated by Eq. 11.

For the estimation of λ in Eq. 9, there are two intuitions: 1) if R_q is more relevant to R_X, it is more likely to represent user's current interests and thus λ should be larger; 2) If X contains more clicks, R_X more covers the current interests and thus λ should be smaller. Accordingly, we estimate λ as follows:

$$\lambda = \sigma_4(w_4[R_X; R_q; |R_X - R_q|; n(X)] + b_4), \tag{15}$$

where $n(X)$ is the size of click sequence X. The σ_4 is a three-layer MLP with the same activation functions of σ_1 in Eq. 8.

4 Experiments

Comparison Solutions. Since interest modelling in hotel search has not been explored, we select SOTA solutions in product recommendation as baselines: 1) Base only uses profile features; 2) DNN encodes all clicks by using a MLP with three layers, and averages the representations of all clicks as the user's interests; 3) DIN [5] first estimates the relevance between each click and the target, and then uses the relevance to sum the representations of all clicks; 4) SIM [8] searches the important clicks and then uses DIN [5] to model the user's interests over the important clicks; 5) DIEN [3] uses an interest extractor layer to capture temporal interests and an interest evolving layer to capture the interest evolving process; 6) SASRec [12] uses the attention mechanism to capture long-term semantic and estimate the relevance of each historical click; 7) DFN [6] uses an internal feedback and external feedback components to capture the fine-grained interactions among behaviors; 8) FMLP [7] is a MLP model with learnable filters and designed for reducing the effect of noise clicks.

Datasets. The *Indu* dataset is sampled from the user behavior logs in Meituan-Hotel. According to the method of sampling [13], we take four weeks' samples as the training data from November 1, 2021 to November 28, 2021, and use the following two days' samples as the evaluation data and testing data, respectively. The dataset contains about 5.5 million users and 1.3 million hotels. It includes $2, 103, 945, 115$ negative examples, $136, 225, 279$ positive examples with click behaviors and $85, 665, 472$ positive examples with order behaviors. The maximum length of behavior sequences is 500 and minimum length is 0. Note that every user is equipped with a behavior sequence of six months.

Evaluation Metrics. For the offline experiments, we use the following metrics:

- AUC. AUC is a widely used metric in IR and data mining tasks.

- QAUC. In hotel domain, we highlight the AUC values of queries, and adapt the user-centric weighted AUC [5] to the query-centric weighted AUC: $QAUC = (\sum_{i=1}^{n} num_i \cdot AUC_i)/ \sum_{i=1}^{n} num_i$ where n is the number of queries, AUC_i is AUC of query i, and num_i is the number of exposed hotels for query i.

- QNDCG [11]. NDCG is a measure of ranking quality considering both the relevance and ranked positions. QNDCG is the average NDCG of all queries.

CTR and CTCVR are taken as online metrics. The significance is tested against TCN by a two-tailed paired t-test, and † denotes the difference at 0.05.

4.1 Experimental Results

Comparison with baselines. We perform all models and report their achieved metrics in Table 1. The notation *Indu without profile features* denotes that the profile features of *Indu* are not applied to all models. It can be seen that 1) DIN and SIM models perform better than other baselines, i.e., DIEN, SASRec, DFN. In hotel search, users' click sequences are so sparse that they can not provide sufficient semantics to model the relations among clicks. So the solutions of modelling the relations among clicks i.e., DIEN, SASRec and DFN, can not make full use of their advantages on the sparse sequences; 2) TCN outperforms all baselines on all metrics. On *Indu*, 0.146% AUC improvement over SIM is very significant to hotel business because it brings 2.17% increase in CTCVR (see online A/B test in Table 3) and an increase of about 5500 orders. These comparisons verify that TCN can model user's interests better than baselines.

In principle, TCN first models the interest focus on each click, and then construct the user's interest representation based on the interest focuses, while baselines don't consider and model the interest focus. Besides, the interest smoothing network in TCN can effectively address the interest gap challenge, while baselines can not well address the challenge. Although DIEN uses an interest evolving layer to capture the interest evolving process, DIEN can not well address the interest

Table 1. Metric comparison. Since AUC is a number not a list, the significance of AUC metrics is not tested.

Model	AUC	QAUC	QNDCG	AUC	QAUC	QNDCG
	Indu			*Indu without profile features*		
Base	0.92373	$0.83687^\dagger$	$0.78718^\dagger$	—	—	—
DNN	0.92522	$0.83912^\dagger$	$0.78855^\dagger$	0.68052	$0.60316^\dagger$	$0.59816^\dagger$
DIN	0.92559	$0.83937^\dagger$	$0.78877^\dagger$	0.69145	$0.61444^\dagger$	$0.61284^\dagger$
SIM	0.92570	$0.84019^\dagger$	$0.78947^\dagger$	0.68995	$0.61018^\dagger$	$0.61261^\dagger$
DIEN	0.92482	$0.83817^\dagger$	$0.78788^\dagger$	0.67976	$0.59894^\dagger$	$0.60376^\dagger$
SASRec	0.92513	$0.83823^\dagger$	$0.78799^\dagger$	0.66862	$0.58795^\dagger$	$0.59711^\dagger$
DFN	0.92492	$0.83842^\dagger$	$0.78807^\dagger$	0.64839	$0.59296^\dagger$	$0.59982^\dagger$
FMLP	0.92517	$0.83861^\dagger$	$0.78798^\dagger$	0.63819	$0.59765^\dagger$	$0.58921^\dagger$
TCN	**0.92716**	**0.84139**	**0.79059**	**0.80165**	**0.66252**	**0.64431**

gap, because in hotel domain, 1) users' click sequences are so short that they can not provide sufficient statistics for interest evolution learning; 2) the time intervals between clicks are so long that they severely weaken the relationships among clicks and hinder the modelling of interest evolution. The experimental results in Table 1 illustrate that TCN can address the interest focus and interest gap challenges more effectively than baseline models.

Table 2. Effectiveness of components in TCN.

Model	AUC	QAUC	QNDCG	AUC	QAUC	QNDCG
	Indu			*Indu without profile features*		
TCN-DCN	0.92635	0.83968	0.78913	0.78395	0.63712	0.62696
TCN-ISN	0.92619	0.83952	0.78906	0.68994	0.60547	0.59930
TCN-IDU	0.92683	0.84053	0.78965	0.74216	0.62953	0.61845
TCN-ISU	0.92695	0.84091	0.78981	0.75674	0.64517	0.62759
TCN	**0.92716**	**0.84139**	**0.79059**	**0.80165**	**0.66252**	**0.64431**

Ablation Study. We conduct an ablation study to test the effectiveness of components in TCN and show the results in Table 2. The notation TCN-DCN, TCN-ISN, TCN-IDU and TCN-ISU denote that the divergence-based calibration network, the interest smoothing network, the interest domination unit, and the interest smoothing unit are not applied to TCN, respectively. In TCN-ISN, we average the embedded representations of the interest focuses of all clicks as the final user's interest representation. From Table 2, we can see that 1) the metrics achieved by TCN-DCN are lower than those achieved by TCN. This illustrates that the divergence-based calibration network is important to interest modelling, and verifies the effectiveness of the divergence-based calibration network; 2) The metrics achieved by TCN-ISN are lower than those achieved by TCN, which verifies the importance of the interest smoothing network. As well, it demonstrates that the interest smoothing network can well bridge the interest gap; 3) The metrics achieved by TCN-DCN are higher than those achieved by TCN-ISN, which illustrates that the interest smoothing network is more important to interest modelling than the divergence-based calibration network; 4) TCN achieves the better metrics than TCN-IDU and TCN-ISU. This verifies the effectiveness of the interest domination unit and interest smoothing unit.

Table 3. Online A/B testing (relative improvements).

Metric	DNN	SIM	TCN
CTCVR	0.00%†	+0.73%†	**+2.90%**
CTR	0.00%†	+0.36%†	**+1.53%**

Online A/B Testing. We perform the A/B tests and show the results in Table 3. Considering the computational cost and economic benefits, we select DNN, SIM and TCN to conduct the A/B tests for two weeks. Each model is assigned with 20% traffic. It can be seen that TCN achieves the best online CTCVR and CTR metrics. Compared to DNN, TCN achieves the +2.91% CTCVR improvement and +1.53% CTR improvement. These metric improvements are much higher than those achieved by SIM, and significant to hotel business.

5 Conclusion

We study the interest modelling task for reranking hotels, and identify two new challenges, i.e., interest focus challenge and interest gap challenge. To capture interest focus, we propose three insights based on search data, and develop a divergence-based calibration network. To bridge the interest gap, we propose an interest smoothing principle, and then develop an interest smoothing network by reusing the query data. By experiments, we find 1) modeling the interest focus on each click is important for constructing the user's interest representation over all clicks; 2) For bridging interest gaps, the combination of the interest domination unit and the interest smoothing unit performs better than any single one.

References

1. Haldar, M., et al.: Improving deep learning for airbnb search. In: SIGKDD, pp. 2822–2830 (2020)
2. Abdool, M., et al.: Managing diversity in airbnb search. In: SIGKDD, pp. 2952–2960 (2020)
3. Zhou, G., et al.: Deep interest evolution network for click-through rate prediction. In: AAAI, pp. 5941–5948 (2019)
4. Zhai, C., et al.: A study of smoothing methods for language models applied to ad hoc information retrieval. In: SIGIR, pp. 334–342 (2001)
5. Zhou, G., et al.: Deep interest network for click-through rate prediction. In: SIGKDD, pp. 1059–1068 (2018)
6. Xie, R., et al.: Deep feedback network for recommendation. In: IJCAI, pp. 2519–2525 (2020)
7. Zhou, K., et al.: Filter-enhanced MLP is all you need for sequential recommendation. In: WWW, pp. 2388–2399 (2022)
8. Pi, Q., et al.: Search-based user interest modeling with lifelong sequential behavior data for click-through rate prediction. In: CIKM, pp. 2685–2692 (2020)
9. Zhang, Z., et al.: Generalized cross entropy loss for training deep neural networks with noisy labels. In: NeurIPS, pp. 8792–8802 (2018)
10. Ma, J., et al.: modeling task relationships in multi-task learning with multi-gate mixture-of-experts. In: SIGKDD, pp. 1930–1939 (2020)
11. Wang, Y., et al.: A theoretical analysis of NDCG type ranking measures. In: The 26th Annual Conference on Learning Theory, pp. 25–54 (2017)
12. Kang, W.-C., et al.: Self-attentive sequential recommendation. In: ICDM, pp. 197–206 (2018)
13. Chen, T., et al.: MIND: a large-scale dataset for news recommendation. In: ACL (2020)

A Novel Deep Learning Framework for Interpretable Drug-Target Interaction Prediction with Attention and Multi-task Mechanism

Yubin Zheng[1], Peng Tang[1], Weidong Qiu[1(✉)], Hao Wang[2], Jie Guo[1], and Zheng Huang[1]

[1] School of Cyber Science and Engineering, Shanghai Jiao Tong University, Shanghai, China
{zybhk21,tangpeng,qiuwd,guojie,huang-zheng}@sjtu.edu.cn
[2] College of Basic Medical Sciences, Shanghai Jiao Tong University, Shanghai, China
angela_wanghao@sjtu.edu.cn

Abstract. The measurement of drug-target interaction(DTI) is a major task in the field of drug discovery, where drugs are typically small molecules and targets are typically proteins. Traditional DTI measurements in the lab are time consuming and expensive. DTI can be predicted through the use of computational methods like ligand similarity comparison and molecular docking simulation. However, these methods strongly rely on domain expertise. Deep learning has recently advanced, and some deep learning techniques are being used to predict DTI. These deep learning ways can extract drug and target features automatically without domain knowledge and produce good results. In this work, we propose an end-to-end deep learning framework to predict DTI. The unsupervised method Mol2Vec with self-attention is used to extract the drug features. To extract the target features, we pre-train a BERT model, which is the state-of-the-art model for many text comprehension tasks in NLP. In order to improve the generalization ability of the model, we introduce a multi-task learning mechanism by using two transformer encoder-decoders. As far as we know, we are the first to apply Mol2Vec, BERT, attention mechanism and multi-task mechanism to one model. The experiment results show that our model outperforms other latest deep learning methods. Finally, we interpret our model through a case study by visualizing the predicted binding sites.

Keywords: drug-traget interaction · BERT · Mol2Vec · attention mechanism · multi-task mechanism · deep learning

1 Introduction

Drug-target interaction plays an important role in drug discovery. The production of a new kind of drug for one type of disease is a long process with lots of difficulties. The identification of drug-target interaction can discover potential effective molecules to a special disease's target. Since wet lab experiments to measure DTI are time consuming and expensive [4, 8], some computational methods for DTI prediction can be efficient and urgently demanded.

Two computation approaches are mainly used to predict drug-target interaction. They are molecular docking and machine learning. Molecular docking can simulate the docking process of the proteins and molecules. The optimal matching mode can be identified during this process, and the matching scores will be computed to evaluate the drug-target interaction. Molecular docking is widely used because of its good performance and interpretability. However, it has two serious shortcomings. The first is that domain knowledge is necessary for the matching process. The second is that large scale simulation is time consuming.

Machine learning approaches have advantages compared with molecule docking. Some machine learning methods can extract the features of drug and target automatically without domain expertise. Additionally, machine learning models can quickly produce a large-scale DTI prediction. DTI prediction can be viewed as a binary classification problem in machine learning methods. The input of the model is a pair of candidate drug and target. The model will output the label which can indicate whether there is a an interaction between the candidate drug and target. Deep learning is a sub-field of machine learning. In recent years, with the rapid development of deep learning, the classification models can achieve better performance with deep learning methods.

In this paper, we propose an end-to-end interpretable deep learning framework to predict DTI. The inputs of the model are molecule SMILES strings and protein sequences, which are the sequence information of the molecules and proteins. The molecules are represented by SMILES strings, which contain the molecule structure information. Mol2Vec [12] is an unsupervised model which learns to represent the substructures of the molecules. Considering the correlations between substructures in different positions, we develop the Mol2Vec by adding a multi-head self-attention [20] layer after we extracted the molecule features by Mol2Vec. The proteins are represented by long amino acid sequences which are similar to the sentences in Natural Language Process(NLP). BERT [5] is a model proposed in NLP which achieves state-of-the-art in many text understanding tasks. We pre-train a BERT model with 700,000 protein sequences to obtain the latent protein representations.

The DTI is the interaction between drug and target. It might be challenging to determine which drugs are likely to interact with which targets. In other words, DTI is a two-way selection procedure. We propose a multi-task learning [15] approach using two transformer encoder-decoders to train the model to learn this bidirectional selection process. With the multi-task learning mechanism, two tasks share the bottom layers [2], which can improve the generalization ability of the model. In conclusion, we make the following contribution:

- We propose a novel deep learning framework to predict drug-target interaction. To the best of our knowledge, we are the first to apply Mol2Vec, BERT, attention mechanism and multi-task mechanism to one model to predict the drug-target interaction.

- We develop the Mol2Vec by adding multi-head self-attention mechanism. A BERT model is pretrained with 700,000 protein sequences in PDB database to obtain the latent protein representation. We also introduce a multi-task mechanism to make model learn the bidirectional selection process of the drug and target.
- We compare our model with other latest deep learning models on four metrics: area under curve(AUC), precision, recall and F1. The experiment results show that our model outperforms other latest deep learning models. What's more, we interpret our model through a case study by visualizing the predicted binding sites.

2   Related Work

By simulating the matching process between small molecules and proteins, molecular docking [7] selects the optimal matching mode between them. The DTI interaction will be evaluated using the optimal matching score. High accuracy and strong interpretability are two strengths of this method, but professional domain knowledge is required. Additionally, large scale molecular docking is time consuming.

Machine learning methods have been applied to predict DTI. Most of them are based on ligands similarity measurement [6]. A new small molecule is compared to a small molecule that is previously known to be effective against the protein target in this method. Some new molecules with similar structures to effective compounds will be selected as candidate molecules for drug development. It is difficult to find efficient small compounds with novel structures using ligand similarity measurements. Faulon et al. [9] improved support vector machine (SVM) to predict DTI. Yamanishi et al. [24] combined chemical, genomic and pharmacological data to predict drug-target binding using ligands similarity measurements.

In recent years, deep learning has developed rapidly, which provides new approaches for DTI prediction. These deep learning methods can extract drug and target features automatically without professional domain knowledge. To make the model work well, the deep learning model is trained on huge database. Many deep learning models have been proposed to predict DTI. Jiang et al. [13] constructed a kind of graph data to represent the relationship between the protein and molecule. The relationship graph is used to predict DTI. Gao et al. [10] used graph CNN to extract molecule information and LSTM to project protein sequences. Karimi et al. [14] proposed a model using recurrent neural network to extract sequence information and convolution neural network to predict DTI. Wang et al. [21] used Restrict Boltzmann machine to represent DTI pairs and trained a deep belief network(DBN) to predict DTI. Zhao et al. [27] proposed AttentionDTA model which associates attention mechanism to predict the binding affinity of DTI. Nguyen et al. [17] proposed GraphDTA model using graph convolution network to extract molecule information. Chen *et al.* [3] constructed special dataset for DTI and proposed a new transformer neural network named transformerCPI.

3   Methodology

There are three modules in the proposed model. The first module consists of the Mol2Vec model and a multi-head self-attention layer. It receives SMILES strings as input and outputs the molecule representation vectors. The second module contains a pre-trained BERT model. The inputs are the protein sequences and the outputs are the protein representation vectors. The third module is a multi-task module including two transformer encoder-decoders. It receives molecule and protein representation vectors. The output vectors are passed to fully connected layers to get two predicted probabilities. Then the average of the two predicted probabilities is calculated to be the final predicted results. Figure 1 shows the architecture of the proposed model.

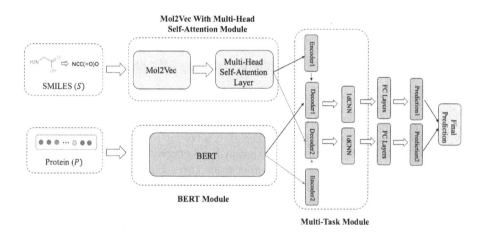

Fig. 1. The architecture of the proposed model.

3.1   Mol2Vec with Multi-Head Self-Attention

SMILES strings [23] are used to represent small drug molecules. A chemical graph structure of a molecule can be reconstructed by structure diagram generation algorithms with SMILES strings. SMILES strings therefore include the molecule's structural information. Mol2Vec [12] is an unsupervised model trained on large molecule database which learns to represent the substructures of a molecule chemistry graph. The Mol2Vec model is trained using the skip-gram

[16] approach. However, to represent a molecule, Mol2Vec add all substructures' embedding representation vectors, which ignores the correlations of the substructures in different positions in a molecule. We develop the Mol2Vec by using the substructures sequence to represent a molecule. Figure 2 shows the substructures sequence of the molecule with SMILES string 'NCC(=O)O'. Through the Mol2Vec model, each substructure can be denoted by a vector. Then the SMILES string of a molecule can be turned into a matrix $D(n, d)$, where n is the number of substructures and d is the d dimensional vector representation of each substructure.

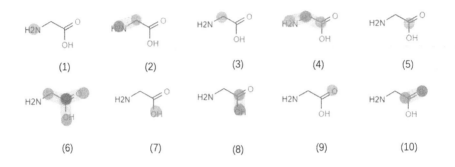

Fig. 2. The substructures sequence of the molecule with SMILES string 'NCC(=O)O'.

Only the relationship between nearby substructures are taken into account by the skip-gram approach in Mol2Vec. We introduce a self-attention mechanism [20] to take into account the relationship between nonadjacent substructures. The SMILES string of a molecule can be turned into a matrix $D(n, d)$ by Mol2Vec model. To calculate the attention between different substructures, we first calculate three matrices Q, K, V:

$$Q = DW_Q \tag{1}$$

$$K = DW_K \tag{2}$$

$$V = DW_V \tag{3}$$

Here, W is the weight matrices. After that, tree matrices $Q, K, V \in R^{nk}$ are used to calculate attention matrix α and new representation $D' \in R^{nk}$:

$$\alpha = softmax(\frac{QK^T}{\sqrt{d}}) \tag{4}$$

$$D' = \alpha V \tag{5}$$

Transformer and BERT both employ multi-head self-attention, which can learn various forms of attention. We use three heads to discover the different relationships between a molecule's substructures. The calculation of each head is the same formula as the attention calculation process above. Then we can obtain D_1', D_2', D_3':

$$D_1' = \alpha_1 V_1 \tag{6}$$

$$D_2' = \alpha_2 V_2 \tag{7}$$

$$D_3' = \alpha_3 V_3 \tag{8}$$

Then three matrices are concatenated and a linear transformation is applied to the concatenated matrix:

$$D_n = Linear(D_1' \oplus D_2' \oplus D_3') \tag{9}$$

where $\oplus$ is the concatenation operator. $D_n \in R^{nk}$ is the new representation of the molecule.

3.2   Bidirectional Encoder Representation from Transformers

A kind of protein is represented by a amino acid sequence $P = (a_1, ..., a_n)$, where a_1 is one of the 23 types of amino acids. The amino acid sequences are similar to the sentences in natural language processing tasks. With the assumption that the amino acid sequences consist of a large amount of protein semantic information, we apply bidirectional encoder representation from transformers(BERT) [5] to learn the protein representation.

BERT is a pre-trained language representation model based on transformer which achieves state-of-the-art in many text understanding tasks. By splitting the amino acid sequences into overlap 3-gram sequentially, such as ['MYOGLO'] to ['MYO', 'YOG', 'OGL', 'GLO'], we build a vocabulary on PDB protein dataset [1]. Then we pre-train the BERT model with MLM task on 700,000 amino acid sequences in PDB protein dataset. If we use less than three amino acids as a group to build a vocabulary, the number of the vocabulary will be small. The protein sequences can not contain much semantic information with small vocabulary. If we use more than three amino acids as a group, the length of the protein sequence will be very short. In MLM task, some words are hidden as MASK labels and the model is pre-trained to learn what are the hidden words. The loss function of MLM task is:

$$F_{loss} = \frac{1}{n} \sum_{i}^{n} H(mask_i, label_i) \tag{10}$$

Here, H is the cross entropy function, $mask_i$ is the model prediction of i-th MASK, and $label_i$ is the true label of the i-th hidden word. In one protein sequence, we randomly choose 15% of the words and hide them as MASK labels. The number of the MASK in the sentence is n. Figure 3 shows the pre-trained BERT model architecture.

In MLM task, the MASK words are predicted according to the words which are not hidden. In NLP, traditional recurrent neural network methods generate the embedding of each word using the information of few words in front of it.

Information of the words which behind the current word and have a long distance in front of the current word are not used. The multi-head attention mechanism in BERT can use the bidirectional and distant words' information to update the current word's embedding. The latent semantic representation of the proteins can be obtained by BERT. Although the protein sequences are similar to the sentences in NLP, the correlation between different protein sequences is not clear. So the NSP(next sentence prediction) pre-train task is not used.

Research [5] shows that it is difficult to achieve better results by using the sentences vectors obtained by pre-trained BERT model directly for classification tasks. The pre-trained model needs to be fine-tuned to adapt to the downstream machine learning tasks. So the parameters in BERT model are not frozen. The downstream DTI classification task will fine-tune the BERT model through error back propagation.

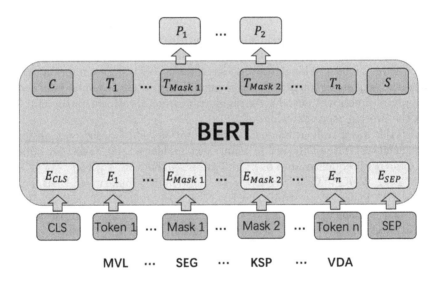

Fig. 3. The pre-trained BERT model architecture. The input masks represent the tokens which are hidden. During the pre-training process, model will output the mask prediction and calculate the loss between prediction and true label of the mask.

3.3 Multi-Task Mechanism

Mol2Vec with self-attention can produce the molecule representation $M(n,k)$. BERT can obtain the protein representation $P(m,k)$. So we can get the good representation of the molecule and protein. Then we introduce a multi-task mechanism to do the downstream DTI classification task by using two transformer encoder-decoders [20]. In the first task, the molecule sequences are encoded and the protein sequences are decoded with multi-head attention mechanism. The second task encodes the protein sequences and decodes the molecule sequences, which is different from the first task. Then we can get the two tasks' interaction vectors $M' \in R^{nk}$ and $P' \in R^{mk}$:

$$\alpha_m = softmax(\frac{Q_p K_m^T}{\sqrt{k}}) \tag{11}$$

$$M' = \alpha_m V_m \tag{12}$$

$$\alpha_p = softmax(\frac{Q_m K_p^T}{\sqrt{k}}) \tag{13}$$

$$P' = \alpha_p V_p \tag{14}$$

Here, Q_m, K_m and V_m are the matrices calculated by molecule matrix M with the formula in Sect. 3.1. Q_p, K_p and V_p are the matrices calculated by protein matrix P. We also introduce the multi-head attention mechanism:

$$M_n = Linear(M_1' \oplus M_2' \oplus M_3') \tag{15}$$

$$P_n = Linear(P_1' \oplus P_2' \oplus P_3') \tag{16}$$

After that, each task's interaction vectors M_n and P_n will be passed into a one dimension convolution neural network(1dCNN) [26] block to predict DTI.

Let $x_{i:i+j}$ be the concatenation of interaction vectors $x_i, x_{i+1}, x_{i+2}, ..., x_{i+j}$. A convolution filter $w \in R^{hk}$ is applied to a window of h vectors to produce a new feature. For example, the i-th new feature c_i is generated from a window of interaction vectors $x_{i:i+h-1}$ by:

$$c_i = f(w \cdot x_{i:i+h-1} + b) \tag{17}$$

Here $b \in R$ is a bias term and f is a non-linear function such as the ReLU function. This filter is applied to each window of the interaction vectors $x_{1:h}, x_{2:h+1}, ..., x_{n-h+1:n}$ to produce a new feature map:

$$c = [c_1, c_2, ..., c_{n-h+1}] \tag{18}$$

A max-over-time pooling operation is applied over the feature map, which takes the maximum value $c' = max\{c\}$. The pooling operation is to capture the most important feature. It can deal a feature map with different lengths. The features obtained by the 1dCNN will be passed to fully connected layers to predict DTI labels.

Figure 4 shows the architecture of two tasks. The pre-trained BERT model will be fine-tuned to adapt to the DTI prediction task during the training process. The multi-task mechanism can make model learn the bidirectional selection process of drug and target. Two tasks share the bottom parameters [2], which will also improve the generalization ability of the model.

Let the first task's prediction be y_1 and the second task's prediction be y_2. The true label is l. The training goal of the model is to minimize the weighted sum of the two tasks' loss functions:

$$loss1 = \frac{1}{n} \sum_{i=1}^{n} l_i log y_{1i} + (1 - l_i) log(1 - y_{1i}) \tag{19}$$

$$loss2 = \frac{1}{n} \sum_{i=1}^{n} l_i log y_{2i} + (1 - l_i) log(1 - y_{2i}) \tag{20}$$

$$loss = \lambda loss1 + (1 - \lambda)loss2 \tag{21}$$

Here, λ is a hyper parameter. It can be adjusted according to the importance of the two tasks.

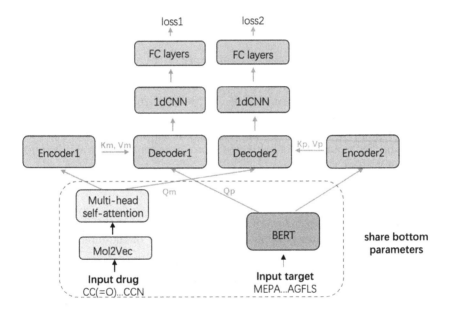

Fig. 4. The proposed model architecture combines two tasks. The training goal is to minimize the weighted sum of two tasks' loss functions.

4   Experiments

4.1   Dataset

Human dataset [19] is a public DTI dataset. Human dataset includes DTI pairs from DrugBank 4.1 and Matador. The negative samples are highly credible as they are obtained from a systematic screening framework. Human dataset contains 3364 positive DTI samples and 2848 negative DTI samples between 1052 compounds and 852 proteins.

BindingDB [11] is a public, web-accessible database of measured binding affinities, focusing chiefly on the interactions of drug-targets proteins and small, drug-like molecules. BindingDB contains 1.3 million data records, each of which contains information such as the identifier records, molecule SMILES strings, protein sequences, IC50 values, etc. By following the activity threshold discussion in [22], record is positive if its IC50 is less than 100nm, negative if IC50 greater than 10000nm. Then we construct a binary classification dataset with 60000 positive samples and 59855 negative samples.

Table 1. Summary of the datasets.

Data	Positive	Negative	Train	Validation	Test
Human	3364	2848	4969	621	622
BindingDB	60000	59855	95884	11985	11986

We split DTI pairs into training, validation, and testing datasets. Statistics of the datasets are shown in Table 1.

4.2 Training Details

We pre-train the BERT model with 700,000 protein sequences from PDB protein dataset for 8 epoch. The batch size of the BERT model is 64. We use Adam gradient descent optimization and the initial learning rate is 0.00001. After pre-training, the embedding representation of each token is a 768 dimensional vector. Through Mol2Vec model with multi-head self-attention mechanism, each substructure of a molecule is represented by a 300 dimensional vector.

The new molecule and protein representation will be passed to the multi-task module with two transformer encoder-decoders. The batch size of each training step is 64. A batch of molecule and protein samples is randomly selected from the training data. We use Adam gradient descent optimization with initial learning rate equals to 0.00005 to train the model. After that, two tasks' interaction vectors are then fed into a convolution neural network with 96 different filters. After max-over-time pooling, two 96 dimensional vectors are passed to a 96-1024-1024-512-2 fully connected network to predict the DTI labels. The training goal is to minimize the weighted sum of two tasks' loss functions. The λ is set to 0.6 according to the experiments.

4.3 Baselines

We compare the proposed model with the following methods:

DeepDTA [18] applies one-dimensional convolution neural network to handle the molecule and protein sequences. The obtained embedding of the molecule and protein will be concatenated together and sent to fully connect layers to predict DTI.

GraphDTA [17] views molecules as a kind of graph data and apply graph convolution network to extract the molecule graph information from SMILES strings. The protein sequences are projected by some one-dimensional convolution layers. Then the molecule vector and the protein vector are concatenated together and sent to fully connected layers to predict DTI labels. GraphDTA combines graph convolution neural network and one dimensional convolution neural network.

E2E [10] includes graph convolution network and LSTM. Graph convolution network updates the embedding representation of the molecule with SMILES strings as inputs. LSTM is used to extract the information from long protein sequences. The weighted sum of each word's vectors is to be the final representation of a molecule or protein. Two representation vectors are concatenated and sent to fully connected layers to predict DTI.

TransformerCPI [3] is based on Transformer, which is a model to deal with machine translation task in nature language process(NLP). In TransformerCPI, protein sequence is the input of encoder, while atomic sequence is the input of decoder. The decoder outputs interaction feature vectors with the same length as atomic sequence. These interaction vectors are passed to fully connected layers to predict DTI labels.

5   Results and Analysis

In this section, we report the experimental results and make further analysis.

5.1 Effectiveness

The experimental results are shown in Table 2. From the results, our model performs better than other baseline models in all evaluation metrics, which proves the effectiveness of the proposed model.

(1) **DeepDTA** outperforms **GraphDTA** and **E2E** which use graph convolution network to obtain the embedding representation of molecule graph. This result may due to the molecule graph is small-scale compared than social media network. The convolution operation on each node of the molecule graph may cause the embedding representation overfitting.

(2) **TransformerCPI** achieve a better performance compared to other baseline models. This indicates that the attention mechanism in Transformer can extract more semantic information from sequence data than LSTM and

TextCNN, which are classical methods to process sequence data. The interaction vectors output by Transformer decoder is a better way to denote drug-target interaction than just concatenating the molecule vector and protein vector together.

(3) In BindingDB and Human dataset, our model **MBMT** exceed all baselines in all evaluation metrics. Our model can learn the relationship between different substructures of molecules with multi-head self-attention and extract deep semantic representation of protein sequences with BERT. A multi-task mechanism is introduced in our model. Through this mechanism, the model can learn the bidirectional selection process of drug and target. This mechanism also shares bottom parameters of the model which can improve the generalization of the model.

(4) In order to verify the effectiveness of multi-task mechanism, we record the evaluation metrics of single task and multi-task on the test dataset during the training process. The model with multi-task mechanism has a better performance than one single task in AUC and F1. Although **MBMT** perform worse than **MBMT**(task1) in Human's Precision and BindingDB's Recall, F1 which is the overall consideration of Precision and Recall is higher in **MBMT** multi-task model. This shows that multi-task mechanism can improve the performance and generalization ability of the model.

Table 2. The average AUC, Precision, Recall and F1 on Human and BindingDB test datasets.

Model	Human				BindingDB			
	AUC	Precision	Recall	F1	AUC	Precision	Recall	F1
DeepDTA	0.9640	0.9135	0.9019	0.9077	0.9221	0.8357	0.8732	0.8540
GraphDTA	0.9603	0.8821	0.9118	0.8967	0.9123	0.8504	0.8184	0.8341
E2E	0.9539	0.8934	0.9098	0.9015	0.9217	0.8456	0.8682	0.8568
TransformerCPI	0.9732	0.9157	0.9254	0.9205	0.9426	0.8676	0.8827	0.8751
MBMT(task1)	0.9829	**0.9228**	0.9462	0.9344	0.9831	0.9292	**0.9572**	0.9430
MBMT(task2)	0.9723	0.9046	0.9304	0.9173	0.9799	0.9228	0.9477	0.9351
MBMT	**0.9834**	0.9205	**0.9525**	**0.9362**	**0.9838**	**0.9362**	0.9540	**0.9450**

5.2 Ablation Study

In order to explore the impact of each component in MBMT, we conduct ablation experiments over three key components in MBMT, including multi-head self-attention mechanism in Mol2Vec, BERT and multi-task mechanism. For ablation experiments, each component is removed while other experiments conditions remain the same. In particular, when the multi-task mechanism is removed, the model can be regarded as two models with each task.

Table 3. Performance(AUC) of MBMT models with different component removing.

Model	MBMT	-attention	-BERT	-task1	-task2
Human	0.9834	0.9807	0.9755	0.9777	0.9824
BindingDB	0.9838	0.9812	0.9762	0.9781	0.9826

From the results in Table 3, we can observe that after removing the BERT, the performance of the MBMT decrease a lot. This is because the embedding outputs of the pre-trained BERT model are better latent representation of proteins, which contain a lot of semantic information of protein. When the self-attention mechanism in Mol2Vec is removed, there is a little performance loss, which indicates that the substructures of a molecule in different positions can really influence each other. As molecules have shorter sequences than proteins do, task 2's interaction vectors have shorter length than task 1's. This will make task 2 harder to train than task 1. So removing task1 will have more performance loss than removing task2.

5.3   Hyper-Parameter Study

λ is a hyper parameter that can be adjusted according to the importance of the two single tasks. The descent direction of the model parameters' gradient is determined by the task with larger weight of the loss function. It means the task with larger weight of loss function should be easier to train by gradient descent. To choose the best parameter λ, we test six different λ values. The performance(AUC) of the model with different λ values is in Fig. 5a. As a result, the classification performance of the model is best when λ is 0.6. This indicates that the first task is easier to train. This can be explained by that the task 2's interaction vectors have shorter length than task 1's, which will make task 2 harder to train than task 1.

To investigate the impact of the number of attention heads, we conduct experiments with number of attention heads from 1 to 6. Since the importance of the two tasks is affected by the number of attention heads set for the two tasks, the number of attention heads set for the two tasks is consistent. Figure 5b shows the performance(AUC) of the model with different number of attention heads in Human and BindingDB datasets. As a result, the classification performance of the model is best when the number of heads is 3. Multi-head attention can make the model learn various forms of attention. So it is no doubt that adding the number of heads will improve the performance of the model. But from the result, we can see it is not the case that the model works well with more heads. The reason is that the number of model parameters increases as the number of heads increases. More parameters will make the model easier to overfitting, which will reduce the generalization ability of the model.

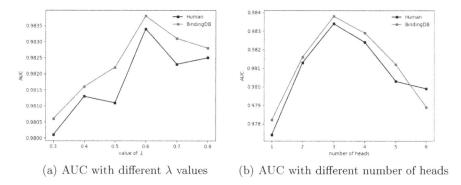

(a) AUC with different λ values   (b) AUC with different number of heads

Fig. 5. The performance of the model with different λ values and number of heads in Human and BindingDB datasets.

5.4 Interpretability Study

One important advantage of our model is the interpretability. To demonstrate that the attention mechanism in our model can make the model interpretable, we visualize the attention scores in protein sequence with heatmap. Figure 7 shows the structure of complex with MDM2 protein and Nutlin-3a small molecule. The binding sites are shown in the bottom bar in Fig. 6. The black regions are the true binding sites in the protein sequence. We visualize the attention scores output by our model in the protein sequence with heatmap, which are shown in the above bar in Fig. 6. The attention scores increase with the darker blue color. By comparison, we can find that many darker places of the two bars correspond to each other. This indicates that the attention scores can well reflect the importance of each part in the protein sequence for DTI prediction.

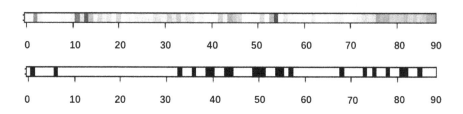

Fig. 6. The prediction of the binding sites with attention scores and the true binding sites in the M2MD protein sequence.

Fig. 7. The complex of M2MD protein and Nutlin-3a small molecule.

6   Conclusion

We have proposed an end-to-end deep learning framework to predict drug-target interactions from low level representations. The proposed model can learn the relationship between different substructures of molecules with multi-head self-attention mechanism. Deep semantic representation of protein sequences can be obtained by BERT. A multi-task mechanism is also introduced to make the model learn the bidirectional selection process of drug and target. Experimental evaluation shows that the proposed model overall outperforms all baselines. The model has a good performance on active class classification which can provide some candidate molecules as reference for new drug. Finally, we prove that our model is interpretable through a case study.

For deep learning models, the models' performance almost depend on the amount of data for training. It is worth noting that federated learning [25] can aggregate data from all data parties and protect the privacy of each party. In future work, we will apply federated learning to our deep learning model.

References

1. Bank, P.D.: Protein data bank. Nat. New Biol. **233**, 223 (1971)
2. Caruna, R.: Multitask learning: a knowledge-based source of inductive bias. In: Machine Learning: Proceedings of the Tenth International Conference, pp. 41–48 (1993)
3. Chen, L., et al.: Transformercpi: improving compound-protein interaction prediction by sequence-based deep learning with self-attention mechanism and label reversal experiments. Bioinformatics **36**(16), 4406–4414 (2020)
4. Chen, X., et al.: Drug-target interaction prediction: databases, web servers and computational models. Briefings Bioinf. **17**(4), 696–712 (2016)
5. Devlin, J., Chang, M.W., Lee, K., Toutanova, K.: Bert: pre-training of deep bidirectional transformers for language understanding. arXiv preprint arXiv:1810.04805 (2018)
6. Ding, H., Takigawa, I., Mamitsuka, H., Zhu, S.: Similarity-based machine learning methods for predicting drug-target interactions: a brief review. Briefings Bioinf. **15**(5), 734–747 (2014)

7. Ewing, T.J., Makino, S., Skillman, A.G., Kuntz, I.D.: Dock 4.0: search strategies for automated molecular docking of flexible molecule databases. J. Comput.-Aided Mol. Des. **15**(5), 411–428 (2001)
8. Ezzat, A., Wu, M., Li, X.L., Kwoh, C.K.: Computational prediction of drug-target interactions using chemogenomic approaches: an empirical survey. Briefings Bioinf. **20**(4), 1337–1357 (2019)
9. Faulon, J.L., Misra, M., Martin, S., Sale, K., Sapra, R.: Genome scale enzyme-metabolite and drug-target interaction predictions using the signature molecular descriptor. Bioinformatics **24**(2), 225–233 (2008)
10. Gao, K.Y., Fokoue, A., Luo, H., Iyengar, A., Dey, S., Zhang, P., et al.: Interpretable drug target prediction using deep neural representation. In: IJCAI, vol. 2018, pp. 3371–3377 (2018)
11. Gilson, M.K., Liu, T., Baitaluk, M., Nicola, G., Hwang, L., Chong, J.: Bindingdb in 2015: a public database for medicinal chemistry, computational chemistry and systems pharmacology. Nucleic Acids Res. **44**(D1), D1045–D1053 (2016)
12. Jaeger, S., Fulle, S., Turk, S.: Mol2vec: unsupervised machine learning approach with chemical intuition. J. Chem. Inf. Model. **58**(1), 27–35 (2018)
13. Jiang, D., et al.: Interactiongraphnet: a novel and efficient deep graph representation learning framework for accurate protein-ligand interaction predictions. J. Med. Chem. **64**(24), 18209–18232 (2021)
14. Karimi, M., Wu, D., Wang, Z., Shen, Y.: Deepaffinity: interpretable deep learning of compound-protein affinity through unified recurrent and convolutional neural networks. Bioinformatics **35**(18), 3329–3338 (2019)
15. Luong, M.T., Le, Q.V., Sutskever, I., Vinyals, O., Kaiser, L.: Multi-task sequence to sequence learning. arXiv preprint arXiv:1511.06114 (2015)
16. Mikolov, T., Chen, K., Corrado, G., Dean, J.: Efficient estimation of word representations in vector space. arXiv preprint arXiv:1301.3781 (2013)
17. Nguyen, T., Le, H., Quinn, T.P., Nguyen, T., Le, T.D., Venkatesh, S.: Graphdta: Predicting drug-target binding affinity with graph neural networks. Bioinformatics **37**(8), 1140–1147 (2021)
18. Öztürk, H., Özgür, A., Ozkirimli, E.: Deepdta: deep drug-target binding affinity prediction. Bioinformatics **34**(17), i821–i829 (2018)
19. Tsubaki, M., Tomii, K., Sese, J.: Compound-protein interaction prediction with end-to-end learning of neural networks for graphs and sequences. Bioinformatics **35**(2), 309–318 (2019)
20. Vaswani, A., et al.: Attention is all you need. In: Advances in Neural Information Processing Systems, vol. 30 (2017)
21. Wang, Y., Zeng, J.: Predicting drug-target interactions using restricted boltzmann machines. Bioinformatics **29**(13), i126–i134 (2013)
22. Wang, Z., Liang, L., Yin, Z., Lin, J.: Improving chemical similarity ensemble approach in target prediction. J. Cheminformatics **8**(1), 1–10 (2016)
23. Weininger, D.: Smiles, a chemical language and information system. 1. introduction to methodology and encoding rules. J. Chem. Inf. Comput. Sci. **28**(1), 31–36 (1988)
24. Yamanishi, Y., Kotera, M., Kanehisa, M., Goto, S.: Drug-target interaction prediction from chemical, genomic and pharmacological data in an integrated framework. Bioinformatics **26**(12), i246–i254 (2010)
25. Zhang, C., Xie, Y., Bai, H., Yu, B., Li, W., Gao, Y.: A survey on federated learning. Knowl.-Based Syst. **216**, 106775 (2021)

26. Zhang, Y., Wallace, B.: A sensitivity analysis of (and practitioners' guide to) convolutional neural networks for sentence classification. arXiv preprint arXiv:1510.03820 (2015)
27. Zhao, Q., Xiao, F., Yang, M., Li, Y., Wang, J.: Attentiondta: prediction of drug-target binding affinity using attention model. In: 2019 IEEE international conference on bioinformatics and biomedicine (BIBM). pp. 64–69. IEEE (2019)

ShapeWordNet: An Interpretable Shapelet Neural Network for Physiological Signal Classification

Wenqiang He[1], Mingyue Cheng[1], Qi Liu[1(✉)], and Zhi Li[2]

[1] Anhui Province Key Laboratory of Big Data Analysis and Application, University of Science and Technology of China, Hefei, China
{wenqianghe,mycheng}@mail.ustc.edu.cn, qiliuql@ustc.edu.cn
[2] Shenzhen International Graduate School, Tsinghua University, Shenzhen, China
zhilizl@sz.tsinghua.edu.cn

Abstract. Physiological signals are high-dimensional time series of great practical values in medical and healthcare applications. However, previous works on its classification fail to obtain promising results due to the intractable data characteristics and the severe label sparsity issues. In this paper, we try to address these challenges by proposing a more effective and interpretable scheme tailored for the physiological signal classification task. Specifically, we exploit the time series shapelets to extract prominent local patterns and perform interpretable sequence discretization to distill the whole-series information. By doing so, the long and continuous raw signals are compressed into short and discrete token sequences, where both local patterns and global contexts are well preserved. Moreover, to alleviate the label sparsity issue, a multi-scale transformation strategy is adaptively designed to augment data and a cross-scale contrastive learning mechanism is accordingly devised to guide the model training. We name our method as ShapeWordNet and conduct extensive experiments on three real-world datasets to investigate its effectiveness. Comparative results show that our proposed scheme remarkably outperforms four categories of cutting-edge approaches. Visualization analysis further witnesses the good interpretability of the sequence discretization idea based on shapelets.

Keywords: Physiological Signal Classification · Shapelet-based Sequence Discretization · Interpretability · Contrastive Learning

1 Introduction

Physiological signal is an invaluable type of medical time series, which has broad applications in the healthcare domains, such as emotion recognition, seizure detection and heartbeat classification [11]. To effectively indicate the health state of human body, relevant information is often recorded simultaneously by multiple sensors through high-frequency and long-time sampling. For example, an

X. Wang et al. (Eds.): DASFAA 2023, LNCS 13946, pp. 353–369, 2023.
https://doi.org/10.1007/978-3-031-30678-5_27

electrocardiogram (ECG) signal recording can be sampled in 12 channels at a frequency of 500 HZ for at least 10 s to be used for diagnosing the cardiovascular condition of a patient. Nowadays, the fast progress of IoT spurs an explosive increase of physiological signals, making the traditional way of manually classifying such high-dimensional data not only costly but also inefficient [21]. Hence, recent research has turned to artificial intelligence and machine learning for technical assistance [6].

Given the temporal data property, the physiological signal classification (PSC) task is often viewed as a typical time series classification (TSC) problem in the machine learning field, where a plentitude of TSC methods have been proposed and can be roughly grouped into two categories: the classical algorithms and the deep learning (DL) based approaches [3,17]. Classical TSC methods focus on explainable feature engineering, where distinguishable features are designed and extracted from various perspectives. For instance, the "golden standard" 1-NN Dynamic Time Warping (DTW) paradigm [10,22] concentrates on comparing the similarity of global patterns, while the shapelet-based approaches [14,23,33] aim at mining discriminative subsequences that maximally represent a class. Nevertheless, despite effective on small-scale and univariate datasets, classical methods do not scale well to the PSC task due to the difficulty of large-space feature selection and the inability to capture multi-variate interaction [31].

In the past few years, deep learning based methods have reported incredible advancements in the TSC field, which avoid handcraft feature design and laborious feature selection by directly learning informative and low-dimensional representations from raw data [17,34]. However, DL approaches require a large amount of labelled data to supervise model training, which is quite limited in the PSC scenario and may lead to performance degeneration. Besides, DL models provide little insight into the decisive factors and such black-box natures would impair their credibility in the healthcare field [1,24]. To overcome the above challenges and better adapt to the PSC task, one natural idea is to make full use of the strengths of both the classical and the deep learning based methods [34].

In this article, we propose a two-stage model named ShapeWordNet to provide a more effective and interpretable solution to the PSC problem. To be specific, we novelly take advantage of the time series shapelets to extract discriminative local patterns as elementary "words" that contain certain class-relevant "semantics" of the original data. Then, based on these explainable words, we discretize the point-wise raw signal into a word-wise token sequence dubbed ShapeSentence for the whole-series information distillation, where we believe both significant local patterns and global contexts are well preserved. Moreover, in order to alleviate the label sparsity issue, the large shapelet candidate space is adaptively leveraged to augment the raw data in a multi-scale transformation way, and a cross-scale contrastive learning mechanism is accordingly constructed as an auxiliary objective to capture the data invariance. Finally, a scale-aware feature integrator is devised to fuse the representations of multi-scale ShapeSentences for class label prediction.

In summary, the main contributions of our work are as follows:

- We propose an effective and interpretable scheme named ShapeWordNet tailored to the physiological signal classification task, which integrates the representation learning strengths of deep neural networks with the interpretability advantages of time series shapelets.
- We design a ShapeWord Discretization strategy to deal with the intractable data properties of physiological signals and devise a cross-scale contrastive learning mechanism to alleviate the label sparsity issues. To the best of our knowledge, this is the first work to utilize shapelets for explainable sequence discretization and deep learning's representational ability promotion.
- We conduct extensive experiments on three real-world datasets to investigate the effectiveness of ShapeWordNet. The comparative results validate the model's outperformance over four categories of TSC methods and the visualization analysis illustrates the good interpretability of the sequence discretization idea based on shapelets.

2   Preliminaries

2.1   Problem Formulation

Given a group of physiological signals $T = \{T_1, T_2, ..., T_m\} \in \mathcal{R}^{m \times d \times n}$ and the corresponding label set $\mathcal{Y} = \{y_1, y_2, ..., y_m\} \in \mathcal{R}^m$, where each sample $T_i \in \mathcal{R}^{d \times n}$ is a d-dimensional sequence of n time steps associated with a label y_i, the goal of physiological signal classification is to train a model $f_\Theta : T \mapsto y$ to predict the class label for a target instance.

2.2   Definitions

Definition 1: Shapelet. A shapelet $\tilde{S} \in \mathcal{R}^l$ $(1 \leq l \leq n)$ is a type of subsequence that well discriminates classes [33]. A good shapelet is supposed to have small $sDist$, i.e. the shapelet distance [4], to instances of one class and have large $sDist$ to those of another. The $sDist$ is defined as the minimum euclidean distance between $\tilde{S}$ and any subseries $w \in W^l$ of a given time series $T \in \mathcal{R}^n$:

$$sDist(\tilde{S}, T) = \min_{w \in W^l} (dist(\tilde{S}, w)). \tag{1}$$

Definition 2: ShapeWord. A ShapeWord is defined as the cluster centroid of a set of similar shapelets, which represents the abstract prototype [20] of their shared local pattern and can be referred to by the cluster label token SW:

$$\begin{aligned} ShapeWord &= ClusterCentroid(\tilde{S}_1, ..., \tilde{S}_v), \\ SW &= ClusterLabel(\tilde{S}_1, ..., \tilde{S}_v). \end{aligned} \tag{2}$$

Definition 3: ShapeSentence. A ShapeSentence SS is the discretized token sequence of the continuous raw series T, where each token SW_i refers to a Shape-Word and s is the length of this ShapeSentence:

$$T = [t_1, ..., t_n] \in \mathcal{R}^{d \times n} \rightarrow SS = [SW_1, ..., SW_s] \in \mathcal{R}^{d \times s}. \tag{3}$$

3   ShapeWordNet

The overall architecture of our proposed ShapeWordNet is shown in Fig. 1, which consists of two stages: the **ShapeWord Discretization** stage and the **Cross-scale Contrastive Learning Assisted Classification** stage.

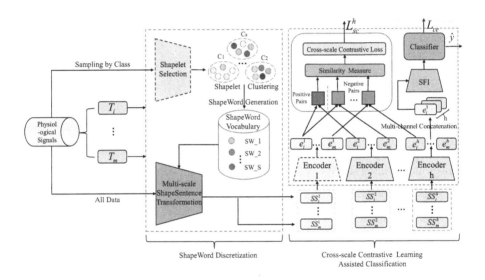

Fig. 1. An overview of the ShapeWordNet model.

3.1   ShapeWord Discretization

The first stage includes three steps: (1) **Shapelet Selection**, (2) **ShapeWord Generation** and (3) **Muti-scale ShapeSentence Transformation**.

Shapelet Selection. Shapelets are discriminative subsequences that can offer explanatory insights into the problem domain [33]. In this paper, we seize on such advantages of shapelets to extract interpretable and prominent local patterns. Traditional way of selecting shapelets is to evaluate the maximum information gain among all possible splits for each shapelet candidate, which would be extremely time-consuming in the physiological signal classification (PSC) scenario. To fast select shapelets, we combine the single-scan shapelet discovery algorithm [23] with a random sampling strategy. Specifically, we first select 10 samples from each class at random. Then we generate shapelet candidates with a sliding window and evaluate each candidate's discrimination ability with the F-statistic measure that assesses the mean $sDist$ distribution differences between classes:

$$F(S) = \frac{\sum_{v=1}^{V} \left(\bar{d}_{S,v} - \bar{d}_S \right)^2 \Big/ (V-1)}{\sum_{v=1}^{V} \sum_{j=1}^{N_v} \left(d_{S,v,j} - \bar{d}_{S,v} \right)^2 \Big/ (N-V)}, \tag{4}$$

where V is the class number, N is the total sample number, N_v is the number of class v, S is the shapelet candidate to be assessed, $\bar{d}_S$ is the mean value of its $sDist$ vector D_S, and $d_{S,v,j}$ is its $sDist$ with the j-th sample of class v.

ShapeWord Generation. Although plenty of shapelets can be easily found, many of them are similar to each other, which brings about feature redundancy and increases computational complexity. To alleviate this issue, we propose to generate the prototypes [20] that contain the key information shared by similar shapelets as the elementary units for sequence discretization. Toward this end, we cluster the selected shapelets with K-means and define the cluster centroids as their prototypes [28], as is suggested in Eq. 2. Those prototypes are named ShapeWord and assigned with numeric cluster label tokens for reference. For the multivariate case, we simply repeat the aforementioned algorithm and generate ShapeWords for each variable. In doing so, we establish a *vocabulary* of ShapeWords which encompasses the significant local patterns over all variable dimensions, as is illustrated in Fig. 1.

Moreover, to validate the discriminative edges of ShapeWords, we conduct a comparison experiment between the selected shapelets and the generated Shape-Words on the Sleep dataset [19]. In our experiment, we first generate shapelet candidates of different lengths from 5 to 200. For each scale, we select the top-100 shapelets to produce corresponding ShapeWords via K-means, where the cluster number is set equal to the class number, i.e. $N_{SW} = 8$. Then, we establish a validation set containing 10 random samples of each class to compare the average F-statistic qualities of the ShapeWords with that of top-100 shapelets. The results in Fig. 2 show that the mean F-statistic scores of the ShapeWords are competitively higher regardless of scales, which concretely demonstrate the effectiveness of ShapeWords in representing the prototypes of similar shapelets.

Multi-scale ShapeSentence Transformation. Traditional shapelet-based methods focus on extracting local patterns for classification, while in the PSC scenario, the global contextual information such as the periodicity and variation of local patterns is also of critical importance. For example, the sinus arrhythmia can be more effectively diagnosed from a periodic perspective. Hence, based on the extracted local patterns represented by ShapeWords, we discretize the entire sequence to further distill the global contexts. Firstly, we segment the original signals into non-overlapping subsequences via a sliding window of the

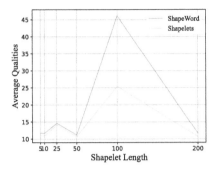

Fig. 2. Results of the average F-statistic quality comparison between ShapeWords and the top-100 Shapelets w.r.t. the shapelet lengths, which is the higher the better.

ShapeWord size. Then, we assign each segment the cluster label token that refers to its nearest ShapeWord according to the Euclidean Distance. In doing so, a long and complex point-wise time series can be interpretably compressed into a much shorter and simpler word-wise token sequence, which preserve the key features both locally and globally and can be more robust to noise disturbance [30]. We call such a token sequence ShapeSentence to suggest it is like a meaningful sentence in the natural language, where the whole sentence and its constituent words contain information at different semantic levels.

In addition, the ShapeSentence is quite scalable and can be adaptively extended as a data augmentation strategy to mitigate data sparsity issue. To be specific, we make the best of the large shapelet candidate space to generate different sizes of ShapeWords and transform each sample into multiple scales of ShapeSentences, which are further utilized to construct self-supervised signals in the contrastive learning mechanism. We dub this data augmentation technique as the **M**ulti-scale **S**hapeSentence **T**ransformation (MST) strategy and summarize its procedure in Algorithm 1. When $max = min + 1$, it is the simplest single-scale ShapeSentence transformation version mentioned above.

3.2   Cross-scale Contrastive Learning Assisted Classification

The second stage adopts the paradigm of multi-task learning, where the model training is assisted by the **Cross-scale Contrastive Learning** and the learnt representations are fused through **Scale-aware Feature Integration** before final classification.

Cross-scale Contrastive Learning. Self-supervised learning has emerged as an alternative paradigm to overcome deep learning's heavy dependence on manual labels by leveraging the input data itself as supervision [25]. In recent years, great breakthroughs in this field has been achieved by the contrastive learning [7], which aims at "learning to compare" through the Noise Contrastive

---

Algorithm 1 Multi-scale ShapeSentence Transformation (MST)

1: **Input:** sample set T, ShapeWord scale range $[min, max]$, ShapeWord vocabulary $SWList$
2: $AugmentedData \leftarrow \emptyset$
3: **for** T_i in T **do**
4:     $MShapeSentences \leftarrow \emptyset$
5:     **for** $l \leftarrow min$ to max **do**
6:         $W_i^l \leftarrow segmentSequences\,(T_i, l)$
7:         **for** subsequence S in W_i^l **do**
8:             $SW_h \leftarrow findClosest\,(S, SWList)$
9:             $MShapeSentences.append\,(SW_h)$
10:         **end for**
11:         $AugmentedData.append\,(MShapeSentences)$
12:     **end for**
13: **end for**
14: **return** $AugmentedData$

Estimation (NCE) [15] or the InfoNCE objectives [27]. In this work, we adpatively design a cross-scale contrastive learning mechanism to alleviate the label sparsity issues of PSC by constructing the self-supervised signals based on the multi-scale transformed ShapeSentences. Since a pair of large-scale and small-scale ShapeSentences of the same sample can be regarded as its observations from multi-scale perspectives, it is safe to hypothesize that there exists latent invariance behind them which we can enable the feature encoders to capture [16]. With this intuition, we encode each scale of ShapeSentence into fixed-size representations and compute the InfoNCE loss by comparing their similarities.

As is illustrated in Fig. 1, for instance, h scales of sample i's ShapeSentences, i.e. $SS_i^1, SS_i^2, ..., SS_i^h$, are first fed into a set of encoders $En^1\,(\cdot), ..., En^h\,(\cdot)$ to obtain their representations, i.e. $e_i^1 = En^1\left(SS_i^1\right), ..., e_i^h = En^h\left(SS_i^h\right)$ (one encoder corresponds to one scale). Then, in order to make the encoders capable of capturing the invariance shared by different scales of ShapeSentences, we define the cross-scale contrastive loss L_{sc}^h as:

$$L_{sc}^h = \frac{1}{\binom{h}{2}} \sum_{u=1}^{h-1} \sum_{v=u+1}^{h} L_{u,v}^{sc}, \tag{5}$$

$$L_{u,v}^{sc} = E_{\left(e_i^u, e_i^v\right) \sim P_{u,v}^i} \left[-\log \frac{f\left(e_i^u, e_i^v\right)}{f\left(e_i^u, e_i^v\right) + \sum_{i \neq j} f\left(e_i^u, e_j^v\right)} \right], \tag{6}$$

$$f\left(e_i^u, e_i^v\right) = \left(e_i^u\right)^T e_i^v \Big/ \tau, \tag{7}$$

where $L_{u,v}^{sc}$ is the contrastive loss between the representations of the u-scale ShapeSentences and the v-scale ShapeSentences, with $P_{u,v}^i$ as their joint sample distribution and $f\,(\cdot)$ being the representation similarity measure. In this article, we simply apply the vector inner product to compute representation similarities, view representation pair (e_i^u, e_i^v) from the same sample as positive, and randomly select $N - 1$ different samples from the marginal distributions of other samples within the same one mini-batch as negative, e.g. the negative e_j^v from the v-scale marginal distribution of another sample j, as is suggested by InfoNCE [27].

In our scheme, we choose the deep dilated causal convolutional neural network [12] as the encoder backbone given its high efficiency and outstanding excellence in capturing long-range dependencies [5]. Besides, to learn the multivariate interactions, we input different variable's ShapeSentence into different channels of the encoder to obtain the representation feature vectors [35].

Scale-aware Feature Integration for Classification. Before classification, we need to integrate the learned representations from different scales of Shape-Sentences at first. In terms of multi-scale feature fusion, general methods like average pooling, maxpooling and direct concatenation [8] are all model-agnostic that do not take the domain particularities into consideration. To adaptively make full use of the multi-scale information in our method, we regard each scale's representation as complementary to the raw data's invariant features and concatenate them by channel. We then input the concatenated representation tensor into the the **S**cale-aware **F**eature **I**ntegrator (SFI) for feature fusion:

$$C_i = SFI\left(E_i\right), \tag{8}$$

where SFI is a single one-dimensional convolution layer with different channels catering to different scales of ShapeSentence representations, $E_i = \left[e_i^1, ..., e_i^h\right] \in \mathcal{R}^{p \times h}$ is sample i's concatenated representation tensor, and $C_i \in \mathcal{R}^q$ is the integrated feature vector.

Finally, we input the fused feature representation C_i into a single linear classifier to obtain the classification outcome $\hat{y}_i$:

$$\hat{y}_i = W_c^\top C_i + W_0, \tag{9}$$

where $W_c \in \mathcal{R}^{c \times q}$ and $W_0 \in \mathcal{R}^c$ are learnable parameters.

Multi-task Optimization. To optimize the whole network, we combine the cross-entropy loss for classification with the cross-scale contrastive loss for self-supervision as a multi-task goal [29] for joint training:

$$
\begin{aligned}
L &= L_{ce} + \lambda L_{sc}^h \\
&= -\sum_c^{|C|} y_c \log\left(\hat{y}_c\right) + \frac{\lambda}{\binom{h}{2}} \sum_{u=1}^{h-1} \sum_{v=u+1}^{h} L_{u,v}^{sc},
\end{aligned} \tag{10}
$$

where λ is used to balance different losses and h is the number of scales.

4 Experiments

4.1 Experimental Setup

Datasets. We conduct experiments on three real-world public datasets from PhysioNet [13], where two datasets are ECG signals used for cardiac disease classification in the 2020 Physionet/Computing Cardiology Challenge [2] and one dataset contains EEG signals popular in sleep-stage classification [19]. In our experiment, we pick out part of the single-label samples and randomly split them into 80%-20% train-test datasets. The statistics of these datasets are summarized in Table 1, where Trainsize/Testsize means the sample number of the training/testing dataset, Dim/Len refers to the variable number and time steps, and Ratio stands for the ratio of one class number to all.

Table 1. Descriptive statistics of three datasets.

Property	Datasets		
	CPSC	Georgia	Sleep
Trainsize	5,123	2,676	12,787
Testsize	1,268	668	1,421
Dim/Len	12/5,000	12/5,000	2/3,000
Category	ECG	ECG	EEG
Multi-class Ratio (%)	[14.07, 24.24, 15.42, 9.53, 12.26, 10.85, 3.03, 2.69, 7.91]	[52.73, 12.97, 7.47, 6.50, 6.69, 3.81, 3.77, 3.51, 2.54]	[48.42, 3.65, 21.40, 4.25, 4.40, 9.93, 0.08, 7.86]
Binary Ratio (%)	[14.07, 85.93]	[52.73, 47.27]	[48.42, 51.58]

Baselines and Variants. We compare three variants of our ShapeWordNet (**SWN**) scheme with four categories of time series classification (TSC) baselines:

(1) **Shapelet-based:** We employ the Shapelet Transformation (**ST**) [23] and Learning Shapelets (**LS**) [14] for the first baseline group. ST searches the shapelets and transforms the original data into distance vectors, while LS learns the shapelets directly by optimizing the goal function of classification.

(2) **Dictionary-based:** We pick out the classical **SAX-VSM** [32] and the recent **WEASEL+MUSE** [31] for sequence discretization comparison, which respectively utilize the SAX words and SFA words for time series discretization and build classifiers based on their frequency patterns.

(3) **SOTA:** This group contains two state-of-the-art TSC methods, including the non-DL model MiniRocket [9] and the DL model TapNet [34].

(4) **CNN-based:** Considering the noticeable achievements of convolutional neural networks (CNN) in sequence modeling [5], we adopt four different architectures of CNN-based TSC approaches as the deep learning baselines, which are **MCNN** [8], **LSTM-FCN** [18], **ResCNN** [36] and **TCN** [5].

(5) **Variants:** We put forward three variants of our method for ablation study: (1) **SWN w/o SD** stands for the ShapeWordNet without ShapeWord Discretization, i.e. the encoder backbone and the **TCN** baseline, (2) **SWN w/o CCLM** represents the ShapeWordNet without the Cross-scale Contrastive Learning Mechanism, and (3) **SWN w/o SFI** means the ShapeWordNet without the Scale-aware Feature Integrator.

Implementation Details. We set the *layer_depth* = 3, *kernel_size* = 3, and *out_channels* = 50 as the default parameters for each of the dilated causal convolutional encoders. In terms of the ShapeWord/shapelet/word number parameters in three SWN variants, Dictionary-based and Shapelet-based baselines, we set them equal to the task class number times the variable dimensions of each dataset, i.e. 8*2 for Sleep and 9*12 for CPSC and Georgia. As for the ShapeWord/shapelet/word lengths, we set 10 for SWN w/o CCLM, SAX-VSM, WEASEL+MUSE, ST and LS, and set [10,25,50] for SWN w/o SFI and SWN. In SWN w/o SFI and SWN, the λ in Eq. 10 to balance loss is set to be 0.5. Besides, we set the *batch_size* = 30, *training_epochs* = 50 and use Adam optimizer with *learning_rate* = 0.001 for all methods.

We evaluate the classification performance with two metrics: **Accuracy (ACC)** and **Macro F1-score (MAF1)**. ACC can measure the overall performance by calculating how many samples are correctly classified in total, while MAF1 can avoid the measurement bias caused by class imbalance and assess a model's discrimination ability more fairly.

We have implemented the proposed method in python 3.7 and run all the experiments on a machine of CentOS 7.9.2009 with 4 T V100 and 2 Intel Xeon Gold 5218 @2.30 GHz.

4.2   Performance Comparison

To comprehensively evaluate the performance of our method, we conduct two experimental tasks of binary classification (BC) and multi-class classification (MC) on each dataset, where all the labels in MC other than the positive ones constitute the negative labels in BC. According to Table 1, the BC class ratios of Georgia and Sleep are more balanced than their MC class ratios, which enables BC tasks to serve as the control experiments concerning the label sparsity issue. Although the BC class ratio of CPSC is less balanced than MC class ratio, CPSC has a more balanced MC ratio than Georgia and Sleep, making it a control dataset to indicate the influence of class balance on model performance. Table 2 reports the ACC and MAF1 of different methods on two tasks times three datasets and denotes the best ACC and MAF1 for each task with boldface. Consistent with our intuition, the major observations are summarized as follows:

(1) We can see that our SWN variants significantly outperform four categories of TSC methods on three datasets with an average improvement of 9.28% on BC tasks and 31.40% on MC tasks. These results strongly testify the superiority of our method in dealing with the PSC problem especially on the condition of severe label sparsity issues.
(2) Our proposed ShapeWordNet surpasses the classical Shapelet-based ST and LS methods by a large margin, which underlines the significance of utilizing deep learning models to capture multivariate interaction and distilling both the local and global information for time series classification.
(3) Compared with SAX-VSM and WEASEL+MUSE, our method wins overwhelmingly on all tasks. We believe this phenomenon concretely demonstrates the advantages of ShapeWords in representing discriminative local

Table 2. The performance comparison between different methods on BC and MC tasks of three datasets with ACC and MAF1 metrics. The best performing methods are boldfaced and the best beaselines are underlined. Improv (%) measures the relative improvements of SWN variants over the best baselines. Note that the improvements are statistically significant with a two-sided t-test $p<0.01$.

Method	CPSC				Georgia				Sleep			
	2 Classes		9 Classes		2 Classes		9 Classes		2 Classes		8 Classes	
	ACC	MAF1	ACC	MAF1	ACC	MAF1	ACC	MAF1	ACC	MAF1	ACC	MAF1
LS	0.8400	0.4579	0.1554	0.0299	0.4970	0.4970	0.2814	0.1112	0.7847	0.7843	0.6946	0.3244
ST	0.6500	0.5911	<u>0.6900</u>	0.1331	0.5100	0.4516	0.2500	<u>0.2319</u>	0.8300	0.8311	0.5600	0.2839
SAX-VSM	0.1617	0.1426	0.1483	0.1141	0.5973	<u>0.5904</u>	0.1272	0.1106	0.6868	0.6666	0.5771	0.2994
WEASEL+MUSE	0.8303	0.4987	0.2505	0.3850	<u>0.6102</u>	0.5561	<u>0.5653</u>	0.1567	0.7136	0.7126	0.5517	0.3334
MiniRocket	0.8450	0.4646	0.0935	0.0220	0.5308	0.3802	0.0793	0.0164	<u>0.9369</u>	<u>0.9363</u>	0.6864	0.4211
TapNet	0.7167	0.5366	0.1372	0.1148	0.5132	0.5123	0.2695	0.1180	0.7910	0.7904	0.5742	0.2775
MCNN	0.8564	0.4613	0.1044	0.0210	0.4850	0.3266	0.4850	0.0726	0.5011	0.3338	0.5039	0.0838
LSTM-FCN	<u>0.8967</u>	<u>0.7865</u>	0.6727	<u>0.6134</u>	0.5150	0.3399	0.1617	0.0607	0.7607	0.7566	0.6144	0.2489
ResCNN	0.8431	0.4574	0.1300	0.0489	0.5225	0.4976	0.1766	0.0417	0.7741	0.7575	0.1323	0.0535
TCN (SWN w/o SD)	0.8904	0.7734	0.6491	0.5423	0.5928	0.5883	0.5045	0.1740	0.8951	0.8946	<u>0.7241</u>	<u>0.4645</u>
SWN w/o CCLM	0.9014	0.7893	0.6924	0.6395	**0.7590**	**0.7590**	0.7246	0.3650	0.9198	0.9192	0.76425	0.5210
SWN w/o SFI	0.9085	0.8059	0.7397	0.6778	0.7350	0.7325	0.7156	**0.4538**	0.9346	0.9343	0.8023	0.5638
SWN	**0.9101**	**0.8212**	**0.7516**	**0.7156**	0.7096	0.7094	**0.7365**	0.4256	**0.9374**	**0.9370**	**0.8093**	**0.5645**
Improv (%)	1.49	4.41	8.93	16.66	19.70	20.58	45.99	83.53	4.73	4.74	11.77	21.53

patterns over the SAX words and SFA words. Besides, it also indicates learning abstract feature representations is more effective than relying on discrete statistical patterns.

(4) In contrast to the CNN-based methods that rank the second in 5 tasks (3 for LSTM-FCN and 2 for TCN), our method works particularly better on three MC tasks by obtaining an average of 19.09% rise in MAF1. Such a noticeable improvement soundly validates the effectiveness of leveraging ShapeWord Discretization and Cross-scale Contrastive Learning in label sparsity mitigation and invariant feature extraction.

4.3 Ablation Study

To evaluate the effectiveness of each component, we investigate the performance of three variants. Firstly, we observe that SWN w/o CCLM substantially surpasses the SWN w/o SD by approximately 17.92% on CPSC MC, 12.16% on Sleep MC and 109.77% on Georgia MC in MAF1, which substantiates the effectiveness of ShapeWord Discretization in reducing noise disturbance and distilling prominent features. In addition, the fact that SWN and SWN w/o SFI defeat SWN w/o CCLM on 10 tasks with at least 5.99% rise of MC MAF1 tellingly verifies the remarkable progress made by the CCLM component in relieving the label sparsity. Moreover, the slight edges of SWN over SWN w/o SFI on 9 tasks indicate that SFI may contribute to promoting performance more or less.

4.4   ShapeWord Discretization Analysis

As the results in Table 2 imply, ShapeWord Discretization plays the most critical role in boosting the model's generalization. Therefore, it is necessary to explain how it really takes effect. To investigate the interpretability and effectiveness of ShapeWord Discretization, we take a close look at its characteristics through two visualization experiments.

First, we conduct a case study to intuitively explore the interpretability of the ShapeWord Discretization. Figure 3 illustrates a case of two atrial fibrillation (AF) samples and two sinus rhythm (NSR) samples, where the four raw signals in blue lines are discretized into four ShapeSentences ([1, 1, 8, 8, 1], [8, 8, 1, 1, 1], [1, 1, 1, 3, 1] and [1, 1, 3, 1, 1]) in red lines respectively. It can be noticed that the pattern of AF, i.e. the disease of sustained tachyarrhythmia commonly seen in clinical practice, seems to be captured by the violently fluctuated ShapeWord SW_8 presented below. In contrast, the peculiarities of NSR, i.e. the normal state, seem to be represented by the combination of the ShapeWord SW_1 and SW_3.

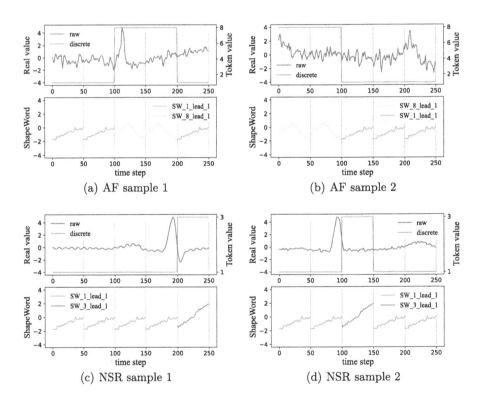

Fig. 3. A case of two AF samples and two NSR samples transformed by ShapeWord Discretization with $window\_size = 50$. The raw signals are in blue lines and their ShapeSentences are in red ones, where each token refers to a ShapeWord shown below. (Color figure online)

Hence, it is reasonable to believe that both the discriminative local patterns and their coherence can be well preserved via ShapeWord Discretization.

Second, to vindicate the contributions of ShapeWord Discretization to DL model's representation learning, we conduct the t-SNE analysis [26] to compare the representations output by SWN w/o CCLM and SWN w/o SD respectively on the Georgia BC task. In Fig. 4(a) and 4(b), the sample representations output by SWN w/o CCLM (i.e. the variant with the ShapeWord Discretization) are more closely grouped with a clearer clustering boundary than those output by the SWN w/o SD, which defensibly testifies the effect of ShapeWord Discretization on relieving the influence of data noise and promoting the representational ability of deep neural networks.

4.5   Hyper-parameter Sensitivity

In this section, we discuss the impact of several key hyper-parameters on the performance of our method, which includ the Scale Number N_s, the Loss Balance Factor λ, the ShapeWord Number N_{SW} and the ShapeWord Length L_{SW}.

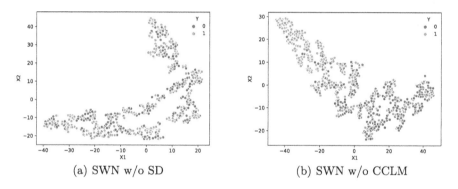

(a) SWN w/o SD   (b) SWN w/o CCLM

Fig. 4. T-SNE visualization of representations produced by (a) SWN w/o SD and (b) SWN w/o CCLM on Georgia BC task (0 indicates positive and 1 means negative).

Performance W.r.t. Scale Number. We display the model performance w.r.t. the scale number N_s in Fig. 5, where the scales for MST are successively picked from the range [5,10,25,50,100], e.g. 2 for [5,10] and 3 for [5,10,25]. Figure 5(a) and Fig. 5(b) show that for different datasets the impact of scale number is different and an ideal interval of this parameter shared by three datasets seems to be around [2,3]. Based on this observation, we implement 3 scales of MST for SWN in our experiments.

Performance W.r.t. Loss Balance Factor. The parameter λ in Eq. 10 is to balance the loss between classification training and contrastive learning. As is illustrated in Figure 5(c) and 5(d), the performance of SWN seems more likely to be affected by λ on MC tasks than on BC tasks, and the overlapping optimal interval of λ for three datasets is suggested to be $[0.3, 0.7]$. In our experiment, we adopt $\lambda = 0.5$ as default given the fact that SWN obtains the best performance on two datasets under this condition.

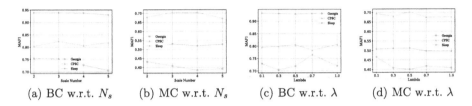

(a) BC w.r.t. N_s (b) MC w.r.t. N_s (c) BC w.r.t. λ (d) MC w.r.t. λ

Fig. 5. SWN's performance of MAF1 w.r.t. the number of scales N_s and the loss balance factor λ on three datasets over BC task and MC task.

Performance W.r.t. ShapeWord Number. Since ShapeWord is defined as the centroid of a cluster of similar shapelets, its number hence depends on the cluster number. In this article, we leverage K-means to generate ShapeWords. To exclude the impact of contrastive learning, we adopt the variant SWN w/o CCLM to conduct the sensitivity experiment regarding the parameter of Shape-Word Number N_{SW}. Graphs in Fig. 6 show that despite the influence of Shape-Word Length L_{SW}, the overlapping optimal interval of three datasets for N_{SW} is approximately $[5, 12]$. In our experiment, we choose the class number as this parameter's default setting for each dataset (i.e. 8 for Sleep, 9 for CPSC and Georgia), which is also consistent with the original definition of shapelets.

Performance W.r.t. ShapeWord Length. It can be observed from Fig. 6 that the performance curve of each dataset shows a trend of first moving up and then going down as L_{SW} increases, which indicates short ShapeWords are more suitable for physiological signal discretization. And the overlapping optimal interval of ShapeWord Length among three datasets is about $[10, 50]$, which is why we choose the three scales $[10, 25, 50]$ for SWN.

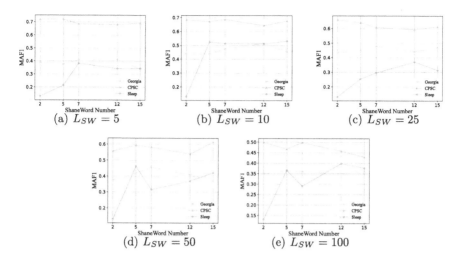

Fig. 6. Performance of SWN w/o CCLM w.r.t. ShapeWord Number in MC tasks under five different settings of ShapeWord Length.

5 Conclusion

In this paper, we proposed ShapeWordNet to deal with physiological signal classification. The uniqueness of our model is to generate prototypes of discriminative local patterns via shapelets and to discretize point-wise raw signals into token sequences of subseries. Given the label sparsity issue, we designed a cross-scale contrastive learning mechanism to assist model optimization, where a multi-scale ShapeSentence transformation strategy was adaptively utilized to augment the data. The experimental results demonstrated both the effectiveness and the interpretability of our method, paving the way for its extension to general time series analysis in the future.

Acknowledgement. This research was partially supported by grant from the National Natural Science Foundation of China (Grant No. 61922073). This work also thanks the support of fundings MAI2022C007 and WK5290000003.

References

1. Ahmad, M.A., Eckert, C., Teredesai, A.: Interpretable machine learning in healthcare. In: Proceedings of the 2018 ACM International Conference on Bioinformatics, Computational Biology, and Health Informatics, pp. 559–560 (2018)
2. Alday, E.A.P., et al.: Classification of 12-lead ecgs: the physionet/computing in cardiology challenge 2020. Physiol. Meas. **41**(12), 124003 (2020)
3. Bagnall, A., Lines, J., Bostrom, A., Large, J., Keogh, E.: The great time series classification bake off: a review and experimental evaluation of recent algorithmic advances. Data Min. Knowl. Disc. **31**(3), 606–660 (2017)

4. Bagnall, A., Lines, J., Hills, J., Bostrom, A.: Time-series classification with cote: the collective of transformation-based ensembles. IEEE Trans. Knowl. Data Eng. **27**(9), 2522–2535 (2015)

5. Bai, S., Kolter, J.Z., Koltun, V.: An empirical evaluation of generic convolutional and recurrent networks for sequence modeling. arXiv:1803.01271 (2018)

6. Che, Z., Purushotham, S., Cho, K., Sontag, D., Liu, Y.: Recurrent neural networks for multivariate time series with missing values. Sci. Rep. **8**(1), 1–12 (2018)

7. Chen, T., Kornblith, S., Norouzi, M., Hinton, G.: A simple framework for contrastive learning of visual representations. In: International Conference on Machine Learning, pp. 1597–1607. PMLR (2020)

8. Cui, Z., Chen, W., Chen, Y.: Multi-scale convolutional neural networks for time series classification. arXiv preprint arXiv:1603.06995 (2016)

9. Dempster, A., Schmidt, D.F., Webb, G.I.: Minirocket: a very fast (almost) deterministic transform for time series classification. In: Proceedings of the 27th ACM SIGKDD Conference on Knowledge Discovery & Data Mining, pp. 248–257 (2021)

10. Ding, H., Trajcevski, G., Scheuermann, P., Wang, X., Keogh, E.: Querying and mining of time series data: experimental comparison of representations and distance measures. Proc. VLDB Endowment **1**(2), 1542–1552 (2008)

11. Faust, O., Hagiwara, Y., Hong, T.J., Lih, O.S., Acharya, U.R.: Deep learning for healthcare applications based on physiological signals: a review. Comput. Methods Programs Biomed. **161**, 1–13 (2018)

12. Franceschi, J.Y., Dieuleveut, A., Jaggi, M.: Unsupervised scalable representation learning for multivariate time series. In: Advances in Neural Information Processing Systems, vol. 32 (2019)

13. Goldberger, A.L., et al.: Physiobank, physiotoolkit, and physionet: components of a new research resource for complex physiologic signals. Circulation **101**(23), e215–e220 (2000)

14. Grabocka, J., Schilling, N., Wistuba, M., Schmidt-Thieme, L.: Learning time-series shapelets. In: Proceedings of the 20th ACM SIGKDD International Conference on Knowledge Discovery and Data Mining, pp. 392–401 (2014)

15. Gutmann, M., Hyvärinen, A.: Noise-contrastive estimation: a new estimation principle for unnormalized statistical models. In: Proceedings of the Thirteenth International Conference on Artificial Intelligence and Statistics, pp. 297–304. JMLR Workshop and Conference Proceedings (2010)

16. Hou, M., et al.: Stock trend prediction with multi-granularity data: a contrastive learning approach with adaptive fusion. In: Proceedings of the 30th ACM International Conference on Information & Knowledge Management, pp. 700–709 (2021)

17. Ismail Fawaz, H., Forestier, G., Weber, J., Idoumghar, L., Muller, P.-A.: Deep learning for time series classification: a review. Data Min. Knowl. Disc. **33**(4), 917–963 (2019). https://doi.org/10.1007/s10618-019-00619-1

18. Karim, F., Majumdar, S., Darabi, H., Chen, S.: LSTM fully convolutional networks for time series classification. IEEE Access **6**, 1662–1669 (2017)

19. Kemp, B., Zwinderman, A., Tuk, B., Kamphuisen, H., Oberye, J.: Analysis of a sleep-dependent neuronal feedback loop: the slow-wave microcontinuity of the EEG. IEEE Trans. Biomed. Eng. **47**(9), 1185–1194 (2000)

20. Kolodner, J.L.: An introduction to case-based reasoning. Artif. Intell. Rev. **6**(1), 3–34 (1992)

21. Krupinski, E.A., Berbaum, K.S., Caldwell, R.T., Schartz, K.M., Kim, J.: Long radiology workdays reduce detection and accommodation accuracy. J. Am. College Radiol. **7**(9), 698–704 (2010)

22. Lines, J., Bagnall, A.: Time series classification with ensembles of elastic distance measures. Data Min. Knowl. Disc. **29**(3), 565–592 (2015)
23. Lines, J., Davis, L.M., Hills, J., Bagnall, A.: A shapelet transform for time series classification. In: Proceedings of the 18th ACM SIGKDD International Conference on Knowledge Discovery and Data Mining, pp. 289–297 (2012)
24. Lipton, Z.C.: The mythos of model interpretability: in machine learning, the concept of interpretability is both important and slippery. Queue **16**(3), 31–57 (2018)
25. Liu, X., et al.: Self-supervised learning: generative or contrastive. IEEE Trans. Knowl. Data Eng. **35**(1), 857–876 (2021)
26. Van der Maaten, L., Hinton, G.: Visualizing data using t-sne. J. Mach. Learn. Res. **9**(11), 2579–2605 (2008)
27. Van den Oord, A., Li, Y., Vinyals, O.: Representation learning with contrastive predictive coding. arXiv e-prints pp. arXiv-1807 (2018)
28. Qiao, F., Wang, P., Wang, W., Wang, B.: An interpretable time series classification approach based on feature clustering. In: International Conference on Database Systems for Advanced Applications, vol. 13246. pp. 664–672. Springer, Cham (2022). https://doi.org/10.1007/978-3-031-00126-0_50
29. Ruder, S.: An overview of multi-task learning in deep neural networks. arXiv preprint arXiv:1706.05098 (2017)
30. Schäfer, P.: The boss is concerned with time series classification in the presence of noise. Data Min. Knowl. Disc. **29**(6), 1505–1530 (2015)
31. Schäfer, P., Leser, U.: Multivariate time series classification with weasel+ muse. arXiv preprint arXiv:1711.11343 (2017)
32. Senin, P., Malinchik, S.: Sax-vsm: interpretable time series classification using sax and vector space model. In: 2013 IEEE 13th International Conference on Data Mining, pp. 1175–1180. IEEE (2013)
33. Ye, L., Keogh, E.: Time series shapelets: a new primitive for data mining. In: Proceedings of the 15th ACM SIGKDD International Conference on Knowledge Discovery and Data Mining, pp. 947–956 (2009)
34. Zhang, X., Gao, Y., Lin, J., Lu, C.T.: Tapnet: multivariate time series classification with attentional prototypical network. In: Proceedings of the AAAI Conference on Artificial Intelligence, vol. 34, pp. 6845–6852 (2020)
35. Zheng, Y., Liu, Q., Chen, E., Ge, Y., Zhao, J.L.: Time series classification using multi-channels deep convolutional neural networks. In: Li, F., Li, G., Hwang, S., Yao, B., Zhang, Z. (eds.) WAIM 2014. LNCS, vol. 8485, pp. 298–310. Springer, Cham (2014). https://doi.org/10.1007/978-3-319-08010-9_33
36. Zou, X., Wang, Z., Li, Q., Sheng, W.: Integration of residual network and convolutional neural network along with various activation functions and global pooling for time series classification. Neurocomputing **367**, 39–45 (2019)

DualFraud: Dual-Target Fraud Detection and Explanation in Supply Chain Finance Across Heterogeneous Graphs

Bin Wu[1], Kuo-Ming Chao[2], and Yinsheng Li[1(✉)]

[1] School of Computer Science, Fudan University, Shanghai, China
{bwu18,liys}@fudan.edu.cn
[2] Department of Computing and Informatics, Bournemouth University, Bournemouth, UK
kchao@bournemouth.ac.uk

Abstract. In supply chain finance, detecting fraudulent enterprises and transactions is crucial to minimize financial loss. Enterprises and transactions have heterogeneous information and fraud labels, thus, leveraging such information well can simultaneously improve fraud detection performance in enterprise and transaction domains. This paper describes our newly proposed multitask learning framework, DualFraud, which detects fraudulent enterprises and transactions with explainability based on heterogeneous graphs in supply chain finance. The main contributions of this work are the proposed framework that can facilitate these two domains to share and enhance learning and modelling capabilities to improve fraud projection. The explainer component is attached to generate rich and meaningful explanations for risk controllers across enterprise and transaction graphs. Experiments on datasets prove the effectiveness of fraud detection and explainability in both enterprises and transactions. The proposed DualFraud outperforms the other methods in the selected criteria.

Keywords: Fraud Detection · Heterogeneous Graph · Graph Neural Network · Multitask · Explainability · Supply Chain

1 Introduction

Supply chain finance builds upon the trust and credit among suppliers, buyers and financiers to reduce financing costs and manage their cash flow effectively. Fraud has a significant negative impact on supply chain participants and financial institutions, but it is not easy to detect. Fraudsters tend to hide them and prevent them from being uncovered. In recent years, graph neural networks have been gaining popularity in financial applications due to their ability to model complex finance networks and capture individual and structural information [13].

Figure 1 illustrates an example of supply chain finance. The fraud enterprise connects to other fraud enterprises through investment relationship and other

© The Author(s), under exclusive license to Springer Nature Switzerland AG 2023
X. Wang et al. (Eds.): DASFAA 2023, LNCS 13946, pp. 370–379, 2023.
https://doi.org/10.1007/978-3-031-30678-5_28

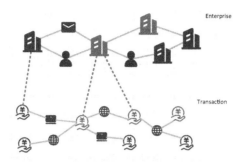

Fig. 1. An example of supply chain finance.

normal enterprises through common persons or emails. The fraud enterprise has conducted several fraud transactions. Transactions are connected through common networks and devices. In this paper, we propose to address three challenges. Firstly, enterprises and transactions have different features and relations. Different graphs need to be designed to capture their characteristics. Besides, business transactions are more dynamic than enterprise records. Decoupling of graphs facilitates the scalability of the subsequent model. Secondly, we need to improve the performance of fraudulent enterprise and transaction detection in the meantime while keeping robustness and flexibility. Fraud detection of enterprises and transactions are two different but related tasks. It is necessary to devise a novel model to improve the performance of both tasks simultaneously by leveraging data richness and diversity in both domains. Also, we need to provide explanations of fraud detection across graphs to facilitate further processes in the business unit. In our multi-graph setting, the model needs to provide explanations from both enterprise and transaction perspectives.

To tackle the above issues, we propose DualFraud, a dual-target fraud detection and explanation framework for enterprises and transactions. DualFraud can effectively predict the legitimacy of enterprises and transactions, and generate explanations on multiple heterogeneous graphs in the multitask setting. We construct two separate heterogeneous graphs of different node types to tackle multi-source and heterogeneous information. We use metapath [11] aggregated GCNs [3] to generate embeddings of enterprises and transactions separately. To enhance both tasks, we consider fraud enterprise and transaction detection together from multitasking learning. We use Att-BLSTM [18] to learn representations of the transaction history. We extend GNNExplainer [15] to multiple graphs by utilizing the attention scores. The contributions of this work are as follows:

1. We present a novel multitask learning framework DualFraud to capture and represent both fraud enterprises and transactions in a unified way.
2. We add explainability into DualFraud to form an end-to-end fraud detection mechanism in supply chain finance.
3. We have conducted experiments on datasets to show its superiority compared with the state-of-the-art models in fraud detection.

2   Literature

For fraud transaction detection, IHGAT [5] constructs a heterogeneous transaction-intention network in e-commerce platforms to leverage the cross-interaction information over transactions and intentions. xFraud [10] constructs a heterogeneous graph to learn expressive representations. For enterprises, ST-GNN [14] addresses the data deficiency problem of financial risk analysis for SMEs by using link prediction and predicts loan default based on a supply chain graph. HAT [17] proposes a heterogeneous-attention-network-based model to facilitate SME bankruptcy prediction. These methods focus on a single task of fraud detection only. For GNN-based fraud detection in the multitask setting, MvMoE [4] proposes a multi-view and multitask learning-based approach to solve credit risk and limit forecasting simultaneously. GraphRfi [16] proposes a framework with GCN and neural random forest to solve robust recommendation and fraudster detection. These methods cannot be used directly for dual-target fraud detection tasks. Also, these methods use only one GNN component, and information cannot be shared and propagated from different perspectives. For explainability, GNNExplainer [15] is an model-agnostic approach for providing interpretable explanations for predictions. xFraud [10] extends it to heterogeneous graphs. Our proposed framework takes this work further to multiple heterogeneous graphs.

3   Methodology

The DualFraud framework is shown in Fig. 2, which consists of two key modules: detector and explainer.

3.1   Detector of DualFraud

The Detector is responsible for learning from labeled data and detecting possible frauds. To fully utilize the labels and features, two heterogeneous graphs of different node types are constructed: a heterogeneous enterprise graph $\mathcal{G}_e = \{\mathcal{V}, E\}$ and a heterogeneous transaction graph $\mathcal{G}_t = \{\mathcal{U}, E'\}$. We adopt metapath and GCN to capture the information of nodes and heterogeneous topological neighborhood structure simultaneously. We assume that initial node representations of each enterprise node $v \in \mathcal{V}$ is h_v^0. The kth layer embedding is:

$$\mathbf{h}_v^k = \sigma \left(\mathbf{W}_k \sum_{u \in N(v) \cup v} \frac{\mathbf{h}_u^{k-1}}{\sqrt{|N(u)||N(v)|}} \right) \tag{1}$$

where $\sigma(\cdot)$ is the activation function, and $\mathbf{W}_k$ is the weight matrix. $N(v)$ is the set of nodes in the neighborhood of node v.

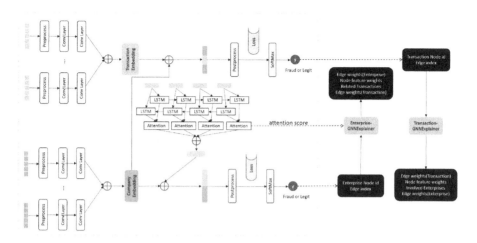

Fig. 2. DualFraud detector and explainer.

We apply two graph convolutional layers, with skip connections, to the node representation. The final representation is $\mathbf{z}_v = \mathbf{h}_v^K$. We select several metapaths manually based on domain experts to capture different relations. We concatenate representations from different metapaths as the embedding of the enterprise node, $\mathbf{z}_v^F = \bigoplus_{i=1}^{M} \mathbf{z}_v^i$, where $\oplus$ denotes the concatenation of two vectors, and M is the number of metapaths. Similarly, we get the embedding $\mathbf{z}_u^F$ of transaction nodes from the heterogeneous transaction graph $\mathcal{G}_t = \{\mathcal{U}, \mathcal{E}'\}$.

We use bidirectional LSTM and Attention to learn the representation of transaction history and concatenate it with node embedding $\mathbf{Z}_v^F$ of the enterprise. The node embeddings of transactions are $\mathbf{Z}_{uT}^F = \{z_{u1}^F, z_{u2}^F, ..., z_{ut}^F\}$ from the transaction graph. For each transaction, the LSTM layer computes the hidden state h_t at time t. The BLSTM contains two LSTM layers in opposite directions. The output of the ith transaction is the element-wise sum of outputs from forward and backward LSTM layers: $h_i = \left[\overrightarrow{h_i} + \overleftarrow{h_i}\right]$. We adopt the attention mechanism to capture transactions that are important to the enterprise. The h_i is fed into a one-layer MLP to get the hidden representation u_i. We can get the importance weight α_i through a Softmax function.

$$u_i = \tanh\left(W_a h_i + b_a\right),$$
$$\alpha_i = \frac{\exp\left(u_i\right)}{\sum_i \exp\left(u_i\right)}, \tag{2}$$
$$\mathbf{z}_v^A = \sum_i \alpha_i h_i$$

The concatenated embeddings of enterprises are defined as: $\mathbf{z}_v^{new} = \mathbf{z}_v^F \oplus \mathbf{z}_v^A$. The attention scores of historical transactions of enterprise v are defined as $\mathbf{Att}_v = \{\alpha_1, \alpha_2, ..., \alpha_n\}$, where n is the number of transactions of enterprise v.

For transactions, we concatenate embeddings of the sender $\mathbf{z}_{vs}^F$ and receiver $\mathbf{z}_{vr}^F$ to get the embeddings: $\mathbf{z}_u^{new} = \mathbf{z}_u^F \oplus \mathbf{z}_{vs}^F \oplus \mathbf{z}_{vr}^F$.

We use the standard supervised cross-entropy loss to train the model. Instead of training these two tasks separately, we combine their losses and jointly minimize the following loss function:

$$\mathcal{L} = \lambda\mathcal{L}_{enterprise} + (1 - \lambda)\mathcal{L}_{transaction} \tag{3}$$

where λ is a hyper-parameter to balance the effect of both parts.

3.2   Explainer of DualFraud

We add explainability to DualFraud by extending GNNExplainer [15] to multiple heterogeneous graphs. The model's prediction of node v is given by $\hat{y} = \Phi(G_c(v), X_c(v), I_c(v))$, where Φ is the model trained in the detection process. The prediction is determined by graph structural information $G_c(v)$, node feature information $X_c(v)$, and information from the other graph $I_c(v)$. It implies that we need to consider these three aspects to explain $\hat{y}$. Formally, the explainer component generates explanations for the prediction

$$\hat{y} = \left(G_{vS}, X_{vS}^F, \bigcup_{i\in k} G_{uSi}, \bigcup_{i\in k} X_{uSi}^F \right) \tag{4}$$

where G_{vS} is a small subgraph of the target node, G_{uS} are subgraphs from the other graph of related nodes. X_{vS} and X_{uS} are the associated feature of G_{vS} and G_{uS}. X_{vS}^F and X_{uS}^F are small subsets of node features (masked out by the mask F, i.e., $X_{vS}^F = \{x_j^F \mid v_j \in G_{vS}\}, X_{uS}^F = \{x_j^F \mid v_j \in G_{uS}\}$).

For enterprises, we choose the top $\mathbf{k}$ transactions of attention scores $\mathbf{Att}_v$ generated by the Att-BLSTM component since attention reflects the importance of this transaction to the enterprise. For transactions, we apply the explainer to the involved enterprises. The subgraphs of enterprises and transactions are linked together as the final explanation result.

4   Experiment

4.1   Datasets and Comparing Methods

This research adopts two different datasets, one synthetic and one real-world. The statistical information describing datasets is shown in Table 1.

Synthetic Dataset. We construct synthetic datasets referring to the design of experimental datasets of GNNExplainer [15]. For enterprise relationships, we construct a Tree-Cycles dataset. It starts with a base 9-level balanced binary tree and 20 six-node cycle motifs, which are attached to random nodes of the base graph. The resulting graph is further perturbed by adding 0.1N random edges. The tree-like structure is similar to the relationships of enterprises in supply

Table 1. Statistical information of datasets.

	Nodes (Fraud%)	#Features	Relation	#Edges
Synthetic Enterprise	1143 (10.5%)	8	C-C	1181
Synthetic Transaction	4575 (10.5%)	8	T-T	4749
Real-world Enterprise	13489 (26.4%)	89	C-I-C C-C C-M-C ALL	53874 15908 139413 209195
Real-world Transaction	50000 (1.2%)	23	T-A-T	206666

chain. Nodes in the base graph are labeled with 0 to represent normal enterprises. The structure of cycle motifs is similar to the relationship in self-financing fraud, one of the common types of supply chain accounts-receivable frauds[1]. Nodes in the cycle motifs are labeled with 1 to represent fraud enterprises. Nodes in the same class have normally distributed feature vectors. Similarly, for transaction relationships, we construct a base 11-level balanced binary tree and 80 six-node cycle motifs. Every fraud transaction is mapped to two random fraud enterprises, while normal transactions are mapped to random enterprises.

Real-World Dataset. We use public real-world datasets of HAT [17] and BankSim [8] to construct our experimental dataset. The SME's Bankruptcy Dataset of HAT contains the board member network and shareholder network for 13489 companies, in which 3563(26.4%) companies go bankrupt. Specifically, 1000 companies are selected firstly, which located in a south-eastern city in China and went bankrupt in 2018. Then, all the shareholders and board members for these enterprises are collected and this process is repeated for the collected enterprises twice. BankSim is based on a sample of aggregated transactional data provided by a bank. Every transaction in the BankSim dataset has the label of fraud(1.2%) or normal(98.8%). We define entities in BankSim with at least one fraudulent transaction as fraudulent and match them with fraudulent enterprises in the HAT data. The rest of the entities are matched with normal enterprises.

We evaluate the performance by comparing with the following methods:

- GraphSage [2]: A popular inductive GNN framework generates embeddings by sampling and aggregating features from a node's local neighborhood.
- GEM [7]: A heterogeneous GNN approach for detecting malicious accounts which adopts attention to learn the importance of different types of nodes.
- SemiGNN [12]: A semi-supervised graph attentive network for financial fraud detection that utilizes the multi-view labeled and unlabeled data.
- GraphConsis [6]: A model that tackles inconsistency problems in applying GNNs in fraud detection problems.
- RioGNN [9]: A SOTA fraud detection model in reinforced, recursive and flexible neighborhood selection guided multi-relational GNN architecture.
- DualFraud-S: Version with DualFraud's embedding sharing module removed.

Due to the class imbalance in fraud detection, we select Macro-F1 and AUC to evaluate the performance of the models.

---
[1] http://www.nifd.cn/ResearchComment/Details/2582.

4.2 Experimental Settings and Implementation

In DualFraud, we set the hidden embedding size to 16, the number of graph convolution layers to 2, the number of recurrent layers to 1, and the dimension of attention layer to 5. For model optimization, we use Adam optimizer and set the learning rate to 0.0003. The codes for GraphSage, GEM, SemiGNN, and GraphConsis are from DGFraud-TF2 [1]. The codes for other models come from respective authors' implementations. For GraphSage which adopts homogeneous graphs, the edges of different types are treated as the same. For the datasets, we distribute them according to the ratio 68:12:20 for training, validation, and test set, respectively. The source code of our model and constructed datasets are available[2]. All experiments are conducted by Python 3.6, NVIDIA GeForce GTX 1080 and AMD Ryzen 5 2600.

4.3 Results

Fraud Detection. We present experimental results in Table 2 of the fraud enterprise and transaction detection tasks on the synthetic and real-world datasets. In both synthetic and real-world datasets, DualFraud outperforms all the compared methods in F1 and AUC. In the real-world dataset, SemiGNN cannot complete the experiment because of out-of-memory. For fraud transaction detection in real-world dataset, multiple methods including GraphSage and GEM cannot perform discrimination effectively due to the severe class imbalance problem. They predict all transactions in the test dataset as normal. While DualFraud-S performs poorly compared with other methods, DualFraud improves the performance by a large margin. It demonstrates the effectiveness of sharing embeddings to exploit both enterprise and transaction information. By sharing information through the multitask framework, information from other networks is used as side information to enhance the features of the current task. This increases the difference of features of samples from different categories, which is more effective in real-world transaction dataset that has severe class imbalance. On the task of fraud enterprise detection in real-world dataset, DualFraud is only marginally better than the baselines compared with other tasks. The results are consistent with the experimental results in the paper [17]. This may be due to the fact that the features and relationships built by the dataset are not effective in classification task.

Parameter Study. For the multitask setting, we study the parameter sensitivity of weights of different tasks in Fig. 3. We run experiments on real-world dataset using different λ parameters in Eq. 3. We can notice that the larger the lambda value,

---

[2] https://github.com/anonymousDualFraud/DualFraud.

Table 2. Comparison of methods on Synthetic(S.) dataset and Real-world(R.) dataset.

	S. Enterprise		S. Transaction		R. Enterprise		R. Transaction	
	F1	AUC	F1	AUC	F1	AUC	F1	AUC
GraphSage	0.6790	0.6250	0.6286	0.6088	0.6465	0.6303	0.4971	0.5000
GEM	0.9069	0.9069	0.8613	0.8947	0.6436	0.6354	0.4971	0.5000
SemiGNN	0.6216	0.7985	0.4281	0.6137	OOM	OOM	OOM	OOM
GraphConsis	0.5398	0.5343	0.5102	0.5150	0.6498	0.6338	0.5107	0.5074
RioGNN	0.6177	0.6524	0.5566	0.5667	0.6659	0.6612	0.5509	0.5707
DualFraud-S	0.8743	0.8468	0.8958	0.8867	0.4595	0.5122	0.8841	0.8581
DualFraud	**0.9758**	**0.9583**	**0.9397**	**0.9458**	**0.6974**	**0.7492**	**0.8891**	**0.8792**

i.e., the higher the weight of the task, the smaller the loss value in task 1. Since the method performs much better on task 2 than on task 1, different lamda parameters have no significant effect on task 2. Thus, the weights can be adjusted according to the performance of different tasks in practical use.

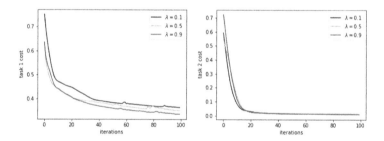

Fig. 3. Training loss in different λ parameter.

Explainability. We present case studies on flagging frauds in Fig. 4. Different shapes and colored edges indicate different types of nodes and relations respectively. The edges' thickness represents the relations' importance in the prediction. The thicker the more important. We use ground truth labels as the color of nodes of enterprises and transactions. If the edge between the red node and the target node is bolded, it indicates that the explainer provides a valid explanation. If all the provided transaction nodes connected to the target enterprise node are red, it indicates that the explainer found the suspicious transactions accurately.

In Fig. 4(a), we present an explanation result for a fraud enterprise(node *e1023*) in the synthetic dataset. The explainer catches most of the critical edges with other fraud enterprises in the Cycle motif of the target node. Also, the explainer identifies the enterprise's fraud transactions(node *t4118* and node *t4200*) with the highest attention score. Then the explainer shows the most suspicious edges from the perspective of transactions. The subgraph provides good explanations from enterprise and transaction perspectives.

In Fig. 4(b), we present an explanation result for a fraud enterprise(node *e12527*) in the real-world dataset. The enterprise *e12527* has the same share-holders as enterprises *e6336* and *e10047*, both of which are fraudulent enterprises. The enterprise *e12527* has the same board members as enterprises *e2934*, *e5440*, *e1668*, *e5915* and *e5495*, in which only enterprise *e1668* is fraudulent. Although enterprise *e5495* is normal, it has the same board members as several fraudulent enterprises including *e2907*, *e7788*, *e11416* and *e8514*. The transaction(node *t4681*) with the highest attention score in the history of transactions of enterprise *e12527* is a fraud. This transaction is related to multiple fraudulent transactions. This demonstrates the ability of DualFraud to identify suspicious enterprises and transactions, even when the proportion of normal and fraud in the neighborhood is similar.

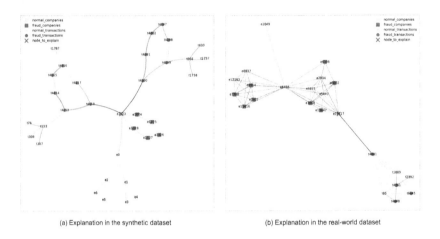

(a) Explanation in the synthetic dataset   (b) Explanation in the real-world dataset

Fig. 4. Explanation of fraud enterprises and transactions.

5   Conclusion

The paper proposes an end-to-end GNN framework called DualFraud in supply chain financial fraud detection. The experimental results demonstrate the effectiveness and practicality of DualFraud on fraud detection and explanation. For future work, when faced with data from more sources, the model needs to handle data and provide explanations from more perspectives. Also, it needs to address conflicts between data from different sources. In transactions, using dynamic graph neural networks to deal with the transaction graph in the model is worthwhile for further investigation.

Acknowledgements. The work has been supported by the National Key Research and Development Program of China (Grant No. 2019YFB1404904).

References

1. Dou, Y., Liu, Z., Sun, L., Deng, Y., Peng, H., Yu, P.S.: Enhancing graph neural network-based fraud detectors against camouflaged fraudsters. In: CIKM, pp. 315–324 (2020). https://doi.org/10.1145/3340531.3411903
2. Hamilton, W.L., Ying, Z., Leskovec, J.: Inductive representation learning on large graphs. In: NeurIPS, pp. 1024–1034 (2017)
3. Kipf, T.N., Welling, M.: Semi-supervised classification with graph convolutional networks. In: ICLR (2017)
4. Liang, T., et al.: Credit risk and limits forecasting in e-commerce consumer lending service via multi-view-aware mixture-of-experts nets. In: WSDM, pp. 229–237 (2021)
5. Liu, C., Sun, L., Ao, X., Feng, J., He, Q., Yang, H.: Intention-aware heterogeneous graph attention networks for fraud transactions detection. In: KDD, pp. 3280–3288 (2021). https://doi.org/10.1145/3447548.3467142
6. Liu, Z., Dou, Y., Yu, P.S., Deng, Y., Peng, H.: Alleviating the inconsistency problem of applying graph neural network to fraud detection. In: SIGIR, pp. 1569–1572 (2020). https://doi.org/10.1145/3397271.3401253
7. Liu, Z., Chen, C., Yang, X., Zhou, J., Li, X., Song, L.: Heterogeneous graph neural networks for malicious account detection. In: CIKM, pp. 2077–2085 (2018). https://doi.org/10.1145/3269206.3272010
8. Lopez-Rojas, E.A., Axelsson, S.: Banksim: A bank payments simulator for fraud detection research. In: EMSS, pp. 144–152 (2014)
9. Peng, H., Zhang, R., Dou, Y., Yang, R., Zhang, J., Yu, P.S.: Reinforced neighborhood selection guided multi-relational graph neural networks. TOIS **40**(4), 1–46 (2021)
10. Rao, S.X., et al.: xFraud: explainable fraud transaction detection. In: Proceedings of VLDB Endow, vol. 15, no. 3, pp. 427–436 (2021). https://doi.org/10.14778/3494124.3494128
11. Sun, Y., Han, J., Yan, X., Yu, P.S., Wu, T.: Pathsim: meta path-based top-k similarity search in heterogeneous information networks. Proc. VLDB Endow. **4**(11), 992–1003 (2011)
12. Wang, D., et al.: A semi-supervised graph attentive network for financial fraud detection. In: ICDM, pp. 598–607 (2019). https://doi.org/10.1109/ICDM.2019.00070
13. Wang, J., Zhang, S., Xiao, Y., Song, R.: A review on graph neural network methods in financial applications. arXiv preprint arXiv:2111.15367 (2021)
14. Yang, S., et al.: Financial risk analysis for smes with graph-based supply chain mining. In: IJCAI, pp. 4661–4667 (2020)
15. Ying, Z., Bourgeois, D., You, J., Zitnik, M., Leskovec, J.: Gnnexplainer: generating explanations for graph neural networks. In: NeurIPS, pp. 9240–9251 (2019)
16. Zhang, S., Yin, H., Chen, T., Nguyen, Q.V.H., Huang, Z., Cui, L.: Gcn-based user representation learning for unifying robust recommendation and fraudster detection. In: SIGIR, pp. 689–698 (2020). https://doi.org/10.1145/3397271.3401165
17. Zheng, Y., Lee, V.C.S., Wu, Z., Pan, S.: Heterogeneous graph attention network for small and medium-sized enterprises bankruptcy prediction. In: KDD, pp. 140–151 (2021). https://doi.org/10.1007/978-3-030-75762-5_12
18. Zhou, P., et al.: Attention-based bidirectional long short-term memory networks for relation classification. In: ACL, pp. 207–212 (2016)

Process Drift Detection in Event Logs with Graph Convolutional Networks

Leilei Lin[1,4], Yumeng Jin[1], Lijie Wen[3,4], Wenlong Chen[2(✉)], Ying Di[1], Yusong Xu[3], and Jianmin Wang[3,4]

[1] School of Management, Capital Normal University, Beijing, China
leilei_lin@cnu.edu.cn
[2] Information Engineering College, Capital Normal University, Beijing, China
chenwenlong@cnu.edu.cn
[3] School of Software, Tsinghua University, Beijing, China
{wenlj,xys20,jimwang}@tsinghua.edu.cn
[4] Beijing Key Laboratory of Industrial Big Data System and Application, Beijing, China

Abstract. The only constant is change. Drift detection in process mining is a family of methods to detect changes by analyzing event logs to ensure the accuracy and reliability of business processes in process-aware information systems (e.g., ERP systems). However, artificial feature selection is still a mountain to climb in existing methods, which requires high domain knowledge for users. In this paper, we propose a novel approach by using **Graph** convolutional networks for **Drifts Detection** in the event log, we name it **GDD**. Specifically, 1) we transform event sequences into two directed graphs by using two consecutive time windows, and construct the line graphs for the directed graphs to capture the orders between different activities; 2) we use graph convolutional networks to capture the features in these graphs, and augment the original graphs with virtual nodes to represent the latent aspects of the graphs; 3) we calculate the distances between virtual nodes, and use the K-means algorithm to find the outliers that are considered as candidate change points. Then, a filter mechanism is used to confirm the actual change points. The experiments on simulated event logs and real-life event logs confirmed the improvements of GDD compared with the baselines.

Keywords: Graph neural networks · Line graph · Process mining · Process changes · Feature propagation

1 Introduction

Process-Aware Information Systems (PAISs) play an important role in business process management (BPM), while process mining serves as a bridge between data mining and business process management [1]. As shown in Fig. 1a, process mining aims to extract valuable information from event logs generated by

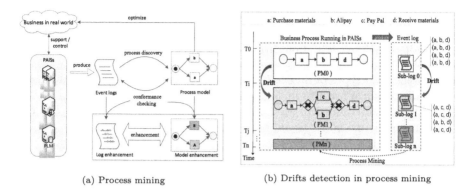

(a) Process mining   (b) Drifts detection in process mining

Fig. 1. An example for illustrating the concept drift in process mining

PAISs for configuring and optimizing the ongoing business process. To achieve the aim of process optimization, many classical techniques in process mining have been proposed to support three scenarios: process discovery [2,3] (i.e., extracting process model from an event log without using any apriori information), conformance checking [4] (i.e., monitoring deviations by comparing models and event logs) and process enhancement [5] (i.e., extending or improving an existing model using additional information in event logs). However, these techniques assume processes to be in a steady state. Bose et al. [6] analyzed processes in more than 100 organizations and found that it is very unrealistic to assume that the process being studied is in a steady state. As we know, business processes are continuously evolving in order to adapt to changing circumstances. Some changes are planned, but others occur unexpectedly (e.g. the COVID-19 outbreak) and may remain unnoticed by a majority of stakeholders.

This motivated academics to devise methods and tools that enable business analysts to identify process changes as early as possible. Business process drift detection [7–12] is a family of methods to detect changes by analyzing the event log to ensure the accuracy and reliability of process mining. Figure 1b uses a procurement process as an example to show us the phenomenon of concept drift in process mining. As the business expands, the process PM_0 evolves into the process PM_1 at the time point T_i. Therefore, there is also a change point T_i in the event log generated by the running process. The goal of drift detection is to accurately identify this change point in the log and then divide the log into sub-logs to ensure the accuracy of the process mining.

Still, state-of-the-art methods in this area are limited by artificial feature selection, which will lead to two problems: 1) increase the burden on users. For example, Bose et al. [6,7] propose global features and local features, for users that are asked to identify the features to be used for drift detection, implying that they have some a-priori knowledge of the possible nature of the drift; 2) low accuracy. When users decide to adopt a certain feature to detect changes in the event log, it also means that changes caused by other features will be ignored. To bridge these gaps, we propose GDD (using **G**raph convolutional networks

[14] for **Drift Detection** in the event log), which is able to implicitly capture the changing features in the event log by multi-layer convolution as much as possible. The main contributions are summarized as follows.

- We design two types of directed graphs, activity graph and order graph, for capturing change features in the logs.
- A virtual node is added to each directed graph for GCN operation. Then the features of virtual nodes would be treated as the features of directed graphs.
- Inspired by the Feature Fusion that can improve classification results, we similarly aggregate the distances of two kinds of virtual nodes (i.e., activity level and order level) for improving drift detection.
- The K-means algorithm is adopted to find the outliers that are considered as candidate change points. Then, a filtering mechanism is used to locate the real change points, which guarantees the robustness of GDD.

In the experiments, we evaluate the performance of GDD on different datasets. Empirical results show that the proposed method achieved state-of-the-art drift detection in process mining.

2   Related Work

Business processes are prone to change in response to various factors. For this reason, before mining the process model, the moments of change must be found in the event log. Drift detection originated in the field of data mining (a.k.a. concept drift) [13], but is a relatively young research topic that has gained popularity in process mining in the last 10 years.

Bose et al. [6,7] used the strategy of statistical hypothesis testing to determine whether there is a difference between two populations in the logs. The authors define two kinds of features for the user to choose: global features and local features, while global features include relation type count (RC) and relation entropy (RE), and local features include window count and J-Measure. Then, the Hotelling T2 test and K-S test are adopted respectively to detect change points in the event log. Instead of defining a large number of features in [6], a *Run* feature is defined in [8]. The *Run* feature captures accurately concurrency relation in the log. The authors also used hypothesis testing to monitor the change in the distribution of *Run* for detecting drift in the logs. Ostovar et al. [9] substitute the *Alpha* feature for the *Run* feature, because *Alpha* not only captures the concurrent relation in the log, but also the loop relation. Kumar et al. [18] proposed an *Event correlation* feature for drift detection, but the accuracy is low. In [10], the authors used A* algorithm to find the change points. However, this strategy requires expert knowledge to specify the editing cost.

Zheng et al. [11] proposed a new approach to detect drifts based on the DBSCAN algorithm. But, this method is sensitive to noise. The bidirectional testing strategy is used to overcome the noise problem in [12]. Carmona et al. [17] constructed the inequality for detecting drifts based on abstract interpretation.

The authors skillfully transformed the drift detection problem into a numerical computation problem, but one drawback is that drift detection is discontinuous.

Lin et al. [16] detected the change points based on local completeness, namely LCDD. The accuracy of this method is very high, but it can not deal with the noise in the log. Maggi et al. [15] mined the declarative process model based on linear temporal logic, and then the sliding window is used to determine whether the model needs to be updated. When the model is updated, a drift point is found. In [19], the authors brilliantly developed a visualization approach to find out the change points in the logs. Unfortunately, the method requires expert knowledge to determine exactly where the drift occurs.

Table 1. Evaluation capabilities of different approaches

Literature	Strategy	Feature	Evaluation Capability		
			No expert	High accuracy	Robustness
[6,7]	Hypothesis testing	Global and local features	✗	✗	✓
[8]	Hypothesis testing	Run	✓	✗	✓
[9]	Hypothesis testing	Alpha	✓	✗	✓
[10]	A* algorithm	Edit Distance	✗	✗	✓
[11]	DBSCAN	Direct succession	✓	✗	✗
[12]	Bidirectional testing	Directly follows	✓	✗	✓
[15]	Lossy counting	Frequency	✓	✗	✗
[16]	local completetness	Direct succession	✓	✓	✗
[17]	Abstract interpretation	Parikh vector	✓	✗	✗
[18]	Hypothesis testing	Event correlation	✓	✗	✗
[19]	visualization+clustering	Declare constraints	✗	✗	✓
This paper	GCN	Implicit in the graphs	✓	✓	✓

As already remarked, the existing methods have some limitations. Hence, we propose a novel approach for detecting change points based on graph convolutional networks, we name it GDD. As shown in Table 1, we highlight the differences between GDD and the other methods.

3 Preliminaries

In this section, a number of basic definitions will be introduced, and these definitions are borrowed from [8,16].

Definition 1 (Event log). *Let $\mathcal{T}$ be a finite set of activities. An event log $\mathcal{L}$ over $\mathcal{T}$ is defined as $\mathcal{L} = (\mathcal{E}, \mathcal{C}, \gamma, \tau, \succ)$, where $\mathcal{E}$ is a set of events, $\mathcal{C}$ is a set of case identifiers, $\gamma : \mathcal{E} \to \mathcal{C}$ a surjective function relating events to cases, $\tau : \mathcal{E} \to \mathcal{T}$ a function relating events to activities and $\succ \subseteq \mathcal{E} \times \mathcal{E}$ a total order on events.*

Definition 2 (Trace). *Let T be a finite set of activities and $\mathcal{L} = (\mathcal{E}, \mathcal{C}, \gamma, \tau, \succ)$ a log over T. For all cases $c \in \mathcal{C}$, we define the trace $\sigma_c = \langle e_1, e_2, \cdots, e_n \rangle$ as an event sequence over the event set $\mathcal{E}(\sigma_c) = \{e \in \mathcal{E} | \gamma(e) = c\}$ as all events relating to case c, where n is the number of events in $\mathcal{E}(\sigma_c)$, $e_i \in \mathcal{E}(\sigma_c)$, $i \in \{1, 2, \cdots, n-1\}$ and if $1 \leq i < n$, it holds $e_i \succ e_{i+1}$. The length of the trace σ_c is defined as the number of elements in $\mathcal{E}(\sigma_c)$, which is denoted as $|\sigma_c|$.*

Definitions 1 and 2 state that an event log consists of a set of traces, each capturing the sequence of events for a given case of the process ordered by timestamp. Sometimes a simple event log will be written as a multi-set of traces, which ignores the orders of traces [1]. However, we do not represent event logs as multisets, because the order of each trace would be treated as a time point in drift detection [6,9,16]. It is reasonable to regard the order of trace as a time point, while each trace in logs has a unique ID and start-time. As shown in Fig. 1b, we assume that an event log $\mathcal{L} = [\sigma_0 = \langle a, b, d \rangle, \sigma_1 = \langle a, b, d \rangle, \sigma_2 = \langle a, b, d \rangle, \sigma_3 = \langle a, b, d \rangle, \sigma_4 = \langle a, c, d \rangle, \sigma_5 = \langle a, c, d \rangle, \sigma_6 = \langle a, b, d \rangle, \sigma_7 = \langle a, c, d \rangle]$ has 8 traces. Then, we want to find out the change point T_i (i.e., σ_4) for dividing the event log $\mathcal{L}$ into two sub-logs: one is $\mathcal{L}_0 = [\sigma_0 = \langle a, b, d \rangle, \sigma_1 = \langle a, b, d \rangle, \sigma_2 = \langle a, b, d \rangle, \sigma_3 = \langle a, b, d \rangle]$ and another is $\mathcal{L}_1 = [\sigma_4 = \langle a, c, d \rangle, \sigma_5 = \langle a, c, d \rangle, \sigma_6 = \langle a, b, d \rangle, \sigma_7 = \langle a, c, d \rangle]$.

4   Methodology

To solve the drift detection issue for event logs, we propose GDD, a fully automated method without artificial feature selection. In this section, we will explain the details of our method.

Figure 2 shows an overview of our proposed method. Initially, we convert the traces in the event log into two directed graphs by using two consecutive time windows that would continuously move through the log. Each time the windows move, the directed graphs are updated. Then, we apply the graph convolutional networks for node embedding. After embedding, the high-dimension vector of each node contains the information of its neighbors. A virtual node is added in each directed graph to capture the features of the entire directed graph by connecting to all nodes. Next, we calculate the distances of virtual nodes representing the activity graph and order graph respectively. Finally, the K-means algorithm and a fliter mechanism are used to comfirm the actual change points.

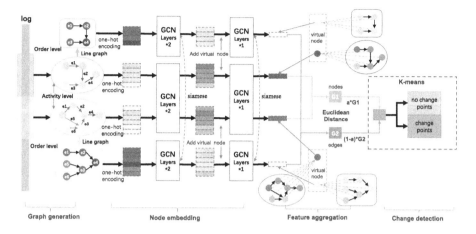

Fig. 2. Overview of our technique named GDD for drift detection

4.1   Graph Generation

From a statistical viewpoint, the problem of business process drift detection can be formulated as follows: *identify a time point when there is a statistically significant difference between the observed process behavior before and after this point* [8]. So, various features are manually selected for hypothesis testing in Fig. 1. However, it is difficult to fully take into account the change features in the event log among these selections. We believe the intuition behind concept drift is that a change happened in a business process must be caused by the changing of activities or the changing of orders among activities. In this setting, a more suitable approach for drift detection is to analyze the differents of the activities and the orders between two sub-logs obtained from consecutive time windows. So, if we can determine that two directed graphs are different where nodes and edges represent activities and orders respectively, a change point can be detected. It is worth emphasizing that this paper, like the references in Table 1, focuses on the control flow perspective while many perspectives can be studied in the business processes (e.g., resource perspective).

Definition 3 (Activity Graph). *Let T be a finite set of activities in a sub-log $\mathcal{L}$. We define a directed graph $\mathcal{AG} = (\mathcal{V}, E)$ be an activity graph, $\mathcal{V}$ represents the set of nodes, E represents the set of ordered pairs of nodes, $E = \mathcal{V} \times \mathcal{V}$, where (1) $\forall n \in T$, $\exists n \in \mathcal{V}$, and (2) if there is an oder $e_i \succ e_{i+1}$ occurring in a trace σ_c, $\tau(e_i) = a, \tau(e_{i+1}) = b$ and $a, b \in \mathcal{V}$ then $\exists (a, b) \in E$.*

Definition 4 (Order Graph). *Let $\mathcal{AG} = (\mathcal{V}, E)$ be an activity graph, we define a line graph $\mathcal{OG} = (\widetilde{\mathcal{V}}, \widetilde{E},)$ be an order graph, where (1) each node $n \in \widetilde{\mathcal{V}}$ represents an edge in E, and (2) each edge $e = (a, b) \in \widetilde{E}$ and $a, b \in \widetilde{\mathcal{V}}$ represents two directed edges $(a, x), (x, b)$ in E and $a, x, b \in \mathcal{V}$.*

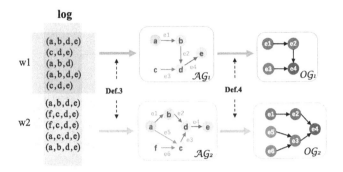

Fig. 3. Activity Graph and Order Graph

To capture the features of activities in the log, we convert the log into a directed graph. Definition 3 states an activity graph $\mathcal{AG}$, where the nodes and edges in $\mathcal{AG}$ represent the activities and orders in the log respectively. Now, the problem of drift detection is transformed into the problem of directed graph convolution. However, if the convolution operation is performed only on the activity graph, it is easy to ignore the information on the edges in $\mathcal{AG}$. Because the graph convolution operation just updates the features of nodes in $\mathcal{AG}$. Inspired by the CensNet [20] that can improve classification performance by convolution with edge-node switching graph. Similarly, we construct a line graph, called $\mathcal{OG}$ in Definition 4, to capture order features in the event log.

An example is given in Fig. 3 to illustrate the conversion process from event logs to $\mathcal{AG}$ and $\mathcal{OG}$. As shown in Fig. 3, we first apply two windows, $w1$ and $w2$, to obtain two sub-logs from the original log. The size of windows is a hyperparameter and $w1 = w2 = 5$ in Fig. 3. Then, the nodes and edges are extracted from the sub-logs to construct two activity graphs based on Definition 3. Specifically, there are five nodes $\mathcal{V} = \{a, b, c, d, e\}$ and four edges $E = \{(a, b), (b, d), (c, d), (d, e)\}$ extracting for the window $w1$ and six nodes $\mathcal{V} = \{a, b, c, d, e, f\}$ and six edges $E = \{(a, b), (b, d), (c, d), (d, e), (a, c), (f, c)\}$ extracting for the window $w2$. Finally, the order graphs are built based on Definition 4, where the nodes in $\mathcal{OG}$ represent the edges in $\mathcal{AG}$.

4.2   Node Embedding

According to Definitions 3 and 4, we have constructed the directed graphs (i.e., $\mathcal{AG}$ and $\mathcal{OG}$) from the event log. After that, some methods have to be used for calculating the differences between these graphs. The traditional graph traversal algorithms, like Depth-First-Search and DeepWalk [21], face the following problems when comparing the difference between two directed graphs: (1) inefficiency. Their efficiency is an obvious drawback when dealing with very complex structures; (2) low accuracy. The traversal algorithms can only compare local features (nodes or second-order proximity) between two directed graphs, they cannot observe the difference between two graphs from the overall topology. At the same time, graph neural networks [14] have shown extraordinary talents in graph classification in recent years. So, we decide to map the features of the directed graph into high dimensional vectors by applying the multi-layer Graph Convolutional Network [14], where the layer-wise propagation rule is following:

$$H^{(l+1)} = \sigma\left(\widetilde{D}^{-\frac{1}{2}} \widetilde{A} \widetilde{D}^{-\frac{1}{2}} H^{(l)} W^{(l)}\right). \tag{1}$$

Here, $\widetilde{A} = A + I_N$ is the adjacency matrix of the directed graph (i.e, $\mathcal{AG}$ or $\mathcal{OG}$) with added self-connections. I_N is the identity matrix and $\widetilde{D}_{ii} = \sum_j \widetilde{A}_{ij}$. $W^{(l)}$ is a layer-specific trainable weight matrix, $\sigma(\cdot)$ represents an activation function defaulted to the $ReLU(\cdot)$. $H^{(l)} \in \mathbb{Z}^{N \times D}$ is the matrix of activations in the lth layer, while $H^{(0)} = X$.

We know that the matrix I_N is to prevent nodes from losing their own information in the process of feature propagation, while the matrix $\widetilde{D}$ can ensure the normalization of features. But, the trainable weight matrix $W^{(l)}$ is randomly generated, which will seriously impress the similarity computation of the two directed graphs. Hence, we apply the **siamese mechanism** [22] to construct two GCN models with tied weights for $\mathcal{AG}$ convolution as well as $\mathcal{OG}$ convolution. For example, there are two GCN models M and N and two weight matrices $W^M =< w_1^M, w_2^M, \cdots, w_m^M >$ and $W^N =< w_1^N, w_2^N, \cdots, w_n^N >$ which are used to perform convolution in two graphs inputs, we solely focus on siamese architectures with tied weights such that:

$$\begin{cases} W \equiv N \\ m \equiv n \\ w_i^M \equiv w_i^N, \forall i \in [1, 2, \cdots, m] \end{cases}$$

Before graph convolution, each node in $\mathcal{AG}$ and $\mathcal{OG}$ will be assigned a high-dimensional vector based on one-hot encoding, **One_Hot** $(v) \in \mathbb{R}^k$ and k represents the number of the set of nodes in two graphs. So, the initial feature of a directed graph is $X = \{v_1, v_2, \cdots, v_k\} = H^{(0)}$. Then, a two-layer GCN is built for node embedding with an adjacency matrix A. We first calculate $\hat{A} = \widetilde{D}^{-\frac{1}{2}} \widetilde{A} \widetilde{D}^{-\frac{1}{2}}$ in a pre-processing step, and the forward model then takes the simple form:

$$Z = f(X, A) = softmax\left(\hat{A} ReLU\left(\hat{A} X W^{(0)}\right) W^{(1)}\right), \tag{2}$$

where $W^{(0)} \in \mathbb{R}^k$ is an input-to-hidden weight matrix for a hidden layer with k feature maps. The softmax activation function, defined as $softmax(x_i) = \frac{1}{\sum_i exp(x_i)} exp(x_i)$, is applied row-wise.

After convolution by two-layer GCNs, the vector of each node in $\mathcal{AG}$ and $\mathcal{OG}$ encompasses the information of its neighboring nodes. In order to accurately portray the topological information of each directed graph, we add a **virtual node** to each directed graph. The idea is inspired by [23] which has shown the virtual node can facilitate graph representation learning. Therefore, four virtual nodes are added to $\mathcal{AG}_1$, $\mathcal{OG}_1$, $\mathcal{AG}_2$ and $\mathcal{OG}_2$ respectively after two-layer GCNs. Then, all existing nodes in each of the graphs will be connected to the virtual node to form an expanded graph. The virtual node represents the latent aspects of the expanded graph, which are not immediately available from the attributes and local connectivity structures. Next, we would use a one-layer GCN for the expanded graphs to obtain the vector of each virtual node. So, The representation of the virtual node is then the representation of the entire graph.

4.3   Feature Aggregation

After graph generation and the node embedding, we now obtain the vectors of four virtual nodes $< v_1^{\mathcal{A}}, v_2^{\mathcal{A}}, v_1^{\mathcal{O}}, v_2^{\mathcal{O}} >, v \in \mathbb{R}^{k+1}$ to represent all the features of the four graphs. Among them, $v_1^{\mathcal{A}}$ and $v_2^{\mathcal{A}}$ represent the features of the activity graphs $\mathcal{AG}_1$ and $\mathcal{AG}_2$, $v_1^{\mathcal{O}}$ and $v_2^{\mathcal{O}}$ represent the features of the order graphs $\mathcal{OG}_1$ and $\mathcal{OG}_2$, respectively.

Next, we use the **Euclidean Distance** to calculate the distance between the two graphs separately. An aggregation factor $\alpha \in [0.0, 1.0]$ is introduced to aggregate the distances of the two type graphs to obtain the final distance of the two sub-logs as follows:

$$d(w1, w2) = \alpha \left\| v_1^{\mathcal{A}} - v_2^{\mathcal{A}} \right\| + (1 - \alpha) \left\| v_1^{\mathcal{O}} - v_2^{\mathcal{O}} \right\|, \qquad (3)$$

where the value of $d(w1, w2)$ is in [0.0, 1.0], because we apply **Min-Max Normalization** to normalize each distance of two type graphs. The aggregation factor α is set to 0.5 in this paper, which is used to balance the influence of nodes and edges on drift detection.

4.4   Change Detection

In this section, we provide details about change detection based on the distance of two sub-logs. As shown in Fig. 4, there are two steps for detecting change points: (1) we use the K-means algorithm to cluster all points into two classes, where the distant ones are the candidate change points, and (2) we use a filter mechanism to find the actual change points from the candidate points.

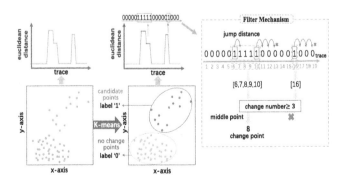

Fig. 4. An example for illustrating the detail of change detection. Two steps are described: K-means algorithm be used to find the candidate points, then filter mechanism be adopted to confirm the actual change points.

K-Means Algorithm. With the sliding of two consecutive windows, many points are obtained from the event log. Each of these points is the intersection of the windows (i.e., the location of the 1st trace in w2). Figure 4 depicts the details of clustering: we first use the K-means algorithm (i.e., K=2) to cluster these points into two classes based on the distance of each point. Second, all points in the class with a large distance will be marked as '1' label, and the points in the other class are marked as '0' label. We believe that changes happened in the event log will result in a larger distance between windows w1 and w2. Therefore, the points with label '1' are considered as candidate change points.

Filter Mechanism. In order to ensure the robustness of our method, we introduce a filter mechanism to improve the performance of GDD. We have found that some situations also cause the distance to become larger rather than just concept drift: (a) Noise traces. This kind of trace is often generated due to employee operation errors or system failures. However, these traces occur less frequently; (b) Execution probability triggered by business branching. There are branching structures such as parallel and alternative in the business process model, and these structures can interfere with the drift judgment. The form triggered by the branch structure is that the labels '0' and '1' appear alternately in the event log. So, we provide a simple-to-implement and efficient mechanism, namely the Filter Mechanism, to reduce the interference of the above situations. The operation steps of the Filter Mechanism are shown below.

The first step is grouping. We use the hyperparameter *jump distance* (abbr., j) to group all the candidate points into different groups. As shown in Fig. 4, assuming $j = 3$, we start from the location of first label '1' and jump three points forward. If there is a point with '1' in the three points, this point will be replaced with the new starting point and continues to jump until no point with label '1' is found. Then, a group is divided successfully. We repeat the above steps until all the points with label '1' are grouped successfully. There are two groups [6,7,8,9,10] and [16] be found in Fig. 4, where each numeral in the groups represents the location of drift.

The second step is filtering. A hyperparameter *change number* (abbr., n) is used to filter out the groups with the number of candidate points less than n. In Fig. 4, the group [16] will be filter out.

The last step is to confirm the actual change points. When the label '1' is no longer appears, it means that the evolution tends to be stable. In other words, a group represents only one evolution. Therefore, we use the middle point to replace all the points in a group. The location of middle point is defined by $middle\ point = group.get(\lfloor \frac{group.size-1}{2} \rfloor)$, where $group.size$ represents the number of candidate points in a group, the function $group.get(i)$ means getting the i-th candidate point from a group and $i = \lfloor \frac{group.size-1}{2} \rfloor$.

Finally, the locations of change points are found (i.e., middle points) through the K-means algorithm and Filter Mechanism. Also, these locations are considered as the point of time when the concept drift occurs in the event log. Using these change points, we are then able to accurately divide the event log into different versions of business processes and provide support for process mining.

5   Experiments

The proposed method (i.e., GDD) is implemented as a stand-alone Python application, and all the code are publicly-available[1].

5.1   Datasets

We conduct experiments on public datasets including PRODRIFT and PTCOST, since they are often used to test the performance of drift detection methods [8,9,12,16].

— PRODRIFT[2], provided by Maaradji et al. [8], is the most cited simulation dataset. This dataset consists of 72 synthetic event logs generated from the loan application process (Fig. 5). The model is composed of different control-flow structures including sequences, loops, alternative and parallel branches. To generate the logs, the base model (Fig. 5) was modified systematically by applying, in turn, one out of 12 simple change patterns described in [8]. These modifications reveal different change patterns, which are categorized into insertion ("I"), resequentialisation ("R") and optionalisation ("O"). Besides, 66 composite change patterns (RIO, ROI, IOR, IRO, OIR, ORI) were created by combining the simple change patterns. Finally, to vary the distance between the drifts, four event logs of sizes 2,500 traces (abbr., 2.5k), 5,000 traces (abbr., 5k), 7,500 traces (abbr., 7.5k) and 10,000 traces (abbr., 10k) were generated for each of the change patterns, and each event log has 9 change points.

— PTCOST[3], provided by Travel Agency, is a real-log, which contains events pertaining to two years (2017–2018) of travel expense claims. The dataset consists of 18246 events, and 2099 traces. As described in the documentation of PTCOST, the processes in 2017 and 2018 are different since 2017 is a pilot year. The information implies that there is a potential drift point for the event log sometime between the end of 2017 and the beginning of 2018.

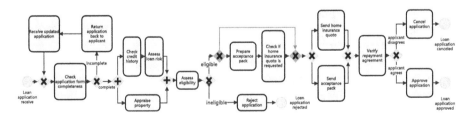

Fig. 5. The process model for the loan application, where the base model does not contain the exclusive gateways(marked as red) but the change pattern CM does. (Color figure online)

---

[1] https://github.com/lll-lin/GDD.

[2] https://data.4tu.nl/articles/dataset/Business_Process_Drift/12712436.

[3] https://data.4tu.nl/articles/dataset/BPI_Challenge_2020_Prepaid_Travel_Costs/12696722?backTo=/collections/__/5065541.

5.2   Baselines and Metrics

Baselines. We know that many drift detection methods being published in the past years, but it is not possible to assess all the methods. So, we decide to compare GDD with some methods that satisfy the following criteria: (i) the methods can handle noise and their implementations are publicly accessible; (ii) if a method has many extended versions, only the best one will be selected; (iii) the input of the method is the set of traces or the smallest unit is a trace, while the output is the set of process drifts. The application of the three criteria resulted in a selection of baseline methods: **Run** [8], **Alpha** [9] and **BM** [12].

Metrics. Standard F-score metric [6–12] for evaluating the performance among GDD, Run, Alpha and BM is used in our evaluation, where $precision = TP/(TP + FP)$, $recall = TP/(TP + FN)$, and $F\text{-}score = (2 * precision * recall)/(precision + recall)$. TP refers to true positive, FN refers to false negative, and FP refers to false positive. To describe the three variables, a lag period (r) is defined:

- TP: a change point t is detected and there is an actual drift in the integer interval $[t - r, t + r]$.
- FP: a change point t is detected and there is not an actual drift in the integer interval $[t - r, t + r]$.
- FN: there is an actual drift t and there is no change point detected in the integer interval $[t - r, t + r]$.

5.3   Evaluation on Different Parameter Settings

In the first experiment, we evaluate the impact of window sizes w, jump distance j and change number n. We choose the PRODRIFT to test the impact of different parameters in GDD, because this dataset contains different change intervals.

First, the impact of w would be discussed, while j and n are fixed at 3. As shown in Fig. 6a, we test a total of 5 different window sizes ranging from 50 to 500. When the window size is set to $w = 100$, the F-score is the highest, followed by $w = 50$. If the window increases to 200, the performance of GDD starts to drop. When $w = 500$, the F-score is the worst. Therefore, the window size should not be too large, and the accuracy will drop when the window size is larger than the minimal distance between two consecutive drfits. This conclusion is almost identical to that of most current window-based drift detection methods.

Next, Fig. 6b shows us the impact of j with 4 different sizes ($w = 100$ and $n = 3$). We can see that F-scores are highest when $j = 3$. But, it does not perform well when $j = 10$, because it tends to lead to missing the actual drift points for the dataset with small change intervals (e.g., 2.5k).

Last, we set the parameter n with 4 different sizes for observing the F-score of GDD, while w and j are fixed at 100 and 3 respectively. Overall, the parameter of the change number is insensitive in the simulation log. When $n = 3$, the F-score is the most stable, while the F-score decreases slightly when $n = 10$. Because

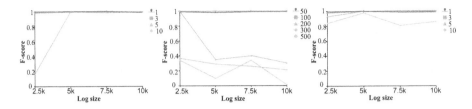

Fig. 6. F-scores under different parameter settings.

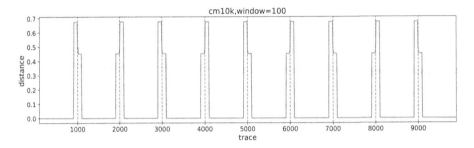

Fig. 7. Visualization of the detection result in the CM10k log and the window size = 100

there are no noise data in the simulation log, it also misses a small number of drift points if the value is too large. In the remaining text, we report the results with $w = 100$, $j = 3$ and $n = 3$.

5.4 Comparing with the Baselines

In the second experiment, the accuracy of our proposed method (GDD) and the baselines (Run, Alpha and BM) are compared for each change pattern. Figure 7 depicts the visualization of the detection result by GDD. We can see that the distances between two sub-logs sampled by two consecutive two windows are very high near the actual change point, which matches our initial conjecture. Then, we start by comparing the four methods on the 10k log to see how each method performs specifically on the 18 change pattern, while the delay period decreases from 100 to 50. We find that GDD has perfect F-scores on each pattern. However, the F-scores of most patterns drop dramatically in Run an Alpha. Such as, Fig. 8b reports that when $r = 50$, the F-scores of LP are zero by using Alpha and Run. At the same time, the F-scores of BM on the change patterns of CM, CB and LP also decrease significantly.

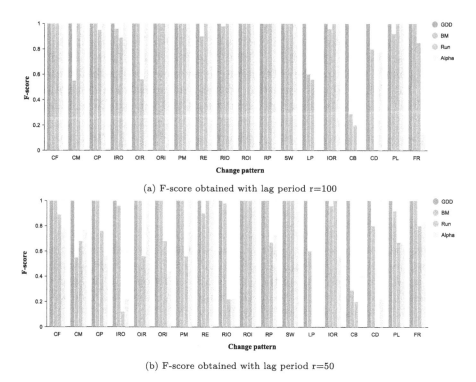

(a) F-score obtained with lag period r=100

(b) F-score obtained with lag period r=50

Fig. 8. Comparison of F-score per change pattern in the 10k log.

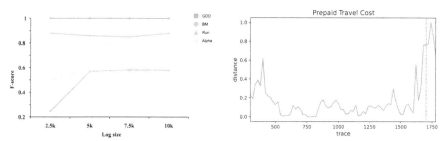

(a) F-scores obtained with differnt log sizes (b) The visualization of GDD on real-life log

Fig. 9. Evaluation on simulated logs and real-life log.

Besides, we test the four methods on four different sizes of logs (2.5k, 5k, 7.5k, 10k), and the F-score for each log is the average of 18 change patterns. Figure 9a shows that GDD keeps the highest F-score of 1, while the BM achieves an average F-score of 0.85, Alpha achieves 0.58 and Run achieves 0.51. Hence, the experimental results mean our method is more accurate.

Last, we evaluate our algorithm on the PTCOST dataset (see Fig. 9b). Since the real data have a large number of infrequent behaviors that can easily interfere with drift detection as noise, we used two sets of parameters:(1) when $w = 100$, $j = 3$ and $n = 3$, we obtain two drift points [325, 1710], of which the drift points 325 is a false positive; (2) when $w = 100$, $j = 3$ and $n = 5$, we obtain only one drift point [1700], and the 1700th trace is the new process that started in January 2018. This result is consistent with the documentation description of the PTCOST. Therefore, the results show that GDD can handle real-life logs very well.

On the aspect of software and hardware, we conduct all tests on an Intel i5-12500H@2.50 GHZ with 16 GB RAM (64 bit), running Windows 10 and python3.7. The average running time on the 10k dataset is 40 s. If you run the test on Goolge Colab, the average time is about 2 min.

5.5 Ablation Study

To clarify the contributions of order graphs, virtual nodes and the filter mechanism, we design $\mathcal{OG}$-$\emptyset$, virtual node-$\emptyset$ and filter mechanism-$\emptyset$. $\mathcal{OG}$-$\emptyset$ and filter mechanism-$\emptyset$ respectively denote the order graphs deletion and the filter mechanism deletion. Virtual node-$\emptyset$ means that we replace the virtual node with the arithmetic mean of each node in the directed graph.

Table 2. Ablation analysis in different logs.

Method	PRODRIFT				PTCOST
	2.5K	5K	7.5K	10K	
$\mathcal{OG}$-$\emptyset$	0.90	0.90	0.92	0.89	0.79
virtual node-$\emptyset$	0.99	0.89	0.91	0.94	0.85
filter mechanism-$\emptyset$	0.99	0.99	0.91	0.90	0.60
GDD	1.0	1.0	1.0	1.0	1.0

As shown in Table 2, the average contributions of order graphs and virtual nodes to the F-score of GDD on the simulation log reach 10%, while the average contribution of the filter mechanism is only 5%. On the contrary, in the real-life log, the contribution of the filtering mechanism is the best, which reaches 40%. Experiments show that the filter mechanism is necessary to improve the robustness of GDD. In addition, the contribution of $\mathcal{OG}$ in the real-log is greater than that in simulation Logs. The essential reason is that the number of edges is far more than the number of nodes in the real business process.

6   Conclusion and Future Work

Drift detection is a valuable topic in process mining, which makes full use of the event log recorded in today's information system to improve business process management. In this paper, we propose a new process drift detection method

(GDD) that can accurately locate the process drift points. Different from previous work, GDD does not rely on artificial feature selection but applies graph convolutional networks to automatically obtain all the change features implied in the logs. We empirically demonstrated that our method outperforms state-of-the-art baselines.

In the future, we will continue to study the following aspects: On the one hand, we aim to propose a way to automatically determine the window size for different logs. On the other hand, we plan to extend the work to support gradual drift detection.

Acknowledgements. The work was supported by the National Key Research and Development Program of China (2018YFB1800403), the general project numbered KM202310028003 of Beijing Municipal Education Commission, the National Natural Science Foundation of China (61872252).

References

1. van der Aalst, W.M.P.: Process Mining: Discovery, Conformance and Enhancement of Business Processes. Springer, New York (2011)
2. Leemans, S.J.J., Fahland, D., van der Aalst, W.M.P.: Discovering block-structured process models from event logs containing infrequent behaviour. In: Lohmann, N., Song, M., Wohed, P. (eds.) BPM 2013. LNBIP, vol. 171, pp. 66–78. Springer, Cham (2014). https://doi.org/10.1007/978-3-319-06257-0_6
3. Guo, Q., Wen, L., Wang, J., Yan, Z., Yu, P.S.: Mining invisible tasks in non-free-choice constructs. In: Motahari-Nezhad, H.R., Recker, J., Weidlich, M. (eds.) BPM 2015. LNCS, vol. 9253, pp. 109–125. Springer, Cham (2015). https://doi.org/10.1007/978-3-319-23063-4_7
4. Sander, J.J., Leemans, A.F.S., Aalst, W.M.P.: Earth moves' stochastic conformance checking. In: International Conference on Business Process Management. Vienna, Austria, pp. 127–143 (2019)
5. Fahland, D., Aalst, W.M.P.: Model repair - aligning process models to reality. Inf. Syst. **47**(1), 220–243 (2015)
6. Bose, R.P.J.C., Aalst, W.M.P., et al.: Dealing with concept drifts in process mining. IEEE Trans. Neural Netw. Learn. Syst. **25**(1), 154–171 (2014)
7. Bose, R.P.J.C., van der Aalst, W.M.P., Žliobaitė, I., Pechenizkiy, M.: Handling concept drift in process mining. In: Mouratidis, H., Rolland, C. (eds.) CAiSE 2011. LNCS, vol. 6741, pp. 391–405. Springer, Heidelberg (2011). https://doi.org/10.1007/978-3-642-21640-4_30
8. Maaradji, A., Dumas, M., La Rosa, M., Ostovar, A.: Fast and accurate business process drift detection. In: Motahari-Nezhad, H.R., Recker, J., Weidlich, M. (eds.) BPM 2015. LNCS, vol. 9253, pp. 406–422. Springer, Cham (2015). https://doi.org/10.1007/978-3-319-23063-4_27
9. Ostovar, A., Maaradji, A., Rosa, M.L., et al.: Detecting drift from event streams of unpredictable business processes. In: International Conference on Conceptual Modeling. Gifu, Japan, pp. 330–346 (2016)
10. Ostovar, A., Leemans, S., Rosa, M.L.: Robust drift characterization from event streams of business processes. ACM Trans. Knowl. Disc. Data **14**(3), 1–57 (2020)

11. Zheng, C., Wen, L., Wang, J.: Detecting process concept drifts from event logs. In: Panetto, H., et al. (eds.) OTM 2017. LNCS, vol. 10573, pp. 524–542. Springer, Cham (2017). https://doi.org/10.1007/978-3-319-69462-7_33

12. Lu, Y., Chen, Q., Poon, S.: A robust and accurate approach to detect process drifts from event streams. In: Polyvyanyy, A., Wynn, M.T., Van Looy, A., Reichert, M. (eds.) BPM 2021. LNCS, vol. 12875, pp. 383–399. Springer, Cham (2021). https://doi.org/10.1007/978-3-030-85469-0_24

13. Widmer, G., Kubat, M.: Learning in the presence of concept drift and hidden contexts. Mach. Learn. **23**(1), 69–101 (1996)

14. Kipf, T.N., Welling, M.: Semi-supervised classification with graph convolutional networks. In: The 5th International Conference on Learning Representations. Toulon, France, pp. 1–14 (2017)

15. Maggi, F.M., Burattin, A., Cimitile, M., Sperduti, A.: Online process discovery to detect concept drifts in LTL-based declarative process models. In: Meersman, R., et al. (eds.) OTM 2013. LNCS, vol. 8185, pp. 94–111. Springer, Heidelberg (2013). https://doi.org/10.1007/978-3-642-41030-7_7

16. Lin, L., Wen, L., Lin, L., Pei, J., Yang, H.: LCDD: detecting business process drifts based on local completeness. IEEE Trans. Serv. Comput. **15**(4), 2086–2099 (2022)

17. Carmona, J., Gavaldà, R.: Online techniques for dealing with concept drift in process mining. In: Hollmén, J., Klawonn, F., Tucker, A. (eds.) IDA 2012. LNCS, vol. 7619, pp. 90–102. Springer, Heidelberg (2012). https://doi.org/10.1007/978-3-642-34156-4_10

18. Kumar, M.V.M., Thomas, L., Annappa, B.: Capturing the sudden concept drift in process mining, pp. 132–143. Algorithms Theories Anal. Event Data, Brussels, Belgium (2015)

19. Yeshchenko, A., Di Ciccio, C., Mendling, J., et al.: Visual drift detection for sequence data analysis of business processes. IEEE Trans. Vis. Comput. Graph. (2021). https://doi.org/10.1109/TVCG.2021.3050071

20. Jiang, X., Ji, P., Li, S.: CensNet: convolution with edge-node switching in graph neural networks. In: International Joint Conferences on Artificial Intelligence (IJCAI), pp. 2656–2662 (2019)

21. Perozzi, B., Al-Rfou, R., Skiena, S.: DeepWalk: online learning of social representations. In: Proceedings of the 20th ACM SIGKDD International Conference on Knowledge Discovery and Data Mining, pp. 701–710 (2014)

22. Mueller, J., Thyagarajan, A.: Siamese recurrent architectures for learning sentence similarity. In: The Association for the Advancement of Artificial Intelligence (AAAI), pp. 2786–2792 (2016)

23. Pham, T., Tran, T., Dam, H., Venkatesh, S.: Graph classification via deep learning with virtual nodes. arXiv:1708.04357 (2017)

Decoupling Graph Neural Network with Contrastive Learning for Fraud Detection

Lin Meng[1(✉)], Yuxiang Ren[1], and Jiawei Zhang[2]

[1] IFM Lab, Department of Computer Science, Florida State University,
Tallahassee, FL, USA
lin@ifmlab.org, renyuxiang931028@gmail.com
[2] IFM Lab, Department of Computer Science, University of California,
Davis, CA, USA
jiawei@ifmlab.org

Abstract. Recently, many fraud detection models introduced graph neural networks (GNNs) to improve the model performance. However, fraudsters often disguise themselves by camouflaging their features or relations. Due to the aggregation nature of GNNs, information from both input features and graph structure will be compressed for representation learning simultaneously. On the one hand, since not all neighbors provide useful information due to camouflage, aggregating information from all neighbors may potentially decrease the model performance. On the other hand, the structure including all neighbors is not reliable due to the relation camouflage. In this paper, we propose to decouple attribute learning and structure learning to avoid the mutual influence of feature and relation camouflage. Therefore, the model first learns its embedding seperately and then combine them together with label-guided contrastive losses to make predictions better. We conduct extensive experiments on two real-world datasets, and the results show the effectiveness of the proposed model.

Keywords: Fraud Detection · Camouflage · Graph Neural Networks

1 Introduction

Fraud detection is an important task in our daily life to fight against malicious actions or intentions. Fraudulent activities exist in many scenarios. For example, opinion fraud in online review platforms affects the behavior of customers [2], fake news in social media [1] misleads the opinion of people, and financial fraud in financial platforms can cause severe financial loss to customers [10,17]. In these applications, the extensive interactions among people in both the online and offline world enable researchers to solve the problem from the graph perspective by treating all people as nodes and interactions as edges. Due to the superior representation power of graph neural networks (GNNs), GNN-based fraud detection has drawn extensive attention in both industry and academia [2,11,12]. Generally, GNNs aggregate all the information from the neighborhood and then update the information for the center node with it via a linear transformation [7,21,26]. Meanwhile, fraudsters camouflage themselves to avoid being detected by the fraud detection systems [5]. Typically, the camouflage behaviors of fraudsters can be categorized into attribute camouflage and relation camouflage. Attribute camouflage [2] refers to fraudsters fabricating their attributes like regular customers, while relation camouflage [2] indicates the relations that misguide the classifier.

The effectiveness of GNNs mainly comes from propagating information from a "homophily" neighborhood, which means the central nodes rely on information propagated from neighboring nodes in the same class. However, when dealing with fraud detection, the camouflage phenomenon violates the assumption since camouflage in GNNs produces noisy information propagated from nodes with different labels [2,12]. Furthermore, attribute and connection are entangled during GNN propagation, making attribute learning and structure learning affect each other, which may exaggerate false information. The influence of connection camouflage can be illustrated by Fig. 1. Therefore, applying GNNs to fraud detection is challenging from three perspectives: (1) feature camouflage makes nodes with similar features have different labels;

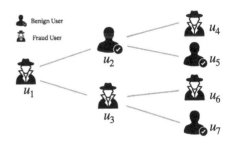

Fig. 1. Example of Introducing Noise When Applying A Two-layer GNN to Fraud Detection. In u_1-u_2-u_4, u_4 provides essential information for u_1, but it receives noise from u_2 unavoidably due to camouflage superposition. In u_1-u_3-u_7 and u_1-u_2-u_5, u_1 receives the noisy information from u_2, u_5 and u_7 due to camouflage on the first or second layer. Aggregated messages mislead GNN, making it predict wrong labels.

(2) connections camouflage leads GNN to aggregate noise information. More noisy information will be learned if with multiple GNN layers; and (3) both types of camouflage are mixed during the propagation at each layer, which further damages the performance of GNN.

To address these aforementioned challenges posed by applying GNNs to fraud detection, we propose a framework that decouples attribute learning and structure learning, named **DC-GNN** (**D**ecoupled **C**ontrastive **G**raph **N**eural **N**etwork). To accomplish it, it contains three modules: (1) individual attribute encoding, which encodes four attributes together and then uses a graph transformer to identify useful neighbors that reveal characteristcs of fraudsters; (2) local structure encoding, which learns the structure feature that well identifies fraudsters; and (3) label-guided contrastive losses enhanced optimization, which contrasts the embeddings of fraudulent nodes and benign nodes so that the model learns more robust node embeddings.

In this paper, we detect fraudsters in multi-relational graphs. Existing models for multi-relational graphs mainly use the node embeddings learned by GNNs where they filter out camouflaged relations and then aggragate node features from "homophily" neighborhood [2,12]. However, camouflaged relations can also reveal characteristics of fraudsters [16]. Thus, we use adjacency matrices to learn the structure characteristics of fraudster so that node features are not required during structure learning. The main contributions are summarized below:

– We propose a novel framework that learns attribute and structure embeddings separately to overcome the mutual effects brought by camouflage in GNNs for fraud detection.
– We propose a label-guided contrastive loss to enhance optimization, which improves the robustness of the model by contrasting fraudsters and benign nodes.
– We conduct extensive experiments on two real-world datasets, and the performance shows the effectiveness of the proposed model.

2   Notations and Problem Definition

Notations. Generally, scalars are denoted as lowercase letters (e.g., x), the lowercase bold faced letters (e.g., $\mathbf{x}$) represent vectors, and capital bold faced letters (e.g., $\mathbf{X}$) to represent matrices. Sets or tensors are denoted as calligraphic letters (e.g., $\mathcal{X}$). $\mathbf{X}(i,:)$ and $\mathbf{X}(:,j)$ denotes i^{th} row and j^{th} column of $\mathbf{X}$, respectively. $\mathbf{X}(i,j)$ denotes the element in i^{th} row and j^{th} column of $\mathbf{X}$. $\| \cdot \|_F$ represents matrix F-norm. $\cup$ and $\odot$ represent concatenation and vector inner product, respectively.

Definition 1 *(Multi-relational Graph). A multi-relation graph can be represented as $\mathcal{G} = (\mathcal{V}, \{\mathcal{E}_r\}|_{r=1}^{R}, \mathcal{X})$, where $\mathcal{V} = \{v_1, \ldots, v_n\}$ denotes the node set, $\mathcal{X} = \{\mathbf{x}_1, \ldots, \mathbf{x}_n\}$ represents a set of all node features. Each node v_i is associated with a d-dimensional feature vector $\mathbf{x}_i \in \mathbb{R}^d$ An edge $e_{ij} = (v_i, v_j, r) \in \mathcal{E}_r$ if v_i and v_j is connected via relation $r \in \{1, \ldots, R\}$. A label set $\mathcal{Y}$ denotes a node label set.*

In this paper, we relax the definition of terminology about "hop", which denotes the nodes within certain distance measures.

Definition 2 *(K-hop Neighbors). Suppose the distance between node v_i to v_j is d_{ij}. K-hop neighbors of node v_i are a set of nodes that $d_{ij} \leq K$ holds. Given a node $v_i \in \mathcal{V}$ in the graph, we use $\mathcal{N}_i^{(K)}$ to denote K-hop neighbors.*

Problem Definition. Given a multi-relational graph $\mathcal{G} = (\mathcal{V}, \{\mathcal{E}_r\}|_{r=1}^{R}, \mathcal{X})$ and the corresponding label set $\mathcal{Y}$ for all nodes (accounts or reviews) in $\mathcal{V}$, we model the fraud detection problem as a binary classification task. Formally, our goal is to find a function $f(\cdot)$, s.t.

$$f(\mathcal{G}) \rightarrow \mathcal{Y}.$$

where $\mathcal{Y} = \{0,1\}_1^n$, and $y_i = 1$ if v_i is fraudulent while $y_i = 0$ if v_i is benign.

3   Proposed Method

To alleviate the effects of camouflage, we decouple the GNN learning process into individual attribute encoding and multi-relational local structure encoding; after that, we fuse attribute embedding and structure embedding as the final node representation. To better assist the learning process, we propose two label-guided contrastive losses. The overall framework is shown in Fig. 2.

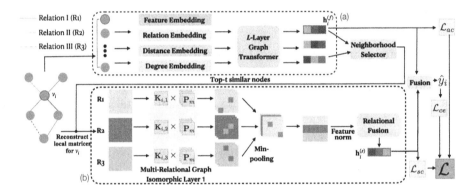

Fig. 2. DC-GNN Architecture. (Module (a): individual attribute encoding with a graph transformer; module (b): multi-relational local structure encoding with multi-relational IsoNN including multi-relational graph isomorphic layer, min-pooling layer, feature normalization (norm), and relational fusion; $\mathcal{L}$: the final loss, which includes $\mathcal{L}_{ce}$ for label prediction, $\mathcal{L}_{ac}$ and $\mathcal{L}_{sc}$ for attribute and structure contrastive losses, respectively.)

3.1   Individual Attribute Encoding

Node Attributes. Since fraudulent nodes often camouflage themselves on node features, we propose to add three more graph related attributes to determine if a node is fraudulent or not. Since nodes are in a multi-relational graph, we choose connected relation types, distances towards the central node, and degrees under each relational graph. The reasons why we choose them as additional attributes are: (1) fraudsters are often involved in many social activities since they want to make profits. For example, in online review graphs, fraudsters often write an untruthful review for many items for better promotion. Therefore, fraudsters would connect to many users who also write reviews for those items; (2) additionally, many fraudsters are also involved in different types of activities, which potentially reveals the activity type of fraudsters, such as writing fake reviews or posting fake reviews at the same time; (3) moreover, the distances towards the given node show the influence of nodes towards the given node.

Formally, given a multi-relational graph $\mathcal{G} = (\mathcal{V}, \{\mathcal{E}_r\}|_{r=1}^{R}, \mathcal{X})$, for a node v_i, we encode the corresponding node feature $\mathbf{x}_i$ as follows

$$\mathbf{x}_i^{(a)} = \text{MLP}(\mathbf{x}_i) \in \mathbb{R}^{d_h \times 1}.$$

Since we are dealing with multi-relations, we use a vector $\mathbf{r}_i \in \{0,1\}^R$, where $\mathbf{r}_i(j) = 1$ indicates the node has a j-th relation type connection. One MLP layer is adopted to get the embedding of the relation vector.

$$\mathbf{x}_i^{(r)} = \text{MLP}(\mathbf{r}_i) \in \mathbb{R}^{d_h \times 1}.$$

We use the postional embedding proposed in [20] to encode the distance $P(v_i)$ of v_i to a node (i.e., the distance is decided by the shortest distance between v_i to the central node in this paper)

$$\mathbf{x}_i^{(d)} = \text{Pos-emb}(P(v_i))$$
$$= \left[\sin\left(\frac{P(v_i)}{10000^{\frac{2l}{d_h}}} \right), \cos\left(\frac{P(v_i)}{10000^{\frac{2l+1}{d_h}}} \right) \right]_{l=0}^{\lfloor \frac{d_h}{2} \rfloor},$$

where $\mathbf{x}_i^{(d)} \in \mathbb{R}^{d_h \times 1}$. The index l iterates throughout all the entries in the above vector to compute the entry values with $sin(\cdot)$ and $cos(\cdot)$ functions for the node based on its distance. For degree d_i, since its number could be extremely large, we first normalize it and use MLP to obtain the information of it.

$$\mathbf{x}_i^{(e)} = \text{MLP}\left(\frac{d_i}{\sum_{r=1}^{R} \|\mathcal{E}_r\|} \right) \in \mathbb{R}^{d_h \times 1}.$$

By adding four attributes together, we obtain the initial node embeddings for all nodes.

$$\mathbf{x}_i' = \mathbf{x}_i^{(a)} + \mathbf{x}_i^{(r)} + \mathbf{x}_i^{(d)} + \mathbf{x}_i^{(e)}.$$

Attribute Encoding with Graph Transformer. Due to the great representation power of transformer [20], we use it as our attribute encoder. In graphs, we regard the context of a node as its neighborhood. Therefore, we choose K-hop neighbors directly as the context of the given node.

For each node, we obtain its K-hop neighbors by preprocessing the graph data. Suppose we have the K-hop neighbor set $\mathcal{N}_i^{(K)}$ for v_i, the corresponding input feature matrix $\mathbf{X}_i \in \mathbb{R}^{(|\mathcal{N}_i^{(K)}|+1) \times h}$ and $\mathbf{X}_i = [\mathbf{x}_i', \mathbf{x}_{i,1}', \cdots, \mathbf{x}_{i,|\mathcal{N}_i^{(K)}|}']$. We set $\mathbf{H}_i^{(0)} = \mathbf{X}_i$ and feed it into the L-layer transformer:

$$\begin{aligned} \mathbf{H}_i^{(l)} &= \text{Transformer}\left(\mathbf{H}_i^{(l-1)}\right) \\ &= \text{Softmax}\left(\frac{\mathbf{Q}\mathbf{K}^\top}{\sqrt{d_h}}\right)\mathbf{V} \end{aligned}, \text{ where } \begin{cases} \mathbf{Q} = \mathbf{H}_i^{(l-1)}\mathbf{W}_Q^{(l)} \\ \mathbf{K} = \mathbf{H}_i^{(l-1)}\mathbf{W}_K^{(l)} \\ \mathbf{V} = \mathbf{H}_i^{(l-1)}\mathbf{W}_V^{(l)} \end{cases}$$

where $\mathbf{H}_i^{(L)} \in \mathbb{R}^{(|\mathcal{N}_i^{(K)}|+1) \times d_h}$, and $\mathbf{W}_Q^{(l)}, \mathbf{W}_K^{(l)}, \mathbf{W}_V^{(l)} \in \mathbb{R}^{d_h \times d_h}$. Therefore, we get the embeddings of $\mathbf{H}_i^{(L)} = [\mathbf{h}_i^{(L)}, \mathbf{h}_{i,1}^{(L)}, \cdots, \mathbf{h}_{i,|\mathcal{N}_i^{(K)}|}^{(L)}]$. Note that due to the feature camouflage, we take the $\mathbf{h}_i^{(L)}$ directly as the final attribute embedding $\mathbf{h}_i^{(f)}$ of node v_i to avoid incorporating embeddings from the benign class.

3.2　Multi-relational Local Structure Encoding

Relations between nodes also reveal important information for detecting fraudsters. As indicated in previous works [16], fraudsters often act collectively, which also means the fraudsters may share similar local structures. However, traditional GNN cannot learn good structural information since fraudsters often hide themselves in connections with benign nodes, which introduce features belongs to benign nodes. To better learn the local structural features, we propose to encode the adjacency matrix solely. Thus, we utilize isomorphic graph neural network [14]. However, IsoNN originally is used for homogeneous graph classification, we extend it to multi-relational graphs to capture the local structure embeddings for each node.

Local Adjacency Matrices Reconstruction. We need to reconstruct the local adjacency matrices for all nodes and then each node learns its local structure representation. However, reconstructing the local adjacency matrices with its K-hop neighborhood is not realistic, since it may involve several nodes, causing high computational costs. Therefore, we select useful nodes from K-hop neighbors. Nodes are selected based on the similarities between embeddings in the output of transformer $\mathbf{H}_i^{(L)}$ of the given node and its neighbors. Here, we calculate the similarity as follows:

$$s_{ij} = \text{sim}(\mathbf{h}_i, \mathbf{h}_{i,j}) \quad \forall v_j \in \mathcal{N}_i^{(K)}$$

Here, Euclidean distance is the similarity measurement. We select Top-t similar nodes $\{v_j, \cdots, v_k\}$ and itself v_i to reconstruct local adjacency matrix. Note that for the rest of the paper, we treat $\{v_j, \cdots, v_k\}$ together with central node v_i as selected t nodes just for simplicity.

With the multi-relational graph $\mathcal{G} = (\mathcal{V}, \{\mathcal{E}_r\}|_{r=1}^R, \mathcal{X})$, and the selected nodes $\{v_i, v_j, \cdots, v_k\}$, we reconstruct local adjacency matrices for all relations, which is denoted as $\mathcal{A}_i = \{\mathbf{A}_{i,r} \in \{0,1\}^{t \times t} | 1 \le r \le R\}$, and the local adjacency matrix $\mathbf{A}_{i,r}$ of node v_i in r-th relation is as follows

$$\mathbf{A}_{i,r}(p,q) = \begin{cases} 1, & e_{pq} \in \mathcal{E}_r, \\ 0, & \text{otherwise.} \end{cases}$$

Note that the order of nodes will not affect the embedding results much since IsoNN can alleviate the node order constraint posed by the adjacency matrix.

Structure Encoding with Multi-relational IsoNN. Graph isomorphic layer in IsoNN adopts learnable kernel variable $\mathbf{K} \in \mathbb{R}^{m \times m}$ to learn the regional structure information. Here, to incorporate with multi-relational graph, we propose a multi-relational Graph Isomorphic Layer. we use relational kernels $\{\mathbf{K}_1, \ldots, \mathbf{K}_R\}$, where $\mathbf{K}_r \in \mathbb{R}^{m \times m}$ is for r^{th} relation. We can have multiple channels, so the learnable kernel for relation r is denoted as $\mathcal{K}_r \in \mathbb{R}^{c \times m \times m}$. Then features can be learned by

$$\mathcal{F}_{i,r}(l,j,s,h) = \|\mathbf{P}_j \mathcal{K}_r(l,:,:) \mathbf{P}_j^\top - \mathbf{A}_{i,r}(s:s+m, h:h+m)\|_F$$

where $\mathbf{P}_j \in \{0,1\}^{m \times m}$ is a permutation matrix from permutation set $\{\mathbf{P}_1, \cdots, \mathbf{P}_{m!}\}$. $\mathcal{F}_{i,r} \in \mathbb{R}^{c \times m! \times (t-m+1) \times (t-m+1)}$ because $\mathbf{A}_{i,r}$ generates $(t-m+1) \times (t-m+1)$ sub-matrices in total. After computing all possible permutations, we need to find the features made by the optimal permutation. Therefore, a min-pooling layer is adopted on dimension caused by permutation matrices to find the minimal value computed by optimal permutation

$$\bar{\mathcal{F}}_{i,r}(l,s,h) = \text{min-pooling}\left(\mathcal{F}_{i,r}(l,:,s,h)\right),$$

where $\bar{\mathcal{F}}_{i,r} \in \mathbf{R}^{c \times (t-m+1) \times (t-m+1)}$. To reduce the number of parameters, we adopt an average pooling layer for $\bar{\mathcal{F}}_{i,r}$ among all channels.

$$\hat{\mathbf{F}}_{i,r}(s,h) = \frac{\sum_{l=1}^c \bar{\mathcal{F}}_{i,r}(l,s,h)}{c},$$

where $\hat{\mathbf{F}}_{i,r} \in \mathbb{R}^{(t-m+1) \times (t-m+1)}$. Since values of the learned features can vary within a large range, we normalize those learned features. Moreover, differentiating which region contributes more can build better structural features. We reshape $\{\hat{\mathbf{F}}_{i,1}, \ldots, \hat{\mathbf{F}}_{i,R}\}$ into vectors and concatenate them as $\hat{\mathbf{F}}_i \in \mathbb{R}^{R \times (t-m+1)^2}$, then normalize it by

$$\mathbf{F}_i(r,s) = 1 - \frac{\exp(\hat{\mathbf{F}}_i(r,s))}{\sum_j \exp(\hat{\mathbf{F}}_i(r,j))}.$$

Here, we use '1' for subtraction because the smaller values show better matching between templates and subgraphs. Therefore, the local structure embedding $\mathbf{h}_i^{(s)}$ is

$$\mathbf{h}_i^{(s)} = \text{ReLU}\left(\text{MLP}\left(\bigcup_{r=1}^R \mathbf{F}_i(r,:)\right)\right).$$

where $\mathbf{h}_i^{(s)} \in \mathbb{R}^{d_h}$. The time cost for learning isomorphic features is $\mathcal{O}(cm!m^3R(t - m + 1)^2)$, where $(t - m + 1)^2$ is the number of submatrices, m^3 corresponds to matrix multiplication time cost, $m!$ is introduced in enumerating permutation matrices. Thus, by choosing small kernel size (small m) and few nodes (small t), the training is feasible.

3.3 Label-Guided Contrastive Loss Enhanced Optimization

Label Prediction. To predict node labels, we concatenate attribute embedding and structure embedding together as the final node embedding. Formally, the prediction for node v_i can be made by

$$\hat{y}_i = \text{MLP}\left(\text{ReLU}\left(\text{MLP}(\mathbf{h}_i^{(f)} \cup \mathbf{h}_i^{(s)})\right)\right).$$

The loss function for label prediction is weighted cross entropy.

$$\mathcal{L}_{ce} = -\sum_{i \in \mathcal{B}}\left(\gamma y_i \log\left(\hat{y}_i\right) + (1 - y_i) \log(\hat{y}_i)\right),$$

where γ is the imbalance ratio of fraud labels ($y_i = 1$) to benign labels ($y_i = 0$), $\mathcal{B}$ is the training batch.

Label-Guided Contrastive Loss. To better assist the optimization, we introduce two label-guided contrastive losses to provide extra guidance on learning node embeddings. As aforementioned, we separate the attribute and the structure since they would affect each other, which might make fraudsters harder to be discovered. Therefore, we contrast the attribute embedding and structure embedding separately. We set that positive pairs are those with the same labels and negative pairs are those with different labels based on the training set. Here, we define the attribute label-guided contrastive loss as:

$$\mathcal{L}_{ac} = -\log \frac{\exp\left(f\left(\mathbf{h}_i^{(f)}, \mathbf{h}_{i+}^{(f)}\right)/\tau\right)}{\exp\left(f\left(\mathbf{h}_i^{(f)}, \mathbf{h}_{i+}^{(f)}\right)/\tau\right) + \sum_{i- \in \mathcal{C}} \exp\left(f\left(\mathbf{h}_i^{(f)}, \mathbf{h}_{i-}^{(f)}\right)/\tau\right)},$$

where i^+ means node v_{i+} has $y_{i+} = y_i$, i^- means node v_{i-} has $y_{i-} \neq y_i$, $\mathcal{C}$ is set containing negative instances, and τ is temperature. To get the negative instance set, we randomly select b nodes with different labels from the training set. Similarly, we can get $\mathcal{L}_{sc}$ with structure embedding. We formulate the objective function of DC-GNN as:

$$\mathcal{L} = \mathcal{L}_{ce} + \lambda_1 * \mathcal{L}_{ac} + \lambda_2 * \mathcal{L}_{sc}$$

where λ_1 and λ_2 control the weight of $\mathcal{L}_{ac}$ and $\mathcal{L}_{sc}$, respectively. The training goal is to minimize $\mathcal{L}$ and the parameters are updated by backpropagation.

4   Experiments

4.1   Experimental Settings

To validate the effectiveness of the proposed model, we use two real-world datasets – Yelp [16] and Amazon (AMZ) [13]. Yelp includes hotel and restaurant reviews filtered (spam) and recommended (legitimate) on Yelp platform. Amazon includes product reviews under the Musical Instruments category.

Table 1. Dataset Statistics

	# Nodes (Frauds%)	Relations		
Yelp	45,954 (14.5%)	R-U-R (49,315)	R-T-R (573,616)	R-S-R (3,402,743)
AMZ	11,944 (9.5%)	U-P-U (175, 608)	U-S-U (3,566,479)	U-V-U (1,036,737)

The number of edges belonging to each relation is shown in Table 1. More information can be found at [2].

Comparison Methods and Evaluation Metrics. Our baselines include GNN models and GNN-based fraud detection models. RGCN [18], Graph-SAGE [4] are two general GNN models. For GNN-based models, we add FdGars [24] which is based on GCN, GraphConsis [12] which aggregates selected neighbors by similarity, DCI [25] which decouples representation learning and classification to enhance the performance, CARE-GNN [2] that uses a reinforcement learning module for neighborhood selection, FRAUDRE [30] which considers four types of inconsistencies in fraud detection.

Fraud detection suffers severe class imbalance. Therefore, AUC and average precision (AP) are our evaluation metrics.

Experimental Setups. We partition the datasets into training, validation, and test data. We choose $p \in \{60\%, 80\%\}$ of a dataset to train the model, and use 20% of the remaining data as the validation set. The rest is test data. We set the embedding size $d_h = 64$ for all comparison methods and set the same random seed to make a fair comparison. Parameters for graph transformer include layer number $L = 2$, and the number of heads $= 8$. In the individual attribute encoding, we set $K = 2$, but $|\mathcal{N}_i^{(2)}|$ is not a fixed number for all nodes. Therefore, we select the same number of neighbors from $\mathcal{N}_i^{(2)}$ to make it easy to train. We set the number of neighborhoods to 201 for Amazon and 61 for Yelp. For structure encoding, we only use 1 multi-relational graph isomorphic layer for simplicity. We set $t = 5$, the kernel size $m = 2$ and channel number $c = 10$ for Amazon, and $t = 20$, $m = 3$ and $c = 10$ for Yelp. In contrastive loss, $\tau = 0.001$, $b = 15$, and $\lambda_1 = 0.001$ and $\lambda_2 = 0$. for Amazon, and $\tau = 0.001$, $b = 5$, $\lambda_1 = 0$ and $\lambda_2 = 0.001$ for Yelp. The learning rate is 0.001 and 4096 is batch size. Adam optimizer is utilized.

Table 2. Fraudulent nodes similarity between its neighborhood. (FeatSim denotes feature similarity and LabelSim denotes label similarity. 2-hop via F/B means the 2-hop neighborhood retrieved via fraudulent nodes/benign nodes.)

datasets	Relation	FeatSim (1-hop)	LabelSim (1-hop)	FeatSim (2-hop via F)	LabelSim (2-hop via F)	FeatSim (2-hop via B)	LabelSim (2-hop via B)
Yelp	R-U-R	0.9906	0.9089	0.9955	0.9817	0.9940	0.5306
	R-T-R	0.9880	0.1764	0.9882	0.1989	0.9882	0.1932
	R-S-R	0.9878	0.1857	0.9879	0.1914	0.9879	0.1910
	ALL	0.9878	0.1838	0.9878	0.1737	0.9876	0.0955
Amazon	U-P-U	0.7107	0.1673	0.7069	0.1687	0.6121	0.0751
	U-S-U	0.6866	0.0558	0.6287	0.0638	0.6254	0.0499
	U-V-U	0.6969	0.0532	0.6239	0.0388	0.6153	0.0308
	ALL	0.6866	0.0722	0.6169	0.0684	0.6331	0.0528

For the baselines' implementation, we follow the setting reported in their papers. We use the codes of DG-Fraud[1] for GraphSAGE, FdGars and GraphConsis, and codes published by the authors for RGCN, CARE-GNN, DCI, FRAUDRE. For DC-GNN, we pre-train the individual attribute encoding module to get the ranking of the neighbhorhood so that we can reconstruct the adjacency matrices before training DC-GNN. All experiments are run on 256GB Linux server.

4.2   Camouflage Evidence in 2-Hop Neighborhood

We use the same equation in [2] to calculate similarity scores. Table 2 shows the feature and label similarity scores among 1-hop and 2-hop neighborhood. According to the similarity scores of 1-hop neighbors, we observe that label similarity is as low as 0.1838 for Yelp and 0.0722 for Amazon, which indicates that fraudsters connect many benign nodes in both datasets, showing that connection camouflage is severe in almost all relations (except R-U-R in Yelp). Additionally, under such a low label similarity score, the feature similarity score is as high as 0.9878 for Yelp and 0.6866 for Amazon, which shows feature camouflage exists since fraudsters connect benign nodes sharing similar features. To have a better understanding of how camouflage affects the GNNs in 2-hop neighborhood. We have two scenarios: if the intermediate node is fraudulent (Via F) or benign (Via B). If the intermediate node is fraudulent, then both feature camouflage and connection camouflage are still severe as the scores are similar to those of 1-hop. If the intermediate node is benign, two scores heavily drop compared with scores of 1-hop. This shows the information transmitted through benign nodes brings much misleading information, which further degrades the performance of traditional GNN models.

---

[1] https://github.com/safe-graph/DGFraud-TF2.

Table 3. Classification Results. (The best scores are bold, the second best is underlined.)

Dataset	Metric	Training Percentage	RGCN	Graph-SAGE	FdGars	Graph-Consis	DCI	CARE-GNN	FRAU-DRE	DC-GNN
Yelp	AUC	60%	61.18	54.31	48.22	85.55	63.41	78.86	<u>86.08</u>	**88.13**
		80%	61.66	51.79	46.82	<u>86.60</u>	63.45	78.34	86.25	**88.18**
	AP	60%	23.46	17.32	14.59	54.18	21.07	42.17	<u>55.85</u>	**58.82**
		80%	24.33	14.87	14.19	56.10	17.53	41.73	<u>57.00</u>	**58.94**
Amazon	AUC	60%	18.84	73.78	40.75	89.37	88.52	92.66	<u>93.15</u>	**94.87**
		80%	18.97	75.88	41.48	90.52	86.04	93.02	<u>94.23</u>	**95.43**
	AP	60%	37.86	24.52	7.94	81.04	53.63	82.69	<u>84.37</u>	**86.95**
		80%	41.51	28.39	9.30	81.80	38.08	83.80	<u>85.47</u>	**87.56**

4.3 Overall Performance

The overall performance is shown in Table 3. RGCN and GraphSAGE and FdGars, are general graph neural networks and perform badly on both datasets. The main reason is traditional GNNs cannot handle the noises introduced by camouflage, especially when severe camouflage that exists in both datasets as illustrated in Subsect. 4.2. Among them, FdGars gets the worst performance since it cannot deal with multi-relations and is unable to filter out any noise brought by camouflage. DCI performs better in AUC metric than traditional GNN since it decouples the end-to-end classification into representation learning and classification, and the self-supervised graph learning in DCI captures the relatively comprehensive information about fraudsters. Moreover, when the number of training data increases, DCI's performance decreases, too. It shows the decoupling strategy of DCI is unable to learn well when facing more instances with camouflage. GraphConsis, CARE-GNN, and FRAUDRE have relatively good performances among all methods. FRADURE has the best performance among them since it considers four types of inconsistencies caused by camouflage and propagates additional inconsistency features. However, the noise brought by dissimilar neighbors is unable to remove. Compared with FRAUDRE, CARE-GNN is worse on two datasets. Especially on Yelp, AUC and AP are much lower. CARE-GNN selects nodes at each layer, which shows that the selection can filter out some neighbors connected by camouflaged edges successfully. However, the aggregation is based on node features. Thus, for Amazon, whose features provide more information than structures, it works well. but for Yelp, whose features do not provide much information (see discussion in Subsect. 4.5), it cannot work well. With more training data, its performance on Yelp is even worse. Graph-Consis has better performance than CARE-GNN, but worse than FRAUDRE, since it can select k-hop neighbors and filter out most of the neighbors who do not share the same class with the central node. More training data brings better performance. DC-GNN outperforms all methods. For Yelp, it has around 2%, 2% improvement over the second-best scores in AUC and AP, respectively. For Amazon, DC-GNN has about 1.5% and 2% improvement than the second best scores in AUC and AP, respectively. It means such a decoupled learning process can minimize the mutual influence between feature camouflage and relation camouflage and learn useful local structure information as much as possible. Overall, the performance shows the effectiveness of DC-GNN.

4.4 Discussion on Reconstructed Matrices

To better understand how the node selection works, we show the average reconstructed local adjacency matrices with 20 selected nodes on Yelp in Fig. 3. Specifically, Fig. 3a–3c show the reconstructed matrices for fraudulent nodes, while Fig. 3d–3f show reconstructed matrices for benign nodes. All reconstructed local adjacency matrices of fraudsters have clear different patterns from those for benign nodes. It illustrates the reconstructed matrices are able

Table 4. Study on Selected Nodes of Yelp.

#Selected Nodes	AUC	AP	Distance	LabelSim
5	84.41	47.33	1.1311	0.2111
10	85.69	52.32	1.1955	0.2033
15	86.57	54.54	1.2197	0.2026
20	88.13	58.82	1.2306	0.2008

to give characteristics of fraudulent nodes. Furthermore, we show the distance and label similarity of selected nodes to central nodes in Table 4. Distance is bigger than 1, which shows selected nodes contain 2-hop neighbors. By selecting 2-hop neighbors, DC-GNN can reach more nodes and obtain richer information. The label similarity is low, which means the most of the selected neighbors are benign nodes. Thus, the reconstructed adjacency matrices are made of camouflaged edges, edges between benign nodes, and edges between fraudulent nodes. The camouflaged edges provide the links to benign nodes, while edges between benign nodes reveal the"community" that the fraudsters want to cheat. Therefore, the performance increases with the number of selected nodes increases. DC-GNN achieves the best performance with 20 nodes.

4.5 Ablation Study

To study the effectiveness of each module, we also conduct an ablation study with 60% training data. Fig. 4 shows the results. We first show how attribute and structure modules contribute seperately. Then, we add contrastive loss (CL) to attribute encoding and structure encoding separately and collectively. As illustrated in Fig. 4a, single attribute embedding or structure embedding cannot provide enough information to detect fraudsters in Amazon. By combining two modules, the performance gets a huge improvement. Interestingly, CL can better assist attributes than structure. It means attributes of fraudulent nodes are different from those of benign nodes. Meanwhile, structures of fraudulent nodes are extremely hard to distinguish from benign nodes, even getting worse performance with CL. As shown in Fig. 4b, combined embedding with CL is also the best performer on Yelp. However, different from Amazon, attribute embedding with CL gets much worse than it without CL, which means the attribute embedding in

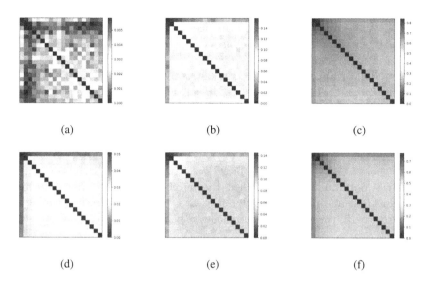

Fig. 3. Reconstructed Local Adjcency Matrices of Yelp. (Fig. 3a–3c for fraudulent nodes and Fig. 3d–3f for benign nodes.)

Yelp is hard to provide the right guidance under CL. Instead, the structure can provide useful guidance under CL to improve the model performance.

Number of Neighborhood. Here, we also discuss how the number of neighbors affects performance. As illustrated in Fig. 5, in both datasets, the difference is not much when choosing different numbers of neighbors for the individual attribute encoding module. Even for Amazon, DC-GNN has relatively high performance with 40 neighbors. Therefore, we can use a small-sized "context" to train the individual attribute encoding module, which can keep the training time within a reasonable range.

4.6 Parameter Analysis

Parameters Analysis in Contrastive Loss. We have four parameters in contrastive loss, including the coefficient of attribute contrastive loss λ_1, coefficient of structure contrastive loss λ_2, temperature τ, and the number of negative instances b. The overall results are shown in Fig. 6.

- λ_1. We show the results that $\lambda_1 \in \{0.0001, 0.001, 0.01, 0.1\}$ in Fig. 6a and 6e. For Yelp, the performance has a sudden drop when $\lambda_1 = 0.01$. Scores of 0.0001 and 0.001 are similar and better than those of 0.1. For Amazon, the performance remains stable under different λ_1.
- λ_2. Similar to λ_1, we show the results that $\lambda_2 \in \{0.0001, 0.001, 0.01, 0.1\}$ in Fig. 6b and 6f. Figures show that the performance affected by structure

information has a clear trend. When λ_2 decreases, the performance decreases as well. While for Yelp, the performance increases at $\lambda_2 = 0.001$ and then decreases, showing the best value for λ_2 is 0.001. For Amazon, the best value is 0.1.

- τ. τ is an important parameter in the contrastive loss. we set $\tau \in \{1 \times 10^{-6}, 1 \times 10^{-5}, 1 \times 10^{-4}, 1 \times 10^{-3}, 1 \times 10^{-2}, 1 \times 10^{-1}\}$. From Fig. 6c, we observe a sharp increase when τ change from 1×10^{-5} to 1×10^{-4}, and have a slight increase afterwards. From Fig. 6g, the performance first goes down at 1×10^{-4} and then reach a peak at 1×10^{-3}. Therefore, a reasonable τ is curial for model performance.

- b. We choose the b from $\{5, 10, 15, 20\}$. As indicated in Fig. 6d and 6h. For Yelp, the best number is 5. When the number increases, the performance decreases first and remains steady afterward. For Amazon, when the number increases to 15, the performance reaches a peak.

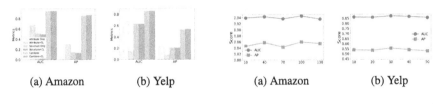

| (a) Amazon | (b) Yelp | (a) Amazon | (b) Yelp |

Fig. 4. Ablation Study   **Fig. 5.** Number of Neighbhors

5   Related Work

5.1   Graph Neural Networks and Graph Contrastive Learning

Graph Neural Networks (GNNs) [4,7,18] bring much easier computation along with better performance for graph-structured data. Generally, GNNs utilize the message-passing framework, which first aggregates all the message coming from the connected neighborhood, and then update the embedding for the central node. Prevailing methods to capture graph properties are in two granularities, including node [7,18], subgraph [6,9,27]. From a node view, GCN [7], Graph-SAGE [4] are some of the earliest works focusing on node classification. In heterogeneous graphs, many researchers use meta-path [3,33] to construct homogeneous graphs, and then apply GNN layers or attention mechanisms on top of them. From the subgraph view, researchers claim subgraph can bring finer information on subgraph level [6,19]. For instance, SubG-Con [6] better node embeddings by constructing node embeddings with subgraph embeddings. When using subgraph embeddings, a graph pooling layer like DGCNN [31], DiffPool [28] SAGPool [8], etc. are added after a GNN layer. These methods learn graph embedding with the help of node features. Different from them, IsoNN [14] can solely be based on the graph structure, but it cannot deal with large subgraphs.

Graph contrastive learning is often used in self-supervised learning, which requires positive samples and negative samples. Many works in literature contrast node embedding with its corresponding graph embedding or subgraph embedding [6,22], some contrast subgraph with subgraph [15,29]. However, in fraud detection, such contrast cannot be held due to camouflage. The neighborhood of fraudsters is noisy by having many benign nodes around. Therefore, previous contrast methods cannot be applied to fraud detection directly.

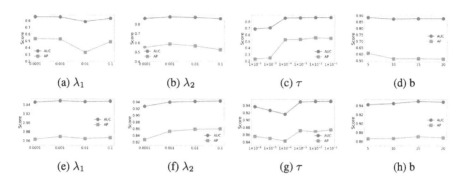

Fig. 6. Contrastive Loss Parameter Analysis Two datasets. Fig. 6a–6d shows the performance on Yelp, Fig. 6e–6h performance on Amazon.

5.2   GNN-Based Fraud Detection

Recent works have made much progress in applying GNN to fraud detection. FdGars [24] is the first paper using GCN [7] to detect fraudsters for online app review systems. Since only a very small portion of data has labels in fraud detection, SemiGNN [23] utilizes both labeled and unlabeled data in the loss function, while Player2Vec [32] construct multi-view networks from abundant information in the heterogeneous graph to enrich information. On the other hand, fraudsters often disguise themselves, therefore, some works like GraphConsis [12] and FRAUDRE [30] consider incorporating inconsistencies of node features and relations when applying GNN. Some other works [2,11,12] choose to filter out neighbors that do not share the same class before aggregation. CARE-GNN [2] utilizes reinforcement learning to distinguish nodes with camouflaged behaviors. Besides, some works still claim other issues, like class imbalance in GNN, incomplete information in the graph, etc. For example, PC-GNN [11] remedies the class imbalance problem by picking and choosing neighbors to aggregate at each layer. DCI [25] decouples the representation learning and classification to obtain performance gain. With many works in fraud detection, but rare papers are about the mutual effects of node features and graph structure under camouflage.

6   Conclusion

We study the fraud detection in graph setting with camouflage behaviors. Since feature camouflage and connection camouflage affect each other in traditional GNN learning. Therefore, we propose DC-GNN that decouples the learning process of traditional GNNs into attribute learning and structure learning, which adopts a graph transformer to learn attribute embedding based on four attributes and then select nodes from K-hop neighbors to reconstruct local adjacency matrices for all nodes. To learn structure embeddings, DC-GNN utilizes the multi-relational IsoNN. DC-GNN predicts labels by combining the attribute embedding and structure embedding. It also uses label-guided contrastive losses to enhance the performance. We conduct extensive experiments on two real-world datasets, and the results demonstrate the effectiveness of DC-GNN.

Acknowledgement. This work is partially supported by NSF through grants IIS-1763365 and IIS-2106972.

References

1. Bian, T., et al.: Rumor detection on social media with bi-directional graph convolutional networks. In: AAAI (2020)
2. Dou, Y., Liu, Z., Sun, L., Deng, Y., Peng, H., Yu, P.S.: Enhancing graph neural network-based fraud detectors against camouflaged fraudsters. In: CIKM, pp. 315–324 (2020)
3. Fu, X., Zhang, J., Meng, Z., King, I.: Magnn: metapath aggregated graph neural network for heterogeneous graph embedding. In: Proceedings of the Web Conference 2020, pp. 2331–2341 (2020)
4. Hamilton, W.L., Ying, R., Leskovec, J.: Inductive representation learning on large graphs. In NeurIPS, pp. 1025–1035 (2017)
5. Hooi, B., Song, H.A., Beutel, A., Shah, N., Shin, K., Faloutsos, C.: Fraudar: bounding graph fraud in the face of camouflage. In: Proceedings of the 22nd ACM SIGKDD International Conference on Knowledge Discovery and Data Mining (2016)
6. Jiao, Y., Xiong, Y., Zhang, J., Zhang, Y., Zhang, T., Zhu, Y.: Sub-graph contrast for scalable self-supervised graph representation learning. In: ICDM, pp. 222–231. IEEE (2020)
7. Kipf, T.N., Welling, M.: Semi-supervised classification with graph convolutional networks. arXiv preprint arXiv:1609.02907 (2016)
8. Lee, J., Lee, I., Kang, J.: Self-attention graph pooling. In: ICML (2019)
9. Li, X., Wei, W., Feng, X., Liu, X., Zheng, Z.: Representation learning of graphs using graph convolutional multilayer networks based on motifs. Neurocomputing **464**, 218–226 (2021)
10. Liang, T., et al.: Credit risk and limits forecasting in e-commerce consumer lending service via multi-view-aware mixture-of-experts nets. In: WSDM, pp. 229–237 (2021)
11. Liu, Y., et al.: Pick and choose: a gnn-based imbalanced learning approach for fraud detection. In: Proceedings of the Web Conference 2021, pp. 3168–3177 (2021)

12. Liu, Z., Dou, Y., Yu, P.S., Deng, Y., Peng, H.: Alleviating the inconsistency problem of applying graph neural network to fraud detection. In: SIGIR, pp. 1569–1572 (2020)
13. McAuley, J.J., Leskovec, J.: From amateurs to connoisseurs: modeling the evolution of user expertise through online reviews. In: WWW, pp. 897–908 (2013)
14. Meng, L., Zhang, J.: Isonn: isomorphic neural network for graph representation learning and classification. arXiv preprint arXiv:1907.09495 (2019)
15. Qiu, J., et al.: GCC: graph contrastive coding for graph neural network pretraining. In: Proceedings of the 26th ACM SIGKDD International Conference on Knowledge Discovery & Data Mining, pp. 1150–1160 (2020)
16. Rayana, S., Akoglu, L.: Collective opinion spam detection: bridging review networks and metadata. In: SIGKDD (2015)
17. Ren, Y., Zhu, H., Zhang, J., Dai, P., Bo, L.: Ensemfdet: an ensemble approach to fraud detection based on bipartite graph. In: ICDE (2021)
18. Schlichtkrull, M., Kipf, T.N., Bloem, P., van den Berg, R., Titov, I., Welling, M.: Modeling relational data with graph convolutional networks. In: Gangemi, A., et al. (eds.) ESWC 2018. LNCS, vol. 10843, pp. 593–607. Springer, Cham (2018). https://doi.org/10.1007/978-3-319-93417-4_38
19. Sun, Q., et al.: Sugar: subgraph neural network with reinforcement pooling and self-supervised mutual information mechanism. In: WWW, pp. 2081–2091 (2021)
20. Vaswani, A., et al.: Attention is all you need. In: Advances in Neural Information Processing Systems, pp. 5998–6008 (2017)
21. Veličković, P., Cucurull, G., Casanova, A., Romero, A., Lio, P., Bengio, Y.: Graph attention networks. arXiv preprint arXiv:1710.10903 (2017)
22. Veličković, P., et al.: Deep graph infomax. arXiv preprint arXiv:1809.10341 (2018)
23. Wang, D., et al.: A semi-supervised graph attentive network for financial fraud detection. In: 2019 IEEE International Conference on Data Mining (ICDM), pp. 598–607. IEEE (2019)
24. Wang, J., Wen, R., Wu, C., Huang, Y., Xion, J.: Fdgars: fraudster detection via graph convolutional networks in online app review system. In: WWW (2019)
25. Wang, Y., Zhang, J., Guo, S., Yin, H., Li, C., Chen, H.: Decoupling representation learning and classification for gnn-based anomaly detection. In: SIGIR, pp. 1239–1248 (2021)
26. Xu, K., Hu, W., Leskovec, J., Jegelka, S.: How powerful are graph neural networks? In: ICLR (2018)
27. Yang, C., Liu, M., Zheng, V.W., Han, J.: Node, motif and subgraph: leveraging network functional blocks through structural convolution. In: 2018 IEEE/ACM International Conference on Advances in Social Networks Analysis and Mining (ASONAM), pp. 47–52. IEEE (2018)
28. Ying, R., You, J., Morris, C., Ren, X., Hamilton, W.L., Leskovec, J.: Hierarchical graph representation learning with differentiable pooling. In: Proceedings of the 32nd International Conference on Neural Information Processing Systems, pp. 4805–4815 (2018)
29. You, Y., Chen, T., Sui, Y., Chen, T., Wang, Z., Shen, Y.: Graph contrastive learning with augmentations. Adv. Neural Inf. Process. Syst. 33, 5812–5823 (2020)
30. Zhang, G., et al.: Fraudre: fraud detection dual-resistant to graph inconsistency and imbalance. In: 2021 IEEE International Conference on Data Mining (ICDM), pp. 867–876. IEEE (2021)
31. Zhang, M., Cui, Z., Neumann, M., Chen, Y.: An end-to-end deep learning architecture for graph classification. In: AAAI (2018)

32. Zhang, Y., Fan, Y., Ye, Y., Zhao, L., Shi, C.: Key player identification in underground forums over attributed heterogeneous information network embedding framework. In: Proceedings of the 28th ACM International Conference on Information and Knowledge Management, pp. 549–558 (2019)
33. Zhao, J., Wang, X., Shi, C., Binbin, H., Song, G., Ye, Y.: Heterogeneous graph structure learning for graph neural networks. In: Proceedings of the AAAI Conference on Artificial Intelligence, vol. 35, pp. 4697–4705 (2021)

TUAF: Triple-Unit-Based Graph-Level Anomaly Detection with Adaptive Fusion Readout

Zhenyang Yu[1], Xinye Wang[1(✉)], Bingzhe Zhang[1], Zhaohang Luo[2], and Lei Duan[1(✉)]

[1] School of Computer Science, Sichuan University, Chengdu, China
{yuzhenyang,wangxinye,zhangbingzhe}@stu.scu.edu.cn, leiduan@scu.edu.cn
[2] Nuclear Power Institute of China, Chengdu, China
luozhaohang@qq.com

Abstract. Graph-level anomaly detection (GAD) has emerged as a significant research direction due to its practical application in diverse domains, such as toxic drug identification and compound activity assay. Existing GAD methods generally regard the node as the basic unit to learn graph representation, thus ignoring the vital information of the triple structure "node-edge-node". Intuitively, the occurrences of anomalous events in form of triples are the primary cause of an abnormal graph. Meanwhile, previous works adopt trivial readout strategies to obtain the graph-level representation without considering the different contributions of nodes. In this paper, we propose a novel GAD method named TUAF, based on triple-unit graphs with adaptive fusion readout. Specifically, we first transform the original graph into the triple-unit graph, and then learn triple representations for capturing abundant information about an edge and its corresponding nodes simultaneously. Furthermore, we design an adaptive fusion readout to obtain a high-quality graph-level representation by adaptively learning the optimal gravity coefficient for each triple. Through extensive experiments, we demonstrate the effectiveness of TUAF on 18 real-world datasets.

Keywords: Graph-level anomaly detection · Triple-unit graphs · Graph convolutional networks · Adaptive fusion readout

1 Introduction

Graph-based anomaly detection has achieved great success in various domains due to the excellent representation abilities of graphs and advanced graph representation learning techniques [1,14]. Existing works mainly focus on detecting abnormal nodes/edges/sub-graphs in a single large graph [4,6,21], while

This work was supported in part by the National Natural Science Foundation of China (61972268), and the Joint Innovation Foundation of Sichuan University and Nuclear Power Institute of China.

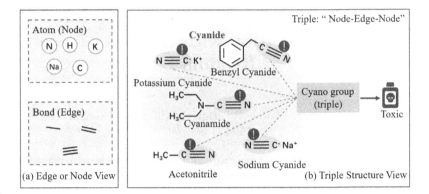

Fig. 1. A real-world example of the importance of triple in the graph-level anomaly detection task. Take the edge or node as the basic unit, we treat the elements of molecules separately. In the perspective of triple, we regard those "node-edge-node" structures as triples.

few studies attempt to explore the abnormal graphs detection from a set of graphs. It is worth noting that graph-level anomaly detection plays a vital role in many practical applications, such as recognizing molecule toxicity and drug side-effects [12], monitoring anomalies of transaction logs [15], and discriminating fake news [5]. With few studies but significant, there are still some challenges:

- **(CH1) How to learn graph representation for capturing triple structure information sufficiently?** Graph convolutional network (GCN) and its variants have shown powerful capability of learning graph representation [23]. However, a majority of graph representation learning techniques typically treat edges [9,22] or nodes [10,19] as the basic unit for calculation and aggregation. Intuitively, abnormal conditions are usually triggered by the triple of "node-edge-node". It means that considering edges or nodes separately is more likely to overlook anomalous information for detecting abnormal graphs. In addition, an edge establishes the tight bond between two nodes, while the two nodes are essential to forming an edge. Therefore, an edge and its corresponding nodes (i.e., a triple) are inseparable. The elaboration is given in Example 1.

Example 1. Toxic molecule detection is one of the important graph-level anomaly detection tasks. Figure 1(a) illustrates that these cyanides consist of five types of atoms and three types of chemical bonds, while in Fig. 1(b) the triple "C≡N" is the toxic cause of cyanides according to [8]. Cyanides are clinically proven to be a kind of toxic molecule in extensive cases, such as benzyl cyanide, cyanamide and acetonitrile. The components of cyanide seem to be normal in isolation (e.g., carbon atom, nitrogen atom and the triple bond). The toxicity of the molecule is caused by the formation of the triple "C≡N", named cyano group.

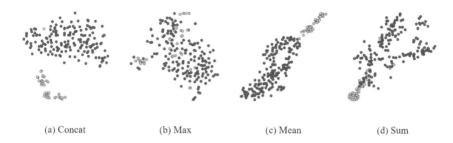

(a) Concat   (b) Max   (c) Mean   (d) Sum

Fig. 2. Illustration of graph-level representations output by four trivial readout strategies on a popular dataset AIDS. Visualization in (a) - (d) is based on t-SNE.

It can be observed that anomalous events usually occur in the form of triples which lead to an abnormal graph. Hence, learning graph representation that captures triple structure information is a crucial factor for detecting graph-level anomalies accurately.

- **(CH2) How to design an appropriate readout to obtain the graph-level representation for anomaly detection?** A proper readout strategy is helpful to generate a high-quality entire graph representation by aggregating the basic unit representations of a graph. Then, the abnormal graph detector can leverage the graph-level representation to calculate the anomaly score. However, the trivial readout strategies generate the suboptimal graph-level representation thus the abnormal graphs cannot be well distinguished in the potential space as shown in Fig. 2. This is because they capture general patterns of the graph on most occasions, while the different contributions of the basic units to the final graph-level representation are neglected.

To address the above challenges, we propose a novel graph-level anomaly detection method based on the **T**riple-**U**nit graph with **A**daptive **F**usion readout (TUAF). Specifically, **to tackle CH1**, we first transform the original graph into a triple-unit graph for capturing sufficient information about the anomalous event. After that, we utilize triples as the basic units for learning and updating the graph representation. **To tackle CH2**, before identifying anomalous graphs, we apply an adaptive fusion readout strategy. To alleviate the flattened results by trivial readout operation, it produces the graph-level representation by making a weighted summation of all triples.

To summarize, the main contributions of this paper are as follows:

- We construct a triple-unit graph that regards the triple structure "node-edge-node" as the basic unit. In this way, the anomalous events of an abnormal graph can be captured to learn graph representation sufficiently.
- We design an adaptive fusion readout strategy to obtain a high-quality graph-level representation. It explores the contribution of each triple to the entire-graph representation by learning gravity coefficients adaptively.
- We propose a novel method for graph-level anomaly detection. Extensive experiments verify the effectiveness of TUAF on 18 real-world datasets.

2   Related Work

We review the existing works on graph-level anomaly detection and graph representation learning which are most relevant to our work.

2.1   Graph-Level Anomaly Detection

Graph-based anomaly detection has received extensive attention on diverse types of graphs (e.g., static graphs, attribute graphs, and dynamic graphs) in recent years [14]. Most works have shown advanced performance on detecting anomalous nodes [4,11], anomalous edges [6,28], and anomalous subgraphs [21,29] in a single large graph. However, graph-level anomaly detection that recognizes the abnormal graphs in a collection of graphs has been explored fairly less.

Zhao et al. [27] introduced OCGIN to tackle the problem in a heterogeneous graph. Through delving into details, it discussed the intriguing "performance flip" phenomenon. Qiu et al. [16] proposed OCGTL to alleviate performance flip and hypersphere collapse by learning one-class term and transformation term simultaneously. Inspired by the knowledge distillation ideology, Ma et al. [13] presented GLocalKD that constructed an initial network and a prediction network separately as well as considered both local and global anomalies. Moreover, the other part of the studies endeavors to conduct complex graphs or generalized graphs. For instance, Nguyen et al. [15] designed CODEtect to address directed labeled graphs with multiple edges by identifying key motifs in transaction streams. Kipf et al. [5] utilized text representation learning techniques and graph neural network to detect fake news on a news propagation graph named UPFD.

Most of the methods we have mentioned above ignore the rich information of the triples, which usually represent anomalous event occurrence. Besides, to obtain entire graph representations, they mainly choose trivial readout strategies such as max or mean that lead to suboptimal results.

2.2   Graph Representation Learning

Graph representation learning plays an important role in graph anomaly detection and an effective graph representation learning technique can greatly improve our task. According to the basic unit of computation and aggregation in a graph, the approaches can be divided into two categories, i.e., edge-based and node-based approaches.

Node-Based Approaches. The original graph neural network (GNN) and its variants learn the graph representation in terms of nodes, e.g., graph convolutional network (GCN) [10], graph attention network (GAT) [19], and graph isomorphic network (GIN) [24]. Wang et al. [20] introduced semiGNN to learn node representations in a semi-supervised manner and it has shown excellent results in node classification. Xu et al. [25] proposed Graph2Seq to aggregate the node representations into a graph sequence representation by sum and max operations.

Edge-Based Approaches. Gong *et al.* [7] employed EGNN to take into account edge features. CensNet which was proposed by Jiang *et al.* [9] learned the graph representation by alternately updating the node and edge representations. Yang *et al.* [26] designed NENN based on GAT, which trained node-level and edge-level attention layers, respectively. Furthermore, similar to CensNet, Wang *et al.* [22] presented EGAT to incorporate edge characteristics into GAT.

3 Preliminaries

In this part, we present key concepts of the undirected labeled graph and triple-unit graph, then we formulate the problem of graph-level anomaly detection.

Definition 1. *Undirected Labeled Graph.* *An undirected labeled graph is $G = \{V, E, L_v, L_e, \phi, \varphi\}$, where V and E represent sets of nodes and edges, respectively. In addition, ϕ and φ denote two corresponding label mapping functions. The former is a node label mapping function $\phi: V \to L_v$ and the latter is an edge label mapping function $\varphi: E \to L_e$, where L_v and L_e are collections of node labels and edge labels accordingly.*

Definition 2. *Triple-Unit Graph.* *A triple-unit graph is $\widetilde{G} = \{T, R, X\}$ containing a triple set T, a relation set R and a triple feature matrix X. In a triple-unit graph, a triple is denoted as $t = \langle u, e, v \rangle$. Among them, $u, v \in V$ are two neighboring nodes of an undirected labeled graph G following the Definition 1, while $e \in E$ represents an edge between node u and v. A relation $r \in R$ connecting two triples indicates that they share the same node. According to functions ϕ and φ, each row of X refers to a triple feature. It can be denoted as $x = [l_u; l_e; l_v]$ by concatenating the labels of u, e and v, where $l_u, l_v \in L_v$ and $l_e \in L_e$.*

Contrary to detecting anomalous nodes, edges, and sub-graphs in a large single graph, we aim to recognize unusual graphs that deviate significantly from others (e.g., identifying toxic drugs from a collection of molecules). Therefore, our problem can be formulated as:

Problem 1. *Graph-level Anomaly Detection.* Given a set of graphs $\mathcal{G} = \{G_i\}_{i=1}^{|\mathcal{G}|}$, where each graph $G \in \mathcal{G}$ is an undirected labeled graph. The goal of graph-level anomaly detection is to learn an anomaly score function $\hat{y}_G = \mathcal{F}(G; \Theta)$, where $\hat{y}_G$ represents the probability that the graph G is anomalous, and Θ is a set of the trainable parameters inside the anomaly score function $\mathcal{F}$.

4 The Design of TUAF

In this section, we first introduce the overall architecture of TUAF. Then, we will discuss three components of TUAF: (i) the triple-unit graph generation which treats the "node-edge-node" structures of the input graph as triples; (ii) triple representations updating through graph encoder and decoder; and (iii) recognizing anomalous graph by training an anomaly detector with adaptive fusion readout. Finally, we describe the joint optimization process in detail.

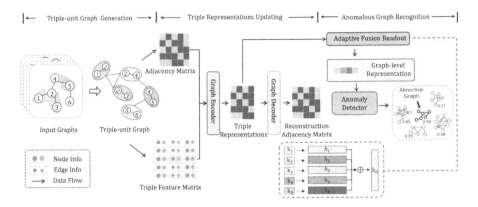

Fig. 3. The architecture of TUAF includes triple-unit graph generation, triple representations updating, the anomalous graph recognition, respectively.

4.1 Architecture Overview

The overall framework of TUAF is illustrated in Fig. 3, we will introduce following the guidance of data flow arrows.

In the triple-unit graph generation stage, it aims to transform the input graphs into triple-unit graphs. Each input is composed of an undirected labeled graph, and the output includes an adjacency matrix of the triple-unit graph and a triple feature matrix.

In the triple representations updating stage, it learns and updates the triple representations through graph encoder and decoder modules. Given an adjacency matrix and a triple feature matrix, a graph encoder module yields the triple representations. Then, leveraging the corresponding triple representations, the graph decoder module reconstructs the triple-unit graph to obtain the reconstruction adjacency matrix. The reconstruction loss is calculated by it and the original adjacency matrix (i.e., $Loss_1$).

In the anomalous graph recognition stage, it first conducts a readout operation to obtain the graph-level representation and then through a one-class classifier so as to recognize the abnormal graph. To provide a high-quality graph-level representation, we design an adaptive fusion readout strategy applied to triple representations aggregation. Subsequently, taking the graph-level representation as input we can calculate the one-class classifier loss (i.e., $Loss_2$).

With the process above-mentioned, the method is jointly optimized based on the losses of reconstruction $Loss_1$ and one-class classifier $Loss_2$, and the detail is given in Sect. 4.5.

4.2 Triple-Unit Graph Generation

In reality, anomalous events usually occur in form of triples, like the connection of carbon and nitrogen with a triple bond (i.e., cyano group "C ≡ N") shown in

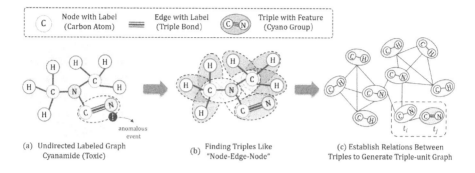

Fig. 4. Triple-unit graph generation process of cyanamide molecule.

Fig. 4(a). In other words, a triple serves as the minimum unit that leads to an abnormal graph. Hence, we propose an approach to transform the original graph G (e.g., an undirected labeled graph) to the triple-unit graph $\widetilde{G}$ for capturing anomalous information adequately.

Concretely, given an undirected labeled graph G with a set of nodes V and a set of edges E, we first construct the triple set $T = \{t_1, t_2, \cdots, t_{|T|}\}$ by finding triples that satisfy the "node-edge-node" structure. Thus, a triple $t = \langle u, e, v \rangle$, where $u, v \in V$ are two neighboring nodes and $e \in E$ represents the edge between them. It can be observed in Fig. 4(b) that each $v \in V$ could correspond to one or multiple triples and the number of edges $|E|$ is identical to the number of triples $|T|$.

Next, to get the relation set R, we must establish the relations among triples according to the link rule. Simply put, there exists a relation between two triples on the condition that they share the same node. For example, as shown in Fig. 4(c), triple t_i and triple t_j have a relationship because they share the same carbon atom. Meanwhile, the adjacency matrix $A \in \mathbb{R}^{|T| \times |T|}$ of the triple-unit graph $\widetilde{G}$ can be obtained in terms of relations, where A_{ij} represents the element of A in row i, column j. Note that $A_{ij} = 1$ indicates that the triple t_i is connected with triple t_j.

In addition, L_v and L_e are a node label set and an edge label set of G, respectively, such as the types of atoms and bonds. Thus, we can get the triple feature $X \in \mathbb{R}^{|T| \times D}$, where the triple feature dimension D is equal to $2|L_v| + |L_e|$. Each row of X represents the feature of a triple by concatenating the node and edge label with one-hot coding which can be denoted as $x = [l_u; l_e; l_v]$.

4.3 Triple Representation Updating

To obtain superior triple representations, the graph encoder takes the triple-unit graph as input to learn the representation of triples in the latent space, and then the graph decoder outputs a new adjacency matrix by reconstruction.

As for graph encoder, it is well known that there are various graph representation learning techniques, such as GCN [10], GAT [19], GIN [24] and their variants. In our work, we pick up the widely used and effective GCN serving as the graph encoder module. Specifically, given an adjacency matrix A and a triple feature matrix X, the graph encoder learns the triple representations Z as follows:

$$Z^{(l+1)} = \sigma \left(D^{-\frac{1}{2}} \widetilde{A} D^{-\frac{1}{2}} Z^{(l)} W_1^{(l)} \right) \tag{1}$$

where $Z^{(l)}$ and $W_1^{(l)}$ are the input and trainable weight matrix of the l-th graph convolutional layer respectively, while $Z^{(l+1)}$ is the output. Note that X is taken as the input of the first layer, i.e., $Z^{(0)} = X$. Here, $\widetilde{A}$ has a slight difference from A which is a sum of the identity matrix and itself, and D denotes the diagonal degree matrix of the triple-unit graph $\widetilde{G}$. Besides, $\sigma(\cdot)$ is an activation function, we generally choose ReLU. Iteratively, we can obtain the final output Z.

With regard to the graph decoder, the triple representations Z is given to compute the similarity among triples by performing the inner product. In fact, the reconstruction adjacency matrix $\hat{A}$ is composed of the similarities of all pairs of triples. The calculation process is described as follows:

$$\hat{A} = \text{Sigmoid} \left(Z Z^\top \right) \tag{2}$$

where Sigmoid is the activation function, and $Z^\top$ refers to the transpose operation of Z.

In order to get the optimal representations, we update the triple representations on the basis of reconstruction loss, which can be defined as:

$$Loss_1 = \left\| A - \hat{A} \right\|_F^2 = \sum_{i,j=1}^{|T|} \left(A_{ij} - \hat{A}_{ij} \right)^2 \tag{3}$$

where $\| \cdot \|_F$ refer to the Frobenius norm, i.e., squaring and re-squaring the absolute values of the matrix elements.

4.4 Anomalous Graph Recognition

Through the above stage, we obtain the triple representations of the triple-unit graph. To tackle the anomalous graph recognition task, it is necessary to fuse triple representations into a graph-level representation at first. Then, the anomaly detector can leverage it to identify whether the graph is abnormal or normal.

Adaptive Fusion Readout. Note that the anomaly detector cannot directly utilize the triple representations for identifying anomalous graphs. To get the graph-level representation, there have been a number of trivial ways to integrate the basic units of a graph. For instance, it can be obtained by various readout strategies, such as mean, sum, max, and concat. However, these readout strategies solely capture the general pattern of the graph and fail to take into account the different contributions of units (e.g., triples).

Hence, to handle the above problem, we design an adaptive fusion readout strategy to produce a high-quality graph-level representation, which is more conducive to the advancement of the graph-level anomaly detection task. To be specific, we first calculate the gravity coefficient to distinguish the importance of each triple as follows:

$$k = \text{LeakyReLU}\left(W_2\left[W_3 h \odot W_4 \hat{h}\right]\right) \tag{4}$$

where k indicates the importance of a triple to the final graph representation. h is a triple representation, while $\hat{h}$ is a cursory representation of the global graph by conducting a mean fusion operation on all triples. Here, $\odot$ denotes the element-wise product. W_2, W_3, and W_4 are the learnable parameters, and LeakyReLU is the activation function that is capable of exploiting both positive and negative signals.

Subsequently, to make it comparable to different triples of a graph, we adopt Softmax function to normalize the gravity coefficient:

$$\hat{k}_i = \text{Softmax}(k) = \frac{\exp(k_i)}{\sum_{i=1}^{|T|} \exp(k_i)} \tag{5}$$

Based on the triple representations and their corresponding gravity coefficients, we can calculate a weighted sum to obtain the graph-level representation as follows:

$$h_G = \sigma\left(\sum_{i=1}^{|T|} \hat{k}_i h_i\right) \tag{6}$$

where $\sigma(\cdot)$ denotes an activation function, and h_G is the final graph representation of the input graph G.

Anomaly Detector Training. To facilitate comparison, we use one of the most popular one-class classifiers deep support vector data description (DeepSVDD) [17] as anomaly detector followed [16,27]. The goal is to learn a hypersphere to enclose normal samples and exclude anomalies from it. Specifically, we learn a hypersphere by minimizing the average distance of all graph-level representations with center c and the loss can be defined as:

$$Loss_2 = \frac{1}{|\mathcal{G}|} \sum_{G \in \mathcal{G}} \|h_G - c\|^2 \tag{7}$$

where h_G represents the graph-level representation of the input graph G, and $|\mathcal{G}|$ indicates the total number of the input graphs. It should be noted that the center c is initialized by calculating the mean of all training graph-level representations according to [17].

Moreover, for a given test graph sample G, the one-class classifier outputs the anomaly score which is defined by the distance of its graph-level representation to the center c. It is formulated as follows:

$$Score\,(h_G^*) = \|h_G^* - c\|^2 \tag{8}$$

where h_G^* denotes the optimal graph-level representation after all modules have completed training and the parameters inside them are fixed.

4.5   Joint Optimization

Based on the above description, there are two essential losses including reconstruction loss $Loss_1$ (Sect. 4.3) and one-class classifier loss $Loss_2$ (Sect. 4.4). To perform graph-level anomaly detection task, the total loss function under joint optimization is defined as:

$$Loss = \sum_{i,j=1}^{|T|} \left(A_{ij} - \hat{A}_{ij}\right)^2 + \frac{1}{|\mathcal{G}|} \sum_{G \in \mathcal{G}}^{|\mathcal{G}|} \|h_G - c\|^2 + \lambda\|\theta\|_2^2 \tag{9}$$

where θ denotes the parameters of TUAF and λ controls the L_2 regularization that is utilized to alleviate overfitting. As a result, TUAF is optimized according to this final loss by backpropagation to train an optimal anomaly detector for recognizing abnormal graphs.

5   Experiments

In this section, we conduct experiments extensively to evaluate TUAF in a comprehensive way. We aim to answer the following research questions:

- **RQ1:** How does TUAF perform vs. the state-of-the-art baselines on graph-level anomaly detection task?
- **RQ2:** What is the effect of triple structures which consider node information and edge information simultaneously?
- **RQ3:** Does each of the key modules makes a contribution to TUAF?
- **RQ4:** How does the proposed readout strategy facilitate TUAF?

5.1   Experimental Setups

Datasets. TUAF is evaluated on 18 real-world datasets, which are collected from graph benchmarks platform[1]. The statistics of these datasets are summarized in Table 1. In order to adapt to the anomaly detection task, we regard the class with minority graphs as anomalies, following [3,13]. Actually, these datasets could prove to be an anomaly semantically, such as the toxic drug and compounds associated with tumors. It is worth noting that one important criterion for selecting these datasets is to satisfy that both edges and nodes have category labels so as to the construction of our triples.

---

[1] https://chrsmrrs.github.io/datasets/docs/datasets/.

Table 1. The statistics of dataset.

Dataset	#Graphs	#V-types	#E-types	#Outliers	Description
PTC_MR	344	18	4	152	Carcinogenic on Male Rats
PTC_FM	349	18	4	143	Carcinogenic on Female Mice
PTC_FR	351	19	4	121	Carcinogenic on Female Rats
AIDS	2000	38	3	400	Anti-HIV Activity
ARE	7953	54	4	1237	Antioxidant Responsive Element
MMP	8099	54	4	1238	Mitochondrial Membrane Potential
HSE	9024	54	4	458	Heat Shock Factor Response Element
AhR	9048	51	4	1051	Aryl Hydrocarbon Receptor
PPAR	9053	53	4	268	Peroxisome Proliferator-activated Receptor
p53	9516	54	4	605	DNA damage p53 pathway
PC-3	27509	45	3	1568	Prostate Tumor
MCF-7	27770	46	3	2294	Breast Tumor
MOLT-4	39765	64	3	3140	Leukemia Tumor
UACC257	39988	64	3	1643	Melanoma Tumor
SN12C	40004	65	3	1955	Renal Tumor
SF-295	40271	65	3	2025	Central Nerv Sys Tumor
SW-620	40532	65	3	2410	Colon Tumor
P388	41472	72	3	2298	Leukemia Tumor

Baselines. To evaluate the performance, we compared TUAF with seven methods including four variants of support vector machines (SVM) with two graph kernels and three recent methods based on GNN. Since the multiple edge graph and text information required by CODEtect [15] and UPFD [5] are different from our objectives, they are not included in the comparison.

- **SVM variants** are traditional methods including $KSVM_{GR}$, $KSVM_{SP}$, $LSVM_{GR}$ and $LSVM_{SP}$. A linear/kernel SVM is utilized as a detector combined with graphlet kernel (GR) [18] and shortest-path kernel (SP) [2].
- **OCGIN** [27] is a GNN-based method that employs a heterogeneous graph network with a one-class classifier SVDD to recognize anomalous graphs.
- **OCGTL** [16] is a GNN-based method by incorporating a one-class classifier and graph transformation learning. Besides, it leverages a new readout approach to pick the most suitable aggregation way.
- **GLocalKD** [13] is a GNN-based method. It constructs a random network and a predictive network based on the knowledge distillation ideology.

Evaluation. Due to the data imbalance, we use area under curve (AUC) which is the less sensitive and popular anomaly detection evaluation metric [3,13]. A higher AUC value implies better performance to identify anomalies. We report results by the mean and standard deviation of AUC (%) based on 5-fold cross-validation for most datasets, apart from ARE, MMP, AhR, PPAR, and p53 that have provided training, validation, and testing splits. For these datasets, we run five times independently with distinct random seeds to obtain the average result.

Implementation Details. TUAF is implemented by the deep learning framework PyTorch. We employed three GCN layers for graph representation learning to stay the same with the other baselines. We set the dimension of the hidden layer to 512 except for the PTC_FM, AhR, ARE, MMP, and PPAR datasets, where the dimension of the hidden layer is set to 256. We adopt SGD optimizer with 5×10^{-4} weight decay recommended by [17]. Besides, the learning rate is selected from $\{10^{-1}, 10^{-2}, \cdots, 10^{-5}\}$ by grid search, and the number of 50 epochs is enough to g et great results. Our source code and datasets are publicly available on GitHub[2]. As for baselines, we execute the open-source codes provided by the authors with their recommended parameter settings.

Table 2. The performance with compared methods.

Dataset	$KSVM_{GR}$	$KSVM_{SP}$	$LSVM_{GR}$	$LSVM_{SP}$	OCGIN	OCGTL	GLocalKD	TUAF
PTC_MR	54.4±0.9	58.2±1.3	54.8±1.0	58.2±1.5	48.2±9.4	62.5±7.6	52.8±9.3	**68.7±3.1**
PTC_FM	55.5±1.4	56.2±1.3	56.1±1.3	57.0±1.1	49.1±9.5	62.0±10.1	59.7±3.5	**67.5±5.8**
PTC_FR	53.1±0.5	59.8±1.5	54.2±1.9	57.7±0.3	48.2±8.4	56.9±8.0	52.5±8.4	**82.7±1.7**
AIDS	82.0±0.5	97.0±0.0	82.4±0.6	97.3±0.1	56.0±18.3	99.1±1.0	99.7±0.6	**99.9±0.0**
ARE	51.0±0.1	52.3±0.4	50.8±0.2	54.3±0.3	54.4±10.9	57.2±3.1	56.5±0.4	**63.2±4.4**
MMP	50.8±0.1	60.9±1.1	50.8±0.3	64.7±0.6	52.8±12.0	57.9±2.5	67.4±0.2	**72.2±1.5**
HSE	56.8±0.2	54.6±0.8	55.2±0.4	57.4±0.8	55.6±11.1	58.7±4.0	58.3±0.2	**61.8±1.0**
AhR	51.7±0.1	52.2±0.1	51.6±0.3	54.3±0.4	58.0±11.4	55.4±2.7	55.3±0.2	**76.6±5.7**
PPAR	54.8±0.1	55.5±0.5	54.6±0.3	55.1±0.4	46.9±10.5	58.4±6.9	64.0±0.2	**66.7±2.8**
p53	50.9±0.1	53.8±0.7	50.7±0.1	53.3±0.6	51.4±11.8	56.6±3.0	63.9±0.2	**69.0±1.5**
PC-3	50.4±0.0	50.6±0.1	50.2±0.0	50.2±0.1	52.5±2.6	65.6±3.1	67.9±1.4	**70.8±0.9**
MCF-7	50.3±0.0	50.9±0.1	50.0±0.0	50.3±0.1	50.1±4.2	64.7±2.1	64.6±0.8	**67.8±0.5**
MOLT-4	–	–	50.1±0.0	50.1±0.0	49.6±5.4	50.9±2.2	63.8±1.3	**64.8±0.7**
UACC257	–	–	50.1±0.0	50.1±0.0	50.5±8.6	50.9±1.7	68.5±1.8	**70.7±0.7**
SN12C	–	–	50.1±0.0	50.1±0.0	51.1±8.3	55.3±6.9	68.1±1.6	**69.8±0.6**
SF-295	–	–	50.2±0.0	50.2±0.0	50.5±9.6	50.3±1.0	66.0±1.1	**70.9±1.4**
SW-620	–	–	50.0±0.0	50.0±0.0	50.3±9.2	50.0±0.3	69.0±1.7	**72.3±1.0**
P388	–	–	53.7±0.1	55.4±0.3	48.9±9.3	51.0±1.7	62.2±1.5	**66.1±0.8**

"–": the data scale is so large that kernel methods fail to get the results.

5.2   Performance Comparisons (RQ1)

We compare TUAF with seven of the state-of-the-art graph-level anomaly detection baselines mentioned in Sect. 5.1 to estimate the performance. The results are listed in Table 2, and the analysis are as follows:

– TUAF attains the best performance over all baselines. Compared to its strongest contenders, TUAF achieves more than 5% improvement on average. Notably, the AUC value has been improved significantly on several datasets such as PTC_FM (22.9%), ARE (6.0%) and AhR (18.6%). The experimental

---

[2] https://github.com/scu-kdde/OAM-TUAF-2023.

results indicate that TUAF has the powerful capability of detecting abnormal graphs because it can capture anomalous information sufficiently based on triple structures. Furthermore, the designed adaptive fusion readout is also helpful to produce a high-quality graph-level representation thus promoting the effectiveness.

– Nearly all GNN-based methods outperform the traditional methods, which reflects that GNN has an excellent ability to learn graph representation to improve anomaly detection. In addition, compared to graph kernel, GNN focuses on the local information such as node and its neighbors which to some extent implies the anomalous causes. Among SVM variants, the SP kernel shows better performance than GR which indicates that it is adaptive on graphs with different distributions.

– Among GNN-based methods, OCGTL and GLocalKD achieve higher AUC than OCGIN on most datasets. This is because OCGTL adopts OCPool to find the best readout strategy, which preserves more critical information to aggregate nodes into the final graph representation. As for GLocalKD, it exploits locally and globally to capture as much anomalous information as possible. This further validates the importance of triple structures which indicates that anomalous events and an appropriate readout strategy can also benefit the performance of anomaly detection.

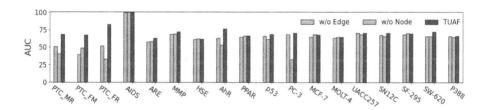

Fig. 5. The effect of triple structure.

5.3 The Effect of Triple Structure (RQ2)

To investigate the effect of triple structure, we remove the edge and node information of the triple, respectively. Specifically, we generate two variants of TUAF that only use node or edge information in triple-unit graphs:

– **TUAF w/o Edge:** TUAF without using edge information.
– **TUAF w/o Node:** TUAF without using node information.

As shown in Fig. 5, the performance of TUAF w/o Edge and TUAF w/o Node is inferior to TUAF on most occasions. Note that TUAF w/o Edge is significantly worse than TUAF w/o Node on some datasets, such as PTC_FM, PPAR, and MCF-7. One possible reason is that the edge information dominates these datasets. Conversely, the reverse occurs in PTC_FR, AhR, and PC-3. We found that TUAF w/o Node obtain the best results on MCF-7 and MOLT-4. We infer that it can distinguish between abnormal and normal graphs by using the node information on these two datasets, while adding edge information may introduce some noise. In a nutshell, a graph-level anomaly detection task can obtain the maximum enhancement generally while using a triple structure, because it contains abundant information that represents the occurrence of anomalous events.

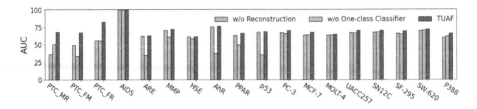

Fig. 6. The ablation of key modules.

5.4 Ablation Study (RQ3)

To verify the effectiveness of two key components separately, we conduct an ablation study of TUAF on two variants as follows:

- **TUAF w/o Reconstruction:** TUAF without graph decoder modules.
- **TUAF w/o One-class Classifier:** TUAF without one-class classifier.

Based on results shown in Fig. 6, we observe that the complete TUAF achieves the best performance. The AUC value of TUAF w/o Reconstruction and TUAF w/o One-class Classifier is lower than TUAF, which demonstrates the contribution of both components of TUAF. On the one hand, TUAF w/o Reconstruction obtains weaker performance than TUAF on PTC_MR, PTC_FM, and P388, which suggests that the reconstruction process can optimize the triple representations to some extent. On the other hand, TUAF w/o One-class Classifier gets even poor performance on a majority of datasets, such as ARE, AhR, PPAR, and p53. It is indicative of the important role of a one-class classifier that serves as the anomaly detector. Hence, the evidence manifests that it is indispensable to adopt two parts of losses to jointly optimize our method.

5.5 Analysis of Readout Strategy (RQ4)

To explore the impact of different readout strategies, we compared the adaptive fusion readout strategy with four popular strategies. As shown in Fig. 7, the proposed readout achieves superior performance on a majority of datasets. The max readout strategy leads to the poorest results. Because the smaller values

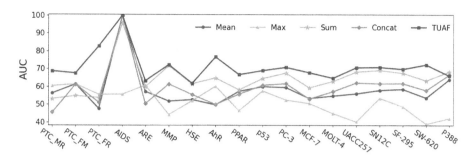

Fig. 7. The comparison of readout strategies.

in some dimensions may also be the key to causing anomalies. The sum and concat readout achieve sub-optimal results on most occasions due to taking all triple representations into account. In addition, the result of the mean readout is unstable and weaker because it flattens the representations of triple thus overlooking the abnormal information. It is observed that the sum readout strategy obtains a better result on P388, the probable reason may be that nearly all triple representations are quite similar and equally important in this dataset. Therefore, the effectiveness of the proposed readout strategy is verified.

6 Conclusion

In this paper, we propose TUAF to tackle graph-level anomaly detection. Through triple-unit graph generation, the anomalous information can be captured sufficiently by considering the triple structure "node-edge-node". After learning and updating triple representation, an adaptive readout strategy is conducive to obtaining high-quality graph-level representation. Extensive experiments have shown that the proposed method achieves superior performance over the state-of-the-art counterparts. Moreover, we find that a suitable graph readout approach is essential to the graph-level anomaly detection task, and experimental results show that the readout strategy we designed achieves the best results on most datasets compared to other popular strategies.

References

1. Akoglu, L., Tong, H., Koutra, D.: Graph based anomaly detection and description: a survey. DMKD **29**(3), 626–688 (2015)
2. Borgwardt, K.M., Kriegel, H.: Shortest-path kernels on graphs. In: ICDM, pp. 74–81 (2005)
3. Campos, G.O.: On the evaluation of unsupervised outlier detection: measures, datasets, and an empirical study. Data Min. Knowl. Disc. **30**(4), 891–927 (2016)
4. Ding, K., Li, J., Bhanushali, R., Liu, H.: Deep anomaly detection on attributed networks. In: SDM, pp. 594–602 (2019)

5. Dou, Y., Shu, K., Xia, C., Yu, P.S., Sun, L.: User preference-aware fake news detection. In: SIGIR, pp. 2051–2055 (2021)
6. Duan, D., Tong, L., Li, Y., Lu, J., Shi, L., Zhang, C.: AANE: anomaly aware network embedding for anomalous link detection. In: ICDM, pp. 1002–1007 (2020)
7. Gong, L., Cheng, Q.: Exploiting edge features for graph neural networks. In: CVPR, pp. 9211–9219 (2019)
8. Hendry-Hofer, T.B., et al.: A review on ingested cyanide: risks, clinical presentation, diagnostics, and treatment challenges. JMT **15**(2), 128–133 (2019)
9. Jiang, X., Ji, P., Li, S.: Censnet: Convolution with edge-node switching in graph neural networks. In: IJCAI, pp. 2656–2662 (2019)
10. Kipf, T.N., Welling, M.: Semi-supervised classification with graph convolutional networks. In: ICLR (2017)
11. Kumagai, A., Iwata, T., Fujiwara, Y.: Semi-supervised anomaly detection on attributed graphs. In: IJCNN, pp. 1–8 (2021)
12. Li, J., Cai, D., He, X.: Learning graph-level representation for drug discovery. CoRR abs/1709.03741 (2017)
13. Ma, R., Pang, G., Chen, L., van den Hengel, A.: Deep graph-level anomaly detection by glocal knowledge distillation. In: WSDM, pp. 704–714 (2022)
14. Ma, X., et al.: A comprehensive survey on graph anomaly detection with deep learning. TKDE (2021)
15. Nguyen, H.T., Liang, P.J., Akoglu, L.: Detecting anomalous graphs in labeled multi-graph databases. TKDD **17**, 1–25 (2022)
16. Qiu, C., Kloft, M., Mandt, S., Rudolph, M.: Raising the bar in graph-level anomaly detection. In: IJCAI, pp. 2196–2203 (2022)
17. Ruff, L., et al.: Deep one-class classification. In: ICML, pp. 4390–4399 (2018)
18. Shervashidze, N., Vishwanathan, S.V.N., Petri, T., Mehlhorn, K., Borgwardt, K.M.: Efficient graphlet kernels for large graph comparison. In: AISTATS, pp. 488–495 (2009)
19. Velickovic, P., Cucurull, G., Casanova, A., Romero, A., Liò, P., Bengio, Y.: Graph attention networks. In: ICLR (2018)
20. Wang, D., et al.: A semi-supervised graph attentive network for financial fraud detection. In: ICDM, pp. 598–607 (2019)
21. Wang, H., Zhou, C., Wu, J., Dang, W., Zhu, X., Wang, J.: Deep structure learning for fraud detection. In: ICDM, pp. 567–576 (2018)
22. Wang, Z., Chen, J., Chen, H.: EGAT: edge-featured graph attention network. In: ICANN, pp. 253–264 (2021)
23. Wu, Z., Pan, S., Chen, F., Long, G., Zhang, C., Yu, P.S.: A comprehensive survey on graph neural networks. TNNLS **32**(1), 4–24 (2021)
24. Xu, K., Hu, W., Leskovec, J., Jegelka, S.: How powerful are graph neural networks? In: ICLR (2019)
25. Xu, K., Wu, L., Wang, Z., Feng, Y., Sheinin, V.: Graph2seq: graph to sequence learning with attention-based neural networks. CoRR abs/1804.00823 (2018)
26. Yang, Y., Li, D.: NENN: incorporate node and edge features in graph neural networks. In: ACML, pp. 593–608 (2020)
27. Zhao, L., Akoglu, L.: On using classification datasets to evaluate graph outlier detection: peculiar observations and new insights. Big Data (2020)
28. Zheng, L., Li, Z., Li, J., Li, Z., Gao, J.: Addgraph: anomaly detection in dynamic graph using attention-based temporal GCN. In: IJCAI, pp. 4419–4425 (2019)
29. Zheng, M., Zhou, C., Wu, J., Pan, S., Shi, J., Guo, L.: Fraudne: a joint embedding approach for fraud detection. In: IJCNN, pp. 1–8 (2018)

Region-Aware Graph Convolutional Network for Traffic Flow Forecasting

Haitao Liang, An Liu$^{(\boxtimes)}$, Jianfeng Qu, Wei Chen, Xiaofang Zhang, and Lei Zhao

School of Computer Science and Technology, Soochow University, Suzhou, China
20204227027@stu.suda.edu.cn,
{anliu,jfqu,robertchen,xfzhang,zhaol}@suda.edu.cn

Abstract. Urban traffic flow prediction is a crucial service in intelligent transportation systems. It is very challenging due to the complex spatiotemporal dependencies and inherent uncertainty caused by dynamic urban traffic conditions. Recent work has focused on designing complex Graph Convolutional Network (GCN) architectures to capture spatial dependencies among segment-level traffic status and achieves state-of-the-art performance. But these GCN based methods has two shortcomings. One on hand, they ignore cross-region movement which reflects traffic flow transfer patterns at the regional level. On the other hand, they fail to capture the long-term temporal dependencies of traffic flows due to its non-linearity and dynamics. In order to address the above-mentioned deficiencies, we propose a novel Region-aware Graph Convolution Networks (RGCN) for traffic forecasting. Specially, a DTW-based pooling layer is introduced to capture the cross-regional spatial correlation, based on which a traffic region graph is constructed from the original traffic network and is employed to model cross-region traffic flow. Besides, a transformer-based temporal module is proposed to model long-term and dynamic temporal dependencies across multiple time steps. The proposed model is evaluated on two public traffic network datasets and the experimental results show that RGCN outperforms the state-of-the-art baselines.

Keywords: Graph Neural Network · Traffic Flow Prediction · Spatio-temporal data mining

1 Introduction

With the development of urbanization and the advancement of information technology, although more and more practical problems have surfaced, corresponding solutions have also emerged. As one of the most important issues, urban transport systems are under enormous pressure from growing populations and vehicles. We need to develop an advanced Intelligent Transportation Systems (ITS) [1,2] to deal with the problem. Currently, traffic flow prediction has become a vital component of advanced ITS. Accurate traffic prediction information can provide traffic managers with a strong basis for traffic decision-making, and at the same time allow drivers to choose a smoother road to travel, thereby avoiding or alleviating traffic congestion. However, prediction is particularly challenging due to the dynamic spatial dependencies of traffic networks and the complex temporal dependencies of traffic flows.

Traffic flow forecasting, which can be considered as a typical spatial and temporal problem, is to predict several road nodes' future traffic flow in an area based on the historical traffic data. In early research, many traditional algorithms were widely used in traffic flow prediction tasks. Recent years have witnessed a great success of deep learning. Recurrent Neural Networks (RNN) and their variants, Long Short Term Memory (LSTM) and Gated Recurrent Units (GRU), were first applied to traffic flow prediction tasks, due to their great success in sequence learning. And then, to make up for their inability to capture spatial dependencies, Convolutional Neural Networks (CNNs) and Graph Convolutional Networks (GCNs) were introduced into this research field and brought some SOTA effects. In particular, the models combined with GCNs provide more accurate prediction results than traditional algorithms in traffic flow prediction tasks on unstructured road networks. However, existing GNN-based models still lack satisfactory progress in traffic prediction, mainly due to the following two difficulties.

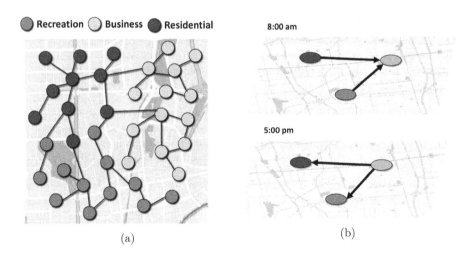

Fig. 1. An illustration of regional information in traffic road network. (a) shows three different kinds of road segments with various functionalities. (b) shows how the trend of traffic flow shifts between traffic regions changes over time.

- **Lack of Mining Regional Information.** Most of the existing work simply takes into account spatial information at the segment level. In fact, the traffic road network is characterized by a hierarchical structure, which is composed of traffic regions with different functionalities. The transfer of traffic flows usually takes place between these regions. As shown in Fig. 1, during the morning rush hour, the traffic flow in the residential region usually has a tendency to shift to the business region. But at noon lunch time, or after working hours, this traffic flow transfer pattern between region may change. The transfer pattern of traffic flow between regions is an important part of traffic flow, but existing methods cannot capture this pattern, which limits the performance of the model.

- **Lack of long-Term Temporal Correlations Extraction.** Forecasting traffic flows over long term is more complex but more critical than in the short term, because it can provide an earlier alert to future traffic congestion problems. Whether the methods are based on RNNs or CNNs, they still struggle in making an accurate long-term forecast. In long sequence modelling tasks, RNNs often suffer from the issue of gradient vanishing. Likewise, CNNs methods cannot capture long-range dependencies due to their limited receptive fields by the convolution kernels. Although TCNs adopt dilated convolutions to achieve exponentially large receptive fields, it still requires stacking several convolutional layers to connect any two positions in the sequence, which impairs its ability to learn long-term dependencies. Therefore, a more effective method is required for more accurate traffic flow prediction.

In this paper, we seek to address several challenges facing the traffic prediction problem mentioned above and propose a new paradigm of Region-aware Graph Convolutional Network (RGCN) for Traffic Flow Forecasting. Our contributions are summarized below.

- A new model named Region-aware Graph Convolutional Network is proposed to capture cross-region traffic flow transfer patterns by a DTW-based pooling method which can mine the spatial correlations of nodes with similar semantic information.
- In contrast to the existing methods, we propose a transformer-based temporal module to model long-range temporal dependencies.
- We carry out extensive experiments on our model and compare with multiple state-of-the-art baselines. Results demonstrate the effectiveness and robustness of our model.

This paper is structured as follows. In Sect. 2, we present technical developments in the field of traffic flow forecasting and the basic concepts of graph neural networks. In Sect. 3, we formally define the traffic flow prediction task. In Sect. 4, we describe the main framework of the model and some important modules. In Sect. 5, extensive experiments are conducted, and the effectiveness of our model is demonstrated. Finally, we conclude our paper and present our future work in Sect. 6.

2   Related Work

2.1   Traffic Flow Forecasting

Traffic flow prediction algorithms have been extensively researched over the last decade. From traditional algorithms to deep learning algorithms, the prediction results of the tasks have been made increasingly accurate. Traditional algorithms include some statistical methods and traditional machine learning methods such as ARIMA [7], SVR [4] and KNN [3]. Some of these models are based on linear time series methods that rely on static assumptions, while others require human-designed feature engineering, which is time-consuming and labor-intensive and cannot handle complex traffic data.

Due to the rapid development of deep neural networks, traditional algorithms have gradually been replaced. Earlier studies have employed RNNs (including LSTM and GRU) that can efficiently process sequential data for traffic flow prediction. Without considering spatial dependencies, the performance of RNNs models cannot be further improved. Therefore some studies treat traffic road networks as grid structures and incorporate CNNs modules into their models [12,13] to capture the spatial characteristics of traffic networks. However, the spatial correlation captured by CNNs is limited, because the traffic road network has a complex topology and is essentially non-Euclidean data. Afterwards, the researchers introduced GCNs to model spatial dependencies. STGCN [10] adopts graph convolution to extract spatial dependencies and gated CNN to mine temporal patterns. Based on this, GWNET [9] and AGCRN [6] propose an adaptive adjacency matrix to learn dynamic spatial correlations. STSGCN [11] synchronously captures the localized spatial-temporal correlations. In order to solve the drawbacks of existing models that only consider the traffic flow prediction at the road segment level, HGCN [5] constructs a two-stream graph network to consider micro and macro traffic information. However, when pooling is performed, HGCN ignores the semantic correlation between nodes, resulting in not being able to obtain better performance.

2.2   Graph Convolution Network

Graph Neural Networks (GNNs) are a class of deep learning methods that perform well on graph data, enabling predictions on nodes [9,10], edges, or graphs [14–16]. With GNN, operations can be achieved that traditional convolution (CNN) cannot, such as capturing the spatial dependencies of unstructured data. To capture these spatial dependencies between data, GNNs use the message-passing approach to aggregate information about the neighbors of entities in the network into the entities themselves. There are two main categories of GNNs based on different convolution operations. One is Spectral-based Graph Convolutional Networks [17] which find the corresponding Fourier basis in spectral domain. The other is Spatial-based Graph Convolutional Networks [18] which define graph convolution based on the spatial relationship with neighbours of nodes. Spatially based graph convolution operations are widely used in a variety of tasks on graphs, such as traffic flow prediction, which is defined as:

$$GCN(\mathcal{X}^l, A) = \sigma(A\mathcal{X}^{l-1}W^l) \tag{1}$$

where l represents the index number of layers of the neural network, $\mathcal{X}$ represents the input signal and W represents the learnable weight matrix. $A \in \mathcal{R}^{N \times N}$ denotes the normalized adjacency matrix with self-loops and is defined as below:

$$A = I + \hat{D}^{-\frac{1}{2}} \hat{A} \hat{D}^{-\frac{1}{2}} \tag{2}$$

where I is the identity matrix, $\hat{A}$ is the adjacency matrix and $\hat{D}_{ij} = \sum_j \hat{A}_{ij}$ is the diagonal matrix of $\hat{A}$.

3 Problem Statement and Preliminaries

In this section, we give some preliminaries and the mathematical definition of the traffic prediction problem.

Preliminaries. We denote a road network as a weighted directed graph $\mathcal{G} = (\mathcal{V}, \mathcal{E}, \mathcal{A})$. Here $\mathcal{V}$ is a set of $N = |\mathcal{V}|$ vertices representing vertices on the road network. $\mathcal{E}$ is a set of edges representing the connectivity between vertices. $\mathcal{A} \in \mathcal{R}^{N \times N}$ is an adjacency matrix recording the weighted connectedness between two vertices. The traffic condition at time step t is represented as $X_t \in \mathcal{R}^{N \times F}$ on Graph $\mathcal{G}$, where F is the recorded time series features on each vertices on road network. $\mathcal{X} = (X_1, X_2, ..., X_\tau)^T \in \mathcal{R}^{N \times F \times \tau}$ denotes the value of all the features of all the nodes over τ timestamps.

Problem Studied. Traffic flow forecasting aims to forecast the most likely future length-J sequence of node features in $\mathcal{G}$ given the previous τ observations $\mathcal{X}$:

$$\hat{X}_{t+1}, ..., \hat{X}_{t+j} = \underset{X_{t+1},...,X_{t+j}}{\arg\max} \; P(X_{t+1}, ..., X_{t+j} | X_{t-\tau+1}, ..., X_t; \mathcal{G}) \tag{3}$$

4 Methodology

The overall framework of RGCN is based on an architecture inspired by HGCN [5], as shown in Fig. 2 (a). The model has a two-stream computation process. The two streams model traffic flow transition patterns at the segment level and at the region level, respectively. During the calculation process, the two kinds of information are interactively fused. Two core technical modules, which are DTW-based pooling layer and transformer-based temporal module contained in this model, along with other components will be described in detail in this section.

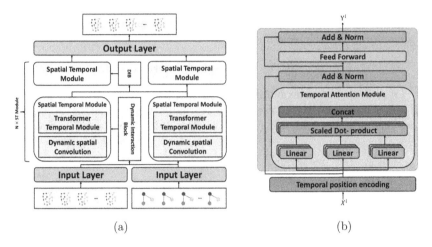

(a)   (b)

Fig. 2. An illustration of the proposed framework. (a) is the framework of our model which has a two-stream computation process. (b) is the detail of our transformer-based temporal module.

4.1   DTW-Based Pooling Layer

In order to better capture the cross-region traffic flow transfer patterns, firstly, we propose to mine the semantic information of traffic segments. With this approach, the distribution of traffic on roads with the same functionality or grade will have a high degree of similarity, and based on this, we will cluster the traffic road network to form a traffic region network to model traffic flow at region level. The processing flow is shown in Fig. 3. First, the Dynamic Time Warping algorithm (DTW) is used to capture the semantic similarity between traffic segments. In several traffic flow prediction works [19,20], DTW is used to construct a graph matrix to further learn the semantic relationship among the traffic segments, and its effectiveness is proved. Here we use it as a necessary preprocessing before the clustering operation. DTW uses the sum of the distances between all similar points, called Warp Path Distance, to measure the similarity between two time series. Given two time series $X = (x_1, x_2, ..., x_n)$ and $Y = (y_1, y_2, ..., y_m)$, firstly, generate the original distance matrix $Dist_{n \times m}$ whose element is $Dist_{i,j} = |x_i - y_j|$. Then, the cost matrix D could be defined:

$$D(i, j) = Dist(i, j) + min[D(i - 1, j), D(i, j - 1), D(i - 1, j - 1)] \quad (4)$$

After the iteration of the algorithm, finally, $D(|X|, |Y|)$ can be described as the similarity between two time series. Based on the DTW algorithm, the similarity relation Graph G_{DTW} between nodes is constructed, whose adjacency matrix is A_{DTW} and the generation process of Warp Path is shown in Fig. 3(a).

To enable modelling of cross-region traffic flow transfer patterns in traffic road networks, we employ pooling operations in GNNs. Pooling operations are important for deep models especially on image tasks, where they help expand the

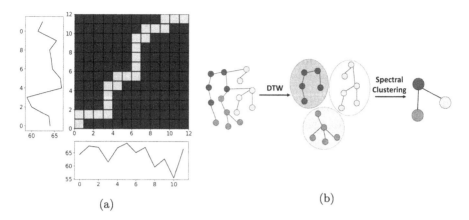

(a)   (b)

Fig. 3. DTW-based pooling processing.(a): The generation process of Warp Path between two time series. (b) shows the execution flow of the DTW-based pooling layer: A new graph is constructed from the original traffic network graph through semantic similarity, and on this basis, a new traffic region graph is clustered by the spectral clustering algorithm.

receptive field and reduce computational cost. Pooling of images is very straightforward, but Graph pooling, which is limited by a non-European data structure, is a separate but important area of research and cannot be manipulated simply. For the sake of simplicity, we employ the spectral clustering algorithm as a pooling operation in our preprocessing stage.

The object of spectral clustering is to cut the graph into K subgraphs that are not connected to each other. In our paper, traffic road network is divided into K traffic regions with different functionality. The choice of K is a trick, and we will discuss it in detail in the experimental section. One of the cores of spectral clustering algorithm is the construction of similarity matrix. And this matrix has been obtained by DTW which is A_{DTW}. It can reflect the similarity between road sections more reasonably than a geographical adjacency matrix, thus making clustering more effective. From this construction, we can get a new traffic region graph $\mathcal{G}^{\mathcal{R}}$, and the value of all the features of all the regions over τ timestamps can be denoted as $\mathcal{X}^{\mathcal{R}} = (X_1^R, X_2^R, ..., X_\tau^R)^T \in \mathcal{R}^{K \times F \times \tau}$. The traffic flow of the region is represented by the average of the traffic flow of the road segment. In the end, the connectivity between regions is determined by whether the road segments within the region are connected.

4.2   Transformer-Based Temporal Module

Although the capture of long-term and dynamic temporal dependencies of traffic flow is difficult, it is very important for accurate prediction. In recent years, time series forecasting models [22,23] that incorporate the transformer structure [21] have had great success. Inspired by this, we design a transformer-based temporal module to capture long-term temporal dependencies in our traffic flow forecasting task, as shown in Fig. 2 and described in detail below.

Temporal Position Encoding. In some Transformer model based tasks, such as NLP, the positional encoding of the token is provided as input, because they are not only part of the grammatical structure of a sentence, but are also important concepts for expressing semantics. Likewise, in traffic flow prediction task, the order information of a time series can characterize the dependencies between time steps. For example, intuitively, time steps that are closer may have a larger impact on each other. Inspired by transformer, the temporal position encoding e_{tp} is defined as:

$$
e_{tp} = \begin{cases} sin(\tau/10000^{2i/d}), & if\ \tau = 0, 2, 4... \\ cos(\tau/10000^{2i/d}), & otherwise \end{cases} \tag{5}
$$

where τ is the index of the element in the input and d is the dimension of model embedding.

Temporal Attention Module. Self-attention mechanism is an improvement of attention mechanism, which reduces the dependence on external information and is better at capturing the internal correlation of data or features. From this, we use Multi-Head Self-Attention to capture the long-term dependencies of traffic flow as well as the dynamic dependencies between time steps. First of all, we employ Scaled Dot-Product Attention as our attention function:

$$
Attention(Q, K, V) = softmax(\frac{QK^T}{\sqrt{d}})V \tag{6}
$$

where d is the dimension of model embedding and the queries $Q \in \mathcal{R}^{T \times d^\tau}$, the keys $K \in \mathcal{R}^{T \times d^\tau}$ and the values $V \in \mathcal{R}^{T \times d^\tau}$ can be calculated as:

$$
\begin{aligned} Q &= X^i W_q \\ K &= X^i W_k \\ V &= X^i W_v \end{aligned} \tag{7}
$$

where W_q, W_k and W_v are the learned linear mappings and X^i is the sequence of road segment i. Based on this, multi-head self-attention linearly projects the queries, keys and values into different representation subspaces and then performs the attention function. The output is concatenated by several heads. Formally,

$$
\begin{aligned} MultiHeadAttention(Q, K, V) &= Concat(head_1, ..., head_h)W^O \\ head_i &= Attention(QW_i^q, KW_i^k, VW_i^v) \end{aligned} \tag{8}
$$

where h is the number of attention heads, W_i^q, W_i^k, W_i^v are the weight matrices for different heads and W^O is the linear output projection. Finally, we achieve the output $H^i \in \mathcal{R}^{T \times d}$, the representation of the node i. Through the multi-headed attention mechanism, we are able to capture complex dynamic correlations from traffic data, enabling accurate long-term predictions to be made.

4.3 Dynamic Spatial Convolution

In Eq. 2, the main calculation of the GCN is revealed. It relies on the spatial relationships defined by the adjacency matrix and is applied to update the representation of the nodes by means of message passing among them. However, since the spatial dependencies between nodes in the graph change over time, neither the geographical proximity nor the adjacency matrix learned using gradient descent [6,9] can represent their complex spatial correlation. Inspired by self-attention mechanism, shown in Eq. 6, we dynamically compute the spatial correlation among the nodes at each input and treat the attention matrix as a new adjacency matrix which is defined as below:

$$A^D = softmax(\frac{\mathcal{X}_{l-1}\mathcal{X}_{l-1}^T}{\sqrt{d}}) \in \mathcal{R}^{N \times N} \tag{9}$$

where $\mathcal{X} \in \mathcal{R}^{N \times d}$ is the output of the previous module and A_{ij}^D is considered as representing the spatial relationship between node i and node j. From this, the new graph convolution can be defined as:

$$GCN^D(\mathcal{X}^l, A) = \sigma((A + A^D)\mathcal{X}^{l-1}W^l) \tag{10}$$

Here we fuse the original static geographic proximity relationship between traffic nodes and the learned dynamic spatial dependencies to capture the transfer patterns of traffic flow more effectively.

4.4 Extra Components

Dynamic Interaction Block. In order to integrate the segment and region information obtained above into our model and guide us for more accurate and efficient training, we follow [5], and employ the dynamic interaction block. It is based on the attention mechanism similar to Eq. 6 and is able to establish messages interaction between traffic nodes and regions.

Input and Output Layer. A fully connected layer is added at the top of the network to transform the input into a high-dimension space, which can improve the representation power of the network. Besides, the output module fuses the outputs of segment level and region level module with Dynamic Interaction Block and consists of two fully connected layers, aggregating the learned deep features for final predictions.

Loss Function. Mean absolute error (MAE) is selected as our loss function to measure the absolute value error between the ground truth Y and the prediction values $\hat{Y}$ of our model.

$$L(Y, \hat{Y}) = \frac{1}{Q} \sum_{i=t+1}^{i=t+j} |Y_i - \hat{Y}_i| \tag{11}$$

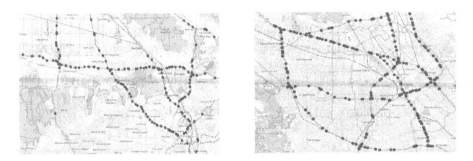

Fig. 4. Distribution of detectors in METR-LA (left, 207 sensors) and PEMS-BAY (right, 325 sensors) datasets.

5   Experiment

To evaluate the performance of our proposed model, we conducted extensive experiments on two real-world datasets. A detailed description is presented below.

5.1   Dataset

RGCN is evaluated on two traffic datasets, METR-LA and PEMS-BAY, which consist of the traffic speed readings collected from loop detectors and aggregated over 5 min intervals, shown in Fig. 4. METR-LA contains 34,272 time steps of 207 sensors collected in Los Angeles County over 4 months. PEMS-BAY contains 52,116 time steps of 325 sensors collected in the Bay Area over 6 months, shown in Table 1.

5.2   Baseline Methods

We compared RGCN with several traditional methods and state-of-the-art prediction models, which are as follows:

- HA: Historical Average method is used to predict the future speed using the average value of historical speed data

Table 1. Result on PEMS dataset

DataSet	Node	Edge	Time Range	Sample	Sample rate
METR-LA	207	1515	2012/03/01–2012/06/27	34272	5 min
PEMS-BAY	325	2369	2017/01/01–2017/06/30	52116	5 min

- ARIMA [7]: Auto-Regressive Integrated Moving Average method with a Kalman filter, which is a typical statistical model in time series field.
- FC-LSTM [8]: It is a RNN-based sequence-to-sequence model that uses fully-connected LSTM hidden layers in encoder and decoder
- STGCN [10]: Spatial-Temporal Graph Convolutional Network adopts graph convolution and gated CNN to extract spatial and temporal patterns.
- GWNET [9]: Graph WaveNet combines gated TCN with spatial GCN and proposes an adaptive adjacency matrix to learn dynamic spatial correlations.
- STSGCN [11]: Spatial-Temporal Synchronous Graph Convolutional Network captures the complex localized spatial-temporal correlations through a spatial-temporal synchronous modeling mechanism.
- HGCN [5]: Hierarchical Graph Convolution Networks constructs a two-stream graph network to consider micro and macro traffic information.

5.3 Experiment Settings

We implemented our model via the PyTorch framework and on the following hardware platform: (CPU: Intel(R) Xeon(R) CPU E5-2650 v4 @ 2.20 GHz, GPU: NVIDIA GTX 1080Ti). In the experiments, datasets are split as 70% training, 20% validation and 10% testing in chronological order. The length of input and output time sequence are both 12 (1 h). The dimension of model embedding is 64 and the number of attention heads h is 4. The Adam optimizer with the initial leaning rate of 10^{-4}, the epoch of 100 and the batch size of 32 are set for model training. For evaluation, mean absolute error (MAE), root mean square error (RMSE) and mean absolute percentage error (MAPE) are used as the evaluation metrics.

5.4 Experimental Results

The results of RGCN compared with other baseline methods for 15 min, 30 min, and 1 h ahead predictions on two datasets are shown in Table 2. We can see that our model achieves the state-of-the-art performance in three evaluation metrics and the improvements are increased with prediction time gets longer. Here, several observations can be get by further analyses. First, among all the methods, the performance of traditional statistical algorithms and the time series model (HA, ARIMA and FC-LSTM) is relatively poor. Because, compared with

Table 2. Performance of RGCN and other models

Datasets	Models	15 min			30 min			60 min		
		MAE	RMSE	MAPE	MAE	RMSE	MAPE	MAE	RMSE	MAPE
PEMS-BAY	HA	2.88	5.59	6.55%	2.88	5.59	6.55%	2.88	5.59	6.55%
	ARIMA	1.62	3.30	3.50%	2.33	4.76	5.40%	3.38	6.50	8.30%
	FC-LSTM	2.05	4.19	4.80%	2.20	4.55	5.20%	2.37	4.96	5.70%
	STGCN	1.41	2.97	3.00%	1.79	4.26	4.15%	2.47	5.66	5.74%
	GWNET	**1.30**	**2.74**	**2.73%**	1.63	3.70	3.67%	1.95	4.52	4.63%
	STSGCN	1.41	2.98	3.00%	1.79	4.01	4.01%	2.15	5.05	5.11%
	HGCN	1.32	2.80	3.10%	1.65	3.76	3.77%	1.95	4.53	4.70%
	RGCN	1.31	2.77	2.75%	**1.62**	**3.65**	**3.60%**	**1.87**	**4.42**	**4.50%**
METR-LA	HA	4.16	7.80	13.02%	4.16	7.80	13.02%	4.16	7.80	13.02%
	ARIMA	3.99	8.21	9.60%	5.15	10.45	12.70%	6.90	13.23	17.4%
	FC-LSTM	3.44	6.30	9.60%	3.77	7.23	10.90%	4.37	8.69	13.2%
	STGCN	2.90	5.86	7.72%	3.50	7.30	10.00%	4.63	9.80	13.00%
	GWNET	2.69	5.15	6.90%	3.07	6.22	8.37%	3.53	7.37	10.01%
	STSGCN	3.25	7.20	7.10%	4.01	9.42	10.10%	4.85	9.98	12.91%
	HGCN	2.72	5.55	7.20%	3.20	6.51	8.50%	3.71	7.51	10.58%
	RGCN	**2.66**	**5.14**	**6.84%**	**3.03**	**6.16**	**8.23%**	**3.46**	**7.30**	**9.72%**

deep learning approaches, they only take temporal correlations into consideration and cannot utilize the spatial dependencies of the spatial-temporal network. In contrast, deep learning models (STGCN, GWNET) that incorporate temporal and spatial dependencies are able to learn more precise intrinsic patterns from the data. In particular, GWNET, with its proposed adaptive adjacency matrix, has led to further improvements in prediction accuracy. However, due to the lack of resolution of cross-regional traffic flows, the predictions do not perform as well as our model. The models (HGCN, STSGCN) use hierarchical or synchronous ideas and achieve the good results. However, they fall short of our model because they do not go deeper into the hierarchy of the traffic network and do not offer a better solution for long-term time prediction. To make the results of our model more intuitive, a visualisation of the prediction results of the RGCN compared to other models is shown in Fig. 7, demonstrating the superior fitting ability of our proposed model.

5.5   Study on DTW-Based Pooling Layer

The original geographical adjacency matrix of the traffic network only reflects spatial correlations and ignores semantic connections. As shown in Fig. 5, some road segments are geographically divided, but they are semantically similar, e.g. they are both in residential areas. They share the similar distributions of traffic flow and experiments have shown this semantic similarity to be a non-negligible

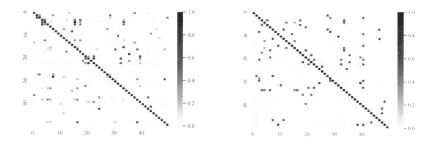

Fig. 5. Comparison of the adjacency matrix on geographic space (left) and the adjacency matrix generated by DTW (right).

element. The choice of K is a trick. We experimented with several values and chose the best one to take. As shown in Fig. 6, we used the MAE on a one-hour forecast as the selection criterion. We can observe that the optimal k for PEMS-BAY is 15, while for METR-LA it is 5.

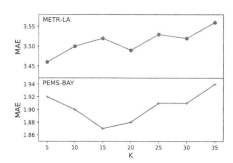

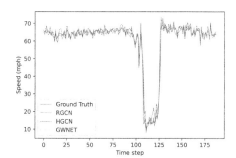

Fig. 6. Experiments on choices of K   **Fig. 7.** Visualisation of predictions

5.6   Ablation Study

To verify the effectiveness of each module in our model, we conduct ablation experiments in this section. Based on original RGCN, we keep part of components to form variants as follows:

- W/O DP: RGCN ignores the regional information of traffic road network with removing DTW-based pooling layer.
- W/O TT: RGCN remove the transformer-based temporal module and replace it with dilated TCN.
- W/O DS: RGCN removes the dynamic spatial convolution module and employs geographic proximity-based convolution.

We conducted ablation experiments on METR-LA, and the results of the comparison are shown in Table 3. It demonstrates that any one component is essential to our results. We observe the importance and effectiveness of the DTW-based pooling layers for our model. It captures regional information about the traffic network, which can help us to mine traffic flow patterns across traffic regions and thus improve the performance of our model. The transformer-based temporal module enhances the modeling ability of the model for long-term prediction, without which the model performance drops significantly. We also did ablation experiments on Dynamic spatial Convolution. Obviously, it helps the model to improve the performance by dynamically capturing the spatial connections between nodes. In the end, our RGCN incorporates the best of all modules to obtain optimal results.

Table 3. Ablation Study on RGCN

Method	15 min			30 min			60 min		
	MAE	RMSE	MAPE	MAE	RMSE	MAPE	MAE	RMSE	MAPE
RGCN W/O DP	2.70	5.18	6.92%	3.07	6.18	8.38%	3.54	7.35	10.0%
RGCN W/O TT	2.68	5.16	6.85%	3.06	6.20	8.34%	3.54	7.37	9.95%
RGCN W/O DS	2.67	5.15	6.90%	3.03	6.18	8.27%	3.48	7.35	9.80%
RGCN	**2.66**	**5.14**	**6.84%**	**3.03**	**6.16**	**8.23%**	**3.46**	**7.30**	**9.72%**

6   Conclusions

In this paper, a Region-aware Graph Convolutional Network for traffic flow forecasting is proposed to predict future traffic conditions based on historical traffic flow data. A DTW-based pooling layer is developed to construct a traffic region network graph from the original traffic network that can mine potential regional attributes in traffic flows. A transformer-based temporal module is designed to capture long-term and dynamic temporal dependencies. And extensive experiments on two real-world datasets show that our model is superior to the existing models. Our future research will focus on mining deeper semantic information about road segments and considering more other factors that affect traffic flow, such as weather and traffic accidents, using datasets with these attributes. In the meantime, we will attempt to apply our model to more spatio-temporal data mining issues.

Acknowledgements. This work is supported by Natural Science Foundation of Jiangsu Province (Grant Nos. BK20211307), and by project Funded by the Priority Academic Program Development of Jiangsu Higher Education Institutions.

References

1. Zhang, J., Wang, F.Y., Wang, K., Lin, W.H., Xu, X., Chen, C.: Data-driven intelligent transportation systems: a survey. IEEE Trans. Intell. Transp. Syst. **12**(4), 1624–1639 (2011)

2. Mori, U., Mendiburu, A., Álvarez, M., Lozano, J.A.: A review of travel time estimation and forecasting for advanced traveller information systems. Transportmetrica A: Transp. Sci. **11**(2), 119–157 (2015)

3. Van Lint, J., Van Hinsbergen, C.: Short-term traffic and travel time prediction models. Artif. Intell. Appl. Crit. Transp. Issues **22**(1), 22–41 (2012)

4. Wu, C.H., Ho, J.M., Lee, D.T.: Travel-time prediction with support vector regression. IEEE Trans. Intell. Transp. Syst. **5**(4), 276–281 (2004)

5. Guo, K., Hu, Y., Sun, Y., Qian, S., Gao, J., Yin, B.: Hierarchical graph convolution network for traffic forecasting. In: Proceedings of the AAAI Conference on Artificial Intelligence, vol. 35, pp. 151–159 (2021)

6. Bai, L., Yao, L., Li, C., Wang, X., Wang, C.: Adaptive graph convolutional recurrent network for traffic forecasting. In: Advances in Neural Information Processing Systems, vol. 33, pp. 17804–17815 (2020)

7. Box, G.E., Pierce, D.A.: Distribution of residual autocorrelations in autoregressive-integrated moving average time series models. J. Am. Stat. Assoc. **65**(332), 1509–1526 (1970)

8. Sutskever, I., Vinyals, O., Le, Q.V.: Sequence to sequence learning with neural networks. In: Advances in Neural Information Processing Systems, vol. 27 (2014)

9. Wu, Z., Pan, S., Long, G., Jiang, J., Zhang, C.: Graph WaveNet for deep spatial-temporal graph modeling. In: Proceedings of the Twenty-Eighth International Joint Conference on Artificial Intelligence, IJCAI 2019, Macao, China, August 10–16, 2019, pp. 1907–1913. ijcai.org (2019)

10. Yu, B., Yin, H., Zhu, Z.: Spatio-temporal graph convolutional networks: a deep learning framework for traffic forecasting. In: Proceedings of the Twenty-Seventh International Joint Conference on Artificial Intelligence, IJCAI 2018, July 13–19, 2018, Stockholm, Sweden, pp. 3634–3640. ijcai.org (2018)

11. Song, C., Lin, Y., Guo, S., Wan, H.: Spatial-temporal synchronous graph convolutional networks: a new framework for spatial-temporal network data forecasting. In: Proceedings of the AAAI Conference on Artificial Intelligence, vol. 34, pp. 914–921 (2020)

12. Zhang, J., Zheng, Y., Qi, D.: Deep spatio-temporal residual networks for citywide crowd flows prediction. In: Thirty-First AAAI Conference on Artificial Intelligence (2017)

13. Yao, H., Tang, X., Wei, H., Zheng, G., Li, Z.: Revisiting spatial-temporal similarity: a deep learning framework for traffic prediction. In: Proceedings of the AAAI Conference on Artificial Intelligence, vol. 33, pp. 5668–5675 (2019)

14. Liu, G., Wang, Y., Orgun, M.A.: Finding k optimal social trust paths for the selection of trustworthy service providers in complex social networks. In: 2011 IEEE International Conference on Web Services, pp. 41–48 (2011)

15. Liu, G., et al.: MCS-GPM: multi-constrained simulation based graph pattern matching in contextual social graphs. IEEE Trans. Knowl. Data Eng. **30**(6), 1050–1064 (2018)

16. Liu, G., et al.: TOSI: a trust-oriented social influence evaluation method in contextual social networks. Neurocomputing **210**, 130–140 (2016)

17. Kipf, T.N., Welling, M.: Semi-supervised classification with graph convolutional networks. In: 5th International Conference on Learning Representations, ICLR 2017, Toulon, France, April 24–26, 2017, Conference Track Proceedings. OpenReview.net (2017)

18. Hamilton, W.L., Ying, Z., Leskovec, J.: Inductive representation learning on large graphs. In: Advances in Neural Information Processing Systems 30: Annual Con-

ference on Neural Information Processing Systems 2017, December 4–9, 2017, Long Beach, CA, USA, pp. 1024–1034 (2017)

19. Li, M., Zhu, Z.: Spatial-temporal fusion graph neural networks for traffic flow forecasting. In: Proceedings of the AAAI Conference on Artificial Intelligence, vol. 35, pp. 4189–4196 (2021)

20. Fang, Z., Long, Q., Song, G., Xie, K.: Spatial-temporal graph ode networks for traffic flow forecasting. In: Proceedings of the 27th ACM SIGKDD Conference on Knowledge Discovery & Data Mining, pp. 364–373 (2021)

21. Vaswani, A., et al.: Attention is all you need. In: Advances in Neural Information Processing Systems, vol. 30 (2017)

22. Zhou, H., et al.: Informer: Beyond efficient transformer for long sequence time-series forecasting. In: Proceedings of the AAAI Conference on Artificial Intelligence, vol. 35, pp. 11106–11115 (2021)

23. Wu, H., Xu, J., Wang, J., Long, M.: Autoformer: decomposition transformers with auto-correlation for long-term series forecasting. In: Advances in Neural Information Processing Systems, vol. 34, pp. 22419–22430 (2021)

SMART: A Decision-Making Framework with Multi-modality Fusion for Autonomous Driving Based on Reinforcement Learning

Yuyang Xia[1], Shuncheng Liu[1], Rui Hu[2], Quanlin Yu[2], Xiushi Feng[3],
Kai Zheng[1,2,3(✉)], and Han Su[1,2(✉)]

[1] School of Computer Science and Engineering, University of Electronic Science and
Technology of China, Chengdu, China
{xiayuyang,liushuncheng}@std.uestc.edu.cn, {zhengkai,hansu}@uestc.edu.cn
[2] Yangtze Delta Region Institute (Quzhou), University of Electronic Science and
Technology of China, Chengdu, China
{hurui03, quanlin.yu}@std.uestc.edu.cn
[3] Shenzhen Institute for Advanced Study, University of Electronic Science and
Technology of China ,Chengdu, China
xiushifeng@std.uestc.edu.cn

Abstract. Decision-making in autonomous driving is an emerging technology that has rapid progress over the last decade. In single-lane scenarios, autonomous vehicles should simultaneously optimize their velocity decisions and steering angle decisions to achieve safety, efficiency, comfort, small impacts on rear vehicles, and small offsets to the lane center line. Previous studies, however, have typically optimized these two decisions separately, ignoring the potential relationship between them. In this work, we propose a decision-making framework, named SMART (deciSion-Making frAmework based on ReinforcemenT learning), to optimize the velocity and steering angle of the autonomous vehicle in parallel. In order for the autonomous vehicle to effectively perceive the curvature of the lane and interactions with other vehicles, we adopt a graph attention mechanism to extract and fuse the features from different modalities (i.e., sensor-collected vehicle states and camera-collected lane information). Then a hybrid reward function takes into account aspects of safety, efficiency, comfort, impact, and lane centering to instruct the autonomous vehicle to make optimal decisions. Furthermore, our framework enables the autonomous vehicle to adaptively choose the duration of an action, which helps the autonomous vehicle pursue higher reward values. Extensive experiments evidence that SMART significantly outperforms the existing methods in multiple metrics.

Keywords: Autonomous driving · Multi-modality fusion ·
Reinforcement learning

Y. Xia and S. Liu—Both authors contribute equally to this paper.

1   Introduction

Automated driving is considered to be one of those technologies that could herald a major shift in transportation [1]. They are expected to bring society a huge number of benefits, e.g., improved safety, increased capacity, and reduced fuel use and emissions. With the advancement of sensors, GPS, and deep neural networks, autonomous vehicles are able to obtain reliable information about surrounding vehicles and lanes [6], which aids in their decision-making. In this work, we propose a holistic decision-making framework for autonomous vehicles in single-lane scenarios.

Decision-making in single-lane scenarios can be divided into velocity decisions and steering angle decisions. For velocity decisions, traditional methods [13,22] use mathematical formulas to model the relationship between two consecutive vehicles. For example, with the help of Adaptive Cruise Control (ACC) [13], autonomous vehicles can maintain a constant gap from the preceding vehicle with safety and efficiency considerations. Due to the potential of reinforcement learning to optimize multiple objectives simultaneously, many researchers adopt reinforcement learning to solve velocity decision problems and outperform traditional models. Specifically, [23] proposes a velocity decision model based on reinforcement learning, which considers not only the safety and efficiency of autonomous vehicles but also improving driving comfort and reducing the impact on rear traffic flows. For steering angle decisions that aim at keeping the autonomous vehicle in the center of the lane, the reinforcement learning approaches [15,20] also achieve minimal offset from the lane center line. However, the previous studies consider these two tasks separately, ignoring the potential relationship between them. Specifically, velocity decision methods do not take into account changes in lane curvature, and steering angle decision methods ignore the influence of other vehicles on the velocity of the autonomous vehicle, where there is always only one vehicle pursuing lane keeping in their traffic environments. It is necessary for autonomous vehicles to optimize their velocity and steering angle simultaneously, based on information about the surrounding traffic environment.

To summarize, our framework takes vehicle states (from onboard sensors) and lane information (from onboard cameras) as input, and then produces velocity decisions and steering angle decisions in parallel. Specifically, we focus on a single-lane environment with one autonomous vehicle and multiple conventional vehicles. Our objective is to enable the autonomous vehicle to be safe, efficient, and comfortable, with small impacts on the rear vehicles and minor offsets from the lane center line. According to our objective, challenges mainly stem from two factors: (1) the autonomous vehicle needs to extract useful information from the multi-modality input, e.g., dynamic vehicle interactions and lane curvatures; (2) there are multiple optimization goals for the velocity and steering angle of the autonomous vehicle. The autonomous vehicle is supposed to adaptively choose stable or fine-grained controls based on changes in environmental information.

To tackle these challenges, we design a decision-making framework based on reinforcement learning, called SMART (deciSion-Making frAmework based on ReinforcemenT learning), to control the velocity and steering angle of the autonomous vehicle. The core of SMART is to allow the free exploration of a reinforcement learning agent, with the goal of optimizing specific metrics. First, we encode the lane information and embed the relative features between the autonomous vehicle and its front and rear vehicles. Then we construct a star graph centered on the autonomous vehicle and use a graph attention mechanism to fuse these features from different modalities, which can help the autonomous vehicle pay more attention to changes in lane curvature and vehicle states. In order to optimize multiple objectives, we design a hybrid reward function to guide the exploration of a reinforcement learning agent. This reward function is comprised of five components that correspond to five optimization goals, i.e., safety, efficiency, comfort, impact, and lane centering. Furthermore, the duration of each decision is variable in this work. As a result, the duration becomes a learnable value, allowing the autonomous vehicle to adaptively choose finer or more stable decisions and thus increase reward values.

The contribution of our paper can be summarized as follows:

- We develop a reinforcement learning-based framework, called SMART, to simultaneously make velocity decisions and steering angle decisions considering multi-modality input.
- We adopt an attention mechanism to aggregate the features from different modalities and design a hybrid reward function to guide the learning process of a policy.
- We design a learnable adaptive duration for each action, allowing the autonomous vehicle to make stable and fine-grained decisions.
- We conduct extensive experiments on a widely-used simulator to evidence the superiority of our framework from multiple metrics.

2   Problem Statement

2.1   Preliminary

In this work, we consider an interaction single-lane traffic environment under the World Coordinate System [18], where there is one autonomous vehicle A and multiple conventional vehicles $\mathbb{C}$. Next, we will explain some definitions and notations used in the rest of this paper.

Lane. A lane L is part of a roadway that is used to guide drivers and reduce traffic conflicts. Following [2], we get the lane information by its lane centerline, which can be represented as a sequence of waypoints, i.e., $L = \langle wp_1, wp_2, ..., wp_n \rangle$, where n is the number of waypoints within a distance threshold D_{max}. These waypoints are sampled with a specific sampling resolution of $1\,\mathrm{m}$, and each waypoint wp consists of its distance dis along the lane and orientation θ relative to A, i.e., $wp = (dis, \theta)$.

Vehicle Feature. We represent the features of the autonomous vehicle as $f_A = (vel, \theta, \mathit{off})$, where $f_A.vel$ denotes the velocity, $f_A.\theta$ denotes the orientation and $f_A.\mathit{off}$ denotes the offset to the lane centerline. For a conventional vehicle C_i, we not only record its velocity, orientation, and lane offset, i.e., $f_{C_i} = (vel, \theta, off)$, but also calculate its relative features to the autonomous vehicle A, i.e., $f(C_i, A) = (dis, vel, \theta, off)$, where $f(C_i, A).dis$ denotes its distance relative to A along the lane [7], $f(C_i, A).vel$ and $f_{C_i}.\theta$ respectively denote its velocity and orientation relative to A, $f(C_i, A).off$ denotes the lane offset. In this work, we denote F as all the vehicle feature of the autonomous vehicle A and its front vehicle C_1 and rear vehicle C_2, i.e., $F = (f_A, f(C_1, A), f(C_2, A))$.

Time Step. We treat the time duration $\mathbb{T}$ of interest as a discretized set of time steps, i.e., $\mathbb{T} = \langle 1, 2, ..., t, t+1, ... \rangle$. In this work, we set the time interval Δt between two consecutive time steps to 0.1 s [23].

Problem Definition. Given the information of vehicles F, and lane L at time step t, our objective is to output an optimal velocity decision and steering angle decision.

Good velocity decisions should meet four requirements: safety, efficiency, comfort, and small impacts on the rear vehicles. Meanwhile, steering angle decisions should keep the autonomous vehicle in the center of the lane as much as possible.

3 Methodology

3.1 Reinforcement Learning Modeling

Based on the preliminaries, the autonomous vehicle will generate velocity decisions and steering angle decisions sequentially. Therefore, we model our problem under the Markov decision process (MDP) [16] and exploit a model-free reinforcement learning method with an adaptive time-step size to solve the MDP. In this work, the agent is the autonomous vehicle A, and the environment is the surrounding traffic scenario of the agent. We derive some definitions as below.

State. The state s^t denotes the multi-modality features that the agent A can collect from the environment at time step t. Therefore, state s^t consists of the vehicle information F and lane information L, i.e., $s^t = [F, L]$.

Action. The action a^t includes the acceleration brake rate abr and steering angle deg of the agent at time step t, i.e., $a^t = [abr, deg]$. abr ranges from thr_b to thr_t, where thr_b is a negative value corresponding to the maximum braking threshold of the agent, and thr_t is a positive value corresponding to the maximum throttle threshold of the agent. The steering angle θ ranges from $-thr_\theta$ to thr_θ, where thr_θ is the maximum steering angle of the agent. In this work, the duration of each action is variable and determined by the agent. We treat the duration as an adaptive time-step size, i.e., $k * \Delta t$, where $k \in \{1, 2, ..., thr_k\}$.

State Transition. While the agent performs action a^t from time step t to $t+k$, the state features obtained from the environment are updated step by step from s^t to s^{t+k}. Among these features, the velocities of all vehicles are clipped with a maximum velocity thr_v.

Reward. After the agent A performs an action a^t at state s^t, we calculate a reward value r to rate the driving performance of this action. Our reward

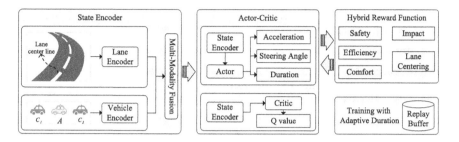

Fig. 1. Framework Overview

function aggregates five rewards from different aspects: safety, efficiency, comfort, impact, and lane centering.

Optimization. The objective of the agent is to choose actions that have great reward values. In this work, we adopt deep determined policy gradient (DDPG) [11] as the reinforcement learning paradigm, which includes a deterministic action policy u and an expected reward function Q. Therefore, the goal of the agent can be converted into finding an optimal policy u with the maximum Q.

3.2 Framework Overview

As shown in Fig. 1, our framework SMART can be divided into three components: state encoder, actor-critic, and hybrid reward function. The state encoder component first encodes lane features and vehicle features, respectively, and then fuses these multi-modality features. Based on the state encoder, the actor-critic component updates the action policy u to output action values and action durations and approximates the Q function. The hybrid reward function component is used to evaluate each action of the agent from five aspects: safety, efficiency, comfort, impact, and lane centering.

3.3 State Encoder

This component first takes the state features from different modalities as input and learns their latent representations, respectively. Then we use a graph attention mechanism to fuse these features and acquire a state encoding from the perspective of the agent.

Lane Encoder. The lane information L provides the geometric features of the lane. Some previous studies [12] rend the lane information as a rasterized image and apply a series of standard convolution layers to process it. However, the rasterized image is an overly complex representation of the lane information and requires significantly more computation to train. Recently, the vectorized lane [2,5] as a more succinct representation has developed rapidly since its strong capability to extract structured lane information. Given the lane information L, we firstly use a linear layer to embed each waypoint inside L and then apply a max-pooling operation over n waypoints, as follows:

$$Enc_L = g_{agg}(\phi_L(L; w_L)), \tag{1}$$

where Enc_L refers to the encoding of the lane information L, g_{agg} refers to the max-pooling operation, the waypoints in L is embedded by a linear layer ϕ_L with parameters w_L.

Vehicle Encoder. The vehicle information F provides the features f of the agent A and its front vehicle C_1 and its rear vehicle C_2. The spatial and dynamic relationships between the agent and other vehicles can serve as a strong cue to the actions of the agent. We encode these vehicle features as follows:

$$Enc_A = \phi_A(f_A; w_A),$$
$$Enc_{C_i} = \phi_C(f(C_i, A); w_C), \tag{2}$$

where Enc_A refers to the encoding of agent A, Enc_{C_i} includes the encoding of C_1 and C_2. For the agent A, we encode its features f_A by a linear layer ϕ_A with parameters w_A. For the conventional vehicles C_1 and C_2, we encode their features by a linear layer ϕ_C with parameters w_C.

Multi-Modality Fusion. After obtaining the encoding of the agent Enc_A, the lane Enc_L, the front vehicle Enc_{C_1} and the rear vehicle Enc_{C_2}, we exploit a graph attention mechanism to aggregate these features from different modalities. Since our goal is to extract useful features that can influence the actions of the agent, we model the agent and other objects with an agent-centric star graph $G = (\mathbb{V}, \mathbb{E})$. As shown in the left side of Fig. 2, the graph has four nodes that respectively denote the agent A, the front vehicle C_1, the rear vehicle C_2 and the lane L, i.e., $\mathbb{V} = [A, C_1, C_2, L]$. We connect the agent node A with other nodes $V' = [C_1, C_2, L]$, to form the star graph. The architecture of the graph attention mechanism is shown on the right side of Fig. 2. With the help of the attention mechanism, the agent can assign different important scores to other objects according to their features (e.g., the lane curvature, the relative distance between the agent and other vehicles). Following the standard process [19], the important score α_x for each node $x \in V'$ are calculated as follows:

$$\alpha_x = \frac{exp(LeakReLU(\phi_2(\phi_1(Enc_A; w_1) \mid\mid \phi_1(Enc_x; w_1)); w_2))}{\sum_{y \in V'} exp(LeakReLU(\phi_2(\phi_1(Enc_A; w_1) \mid\mid \phi_1(Enc_y; w_1)); w_2))}, \tag{3}$$

where ϕ_1 is a linear layer with parameters w_1, $\mid\mid$ is the concatenation operation, and ϕ_2 is a linear layer with parameters w_2 for computing an attention coefficient. Then we normalize the important scores by *Softmax* and acquire the state encoding SE by computing a linear combination between the encoding of each $x \in V'$ and their corresponding important score α_x. The calculation method is as follows:

$$SE = \sum_{x \in V'} \alpha_x \phi_3(Enc_x; w_3), \tag{4}$$

where ϕ_3 is a linear layer with parameters w_3.

3.4   Actor-Critic

The actor network is employed to generate the actions of the agent, and the critic network is employed to compute the Q value.

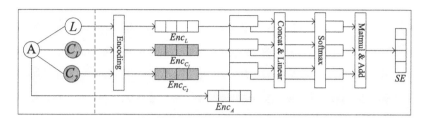

Fig. 2. Example of agent-centric star graph and graph attention mechanism

Actor. After acquiring the state encoding SE, we calculate an action a of the agent, including an acceleration brake rate abr and a steering angle θ. Different from the previous actor network [24], we also compute the duration of this action, i.e., $k * \Delta t$, where k is the time-step size. Therefore, the actor network output three values (i.e., o_1, o_2 and o_3), which are calculated as follows:

$$o_1, o_2, o_3 = Tanh(\phi_a(SE, w_a)), \tag{5}$$

where w_a denotes the parameters of the linear layer ϕ_a. After the activation $Tanh$, o_1, o_2 and o_3 all range from -1 to 1. Next, we need to respectively map these three values to the acceleration brake rate abr, the steering angle θ, and the time-step size k, as follows:

$$abr = \begin{cases} o_1 * thr_t, & o_1 > 0, \\ o_1 * thr_b, & o_1 < 0, \end{cases}$$

$$\theta = o_2 * thr_\theta, \tag{6}$$

$$k = i, \quad o_3 \in \left[-1 + (i-1) * \frac{2}{thr_k}, -1 + i * \frac{2}{thr_k} \right),$$

After the above operation, we get the acceleration brake rate abr from its minimal value thr_b to its maximum value thr_t, and the steering angle θ from its minimum value $-thr_\theta$ to its maximum value thr_θ, and the time-step size k in $\{1, 2, ..., thr_k\}$. For the autonomous vehicle, actions with small time-step sizes indicate a fine-grained vehicle control, and actions with large time-step sizes indicate a stable vehicle state.

Critic. After acquiring the state encoding SE, acceleration brake rate abr, steering angle θ and time-step size k, we can calculate the Q value as follows:

$$Q = \phi_{c_2}(g_{concat}(SE, \phi_{c_1}([abr, \theta, k], w_{c_1})), w_{c_2}), \tag{7}$$

where w_{c_1} and w_{c_2} are the parameters of two linear layers ϕ_{c_1} and ϕ_{c_2}. We first concatenate the state encoding SE and the output of ϕ_{c_1}, and then feed it into ϕ_{c_2} to calculate Q.

3.5   Hybrid Reward Function

The reward value serves as an exploration signal to teach the agent A to learn an optimal action policy u. Our reward function consists of five components as follows.

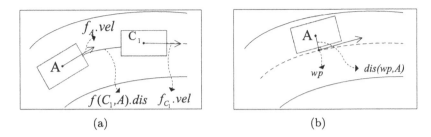

Fig. 3. (a) Example of TTC calculation. (b) Example of offset calculation.

Safety. Time-to-collision (TTC), as a widely-used safety indicator, represents the time span left before a collision if two vehicles maintain their current velocities [4]. A small TTC value indicates a high crash risk. Then we can calculate the TTC value $ttc(C_1, A)$ between agent A and vehicle C_1 as follows:

$$ttc(C_1, A) = \frac{f(C_1, A).dis}{f(C_1, A).vel}, \tag{8}$$

where $f(C_1, A).dis$ denotes the relative distance between agent A and the front vehicle C_1 along the lane, and $f(C_1, A).vel$ denotes the relative velocity. It works only when $f_A.vel$ is larger than $f_{C_1}.vel$, and the agent is supposed to receive a penalty for a small $ttc(C_1, A)$. The safety reward function r_1 is calculated as follows:

$$r_1 = \begin{cases} -3, & collision, \\ log(\frac{ttc(C_1,A)}{thr_{ttc}}), & 0 < ttc(C_1, A) < thr_{ttc}, \\ 0, & otherwise, \end{cases} \tag{9}$$

where thr_{ttc} refers to a bound of TTC. If $ttc(C_1, A)$ is smaller than thr_{ttc}, r_1 will be negative. As $ttc(C_1, A)$ approaches zero, r_1 will be close to negative infinity according to the log operation, which means that the agent will receive a severe punishment to near-crash situations. Then we clip $log(\frac{ttc(C_1,A)}{thr_{ttc}})$ to $[-3, 0)$ because the agent usually causes a collision when $log(\frac{ttc(C_1,A)}{thr_{ttc}}) \leq -3$, and thus $r_1 \in [-3, 0]$.

Efficiency. The velocity of the agent reflects its driving efficiency. Therefore, we calculate the efficiency reward function r_2 as follows:

$$r_2 = \frac{f_A.vel}{thr_v} - 1 \tag{10}$$

where $f_A.vel$ denotes the velocity of the agent, thr_v denotes the maximum velocity, and thus $r_2 \in [-1, 0]$. A larger velocity corresponds to a higher efficiency reward value for the agent.

Comfort. Jerk, defined as the acceleration change rate, is used to evaluate the driving comfort because it has a strong influence on the passenger [9], and a larger JERK corresponds to less comfortable driving of the agent and vice versa. We calculate the JERK value $jerk$ of the agent as follows:

$$jerk = (\frac{acc^t - acc^{t-1}}{\Delta t})^2, \tag{11}$$

where acc^t and acc^{t-1} refer to the acceleration of agent A at the current time step t and last time step $t-1$, respectively, and Δt refers to the time interval. As a result, the comfort reward function r_3 is calculated as follows:

$$r_3 = -\frac{jerk}{thr_j}, \tag{12}$$

where thr_j is the threshold jerk value, and thus $r_3 \in [-1,0]$. The '-' indicates that r_3 is a penalty for a large acceleration change rate.

Impact. The above rewards mainly consider the performance of the agent itself but do not consider its possible impact on the rear vehicle. For example, after the agent has a harsh braking behavior, its rear vehicle will follow this harsh action to maintain a safe distance. As a result, it will produce long-term velocity fluctuations in the rear traffic flow. To solve this issue, we will penalize the agent when its rear vehicle is affected by its behaviors. Following [21], after the agent performs an action at time step t, we record the velocity change of its rear vehicle at next time step $t+1$, and the impact reward function r_4 is calculated as follows:

$$r_4 = \frac{f_{C_2}^{t+1}.vel - f_{C_2}^{t}.vel}{thr_b * \Delta t}, \tag{13}$$

where vehicle C_2 refers to the rear vehicle of agent A, $f_{C_2}^{t+1}.vel$ and $f_{C_2}^{t}.vel$ respectively denote the velocity of C_2 at time step $t+1$ and t, thr_b is the braking threshold, and thus $r_4 \in [-1,0]$. It works when $f_{C_2}^{t}.vel - f_{C_2}^{t+1}.vel > imp_{thr}$, where imp_{thr} is a velocity change threshold. '-' indicates that r_4 serves as a punishment, prompting the agent to reduce and even eliminate its negative impact on the rear vehicle.

Lane Centering. To guide the agent to drive in the center of the lane, we penalize the agent according to its offset relative to the center line of the lane. We first find the corresponding waypoint wp of the agent A, at the lane center line by the interfaces in CARLA [3]. Then we compute the euclidean distance $dis(wp, A)$ between waypoint wp and agent A as shown in Fig. 3b. The lane centering reward function r_5 is calculated as follows:

$$r_5 = -\frac{dis(wp, A)}{wid} \tag{14}$$

where wid is half of the width of the lane, and thus $r_5 \in [-1,0]$. The '-' indicates r_5 serves as a punishment, prompting the agent to follow the curve of the lane.

Among these reward items, the efficiency reward is utilized to encourage the agent to pursue higher driving efficiency, while other rewards are utilized to penalize the agent if an action is dangerous, discomfort, strong-impact, or off-center, respectively. For a given state s^t and action a^t at time step t, we respectively calculate these five reward values and integrate them to get the final reward function, i.e., $r(s^t, a^t) = \sum_{i=1}^{5} w_i r_i$, where w_i is a tunable coefficient of r_i. We search different weighting coefficients to find the best combination that generates a good driving performance.

3.6   Training with Adaptive Duration

After the agent performs an action a^t with a time-step size k, we store the current state s^t, action a^t, k rewards $r(s^{t+i}, a^t)$ ($i \in \{0, 1, ..., k-1\}$), next state s^{t+k} as an experience $e = \left[s^t, a^t, r(s^t, a^t), r(s^{t+1}, a^t), ..., r(s^{t+k-1}, a^t), s^{t+k} \right]$. We store the experience in a replay buffer $D = \langle e_1, ..., e_M \rangle$ with a maximum capacity M, and randomly sample a minibatch of N experiences for network update. The actor network learns to optimize actions and action durations to achieve larger Q values, and the critic network learns to approximate the Q function calculated by the Bellman equation in Q-learning [14].

  Based on the Bellman equation, the Q function $Q(s^t, a^t)$ is equal to the reward value of action a^t plus the discounted future reward value. However, actions in this work may have different durations. Therefore, in order to compare their reward values on a consistent time-step size, we calculate the average reward value for each action over its duration (i.e., $k * \Delta t$). Finally, $Q(s^t, a^t)$ is calculated as follows:

$$Q(s^t, a^t) = \frac{\sum_{i=0}^{k-1} r(s^{t+i}, a^t)}{k} + \gamma Q^{'}(s^{t+k}, u^{'}(s^{t+k})), \qquad (15)$$

where $\sum_{i=0}^{k-1} r(s^{t+i}, a^t)$ is the sum of rewards for action a^t over k time steps, γ is a discount factor, $Q^{'}$ is the future reward calculated by a target critic network and $u^{'}$ is the action policy of a target action network. In DDPG [11], two target networks of the main actor network and the critic network are used to avoid the divergence of the algorithm.

4   Experiments

4.1   Experimental Setup

Dataset. In this work, we conduct our experiments on the CARLA simulator [3]. CARLA is a widely-used project focused on creating a publicly available virtual environment for the autonomous driving industry. We use Town01 and Town02 as the experimental single-lane environments. For traffic flow data, we initialize 1,000 different traffic flows with densities ranging from 10 $vehicles/km$ to 40 $vehicles/km$ as in [21]. Then we split these traffic flows into a training set and a test set with a ratio of 4: 1. Both in the training and the test phases, there are three termination conditions of the autonomous vehicle in an episode: 1) The driving distance reaches 1 km; 2) Collision with other vehicles; 3) Out of the lane. Except that the autonomous vehicle, all conventional vehicles are set to autopilot mode that mimics human drivers in CARLA.

Implementation Details. All the experiments are run on an Intel(R) Xeon(R) Silver 4214 CPU @ 2.20 GHz, and NVidia GeForce RTX 3080 GPU. For the vehicle restrictions, both the acceleration brake rate value and steering angle value range from -1 to 1 (i.e., $thr_b = -1$, $thr_t = 1$, and $thr_\theta = 1$) in CARLA, and the maximum time-step size $thr_\theta = 4$. For the reward function section, the bound value of time-to-collision thr_{ttc} is 4 s, and the maximum velocity thr_v is 25 m/s and the threshold jerk value thr_j is 3600 [23]. The velocity change imp_{thr}

is set to $0.1\,\mathrm{m/s}$, which means the rear vehicle is affected by the autonomous vehicle. The lane width is $4\,\mathrm{m}$ in CARLA and thus wid is $2\,\mathrm{m}$. For the lane information, the distance threshold D_{max} is $10\,\mathrm{m}$ and thus the number n of waypoints is 10.

The state encoder section involves six linear layers ϕ_L, ϕ_A, ϕ_C, ϕ_1, ϕ_2 and ϕ_3, each with 32 units. The actor-critic section involves three linear layers ϕ_a, ϕ_{c_1}, ϕ_{c_2}, each with 64 units. Finally, we train the network using the Adam optimizer [10] with a scheduled learning rate of 0.001 and a batch size N of 64. In DDPG, the size M of the experience replay buffer is 20000, the reward discount factor γ in the Bellman function is 0.9, and the target actor and critic use a soft update mechanism with an updated ratio of 0.01. Finally, our model was trained on the training set for a total of 10 epochs.

Evaluation Metrics. There are 200 episodes for all compared methods in the test phase. Each episode corresponds to a different traffic flow and the autonomous vehicle needs to travel $1\,\mathrm{km}$. To evaluate the performance of velocity decisions, we use three metrics to test the efficiency, comfort, and impact of the autonomous vehicle based on our reward function.

- Average time-to-collision (Avg-ttc). We record the total time when TTC of the autonomous vehicle is less than $2\,\mathrm{s}$. The larger the Avg-ttc, the higher the danger of the autonomous vehicle.
- Average velocity (Avg-vel). We record the average velocity of the autonomous vehicle. A larger Avg-vel indicates higher efficiency.
- Average jerk (Avg-jerk). We record the average acceleration change of the autonomous vehicle. A smaller Avg-jerk indicates a more comfortable driving experience.
- Average impact (Avg-imp). We record the absolute value of the average deceleration of the rear vehicle when the autonomous vehicle has braking behaviors. A small Avg-imp means the autonomous vehicle has a small impact on its rear vehicle.

To evaluate the performance of steering angle decisions, we record the average offset (Avg-dev) of the autonomous vehicle related to the lane center line. A smaller Avg-dev indicates a better performance of lane centering. In our experimental environments, the ratio of the straight section to the curved section is 10: 1. We will compare this metric on the straight section and the curved section, respectively.

- Avg-dev-str. The average offset on the straight section.
- Avg-dev-cur. The average offset on the curved section.

4.2   Overall Performance

In this section, we compare our framework SMART with three baselines.

- Autopilot. This method is the default control method for automatic vehicles in CARLA. This method first gets the next waypoint for the autonomous

vehicle and then uses a PID controller [17] to generate acceleration brake
rates and steering angles.

- DNN [8]. This method outputs acceleration brake rates and steering angles
  of the autonomous vehicle, by using deep learning methods to mimic real-life
  human driving behaviors.
- DQN [15]. This method uses deep Q-learning as the reinforcement learning
  model, which has discrete action space. It discretizes the range of steering
  angle into 9 values and discretizes the range of acceleration brake rates into 3
  values. Therefore, there are 27 action combinations in total. Since this method
  lacks consideration of the front and rear vehicles of the autonomous vehicle,
  we use the same state encoder as our model.

Table 1. Quantitative results of baselines and our model

Methods	Velocity				Steering angle	
	Avg-ttc	Avg-vel	Avg-jerk	Avg-imp	Avg-dev-str	Avg-dev-cur
Autopilot	3.09	15.76	1.26	0.48	**0.02**	0.52
DNN	3.74	15.47	1.11	0.46	0.24	0.65
DQN	2.81	15.83	1.15	0.45	0.17	0.56
SMART	**2.64**	**16.12**	**0.96**	**0.41**	0.12	**0.44**

Evaluation of Velocity Decision. The left side of Table 1 reveals that our
framework SMART achieves superior velocity decision performance in terms of
safety (Avg-ttc), efficiency (Avg-vel), comfort (Avg-jerk), and impact (Avg-imp).
As shown, both DQN and our model that use reinforcement learning achieve
better performance than the traditional method Autopilot and the deep learning method DNN. The reason lies in the strengths of reinforcement learning in
optimizing multiple objectives simultaneously. Furthermore, Our model SMART
with a continuous action space outperforms the DQN with a discrete action
space, because the continuous action space of SMART allows for finer control
and thus improves the velocity performance of autonomous vehicles.

Evaluation of Steering Angle Decision. The right side of Table 1 shows the
average offset of all methods at the straight lane (Avg-dev-str) and the curved lane
(Avg-dev-cur), respectively. A good steering angle decision is supposed to keep the
autonomous vehicle as centered in the lane as possible based on the current lane
curvature and its velocity. As shown, our framework SMART can achieve the smallest offset (Avg-dev-cur) at the curved lane. This is because our state encoder can
effectively capture the upcoming lane curvature and the reward function is capable of instructing the autonomous vehicle to output good steering angle decisions
to follow the current lane. For the steering angle decisions on the straight lane,
except for the DNN method that imitates human driving behaviors, other methods all have a small average offset relative to the lane center line.

Inference Time. In this framework, the time interval between two consecutive
actions is from 0.1 s to 0.4 s. For the calculation of each action, the mean inference
time of our model is 0.007 s. Therefore, compared to the time interval, our model
can output actions for the autonomous vehicle efficiently.

Reward Shaping. To achieve better driving performance, we adopt the grid search [23] to determine coefficients of the reward function in terms of safety, efficiency, comfort, impact, and lane centering, respectively. Finally, we set the coefficients as $w_1 = 1.2$, $w_2 = 0.7$, $w_3 = 1.6$, $w_4 = 0.2$, and $w_5 = 0.9$. Our model under these coefficients has the best performance in velocity decisions and steering angle decisions in the test phase.

4.3 Ablation Experiment

In this section, we evaluate our framework SMART through two ablation studies. We consider two variants of our model, as follows:

– SMART-NoAttn. Compared to our model, we remove the attention model in the multi-modality fusion section. After deleting it, we first concatenate the lane encoding and vehicle encoding of all vehicles and then use a linear layer with 64 units to fuse these features.
– SMART-NoAda. Compared to our model, we remove the adaptive durations of actions when the actor network outputs actions. Instead, the time-step size of each action is fixed at one time step.

Table 2. Quantitative results of variants and our model

Methods	Velocity				Steering angle	
	Avg-ttc	Avg-vel	Avg-jerk	Avg-imp	Avg-dev-str	Avg-dev-str
SMART-NoAttn	2.73	15.93	1.05	0.42	0.15	0.52
SMART-NoAda	2.78	16.06	1.08	0.44	0.14	0.48
SMART	**2.64**	**16.12**	**0.96**	**0.41**	0.12	**0.44**

Effectiveness of Attention Mechanism. To study the effectiveness of the attention mechanism in the multi-modality fusion section, we compare the complete model SMART with the variant SMART-NoAttn in Table 2. As shown, all metrics get worse after removing the attention mechanism. It demonstrates that the attention mechanism can help the autonomous vehicle effectively capture changes in environmental features and output action decisions accordingly. In Fig. 4a, we visualize the important scores of the lane information, front vehicle C_1, and rear vehicle C_2. At time step t_1, the autonomous vehicle drives on curved lanes, and thus the lane information has the highest importance score. At time step t_2 and t_3, the autonomous vehicle drives on straight lanes, and thus the importance scores of vehicles are generally higher than the importance score of the lane information.

Effectiveness of Adaptive Durations of Actions. To study the effectiveness of the adaptive durations of actions, we compare the complete model SMART with the variant SMART-NoAda in Table 2. As shown, all metrics of SMART are better than the variant. It is because the adaptive durations of actions enable the autonomous vehicle to adjust the time step sizes of actions as the changes in environmental information, thus achieving higher reward values. Specifically, actions with long durations generally have higher reward values in comfort, and actions with short durations generally have higher reward values in lane centering, In Fig. 4b, we provide the average reward values of different maximum time-step sizes. We can see that the average reward value obtained by the autonomous vehicle is highest when the maximum time-step size equals 2.

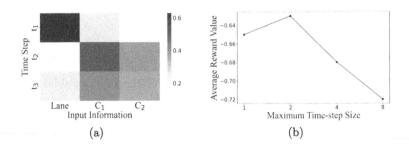

(a)   (b)

Fig. 4. (a) The heatmap of important scores. (b) Average reward values of different maximum time-step sizes.

5   Related Work

Velocity Decision. The traditional velocity decision models consider the relative information between two consecutive vehicles, and then use specific mathematical functions to map the information into a velocity decision. However, these traditional methods are difficult to comprehensively consider multiple optimization objectives. In recent years, many researchers adopt reinforcement learning methods to deal with the velocity decision problem. The key idea of reinforcement learning is to enable the free exploration of an agent to form a driving pattern that targets at optimizing multiple goals simultaneously. For example, [24] has trained a velocity decision model in a single-lane road, which achieves better performance in terms of safety, efficiency, and comfort than traditional methods. Furthermore, [23] considers reducing the impact on the rear vehicles, which can improve the stability and efficiency of the entire traffic flow.

Steering Angle Decision. Traditionally, the PID controller [17] is used to generate the steering angle of autonomous vehicles, which takes as input the position information of an autonomous vehicle and its next waypoint, and then uses kinetics formulas to compute the steering angle. With the rapid development of deep learning techniques, [20] tries to mimic human driving patterns from massive real-world data, to control the steering angle of an autonomous vehicle. However, due to the limited information processing capability, long reaction time, etc., human drivers themselves inevitably have inappropriate behaviors, which may affect the stability of traffic flow. To surpass the lane-following performance of human drivers, [15] adopt a reinforcement learning method that can learn a policy for lane-keeping. This method discretizes the action space of the steering angle and outperforms human drivers after trial and error.

In summary, reinforcement learning methods have achieved the best performance in both velocity decision and steering angle decision. However, these methods do not consider optimizing both tasks simultaneously. Specifically, velocity decision methods do not consider the controlling of steering angle, and steering angle decision methods ignore the influence of other vehicles on the velocity of autonomous vehicles. To the best of our knowledge, this paper pioneers in optimizing the velocity decision and steering angle decision in parallel.

6   Conclusion

In this paper, we focus on optimizing the velocity decisions and steering angle decisions simultaneously with multi-modality input in single-lane scenarios. We adopt a graph attention mechanism to fuse the features from different modalities and design a hybrid reward function to rate the performance of each action. In addition, we enable the autonomous vehicle to adaptively choose the duration of action, which can increase reward values. Extensive experiments evidence that our framework SMART outperforms the existing methods in multiple metrics.

Acknowledgment. This work is partially supported by NSFC (No. 61972069, 61836007, 61832017, 62272086), Shenzhen Municipal Science and Technology R&D Funding Basic Research Program (JCYJ20210324133607021), and Municipal Government of Quzhou under Grant (No. 2022D037, 2021D022).

References

1. Brenner, W., Herrmann, A.: An overview of technology, benefits and impact of automated and autonomous driving on the automotive industry. Digit. Marketplaces Unleashed, 427–442 (2018)
2. Chang, M.F., et al.: Argoverse: 3D tracking and forecasting with rich maps. In: Proceedings of the IEEE/CVF Conference on Computer Vision and Pattern Recognition, pp. 8748–8757 (2019)
3. Dosovitskiy, A., Ros, G., Codevilla, F., Lopez, A., Koltun, V.: CARLA: an open urban driving simulator. In: Proceedings of the 1st Annual Conference on Robot Learning, pp. 1–16 (2017)
4. Evans, L.: Traffic Safety and the Driver. Science Serving Society (1991)
5. Gao, J., et al.: VectorNet: encoding HD maps and agent dynamics from vectorized representation. In: Proceedings of the IEEE/CVF Conference on Computer Vision and Pattern Recognition, pp. 11525–11533 (2020)
6. Gupta, A., Anpalagan, A., Guan, L., Khwaja, A.S.: Deep learning for object detection and scene perception in self-driving cars: survey, challenges, and open issues. Array **10**, 100057 (2021)
7. Gutiérrez, R., et al.: A waypoint tracking controller for autonomous road vehicles using ROS framework. Sensors **20**(14), 4062 (2020)
8. Hsu, T.M., Wang, C.H., Chen, Y.R.: End-to-end deep learning for autonomous longitudinal and lateral control based on vehicle dynamics. In: Proceedings of the 2018 International Conference on Artificial Intelligence and Virtual Reality, pp. 111–114 (2018)
9. Jacobson, I., Richards, L., Kuhlthau, A.: Models of human comfort in vehicle environments. HUMAN FACTORS IN TRANSPORT RESEARCH EDITED BY DJ OBORNE, JA LEVIS 2 (1980)

10. Kingma, D.P., Ba, J.: Adam: a method for stochastic optimization. arXiv preprint arXiv:1412.6980 (2014)

11. Lillicrap, T.P., et al.: Continuous control with deep reinforcement learning. arXiv preprint arXiv:1509.02971 (2015)

12. Luo, W., Yang, B., Urtasun, R.: Fast and furious: real time end-to-end 3D detection, tracking and motion forecasting with a single convolutional net. In: Proceedings of the IEEE Conference on Computer Vision and Pattern Recognition, pp. 3569–3577 (2018)

13. Milanés, V., Shladover, S.E.: Modeling cooperative and autonomous adaptive cruise control dynamic responses using experimental data. Transp. Res. Part C Emerg. Technol. **48**, 285–300 (2014)

14. Mnih, V., et al.: Human-level control through deep reinforcement learning. Nature **518**(7540), 529–533 (2015)

15. Pérez-Gil, Ó., et al.: DQN-based deep reinforcement learning for autonomous driving. In: Bergasa, L.M., Ocaña, M., Barea, R., López-Guillén, E., Revenga, P. (eds.) WAF 2020. AISC, vol. 1285, pp. 60–76. Springer, Cham (2021). https://doi.org/10.1007/978-3-030-62579-5_5

16. Puterman, M.L.: Markov Decision Processes: Discrete Stochastic Dynamic Programming. Wiley, Hoboken (2014)

17. Salfer-Hobbs, M.B.: Acceleration, Braking, and Steering Controller for Polaris GEM e2. Ph.D. thesis (2019)

18. Sun, L., Yan, Z., Mellado, S.M., Hanheide, M., Duckett, T.: 3D of pedestrian trajectory prediction learned from long-term autonomous mobile robot deployment data. In: 2018 IEEE International Conference on Robotics and Automation (ICRA), pp. 5942–5948. IEEE (2018)

19. Veličković, P., Cucurull, G., Casanova, A., Romero, A., Lio, P., Bengio, Y.: Graph attention networks. arXiv preprint arXiv:1710.10903 (2017)

20. Wolf, P., et al.: Learning how to drive in a real world simulation with deep q-networks. In: 2017 IEEE Intelligent Vehicles Symposium (IV), pp. 244–250. IEEE (2017)

21. Xia, Y., Liu, S., Chen, X., Xu, Z., Zheng, K., Su, H.: RISE: a velocity control framework with minimal impacts based on reinforcement learning. In: CIKM 2022, New York (2022)

22. Xiao, L., Wang, M., Van Arem, B.: Realistic car-following models for microscopic simulation of adaptive and cooperative adaptive cruise control vehicles. Transp. Res. Rec. **2623**(1), 1–9 (2017)

23. Xu, Z., et al.: PATROL: a velocity control framework for autonomous vehicle via spatial-temporal reinforcement learning. In: Proceedings of the 30th ACM International Conference on Information & Knowledge Management, pp. 2271–2280 (2021)

24. Zhu, M., Wang, Y., Pu, Z., Hu, J., Wang, X., Ke, R.: Safe, efficient, and comfortable velocity control based on reinforcement learning for autonomous driving. Transp. Res. Part C Emerg. Technol. **117**, 102662 (2020)

Industry Papers

GIPA: A General Information Propagation Algorithm for Graph Learning

Houyi Li[1], Zhihong Chen[2], Zhao Li[3,6], Qinkai Zheng[4], Peng Zhang[5], and Shuigeng Zhou[1(✉)]

[1] School of Computer Science, Fudan University, Shanghai, China
hyli22@m.fudan.edu.cn, sgzhou@fudan.edu.cn
[2] Alibaba Group, Hangzhou, China
jhon.czh@alibaba-inc.com
[3] Zhejiang University, Hangzhou, China
[4] Tsinghua University, Beijing, China
[5] Cyberspace Institute of Advanced Technology,
Guangzhou University, Guangzhou, China
p.zhang@gzhu.edu.cn
[6] Link2Do Technology, Hangzhou, China

Abstract. Graph neural networks (GNNs) have been widely used in graph-structured data computation, showing promising performance in various applications such as node classification, link prediction, and network recommendation. Existing works mainly focus on node-wise correlation when doing weighted aggregation of neighboring nodes based on attention, such as dot product by the dense vectors of two nodes. This may cause conflicting noise in nodes to be propagated when doing information propagation. To solve this problem, we propose a General Information Propagation Algorithm (GIPA), which exploits more fine-grained information fusion including bit-wise and feature-wise correlations based on edge features in their propagation. Specifically, the bit-wise correlation calculates the element-wise attention weights through a multi-layer perceptron (MLP) based on the dense representations of two nodes and their edge; The feature-wise correlation is based on the one-hot representations of node attribute features for feature selection. We evaluate the performance of GIPA on the Open Graph Benchmark proteins (OGBN-proteins) dataset and the Alipay dataset of Alibaba Group. Experimental results reveal that GIPA outperforms the state-of-the-art models in terms of prediction accuracy, e.g., GIPA achieves an average ROC-AUC of 0.8917 ± 0.0007, which is better than that of all the existing methods listed in the OGBN-proteins leaderboard.

Keywords: Graph neural networks · Fine-grained information fusion · Bit-wise and feature-wise attention

1 Introduction

Graph representation learning typically aims to learn an informative embedding for each graph node based on the graph topology (link) information.

H. Li—Contributed to this work when the author worked in Ant and Alibaba Group.

Generally, the embedding of a node is represented as a low-dimensional feature vector, which can be used to facilitate downstream applications. This research focuses on homogeneous graphs that have only one type of nodes and one type of edges. The purpose is to learn node representations from the graph topology [7,8,25]. Specifically, given a node u, either breadth-first search, depth-first search or random walk is used to identify a set of neighboring nodes. Then, u's embedding is learnt by maximizing the co-occurrence probability of u and its neighbors. Early studies on graph embedding have limited capability to capture neighboring information from a graph because they are based on shallow learning models such as SkipGram [21]. Moreover, transductive learning is used in these graph embedding methods, which cannot be generalized to new nodes that are absent in the training graph.

Graph neural networks [9,14,30] are proposed to overcome the limitations of traditional graph embedding models. GNNs employ deep neural networks to aggregate feature information from neighboring nodes and thereby have the potential to gain better aggregated embeddings. GNNs can support inductive learning and infer the class labels of unseen nodes during prediction [9,30]. The success of GNNs is mainly due to the neighborhood information aggregation. However, GNNs face two challenges: *which neighboring nodes of a target node are involved in message passing? and how much contribution each neighboring node makes to the aggregated embedding?*. For the former question, neighborhood sampling [4,9,12,13,38,41] is proposed for large dense or power-law graphs. For the latter, neighbor importance estimation is used to attach different weights to different neighboring nodes during feature propagation. Importance sampling [4,13,41] and attention [11,19,30,33,39] are two popular techniques. Importance sampling is a special case of neighborhood sampling, where the importance weight of a neighboring node is drawn from a distribution over nodes. This distribution can be derived from normalized Laplacian matrices [4,41] or jointly learned with GNNs [13]. With this distribution, at each step a subset of neighbors is sampled, and aggregated with the importance weights. Similar to importance sampling, attention also attaches importance weights to neighbors. Nevertheless, attention differs from importance sampling. Attention is represented as a neural network and is always learned as a part of a GNN model. In contrast, importance sampling algorithms use statistical models without trainable parameters.

Existing attention mechanisms consider only the correlation of node-wise, ignoring the suppression of noise information in transmission, and the information of edge features. In real world applications, only partial users authorize the system to collect theirs profiles. The model cannot learn the node-wise correlation between a profiled user and a user we know nothing about. Therefore, existing models will spread noise information, resulting in inaccurate node representations. However, two users who often transfer money to each other and two users who only have a few conversations have different correlation.

In this paper, to solve the problems mentioned above, we present a new graph neural network attention model, namely General Information Propagation

Algorithm (GIPA). We design a bit-wise correlation module and a feature-wise correlation module. Specifically, we believe that each dimension of the dense vector represents a feature of the node. Therefore, the bit-wise correlation module filters at the dense representation level. The dimension of attention weights is equal to that of density vector. In addition, we represent each attribute feature of the node as a one-hot vector. The feature-wise correlation module performs feature selection by outputting the attention weights of similar dimensionality and attribute features. It is worth mentioning that to enable the model to extract better attention weights, edge features that measure the correlation between nodes are also included in the calculation of attention. Finally, GIPA inputs sparse embedding and dense embedding into the wide end and deep end of the deep neural network for learning specific tasks, respectively. Our contributions are summarized as follows:

1) We design the bit-wise correlation module and the feature-wise correlation module to perform more refined information weighted aggregation from the element level and the feature level, and utilize the edge information.
2) Based on the wide & deep architecture [6], we use dense feature representation and sparse feature representation to extract deep information and retain shallow original information respectively, which provides more comprehensive information for the downstream tasks.
3) Experiments on the Open Graph Benchmark (OGB) [10] proteins dataset (OGBN-proteins) demonstrate that GIPA achieves better accuracy with an average ROC-AUC of 0.8917 ± 0.0007[1] than the state-of-the-art methods listed in the OGBN-proteins leaderboard[2]. In addition, GIPA has been tested on billion-scale industrial Alipay dataset.

2   Related Work

In this section, we review existing attention primitive implementations in brief. [2] proposes an additive attention that calculates the attention alignment score using a simple feed-forward neural network with only one hidden layer. The alignment score $score(q, k)$ between two vectors q and k is defined as $score(q, k) = u^T \tanh(W[q; k])$, where u is an attention vector and the attention weight $\alpha_{q,k}$ is computed by normalizing $scor\tilde{e}_{q,k}$ over all q, k values with the softmax function. The core of the additive attention lies in the use of attention vector u. This idea has been widely adopted by several algorithms [24,37] for natural language processing. [20] introduces a global attention and a local attention. In global attention, the alignment score can be computed by three alternatives: dot-product $(q^T k)$, general $(q^T W k)$ and concat $(W[q; k])$. In contrast, local attention computes the alignment score solely from a vector (Wq). Likewise, both global and local attention normalize the alignment scores with the *softmax* function. [29] proposes a self-attention mechanism based on scaled dot-product.

---

[1] The reproducible code is open source: https://github.com/houyili/gipa_wide_deep.
[2] https://ogb.stanford.edu/docs/leader_nodeprop/#ogbn-proteins.

This self-attention computes the alignment score between any q and k as follows: $scor\tilde{e}_{q,k} = q^T k / \sqrt{d_k}$. This attention differs from the dot-product attention [20] by only a scaling factor of $\frac{1}{\sqrt{d_k}}$. The scaling factor is used because the authors of [29] suspected that for large values of d_k, dot-product is large in magnitude, which thereby pushes the *softmax* function into regions where it has extremely small gradients. More attention mechanisms, like feature-wise attention [26] and multi-head attention [29], can be referred to some surveys [3,22].

In addition, these aforementioned attention primitives have been extended to heterogeneous graphs. HAN [33] uses a two-level hierarchical attention that consists of a node-level attention and a semantic-level attention. In HAN, the node-level attention learns the importance of a node to any other node in a meta-path, while the semantic-level one weighs all meta-paths. HGT [40] weakens the dependency on meta-paths and instead uses meta-relation triplets as basic units. HGT uses node-type-aware feature transformation functions and edge-type-aware multi-head attention to compute the importance of each edge to a target node. It is worthy of mentioning that heterogeneous models must not always be superior to homogeneous ones, and vice versa. In this paper, unlike the node-wise attention in existing methods that ignores noise propagation, our proposed GIPA introduces two MLP-based correlation modules, the bit-wise correlation module and the feature-wise correlation module, to achieve fine-grained selective information propagation and utilize the edge information.

3   Methodology

3.1   Preliminaries

Graph Neural Networks. Consider an attributed graph $\mathcal{G} = \{\mathcal{V}, \mathcal{E}\}$, where $\mathcal{V}$ is the set of nodes and $\mathcal{E}$ is the set of edges. GNNs use the same model framework as follows [9,35]:

$$\tilde{h}_i^k = \phi(\tilde{h}_i^{k-1}, \mathcal{F}_{agg}(\{\tilde{h}_j^{k-1}, j \in \mathcal{N}(i)\})) \tag{1}$$

and where $\mathcal{F}_{agg}$ represents an aggregation function, ϕ represents an update function, $\tilde{h}_i^k$ represents the node embedding based on node feature x_i^k, and $\mathcal{N}(i)$ represents the neighbor node set. The objective of GNNs is to update the embedding of each node by aggregating the information from its neighbor nodes and the connections between them.

3.2   Model Architecture

In this section, we present the architecture of GIPA in Fig. 1, which extracts node features and edge features in a more general way. Consider a node i with feature embedding $\tilde{h}_i$ and its neighbor nodes $j \in \mathcal{N}(i)$ with feature embedding $\tilde{h}_j$. $\tilde{e}_{i,j}$ represents the edge feature between node i and j. The problem is how to generate an expected embedding for node i from its own node feature embedding $\tilde{h}_i$, its neighbors' node features embedding $\{\tilde{h}_j\}$ and the related edge features

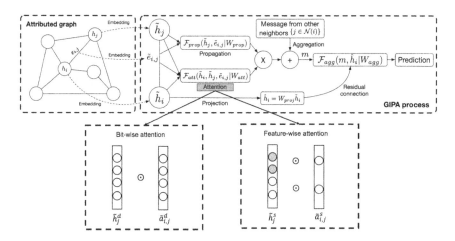

Fig. 1. The architecture of GIPA, which consists of *attention, propagation,* and *aggregation* modules (or processes). The red shadow indicates the bit-wise module and feature-wise module, which extract neighbor information more relevant to the current node through more fine-grained selective information fusion.

$\{\tilde{e}_{i,j}\}$. The workflow of GIPA consists of three major modules (or processes): *attention, propagation* and *aggregation*. First, GIPA computes the dense embedding $\tilde{h}_i^d$, $\tilde{h}_j^d$, and $\tilde{e}_{i,j}$, and sparse embedding $\tilde{h}_i^s$, $\tilde{h}_j^s$ from raw features. Then, the *attention* process calculates the bit-wise attention weights by using fully-connected layers on $\tilde{h}_i^d$, $\tilde{h}_j^d$, and $\tilde{e}_{i,j}$ and feature-wise by using fully-connected layers on $\tilde{h}_i^s$, $\tilde{h}_j^s$, and $\tilde{e}_{i,j}$. Following that, the *propagation* process focuses on propagating information of each neighbor node j by combining j's node embedding $\tilde{h}_j$ with the associated edge embedding $\tilde{e}_{i,j}$. Finally, the *aggregation* process aggregates all messages from neighbors to update the embedding of i. The following subsections introduce the details of each process in one layer.

Embedding Layer. The GIPA (wide & deep) is more suitable for these scenarios: each node in GNN is not composed of text or images, but represents objects such as users, products, and proteins, whose features are composed of category features and statistical features. For example, the category feature in *ogbn-proteins* dataset is what species the proteins come from. And the statistical features in *Alipay* dataset are similar to the total consumption of users in a year.

For dense embedding, each integer number can express a category, each floating-point number can express a statistical value, thus a one-dimensional embedding can represent one feature. However, the sparse embedding of one feature requires more dimensions. For each category feature, one-hot encoding is required. For example, a certain "category feature" has a total of K possible categories, and an $K + 1$ dimensional vector is required to represent the feature. And each "statistical feature" is cut into K categories (using equal-frequency or equal-width cutting method), and then $K + 1$ dimensional one-hot encoding

---

Algorithm 1. The optimization strategy of GIPA

1: **Input:** Graph $\mathcal{G} = \{\mathcal{V}, \mathcal{E}\}$; input features $\{x_v, \forall v \in V\}$; Number of layer K
2: **Note:** $\tilde{h}_v^* \in \{\tilde{h}_v^s, \tilde{h}_v^d\}$
3: $\tilde{h}_v^{d^0} \leftarrow DenseEmb(x_v), \forall v \in V$
4: $\tilde{h}_v^{s^0} \leftarrow SparseEmb(x_v), \forall v \in V$
5: **while** not end of epoch **do**
6:     Select a subgraph $\mathcal{G}_t = \{\mathcal{V}_t, \mathcal{E}_t\} \in \mathcal{G}$
7:     **for** each $k \in [1, K]$ **do**
8:         **if** $k > 1$
9:             $\tilde{h}_i^{*k-1} \leftarrow o_i^{*k-1}, \forall v_i \in V_t$
10:             $\tilde{a}_{i,j}^{*k} \leftarrow \mathcal{F}_{act}(\mathcal{F}_{att}^*(\tilde{h}_i^{*k-1}, \tilde{h}_j^{*k-1}, \tilde{e}_{i,j}|W_{att}^*)), \forall v_i \in V_t$
11:             $m_{i,j}^{*k} \leftarrow \tilde{a}_{i,j}^{*k} * \mathcal{F}_{prop}^*(\tilde{h}_j^{*k-1}, \tilde{e}_{i,j}|W_{prop}^*), \forall v_i \in V_t$
12:             $o_i^{*k} \leftarrow \mathcal{F}_{agg}^*(\sum_{j \in \mathcal{N}(i)} m_{i,j}^{*k}, \hat{h}_{i,j}^{*k-1}|W_{agg}^*), \forall v_i \in V_t$
13:         $\hat{y}_i = Deep(o_i^{dk}) + W_{wide}(o_i^{sk})$
14:         $L \leftarrow \sum_{\forall v_i \in V_t} l(y_i, \hat{y}_i)$
15:         Update all parameters by gradient of L

---

is performed. Thus, the concatenations of dense and sparse embeddings are the inputs of deep part and wide part respectively.

Attention Process. Different from the existing attention mechanisms like self-attention or scaled dot-product attention, we use MLP to realize a bit-wise attention mechanism and a feature-wise attention mechanism. The bit-wise and feature-wise *attention* process of GIPA can be formulated as follows:

$$a_{i,j}^d = \mathcal{F}_{att}^d(\tilde{h}_i^d, \tilde{h}_j^d, \tilde{e}_{i,j}|W_{att}^d) = \text{MLP}([\tilde{h}_i^d||\tilde{h}_j^d||\tilde{e}_{i,j}]|W_{att}^d) \tag{2}$$

$$a_{i,j}^s = \mathcal{F}_{att}^s(\tilde{h}_i^s, \tilde{h}_j^s, \tilde{e}_{i,j}|W_{att}^s) = \text{MLP}([\tilde{h}_i^s||\tilde{h}_j^s||\tilde{e}_{i,j}]|W_{att}^s) \tag{3}$$

where $a_{i,j}^d \in \mathcal{R}^n$ is bit-wise attention weight and its dimension is the same as that of $\tilde{h}_i^d$, $a_{i,j}^s \in \mathcal{R}^m$ is feature-wise attention weight and its dimension is the same as the number of node features, the attention function $\mathcal{F}_{att}^*$ is realized by an *MLP* with learnable weights W_{att}^* (without bias). Its input is the concatenation of the node embeddings $\tilde{h}_i^*$ and $\tilde{h}_j^*$ as well as the edge embedding $\tilde{e}_{i,j}$. As the edge features $\tilde{e}_{i,j}$ measuring the correlation between nodes are input into *MLP*, this attention mechanism could be more representative than previous ones simply based on dot-product. The final attention weight is calculated by an activation function for bit-wise part and feature-wise part:

$$\tilde{a}_{i,j}^* = \mathcal{F}_{act}(a_{i,j}^*) \tag{4}$$

where $\mathcal{F}_{act}$ represents the activation function, such as *softmax, leaky-relu, softplus*, etc. Based on the experimental results, we finally define the activation function as *softplus*. Details can be seen in Sect. 4.3.

Propagation Process. Unlike GAT [30] that considers only the node feature of neighbors, GIPA incorporates both node and edge embeddings during the *propagation* process:

$$p_{i,j}^d = \mathcal{F}_{prop}^d(\tilde{h}_j^d, \tilde{e}_{i,j}|W_{prop}^d) = \text{MLP}([\tilde{h}_j^d||\tilde{e}_{i,j}]|W_{prop}^d), \qquad (5)$$

$$p_{i,j}^s = \mathcal{F}_{prop}^s(\tilde{h}_j^s, \tilde{e}_{i,j}|W_{prop}^s) = \text{MLP}([\tilde{h}_j^s||\tilde{e}_{i,j}]|W_{prop}^s), \qquad (6)$$

where the propagation function $\mathcal{F}_{prop}^*$ is also realized by an MLP with learnable weights W_{prop}^*. Its input is the concatenation of a neighbor node dense and sparse embeddings $\tilde{h}_j^*$ and the related edge embedding $\tilde{e}_{i,j}$. Thus, the *propagation* is done bit-wise and feature-wise rather than node-wise.

Combining the results by *attention* and *propagation* by bit-wise and feature-wise multiplication, GIPA gets the message $m_{i,j}^d$ and $m_{i,j}^s$ of node i from j:

$$m_{i,j}^d = \tilde{a}_{i,j}^d * p_{i,j}^d \quad m_{i,j}^s = \tilde{a}_{i,j}^s \otimes p_{i,j}^s \qquad (7)$$

Aggregation Process. For each node i, GIPA repeats previous processes to get messages from its neighbors. The *aggregation* process first gathers all these messages by a reduce function, summation for example:

$$m_i^* = \sum_{j \in \mathcal{N}(i)} m_{i,j}^* \qquad (8)$$

Then, a residual connection between the linear projection $\hat{h}_i^*$ and the message of m_i is added through concatenation:

$$\hat{h}_i^d = W_{proj}^d \tilde{h}_i^d \quad \hat{h}_i^s = W_{proj}^s \tilde{h}_i^s \qquad (9)$$

$$o_i^d = \mathcal{F}_{agg}^d(m_i^d, \tilde{h}_i^d|W_{agg}^d) = \text{MLP}([m_i^d||\tilde{h}_i^d]|W_{agg}^d) \oplus \hat{h}_i^d \qquad (10)$$

$$o_i^s = \mathcal{F}_{agg}^s(m_i^s, \tilde{h}_i^s|W_{agg}^s) = \text{MLP}([m_i^s||\tilde{h}_i^s]|W_{agg}^s) \oplus \hat{h}_i^s \qquad (11)$$

where an *MLP* with learnable weights W_{agg}^* is applied to get the final dense output o_i^d and sparse output o_i^s . Finally, we would like to emphasize that the process of GIPA can be easily extended to multi-layer variants by stacking the process multiple times. After we get the aggregated output of the node, o_i^d and o_i^s respectively input the depth side and wide side of the Deep&Wide architecture for downstream tasks. See Algorithm 1 for details.

4   Experiments

4.1   Datasets and Settings

Datasets. In our experiments, we choose two edge-attribute dataset: the *ogbn-proteins* dataset from OGB [10] and *Alipay* dataset [17]. The *ogbn-proteins* dataset is an undirected and weighted graph, containing 132,534 nodes of 8 different species and 79,122,504 edges with 8-dimensional features. The task is a multi-label binary classification problem with 112 classes representing different protein functions. The *Alipay* dataset is an edge attributed graph, containing 1.40 billion nodes with 575 features and 4.14 billion edges with 57 features. The task is a multi-label binary classification problem. It is worth noting that due to the high cost of training on *Alipay* data set, we only conduct ablation experiments on the input features of $\mathcal{F}_{att}$ and $\mathcal{F}_{prop}$ in the industrial data set, as Table 3.

Baselines. Several representative GNNs including SOTA GNNs are used as baselines. For semi-supervised node classification, we utilize GCN [14], Graph-SAGE [9], GAT [30], MixHop [1], JKNet [36], DeeperGCN [16], GCNII [5], DAGNN [18], MAGNA [31], UniMP [27], GAT+BoT [34], RevGNN [15] and AGDN [28]. Note that DeeperGCN, UniMP, RevGNN, and AGDN are implemented with random partition. GAT is implemented with neighbor sampling. Except for our GIPA, results of other methods are from their papers or the OGB leaderboard.

Evaluation metric. The performance is measured by the average ROC-AUC scores. We follow the dataset splitting settings as recommended in OGB and test the performance of 10 different trained models with different random seeds.

Hyperparameters. For the number of layers K, we search the best value from 1 to 6. As for the activation function of attention process, we consider common activation functions. For details, please refer to Sect. 4.3.

Running environment. For *ogbn-proteins* dataset, GIPA is implemented in Deep Graph Library (DGL) [32] with Pytorch [23] as the backend. Experiments are done in a platform with Tesla V100 (32G RAM). For *Alipay* dataset, GIPA is implemented in *GraphTheta* [17], and runs on private cloud of Alibaba Group.

4.2   Performance Comparison

Table 1 shows the average ROC-AUC and the standard deviation for the test and validation set. The results of the baselines are retrieved from the ogbn-proteins leaderboard (See footnote 2). Our GIPA outperforms all previous methods in the leaderboard and reaches an average ROC-AUC higher than 0.89 for the first time. Furthermore, GIPA only with 3 layer achieved the *SOTA* performance on *ogbn-proteins* dataset. This result shows the effectiveness of our proposed bit-wise and feature-wise correlation modules, which can leverage the edge features to improve the performance by fine-grained information fusion and noise suppression.

To further investigate the impact of each component in our proposed GIPA, we conduct the ablation study on the ogbn-proteins dataset. As shown in Table 2, compare with GIPA w/o bit-wise module, GIPA w/o feature-wise module.

Table 1. Test and validation performance results (ROC-AUC) on the *ogbn-proteins* dataset. The improvements over comparison methods are statistically significant at 0.05 level.

Method	Test ROC-AUC	Validation ROC-AUC
GCN	0.7251 ± 0.0035	0.7921 ± 0.0018
GraphSAGE	0.7768 ± 0.0020	0.8334 ± 0.0013
DeeperGCN	0.8580 ± 0.0028	0.9106 ± 0.0011
GAT	0.8682 ± 0.0018	0.9194 ± 0.0003
UniMP	0.8642 ± 0.0008	0.9175 ± 0.0006
GAT+BoT	0.8765 ± 0.0008	0.9280 ± 0.0008
RevGNN-deep	0.8774 ± 0.0013	0.9326 ± 0.0006
RevGNN-wide	0.8824 ± 0.0015	0.9450 ± 0.0008
AGDN	0.8865 ± 0.0013	0.9418 ± 0.0005
GIPA-3Layer	0.8877 ± 0.0011	0.9415 ± 0.0023
GIPA-6Layer	$\mathbf{0.8917 \pm 0.0007}$	$\mathbf{0.9472 \pm 0.0020}$

Table 2. Ablation study on the *ogbn-proteins* dataset.

Method	Test ROC-AUC	Validation ROC-AUC
GIPA w/o bit-wise module	0.8813 ± 0.0011	0.9332 ± 0.0009
GIPA w/o feature-wise module	0.8701 ± 0.0021	0.9320 ± 0.0007
GIPA w/o edge feature	0.8599 ± 0.0047	0.9204 ± 0.0038
GIPA	$\mathbf{0.8917 \pm 0.0007}$	$\mathbf{0.9472 \pm 0.0020}$

And the combination of these components (i.e., GIPA) yields the best performance, which indicates the necessity of bit-wise and feature-wise correlation modules.

The average training times per epoch on *ogbn-proteins* with GIPA, 3-Layer GIPA, AGDN are 8.2 s (s), 5.9s, and 4.9s, respectively. The average inference times on whole graph of *ogbn-proteins* with GIPA, 3-Layer GIPA, AGDN are 11.1s, 9.3s, and 10.7s, respectively. Compared with ADGN, GIPA is slower in training speed, but has advantages in inference speed.

4.3   Hyperparameter Analysis and Ablation Study

Effect of the Number of Layers K. To study the impact of the number of layers K on performance, we vary its value to $\{1, 2, 3, 4, 5, 6\}$. As shown in Fig. 2a, with the increase of layers, the performance of the model on the test set gradually converges. However, in the comparison between the five layers and the six layers, the increase in the performance of the model on the verification set is far greater than that on the test set. This result shows that with the increase of complexity, the performance of the model can be improved, but there will be an over-fitting problem at a bottleneck. Therefore, we finally set K to 6.

Analysis on Attention Process with Correlation Module. Because the bit-wise correlation module and the feature-wise correlation module are the most important parts of GIPA, we make a detailed analysis of their architecture, including activation function and edge feature.

Table 3. Ablation on *Alipay* dataset. 'S' and 'D' are primary and neighbor nodes.

$\mathcal{F}_{att}$ input	$\mathcal{F}_{prop}$ input	Feature-wise module	ROC-AUC	F1
Edge	D. Node	None	0.8784	0.1399
Edge & D. Node	Edge & D. Node	None	0.8916	0.1571
Edge & D.+ S. Nodes	Edge & D. Node	None	0.8943	0.1611
Edge & D.+ S. Nodes	Edge & D. Node	Yes	0.8961	0.1623

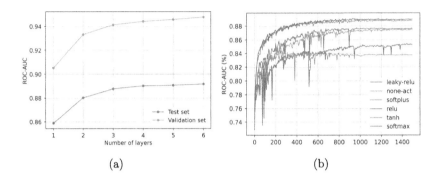

(a)   (b)

Fig. 2. Effect of hyperparameters. (a) ROC-AUC vs. the number of layers K. (b) Convergence on test set vs. activation function of attention.

1) Activation function: as shown in Fig. 2b, by comparing these activation functions: the performance of *tanh* and *relu* are even worse than the one without *activator*. The model using *softmax* as the activation function achieves the best performance, but the performance of model using *softplus* is close to it. This phenomenon indicates that the attention module of GIPA can adaptively learn the correlation weight between nodes. Cause $\sum_{j \in N(i)} a_{ij}$ is needed to be calculated firstly by aggregation from all neighbors, the normalization operation of *softmax* is a costly operation in GNN. On the other hand, GIPA with *softplus* trains faster than that with *softmax* on *ogbn-proteins*. Considering the trade-off between cost and performance, *softplus* is adopted on billion-scale *Alipay* dataset.

2) Edge feature: As shown in Table 2, we find that the performance of GIPA is better than that of GIPA w/o edge feature, which means that the edge feature contains the correlation information between two nodes. Therefore, it is an indispensable feature in attention processing and can help the two correlation modules get more realistic attention weights.

5   Conclusion

We have presented GIPA, a new graph attention network architecture for graph data learning. GIPA consists of a bit-wise correlation module and a feature-wise correlation module, to leverage edge information and realize the fine granularity information propagation and noise filtering. Performance evaluation on the ogbn-proteins dataset has shown that our method outperforms the state-of-the-art methods listed in the ogbn-proteins leaderboard. And it has been tested on the billion-scale industrial dataset.

References

1. Abu-El-Haija, S., et al.: MixHop: higher-order graph convolutional architectures via sparsified neighborhood mixing. In: ICML (2019)
2. Bahdanau, D., Cho, K., Bengio, Y.: Neural machine translation by jointly learning to align and translate. In: ICLR (2015)
3. Brauwers, G., Frasincar, F.: A general survey on attention mechanisms in deep learning. IEEE Trans. Knowl. Data Eng. **35**, 3279–3298 (2021)
4. Chen, J., Ma, T., Xiao, C.: FastGCN: fast learning with graph convolutional networks via importance sampling. In: ICLR (2018)
5. Chen, M., Wei, Z., Huang, Z., Ding, B., Li, Y.: Simple and deep graph convolutional networks. In: ICML (2020)
6. Cheng, H.T., et al.: Wide & deep learning for recommender systems. In: DLRS (2016)
7. Dai, H., Dai, B., Song, L.: Discriminative embeddings of latent variable models for structured data. In: ICML (2016)
8. Grover, A., Leskovec, J.: node2vec: Scalable feature learning for networks. In: SIGKDD (2016)
9. Hamilton, W., Ying, Z., Leskovec, J.: Inductive representation learning on large graphs. In: NeurIPS (2017)
10. Hu, W., et al.: Open graph benchmark: datasets for machine learning on graphs. In: NeurIPS (2020)
11. Hu, Z., Dong, Y., Wang, K., Sun, Y.: Heterogeneous graph transformer. In: WWW (2020)
12. Huang, W., Zhang, T., Rong, Y., Huang, J.: Adaptive sampling towards fast graph representation learning. In: NeurIPS (2018)
13. Ji, Y., et al.: Accelerating large-scale heterogeneous interaction graph embedding learning via importance sampling. TKDD **15**, 1–23 (2020)
14. Kipf, T.N., Welling, M.: Semi-supervised classification with graph convolutional networks. In: ICLR (2016)
15. Li, G., Müller, M., Ghanem, B., Koltun, V.: Training graph neural networks with 1000 layers. In: ICML (2021)
16. Li, G., Xiong, C., Thabet, A., Ghanem, B.: DeeperGCN: all you need to train deeper GCNs. arXiv preprint arXiv:2006.07739 (2020)
17. Li, H., et al.: GraphTheta: a distributed graph neural network learning system with flexible training strategy (2021). https://arxiv.org/abs/2104.10569
18. Liu, M., Gao, H., Ji, S.: Towards deeper graph neural networks. In: SIGKDD (2020)
19. Liu, Z., et al.: GeniePath: graph neural networks with adaptive receptive paths. In: AAAI (2019)
20. Luong, M.T., Pham, H., Manning, C.D.: Effective approaches to attention-based neural machine translation. In: EMNLP (2015)
21. Mikolov, T., Sutskever, I., Chen, K., Corrado, G.S., Dean, J.: Distributed representations of words and phrases and their compositionality. In: Advances in Neural Information Processing Systems, vol. 26, pp. 3111–3119 (2013)

22. Niu, Z., Zhong, G., Yu, H.: A review on the attention mechanism of deep learning. Neurocomputing **452**, 48–62 (2021)
23. Paszke, A., et al.: PyTorch: an imperative style, high-performance deep learning library. In: NeurIPS (2019)
24. Pavlopoulos, J., Malakasiotis, P., Androutsopoulos, I.: Deeper attention to abusive user content moderation. In: EMNLP (2017)
25. Perozzi, B., Al-Rfou, R., Skiena, S.: DeepWalk: online learning of social representations. In: SIGKDD (2014)
26. Shen, T., Jiang, J., Zhou, T., Pan, S., Long, G., Zhang, C.: DiSAN: directional self-attention network for RNN/CNN-free language understanding. In: AAAI (2018)
27. Shi, Y., Huang, Z., Feng, S., Zhong, H., Wang, W., Sun, Y.: Masked label prediction: unified message passing model for semi-supervised classification. arXiv preprint arXiv:2009.03509 (2020)
28. Sun, C., Wu, G.: Adaptive graph diffusion networks with hop-wise attention. arXiv preprint arXiv:2012.15024 (2020)
29. Vaswani, A., et al.: Attention is all you need. In: Advances in Neural Information Processing Systems (2017)
30. Veličković, P., Cucurull, G., Casanova, A., Romero, A., Lio, P., Bengio, Y.: Graph attention networks. In: ICLR (2018)
31. Wang, G., Ying, R., Huang, J., Leskovec, J.: Direct multi-hop attention based graph neural network. arXiv preprint arXiv:2009.14332 (2020)
32. Wang, M., et al.: Deep graph library: towards efficient and scalable deep learning on graphs. In: ICLR (2019)
33. Wang, X., et al.: Heterogeneous graph attention network. In: WWW (2019)
34. Wang, Y.: Bag of tricks of semi-supervised classification with graph neural networks. arXiv preprint arXiv:2103.13355 (2021)
35. Xu, K., Hu, W., Leskovec, J., Jegelka, S.: How powerful are graph neural networks? In: ICLR (2018)
36. Xu, K., Li, C., Tian, Y., Sonobe, T., Kawarabayashi, K.I., Jegelka, S.: Representation learning on graphs with jumping knowledge networks. In: ICML (2018)
37. Yang, Z., Yang, D., Dyer, C., He, X., Smola, A., Hovy, E.: Hierarchical attention networks for document classification. In: Proceedings of the 2016 conference of the North American chapter of the association for computational linguistics: human language technologies, pp. 1480–1489 (2016)
38. Ying, R., He, R., Chen, K., Eksombatchai, P., Hamilton, W.L., Leskovec, J.: Graph convolutional neural networks for web-scale recommender systems. In: SIGKDD, pp. 974–983 (2018)
39. Yun, S., Jeong, M., Kim, R., Kang, J., Kim, H.J.: Graph transformer networks. In: NeurIPS (2019)
40. Zhang, C., Song, D., Huang, C., Swami, A., Chawla, N.V.: Heterogeneous graph neural network. In: SIGKDD (2019)
41. Zou, D., Hu, Z., Wang, Y., Jiang, S., Sun, Y., Gu, Q.: Layer-dependent importance sampling for training deep and large graph convolutional networks. In: NeurIPS (2019)

AntTune: An Efficient Distributed Hyperparameter Optimization System for Large-Scale Data

Jun Zhou[1,2(✉)], Qitao Shi[2], Yi Ding[2], Lin Wang[2], Longfei Li[2], and Feng Zhu[2]

[1] College of Computer Science and Technology,
Zhejiang University, Hangzhou, China
[2] Ant Group, Hangzhou, China
{jun.zhoujun,qitao.sqt,christopher.dy,
fred.wl,longyao.llf,zhufeng.zhu}@antgroup.com

Abstract. Selecting the best hyperparameter configuration is crucial for the performance of machine learning models over large-scale data. To this end, the automation of hyperparameter optimization (HPO) has been widely applied in many automated machine learning (AutoML) frameworks. However, without the effective mechanisms of early stopping and prior knowledge leveraging, such automation is often time-consuming and even inefficient. To improve efficiency, we introduce AntTune, a distributed HPO system that includes parallel optimization, distributed evaluation, tensor cache, etc. Specifically, in AntTune, a time-saving and lightweight mechanism of early stopping is designed to process multiple trials simultaneously. Also, a tree-based meta-learning approach is developed to leverage knowledge from prior tasks and thus it can speed up current HPO tasks. The extensive experiments on both public and industrial datasets demonstrate that our AntTune can improve the state-of-the-art HPO platforms by an average of 3.26% in terms of the effectiveness metrics and 26.25% in terms of tuning time.

Keywords: Hyperparameter Optimization · Automated Machine Learning · Search Space

1 Introduction

Large-scale Machine Learning (ML) has achieved dramatic success recently in various fields. However, building a high-quality ML application heavily relies on tuning hyper parameters by specialized AI scientists and domain experts. To reduce the human effort for frequently tuning hyperparameters, hyperparameter optimization has emerged as a trending topic in both academia and industry. Countless AI projects has been carried out via automated ML (AutoML) due to rapid developments and big breakthrough, but it is still not easy to apply HPO techniques widely in the industry for the following reasons [10]:

Challenge 1. Search Space can be Huge. With AI model size exploding, to find better hyperparameters and fully explore the search space, one may take dozens of trials to get a feasible parameter using massive computing resources.

© The Author(s), under exclusive license to Springer Nature Switzerland AG 2023
X. Wang et al. (Eds.): DASFAA 2023, LNCS 13946, pp. 477–489, 2023.
https://doi.org/10.1007/978-3-031-30678-5_35

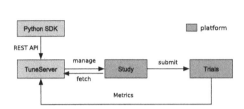

Fig. 1. The high-level architecture of AntTune.   **Fig. 2.** An example of AntTune.

Challenge 2. Model Exploration is Expensive. Though the growth of training data and the complexity of models boost the performance of learning tasks, it could take several days, even weeks to tune models on CPU or specialized high-performance hardware.

Challenge 3. How can Previous Knowledge Help HPO in New Training Datasets? More efficient hyperparameters will be collected on various algorithms when AI approaches are employed frequently. It would be a waste of computing resources if the hyperparameters are tuned from scratch without leveraging knowledge from previous HPO runs on other datasets.

As for **Challenges 1&2**, the distributed HPO and the mechanism of early stopping may be a promising solution. Also, for **Challenge 3**, the idea of meta learning can help to learn prior HPO knowledge. Therefore, in this paper, we propose an industrial-level HPO system, i.e., AntTune, which is efficient and effective for large-scale ML tasks. Our contributions can be summarized as follows:

- To target **Challenges 1&2**, we design and implement our AntTune distributed framework, including six core components, i.e., *TuneServer*, *StudyRunner*, *Distributed TrialRunner*, *Pub/Sub Scheduler*, *Dashboard*, and *TensorCache*, which can run HPO applications simultaneously in distributed platforms. Also, in AntTune, we propose a time-saving and lightweight mechanism to early stop some unpromising trials.
- To target **Challenge 3**, we propose a model warehouse for managing the meta-knowledge of optimization tasks. This makes it possible to learn meta-information or experience from a range of finished tasks and distinguishes AntTune from other HPO systems. Based on the meta-learning approach, optimizing new tasks is much easier and faster than current platforms.
- We conduct extensive experiments on seven public datasets and four industrial datasets to verify the claims mentioned above. The experiments demonstrate that our AntTune can improve the state-of-the-art HPO platforms by an average of 3.26% in terms of the effectiveness metrics, i.e., AUC and RMSE, and 26.25% in terms of tuning time over all datasets.

2   Background

In this section, we first give a few key concepts of HPO. Then two kinds of typical work are discussed.

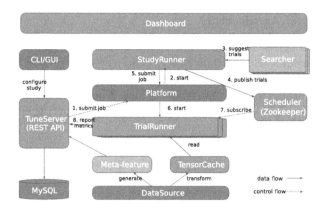

Fig. 3. The detailed architecture and key steps of AntTune.

Definitions. An *Objective function* $f(x)$ attempts to maximize or minimize losses. A *Trial* is a list of hyperparameter values x, which results in an evaluation of $f(x)$. A *Study* represents a process of optimization. Each study contains a collection of trials and configurations related to the optimization. *Search space* is the feasible region for optimizing the study. *Searcher* is the algorithm or tool for suggesting trials to run. *Platform* is the environment where trials are running. It could be a local machine or a cluster (e.g., YARN, Kubernetes).

Hyperparameter Tuning. The automation of hyperparameter optimization has been extensively studied in the literature. SMAC implemented sequential model-based algorithm configuration [6]. TPOT [9] optimized ML pipelines using genetic programming. Tree of Parzen Estimators (TPE) was integrated into HyperOpt [1] and Dragonfly [8] was to perform Bayesian Optimization.

Models Exploration. Many AutoML libraries have been proposed, but most of them tuned models locally or tied to specific AI systems. AutoWEKA [14] and Auto-Sklearn [5] were widely used, combined with the WEKA, and sckit-learn package, and automatically yielded good models for a variety of datasets. However, they are better suited for running locally due to lack of distributed execution. Another type of framework with distributed optimization only focused on specific frameworks, including Katib [16], Auto-Keras [7], Auto-PyTorch [17], Auto-Gluon [4] and etc. Prior open-source softwares of NNI [12] and TransmogrifAI [13] were similar to our system in terms of HPO. NNI was a lightweight toolkit which contains automated feature engineering, hyperparameter optimization, and neural architecture search. TransmogrifAI was an AutoML library with a focus on accelerating ML developer productivity. However, learning meta-knowledge from prior tasks to accelerate the new task is rarely considered in NNI and TransmogrifAI.

3   Overview of AntTune

In this section, we introduce the proposed system, AntTune. Firstly, we give an overview of AntTune. Secondly, we explain the architecture and interface of the system. Finally, we show how to integrate AntTune with an internal AI platform.

3.1   High-Level Architecture

Fig. 1 shows the high-level architecture of AntTune. Firstly, ML developers define a study and launch it with Python SDK. When received the *StartStudy* request, TuneServer submits a job to the platform. The job then leverages the built-in searcher of AntTune to generate trials to run. Finally, those trials are being trained distributedly with PyTorch or TensorFlow and report their intermediate results back to the TuneServer.

3.2   REST API

The TuneServer provides powerful REST API, which makes it lightweight and easy to incorporate with other ML frameworks. The most important APIs are: 1) *CreateStudy*. Create a study that contains a name, an owner, a search space, an optimization goal in {MAXIMIZE, MINIMIZE}, an objective function with a version number, and a trainer. 2) *StartStudy*. It submits a job which requests resources with exactly one container. Once a job successfully runs on the platform, it loads and optimizes the study. 3) *StartTrial*. Start to run a trial which may be terminated early if the metrics calculated by the evaluation are not good. 4) *AddMetric*. Add the metrics of early stopping during the evaluation of a trial. 5) *CreateModule*. Create a module which represents a trainable ML algorithm, e.g., DNNClassifier and XGBoost [2]. 6) *CreateModel*. Create a *Model* that contains the information for building the model, including training/evaluation datasets and the corresponding module.

3.3   Core Components

Figure 3 shows the architecture of AntTune and the key steps for optimizing a study. The main components of AntTune are: 1) *TuneServer* provides the REST API as discussed above. 2) *StudyRunner* manages the parallel optimization of a study. 3) *TrialRunner* is responsible for evaluations of trials. *StudyRunner* maintains a set of *TrialRunner*. 4) *Scheduler* coordinates the distributed execution of each trial. 5) *Dashboard* is a WebUI used for tracking and managing studies and trials. 6) *TensorCache* is a data IO tool optimized for AntTune. Each component will be discussed in detail.

TuneServer. It provides the REST API for communication between different components. For managing and analyzing the studies, TuneServer persists the configuration and current state of all studies and trials to a database management system (e.g., MySQL).

StudyRunner: Parallel Optimization of a Study. When received a StartStudy request, TuneServer submits a job that requests one container with specific memory/CPU/GPU capacities to the platform. When the container is allocated, the study is loaded and then starts to optimize the objective function $f(x)$. Let α be the maximum number of concurrent evaluations of trials. Let β be the total number of trials defined by the study. The key process of *StudyRunner* is: 1) Call

Searcher (e.g., GridSearcher, GeneticSearcher) to suggest trials to run. 2) Publish trials to Zookeeper for fault tolerance. 3) Submit a job to start a *TrialRunner* if there is no idle *TrialRunner* and the number of current running *TrialRunner* is less than α. When received the request, the platform starts a new job to run the *TrialRunner*. 4) Wait for the intermediate result of running trials until at least one metric is reported from *TrialRunner*. 5) Report the result to *Scheduler* and apply *early stopping rules*. The corresponding trial may be terminated early if *early stopping rules* determine that the trial is unpromising. 6) Repeat steps 1 to 5 until the number of running trials reaches β.

TrialRunner: Distributed Evaluation of Trials. Each TrialRunner is a distributed evaluation service, that performs an evaluation of $f(x)$ and evaluates β/α trials on average during the optimization. Each TrialRunner consists of one chief worker and other workers. The main process of the TrialRunner is: 1) Prepare the environment for running an evaluation, i.e., creates an isolated working directory and download the module's code, etc. 2) The chief picks up one trial from a distributed queue and dispatches the trial to workers. 3) When received the trial, all workers start to train the model using data or model parallelism strategies. When the training step reaches evaluation_steps or checkpoint_steps, the workers export the current model M to a distributed file system. 4) The chief starts to evaluate the model once a new checkpoint is exported to the distributed file system. When the evaluation is completed, the chief reports the metric to TuneServer. The trial will be stopped if the call should_stop API returns True. 5) When all workers are completed, repeat the steps from 1 to 4. In contrast to Katib/TransmogrifAI, which submits one job each time when evaluating a trial, AntTune starts at most α jobs that reduce the total count of job submission from β to α, thus decreasing the waiting time for evaluations of trials.

Pub/Sub Scheduler and Fault Tolerant Design. AntTune employs a Pub/Sub ((Publisher/Subscriber) approach to implement the scheduler. It uses Zookeeper to coordinate the execution of trials and send notifications to workers as follows: 1) When a new trial is generated, the scheduler publishes a new node to Zookeeper. All TrialRunners subscribe to the new_trial message. When there is an idle TrialRunner, it takes the node and starts the trial's evaluation. 2) If a TrialRunner consists of more than one worker, it needs to coordinate the distributed execution among the workers. The chief publishes the trial, and other workers subscribe to it. Meanwhile, workers publish their running states, and the chief listens to the states' changes. With the Pub/Sub design, AntTune easily achieves fault tolerance and scalability. The fault tolerance is categorized as two types, i.e., StudyRunner's and TrialRunner's , both of which rely on the Zookeeper Watcher mechanism. For StudyRunner's fault tolerance, when a StudyRunner starts, it registers a temporary node in Zookeeper. TuneServer watches the directories of StudyRunners. If a NodeDeleted event is received, TuneServer restarts the study from the stored state. For TrialRunner's fault tolerance, StudyRunner monitors the Zookeeper directories of its TrialRunners.

TensorCache. In practical scenarios, ML engineers usually build complex pipelines to pre-process input data when developing ML models, e.g., data format conversion and feature engineering. The time cost of exploring a study may be expensive if the pipeline is complicated and time-consuming because each trial must do the same data preparation tasks. TensorCache is developed to handle this problem. An input function that returns datasets with compatible data types for the corresponding ML framework (e.g., tf.data.Dataset for TensorFlow) is specified. AntTune calls the input_function in advance and save the result as tensors to a distributed storage. Trials can train the model by reading the tensors from the distributed storage. The duplicated data preparation tasks are skipped to improve efficiency during the evaluation of trials.

Integration with ML Workflows. AIStudio is an all-in-one AI platform, which is widely used for data pre-processing, feature engineering, model training, and model evaluation in Ant Group. AntTune has been integrated into AIStudio to automate the ML pipeline for improving developer productivity. AIStudio allows AI engineers to automatically create and deploy AI pipelines as follows. Firstly, drag and drop the components from the panel to build a pipeline, as shown in Fig. 2. Secondly, set the configuration for each node. Finally, select a set of nodes as an AutoML-Group and configure an HPO study.

4 Key Algorithms of AntTune

4.1 Automated Asynchronous Early Stopping Rules

The strategy for the termination of unpromising trials is often referred as to automated early stopping. AntTune has out-of-the-box support for a variety of popular early stopping rules, such as the Median Stopping Rule, Population-based Training, Hyperband [11], and its variant [10]. Previous early-stopping algorithms require the ranking of intermediate results to choose the top-performing trials. However, the limited computation resources cannot execute all trials of a study concurrently, trials must wait for the other trials' intermediate results. To tackle the problem, AntTune designs a lightweight resuming mechanism to enable asynchronous early stopping, which processes multiple trials simultaneously. It can be described in two phases: 1) each running trial persists a snapshot of the current training state (e.g., model weights) to a distributed storage when reporting metrics; 2) *StudyRunner* calls the *should_stop* method from *EarlyStoppingRule* to determine whether continue the trial based on the intermediate results. If the results are promising, *StudyRunner* resumes the trial from the saved checkpoint. For example, the asynchronous Hyperband algorithm is implemented using this mechanism, which we call AsyncHyperband.

4.2 Tree-Based Meta-Learning Mechanism

Meta-learning studies how learning systems improve efficiency through experience. It can be employed when knowledge about prior AutoML/HPO systems runs on other datasets is available. We observe that many ML tasks are similar

```
1 create_module("DNNClassifier",type="tf_estimator",package="code.tar.gz")
2 create_model("foo.DNNModel", module="DNNClassifier")
3 insert_model("foo.DNNModel",instance_id="dt=22-05-10",
4  train_in="train_table",eval_in="eval_table",valid_in="valid_input")
5 objective=from_warehouse("foo.DNNModel",instance_id="dt=22-05-10")
6 study=create_study(...); study.optimize(objective)
```

Fig. 4. An example of Model Warehouse.

(e.g., using the same algorithms or a similar dateset) by analyzing the history of our internal AI jobs. However, a lot of HPO tasks are executed from the scratch, which takes time and is expensive. Warm starting the hyperparameter search process by using the best results from the closest seen datasets is a viable strategy. To solve this problem, AntTune proposes a tree-based meta-learning algorithm that leverages knowledge from prior tasks to guide and accelerate the current task. The implementation of meta-learning in AntTune is relatively simple yet efficient for easy deployment. As depicted in previous work, the design of the meta-learning approach depends on meta-data, meta-feature, and meta-models. Furthermore, as to AntTune, a storage system to manage the information about the tasks called *Model Warehouse* is proposed. And optimization tasks are categorized into different groups by the type of ML algorithms. Instead of building a global meta-model, AntTune builds a meta-model per type of ML algorithm.

Meta-Feature. The meta-feature is a collection of measurable properties of the task. It should be interpretable and easy to calculate. We design a set of meta-features by lots of experiments: *General meta-features*, such as number of classes, number of samples, and number of features, etc. *Statistical meta-features*, such as skewness, kurtosis, and number of negative labels over a number of positive labels, etc. *Information-theoretic meta-features*, such as normalized class entropy, and normalized attribute entropy, etc.

Meta-Model. Meta-model learns the patterns among datasets, parameters, and learning performance, i.e., f(meta-features)$\rightarrow$hyperparameters. We do not put much effort into the choice of meta-models since meta-models' performance is often dominated by the meta-feature. And the meta-models should have good generalization abilities to handle bad cases. It is well known that bagging and boosting have been proven successful in real-world applications. Thus AntTune chooses Random Forest as the default meta-model to predict the performance of a hyperparameter configuration on a given task.

Meta-Data and Model Warehouse. As more tasks are completed, more meta-data is gathered, and more prior experience can be exploited. We borrow the term *Data Warehouse* that collects and manages data from varied sources to

---

Algorithm 1: Meta-learning algorithm.

Input: $\mathcal{M}_i$, D_i, T_i, K_i, $study$
Output: A set of hyperparameters
1 $f \leftarrow$ get_meta_feature(D_i); $\kappa \leftarrow$ find_similar_knowledge(K_i, f)
2 **if** $\kappa \neq NULL$ **then**
3 | $S \leftarrow$ the hyperparameter setting of κ
4 **else**
5 | $P \leftarrow$ find_predictor(T_i)
6 | **if** $P = NULL$ and $K_i \neq \emptyset$ **then**
7 | | $P \leftarrow$ build_random_forest_predictor(K_i)
8 | **end**
9 | **if** $P \neq NULL$ **then**
10 | | $S \leftarrow \emptyset$
11 | | **foreach** $n, p \in P$ **do**
12 | | | S.put(n, predict(p, f))
13 | | **end**
14 | **else**
15 | | $S \leftarrow study$.optimize()
16 | | K_i.append((f, S)) // Update the Knowledge Repo
17 | **end**
18 **end**
19 **return** S

---

provide meaningful business insights. Following the definition, we use the term *Model Warehouse* to collect and manage data from various tasks. The *Model Warehouse* contains the following objects: **Module:** Each module represents a kind of AI algorithm, and contains a globally unique name, a version number, an owner, a model type, and a package with training codes compressed into a single tar file. **Model:** Each model contains a globally unique name, an owner, a module name, and a directory which saves model parameters. **Model Instance:** Each model can have one or more model instances with an objective function. It contains an identifier *instance_id* and input datasets. **Knowledge Repository:** It stores the knowledge for training the meta-model. Figure 4 is an example of using Model Warehouse in AntTune. Engineers first define an objective function written in Python and then upload the code to create a module via REST API (line 1). Then engineers insert a trainable model instance (lines 3–5). In line 6, it starts to tune the model. Each call of model tuning triggers a job, and inserts one record to the Knowledge Repository when finished. Therefore, the size of the Knowledge Repository grows gradually, and the performance of meta-models may improve.

Meta-Learning Algorithm. The proposed meta-learning method is summarized in Algorithm 1. The model instance $\mathcal{M}_i$ has the properties (D_i, T_i), where D_i and T_i are the dataset and the module of training $\mathcal{M}_i$, respectively. After F_i (meta-feature) of D_i is extracted by data pre-processing, we try to find similar knowledge from the Knowledge Repository $\mathcal{K}_i$. The similarity of two meta-features F_i, F_j is measured by the Euclidean distance. AntTune finds a predictor from the knowledge that belongs to the module T_i. Each predictor consists of a list of Random Forest regressors, which are trained to predict the performance of a hyperparameter. If appropriate knowledge is not found, AntTune starts a new study to find the best configuration C_i. When the study is finished, it appends a record (F_i, C_i) to the corresponding Knowledge Repository K_i. As shown in line 12, the potentially optimal hyperparameter setting of a study can be derived directly from the corresponding meta-model in several seconds without starting trials. Therefore, it is possible to save a huge amount of computing resources.

5   Experiment

In this section, we first compare our AntTune with two state-of-the-art open-source platforms, i.e., TransmogrifAI [13] and NNI [12], to validate the performance of our AntTune. Next, we conduct ablation studies to demonstrate the detailed contributions of the two main components of our AntTune, i.e., Early Stopping and Meta Learning.

Table 1. The performance (AUC/RMSE/Time(seconds)) on a single machine (**bold**: best among the methods).

Public Dataset	Tasks	TransmogrifAI	NNI	AntTune
Titanic	classification	0.8231/−/63	**0.8274**/−/45.3	0.8273/−/**45**
Wholesale customers	classification	0.8201/−/29.4	0.8232/−/24.1	**0.8237**/−/**23.4**
Tic-Tac-Toe Endgame	classification	0.9478/−/70.6	0.9588/−/49.9	**0.9672**/−/**47.1**
Raisin	classification	0.9036/−/65.2	0.9151/−/47.6	**0.9161**/−/**46.7**
Boston Housing	regression	−/4.294/178	-/**4.2126**/140	−/4.2256/**135**
Forest Fires	regression	−/21.4532/180	−/20.0156/**135**	−/**19.6512**/137
Las Vegas Strip	regression	−/1.4802/198	−/1.3701/161	−/**1.3613**/**152**

Table 2. The performance (AUC/RMSE/Time(seconds)) on a distributed cluster with 48 workers (**bold**: best among the methods).

Industrial Dataset	Tasks	TransmogrifAI	NNI	AntTune
Dist_TaskA1	classification	0.7131/−/13,643	0.7211/−/9,454	**0.7243**/−/**8,595**
Dist_TaskA2	classification	0.7922/−/66,578	0.8014/−/43,678	**0.8048**/−/**39,022**
Dist_TaskB1	regression	−/**115.94**/19,924	−/115.95/11,853	−/115.99/**10,958**
Dist_TaskB2	regression	−/**286.35**/131,498	−/286.36/73,488	-/286.35/**65,739**

5.1   Comparison with TransmogrifAI and NNI

Experimental Setting. To validate the performance of our AntTune, we compare AntTune with the open-source TransmogrifAI and NNI. Specifically, we conduct extensive experiments on a single machine for seven public tasks via using UCI[1] machine learning repository (four binary classification tasks and three

---

[1] UCI URL: https://archive.ics.uci.edu/ml/index.php.

Table 3. Search space of XGBoost for the ablation study of early stopping.

Hyperparameter	max_depth	colsample_bytree	eta	gamma	min_child_weight
Type	discrete	continuous	continuous	continuous	discrete
Range	[4,8,12,16,20,24]	0.5~1.0	0.1~1.0	0.1~1.0	[1,2,3]

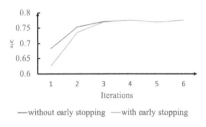

—without early stopping   —with early stopping

■Number of Knowledge —Percentage of meta-learning

Fig. 5. Tuning XGBoost via early stopping rules.

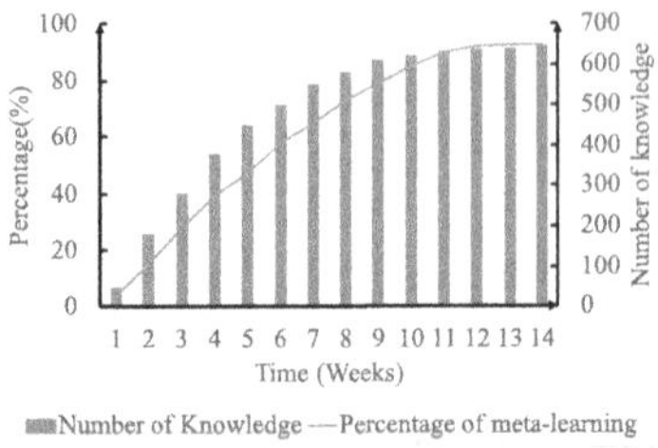

Fig. 6. Meta-learning in real-world scenarios.

regression tasks) and on a distributed cluster with 48 workers for four industrial tasks in our Alipay platform (two binary classification tasks and two regression tasks). Due to space limitations, we list these datasets and tasks in Tables 1 and 2, respectively. As for the four industrial datasets, i.e., Dist_TaskA1/TaskA2 and Dist_TaskB1/TaskB2, we collect them from a classification scenario with 2 million samples (Dist_TaskA1, 72 features per sample) and 10 million samples (Dist_TaskA2, 72 features per sample) and a regression scenario with 1.5 million samples (Dist_TaskB1, 66 features per sample) and 10 million samples (Dist_TaskB2, 66 features per sample) in our AliPay platform. Note that we run each experiment 5 times and then report the average results.

Result and Analysis. Table 1 shows the results of running on a single machine. It demonstrates that our AntTune can improve TransmogrifAI by an average of 3.20% in terms of the effectiveness metrics, i.e., Area Under a ROC (Receiver operating characteristic) Curve (AUC) for classification tasks and root mean square error (RMSE) for regression tasks, and an average of 25.99% in term of tuning time. Also, our AntTune can outperform NNI in most cases. Table 2 shows the results of running on a distributed cluster with 48 workers. Compared with TransmogrifAI and NNI, our AntTune has a greater advantage in the distributed environment. Specifically, our AntTune can improve TransmogrifAI and NNI by an average of 0.80% and 0.22% respectively in terms of the effectiveness metrics and an average of 43.35% and 9.46% respectively in terms of tuning time. Through the data analysis, we find that the efficiency improvement is due to the time saving brought by the following three core components of our AntTune: (1) TrialRunner's distributed method reduces the waiting time by 10%, (2) Pub/Sub Scheduler's fault tolerance mechanism makes the task tolerant to errors and thus reduces the recovery time of tasks, and (3) TensorCache decreases the duplicated IO time by about 10%.

5.2 Early Stopping

Experimental Setting. To demonstrate the detailed contribution of the mechanism of early stopping, we optimize XGBoost's hyperparameters with and without the mechanism of early stopping on a large-scale classification dataset. The dataset contains 40 million samples (80% for training) and each sample consists of 1500 features. Each optimization of the task requires 45 workers, and each worker applies 30G memory. Specifically, we select five hyperparameters as the search space (as listed in Table 3). We set the maximization of AUC as the optimization goal, RACOS [15] as the searcher, and AsyncHyperband (configuration: $R = 45$, $r = 5$, $eta = 3$, rounds $= 3$, iterations $= 6$, and max trials $= 68$) as the early stopping rule. We treat training data sampling as budgets in AsyncHyperband.

Result and Analysis. Figure 5 shows that although data sampling causes performance to decline in the initial iterations, it eventually catches up and produces a comparable outcome, i.e., AUC: 0.773. By sampling the training data and the termination of unpromising trials, the total computation cost is cut by a factor of 46% in this scenario.

5.3 Meta-Learning

Experimental Setting. To demonstrate the detailed contribution of the meta-learning in our AntTune, we choose to optimize the Wide & Deep [3] classification model and observe the percentage of the current tasks that have leveraged the prior knowledge learned from prior tasks over time. In these Wide & Deep tasks, there are 25 hyperparameters and the scale of the dataset ranges from 1 million to 15 million.

Result and Analysis. As shown in Fig. 6, the left vertical axis represents the proportion of current optimization tasks that have leveraged the prior knowledge, the right vertical axis represents the number of knowledge (e.g., prior tasks), and the horizontal axis represents the time. Figure 6 demonstrates that as the amount of knowledge (prior tasks) grows over time, the meta-learning algorithm can help with a variety of optimization tasks. Only 4% of tasks use the meta-learning mechanism's predictions in the first week, but 90% of tasks use the preferable hyperparameter settings provided by meta-models in the eleventh week. The percentage finally converges at about 97%. A significant amount of computational resources are saved since the optimization tasks which can acquire predictions from the meta-model consume only a few seconds as opposed to several days (i.e., dozens of expensive trials). In terms of model results, for instance, meta-learning is deployed in a new coupon push scenario, which also shows a significant advantage over hand-tuning, i.e., CTR (Click-through rate): 0.368 vs. 0.267.

6   Deployment

AntTune has been deployed in production at Ant Group on June 2018. Since then, it has executed a total of 309,423 tasks from 379 real-world scenarios. On average, each task performed 12 trials on 5 million data samples with a running time of 216 min. Break down by scenarios, we observed a 2% to 43% AUC improvement by tuning hyperparameters with AntTune, compared to models with hand-tuned parameters. In addition to improved model performance, we observed a reduction in development time in building an HPO pipeline from around a week to several hours. These amount to the overall improvement of productivity in Ant Group.

7   Conclusion

We introduce an efficient distributed HPO system: AntTune, which is easily integrated with ML frameworks for better AI productivity. Moreover, we discuss the design of AntTune that supports parallel optimization of studies and distributed evaluation of trials. With the help of *Model Warehouse*, automated asynchronous early-stopping algorithms and tree-based meta-learning methods are offered by AntTune. Based on the above design, AntTune has been proven as a powerful and easy-to-use tool for building high-quality ML tasks.

References

1. Bergstra, J., Yamins, D., Cox, D.: Making a science of model search: Hyperparameter optimization in hundreds of dimensions for vision architectures. In: ICML, pp. 115–123. PMLR (2013)
2. Chen, T., Guestrin, C.: Xgboost: a scalable tree boosting system. In: SIGKDD, pp. 785–794 (2016)
3. Cheng, H.T., Koc, L., Harmsen, J., et al.: Wide & deep learning for recommender systems. In: DLRS, pp. 7–10 (2016)
4. Erickson, N., Mueller, J., Shirkov, A., et al.: Autogluon-tabular: robust and accurate automl for structured data. arXiv preprint arXiv:2003.06505 (2020)
5. Feurer, M., Klein, A., Eggensperger, K., Springenberg, J.T., Blum, M., Hutter, F.: Auto-sklearn: efficient and robust automated machine learning. In: Hutter, F., Kotthoff, L., Vanschoren, J. (eds.) Automated Machine Learning. TSSCML, pp. 113–134. Springer, Cham (2019). https://doi.org/10.1007/978-3-030-05318-5_6
6. Hutter, F., Hoos, H.H., Leyton-Brown, K.: Sequential model-based optimization for general algorithm configuration. In: Coello, C.A.C. (ed.) LION 2011. LNCS, vol. 6683, pp. 507–523. Springer, Heidelberg (2011). https://doi.org/10.1007/978-3-642-25566-3_40
7. Jin, H., Song, Q., Hu, X.: Auto-keras: an efficient neural architecture search system. In: SIGK, pp. 1946–1956 (2019)
8. Kandasamy, K., Vysyaraju, K.R., Neiswanger, W., et al.: Tuning hyperparameters without grad students: scalable and robust bayesian optimisation with dragonfly. arXiv preprint arXiv:1903.06694 (2019)

9. Le, T.T., Fu, W., Moore, J.H.: Scaling tree-based automated machine learning to biomedical big data with a feature set selector. Bioinformatics **36**(1), 250–256 (2020)
10. Li, L., Jamieson, K., Rostamizadeh, A., et al.: A system for massively parallel hyperparameter tuning. MLSys **2**, 230–246 (2020)
11. Li, L., Jamieson, K., DeSalvo, G., Rostamizadeh, A., et al.: Hyperband: a novel bandit-based approach to hyperparameter optimization. JMLR **18**(1), 6765–6816 (2017)
12. Microsoft: Neural Network Intelligence. https://github.com/microsoft/nni, Accessed 30 July 2022
13. Salesforce: Transmogrifai's documentation. https://transmogrif.ai/. Accessed 8 July 2022
14. Thornton, C., Hutter, F., Hoos, H.H., et al.: Auto-weka: combined selection and hyperparameter optimization of classification algorithms. In: SIGKDD, pp. 847–855 (2013)
15. Yu, Y., Qian, H., Hu, Y.Q.: Derivative-free optimization via classification. In: AAAI (2016)
16. Zhou, J., Velichkevich, A., Prosvirov, K., Garg, A., Oshima, Y., Dutta, D.: Katib: a distributed general automl platform on kubernetes. In: OpML, pp. 55–57 (2019)
17. Zimmer, L., Lindauer, M., Hutter, F.: Auto-pytorch: multi-fidelity metalearning for efficient and robust autodl. IEEE TPAMI **43**(9), 3079–3090 (2021)

LogLG: Weakly Supervised Log Anomaly Detection via Log-Event Graph Construction

Hongcheng Guo[1], Yuhui Guo[3], Jian Yang[1], Jiaheng Liu[1(✉)], Zhoujun Li[1], Tieqiao Zheng[2], Liangfan Zheng[2], Weichao Hou[2], and Bo Zhang[2]

[1] Beihang University, Beijing, China
{hongchengguo,jiaya,lizj,liujiaheng}@buaa.edu.cn
[2] Cloudwise Research, Beijing, China
{steven.zheng,leven.zheng,william.hou,bowen.zhang}@cloudwise.com
[3] Renmin University of China, Beijing, China
yhguo@ruc.edu.cn

Abstract. Fully supervised log anomaly detection methods suffer the heavy burden of annotating massive unlabeled log data. Recently, many semi-supervised methods have been proposed to reduce annotation costs with the help of parsed templates. However, these methods consider each keyword independently, which disregards the correlation between keywords and the contextual relationships among log sequences. In this paper, we propose a novel weakly supervised log anomaly detection framework, named LogLG, to explore the semantic connections among keywords from sequences. Specifically, we design an end-to-end iterative process, where the keywords of unlabeled logs are first extracted to construct a log-event graph. Then, we build a subgraph annotator to generate pseudo labels for unlabeled log sequences. To ameliorate the annotation quality, we adopt a self-supervised task to pre-train a subgraph annotator. After that, a detection model is trained with the generated pseudo labels. Conditioned on the classification results, we re-extract the keywords from the log sequences and update the log-event graph for the next iteration. Experiments on five benchmarks validate the effectiveness of LogLG for detecting anomalies on unlabeled log data and demonstrate that LogLG, as the state-of-the-art weakly supervised method, achieves significant performance improvements compared to existing methods.

Keywords: Data Security · Log Analysis · Graph Neural Network

1 Introduction

Due to the recent advancement of the IT industry, software systems have become larger and more complex, which brings huge challenges for maintaining systems [1]. To handle these issues, log analysis [2] is adopted as the main technique to identify faults and capture potential risks. With the development of deep learning, fully supervised log anomaly detection approaches [20] have achieved

X. Wang et al. (Eds.): DASFAA 2023, LNCS 13946, pp. 490–501, 2023.
https://doi.org/10.1007/978-3-031-30678-5_36

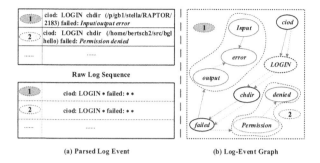

(a) Parsed Log Event   (b) Log-Event Graph

Fig. 1. (a) Number 1 represents log event from normal data while number 2 represents log event from abnormal data. After parsing, they get the same result. (b) Our method exploits the semantic correlation among keywords from different log events by graph. Each event has their corresponding subgraphs, thus we can identify anomaly easily without a parser.

promising successes, which need a large amount of labeled training data consisting of normal and anomalous log sequences. However, labeling massive unlabeled log data is time-consuming and expensive. Thus, semi-supervised learning has attracted increasing interest, resulting in various semi-supervised log anomaly detection methods [9] are proposed. Such methods generally first employ the log parser to generate templates for labeled normal logs and then detect the anomaly by comparing these templates. So they rely heavily on the log parsing stage [7], which means if the log parser does not convert raw logs to the right templates, the detection performance will drop a lot [9]. Besides, in real industrial scenes, labeled normal data is not easily available due to the evolution of the system. In Fig. 1, two log events have the same template (**ciod: LOGIN** ∗ **failed:** ∗∗) while one of the two is the anomaly data. This is because valuable tokens (**Permission denied** and **Input/output error**) are lost after parsing.

In this paper, to address above shortcomings, motivated by WCKG [19], we propose a similar weakly supervised framework for **LOG** anomaly detection via Log-event **G**raph construction, called LoGLG. which is an end-to-end iterative method without utilizing both log parsing and labeled normal log data. How can we perform log anomaly detection without massive labeled normal log data? Naturally, we turn our attention to the keywords in the logs at a finer granularity. With the help of Graph Neural Networks [16], we treat each token as a node in the graph, and by modeling the semantic relationship between nodes, we can obtain the semantics of the whole system. Thus we get rid of the dependence on large-scale normal log sequences. For log parsing, We discard this step, instead, we first sub-word the raw logs and let the model learn to establish connections between important keywords on its own without fixed templates, thus LoGLG is the first end-to-end weakly supervised framework.

Specifically, In each iteration of our model, LoGLG first constructs a log-event graph G with all extracted keywords from log sequences as nodes and each keyword node updates itself via its neighbors. With G, each unlabeled log event L corresponds to one subgraph G_L of G, so we propose a subgraph annotator to assign a pseudo label to each log event L. Besides, a self-supervised method is designed to generate more accurate pseudo-labels. Then, we train a classifier to classify all the unlabeled logs with generated pseudo labels. Based on the classification results, we adopt a simple TF-IDF algorithm to extract more discriminative keywords, where more accurate pseudo labels can be inferred in the next iteration. Extensive experiments on five benchmarks demonstrate that LoGLG effectively detects log anomaly for massive unlabeled log data through a weakly supervised way, and outperforms state-of-the-art methods.

The main contributions of this work are as follows.

- We propose a novel weakly supervised log anomaly detection framework, called LoGLG, which explores the semantic correlation among log data via log-event graph without utilizing a parser and massive normal log sequences.
- An end-to-end iterative process is built to ensure the accuracy of both keywords and pseudo labels, where we design a subgraph annotator to generate pseudo labels for corresponding log events without any labeled log data, thus reducing dependency on log parsing.
- We conduct extensive experiments on five benchmark datasets, and the results demonstrate that our LoGLG gains the state-of-the-art performance compared with existing methods.

2 Related Work

2.1 Log Anomaly Detection

Machine learning based log anomaly detection methods generally contain the following two steps [11] : (i) Preprocessing log messages, (ii) Anomaly detection. These approaches all require a log parser as a preprocessing operation to extract log templates from log messages. Then, a machine learning model is constructed to detect anomalies. In recent years, deep learning methods have been proposed to improve detection performance [4,5,14,18]. Due to the imperfection of log parsing, above methods may lose semantic meaning of log messages, thus leading to inaccurate detection results. Besides, these methods rely heavily on labeled normal data, which is hard to achieve in real world.

2.2 Pre-trained Model

BERT [3] is a model based on transformer network [12], which achieved outstanding performance in various natural language processing tasks [6,13,17]. One of the major characteristics of BERT is that using two unsupervised learning methods containing masked language modeling (MLM) and next sentence prediction (NSP) to perform the pre-training. The system log can be deemed as sequence data because it is a dataset with an order. In this paper, we use the BERT as the log classifier, and our method directly uses unlabeled raw log data to detect anomalies, which follows an iterative process: generating pseudo labels of log sequences using a subgraph annotator, building a log classifier, and updating the keywords based on the classification results. Among them, the most critical step is generating pseudo-labels for unlabeled raw logs. Then, we employ the raw log with pseudo-labels to train the BERT-based log classifier.

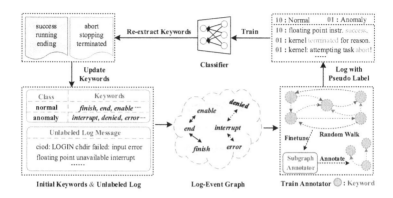

Fig. 2. Overview of LOGLG. Our method follows an iterative paradigm. We first build a log-event graph G, where unlabeled log sequences correspond to log-event subgraphs of G. Then we design a self-supervised task to train the subgraph annotator for pseudo labels, where the random walk is used to generate subgraph as the input of the annotator (i.e., the subgraph formed by pink edges). Then, a log classifier is trained with pseudo labels. Based on the classification, the keywords of log sequences are re-extracted and updated for the next iteration.

3   Proposed Method

3.1   Overview of LOGLG

In Fig. 2, our method is an iterative process, it aims to build a weakly supervised log anomaly detection framework by assigning pseudo labels to unlabeled logs X. The input log data of the model contains two parts: (i) A set of initial keywords $L = \{L_0, L_1\}$ extracted by TF-IDF for unlabeled logs, which are randomly initialed into two categories: normal and anomaly, where $L_i = \{w_1^i, w_2^i, \cdots, w_{k_i}^i\}$ denotes the k_i keywords of the category i, where $i \in \{0, 1\}$. (ii) n unlabeled raw logs $X = \{X_1, X_2, \cdots, X_n\}$ from normal and anomaly data. In each iteration, we first build a log-event graph G based on the keywords from log sequences. To ensure the high-quality of subgraph annotator A, we design a self-supervised task to train and fine-tune the annotator A with noisy labels. After that, pseudo-labels are generated by A. With the synthetic labels, we train a log classifier. After getting the classification results, keywords L are re-extracted and updated for the next iteration, where the training stage ends when there is no change in the keywords at all.

3.2   Log-event Graph Construction

We first exploit the extracted initial keywords of raw log sequences to construct a log-event graph, denoted as $G = (V, E)$, where the keywords as the vertices V and the co-occurrences between keywords as edges E. In each iteration, keywords are re-extract by TF-IDF method based on the results of the classifier. The score S of word w_m is:

$$S(w_m, C_t) = TF(w_m, C_t) \bullet IDF(w_m)^M \tag{1}$$

where M is a hyper-parameter, C_t is two classes, $t \in \{0, 1\}$. Based on the S, the top Z words are selected as the keywords in two classes for the next iteration. we represent each node x_v with the combination of category and index by leveraging the one-hot embedding, thus each node $x_v \in R^{2+|V|}$. Then, the semantic correlation knowledge among raw log sequences can be propagated and aggregated by GNN over the log-event graph G.

3.3   Pseudo Label Generation

After unlabeled log sequences are represented as the subgraphs, the detection of log anomalies is defined as a graph-level classification problem. The keywords of subgraphs belong to a set of vertices in the log-event graph G and the edges among keywords are contained in the log-event graph G. Then, we employ Graph Isomorphism Network (GIN) [16] as the subgraph annotator, which is described as:

$$R_v^{(k)} = Liner^{(k)} \Big(\sum_{u \in N(v)} R_u^{(k-1)} + (1 + \varepsilon^k) \cdot R_v^{(k-1)} \Big) \tag{2}$$

where $R_v^{(k)}$ denotes the representation of node v after the kth update. $Liner^{(k)}$ is a multi-layer perceptron in the kth layer. ε is a learnable parameter. $N(v)$ denotes all the neighbors of node v. Then, the generated subgraph R_G is represented as follows:

$$R_G = concat[\sum_{v \in G} (R_v^{(k)}) | k = 0 \cdots K] \tag{3}$$

where $\sum_{v \in G}(\cdot)$ denotes the aggregation of all node features from the same layer. GIN concatenates the features from all layers as the subgraph representations.

Pre-training on Subgraph Annotator. The pre-training of the annotator is performed on the log-event subgraph. Our self-supervised pre-training process is shown in Algorithm 1, the subgraph derived from the random walk is similar to the subgraph generated by an unlabeled log sequence, where the subgraph annotator A is trained to predict the class of the start point of a random walk.

First, we randomly sample a keyword w_t from one class C_t. This step is the beginning of a random walk. The process of the random walk follows the

---

Algorithm 1. Pre-training of Annotator on Log-event Graph

Require:

    log-event graph G, unlabeled log sequences X, Gaussian parameters μ_m, σ_m^2, edge probability p_{ij}

Ensure:

    pre-trained subgraph annotator A;

1: **repeat**
2:      Randomly sample a class C_t, $t \in \{0, 1\}$;
3:      In class C_t, randomly sample a word ω_t;
4:      Sample L from distribution $N(\mu_m, \sigma_m^2)$;
5:      Perform a random walk on G, where ω_t is the beginning, L is set to be the length, p_{ab} is the probability, then a subgraph G_t is generated;
6:      Input of A is G_t, C_t is the target, loss calculation;
7:      Calculate the gradient and update parameters of A;
8: **until** convergence

Gaussian distribution $N(\mu_m, \sigma_m^2)$, **where the number of random walk steps is same as that of the number of keywords contained in an unlabeled log sequence** X_i. The parameters of the Gaussian distribution $N(\mu_m, \sigma_m^2)$ are estimated with X_i:

$$\mu_m = \frac{1}{n} \sum_i^n tf(X_i) \tag{4}$$

$$\sigma_m^2 = \frac{1}{n-1} \sum_i^n [tf(X_i) - \mu_m]^2 \tag{5}$$

where $tf(X_i)$ denotes the number of keywords in a log sequence X_i. L is the length of random walk from distribution $N(\mu_m, \sigma_m^2)$. Suppose we have two nodes w_a and w_b. The probability of the random walk from node w_a to w_b is defined as follows:

$$p_{ab} = \frac{F_{ab}}{\sum_{\omega_t \in N_{\omega_i}} F_{at}} \tag{6}$$

where F_{at} is the co-occurrence frequency of w_a followed by ω_t. $N(w_a)$ is the neighbors set of w_a.

Then, we set the node ω_t as the starting node to perform L-step random walk. In each step, p_{ab} determines the probability of walking from w_a to neighbor w_b. Thus when the random walk finishs, a subgraph G_t is achieved by us, which is the induced subgraph in the log-event graph G.

In self-supervised pre-training process, we feed the induced subgraph G_t into A, where the annotator A learns to predict the class of start point ω_t. The training objective is defined as the negative log likelihood of C_t:

Throughout the self-supervised pre-training process, we feed the induced subgraph G_t into the annotator A, where the annotator A learns to predict the class of the starting point ω_t. Obviously, there are only two categories on this side, and the training objective is defined as the negative log likelihood of C_t.

$$L_{self-sup} = - \sum_{r \in rand} C_t log(A(G_t)) \tag{7}$$

Improvement Strategy on Annotator. After pre-training the subgraph annotator A, we design an enhanced strategy to fine-tune it by the voting to generate the labels as follows:

$$\hat{y}_a = argmax_k \{\sum_b tf(\omega_b, X_a) | \forall (\omega_b \in L_k)\} \tag{8}$$

where $tf(\omega_b, X_a)$ denotes the term-frequency (TF) of keyword ω_b in log X_a. The loss function is defined as:

$$L_{TF} = - \sum_{a=1}^{n} (kf(X_i) > 0) \hat{y}_a log(A(G_a)) \tag{9}$$

where G_a is the subgraph of log sequence U_a.

3.4 Training Objective on Log Detection

After training the subgraph annotator A, we use it to generate pseudo labels and annotate all unlabeled logs X, which are used to train a log classifier. Our framework can be extended to any log classifier. Generally, the pre-trained BERT [3] is applied to the fields using sequence data, and the log can be deemed as sequence data. Therefore, we use the BERT as the log classifier. Following the previous works [19], we train the classifier on raw logs with pseudo-labels, the training objective of the classifier is defined as:

$$L_{cls} = - \sum_{i=1}^{N} y_i \times log y_i' \tag{10}$$

where y_i is the one-hot distribution of the real category. Only the probability of the real category is 1, the others are 0. y_i' is the distribution after $softmax$.

The predicted labels for all unlabeled logs by the log classifier are used to re-extract keywords of log sequences. To decide whether the model has converged, the change of keywords is defined as:

$$\gamma = \frac{|L^{E_k} - L^{E_k} \cap L^{E_{k-1}}|}{|L^{E_k}|} \tag{11}$$

where L^{E_k} is the keywords set of the kth iteration, ϵ is a hyper-parameter. When $\gamma < \epsilon$, the iteration finishes.

4   Experiment

4.1   Datasets and Evaluation Metrics

We validate the effectiveness of LogLG on five public datasets from LogHub [8][1]. Table 1 shows the details of datasets. For each dataset, considering that logs evolve over time, we select the first 80% (according to the timestamps of logs) log sequences for training and the rest 20% for testing. This setting is the same as the previous work. Since each dataset comes from different systems, thus they require the corresponding preprocessing. We extract the log sequences consistently with the previous works [15,18]. For **HDFS**, we generate log sequences according to the *block_id*. For other datasets, we utilize the **Sliding Window** (size of 20) without overlap to generate log sequences in chronological order. To prevent the model from incorrectly presenting consecutive strings of numbers as keywords, we use a special token [**Num**] to replace consecutive strings of numbers longer than 4 in each log, while strings of numbers less than or equal to 4 in length are retained. In our experiments, we follow the previous work and adopt evaluation metrics: Precision ($\frac{TP}{TP+FP}$), Recall ($\frac{TP}{TP+FN}$) and F_1 Score ($\frac{2*Precision*Recall}{Precision+Recall}$).

Table 1. A summary of the datasets used in this work. Log Messeages are the raw log strings. Log sequences are extracted by ID or sliding window method.

Dataset	Category	#Messages	#Anomaly	#Templates
HDFS	Distributed	11M	17k	52
Hadoop	Distributed	25k	368k	74
OpenStack	Distributed	189k	18k	43
BGL	Supercomputer	5M	20k	450
Thunderbird	Supercomputer	10M	123k	1108

4.2   Implementation Details

The training and evaluation are performed on NVIDIA RTX 2080Ti. Subgraph annotator is a three-layer GIN [16]. The training epoch for the annotator is 30. We set the batch size of self-supervision to 20. For the classifier, we set the batch size to 5 for log sequences. Both the subgraph annotator and the classifier use Adam as optimizer. Their learning rates are 1e–4 and 2e–6, respectively. The classifier uses BERT-base-uncased for initialization. For keyword extraction, we select top 100 keywords respectively for abnormal and normal data. The hyper-parameter M is set to 4.

4.3   Comparison with State-of-the-art Methods

Performance Comparison. In Table 2, LogLG achieves the competitive performance compared with baselines in terms of F_1 score. Labeled positive data

---

[1] https://github.com/logpai/loghub.

Table 2. Performance of our model compared with baselines. LoGLG obtains state-of-the-art results among five datasets.

Model	Type	Log Parser	HDFS F_1 Score	BGL F_1 Score	Thunderbird F_1 Score	Hadoop F_1 Score	OpenStack F_1 Score
DeepLog [4]	semi-supervised	✓	0.875	0.227	0.823	0.742	0.857
	semi-supervised	✗	0.221	0.088	0.172	0.102	0.207
LogAnomaly [14]	semi-supervised	✓	0.878	0.303	0.843	0.759	0.851
	semi-supervised	✗	0.268	0.101	0.231	0.119	0.198
A2log [15]	semi-supervised	✓	–	0.320	0.940	–	–
LAnoBERT [10]	semi-supervised	✗	0.954	0.874	–	–	–
PLELog [18]	semi-supervised	✓	**0.957**	**0.976**	0.949	0.841	0.866
	semi-supervised	✗	0.264	0.223	0.213	0.247	0.195
LoGLG	weakly supervised	✗	0.955	0.963	**0.968**	**0.875**	**0.912**

provides more distribution cues for semi-supervised methods to detect anomalies. We find that some baselines perform worse on BGL like DeepLog and LogAnomaly. This is because of the diversity of log events of BGL. Faced with such erratic log data, LoGLG still performs well. Although there are new log events in the test set, most of the keywords in these events have appeared in the training set. LoGLG seeks keyword associations at a finer granularity, which enhances the performance for detecting anomalies.

Compared with PLElog, LoGLG achieves significant improvements on these benchmark datasets without using the log parser. LoGLG requires no labeled data during the training stage while PLElog relies on massive labeled normal data, which is costly and impractical in industry scenes. Besides, PLElog requires a parser to pre-process the raw log data while LoGLG is an end-to-end framework without a parser. Besides, other baselines perform badly without a parser too, confirming that for template-based methods, the parser stage is heavily needed. For example, Deeplog drops from 0.875/0.823 to 0.221/0.172 on HDFS/Thunderbird.

4.4   Ablation Study

In this section, we mainly validate the effects of key components and parameters in our framework. Experiments are conducted on the Thunderbird.

Effectiveness of Subgraph Annotator. To demonstrate the effect of the annotator, we compare the results with/without the subgraph annotator and with/without self-supervision (SS) on Thunderbird. For the case without a subgraph annotator, we determine to generate pseudo-labels through keyword counting, which is widely operated under weak supervision tasks. For the case without self-supervision, we directly fine-tune the subgraph annotator.

In Fig. 3, we can see that 1) our method with all components of the annotator obtains the highest performance, proving the effectiveness of our subgraph annotator. 2) For the case using keyword counting, the quality of pseudo labels is the

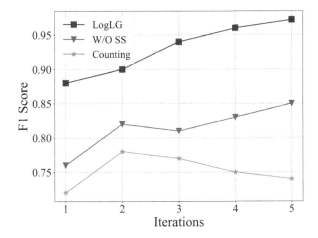

Fig. 3. Results on Thunderbird. where *W/O SS* represents the case without self-supervision and *Counting* represents the case leveraging keyword counting method to generate pseudo-labels.

worst, which leads to the worst classification performance. 3) For the case with fine-tuning but no self-supervised learning, it outperforms the keyword counting by 11% on F_1 score. 4) Self-supervised learning task can boost the performance, exceeding the case without SS by a large margin of 12% on F_1 score.

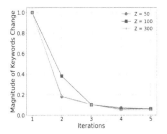
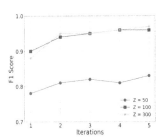

Fig. 4. Effect of keywords number Z on the change of keywords (at left) and the detection results (at right).

Effect of the Number of Keywords. Here, we validate the effect of the number Z of extracted keywords in Fig. 4. We can see that (i) The change of keywords falls below 0.1 in the 3rd update for all three number settings. (ii) Increasing the number of keywords from 50 to 100 brings a significant performance improvement, while more keywords ($Z = 300$) make little change.

4.5 Case Study

Here, we present a case study to show the power of our framework. In the beginning, we take "failed" as the initial keyword by TF-IDF. After two iterations, the keywords are updated and the top 12 keywords are presented in Table 3. Obviously, top 12 keywords extracted are correspond to the exceptions, belonging to "Anomaly". By comparing the keywords between the 1st and the 2nd, we discover that our method has the ability to seek more accurate keywords during iteration. Besides, since we kept the numbers in messages, we could find some combinations of keywords and numbers in the results. For example, 'infinihost0' represents the first host. After analysing, we are surprised to find that these combinations often represent a specific process or host. Exception is injected and passes through this host, which means our model is equipped with the ability of localizing anomalies. Based on this finding, it is possible for LOGLG to trace and conduct root analysis on anomalies, which is far from what those template-parsing based methods can do.

Table 3. Top 12 keywords. Top 12 keywords belonging to *Anomaly* class are list out in the first two iteration. We select *"failed"* as the initial keyword by TF-IDF.

Iteration	Keywords
0	failed
1	denied, failed, ignoring, obj_host_amd64_custom1_rhel4, error, append
	errbuf, tavor_mad, unexpected, get_fatal_err_syndrome, ram0, infinihost0
2	denied, ignoring, infinihost0, failed, error, errbuf
	unexpected, null, get_fatal_err_syndrome, unconfined, append, obj_host_amd64_custom1_rhel4

5 Conclusion

In this paper, we propose a novel end-to-end LOGLG method for weakly supervised log anomaly detection via log-event graph construction. In each iteration, we first construct a log-event graph and transform every unlabeled log sequence into a subgraph. To accurately annotate the subgraphs, we first pre-train the subgraph annotator with a designed self-supervised task and then fine-tune it. Under the generated pseudo labels, we train a detection model to classify the unlabeled data. Then we re-extract keywords from log sequences to update the log-event graph. Extensive experiments on benchmarks demonstrate that LOGLG method without log parsing significantly outperforms existing semi-supervised methods for detecting anomalies on unlabeled log data.

References

1. Chen, J., et al.: An empirical investigation of incident triage for online service systems. In: ICSE 2019 (2019)

2. Chen, Y., et al.: Identifying linked incidents in large-scale online service systems. In: ESEC/FSE 2020 (2020)
3. Devlin, J., Chang, M., Lee, K., Toutanova, K.: BERT: pre-training of deep bidirectional transformers for language understanding. In: NAACL-HLT 2019 (2019)
4. Du, M., Li, F., Zheng, G., Srikumar, V.: Deeplog: anomaly detection and diagnosis from system logs through deep learning. In: CCS 2017 (2017)
5. Guo, H., Lin, X., Yang, J., Liu, J., Zhuang, Y., Bai, J., Zheng, T., Zhang, B., Li, Z.: Translog: A unified transformer-based framework for log anomaly detection. CoRR (2022)
6. Guo, H., et al.: Lvp-m3: language-aware visual prompt for multilingual multimodal machine translation. In: EMNLP (2022)
7. He, P., Zhu, J., Zheng, Z., Lyu, M.R.: Drain: an online log parsing approach with fixed depth tree. In: ICWS 2017 (2017)
8. He, S., Zhu, J., He, P., Lyu, M.R.: Loghub: a large collection of system log datasets towards automated log analytics. CoRR abs/2008.06448 (2020)
9. Le, V., Zhang, H.: Log-based anomaly detection without log parsing. In: ASE 2021 (2021)
10. Lee, Y., Kim, J., Kang, P.: Lanobert: system log anomaly detection based on BERT masked language model. CoRR (2021)
11. Liang, Y., Zhang, Y., Xiong, H., Sahoo, R.K.: Failure prediction in IBM bluegene/l event logs. In: ICDM 2007 (2007)
12. Liu, J., Guo, J., Xu, D.: Geometrymotion-transformer: an end-to-end framework for 3d action recognition. IEEE TMM (2022)
13. Liu, J., Yu, T., Peng, H., Sun, M., Li, P.: Cross-lingual cross-modal consolidation for effective multilingual video corpus moment retrieval. In: NAACL 2022
14. Meng, W., et al.: Loganomaly: unsupervised detection of sequential and quantitative anomalies in unstructured logs. In: IJCAI 2019 (2019)
15. Wittkopp, T., et al.: A2log: attentive augmented log anomaly detection. In: HICSS 2022 (2022)
16. Xu, K., Hu, W., Leskovec, J., Jegelka, S.: How powerful are graph neural networks? In: ICLR 2019 (2019)
17. Yang, J., et al.: UM4: unified multilingual multiple teacher-student model for zero-resource neural machine translation. In: IJCAI 2022, pp. 4454–4460 (2022)
18. Yang, L., et al.: Semi-supervised log-based anomaly detection via probabilistic label estimation. In: ICSE 2021 (2021)
19. Zhang, L., Ding, J., Xu, Y., Liu, Y., Zhou, S.: Weakly-supervised text classification based on keyword graph. In: EMNLP 2021 (2021)
20. Zhang, X., et al.: Robust log-based anomaly detection on unstable log data. In: ESEC/SIGSOFT FSE (2019)

FAN: Fatigue-Aware Network
for Click-Through Rate Prediction
in E-commerce Recommendation

Ming Li, Naiyin Liu, Xiaofeng Pan[(⊠)], Yang Huang, Ningning Li, Yingmin Su,
Chengjun Mao, and Bo Cao

Alibaba Group, Hangzhou, China
{hongming.lm,lily.lnn,yingmin.sym,chengjun.mcj,
zhizhao.cb}@alibaba-inc.com, liuny5@mail2.sysu.edu.cn,
pxfvintage@163.com, 21631140@zju.edu.cn

Abstract. Since *clicks* usually contain heavy noise, increasing research
efforts have been devoted to modeling implicit negative user behaviors
(*i.e.*, *non-clicks*). However, they either rely on explicit negative user
behaviors (*e.g.*, *dislikes*) or simply treat *non-clicks* as negative feedback, failing to learn negative user interests comprehensively. In such
situations, users may experience fatigue because of seeing too many similar recommendations. In this paper, we propose Fatigue-Aware Network
(FAN), a novel CTR model that directly perceives user fatigue from *non-
clicks*. Specifically, we first apply Fourier Transformation to the time
series generated from *non-clicks*, obtaining its frequency spectrum which
contains comprehensive information about user fatigue. Then the frequency spectrum is modulated by category information of the target
item to model the bias that both the upper bound of fatigue and users'
patience is different for different categories. Moreover, a gating network
is adopted to model the confidence of user fatigue and an auxiliary task
is designed to guide the learning of user fatigue, so we can obtain a
well-learned fatigue representation and combine it with user interests for
the final CTR prediction. Experimental results on real-world datasets
validate the superiority of FAN and online A/B tests also show FAN
outperforms representative CTR models significantly.

Keywords: Recommender system · Click-Through Rate Prediction ·
User Fatigue

1 Introduction

Recommender Systems (RS) are becoming increasingly indispensable to help
users discover their preferred items in situations of information overload, therefore improving the user experience and delivering new business value [21,24].
Typically, an industrial e-commerce recommender system consists of matching

---

M. Li and N. Liu–Both authors contributed equally.

X. Wang et al. (Eds.): DASFAA 2023, LNCS 13946, pp. 502–514, 2023.
https://doi.org/10.1007/978-3-031-30678-5_37

and ranking. The matching stage aims to retrieve candidate items related to user interests [5,13], after which the ranking stage predicts precise probabilities of users interacting with these candidate items, *e.g.*, Click-Through Rate (CTR) [26] and Conversion Rate (CVR) [12,17]. In this paper, we focus on the Click-Through Rate (CTR) prediction task of the ranking stage.

Most existing CTR methods [1,7,26] in e-commerce recommendation mainly rely on implicit positive feedback (*i.e.*, *clicks*) as a positive label and infer users' current interests since clicks can be easily collected in practice. However, click behaviors usually contain heavy noise [16,18] since there are gaps between *clicks* and users' real preferences, and outdated interests may exist in historical user behaviors [9]. Moreover, positive feedback is biased toward the choices that the RS offers to its users, as clicks can only be done on items that are exposed to users. As a result, simply focusing on implicit positive feedback will lead to biased modeling of user interests and homogeneous and myopic recommendations, which may harm user experiences [25].

Recently, several researchers [6,10] notice the drawbacks of merely relying on implicit positive feedback and attempt to leverage the more abundant implicit negative feedback (*i.e.*, *non-clicks*) to learn negative user interests. The key idea is to introduce regularization in the loss function to enforce that the representation of positive feedback should be far away from the *non-clicks*. However, a user doesn't click an item does not necessarily mean the user doesn't like the item. Maybe some similar items are displayed nearby, or maybe the exposed items are simply not well noticed. Therefore, ignoring noise in *non-clicks* may lead to conflicts when modeling user interests and result in inaccurate recommendations. Another line of works [3,19,20] tries to make use of explicit negative feedback (*e.g.*, *dislikes*) to distill negative user interests from implicit feedback (*i.e.*, *non-clicks* and *clicks*). However, explicit negative user behaviors are extremely scarce in e-commerce RS. Less than 0.01% of impressions will result in *dislikes*, which are less than one-tenth of *purchases*, according to statistics of our e-commerce platform.

In such situations, models are incapable of learning negative user interests comprehensively and users may experience fatigue due to seeing too many similar recommendations. One way to handle fatigue is to recall more new items at the matching stage, which usually doesn't take effect directly because most existing CTR methods are not friendly to items that users didn't see before. Another common solution is the explore and exploit paradigm [2,8] that considers multiple factors including relevance, novelty, and fatigue [4,11]. However, they simply model user fatigue by statistics such as the number of similar items shown before but ignore the time-frequency distribution of similar recommendations, which causes the loss of information. Besides, it is believable that user interests and user fatigue can affect each other, so the trade-off between exploration and exploitation may not be an optimal choice.

Based on these observations, we propose a novel CTR model: Fatigue-Aware Network (FAN) which can perceive user fatigue from *non-clicks* more comprehensively, therefore achieving improvements in both CTR prediction and user experience. Naturally, we believe that user fatigue should be modeled at the category level because the item level is too fine-grained and the e-commerce RS usually avoids recommending items recommended recently. Given a pair of user and item, we extract *non-clicks* of the same category in recent days and compute the number of this category in each recommendation request, generating a time series that contains rich information about user fatigue. Then we devise a Fatigue Representation Module (FRM) which applies Fast Fourier Transformation (FFT) [14] to the time series to obtain its frequency spectrum, which contains comprehensive information about time-frequency distribution of similar recommendations. Considering the bias that both the upper bounds of fatigue and users' patience are different for different categories (*e.g.*, digital products and clothes), we propose to modulate the frequency spectrum by category information. Moreover, a gating network is adopted to model the confidence of user fatigue according to user activeness and an auxiliary task is designed to guide the learning of FRM. In this way, we can obtain a well-learned fatigue representation. At last, user fatigue is incorporated with user interests to make the final CTR prediction so we can avoid too many similar recommendations and achieve more accurate predictions. Our main contributions are summarized as follows:

- We investigate the difficulties of modeling implicit negative user behaviors (*i.e.*, *non-clicks*) for e-commerce recommendation, and propose to model user fatigue explicitly in CTR prediction to avoid users from seeing too many similar recommendations.
- We propose a novel FAN model that directly perceives user fatigue from *non-clicks*. Benefiting from the frequency-domain representation and category modulation, we are capable of modeling user fatigue comprehensively and accurately. Besides, with an elaborated auxiliary task, FAN can pay attention well to what users are not interested in.
- Experiments on real-world datasets demonstrate the superiority of our FAN model over representative methods, and online tests further show that FAN not only improves model performance but also brings better user experiences. We also conduct extensive analyses to confirm the effectiveness of our design for modeling user fatigue. The code is publicly available[1].

2   Proposed Method

2.1   Model Input

In CTR prediction, the model takes input as $(x, y) \sim (X, Y)$, where x is the feature and $y \in \{0, 1\}$ is the click label. Specifically, the features in this work consist of five parts: 1) user behavior sequence x^{seq}; 2) user features x^u including the user profile and user statistic features; 3) item features x^i such as item id, category, brand, and related statistic features; 4) context features x^c such as position and time information; 5) fatigue time series S, *i.e.*, a sequence of

---

[1] https://github.com/AaronPanXiaoFeng/FAN.

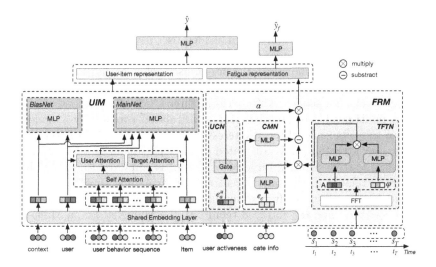

Fig. 1. Framework of the proposed Fatigue-Aware Network (FAN), which consists of the User Interest Module (UIM) and Fatigue Representation Module (FRM). The FRM includes three key components: Time-Frequency Transformation Net (TFTN), Category Modulation Net (CMN), and User Confidence Net (UCN).

statistics of *non-clicks* on target category, which is first proposed in this work and will be detailed below.

Now we describe the extraction process of fatigue time series. Given a pair of user and item, we first retrieve recommendation requests of the user in recent days and order them by time. For each request, we compute the number of *non-clicks* which has the same category as the target item, obtaining a time series $S = \{s_1, s_2, ..., s_T\}$, where s_T denotes the number of *non-clicks* on the target category in T-th request and T is the number of requests. Elements in S can be regarded as raw indicators of fatigue in the time domain. To ensure the correlation between S and fatigue, we adopt the following rules for special cases in the extraction process: 1) If there are clicks on the target category in the i-th request, we set s_i to 0 because it's more related to the user's positive interest and contributes less to fatigue; 2) If there is no item of the target category in a request, we ignore this request because it is not relevant to the target fatigue. By aggregating *non-clicks* on request granularity and organizing statistics in order of time, the raw information about fatigue is more confident and comprehensive.

2.2 User Interest Module

As shown in Fig. 1, the input features of UIM consist of $\boldsymbol{x}^u$, $\boldsymbol{x}^i$, $\boldsymbol{x}^c$ and $\boldsymbol{x}^{seq}$, which are detailed in Sect. 2.1. They can be further grouped into two kinds of features: categorical features and numerical features. We discrete the numerical features based on their boundary values and transform them into the categorical type. Then each categorical feature is encoded as a one-hot vector. Due to the sparseness nature of one-hot encoding, they are further processed by a Shared Embedding Layer so we obtain the embedded user features, item features, context features, and user behavior sequence, *i.e.*, e^u, e^i, e^c and $e^{seq} = \{e^i_1, ..., e^i_l\}$, where e^i_l denotes the item embedding of l-th user behavior and l is the sequence length.

For user behavior sequence, we perform three kinds of attention calculation. Firstly a multi-head self-attention [15] is calculated over e^{seq} to model user preference from multiple views of interest and $\hat{e}^{seq} = \{\hat{e}^i_1, ..., \hat{e}^i_l\}$ is the output. Secondly, user attention a^u is calculated to mine personalized information with e^u attending to $\hat{e}^{seq}$. Thirdly, target attention a^i is employed to activate historical interests related to the target item with e^i attending to $\hat{e}^{seq}$. To preserve the original information for further learning of interactions, we concatenate e^i, e^u and e^c with a^u and a^i, and feed them into the MainNet, *i.e.*, a Multi-Layer Perception (MLP). Meanwhile, we feed e^u and e^c into another MLP (*i.e.*, the BiasNet) to model the bias that different users in different contexts usually behave differently even to similar items. Finally, the outputs of MainNet and BiasNet are concatenated to obtain the user-item representation r_{ui}.

2.3   Fatigue Representation Module

Time-Frequency Transformation Net. Given the fatigue time series S, we argue that it's more beneficial to model user fatigue in the frequency domain than the time domain for two reasons: 1) S only contains magnitude information of *non-clicks* at each request, while in the frequency domain, we can observe both amplitude (related to magnitude) and phase (related to position) for each frequency component, which is more beneficial to capture the periodic evolution of fatigue; 2) In the frequency domain, we can conveniently distinguish the influence of different frequency components for more elaborate modeling.

Motivated by this, we perform N-point FFT to transform S into the frequency domain:

$$
\begin{aligned}
A, \varphi &= FFT(S), \\
A &= [A_1, ..., A_k, ..., A_N] \in \mathbb{R}^N, \\
\varphi &= [\varphi_1, ..., \varphi_k, ..., \varphi_N] \in \mathbb{R}^N,
\end{aligned} \tag{1}
$$

where A and φ are the amplitude and phase vectors respectively.

To capture the different influences of different frequency components adaptively and obtain a high-order representation of amplitude and phase, we feed A and φ into a two-layer MLP respectively. Then, we perform an element-wise multiplication of the results to model the interaction of amplitude and phase and obtain a combination representation:

$$
\begin{aligned}
A' &= MLP_A(A) \in \mathbb{R}^N, \\
\varphi' &= MLP_\varphi(\varphi) \in \mathbb{R}^N, \\
F &= A' \otimes \varphi' \in \mathbb{R}^N,
\end{aligned} \tag{2}
$$

F contains comprehensive information about user fatigue, which is considered as a raw fatigue representation. It's noteworthy that MLP_A and MLP_φ are used to learn that influence of different frequency components is different. For the amplitude, components at low frequencies are more robust, while components at high frequencies may contain more noise and should be attenuated. For the phase, components at low frequencies are more important than those at high frequencies, because the position of high-frequency signals is less sensitive to phase than that of low-frequency signals.

Category Modulation Net. Intuitively, both the upper bound of fatigue and users' patience is different for different categories. Taking inspiration from the spirit of APG [23], we propose to generate model parameters dynamically based on different instances, which helps to capture custom patterns and enhance the model capacity. Thus, we can model user fatigue adaptively for different categories. Specifically, for each category, we formalize the upper bound as a bias vector b_c, and the users' patience as a weight vector w_c, which are generated from category information of the target item respectively:

$$
\begin{aligned}
b_c &= MLP_b(e_c) \in \mathbb{R}^N, \\
w_c &= MLP_\alpha(e_c) \in \mathbb{R}^N,
\end{aligned}
\tag{3}
$$

where e_c is the embedding of category features, including category ID and the corresponding statistical features which can be obtained from item features x^i. To model our intuition, we modulate the raw fatigue representation F as:

$$
F_c = b_c - w_c \otimes F \in \mathbb{R}^N,
\tag{4}
$$

Compared with F, F_c considers the discrepancy of fatigue across different categories and therefore represents fatigue more accurately.

User Confidence Net. Naturally, users with high activeness generate more feedback, which makes modeling user fatigue more confident, and vice versa. Therefore, we employ a gating network to model the confidence of user fatigue according to user activeness. Specifically, we pass e_a^u, the embedded user activeness generated from user features x^u, to a MLP to produce the confidence factor α_u, after which element-wise multiplication is performed between α_u and F_c, i.e.,

$$
\begin{aligned}
\alpha_u &= MLP_u(e_a^u) \in \mathbb{R}^N, \\
F_{uc} &= \alpha_u \otimes F_c \in \mathbb{R}^N,
\end{aligned}
\tag{5}
$$

Through the aforementioned operations, we obtain a fine-tuned fatigue representation F_{uc} which considers the category and the user biases simultaneously.

2.4   Training

On the top of the UIM and FRM, the user-item representation r_{ui} and the fatigue representation F_{uc} are combined by concatenation and passed to a MLP to make the final CTR prediction:

$$
\hat{y} = f_\theta(x) = MLP_o(concat(r_{ui}, F_{uc})),
\tag{6}
$$

The last layer of MLP_o uses *Sigmoid* as activation function to project the prediction to the click probability. We adopt the widely-used logloss as the main loss, which is calculated as follows:

$$L_m = -\frac{1}{|\mathcal{D}|} \sum_{(\boldsymbol{x},y)\in\mathcal{D}} (y \, log \, \hat{y} + (1-y) \, log(1-\hat{y})), \tag{7}$$

where $\mathcal{D}$ denotes training set and $|\mathcal{D}|$ denotes the number of samples in $\mathcal{D}$.

Moreover, to guide the learning of FRM, we design an additional auxiliary task, *i.e.*, predicting the degree of fatigue:

$$\hat{y}_f = MLP_f(\boldsymbol{F_{uc}}), \tag{8}$$

where $MLP_f(\cdot)$ is a 3-layer MLP of which the last layer uses *Sigmoid* as activation function. To find a confident fatigue label for $\hat{y}_f$, we take a user's behaviors in the next three days into consideration. If a user clicks on the target category, we assume that the user has not been over-exposed and mark the fatigue label as negative. If the user doesn't click the target category after a certain number of exposures, we mark the fatigue label as positive. Training samples in other situations are not used in the fatigue prediction task. By aggregating user behaviors over an appropriate time window, we can obtain a relatively stable and confident fatigue label y_f on the target category. Then we calculate logloss between $\hat{y}_f$ and y_f, and formulate the final loss as follows:

$$L_f = -\frac{1}{|\mathcal{D}|} \sum_{(\boldsymbol{x},y_f)\in\mathcal{D}} (y_f \, log \, \hat{y}_f + (1-y_f) \, log(1-\hat{y}_f)),$$
$$L = L_m + \beta L_f. \tag{9}$$

where β is a scaling hyperparameter that gradually increases during the training process, and the optimal maximum value of β is determined by experiments. With the auxiliary loss L_f, the FRM is guided to pay attention to what users are not interested in, therefore learning user fatigue better.

3   Experiments

3.1   Experimental Setup

Datasets. We collect and sample online service logs[2] from the recommendation scenarios in Tmall Mobile between 2022/08/24 and 2022/09/26 as our experimental datasets. Then we split the data into two non-overlapped parts. The data between 2022/08/24 and 2022/09/25 is used for training while the data in 2022/09/26 is collected for testing. Table 1 summarizes the detailed statistics of our datasets.

---

[2] The data collection is under the application's user service agreement and users' private information is protected.

Table 1. Statistics of the established dataset.

#Dataset	#Users	#Items	#Exposures	#Clicks	#Purchases
train	9.40M	7.39M	835.76M	66.62M	277.22K
test	842.71K	2.70M	26.62M	2.15M	10.82K

Evaluation Metrics. For the offline comparison, we use Area Under ROC Curve (AUC) as the evaluation metric. As for the online A/B testing, we use $PCTR = p(click|impression)$ and the average number of user clicks (IPV), which are widely adopted in industrial recommender systems. Moreover, we use Leaf Categories Exposed Number per user (LCEN) and Leaf Categories Clicked Number per user (LCCN) to measure the diversity of recommendations.

Competitors. As a representative in CTR prediction, **DIN** [26] is chosen to be the base model. Besides, we compare the performance of our proposed FAN model with a series of state-of-the-art methods that model both *clicks* and *non-clicks*, *i.e.*, **DFN** [20], **DUMN** [3] and **Gama** [22]. Additionally, to demonstrate the effectiveness of our designed structure in FAN, we also conduct several ablation experiments:

- **FAN_w/o_FRM (UIM)**: As a substructure of FAN, UIM can be used as a deep CTR model by adding prediction layers, which adopts the attention mechanism [15] to model user positive behavior sequence.
- **FAN_w/o_TFTN**: To prove the necessity of the TFTN, we remove the TFTN module and directly make use of the input fatigue time series instead.
- **FAN_w/o_CMN**: In order to verify the gain of the Category Modulation Net(CMN) of FRM to the FAN, we remove the CMN and directly process the output of TFTN via element-wise product with the output of UCN.
- **FAN_w/o_UCN**: Similar to FAN_w/o_CMN, we test the performance of UCN by removing it from FRM.

Implementation Details. All models share the same features, except that DIN ignores *non-clicks* and DFN and DUMN additionally adopt explicit negative feedback of users. All the models are implemented in distributed Tensorflow 1.4 and trained with 10 parameter servers and 4 Nvidia Tesla V100 16GB GPUs. Item ID has an embedding size of 64, category ID and brand ID have an embedding size of 32 while 8 for the other categorical features. We use 8-head attention structures in UIM with a hidden size of 128 and 32-point FFT in FRM. Adagrad optimizer with a learning rate of 0.01 and a mini-batch size of 1024 is used for training. During training, β increases linearly from 0.01 to 0.5 with training steps increasing. We report the results of each method under its empirically optimal hyper-parameters settings.

3.2 Overall Results

For offline evaluation, each model has repeated five times and the best version of each model is selected for online A/B tests, which lasted 3 days from 2022/10/05

Table 2. Offline and online results of comparison experiments

Model	Offline	Online Gain			
	AUC (mean±std.)	PCTR	IPV	LCEN	LCCN
DIN(Base Model)	0.7178 ± 0.00247	0.00%	0.00%	0.00%	0.00%
DFN	0.7194 ± 0.00585	−0.30%	-0.31%	+0.85%	+0.53%
DUMN	0.7218 ± 0.00429	+0.17%	-0.74%	+0.59%	+1.15%
Gama	0.7225 ± 0.00472	+0.89%	+0.42%	−0.26%	+0.57%
FAN_w/o_FRM(UIM)	0.7193 ± 0.00238	−0.85%	+0.79%	+2.52%	+1.41%
FAN(ours)	**0.7249±0.00176**	**+1.63%**	**+1.09%**	**+11.13%**	**+3.29%**

to 2022/10/08. The offline and online comparison results are presented in Table 2 and the major observations can be summarized as follows:

1. Compared with DIN, the UIM model performs better in the offline evaluation. Although the PCTR of UIM is slightly worse, it can recommend more abundant categories than DIN and attract more IPV, implying that our design for modeling positive user interests is effective.
2. The DFN and DUMN perform better than DIN on the offline AUC metric by distilling implicit negative feedback via explicit negative feedback. However, users' explicit negative feedback are extremely scarce in e-commerce, which is insufficient for learning negative user interests comprehensively. As a result, DFN and DUMN can't perform well in online recommendation scenarios.
3. The Gama model slightly outperforms DIN, DFN, and DUMN on AUC, PCTR, and IPV by denoising implicit negative feedback in the frequency domain. However, it still focuses on modeling what users are interested in, so it's incapable of improving the diversity of recommendations. In such situations, users may be overexposed to too many similar recommendations.
4. For both offline and online, our FAN model yields the best performance. It's noteworthy that FAN not only improves the efficiency of online traffic (*i.e.*, PCTR, and IPV) but also achieves impressive gains in diversity. Benefiting from the frequency-domain representation and category modulation, our modeling of fatigue is much more comprehensive and accurate, which helps users avoid seeing too many similar recommendations and explore more new items. Besides, with an elaborated auxiliary task, FAN pays more attention to what users are not interested in than the Gama model.

3.3   Ablation Study

To demonstrate the effectiveness of the designed structure in FAN, we also conduct a series of ablation experiments as detailed in Sect. 3.1. The results are detailed in Table 3. Totally speaking, removing FRM or key substructures of FRM from FAN leads to the decline of AUC, illustrating the effectiveness of our model design. The other supplementary conclusions are summarized as follows:

1. The comparison between FAN_w/o_TFTN and FAN shows that TFTN can achieve better use of the fatigue time series. With the frequency spectrum extracted by FFT, TFTN further models different impacts of different frequency components, helping to learn a better fatigue representation for the final prediction.
2. With CMN removed, FAN_w/o_CMN suffers a significant performance degradation that is almost comparable to removing the whole FRM. This observation proves the correctness of our intuition that both the upper bounds of fatigue and users' patience are different for different categories, and confirms the necessity of category modulation while modeling user fatigue.
3. Comparing FAN_w/o_UCN with FAN, a degradation of AUC is observed, implying that considering user confidence is beneficial to the learning of fatigue representation.

Table 3. Offline results of ablation experiments

Model	AUC (mean±std.)
FAN_w/o_FRM(UIM)	0.7193 ± 0.00238
FAN_w/o_TFTN	0.7225 ± 0.00135
FAN_w/o_CMN	0.7199 ± 0.00147
FAN_w/o_UCN	0.7237 ± 0.00187
FAN	**0.7249 ± 0.00176**

3.4 Effectiveness Analysis

To further analyze how TFTN and FRM work, we visualize the output of the TFTN and FRM. Specifically, we randomly select 1000 samples and feed them to FAN, extracting F and F_{uc} for each sample. Then, we draw mean and standard deviation diagrams for F and F_{uc} respectively, as shown in Fig. 2. Since the symmetry characteristic of FFT, Fig. 2(a) shows an approximately symmetrical structure. Output values between different dimensions are quite different. Frequency components around the 6-th and 15-th dimensions have weaker responses while the other has a higher response, implying that our TFTN act as a second-order bandstop filter in FRM. From Fig. 2(b), we can observe that the symmetrical structure no longer exists, and different frequency components are further activated. This observation implies that our devised CMN and UCN achieve the ability of frequency selection. Moreover, the standard deviation of some frequency components becomes larger while some become smaller, indicating that FRM further weakens the unimportant components and strengthens important components.

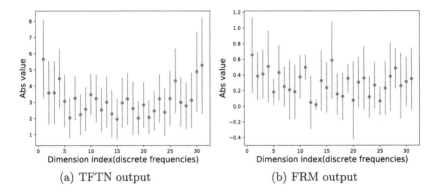

(a) TFTN output   (b) FRM output

Fig. 2. Mean and standard deviation of TFTN and FRM output

4   Conclusion

In this paper, we investigate the difficulties of modeling implicit negative user behaviors (*i.e.*, *non-clicks*) in e-commerce and propose a novel model named FAN which captures the negative user interests in the target category from a perspective of fatigue modeling. In FAN, we apply FFT to obtain time-frequency information from the corresponding sequence of *non-clicks*, which is then used to learn the high-order fatigue representation with category bias and user confidence considered. Moreover, an auxiliary task is elaborately designed to guide the learning of the FRM. In this way, a high-quality fatigue representation can be learned to facilitate improving both the CTR performance and user experience, which is validated by real-world offline datasets as well as online A/B testing. The extensive analysis further confirms the effectiveness of our model design.

References

1. Ai, Z., Wang, S., Jia, S., Guo, S.: Core interests focused self-attention for sequential recommendation. In: Database Systems for Advanced Applications, vol. 13246. pp. 306–314. Springer, Cham (2022). https://doi.org/10.1007/978-3-031-00126-0_23
2. Auer, P.: Using confidence bounds for exploitation-exploration trade-offs. J. Mach. Learn. Res. **3**, 397–422 (2002)
3. Bian, Z., et al.: Denoising user-aware memory network for recommendation. In: Fifteenth ACM Conference on Recommender Systems, pp. 400–410 (2021)
4. Cao, J., Sun, W., Shen, Z.J.M., Ettl, M.: Fatigue-aware bandits for dependent click models. In: Proceedings of the AAAI Conference on Artificial Intelligence, vol. 34, pp. 3341–3348 (2020)
5. Covington, P., Adams, J., Sargin, E.: Deep neural networks for youtube recommendations. In: Proceedings of the 10th ACM Conference on Recommender Systems, pp. 191–198 (2016)
6. Gong, S., Zhu, K.Q.: Positive, negative and neutral: modeling implicit feedback in session-based news recommendation. arXiv preprint arXiv:2205.06058 (2022)

7. Guo, H., Tang, R., Ye, Y., Li, Z., He, X.: Deepfm: a factorization-machine based neural network for CTR prediction. arXiv preprint arXiv:1703.04247 (2017)
8. Li, L., Chu, W., Langford, J., Schapire, R.E.: A contextual-bandit approach to personalized news article recommendation. In: Proceedings of the 19th International Conference on World Wide Web, pp. 661–670 (2010)
9. Li, X., Wang, C., Tong, B., Tan, J., Zeng, X., Zhuang, T.: Deep time-aware item evolution network for click-through rate prediction. In: Proceedings of the 29th ACM International Conference on Information & Knowledge Management, pp. 785–794 (2020)
10. Lv, F., et al.: Xdm: improving sequential deep matching with unclicked user behaviors for recommender system. In: International Conference on Database Systems for Advanced Applications, vol. 13247, pp. 364–376. Springer, Cham (2022). https://doi.org/10.1007/978-3-031-00129-1_31
11. Ma, H., Liu, X., Shen, Z.: User fatigue in online news recommendation. In: Proceedings of the 25th International Conference on World Wide Web, pp. 1363–1372 (2016)
12. Pan, X., et al.: Metacvr: Conversion rate prediction via meta learning in small-scale recommendation scenarios. In: Proceedings of the 45th International ACM SIGIR Conference on Research and Development in Information Retrieval, pp. 2110–2114 (2022)
13. Sarwar, B., Karypis, G., Konstan, J., Riedl, J.: Item-based collaborative filtering recommendation algorithms. In: Proceedings of the 10th international conference on World Wide Web, pp. 285–295 (2001)
14. Soliman, S.S., Srinath, M.D.: Continuous and discrete signals and systems. Englewood Cliffs (1990)
15. Vaswani, A., et al.: Attention is all you need. In: Advances in Neural Information Processing Systems, pp. 5998–6008 (2017)
16. Wang, M., Gong, M., Zheng, X., Zhang, K.: Modeling dynamic missingness of implicit feedback for recommendation. In: Advances in Neural Information Processing Systems, vol. 31 (2018)
17. Wen, H., et al: Entire space multi-task modeling via post-click behavior decomposition for conversion rate prediction. In: Proceedings of the 43rd International ACM SIGIR Conference on Research and Development in Information Retrieval, pp. 2377–2386 (2020)
18. Wen, H., Yang, L., Estrin, D.: Leveraging post-click feedback for content recommendations. In: Proceedings of the 13th ACM Conference on Recommender Systems, pp. 278–286 (2019)
19. Wu, C., et al.: Feedrec: news feed recommendation with various user feedbacks. In: Proceedings of the ACM Web Conference 2022, pp. 2088–2097 (2022)
20. Xie, R., Ling, C., Wang, Y., Wang, R., Xia, F., Lin, L.: Deep feedback network for recommendation. In: Proceedings of the Twenty-Ninth International Conference on International Joint Conferences on Artificial Intelligence, pp. 2519–2525 (2021)
21. Xu, C., Peak, D., Prybutok, V.: A customer value, satisfaction, and loyalty perspective of mobile application recommendations. Decis. Support Syst. **79**, 171–183 (2015)
22. Xu, X., et al.: Gating-adapted wavelet multiresolution analysis for exposure sequence modeling in ctr prediction. In: Proceedings of the 45th International ACM SIGIR Conference on Research and Development in Information Retrieval, pp. 1890–1894 (2022)
23. Yan, B., et al.: APG: adaptive parameter generation network for click-through rate prediction. In: Advances in Neural Information Processing Systems (2022)

24. Zhang, J., Tao, D.: Empowering things with intelligence: a survey of the progress, challenges, and opportunities in artificial intelligence of things. IEEE Internet Things J. **8**(10), 7789–7817 (2020)
25. Zhao, Q., Harper, F.M., Adomavicius, G., Konstan, J.A.: Explicit or implicit feedback? engagement or satisfaction? a field experiment on machine-learning-based recommender systems. In: Proceedings of the 33rd Annual ACM Symposium on Applied Computing, pp. 1331–1340 (2018)
26. Zhou, G., et al.: Deep interest network for click-through rate prediction. In: Proceedings of the 24th ACM SIGKDD International Conference on Knowledge Discovery & Data Mining, pp. 1059–1068 (2018)

Global-Aware Model-Free Self-distillation for Recommendation System

Ang Li[1], Jian Hu[2], Wei Lu[1], Ke Ding[1], Xiaolu Zhang[1], Jun Zhou[1($\boxtimes$)],
Yong He[1], Liang Zhang[1], and Lihong Gu[1]

[1] Ant Group, Hangzhou 310023, China
{liang268038,dingke.dk,heyong.h,zhuyue.zl,lihong.glh}@antgroup.com,
xiaobo.lw@alibaba-inc.com, {yueyin.zxl,jun.zhoujun}@antfin.com
[2] Queen Mary University of London, London, England
jian.hu@qmul.ac.uk

Abstract. The recommendation performance in our Alipay advertising system may suffer from label noise in training data. Earlier approaches that relied on soft targets typically neglected global similarities. Here, we introduce a novel algorithm called Global-aware Model-free Self-Distillation (GMSD) to create soft targets using the information at the global scale. Specifically, we propose calculating the similarities between the target sample and cluster centers produced by clustering the training dataset. The direct calculation of global similarities requires computation across the full dataset, which is prohibitively expensive. Additionally, we develop a contrastive cluster loss (CCLoss) for limiting the distance between the data of the intra-class to be lower than that of the inter-class to promote samples to be drawn to accurate cluster clusters. Extensive experiments on public and industry datasets demonstrate that GMSD outperforms state-of-the-art self-distillation methods in efficiency and effectiveness.

Keywords: Self-distillation · Label-noise · Recommendation system · Clustering

1 Introduction

Deep Neural Networks (DNN) have recently achieved remarkable success in industrial recommendation systems owing to sufficient labeled data. However, it is common for users who click on something they are not interested in or are fond of the items but do not click [3]. This behavior can lead to many noisy labels and deteriorate the model performance. Numerous algorithms have been developed to train on soft targets instead of the original binary labels. Knowledge distillation is introduced to learn multi-view features in the teacher-student format for generating soft targets. Despite its success, it suffers from heavy parameters and computations. BAKE [2] acquires soft ensemble knowledge within the batch

---

A. Li and J. Hu—equal contribution.

X. Wang et al. (Eds.): DASFAA 2023, LNCS 13946, pp. 515–518, 2023.
https://doi.org/10.1007/978-3-031-30678-5_38

affinities to reduce computational complexity. However, some drawbacks are still inevitable. The performance of BAKE significantly depends on the existence of similar samples in the batch. In contrast, in CTR prediction and recommendation, the positive examples are highly sparse, resulting in very few similar positive examples in the batch. So it is necessary to compare the single sample with the entire dataset to calculate the global class similarities in recommendation systems. Assuming that the scale of CTR prediction is N, which is often above millions, the resources and time required are unacceptable.

To tackle these issues, we introduce the **Global-aware Model-free Self-Distillation (GMSD)**, which generates ideal soft targets efficiently based on the similarity of global-level features. Specifically, we leverage clustering to compress the entire dataset into K categories ($K \ll N$). We only need to calculate the similarity with K cluster centers when utilizing global information. Our method reduces the computational complexity from $O(N)$ to $O(K)$. However, due to the highly imbalanced distribution between positive and negative samples, it is easy to mix the positive and negative examples, resulting in poor clustering. To address this issue, we propose to cluster the positive and negative samples independently. Furthermore, we introduce a novel contrastive cluster loss (CCLoss) to facilitate similar examples to be clustered more closely while the decision boundaries between samples from diverse categories to be clearer. Our proposed GMSD outperforms State-of-The-Art self-distillation algorithms in public and industrial datasets.

2   Global-Aware Model-Free Self-distillation

Our proposed Global-aware Model-free Self-distillation (GMSD) is shown in Fig. 1. Firstly, we generate the K clustering centers C^p and C^n of positive and negative samples independently following [4]. Then given a sample x^{anchor}, we can get the intermediate features by the feature extractor f. Finally, we can produce the soft labels by the cosine similarity between the anchor and centers C, which are normed by l_2 function σ.

$$q^{anchor} = \sum_{i=1}^{2K} \sigma(f(x^{anchor}))\sigma(C_i) * g(C_i) \tag{1}$$

Furthermore, To maintain the density of the data inside the same center, we present a novel Contrastive cluster Loss (CCLoss), which constrains samples of the same center to be closer than those from different categories.

$$L_{CC} = \sum_{i=1}^{N} [||f(x_i) - f(x_s)||_2^2 - ||f(x_i) - f(x_d)||_2^2]_+ \tag{2}$$

where x_s represents the sample belongs to the same center with x_i, x_d pertains to the different.

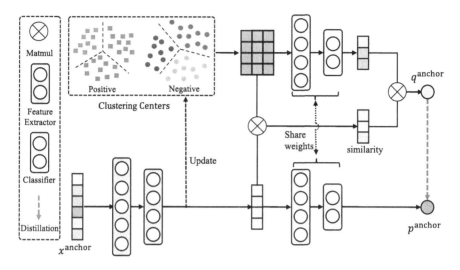

Fig. 1. Architecture of our proposed GMSD. The network consists of feature extractor f, classifier g and cluster C.

3 Experiments

We conduct experiments using the KKBox dataset (Public) and Our Alipay advertising Dataset (Industry). To demonstrate the superiority of our techniques, label-noise learning [1] and self-distillation approaches [2] are chosen for comparison. The results are shown in Table 1. We repeat all experiments ten times and report the average and standard deviation results. According to Table 1, label-noise learning based SEAL [1] and self-distillation based BAKE [2] obtain approximate results but are not comparable to our GMSD in either dataset. GMSD achieves the best performance, demonstrating that global similarities are more valuable than batch affinities, proving the capability of our proposed GMSD. Besides, we compare the inference time of the batch-level algorithm BAKE [2], hand-crafted global-level algorithms, and our cluster-based global-level algorithm. Because the datasets in CTR are over millions or billion, we cannot calculate the similarity between samples and the whole dataset in actual cases. Thus we choose to extend the range of similarity calculation to simulate the global-level algorithm, which is named the hand-crafted global-level algorithm. The computation time of BAKE [2] and the hand-crafted global-level algorithm is 38.13 ms and 85.68 ms. Our GMSD only needs 9.19 ms, saving 75.90% of time compared with BAKE.

3.1 Ablation Study

Direct clustering will result in negative samples covering the positive ones since the number of positive samples is much smaller than that of the negative. Therefore, we experiment with the cluster strategy. In Table 2, *no cluster* is a baseline

Table 1. Comparisons with existing label-noise learning and self-distillation models.

Model	KKBox		Industry	
	AUC	logloss	AUC	logloss
baseline	0.7494 ± 0.0015	0.6113 ± 0.0008	0.7005 ± 0.0011	0.2958 ± 0.0012
SEAL [1]	0.7665 ± 0.0004	0.6025 ± 0.0005	0.7059 ± 0.0005	0.2918 ± 0.0004
BAKE [2]	0.7658 ± 0.0006	0.6023 ± 0.0004	0.7061 ± 0.0006	0.2923 ± 0.0003
GMSD	$\mathbf{0.7757} \pm 0.0003$	$\mathbf{0.5970} \pm 0.0002$	$\mathbf{0.7113} \pm 0.0002$	$\mathbf{0.2893} \pm 0.0001$

Table 2. Ablation study to evaluate different methods of clustering.

Model	KKBox		Industry	
	AUC	logloss	AUC	logloss
no cluster	0.7494 ± 0.0015	0.6113 ± 0.0008	0.7005 ± 0.0011	0.2958 ± 0.0012
fuse	0.7735 ± 0.0004	0.5985 ± 0.0002	0.7101 ± 0.0003	0.2900 ± 0.0001
positive	0.7698 ± 0.0006	0.6002 ± 0.0005	0.7089 ± 0.0005	0.2912 ± 0.0004
negative	0.7715 ± 0.0004	0.5993 ± 0.0003	0.7096 ± 0.0003	0.2908 ± 0.0002
positive + negative	$\mathbf{0.7757} \pm 0.0003$	$\mathbf{0.5970} \pm 0.0002$	$\mathbf{0.7113} \pm 0.0002$	$\mathbf{0.2893} \pm 0.0001$

without self-distillation. *fuse* refers to clustering without distinguishing between positive and negative samples. *positive* refers to only cluster positive examples, while *negative* refers to only cluster negative examples. *positive + negative* refers to separate cluster positive or negative examples to K centers. As shown in Table 2, *positive + negative* achieves the best performance, indicating the effectiveness of GMSD.

4   Online Deployment

We deployed GMSD in real advertising systems and conducted an experiment on A/B test framework for two weeks. GMSD increases online click-through rate (CTR) and Conversion Rate (CVR) by 3.41% and 0.97% compared with BAKE, which illustrates the positive effects of GMSD in a real recommendation system.

References

1. Chen, P., Ye, J., Chen, G., Zhao, J., Heng, P.A.: Beyond class-conditional assumption: a primary attempt to combat instance-dependent label noise. In: AAAI 2021, pp. 11442–11450
2. Ge, Y., et al.: Self-distillation with batch knowledge ensembling improves ImageNet classification. arXiv preprint arXiv:2104.13298
3. Zhao, P., et al.: RLNF: reinforcement learning based noise filtering for click-through rate prediction. In: SIGIR, pp. 2268–2272 (2021)
4. Huan, Z., et al.: Learning to select instance: simultaneous transfer learning and clustering. In: SIGIR 2021, pp. 1950–1954

A Scalable Social Recommendation Framework with Decoupled Graph Neural Network

Ke Tu[1], Zhengwei Wu[1], Binbin Hu[1], Zhiqiang Zhang[1], Peng Cui[2], Xiaolong Li[3], and Jun Zhou[1(✉)]

[1] Ant Group, Hangzhou, China
{tuke.tk,zejun.wzw,bin.hbb,lingyao.zzq,jun.zhoujun}@antgroup.com
[2] Tsinghua University, Beijing, China
cuip@tsinghua.edu.cn
[3] Alibaba Group, Hangzhou, China
xl.li@alibaba-inc.com

Abstract. Social relationships are usually used to improve recommendation quality, especially when users' behavior is very sparse in recommender systems. Most existing social recommendation methods apply Graph Neural Networks (GNN) to capture users' social structure information and user-item interaction information. However, the GNNs need to conduct expensive neighborhood propagation, leading to scalability issues. Some recent works pointed out that the GNNs can be simplified via decoupling. Therefore, we propose a scalable framework for social recommendation to decouple the model into two stages, Gumbel-based feature propagation and self-supervised multi-representation fusion. In the first stage, since the similarity between friends will not change frequently, we pre-train a Gumbel sampling-based attention model offline to learn the importance of each social relation and use the importance as the weight to aggregate the feature during pre-computation. Due to the diversity of user interests, the features are propagated upon different propagation layers to capture information with different aspects. In the second stage, we use the aggregated representations as inputs and fuse the inputs by an attention mechanism to obtain comprehensive embeddings on the online mode to update daily. Besides, we use a contrastive learning way to enrich users' information. Moreover, extensive experimental results demonstrate the scalability and effectiveness of our framework over state-of-the-art algorithms.

Keywords: Decoupled GNN · Social Recommendation · Scalability

1 Introduction

Nowadays, recommender systems are widely used in the industry, such as e-commerce, advertising, and search engines. Due to the widespread existence of data sparsity and cold start issues, recent studies seek to introduce social networks into recommender systems, known as *Social Recommendation*, for enriching "cold" users' information via their neighbors, guided by the principle of homophily [10] that friends tend to share similar interests. Meanwhile, *Graph Neural Networks* (GNNs) [4] has become newly state-of-the-art for modeling network structure, and its effectiveness in facilitating social recommendation has been well studied [19]. Roughly speaking, existing works mainly extend prevailing GNN architectures to adapt for various social recommendation scenarios through social diffusion [15], adversarial training based denoising [17,19] and self-supervised learning [6,16]. Notwithstanding the promising performance in benchmarks, we are crucial about following fundamental questions, aiming at the industrial settings of social recommender systems based on GNNs.

- *Is the recently emerging GNN-based social recommendation approaches suitable for real-world industrial applications?* Current GNN based social recommender systems, such as Diffnet++ [15] and MHCN [20], follow the typical GNN architecture, which recursively obtains user/item representations via nonlinear transformation of its neighbors. Intuitively, the heavy design causes both high computational cost and large space requirement, which severely threatens the scalability of the industrial recommender system. Moreover, the scale of recommender systems, as well as the social networks, is enormous in real-world business and is expected to be exploited for training and inference in a tolerable delay. Such a dilemma highlights an inevitable bottleneck to incorporating GNNs into social recommendation in practice.
- *Is social/interaction relations in reality reliable enough to unfold the strength of GNNs?* Unfortunately, real-world relational data is not only noisy (i.e. users may misclick some unwanted items) but also unreliable (i.e. the strength of social ties is uncertain) in most cases. Due to the recursive embedding propagation of GNNs, the capability of GNN-based social recommendation clearly deteriorates by passing unsatisfying or even harmful information.
- *How to address the problem of the information asymmetry among different views in the social recommendation?* By incorporating social relations, users may naturally have multiple views of neighbors such as friends and clicked items. Due to the sparsity issue, the user may have very few neighbors in some views. The final representations may be mainly influenced by the rich information view by the uniform neighbor aggregation of GNNs. It is important to enrich the sparse context for balancing different views to solve the information asymmetry issue.

In this paper, we borrow the basic idea of GNN decoupling and propose a **S**calable **S**ocial **R**ecommendation framework named **SSR** for the online recommendation system. We decouple our model into two-stage, Gumbel-based [7] feature propagation and self-supervised multi-representation fusion. In the first stage, we pre-process the feature propagation offline. To deal with the noisy and unreliable social relations, we pre-train a Gumbel-attention based GNN to choose the most critical social neighbors. Then we propagate features among the selected neighbors upon different propagation layers to keep the diversity of node representations. In the second stage, we use the pre-processed propagated features as input and apply an attentive aggregation to summarize representations from different propagation layers for users and items on the online daily updated datasets. In terms of the information asymmetry issue among different views, we come up with the contrastive learning [5] module to learn their representations. As the feature propagation in the first stage is pre-processed only once offline, and the online deep neural network in the second stage is efficient, our model can be scaled to large network datasets.

It is worthwhile to highlight the following contributions of this paper:

- We highlight the importance of the scalability problem in the social recommendation and propose a novel model named **S**calable **S**ocial **R**ecommendation framework (**SSR**) for the recommendation system which is scalable for massive industry scenarios. To the best of our knowledge, our model is the first work to introduce decoupled GNN into social recommendation.
- We decouple our model into the offline and online stage. In the offline pre-train stage, we use Gumbel attention to obtain the most useful social friends for solving unreliability issue of social relations. And in the online daily updated stage, we design a model to merge representations from different path-guided views and enrich the interest information for all users by contrastive learning to deal with information asymmetry issue.
- Extensive experimental results, including three common-used public datasets and a large-scale industry dataset, demonstrate the scalability and effectiveness of our framework over the state-of-the-art algorithms.

2 Related Work

2.1 Graph Neural Network

The Graph Neural Networks(GNNs) [22] has been proposed to deal with network data in an end-to-end manner. [1] introduces the graph signal process and defines the convolution in the Fourier spectral domain. GAT [14] applies a self-attention strategy to aggregate nodes in neighborhoods with different weights. Benefit from the strong network representation ability, GNNs are widely used in industry scenarios. Despite their effectiveness, the graph neural networks used in previous works lead to scalability issues in real industry massive scenarios. It is worth mentioning that recently deep insights into GNNs have shed some light on the design of scalable GNNs in practice, which performs simplification via decoupling [13,21]. They decouple GNNs into two-stage, feature propagation, and non-linear mapping. For example, SIGN [13] proposes to delete the nonlinear operators and simplify GNNs into an MLP operating on pre-computed concatenated multi-hop averaged features. Moreover, NARS [21] extends them into a heterogeneous version. However, its effectiveness has not been explored for social recommendation. Most notably, the unreliability of relations could lead to terrible degradation in recommendation performance since neighbor averaging-based strategies are commonly adopted for decoupling.

2.2   Social Recommendation

Due to the social network principle of homophily, users' interests are often influenced by their friends. Recent works aim to introduce social networks into the recommender system to alleviate the issue of data sparsity. The early social recommendation models can be categorized into two groups, factorization methods, and regularization methods. The factorization-based methods, such as SoRec [8], factorize the user-item rating matrices and user-user social matrices and map the factorized user and item representations into the same latent space. As for the regularization methods [9], they add a social regularization loss term, which guides the users' representations are closed to the representations of their friends, to the user preference ranking loss term. Based on these ideas, recent works introduce graph neural networks to describe the social network and preference network. Diffnet++ [15] model the recursive dynamic social diffusion to apply information from social neighbors. ESRF [19] and RSGAN [17] use adversarial training to generate reliable friends for denoising social relations. S^2-MHCN [20] apply hypergraph to capture motif structures to solve the multi-faceted social relations. However, in the real world daily updated recommender systems, none of them can well solve the scalability, unreliability, and information asymmetry issues. In light of the above considerations, we develop a GNN-based social recommender system by graph decoupling to solve all the previous issues.

3   The Proposed Framework

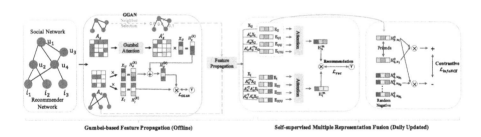

Fig. 1. The Framework of Our Proposed Model SSR.

In this section, we will introduce our proposed framework named **S**calable **S**ocial **R**ecommendations Framework (**SSR**) shown in Fig. 1. Our framework consists of two parts, Gumbel-based feature propagation and self-supervised multi-representation fusion.

3.1   Notations

Let $U = \{u_1, u_2, ..., u_m\}$ and $I = \{i_1, i_2, ..., i_n\}$ denotes the sets of users and items in a traditional recommendation scenario, where m is the number of users and n is the number of items. We define the user feature matrix and item feature matrix as $\mathbf{X_U} \in \mathbb{R}^{m \times d}$ and $\mathbf{X_I} \in \mathbb{R}^{n \times d}$. In social recommendation, there are two networks, user-item interaction network $G_o = \{V_o = \{U, I\}, \mathbf{A}_o\}$ and social network $G_s = \{V_s = U, \mathbf{A}_s\}$ where V_o, V_s denote the node sets and $\mathbf{A}_o$, $\mathbf{A}_s$ denote the adjacency matrix. If user u has interaction with item i, such as clicked and purchased, $\mathbf{A}_{o,ui}$ equals 1 otherwise 0. $\mathbf{A}_{s,u_1u_2} = 1$ denotes user u_1 and u_2 have social relations. $\mathcal{N}_o(u)$ and $\mathcal{N}_s(u)$ are user u's neighbors in network G_o and G_s, respectively. $\mathbf{Y} \in \mathbb{R}^{m \times n}$ is the feedback matrix and y_{ui} which user u's feedback on the item i is either 1 (positive) or 0 (negative or unknown). In our setting, there may be two user-item interaction networks G_o and G_o^m. G_o is historical behavior network for first stage pre-training. And G_o^m is the recent daily updated user-item interaction network. We give the definition of the task Social Recommendation as follows:

Definition 1 (Social Recommendation). *Given the social network $G_s = \{V_s = U, \mathbf{A}_s\}$, user-item interaction network $G_o = \{V_o = \{U, I\}, \mathbf{A}_o\}$ and the features of users $\mathbf{X}_U$ and items $\mathbf{X}_I$, the social recommendation aims to learn a predictive model that effectively forecasts the future user-item interaction.*

3.2 Graph Model Decoupling

Most existing graph neural networks are composed of feature propagation and non-linear mapping. In the model, a non-linear mapping follows a feature propagation step as follows:

$$\mathbf{T}^{(k)} = \mathbf{A}\mathbf{H}^{(k)}, \tag{1}$$

$$\mathbf{H}^{(k+1)} = \sigma_k(\mathbf{T}^{(k)}\mathbf{W}^{(k)}), \tag{2}$$

where $\mathbf{A}$ is sparse adjacency matrix, $\{\mathbf{H}^{(k)}, k = 1, 2, ..., K\}$ is the hidden representations and we set $\mathbf{H}^{(0)}$ equals feature matrix $\mathbf{X}$ and $\mathbf{W}^{(k)}$ is the weight matrix in layer k. The σ_k is a non-linear function. For different graph neural networks, the weighted adjacency matrix $\mathbf{A}$ may be different. For example, GCN [4] and GAT [14] re-weight the adjacency matrix by degree and attention matrix respectively. Since there is a massive sparse-dense matrix multiplication in Eq. 1, it is very time-consuming. As LightGCN [3] pointed out, the primary role of GNNs is based on the message-passing layer instead of the non-linear mapping layer. After deleting the non-linear mapping between layers, the GNNs can be simplified as: $\mathbf{H}_{LightGCN}^{(K)} = \sigma((\alpha_1\mathbf{A}\mathbf{X} + \alpha_2\mathbf{A}^2\mathbf{X} + ... + \alpha_K\mathbf{A}^K\mathbf{X})\mathbf{W})$. In such a situation, the GNNs can be decoupled as feature propagation and non-linear mapping. In the feature propagation step, we can pre-compute all the feature propagations such as $\mathbf{A}\mathbf{X}, \mathbf{A}^2\mathbf{X}, ..., \mathbf{A}^K\mathbf{X}$ since there are no trainable variables. This process runs only once. In the non-linear mapping step, we can merge the propagated features and apply a deep neural network to learn weights $\mathbf{W}$. It can be efficient and scalable. To introduce decoupled GNNs into social recommendation, we enrich these two steps into Gumbel-based feature propagation and self-supervised multi-representation fusion. The first step of our model is

pre-trained offline and runs only once to obtain stable social relationships. And
the second step updates daily online to use the stable social relationships and
the new arrival datasets to forecast the future user-item interactions.

3.3 Gumbel-Based Feature Propagation

Graph Gumbel Attention Network. Different from the traditional recommendation problems, there are social relations in our setting. In real-world social
networks, the relations between users may be noisy and unreliable. However, the
sparse weighted adjacency matrix $\mathbf{A}$ used in previous works is usually constant
and human-designed. To find the useful neighbors who contribute most to the
task, we need sparse attention to sparsify the social networks to obtain a sparse
and most useful weighted adjacency matrix $\mathbf{A}'$. We propose a pre-trained model
named Graph Gumbel Attention Network (GGAN) to obtain the learned adjacency matrix $\mathbf{A}'$. Since the social relations between users are stable in the middle
period and users' interaction behaviors with items change daily, the goal of the
pre-train model is to obtain the denoised and sparse social adjacency matrix. To
make the discrete neighbor selection differentiable, we use Gumbel sampling [7]
along with the reparametrization trick to produce a relaxation of the one-hot
vector to represent the selected new neighbor. For a user u, we first calculate
the edge attentions of each neighbor as follows:

$$\alpha_{uv} = \text{softmax}_{v \in \mathcal{N}_s(u)}(\text{LeakyReLU}(\mathbf{W}_a^T[\mathbf{X}_u \| \mathbf{X}_v])), \tag{3}$$

where $\mathbf{W}_a$ is trainable weight matrix, $\|$ is the concatenate function and
LeakyReLU is a non-linear activation. We denote the user u's attention vec-
tor as $\boldsymbol{\alpha}_u = [\alpha_{u1} \| \alpha_{u2} \| ... \| \alpha_{um}]$. Then the Gumbel-sampled relaxation of one-hot
vector to select the important neighbor is as follows:

$$\mathbf{v}_u = \frac{\exp\left((\log \boldsymbol{\alpha}_u + \mathbf{g})/\tau\right)}{\sum_{v \in \mathcal{N}(u)} \exp\left((\log \alpha_{uv} + \mathbf{g}_v)/\tau\right)}, \tag{4}$$

where $\mathbf{g}$ follows the Gumbel(0, 1) distribution[1]. $\tau \in (0, +\infty)$ is the temperature
hyper-parameter and it controls the sparsity of the one-hot vector. When τ is
smaller, $\mathbf{v}_u$ is closer to a one-hot vector. We conduct Gumbel sampling for T
times in Eq. 4 to obtain the Gumbel attention-based adjacency matrix:

$$\mathbf{A}'_s = \mathbf{A}_s \odot \sum_{i=0}^{T-1} \mathbf{V}^{(i)}/T, \tag{5}$$

where $\odot$ is the dot product, $\mathbf{V}$ is a matrix of all Gumbel attentions, and $\mathbf{A}_s$ is
the sparse adjacency matrix of social network G_s. After that, we aggregate social
neighbors' information like Eq. 1 and 2 to obtain social-based hidden representation $\mathbf{H}_s^{(k+1)}$. For the user-item network, we use the origin sparse adjacency

---

[1] Gumbel(0, 1) can be transformed by uniform distribution, $g = -log(-log(u)), u \in$
Uniform$(0, 1)$.

matrix $\mathbf{A}'_o = \mathbf{A}_o$ to aggregate graph information to obtain interest-based hidden representation $\mathbf{H}_o^{(k+1)}$ in the same way. We combine the final $\mathbf{H}_s^{(K)}$ and $\mathbf{H}_o^{(K)}$ followed an MLP layer to obtain the final user representation $\mathbf{H}_U^{(K)}$. Then we use the inner product of the final user and item representation to predict the feedback labels:

$$\mathcal{L}_{GGAN} = \sum_{(u,i)\in\mathcal{O}} \|\mathbf{H}_u^{(K)}\mathbf{H}_i^{(K)T} - \mathbf{Y}_{ui}\|^2, \tag{6}$$

where $\mathcal{O}$ is the set of all samples including positive samples and randomly sampled negative samples. After optimizing GGAN by Adam, we can obtain the learned sparse social adjacency matrix $\mathbf{A}'_s$.

3.4   Social Representations with Multiple Path-Guided Views

To better keep the diversity of node representations, we use multiple parameter-free propagation layers to generate multiple representations as the input of the second stage. To keep users' short real-time interests, we use the recent daily updated user-item interaction network $\mathbf{A}_{o'}$ along with the previous learned social weighted adjacency matrix $\mathbf{A}_{s'}$ to aggregate neighbors. We aggregate the feature in the way of propagation path sets $M_U = \{U, UI, UU, UIU\}$ and $M_I = \{I, IU, IUI, IUU\}$ as follows:

$$\mathbf{E}_U = \mathbf{X}_U; \mathbf{E}_I = \mathbf{X}_I; \mathbf{E}_{UI} = \mathbf{A}'_o\mathbf{E}_I; \mathbf{E}_{IU} = \mathbf{A}'^T_o\mathbf{E}_U;$$
$$\mathbf{E}_{UU} = \mathbf{A}'_u\mathbf{E}_U; \mathbf{E}_{IUU} = \mathbf{A}'_o\mathbf{E}_{UU}; \mathbf{E}_{UIU} = \mathbf{A}'_o\mathbf{E}_{IU}; \mathbf{E}_{IUI} = \mathbf{A}'^T_o\mathbf{E}_{UI}. \tag{7}$$

The propagation paths can be categorized into four overlapped groups according to the implicit semantics. U and I are the origin feature of users and items. UI and IU contains users' behavior information. UIU and IUI apply high-order network information. UU and IUU rich users' information by social relations. Since the propagation process is parameter-free, this generation can be pre-processed before training.

3.5   Self-supervised Multi-representation Fusion

In the second stage, we fuse the pre-processed features $\{\mathbf{E}_p, p \in M_U \cup M_I\}$ and capture the users' interests in an end-to-end manner. These representations cover different aspects of users and items. We use an attention mechanism to obtain the cross information. The attention aims to learn the importance of each propagated representation of p. For simplicity, here we only present the attentive aggregator for users, and the same aggregator will be used for items:

$$\mathbf{F}_p^0 = \text{LeakyReLU}(\mathbf{E}_p\mathbf{W}_p^0 + \mathbf{b}_p^0), \hat{\beta}_p = \text{LeakyReLU}(\mathbf{F}_p^0\mathbf{W}_p^a + \mathbf{b}_p^a),$$
$$\beta_p = \frac{\exp(\hat{\beta}_p)}{\sum_{p\in M_U}\exp(\hat{\beta}_p)}, \mathbf{F}_U^m = \sum_{p\in M_U}\beta_p * \mathbf{F}_p^0, \tag{8}$$

where $\mathbf{W}_p^0, \mathbf{W}_p^a, \mathbf{b}_p^0, \mathbf{b}_p^a$ are trainable weight matrix, $\mathbf{F}_p^0$ is the representation of propagation path p and $\mathbf{F}_U^m$ is final user representation. To learn the users' interests, we optimize the model with the loss:

$$\mathcal{L}_{rec} = \sum_{(u,i) \in \mathcal{O}} \|\mathbf{F}_u^m \mathbf{F}_i^{mT} - \mathbf{Y}_{ui}^m\|^2, \tag{9}$$

where $\mathbf{Y}_{ui}^m$ is recent rating matrix from daily user-item behavior network G_o^m.

Besides, due to the information asymmetry issue, we introduce contrastive learning as a regularizer to enrich users' propagated path representations. For this purpose, we maximize the mutual information between one user's path representation $\mathbf{F}_{p,u_i}^0$ and one of his randomly sampled neighbor's path representations $\mathbf{F}_{q,u_j}^0$. In such a way, the path representations in sparse domains can be enriched. In particular, we set as the embedding pair $(\mathbf{F}_{p,u_i}^0, \mathbf{F}_{p,u_j}^0)$ with $(u_i, u_j) \in G_s$ where u_i and u_j are reliable neighbors selected by Gumbel attention in Eq. 5 as positive sample. And we generate negative sample $(\mathbf{F}_{p,u_i}^0, \mathbf{F}_{q,u_k}^0)$ by randomly selecting L users $\{u_{k_c}, c = 0, 1, ..., L - 1\}$ with $(u_i, u_{k_c}) \notin G_s$. Then we define the mutual information by InfoNCE [5] is as follows:

$$\mathcal{L}_{infoNCE} = \sum_{(u_i, u_j) \in G_s} - \log \frac{\exp\left(\mathbf{F}_{p,u_i}^0 \mathbf{F}_{q,u_j}^{0T}\right)}{\exp\left(\mathbf{F}_{p,u_i}^0 \mathbf{F}_{q,u_j}^{0T}\right) + \sum_{(u_i, u_{k_c}) \notin G_s} \exp\left(\mathbf{F}_{p,u_i}^0 \mathbf{F}_{q,u_{k_c}}^{0T}\right)}. \tag{10}$$

Then the total loss of the second stage is the combination of two losses,

$$\mathcal{L}_2 = \mathcal{L}_{rec} + \gamma \mathcal{L}_{infoNCE}, \tag{11}$$

where γ is the coefficient to control the balance of the two losses.

3.6   Complexity Analysis

In this section, we discuss the complexity of our model. In the first stage, the most time-consuming part is graph convolution and Gumbel attention. The time complexity is $\mathcal{O}((|\mathbf{A}_s| + |\mathbf{A}_o|)dK + |\mathbf{Y}_{UI}|d)$. It is easy to see that the first stage is linear to the number of all edges which is the same as traditional GNN models. However, it can be pre-trained only once so the online training process will not have this time cost. So we focus more on the second online daily update stage. There is no graph propagation process which is very time-consuming in the second stage. The training time complexity is $\mathcal{O}(|\mathbf{Y}_{UI}^m|Pd)$ where $|\mathbf{Y}_{UI}^m|$ denotes the number of training user-item pairs in the second stage. And P means the number of paths which is a constant. Therefore, the time complexity of the second online stage is close to a multi-layer MLP model and much lower than that of previous GNN-based social recommendation methods.

4   Experiment

In this section, we conduct experiments to evaluate the performance of SSR. Specifically, we aim to answer the following research questions:

- **RQ1**: How does SSR perform compared to the state-of-the-art baselines?
- **RQ2**: How time efficient is the proposed scalable two-stage framework?
- **RQ3**: Can the proposed SSR be scale to real industry massive datasets?
- **RQ4**: How does each of the key components of SSR affect the model?

Table 1. Dataset Statistics

Dataset	#User	#Item	#Feedback	#Density	#Relation
LastFM	1,892	17,632	92,834	0.28%	25,434
Yelp	17,237	38,342	204,448	0.04%	143,765
Flick	8,358	82,120	314,809	0.05%	187,273
Industry	5,183,534	5,433	22,914,019	0.08%	366,640,997

4.1 Datasets and Baselines

In order to comprehensively evaluate the effectiveness of our proposed method, we evaluate the proposed SSR on three common-used public datasets (i.e., Last.fm[2], Yelp[3], Flickr[4]) and one massive real-world industry recommendation dataset(i.e., Industry). Since there is no time logs in the three public datasets, we use the same user-item interaction in both offline and online stage. In this industry setting, we use user-item interaction of the past month to train the first pre-train stage and user-item interaction of one recent week to train the second online daily-updated stage. The detailed descriptions of the four datasets are summarized in Table 1. Following [15], to perform the evaluation, for each user, we randomly select 1000 unrated items that a user has not interacted with as negative samples, followed by the ranking procedure with the positive samples among 1000 negative samples. Two relevancy-based metrics (i.e., *Precision@15* and *Recall@15*) and one ranking-based metric (i.e., *NDCG@15*) are used to evaluate the performance of all methods. To reduce the uncertainty in this process, we repeat this procedure 5 times and report the average results.

We compare SSR with a set of commonly-used social recommendation baselines, including MF-based and GNN-based models. In summary, **BPR** [12] and **FM** [11] are traditional recommendation methods. **LightGCN** [3] is a fast GNN model. **DiffNet++** [15], **ESRF** [19], **MHCN** [20], **SEPT** [18] are GNN-based social recommendation models. For the general settings of all the methods, we empirically set the dimension of latent embeddings to 64, the balance coefficient γ to 0.01, and the batch size to 2000. We use the Adam optimizer for all these models with an initial learning rate of 0.001. For GNN-based models, the number of graph convolutional layers are set as 2.

---

[2] http://files.grouplens.org/datasets/hetrec2011/.
[3] https://www.yelp.com/dataset/challenge.
[4] https://www.flickr.com/.

4.2   Overall Performance Comparison (RQ1)

Since most of the datasets would take too much time on the largest dataset Industry, we conduct the performance comparison over all baselines on public benchmarks, i.e., Last.fm, Yelp, and Flickr. The results are presented in Table 2. From the results, we have the following observations:

- It is easy to see that our proposed SSR achieves significant improvements over the baselines on all datasets for most cases. Even though our method gets some little negative gains on some datasets with MHCN, the MHCN is much more time-consuming and can not scale to massive industry datasets. The actual training time comparison can be found in the next section.

Table 2. Recommendation Performance Comparison

Dataset	Metric	BPR	FM	LightGCN	DiffNet++	ESRF	MHCN	SEPT	SSR
	Prec@15	0.328	0.312	0.309	0.338	0.339	**0.361**	0.327	<u>0.353</u>
Last.fm	Recall@15	0.499	0.475	0.469	0.514	0.516	**0.549**	0.497	<u>0.536</u>
	NDCG@15	0.494	0.480	0.476	0.514	0.518	**0.548**	0.479	<u>0.536</u>
	Prec@15	0.041	0.040	0.038	<u>0.045</u>	0.045	0.045	0.042	**0.047**
Yelp	Recall@15	0.372	0.362	0.345	<u>0.412</u>	0.410	0.410	0.377	**0.426**
	NDCG@15	0.208	0.202	0.188	<u>0.226</u>	0.226	0.224	0.206	**0.236**
	Prec@15	0.052	0.051	0.054	0.054	0.052	<u>0.060</u>	0.055	**0.060**
Flickr	Recall@15	0.183	0.174	0.173	0.186	0.186	**0.223**	0.201	<u>0.221</u>
	NDCG@15	0.132	0.128	0.126	0.133	0.133	**0.159**	0.143	<u>0.155</u>

- The performance of our SSR is much better in the more sparse datasets like Yelp and Flickr. It is reasonable because our SSR enriches the sparse domain by self-supervised learning.

4.3   Efficiency Analysis on Benchmarks (RQ2)

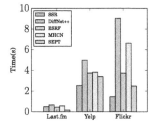
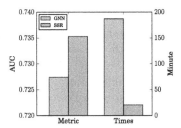

Fig. 2. Left: the training time of social recommendation methods on public benchmarks; Right: the performance and training time on large-scale industry dataset.

We plot the training time per epoch in Fig. 2 left, where only the performance of social recommendation models are shown since only these model uses both social network and recommendation network. In the smallest dataset Last.fm, the training time of all methods including our model and baselines are very close. However, our training time is significantly reduced when the datasets become larger (i.e., Yelp and Flickr). And Diffnet++ and MHCN are two of the most time-consuming methods. It demonstrates the scalability of SSR.

4.4 Performance and Efficiency Analysis on Industry Dataset (RQ3)

In this section, we will demonstrate SSR's ability to handle real industry datasets Industry with ten million scale. For the time limitation, we only run a base graph attention-based GNN on a social recommendation as baselines. To make the GNN runnable on this dataset, we randomly sample ten neighbors for each node like GraphSage [2]. Since the first part of our model is pre-processed offline and only the second part of our model is updated daily, we only count the time of the online part of our model. The results is shown in Fig. 2 right. We can see SSR achieves better performance than the baseline GNN in the left bar graph. Besides, the GNN takes much more time (about 9 times) than SSR in the right bar graph.

4.5 Ablation Analysis (RQ4)

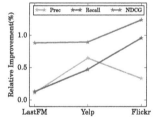
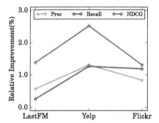

Fig. 3. Left: the relative improvement of SSR over SSR w/o Gumbel attention; Right: the relative improvement of SSR over SSR w/o self-supervised part.

Impact of Social Selection by Gumbel Attention. The effect of social selection by Gumbel attention is shown in Fig. 3 left. We change SSR into a variant, **Adj**, which changes the Gumbel attention into a normal adjacency matrix in the first stage. It measures the effect of denoising in social networks. In the figures, we plot the relative improvement over the variants. We can see that SSR performs the best among the variants on all datasets. It demonstrates that the selection of noisy social relations is necessary for social recommendations.

Impact of the Self-supervised Module. The effect of the self-supervised module is shown in Fig. 3 right. We change SSR into a variant, **-InfoNCE** which deletes the infoNCE loss in the second stage. It measure the importance of enriching the sparse domain by self-supervised contrastive learning. The SSR performs consistently better than the variant. It also proves the importance of the self-supervised module.

5 Conclusion

In this paper, we propose a scalable two-stage social recommendation framework named SSR. We decouple our model into two stages, Gumbel-based feature propagation and self-supervised multi-representation, for offline pre-training and online daily updating respectively. In the first stage, we pre-train a model by Gumbel attention to learn the importance of neighbors and propagate the features with multiple propagation layers. In the second stage, we fuse the preprocessed features and use contrastive learning to enrich the representability. Extensive experimental results demonstrate the scalability and effectiveness of our framework.

References

1. Defferrard, M., Bresson, X., Vandergheynst, P.: Convolutional neural networks on graphs with fast localized spectral filtering. In: NIPS, vol. 29 (2016)
2. Hamilton, W., Ying, Z., Leskovec, J.: Inductive representation learning on large graphs. Adv. Neural Inf. Process. Syst. **30** (2017)
3. He, X., Deng, K., Wang, X., Li, Y., Zhang, Y., Wang, M.: LightGCN: simplifying and powering graph convolution network for recommendation. In: Proceedings of the 43rd International ACM SIGIR Conference, pp. 639–648 (2020)
4. Kipf, T.N., Welling, M.: Semi-supervised classification with graph convolutional networks. In: International Conference on Learning Representations (ICLR) (2017)
5. Liu, X., et al.: Self-supervised learning: generative or contrastive. IEEE TKDE (2021)
6. Long, X., et al.: Social recommendation with self-supervised metagraph informax network. In: Proceedings of the 30th CIKM, pp. 1160–1169 (2021)
7. Lorberbom, G., Johnson, D., Maddison, C.J., Tarlow, D., Hazan, T.: Learning generalized Gumbel-max causal mechanisms. In: NIPS, vol. 34 (2021)
8. Ma, H., Yang, H., Lyu, M.R., King, I.: SoRec: social recommendation using probabilistic matrix factorization. In: CIKM, pp. 931–940 (2008)
9. Ma, H., Zhou, D., Liu, C., Lyu, M.R., King, I.: Recommender systems with social regularization. In: WSDM, pp. 287–296 (2011)
10. McPherson, M., Smith-Lovin, L., Cook, J.M.: Birds of a feather: homophily in social networks. Ann. Rev. Sociol. **27**(1), 415–444 (2001)
11. Rendle, S.: Factorization machines. In: 2010 IEEE International Conference on Data Mining, pp. 995–1000. IEEE (2010)
12. Rendle, S., Freudenthaler, C., Gantner, Z., Schmidt-Thieme, L.: BPR: Bayesian personalized ranking from implicit feedback. arXiv preprint arXiv:1205.2618 (2012)

13. Rossi, E., Frasca, F., Chamberlain, B., Eynard, D., Bronstein, M., Monti, F.: Sign: scalable inception graph neural networks. arXiv preprint arXiv:2004.11198 (2020)
14. Veličković, P., Cucurull, G., Casanova, A., Romero, A., Liò, P., Bengio, Y.: Graph attention networks. In: International Conference on Learning Representations (2018)
15. Wu, L., Li, J., Sun, P., Hong, R., Ge, Y., Wang, M.: DiffNet++: a neural influence and interest diffusion network for social recommendation. In: TKDE (2020)
16. Xia, X., Yin, H., Yu, J., Shao, Y., Cui, L.: Self-supervised graph co-training for session-based recommendation. In: Proceedings of the 30th ACM International Conference on Information and Knowledge Management, pp. 2180–2190 (2021)
17. Yu, J., Gao, M., Yin, H., Li, J., Gao, C., Wang, Q.: Generating reliable friends via adversarial training to improve social recommendation. In: ICDM (2019)
18. Yu, J., Yin, H., Gao, M., Xia, X., Zhang, X., Viet Hung, N.Q.: Socially-aware self-supervised tri-training for recommendation. In: SIGKDD, pp. 2084–2092 (2021)
19. Yu, J., Yin, H., Li, J., Gao, M., Huang, Z., Cui, L.: Enhance social recommendation with adversarial graph convolutional networks. In: TKDE (2020)
20. Yu, J., Yin, H., Li, J., Wang, Q., Hung, N.Q.V., Zhang, X.: Self-supervised multi-channel hypergraph convolutional network for social recommendation. In: Proceedings of the Web Conference 2021, pp. 413–424 (2021)
21. Yu, L., Shen, J., Li, J., Lerer, A.: Scalable graph neural networks for heterogeneous graphs. arXiv preprint arXiv:2011.09679 (2020)
22. Zhang, Z., Cui, P., Zhu, W.: Deep learning on graphs: a survey. IEEE Trans. Knowl. Data Eng. (2020)

Cold & Warm Net: Addressing Cold-Start Users in Recommender Systems

Xiangyu Zhang[1], Zongqiang Kuang[1], Zehao Zhang[1,2], Fan Huang[1(✉)], and Xianfeng Tan[3]

[1] Tencent, Shenzhen, China
{altairzhang,devinkuang,sinohuang}@tencent.com
[2] Tsinghua University, Beijing, China
zhangzeh20@mails.tsinghua.edu.cn
[3] Tencent, Beijing, China
victan@tencent.com

Abstract. Cold-start recommendation is one of the major challenges faced by recommender systems (RS). Herein, we focus on the user cold-start problem. Recently, methods utilizing side information or meta-learning have been used to model cold-start users. However, it is difficult to deploy these methods to industrial RS. There has not been much research that pays attention to the user cold-start problem in the matching stage. In this paper, we propose Cold & Warm Net based on expert models who are responsible for modeling cold-start and warm-up users respectively. A gate network is applied to incorporate the results from two experts. Furthermore, dynamic knowledge distillation acting as a teacher selector is introduced to assist experts in better learning user representation. With comprehensive mutual information, features highly relevant to user behavior are selected for the bias net which explicitly models user behavior bias. Finally, we evaluate our Cold & Warm Net on public datasets in comparison to models commonly applied in the matching stage and it outperforms other models on all user types. The proposed model has also been deployed on an industrial short video platform and achieves a significant increase in app dwell time and user retention rate.

Keywords: recommender systems · cold-start · Cold & Warm Net

1 Introduction

As online information in social media and e-commerce platforms grows explosively, large-scale recommender systems (RS) [11] play an important role in solving the problem of information overload for users. Industrial RS [12] typically contains two stages: matching and ranking. In the matching stage, thousands of items potentially relevant to the user's interests are retrieved from a large-scale candidate pool, which is required to quickly find as many items that satisfy the

---

X. Zhang and Z. Kuang—Equal contributions.

X. Wang et al. (Eds.): DASFAA 2023, LNCS 13946, pp. 532–543, 2023.
https://doi.org/10.1007/978-3-031-30678-5_40

user's interests as possible. After that, the ranking stage is performed to precisely predict the probability of a user interacting with an item.

Recently, the matching stage in recommenders [3] has been paid increasing attention. Various methods have been applied in the matching stage. Conventional collaborative filtering (CF) method [11] depends on the similarity of interacted items between users for the recommendation. State-of-the-art methods based on reinforcement learning [14], graph network [15] and Multi-Interest network [1] focus on user behavior sequence representation owned solely by active users with much interaction behavior. However, these models fail to learn high-quality embeddings for the cold-start users with sparse interaction behavior.

Faced with the cold-start problem, side information [16] has been used to provide a better recommendation. However, methods utilizing side information can only benefit part of the users. There are some attempts [4] to introduce meta-learning into recommender systems, which requires the computation of second-order gradients. Therefore, it cannot meet the scalability required by the matching stage of real-world recommendation scenarios. Scalability is the ability to process large-scale information efficiently.

In this paper, modeling cold-start users in the matching stage is our purpose. The core mission of modeling cold-start users is to learn collaborative information between old and cold-start users and train models effectively. Herein, we propose an implicit embedding net based on cold-start and warm-up experts which solves the problems mentioned above efficiently. According to the frequency of interaction, users can be briefly divided into three categories: cold-start users, warm-up users, and active users. The category of users is dynamically changing with the accumulation of interests and behavior, so it is not suitable to use the same strategies for different types of users. Our embedding net based on cold-start and warm-up experts models the dynamic process of cold-start users towards warm-up and active users without compulsory strategies. Overall, the main contributions of this work can be concluded as follows.

- Dynamic handling of samples. With the division of cold-start and warm-up experts, our Cold & Warm Net can dynamically represent the users' interest in cold-start and warm-up phases. Through the gate network, the net can automatically incorporate the results from two experts according to user type and user state. Cold-start and warm-up experts can learn the differences between samples.
- Flexible teacher selector. Dynamic knowledge distillation is applied to cold-start and warm-up experts using a teacher selector. The selector chooses the right teacher for the cold-start expert according to prediction accuracy. By applying dynamic knowledge distillation, it avoids the underfitting of the cold-start expert while preventing the assimilation of two experts after training, which enables learning sufficient information from cold-start users.
- Explicit modeling of behavior bias. Using a bias net to model the behavior bias of cold-start users explicitly. By utilizing mutual information, user features highly relevant to user behavior are selected. With the combination of the

similarity score from the original net and the bias score from the bias net, information hidden behind user behavior is thoroughly considered.

2   Related Work

In this section, we review the two-tower models based on embedding which are applied in the matching stage and models targeting the cold-start problem.

2.1   Two-Tower Models in the Matching Stage

One of the challenges faced by RS in the matching stage is that the representations of users and items are not in the same latent space. Models based on embedding learn how to map the sparse user and item vectors in high-dimensional space into dense vectors in low-dimensional space and calculate the inner product or cosine similarity between user and item vectors to obtain a relevance score. The idea of deep learning has been applied in the two-tower models. DSSM [8] is a well-known two-tower model utilizing two deep neural networks that map queries and documents into a common space to achieve better search satisfaction. YouTube [3] proposes a deep candidate generation model which can effectively learn the embedding of user and item features. [1] exploits both user profile and behavior information for candidate matching. To tackle sample bias in the matching stage. [7] uses random sampling to acquire negative samples, which successfully bridges the gap in data distribution between training and testing. However, these two-tower models all require close user-item interaction and are incompetent to model cold-start users with rare interaction behavior.

2.2   Cold-Start Problem

The cold-start problem has been one of the long-standing challenges faced by RS. Traditional methods rely on side information to alleviate the cold-start problem, e.g. utilizing social networks among users [16]. Transfer learning-based method [6] is also used to deal with the cold-start problem. [4] uses meta-learner to generate cold-start user embedding. However, most of the existing works focus on the cold-start problem in the ranking stage. [16] applies an attention mechanism in multi-channel matching to extract useful feature interactions. [2] uses an extra adversarial network to generate cold-start item embedding. [5] simulates the cold-start scenarios from the users/items with sufficient interactions and takes the embedding reconstruction as the pretext task. To our knowledge, we are the first to propose an embedding net that dynamically models different types of users in the matching stage.

3   Method

3.1   Problem Description

The objective of the matching stage for RS is to retrieve Top K relevant items from a large-scale candidate pool $\mathcal{I}$ for each user $u \in \mathcal{U}$. To achieve this target, a matching model is built. The input of model is a tuple $(\mathcal{X}_u, \mathcal{X}_i)$, where $\mathcal{X}_u$ denotes user features and $\mathcal{X}_i$ denotes item features. Modeling cold-start users is tough due to the lack of behavior. The core task of Cold & Warm Net is to learn a function that can map original features into user representations. The function can be formulated as:

$$\vec{e}_u = f_{user}(\mathcal{X}_u) \tag{1}$$

where $\vec{e}_u \in \mathbb{R}^{1 \times d}$ denotes the representation vector of user u, d the dimensionality. In addition, the representation vector of target item i is obtained by a function:

$$\vec{e}_i = f_{item}(\mathcal{X}_i) \tag{2}$$

where $\vec{e}_i \in \mathbb{R}^{1 \times d}$ denotes the representation vector of item i. Finally, The top K relevant items are retrieved according to the scoring function:

$$f_{score}(\vec{e}_u, \vec{e}_i) = \vec{e}_u \cdot \vec{e}_i \tag{3}$$

3.2 Cold & Warm Net

As shown in Fig. 1, our Cold & Warm Net consists of two subnets: original cold & warm net and bias net. The original cold & warm net uses user features $\mathcal{X}_u$ and item features $\mathcal{X}_i$ as input while the bias net takes bias features $\mathcal{X}_b$ as input. We divide user features into two categories: user profile features $\mathcal{X}_{up}$(e.g., gender and age) and user action features $\mathcal{X}_{ua}$(also called user behavior). Taking different user features as input, we attain output embedding: user profile embedding $\vec{e}_{up} \in \mathbb{R}^{1 \times d}$ and user action embedding $\vec{e}_{ua} \in \mathbb{R}^{1 \times d}$. Besides, user group embedding $E_{ug} \in \mathbb{R}^{m \times d}$ is provided as prior information for all users. Firstly, We use a pre-trained model for getting all active-user embeddings. Then, taking all active-user embeddings as input, the k-means algorithm is used to attain m clusters. Finally, we use average pooling to aggregate the embeddings of each cluster for generating E_{ug}. The three parts are defined as U_a, U_b and U_c, which are fed into user cold & warm embedding layer to generate user embedding $\vec{e}_u$. Along with item embedding $\vec{e}_i$, similarity score y_{sim_score} is calculated as follows.

$$y_{sim\_score} = \frac{\vec{e}_u \cdot \vec{e}_i}{\|\vec{e}_u\| \|\vec{e}_i\|} \tag{4}$$

The similarity score y_{sim_score} from the original cold & warm net and the bias score y_{bias_score} from the bias net constitute our final output:

$$y = sigmoid(y_{sim\_score} + y_{bias\_score}) \tag{5}$$

User Cold & Warm Embedding Layer. As shown in Fig. 2, our user cold & warm embedding layer is mainly composed of two experts: the cold-start expert and the warm-up expert. To extract and learn valid information that matches the current user, we use the attention mechanism to retrieve the prior information from E_{ug} which contains all user group pre-trained embedding. Cold-start expert

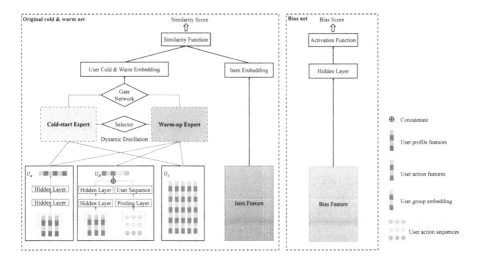

Fig. 1. Cold & Warm Net.

takes input from U_a and U_c which contain user profile information and user group information. Attention embedding takes prior user group information to assist in modeling cold-start users, which can be formulated as:

$$\overrightarrow{e}_{cold}^{a} = softmax(\frac{\overrightarrow{e}_{up}E_{ug}^T}{\sqrt{d}})E_{ug} \tag{6}$$

where $\overrightarrow{e}_{cold}^{a} \in \mathbb{R}^{1 \times d}$ means pre-trained embedding retrieved from E_{ug} using attention mechanism [9]. The output embedding of cold-start expert $\overrightarrow{e}_{cold}$ is:

$$\overrightarrow{e}_{cold} = mlp(\overrightarrow{e}_{up}; \overrightarrow{e}_{cold}^{a}) \tag{7}$$

where $\overrightarrow{e}_{cold} \in \mathbb{R}^{1 \times d}$. The warm-up expert takes input from U_a, U_b and U_c. It is designed for users who possess user profile features $\mathcal{X}_{up}$ and user action features $\mathcal{X}_{ua}$. Taking $\mathcal{X}_{up}$ and $\mathcal{X}_{ua}$ as input, we attain embedding $\overrightarrow{e}_{ut} \in \mathbb{R}^{1 \times d}$. Through the assistance of attention anchor embedding, the output embedding of warm-up expert $\overrightarrow{e}_{warm}$ is defined as follows.

$$\overrightarrow{e}_{warm}^{a} = softmax(\frac{\overrightarrow{e}_{ut}E_{ug}^T}{\sqrt{d}})E_{ug} \tag{8}$$

$$\overrightarrow{e}_{warm} = mlp(\overrightarrow{e}_{ut}; \overrightarrow{e}_{warm}^{a}) \tag{9}$$

where $\overrightarrow{e}_{warm} \in \mathbb{R}^{1 \times d}$. A gate network is used to produce weights for experts.

$$w_{cold}, w_{warm} = f_{weight}(\mathcal{X}_{us}) \tag{10}$$

where $\mathcal{X}_{us}$(such as login state and active degree) denotes state feature. The output user cold & warm embedding is:

$$\overrightarrow{e}_u = w_{cold} \cdot \overrightarrow{e}_{cold} + w_{warm} \cdot \overrightarrow{e}_{warm} \tag{11}$$

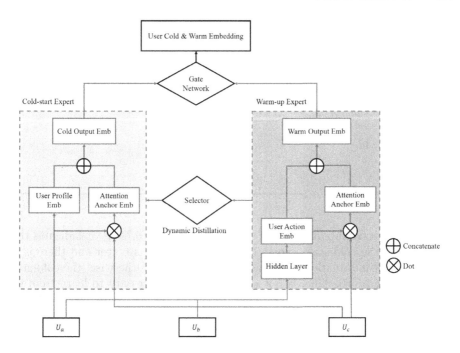

Fig. 2. User cold & warm embedding layer.

The above expression can be understood as a weighted summation of cold-start expert and warm-up expert.

Dynamic Knowledge Distillation. With a mix of experts, cold-start expert suffers from underfitting due to limited information from cold-start users. The reason is that the warm-up expert learns better for active users which own rich behavior features. To avoid underfitting of the cold-start expert, we invent dynamic knowledge distillation(DKD) to distill information from the warm-up expert to the cold-start expert. Binary cross entropy is selected as our loss function. The major loss function L:

$$L = -\frac{1}{N} \sum_{i=1}^{N} (y_i \log \hat{y}_i + (1 - y_i) \log (1 - \hat{y}_i)) \tag{12}$$

where N denotes the number of samples, y_i is the label for each sample. $\hat{y}_i$ denotes the predicted result. L is the loss function except for the cold-start expert. Besides, The auxiliary loss function L_d from dynamic knowledge distillation:

$$L_d = -\frac{1}{N} \sum_{i=1}^{N} l_d \tag{13}$$

where l_d is the loss function of DKD for each sample, which is shown in Algorithm 1. For each sample, we compare the cross entropy loss of cold-start expert

---

Algorithm 1: Dynamic knowledge distillation.

---

1 **foreach** *sample from batch samples* **do**

2 | Calculate $l(\hat{y_i}^c, y_i)$, $l(\hat{y_i}^w, y_i)$

3 | **if** $l(\hat{y_i}^c, y_i) \leq l(\hat{y_i}^w, y_i)$ **then**

4 | | $l_d = l(\hat{y_i}^c, \hat{y_i}^w)$;

5 | **else**

6 | | $l_d = 0$;

7 | **end**

8 **end**

$l(\hat{y_i}^c, y_i)$ and warm-up expert $l(\hat{y_i}^w, y_i)$. If $l(\hat{y_i}^c, y_i) \leq l(\hat{y_i}^w, y_i)$, it indicates that output from the cold-start expert taught by the label is better and there is no need to distill information from the warm-up expert. Otherwise, the cold-start expert taught by the label is not enough and it is necessary to learn from the warm-up expert. $\hat{y_i}^c$ denotes the predicted label for the cold-start expert and $\hat{y_i}^w$ is the predicted label for the warm-up expert. The teacher for knowledge distillation is dynamic to make sure the cold-start expert could learn effective information from the teacher. The loss function of the cold-start expert is defined as L_o.

$$L_o = L + \alpha \cdot L_d \qquad (14)$$

where α is hyperparameter. α determines the strength of distillation from the warm-up expert.

Bias Net. To solve the behavior bias when modeling cold-start users, an additional bias net is applied. The reason why bias net is effective is that the behavior bias is large between cold-start users and active users in real-world recommendation scenarios. For example, active users have several times more click rates than new users, so a bias net is indispensable for describing the bias. We aim to find a set of user features $\mathcal{X}_b$ which are highly relevant to user behavior and feed them into the bias net to get y_{bias_score}. To mine the key features of the behavior bias, mutual information is used to measure the relevance between user features and behaviors. We select the top β relevant features as our bias features $\mathcal{X}_b$. The output bias score y_{bias_score} is:

$$y_{bias\_score} = f_{bias}(\mathcal{X}_b) \qquad (15)$$

y_{bias_score} is used to characterize the bias of target behaviors of people.

4 Experiments

4.1 Offline Evaluation

In this section, we compare our Cold & Warm Net with existing methods applied in the matching stage in terms of recommendation accuracy on two datasets.

Table 1. Statistics of the datasets.

Dataset	# User	# Items	# Interaction
MovieLens 1M	6040	3706	1,000,209
Little-World	433,549	406,140	15,200,286

Datasets and Experimental Setup. Two datasets are chosen to evaluate the recommendation performance in the matching stage. One is MovieLens 1M[1], which is one of the most common datasets used for recommendations. Also, we collect a real-world large-scale dataset from the Little-World[2]. The statistics of datasets are shown in Table 1. Hit rate (HR) and Normalized discounted cumulative gain (NDCG) are adopted as the main metric to evaluate the performance of models in the matching stage, define as:

$$HitRate@K = \sum_{(u,i) \in \mathscr{T}} \frac{I(target\,items\,occur\,in\,topK)}{|\mathscr{T}|} \tag{16}$$

$$NDCG@K = \frac{1}{|U|} \sum_{u \in U} \frac{DCG_k^u}{IDCG_k^u} \tag{17}$$

$$DCG_k^u = \sum_{r=1}^{k} \frac{2^{R_{ur}} - 1}{\log_2(1+r)} \tag{18}$$

where $\mathscr{T}$ denotes the test set containing pair of user and item and I denotes the indicator function. R_{ur}, U, and $IDCG_k^u$ are the real rating of user u for the r-th ranked item, a set of users in the test data and the best possible DCG_k^u for user u, respectively. Specially, in the matching stage, $R_{ir} \in \{0, 1\}$.

Comparing Methods. The following methods widely applied in the matching stage in industry RS are used to compare with our Cold & Warm Net.

- FM [13] A model that utilizes the feature vectors of query and item and feeds them into FM layer.
- YouTubeDNN [3] One of the most commonly used models in the recommendation industry which applies deep neural network to generate item and user embedding.
- DSSM [8] A popular model applied in the matching stage which makes use of rich content features of user and item.

---

[1] https://grouplens.org/datasets/movielens/1m/.

[2] Little-World is a short video platform in QQ, which allows users to create and share micro-videos. Note that we anonymize the data and conduct strict desensitization processing. The data may be made public in the future.

Table 2. Performance comparison of different models in terms of HR and NDCG

(a)Results on full users

Models	MovieLens 1M				Little-World			
	HR@50	HR@100	NDCG@10	NDCG@50	HR@50	HR@100	NDCG@10	NDCG@50
FM	0.0969	0.1922	0.0099	0.0262	0.0513	0.0754	0.0100	0.0173
YouTubeDNN	0.1399	0.2548	0.0153	0.0378	0.0862	0.1461	0.0110	0.0245
DSSM	0.2013	0.3151	0.0226	0.0520	0.0913	0.1511	0.0116	0.0260
Mind	0.2019	0.3322	0.0238	0.0612	0.0917	0.1530	0.0118	0.0262
UMI	0.2348*	0.3697*	0.0305*	0.0664*	0.0920*	0.1546*	0.0119*	0.0270*
Cold & Warm	**0.2556**	**0.3932**	**0.0369**	**0.0750**	**0.1122**	**0.1792**	**0.0155**	**0.0325**
%improve.	8.86%	6.35%	20.98%	12.95%	21.95%	15.91%	30.25%	20.37%

(b)Results on cold-start users

Models	MovieLens 1M				Little-World			
	HR@50	HR@100	NDCG@10	NDCG@50	HR@50	HR@100	NDCG@10	NDCG@50
FM	0.1568	0.2953	0.0211	0.0461	0.0710	0.1047	0.0147	0.0242
YouTubeDNN	0.2444	0.3849	0.0236	0.0639	0.1088	0.1768	0.0138	0.0311
DSSM	0.3666*	0.5356*	0.0657*	0.1173*	0.1109*	0.1775*	0.0159*	0.0326*
Mind	0.3259	0.4807	0.0485	0.1008	0.1074	0.1771	0.0132	0.0300
UMI	0.3360	0.4705	0.0493	0.1002	0.1103	0.1671	0.0143	0.0281
Cold & Warm	**0.4094**	**0.5866**	**0.0678**	**0.1265**	**0.1435**	**0.2200**	**0.0215**	**0.0418**
%improve.	11.67%	9.52%	3.19%	7.84%	29.39%	23.94%	35.22%	28.22%

- Mind [10] the first attempt in representing a user with multiple interest vectors via deep neural network structures.
- UMI [1] State-of-the-art model that relies on multiple user interest representations to achieve superior recommendation accuracy.

The above models are implemented by Tensorflow and Faiss is used to retrieve the top K items from the item pool. The embedding dimension and batch size are set to 32 and 256 respectively for all models. To ensure a fair comparison, for each model, hyperparameters are tuned to achieve the best performance. For Cold & Warm Net, hyperparameters α, β are set to 5e-2 and 10 respectively.

Experimental Results. Table 2 summarizes the performance of Cold & Warm Net in comparison with different models applied in the matching stage in terms of HR@K($K = 50$, 100) and NDCG@K($K = 10$, 50). Obviously, Cold & Warm Net achieves the best recommendation performance among all models on different user categories. FM performs worst among all models revealing the power of deep learning. UMI and Mind which utilize multiple interest representations of user generally performs better than YouTubeDNN which only uses single-user interest representation. UMI performs better than Mind due to exploiting both user profile and behavior information for candidate matching. However, Mind and UMI perform worse than the DSSM model for cold-start users, which may be because cold-start users lack abundant interests. Results on two types of users justify the performance of Cold & Warm Net on different types of users, effectively solving the user cold-start problem faced in recommender systems.

Table 3. Ablation study of Cold & Warm Net.

Models	MovieLens 1M				Little-World			
	Full users		Cold-start users		Full users		Cold-start users	
	HR@100	NDCG@10	HR@100	NDCG@10	HR@100	NDCG@10	HR@100	NDCG@10
Cold & Warm	**0.3932**	**0.0369**	**0.5866**	**0.0678**	**0.1792**	**0.0155**	**0.2200**	**0.0215**
w/o DKD	0.3869	0.0318	0.5540	0.0581	0.1703	0.0144	0.1987	0.0190
w/o Bias Net	0.3930	0.0367	0.5682	0.0632	0.1761	0.0147	0.2122	0.0203

Table 4. Influence of Dynamic Knowledge Distillation on weights (w_{cold}, w_{warm}).

Metrics	Cold-start expert	Warm-up expert
Weights (w/o DKD)	0.0410	0.9590
Weights (DKD)	0.3140	0.6860

Table 3 summarizes the result of the ablation study. It is conducted to evaluate the contribution of the dynamic knowledge distillation(DKD) module and bias net. DKD and bias net designed for cold-start users contribute mainly to solving the problem of modeling cold-start users. For cold-start users, applying DKD brings an increase of 5.88% and 10.72% in HitRate@100 on two datasets while adding bias net brings an increase of 3.24% and 3.68% in HitRate@100. The major boost from DKD may be due to the fact that the cold-start expert learns better user representation with the assistance of the warm-up expert.

4.2 Analysis of Dynamic Knowledge Distillation

In this section, the influence of DKD has been analyzed based on the Little-World dataset. AUC is chosen as the evaluation metric. As shown in Table 4, by applying dynamic knowledge distillation, it greatly improves the weight of cold-start expert w_{cold} from 0.0410 to 0.3140, which allows the cold-start expert to learn sufficient information either from warm-up expert or label and avoids underfitting of cold-start expert. It can be seen from Table 5 that after applying DKD, the AUC of the cold-start expert increases obviously, which demonstrates the effect of DKD on enabling sufficient training. The AUC of the warm-up expert decreases on the train set because DKD reduces losses flowing to the warm-up expert. Therefore, the major contribution of AUC comes from sufficient learning of the cold-start expert. Improved AUC of the whole model on the test set also justifies the influence of DKD on cold-start users.

4.3 Online Experiment

Finally, we deploy Cold & Warm Net in the real-world recommending scenario of Little-World. User retention rate (URR) and app dwell time (APT) are used as the main metrics for cold-start users. All online experimental results are averaged over a week's A/B test on Little-World. The result shows that Cold & Warm

Table 5. Influence of Dynamic Knowledge Distillation on AUC.

Metrics	w/o DKD			DKD		
	Cold-start expert	Warm-up expert	Whole	Cold-start expert	Warm-up expert	Whole
full users	0.5770	0.9255	0.9267	**0.8772**	0.8993	**0.9279**
cold-start users	0.5675	0.7279	0.7281	**0.7384**	0.7434	**0.7548**

Table 6. Online experimental results. Cold & Warm Net performs better in terms of VPI and VSR on cold-start users. DSSM model as the baseline.

Models	VPI	VSR
Cold & Warm	**+23.34%**	**−14.30%**
Mind	−2.05%	+2.76%

Net brings an increase of 3.27% in APT and 1.01% in URR for cold-start users. Meanwhile, we compare Mind and Cold & Warm Net using the DSSM model as the baseline. Video play integrity (VPI) and video skip rate (VSR) are used as the main metrics for user satisfaction evaluation. The result in Table 6 shows that Cold & Warm Net outperforms the DSSM model on both VPI and VSR, which indicates successfully modeling cold-start users and improving user satisfaction.

5   Conclusion

User cold-start problem in the matching stage is a critical challenge faced by RS. However, the solutions are rare both in academia and industry. In this paper, we propose Cold & Warm Net which effectively solves the problem for cold-start users, while in the meantime satisfying the scalability required by the billion-scale matching stage. We first construct our network with two experts and incorporate a gate network to combine results according to the user state. Bias net and DKD module responsible for modeling cold-start users are incorporated. Finally, we evaluate our model through offline and online experiments and it achieves an obvious increase in recommendation performance.

References

1. Chai, Z., et al.: User-aware multi-interest learning for candidate matching in recommenders. In: Proceedings of the 45th International ACM SIGIR Conference on Research and Development in Information Retrieval, pp. 1326–1335 (2022)
2. Chen, H., et al.: Generative adversarial framework for cold-start item recommendation. In: Proceedings of the 45th International ACM SIGIR Conference on Research and Development in Information Retrieval, pp. 2565–2571 (2022)
3. Covington, P., Adams, J., Sargin, E.: Deep neural networks for YouTube recommendations. In: Proceedings of the 10th ACM Conference on Recommender Systems, pp. 191–198 (2016)

4. Dong, M., Yuan, F., Yao, L., Xu, X., Zhu, L.: Mamo: memory-augmented meta-optimization for cold-start recommendation. In: Proceedings of the 26th ACM SIGKDD International Conference on Knowledge Discovery and Data Mining, pp. 688–697 (2020)
5. Hao, B., Zhang, J., Yin, H., Li, C., Chen, H.: Pre-training graph neural networks for cold-start users and items representation. In: Proceedings of the 14th ACM International Conference on Web Search and Data Mining, pp. 265–273 (2021)
6. Hu, G., Zhang, Y., Yang, Q.: CoNet: collaborative cross networks for cross-domain recommendation. In: Proceedings of the 27th ACM International Conference on Information and Knowledge Management, pp. 667–676 (2018)
7. Huang, J.T., et al.: Embedding-based retrieval in Facebook search. In: Proceedings of the 26th ACM SIGKDD International Conference on Knowledge Discovery and Data Mining, pp. 2553–2561 (2020)
8. Huang, P.S., He, X., Gao, J., Deng, L., Acero, A., Heck, L.: Learning deep structured semantic models for web search using clickthrough data. In: Proceedings of the 22nd ACM International Conference on Information and Knowledge Management, pp. 2333–2338 (2013)
9. Kang, W.C., McAuley, J.: Self-attentive sequential recommendation. In: 2018 IEEE International Conference on Data Mining (ICDM), pp. 197–206. IEEE (2018)
10. Li, C., et al.: Multi-interest network with dynamic routing for recommendation at Tmall. In: Proceedings of the 28th ACM International Conference on Information and Knowledge Management, pp. 2615–2623 (2019)
11. Linden, G., Smith, B., York, J.: Amazon.com recommendations: item-to-item collaborative filtering. IEEE Internet Comput. **7**(1), 76–80 (2003)
12. Lv, F., et al.: SDM: sequential deep matching model for online large-scale recommender system. In: Proceedings of the 28th ACM International Conference on Information and Knowledge Management, pp. 2635–2643 (2019)
13. Rendle, S.: Factorization machines. In: 2010 IEEE International Conference on Data Mining, pp. 995–1000. IEEE (2010)
14. Wang, P., Fan, Y., Xia, L., Zhao, W.X., Niu, S., Huang, J.: KERL: a knowledge-guided reinforcement learning model for sequential recommendation. In: Proceedings of the 43rd International ACM SIGIR Conference on Research and Development in Information Retrieval, pp. 209–218 (2020)
15. Ying, R., He, R., Chen, K., Eksombatchai, P., Hamilton, W.L., Leskovec, J.: Graph convolutional neural networks for web-scale recommender systems. In: Proceedings of the 24th ACM SIGKDD International Conference on Knowledge Discovery and Data Mining, pp. 974–983 (2018)
16. Zhang, Y., Shi, Z., Zuo, W., Yue, L., Liang, S., Li, X.: Joint personalized Markov chains with social network embedding for cold-start recommendation. Neurocomputing **386**, 208–220 (2020)

Unsupervised Fraud Transaction Detection on Dynamic Attributed Networks

Yangyang Hou, Daixin Wang, Binbin Hu, Ruoyu Zhuang, Zhiqiang Zhang, Jun Zhou[✉], Feng Zhao, Yulin Kang, and Zhanwen Qiao

Ant Group, Hangzhou, China
{yangyanghou.hyy,daixin.wdx,bin.hbb,zhuangruoyu.zry,lingyao.zzq,
jun.zhoujun,zhaofeng.zhf,yulin.kyl,zhanwen.qiaozw}@antgroup.com

Abstract. Fraud transaction detection is a pressing need in industrial applications, aiming to detect the fraud for a transaction involving the buyer and the seller. Due to the prohibitive cost of accessing appropriate labels for the task in a supervised fashion, unsupervised anomaly detection has become an alternative solution. However, previous methods mainly handcraft some features to detect the fraud on a single entity, which neglects the dynamic and topological nature between the buyer and the seller within the transaction. In this paper, we propose a novel Temporal Structure Augmented Gaussian Mixture Model (**TSAGMM**) for unsupervised fraud transaction detection on dynamic attributed networks. Specifically, we propose a time-encoded graph autoencoder to utilize both the topological structure and temporal information within the dynamic transaction graph to reconstruct the node attributes and graph topology. The learned latent representations as well as reconstruction errors are combined and fed into a density-based model for unsupervised fraud detection. Experimental results on the real-world transaction dataset from Alipay show the superiority of our proposed method among the state-of-the-art methods.

Keywords: Unsupervised Anomaly Detection · Graph Neural Networks · Density Estimation

1 Introduction

In recent years, convenience facilitates the explosive growth of e-commerce and the booming of e-payment, while the underlying issue of the fraud transaction is not negligible. Indeed, the health development of online financial service is greatly threatened by various kinds of fraud transactions, ranging from cash-out fraud transaction [9] to malicious default fraud. In order to alleviate the negative impacts (i.e., incalculable risk-related damages and losses) on individuals and enterprises, *fraud transaction detection* has been an increasingly emerging topic in industrial applications, aiming at safeguarding the capital security in the face of fraudulent behaviors.

© The Author(s), under exclusive license to Springer Nature Switzerland AG 2023
X. Wang et al. (Eds.): DASFAA 2023, LNCS 13946, pp. 544–555, 2023.
https://doi.org/10.1007/978-3-031-30678-5_41

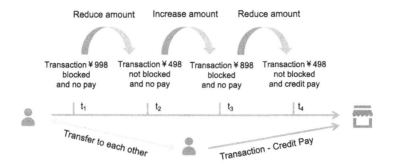

Fig. 1. A real showcase of fraud transactions for aiming at cash out. The two cash-out buyers have created multiple transaction interactions with particular suspicious merchant in a very close period. And the two buyers also have transfer relations before. The two buyers and the seller are in a group aiming to do cash-out.

As the core component of ensuring a healthy environment of online financial services, recent years have witnessed a fruitful line of research in the field of the fraud transaction detection, and attain considerable success [1,3,17,18]. Earlier works mainly focus on the exploration and exploitation of numerous rules summarized by fraud analysts. Unfortunately, the rapid change of fraud patterns is inherently difficult to be fitted with pure rules, hindering the effectiveness. Subsequently, attentions for the fraud transaction detection have been gradually shifted towards machine learning based methods, which could be roughly categorized as tree-based models, deep learning based models or graph-based models [1,10,12,18]. Notwithstanding the competitive performance for automatically uncovering fraud patterns from the data, they still face the following two unresolved limitations :

– **Individual-level detection with static structure.** Previous works mainly perform the transaction fraud detection in the individual level (i.e. a buyer or a seller), which commonly ignore the fact that transactions involve both buyers and sellers. On the other hand, fraud transactions intuitively reveals abnormal patterns in the temporal perspective (e.g., transactions with high frequencies), as shown in Fig. 1. Therefore, the temporal structure associated with a transaction could be a highly discriminative signal for suspicious behaviors.
– **Supervised detection paradigm.** Most of current methods follows the supervised learning paradigm, whose success greatly hinges on large amounts of labeled data. However, in practical scenarios, fraud labels are usually difficult to obtain and insufficient training data also result in a serious data noise problem. Such an inevitable dilemma severely restricts the performance of detection methods with the supervised paradigm.

To address the challenges discussed above, we strive to frame the fraud transaction detection in the setting of unsupervised anomaly detection problem with dynamic attributed graphs. In particular, we propose a Temporal Structure Augmented Gaussian Mixture Model (**TSAGMM** for short) to comprehensively extract the temporal and structural nature of the dynamic transaction graph to detect the transaction-level frauds. In detail, we build TSAGMM upon the general encoder-decoder framework, where a graph neural network encoder with a temporal component explicitly characterizes temporal and topological structure, while a graph reconstruction decoder further aims at the reconstruction of the both topological structure and node attributes. Subsequently, we fuse obtained representations of sellers and buyers, coupled with reconstruction errors into an unified transaction-level representation, and then feed it into a density estimation model for unsupervised fraud detection.

In summary, we highlight the main contributions of this paper as follows:

- **Problem**: As far as we know, we are the first to investigate unsupervised fraud transaction detection in transaction level under dynamic attributed graphs.
- **Model**: We propose TSAGMM, a novel unsupervised graph model to detect the complex fraud patterns based on the temporal and topological transaction behaviors.
- **Evaluation**: Experimental results on a real-world transaction dataset in Alipay prove the effectiveness of our proposed model.

2   Related Work

2.1   Fraud Transaction Detection

In the domain of fraud transaction detection, previous methods usually treat each transaction independently and train supervised models like support vector machines, random forests, etc. [1,18]. Structural information in these methods if applicable, are usually incorporated as hand-crafted features, which are difficult to model subtle interaction information effectively. [12] proposes to use the transaction-intention network to capture the information over transactions and intentions with additional user behaviour sequence data. However, the method is in a supervised fashion to detect some specific patterns of fraud. Furthermore, it performs fraud detection on buyers or sellers, which overlook their coupling effects within the transactions.

2.2   Unsupervised Anomaly Detection

Anomaly detection is one of the common anti-fraud approaches in data science. Tremendous effort has been devoted to unsupervised anomaly detection [3] for tabular data, such as statistical techniques, density-based methods, clustering based methods and so on. Popular used techniques are local outlier factor [2], isolation forest [13], one-class support vector machine [4] etc. Recently, deep learning approaches [16] usually outperform traditional methods for multivariate and high dimensional data. These methods typically can be categorized as a family of encoder-decoder models. The representative is DAGMM [25]. However, all of these methods do not consider the complex interaction patterns for the transaction scenario between the buyers and sellers.

2.3 Graph Based Anomaly Detection

Recent years have seen significant developments in graph neural networks (GNNs) and GNN-based methods are applied to the anomaly detection field [14]. Most of these methods focus on node fraud detection [5,22,24]. Only a few methods focus on edge fraud detection. For example, [6,15,22] focus on the edge fraud detection on static networks. [21,23] are supervised anomaly edge detection on dynamic networks. In our setting, we treat transaction-level fraud detection as an anomalous edge detection problem without any supervision in the dynamic attributed graphs, which is rarely explored before.

3 Preliminary

We first define the dynamic attributed network in the following ways:

Definition 1. *A dynamic attributed network* $\mathbf{G} = (\mathbf{V}, \mathbf{E}, \mathbf{X}, \mathbf{H})$ *consists of: (1) the set of nodes* $\mathbf{V} = \{v_i\}_{i=1}^{N}$ *including the buyers and sellers; (2) the set of edges* $\mathbf{E} = \{e_{ij}\}$ *denoting the relation between node i and node j with the timestamp* $t = t_{ij}$. *Here the relation in our problem contains the transfer relation between buyers and trade relation between the buyer and the seller; (3) the node feature matrix* $\mathbf{X}$ *where the* i^{th} *row vector* $\mathbf{X}_{i.}$ *denotes the attribute information for the* i^{th} *node; and (4) the edge feature matrix* $\mathbf{H}$, *where each element* $\mathbf{H}_{ij}$ *denotes the features of the edge* e_{ij}.

It is worth noting that there may be multiple edges between two nodes in dynamic attributed networks, indicating there are multiple transaction or transfer events occurring between nodes. Then the topological structure of dynamic attributed network $\mathbf{G}$ can be represented by an adjacency matrix $\mathbf{A}$, where $\mathbf{A}_{ij} = k$ if there are k edge events occurring between node v_i and node v_j. Otherwise $\mathbf{A}_{ij} = 0$ if there is no edge between node v_i and v_j.

Then our problem can be defined here:

Definition 2. *Unsupervised fraud transaction detection on dynamic attributed networks: Given the dynamic attributed network* $\mathbf{G}$, *a transaction involving a buyer, denoted as* $\mathcal{B}$, *and a seller denoted as* $\mathcal{S}$, *unsupervised fraud transaction detection aim to predict the fraud score* $s_{\mathcal{B},\mathcal{S}}(\mathbf{G})$ *only based on* $\mathbf{G}$.

4 The Proposed Model

In this section, we introduce our proposed method **TSAGMM** in detail. As shown in Fig. 2, our model consists of three components: (1) The time-encoded graph encoder to model the temporal-structural information within the dynamic attributed graph. (2) The graph reconstruction part to restore the node attributes and graph structure for unsupervised graph learning and (3) The gaussian mixture model to do density-based fraud detection. Since the learning process of graph autoencoders for buyers and sellers are quite similar, we then mainly introduce buyers' as an illustration for space saving.

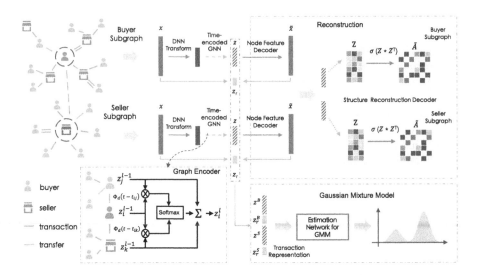

Fig. 2. The overall architecture of the Temporal Structure Augmented Gaussian Mixture Model (TSAGMM) for unsupervised fraud transaction detection.

4.1   Time-encoded Graph Encoder

We propose a time-encoded graph encoder with the attention mechanism to combine the temporal and structural information to learn node representations in a GNN manner. Supposing v_i is the target node at time t, it will aggregate the information from neighboring nodes via the following forms:

$$\mathbf{z}_i^l(t) = \sum_{j \in \mathcal{N}_i(t)} \gamma_{i,j}(t)\mathbf{z}_j^{l-1}(t), \tag{1}$$

where $\mathbf{z}_i^l(t)$ denotes the node embedding at layer l for node v_i, and $\mathbf{z}_i^0 = f(\mathbf{X}_{i.})$, where $f(\cdot)$ denotes a DNN-based model to compress the original node features. $\mathcal{N}_i(t)$ denotes the neighbors of node v_i, whose interaction with v_i takes place at time prior to t. $\gamma_{i,j}(t)$ denotes the attention value to aggregate the representations from neighbor v_j to v_i, whose calculation process will be introduced later.

In particular, our proposed aggregation process not only considers the neighbors' features and edge features as common GNN models do, but more importantly considers the temporal information on each edge. To achieve the idea, we define a functional time encoding technique to represent the time as a combination of several periodic functions [20]:

$$\Phi_d(t) = \sqrt{\frac{1}{d}}[\cos(\omega_1 t), \sin(\omega_1 t), \ldots, \cos(\omega_d t), \sin(\omega_d t)] \tag{2}$$

parameterized by the frequency set $\{\omega_1, \ldots, \omega_d\}$.

Then the attention coefficient between target node i and neighbor node j is computed as follows:

$$\alpha_{i,j}(t) = \text{attn}(\mathbf{q}_i(t), \mathbf{k}_j(t)) = \sigma(\mathbf{a}^T[\mathbf{W}_s\mathbf{q}_i(t) + \mathbf{W}_d\mathbf{k}_j(t)]),$$
$$\mathbf{q}_i(t) = [\mathbf{z}_i^{l-1}(t) \parallel \mathbf{0} \parallel \varPhi_d(0)], \qquad (3)$$
$$\mathbf{k}_j(t) = [\mathbf{z}_j^{l-1}(t) \parallel \mathbf{H}_{ij} \parallel \varPhi_d(t - t_{ij})]$$

where $\parallel$ denotes the concatenate operation, $\sigma(\cdot)$ is the tanh activation function, $\text{attn}(\cdot)$ denotes the attention function parameterized by a, $\mathbf{W}_s$ and $\mathbf{W}_d$. Then, the attention value $\gamma_{i,j}$ can be achieved through the softmax function:

$$\gamma_{i,j}(t) = \frac{\exp(\alpha_{i,j}(t))}{\sum_{k \in \mathcal{N}_i(t)} \exp(\alpha_{i,k}(t))} \qquad (4)$$

In our model, we use a two-layer time-encoded graph encoder to obtain the node embedding $\mathbf{Z} = \{\mathbf{z}_i^2(t_i)\}_{i=1}^N$, which aggregates neighbors' information within two hops.

4.2 Graph Reconstruction

Then we use the learned node embedding $\mathbf{Z}$ to reconstruct both the original node features and the adjacency matrix, aiming to make the model preserve the node attribute information and the graph topology.

In detail, we first use a DNN transform $g(\cdot)$ to project the node embedding $\mathbf{Z}$ into the feature reconstruction space $\tilde{\mathbf{X}} = g(\mathbf{Z}; \theta_g)$. Then given original node feature $\mathbf{X}$ and reconstruction node feature $\tilde{\mathbf{X}}$, we define the feature reconstruction error as: $\boldsymbol{\Delta}_f = [d_e, d_c]$ where d_e and d_c are the Euclidean distance and cosine similarity respectively between $\mathbf{X}$ and $\tilde{\mathbf{X}}$.

Another part aims to reconstruct the original network topology. The reconstruction adjacency matrix is calculated as the inner product between two node embeddings $\tilde{\mathbf{A}} = \sigma(\mathbf{Z}\mathbf{Z}^T)$, where $\sigma(\cdot)$ is the sigmoid function. Then the structure reconstruction error is given by $d_s = \|\bar{\mathbf{A}}_{i\cdot} - \tilde{\mathbf{A}}_{i\cdot}\|_2$, where $\bar{\mathbf{A}}$ is the row-normalization of $\mathbf{A}$.

The overall loss function for graph reconstruction is given by:

$$\mathcal{L}_{recon} = \frac{1}{N} \sum_{i=1}^N \left(d_e\left(\mathbf{X}_{i\cdot}, \tilde{\mathbf{X}}_{i\cdot}\right) + d_c\left(\mathbf{X}_{i\cdot}, \tilde{\mathbf{X}}_{i\cdot}\right) + d_s\left(\bar{\mathbf{A}}_{i\cdot}, \tilde{\mathbf{A}}_{i\cdot}\right) \right) \qquad (5)$$

4.3 Density Estimation

Fraud transactions often consist of abnormal information regarding its buyer and the seller. Therefore, to detect fraud transaction, we obtain the transaction representation by combining the obtained node embedding and reconstruction errors from both the buyer and the seller: $\mathbf{z}^T = [\mathbf{z}^B, \boldsymbol{\Delta}_f^B, \mathbf{z}^S, \boldsymbol{\Delta}_f^S]$. It is worth noting that feature reconstruction errors for the buyer and seller can characterize their own anomaly scores, because a good reconstruction-based model will focus on reconstructing the normal patterns, resulting in a larger reconstruction errors for anomaly data. Therefore, we also combine the feature reconstruction errors with the node embedding for the buyer and seller for fraud detection.

Without labels, we consider using the density-based methods for fraud detection. We assume that normal transactions can be modelled by a mixture of gaussian models, while the fraud transactions will be far-away from the combination of these gaussian distributions. Based on the assumption, we first predict its soft mixture-component membership prediction given the number of mixture components K: $\hat{\gamma} = \text{softmax}(h(\mathbf{z}^T; \theta_h))$, where $h(\cdot)$ is a multi-layer neural network parameterized by θ_h. By traversing all the samples, we can estimate the mixture probability $\hat{\phi}_k$, the mean value $\hat{\mu}_k$, the covariance matrix $\hat{\Sigma}_k$ for each component k in GMM respectively [25]. The sample energy can be inferred by:

$$E(\mathbf{z}^T) = -\log\Big(\sum_{k=1}^{K} \hat{\phi}_k \frac{\exp(-\frac{1}{2}(\mathbf{z}^T - \hat{\mu}_k)^T \hat{\Sigma}_k^{-1}(\mathbf{z}^T - \hat{\mu}_k))}{\sqrt{|2\pi\hat{\Sigma}_k|}} \Big) \qquad (6)$$

4.4   Model Learning and Fraud Detection

To jointly learn the reconstruction errors as well as GMM estimation, given a batch of N transaction data, the training objective function of our proposed model can be formulated as:

$$\mathcal{L} = (\mathcal{L}_{recon}^{\mathcal{B}} + \mathcal{L}_{recon}^{\mathcal{S}}) + \frac{\lambda_1}{N} \sum_{i=1}^{N} E(\mathbf{z}_i^T) + \lambda_2 P(\hat{\Sigma}) \qquad (7)$$

This objective function includes three components: (1) the first two terms denote the graph reconstruction errors for the buyer and seller. (2) $E(\mathbf{z}_i^T)$ is the energy defined in Eq. (6). It describes how possible we could see the transaction samples in the whole training dataset. (3) To avoid trivial solutions when the diagonal entries of covariance matrices degenerate to 0, we penalize small values of the diagonals by the fourth component $P(\hat{\Sigma}) = \sum_{k=1}^{K} \sum_j \frac{1}{\hat{\Sigma}_{kjj}}$ as a regularizer.

In the prediction phase, the sample energy is then employed to assess the abnormality of the transaction data.

5   Experiments

To demonstrate the effectiveness of the proposed method for fraud transaction detection, we conduct comprehensive experiments and present the result.

5.1   Experiments Setup

Evaluation Dataset. With the real-world transaction datasets from Alipay, we sample about 30 million transaction data completed by credit pay in one month for training, and the next month for evaluation. The dynamic graph which contains fund transfer relations between buyers and transaction relations (including credit pays and non-credit pays) between buyers and sellers, has around 70 million interactions per day. For each transaction, we sample two-hops dynamic attributed subgraphs centered by both the buyer and the seller respectively, within 7 days before the creation time of the transaction. We extract 43 features for each user, including user profile, credit history, platform behaviors.[1]

Evaluation Metrics. Since there is no direct labels for the transactions, we collect the evaluation labels from business expert and from buyer and sellers' behaviors a few months later. In all, about 0.6% of the transactions are fraudulent for the evaluation. We select Precision, Recall, F1 score and the LIFT@k% as the evaluation metric. The LIFT@k% measure means the ratio between the bad rate of the top k% transactions with the average bad rate of all transactions. We use this metric because fraud detection usually focuses more on the results ranking ahead.

Comparing Methods. To show the performance, we compare our proposed model **TSAGMM** with three lines of unsupervised methods: popular traditional methods on anomaly detection, NN-based methods and GNN-based methods. The first line contains **LOF** (Local Outlier Factor [2]), **OC-SVM** (One-class support vector machine [4]) and **iForest** (Isolation Forest [13]). NN-based methods contains reconstruction-based method **DAE** [19] and density-based method **DAGMM** [25]. Note that we choose these five methods as representatives because according to [8], these five methods perform good and stable. To make fair comparisons, we extract topology features like degree and the average of neighbors' features and combine them with the basic node features as the node features for these five methods. For graph-based baseline methods, since there is no existing unsupervised graph-based methods for edge-level fraud detection, we combine Graph Autoencoder [7,11] and our temporal encoder to form **TGAE**.

Since LOF and OC-SVM don't scale well, we sample 1% from training dataset for these two methods. We set n_neighbors=10 for LOF and use RBF kernel in OC-SVM. For iForest, we use 100 trees to train. For all deep autoencoding instances, the embedding dimension is set as 8 with three hidden layers (32-unit, 16-unit and 8-unit, respectively). Moreover, we set the number of GMM components is 3 and set λ_1 as 0.1 and λ_2 as 0.001 as they render desirable results. In addition, two time-encoded graph attention layers are employed for our proposed models. We use Adam algorithm to optimize the loss function with the learning rate 0.001 and set batch size to 1024.

---

[1] Please note that the data set does not contain any Personal Identifiable Information (PII). The data set is desensitized and encrypted. Adequate data protection was carried out during the experiment to prevent the risk of data copy leakage, and the data set was destroyed after the experiment. The data set is only used for academic research, it does not represent any real business situation.

Table 1. The performance in the evaluation dataset of credit transactions (best F1 score with corresponding Precision and Recall, as well as LIFT metrics). We also report the relative improvement ratio of our proposed model over all baselines.

Methods	Precision	Recall	F1 score	Lift on top1%	Lift on top2%
LOF	1.26%	6.17%	2.09% (+276%)	2.49 (+318%)	2.13 (+267%)
OC-SVM	2.59%	8.45%	3.96% (+99%)	4.72 (+121%)	4.23 (+85%)
iForest	3.65%	12.02%	5.60% (+40%)	6.80 (+53%)	6.02 (+30%)
DAE	1.02%	11.71%	1.87% (+321%)	1.68 (+520%)	1.70 (+360%)
DAGMM	4.12%	**13.58%**	6.32% (+24%)	7.59 (+37%)	6.75 (+16%)
TGAE	2.70%	8.91%	4.15% (+90%)	5.28 (+96%)	4.45 (+73%)
TSAGMM	**6.31%**	10.41%	**7.86%**	**10.41**	**7.82**

5.2 Offline Results

We present the results for each method in Table 1. We report the best F1 score with corresponding Precision and Recall, as well as the top 1% and 2% LIFT metric. The major results are summarized: It is worth noting that because of high class balance, the absolute values of precision, recall and F1 are not very high. But the F1 score of our proposed method still outperforms the comparing methods by at least 24%, which demonstrates the superiority of the overall performance of our method on fraud detection. Furthermore, we further find that our method achieves at least 35% and 15% improvement on Lift@1% and Lift@2% compared with all baseline methods. It is very important for real-world scenarios because fraud detection usually cares more about the entities ranking very ahead. In addition, we note that although TGAE utilizes the graph data, it performs not good. It demonstrates that designing a sophisticated component for unsupervised fraud detection is very important. In summary, the performance improvement demonstrates our model is able to capturing temporal and structural information effectively for fraud transaction detection.

5.3 Ablation Studies

We conduct the ablation studies to explore the impact of the main components of TSAGMM. Comparing methods are shown as follows:

Table 2. Ablation studies in the evaluation dataset of credit transactions.

Methods	Precision	Recall	F_1	Lift on top1%	Lift on top2%
TSAGMM	**6.312%**	**10.410%**	**7.859%**	**10.41**	**7.82**
TSAGMM_sd	5.733%	9.457%	7.139 %	9.46	6.88
TSAGMM_te	5.610 %	9.254%	6.985%	9.25	7.23
TGAE	2.701%	8.910 %	4.145%	5.28	4.45

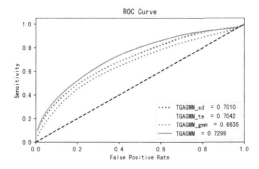

Fig. 3. ROC curves and AUC scores for ablation studies.

- **TSAGMM_sd** removes the graph structure decoder reconstruction module.
- **TSAGMM_te** removes the time encoder part in the encoder function.

We report the results of ablation studies in Table 2 and Fig. 3. Specifically, the performance of TSAGMM_sd and TSAGMM_te drops because the effects of the structural and temporal information are not fully exploited. For TGAE, the result shows the density estimation part is critical for fraud detection. Overall, we can clearly observe that full TSAGMM achieves the best result.

5.4   Performance in Online Environment

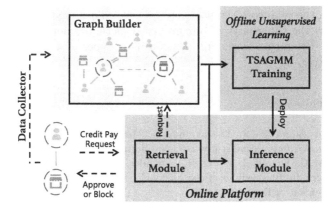

Fig. 4. The deployment of TSAGMM for the online credit transaction service in Alipay.

We deploy the proposed TSAGMM into the online environment of Alipay and report the online results. As shown in Fig. 4, TSAGMM is trained offline every month and then deployed into online environment for serving. Then once a buyer issue a new credit pay request with a seller in the online serving, the retrieval module are employed to extract node features and subgraphs associated with the request efficiently. Then TSAGMM predicts the anomaly score for the request. Together with other online strategies, the system would return the final decision, i.e. approve or block, to the transaction for the credit pay request.

Since we block high-risky transactions in the online environment, we use the repay rate, RP rate for short, as an important online metric. Once the system denies the credit pay request, risky buyers usually close the transactions, while normal buyers would continue to complete the transactions by other means of payments, e.g. debit cards. Therefore, the lower RP rate is, the more risky the transactions are. The average RP rate of online baseline strategy is 27.3%, while our proposed TSAGMM can detect additional highly risky transactions with RP rate 7.8%. Moreover, we randomly extract 100 samples from top ranked transactions and have a check by business experts. The results are as follows: 77% are abnormal (aiming to cash out or fraudulent), 10% are suspicious and 13% are misjudgements. Figure 1 shows a real abnormal transaction example detected by TSAGMM model for cashing out.

6   Conclusion

In this paper, we propose a novel time-encoded graph autoencoding gaussian mixture model for unsupervised fraud transaction detection on dynamic attributed networks. Specifically, we propose a time-encoded graph autoencoder to model the topological structure and temporal information within the dynamic transaction graph. The learned node representations and reconstruction errors are combined for density estimation to perform the fraud detection. Experimental results on the real-world transaction dataset from a credit payment service institute show the superiority of our proposed method among the state-of-the-art methods. The future work will focus on more types of nodes like devices to link users better.

References

1. Bhattacharyya, S., Jha, S., Tharakunnel, K., Westland, J.C.: Data mining for credit card fraud: A comparative study. Decision Support Systems (2011)
2. Breunig, M.M., Kriegel, H.P., Ng, R.T., Sander, J.: LOF: identifying density-based local outliers. In: Proceedings of the 2000 ACM SIGMOD International Conference on Management of Data (2000)
3. Chandola, V., Banerjee, A., Kumar, V.: Anomaly detection: A survey. ACM Computing Surveys (2009)
4. Chen, Y., Zhou, X.S., Huang, T.: One-class SVM for learning in image retrieval. In: Proceedings 2001 International Conference on Image Processing (2001)
5. Ding, K., Zhou, Q., Tong, H., Liu, H.: Few-shot network anomaly detection via cross-network meta-learning. In: Proceedings of the Web Conference (2021)
6. Duan, D., Tong, L., Li, Y., Lu, J., Shi, L., Zhang, C.: Aane: anomaly aware network embedding for anomalous link detection. In: 2020 IEEE International Conference on Data Mining (ICDM), pp. 1002–1007. IEEE (2020)

7. Fan, H., Zhang, F., Li, Z.: Anomalydae: dual autoencoder for anomaly detection on attributed networks. In: ICASSP 2020–2020 IEEE International Conference on Acoustics, Speech and Signal Processing (ICASSP), pp. 5685–5689. IEEE (2020)
8. Han, S., Hu, X., Huang, H., Jiang, M., Zhao, Y.: ADBench: anomaly detection benchmark. arXiv preprint arXiv:2206.09426 (2022)
9. Hu, B., Zhang, Z., Shi, C., Zhou, J., Li, X., Qi, Y.: Cash-out user detection based on attributed heterogeneous information network with a hierarchical attention mechanism. In: Proceedings of the 33rd AAAI Conference on Artificial Intelligence (2019)
10. Huang, D., Mu, D., Yang, L., Cai, X.: Codetect: financial fraud detection with anomaly feature detection. IEEE Access (2018)
11. Kipf, T.N., Welling, M.: Variational graph auto-encoders. arXiv preprint arXiv:1611.07308 (2016)
12. Liu, C., Sun, L., Ao, X., Feng, J., He, Q., Yang, H.: Intention-aware heterogeneous graph attention networks for fraud transactions detection. In: Proceedings of the 27th ACM SIGKDD Conference (2021)
13. Liu, F.T., Ting, K.M., Zhou, Z.H.: Isolation forest. In: 2008 Eighth IEEE International Conference on Data Mining (2008)
14. Ma, X., et al.: A comprehensive survey on graph anomaly detection with deep learning. IEEE Trans. Knowl. Data Eng. (2021)
15. Ouyang, L., Zhang, Y., Wang, Y.: Unified graph embedding-based anomalous edge detection. In: 2020 International Joint Conference on Neural Networks (2020)
16. Pang, G., Shen, C., Cao, L., Hengel, A.V.D.: Deep learning for anomaly detection: a review. ACM Comput. Surv. 54, 1–38 (2021)
17. Phua, C., Lee, V., Smith, K., Gayler, R.: A comprehensive survey of data mining-based fraud detection research. arXiv preprint arXiv:1009.6119 (2010)
18. Prusti, D., Rath, S.K.: Fraudulent transaction detection in credit card by applying ensemble machine learning techniques. In: 2019 10th International Conference on Computing, Communication and Networking Technologies (2019)
19. Sakurada, M., Yairi, T.: Anomaly detection using autoencoders with nonlinear dimensionality reduction. In: Proceedings of the MLSDA 2014 2nd Workshop on Machine Learning for Sensory Data Analysis (2014)
20. Xu, D., Ruan, C., Korpeoglu, E., Kumar, S., Achan, K.: Inductive representation learning on temporal graphs. In: International Conference on Learning Representations (ICLR) (2020)
21. Zhang, G., Li, Z., Huang, J., Wu, J., Zhou, C., Yang, J., Gao, J.: efraudcom: an e-commerce fraud detection system via competitive graph neural networks. ACM Trans. Inf. Syst. (TOIS) 40, 1–23 (2022)
22. Zhang, G., et al.: Fraudre: fraud detection dual-resistant to graph inconsistency and imbalance. In: 2021 IEEE International Conference on Data Mining (ICDM) (2021)
23. Zheng, L., Li, Z., Li, J., Li, Z., Gao, J.: Addgraph: anomaly detection in dynamic graph using attention-based temporal GCN. In: Proceedings of the Twenty-Eighth International Joint Conference on Artificial Intelligence (2019)
24. Zheng, P., Yuan, S., Wu, X., Li, J., Lu, A.: One-class adversarial nets for fraud detection. In: Proceedings of the AAAI Conference on Artificial Intelligence (2019)
25. Zong, B., et al.: Deep autoencoding gaussian mixture model for unsupervised anomaly detection. In: International Conference on Learning Representations (2018)

Personalized Dissatisfied Users Prediction in Mobile Communication Service

Yunong Chen[1], Yuying Lin[1], Bojian Zhang[1], Dongming Zhao[2], Haiwei Zhang[1], and Yanlong Wen[1(✉)]

[1] College of Computer Science, Nankai University, Tianjin, China
{chenyunong,zhangbojian}@dbis.nankai.edu.cn, 2012174@mail.nankai.edu.cn,
{zhhaiwei,wenyl}@nankai.edu.cn
[2] Artificial Intelligence Laboratory, China Mobile Communication Group Tianjin Co., Ltd., Tianjin 300020, China
waitman_840602@163.com

Abstract. In the mobile communication industry, users are always concerned with the services provided by the operators. Users who are dissatisfied with the services will probably change their mobile network operators. Therefore, mobile network operators desire to predict whether users will be dissatisfied with the services by analyzing users' events. Then they stand a chance to timely remedy the services for potential dissatisfied users. Though many existing classification methods are available, they cannot leverage user attribute information well in this task. To address the problem, we propose a Personalized Attention-based Long Short-Term Memory (PA-LSTM) model, consisting of events feature extraction module, user feature extraction module, and personalized prediction module. PA-LSTM makes personalized predictions of dissatisfied users based on both user events and user attributes. Furthermore, PA-LSTM considers the satisfaction tendency of different user groups. Extensive experiments on the industry dataset show that our model performs better than other solutions, verifying the effectiveness of our model.

Keywords: Mobile service · Event analysis · Dissatisfied user prediction

1 Introduction

With the rapid development of mobile communication technology, people can contact each other and access the Internet conveniently with their mobile terminals. Further, mobile service plays an essential role in everyone's life with the widespread popularity of mobile payment service. Under the circumstances, mobile network operators are required to provide high-quality mobile services for massive users. A lot of mobile service event data are logged during the service process, such as monthly bill queries, package information queries, and recharge payments. Mobile service events analysis enables operators to better understand user behavior and distinguish users who are potentially dissatisfied

X. Wang et al. (Eds.): DASFAA 2023, LNCS 13946, pp. 556–567, 2023.
https://doi.org/10.1007/978-3-031-30678-5_42

with their services. Through timely intervention (e.g., recommending packages and preferential activities) to potentially dissatisfied users, the possibility of user complaints or even changing the mobile network operator could be effectively reduced, leading to a more stable user group and greater revenue for mobile network operators. Therefore, the dissatisfied users prediction is a significant task for mobile network operators.

Predicting dissatisfied users is essentially a classification task, which is to classify users into two categories: satisfied users and dissatisfied users. It is similar to the time series classification task [7] since the goal is achieved by analyzing the users' event sequences with timestamps. The current work of time series classification generally uses recurrent neural networks (RNNs) [11] or convolutional neural networks (CNNs) [10] to learn the sequential information of data. Dempster et al. [6] proposed to apply random convolution kernels to extract the features of the original time series. However, the user event in our data is not a continuous variable. The traditional dynamic time warping and kernel methods cannot be applied, and the traditional RNN or CNN should be modified to suit the specific task.

There are also some works on the representation of event sequences. Babaev et al. [3] applied gate recurrent unit (GRU) on user consumption events to credit scoring of retail customers. Feng et al. [8] designed a context-smoothing convolution to represent the learning activities of massive open online course (MOOC) users. These methods are mainly designed for the fields with large-scale long sequential data, such as finance [3], transportation, recommendation, and MOOC [8]. However, mobile users are not likely to interact with mobile network operators frequently. Events generally happen due to the users' specific requirements (e.g., business handling, package query). The event sequence of a user is always weakly sequential and sparse. Additionally, the user attributes are special supplementary data that should be considered. Therefore, existing event modeling methods are not suitable for the dataset in our task.

With weakly sequential and sparse events data in mobile service, existing methods can hardly identify the satisfaction of users. To address that problem, we propose personalized attention-based long short-term memory (PA-LSTM). PA-LSTM has three components: events feature extraction module, user feature extraction module, and personalized prediction module. In events feature extraction, we apply LSTM to extract the features of user event sequences. In user feature extraction, we implement a temporal convolutional network (TCN) on user attributes (e.g. service time, talk time, mobile data usage) to extract user features. Finally, for personalized prediction, we apply the attention mechanism to analyze the important events for specific user, and we further involve the dissatisfaction factor considering the satisfaction tendency of user's group. To sum up, our main contributions are as follows:

1. We notice that the user events data in the field of mobile communication service is weakly sequential and sparse. With the special events data, we propose PA-LSTM model to address the problem of dissatisfied users prediction.

2. We innovatively leverage user attribute information for personalized prediction. We introduce the attention mechanism to guide the representation of the user event sequence. Furthermore, we predict dissatisfied users considering the satisfaction tendency of users' groups.

3. Extensive experiments on the real-world dataset from China Mobile Communications Corporation (CMCC) demonstrate the effectiveness of the proposed PA-LSTM, which provides a reliable reference for the timely intervention of potentially dissatisfied users.

2 Related Work

Classical Supervised Classification. Supervised classification is one of the basic tasks of machine learning with various classical methods, including generative models (e.g., K-Nearest Neighbor, Naive Bayes, and Maximum Entropy Classifier), discriminative models (e.g., Support Vector Machine, Logistic Regression, and Decision Tree) and ensemble models (e.g., Boosting, Bagging and Random Forests). However, traditional supervised classification models lack representation ability for data with complex structures, such as sequential data.

Time Series Classification. Time series classification (TSC) algorithms are often used to classify a new time series, given a set of time series with class labels. Traditional TSC methods include KNN with dynamic time warping, TimeSeries-Forest and Shapelet Transform Classifier, etc. [12]. Deep learning methods have been developed over the years. Wang [16] first proposed Fully Convolutional Networks (FCNs) and later proposed ResNet. Serr à et al. proposed the Encoder [13] based on FCN, replacing the GAP layer with the attention layer. LSTM-FCN [11] added LSTM block parallel with FCN. Dempster et al. [6] innovatively applied a series random convolution kernel to covert feature spaces. TSC algorithms are often designed for long-time and strongly time-sequential inputs such as signal sequences. However, in the dissatisfied user prediction task, the value of the event sequence is independent and discrete. The length of the event sequence can be extremely varied. Therefore, proposing a new method for such weakly-sequential data is necessary.

Event Sequence Representation. The methods of event sequence representation encode the input sequence composed of various events and tags to a dense vector used for downstream tasks. In the neural point process, the representation of event information is based on RNNs and their variants. Many works [5,15,18] applied the attention mechanism to RNN to improve the interpretability of the model. Arjovsky et al. [1] proposed WGAN based on generative adversarial network (GAN) to achieve better performance. Wu et al. [17] applied reinforcement learning to cluster the sequences with different temporal patterns. Babaev et al. [2] proposed CoLES, a self-supervised learning method with contrastive learning for event sequences. However, in the task of dissatisfied user prediction with various data structures, apart from event sequences, additional specific features should also be considered, such as user attributes and the tendency of user groups. As a result, we propose a new method for the particular task.

3 Problem Definition

In this section, we define the terms in the mobile services domain. The data collected in the real world will be described for later use. Finally, we will give a formal definition of the dissatisfied user prediction task.

Definition 1. User Event. A user $u \in \mathbf{U}$ may interact with the mobile network operator when accepting mobile services. We denote a single interaction (monthly bill query, package processing, etc.) as a user event $e_u \in \mathbf{E}$ with timestamp t_{e_u}. Each user event e_u belongs to a certain event category $c_u^e \in \mathbf{C}$, such as complaint event and broadband event. There are $|\mathbf{C}|$ event categories in total.

Definition 2. User Event Sequence. The events e_u of a single user u are temporally ordered by the timestamp t_{e_u}, constituting the user event sequence $S_u = [e_u^1, e_u^2, e_u^3, \cdots, e_u^{|S_u|}]$. The length of the user event sequence $|S_u|$ is determined by the total number of user events of u within the time period.

Definition 3. User Attribute. In addition to the explicit interaction events between the user and the mobile network operator, the user's attribute information a_u^t will also be recorded. The user attribute a_u^t is the monthly statistical information of the user, such as the number of calls, the data usage, the monthly package fee, etc.

Definition 4. User Attribute Sequence. Similar to the user event sequence, the user attribute sequence is composed of temporally ordered attributes. We denote it as $A_u = [a_u^1, e_u^2, e_u^3, \cdots, a_u^{|A_u|}]$, where $|A_u|$ is equal to the number of months in the time period.

Definition 5. User Satisfaction. The mobile network operator regularly conducts telephone surveys on users' satisfaction with the service. The surveyed users give a rating on a scale from 0 to 10, from lowest to the highest satisfaction. A user u with a rating less than 6 is considered a dissatisfied user, with label $L_u = 1$; otherwise, the user is considered a satisfied user, with label $L_u = 0$. Dissatisfied users are more likely to complain about the mobile service and with greater possibilities to change their mobile network operators.

Definition 6. Dissatisfied User Prediction Task. For a user u with user event sequence S_u and user attribute sequence A_u, the goal of the dissatisfied user prediction task is to predict whether the user is a dissatisfied user or not, namely to predict the user's label L_u.

4   Proposed Model

The framework of PA-LSTM is presented in Fig. 1. As shown in the figure, PA-LSTM comprises three parts: events feature extraction, user feature extraction and personalized prediction. We'll introduce the details of these three parts in the following.

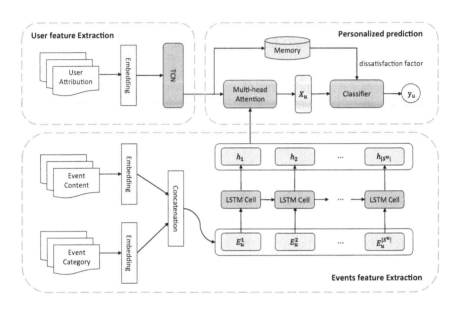

Fig. 1. The framework of PA-LSTM

4.1   Events Feature Extraction

Each event contains three important types of information: event category, event content, and timestamp. The event category involves the coarse-grained semantic information of the event, while the event content involves the details and fine-grained semantics of the event. And the timestamp determines the event position in the sequence. Given the user event sequence $S_u = [e_u^1, e_u^2, e_u^3, \cdots, e_u^{|S_u|}]$, we get corresponding event category sequence $C_u = [c_u^1, c_u^2, c_u^3, \cdots, c_u^{|S_u|}]$. Because both sequences are helpful, we concatenate the embedding of sequential events with their corresponding categories to represent the final user sequence E_u:

$$E_u = [e_u^1 \oplus c_u^1, e_u^2 \oplus c_u^2, e_u^3 \oplus c_u^3, \cdots, e_u^{|S_u|} \oplus c_u^{|S_u|}] \tag{1}$$

Recurrent Neural Networks (RNN) are suitable for modeling sequence data. The Long Short-term Memory network (LSTM) [9] addressed the problem of gradient vanishing and gradient explosion in traditional RNN, achieving outstanding performance in many sequence modeling tasks. So we apply LSTM to represent the user sequence E_u of each time step. For time step t, LSTM calculates the hidden vector h and memory vector m as follows:

$$
\begin{aligned}
g^u &= \sigma \left(W^u h_{t-1} + I^u (e_u^t \oplus c_u^t) \right) \\
g^f &= \sigma \left(W^f h_{t-1} + I^f (e_u^t \oplus c_u^t) \right) \\
g^o &= \sigma \left(W^o h_{t-1} + I^o (e_u^t \oplus c_u^t) \right) \\
g^c &= \tanh \left(W^c h_{t-1} + I^c (e_u^t \oplus c_u^t) \right) \\
m_t &= g^f \odot m_{t-1} + g^u \odot g^c \\
h_t &= \tanh \left(g^o \odot m_t \right)
\end{aligned}
\tag{2}
$$

where σ is the sigmoid activation function, $\odot$ denotes element-wise multiplication, W^u, W^f, W^o, W^c are weight matrices, I^u, I^f, I^o, I^c are projection matrices. We regard the hidden vector h_t as the sequence representation in time step t.

4.2   User Feature Extraction

In addition to the explicit interaction events between users and the mobile network operator, the monthly user attribute is also helpful. As shown in table 1, most items of user attributes are numerical. The user attributes of different months are relatively time-independent, without significant change. Therefore, we use CNN to extract user features in attribute sequences. However, the attribute for a month is only affected by the attributes for the previous month. The traditional CNN involves the information of future months, leading to the information "leakage" [4]. So we adopt the Temporal Convolutional Network [4] instead of traditional CNN. Given the user attribute sequence $A_u = [a_u^1, a_u^2, a_u^3, \cdots, a_u^{|A_u|}]$, the convolution at a_u^t is:

$$
(F * A_u)_{(a_u^t)} = \sum_{k=1}^{K} f_k a_u^{t-K+k}
\tag{3}
$$

where $F = (f_1, f_2, \ldots, f_K)$ contains K filters $f_k \in \mathbb{R}$. Since the user attribute sequence length is relatively short, we abort the dilated convolution strategy in traditional TCN. After the convolution with Eq. 3 for each attribute in A_u, we obtain the final user feature representation E_{A_u}.

4.3   Personalized Prediction

Personalized Attention. Different users may be concerned about the different parts of the event sequence. For example, users using more mobile data may focus on network events, and those with longer talk time are more concerned about call events. Therefore, we use the attention mechanism [14] to represent a personalized sequence. Specifically, we consider the user feature E_{A_u} as the query. The hidden vectors of LSTM at each time step $H_u = [h_1, h_2, h_3, \cdots, h_{|S_u|}]$ are

Table 1. The attributes of a user

Recorded month	Cumulative service time	Monthly talk time	Monthly mobile data usage	Monthly fee	...
4	(1, 2] years	(10,50] minutes	(20,50] G	(50,80] yuan	
5	(1, 2] years	(10,50] minutes	(20,50] G	(50,80] yuan	
6	(1, 2] years	(50,100] minutes	(20,50] G	(50,80] yuan	...

considered as keys and values. Then the personalized representation of the user event sequence is calculated as follows:

$$
\begin{aligned}
Q_u &= W_q E_{A_u} \\
K_u &= W_k H_u \\
V_u &= W_v H_u \\
Z_u &= \mathrm{softmax}\left(\frac{Q_u \times K_u^{\mathrm{T}}}{\sqrt{d_k}}\right) V_u
\end{aligned}
\tag{4}
$$

where W_q, W_k, W_v are weight matrices, d_k denotes the dimension of K_u, softmax($\cdot$) is applied to normalize the attention weight. To further enhance the ability of representation, we apply the multi-head attention mechanism [14], which has multiple groups of weight matrices W_q, W_k, W_v, mentioned in the Eq. 4. Each group of weight matrices is randomly initialized, projecting the input to different representation subspaces. Finally, the representation of the user event sequence with attention weight is obtained:

$$
X_u = \mathrm{Concat}\left(Z_u^1, Z_u^2, \ldots, Z_u^l\right) W_z
\tag{5}
$$

where W_z is the weight matrix, l is the number of attention heads.

Dissatisfaction Factor. The users can be divided into different groups based on their attributes. The users in the same group with similar attributes always show similar behavior patterns and satisfaction tendencies. Therefore, we calculate the dissatisfaction factor for each user according to the labels of similar users. Specifically, we set memory M to store the user feature representation E_{A_u} learned in the training stage. For a user u in the test stage, we select K users from M, whose user feature representation has nearest Euclidean distance with E_{A_u}:

$$
U_K = \mathrm{K\text{-}nearest}(E_{A_u}, M)
\tag{6}
$$

Then we count the proportion of dissatisfied users in U_K, denoted as p. We define the dissatisfaction factor of the user as follows:

$$
r_u = 1 + (p - \alpha)
\tag{7}
$$

where α is the total proportion of dissatisfied users counted during the training stage. The dissatisfaction factor reflects the overall satisfaction tendency for the user's group.

Model Prediction. In the model prediction phase, we apply a linear classifier taking the user sequence representation X_u as the input to get the score of user dissatisfaction:

$$O_u = \text{sigmoid}\,(W_x X_u + b_x) \tag{8}$$

where W_x is the weight matrix, b_x is the offset, sigmoid($\cdot$) is the activation function. Then, we get the final user dissatisfaction score by integrating user dissatisfaction factor r_u:

$$O'_u = r_u O_u \tag{9}$$

Finally, we classify the users according to their dissatisfaction scores:

$$y_u = \begin{cases} 1, & \text{if } O'_u \geq 0.5 \\ 0, & \text{if } O'_u < 0.5 \end{cases} \tag{10}$$

5   Experiments

5.1   Dataset

The dataset for dissatisfied user prediction is collected from the real-world services of CMCC. The dataset contains 1.4 million events from April to September 2022, generated by 130 thousand users. The events can be divided into 7 categories. In the dataset, 22 thousand users are dissatisfied (positive samples), and 108 thousand users are satisfied (negative samples). Besides, user attributes are recorded each month. The length of user attributes is equal to 6 since there are 6 months recorded. We randomly select 80% of users as the training set and the remaining users as the test set.

5.2   Experiment Setup

Baseline Methods. According to Sects. 1 and 2, the dissatisfied user prediction is a classification task with time series events. Hence, we choose some traditional classification methods and time series classification methods as baselines:

Traditional classification method

- Decision Tree: The decision tree adopts the greedy search strategy forming a tree structure to distinguish different samples by top-down regression.
- Random Forest: A random forest is a classifier containing multiple decision trees.
- SVM: Support vector machine is a binary linear margin maximization classifier.

Deep learning method

- LSTM: Long Short-term Memory network [9] is an RNN model that proved effective for modeling time serial data.
- TCN: Temporal Convolutional Network [4] is a designed convolutional network for time series data.
- LSTM-FCN:LSTM fully convolutional network [11] combines LSTM and TCN for time series classification tasks.
- Self Attention: The method utilizes the self-attention mechanism [14] to represent event sequences.
- ALSTM: Attention-Based Long Short-Term Memory [18] attaches an attention layer to the hidden layer of LSTM.

Evaluation Metrics. Accuracy is a common metric for classification problems. However, accuracy is insufficient for the unbalanced CMCC dataset, where the number of satisfied users is far more than that of the dissatisfied user. Therefore, we use precision, recall, and F1-score of dissatisfied users to evaluate model performance. In practice, the precision is related to the additional labor costs of mobile network operators for dissatisfaction user intervention, and the recall is related to the range of potential dissatisfied users the intervention involves.

Implementation Details. For the proposed PA-LSTM, we train the model for 10 epochs with the batch size set to 128. The learning rate of the optimizer Adam is set to 0.001. The dropout is set to 0.5 to prevent overfitting. In the events feature extraction module, we apply a two-layer LSTM, and the dimension of the hidden layer is set to 150. In the user attribute extraction module, a 4-layer TCN is used. The convolution kernel size is set to 3, and the dilatation rate is set to 1. In the personalized prediction module, we employ 8-head attention, and the dimensions of Q, K and V matrices are all equal to 300. For LSTM, TCN or Attention applied in other methods, we use the same settings as PA-LSTM for a fair comparison.

5.3   Performance Comparison

The performances of PA-LSTM and other baseline methods are shown in table 2. It can be concluded that deep learning methods are significantly more efficient than traditional classification methods. PA-LSTM achieves a 3.6% absolute gain on Fl-score than the best baseline ALSTM. Moreover, PA-LSTM highly surpasses other methods on recall metric. Recall indicates the range of potential dissatisfied users the model is able to detect, which is important in practice. LSTM and TCN have comparable performance in the task, both of which can extract the sequential features of the events. However, due to the lack of extracting user attributes, they are inferior to PA-LSTM in the results. It's worth noting that simply using self-attention gets poor performance, probably because of weak time dependency between events. Due to the data imbalance, the gap between ACCs of different methods is relatively small, but PA-LSTM also performs best on ACC.

Table 2. Performance comparison among different methods

Method	ACC	Precision	Recall	F1
Decision Tree	0.921	0.773	0.746	0.759
Random Forest	0.932	0.822	0.754	0.786
SVM	0.934	0.833	0.753	0.791
LSTM	0.941	0.796	0.864	0.829
TCN	0.939	0.793	0.855	0.823
LSTM-FCN	0.939	0.781	0.878	0.827
Self-Attention	0.937	0.792	0.839	0.815
ALSTM	0.942	0.793	0.878	0.833
PA-LSTM	**0.953**	**0.835**	**0.905**	**0.869**

5.4 Ablation Study

To analyze the effectiveness of different modules in PA-LSTM, we conduct ablation experiments on TCN, Memory, and Multi-head Attention. The PA-LSTM methods without these modules are denoted as follows:

- w/o TCN: PA-LSTM without TCN, where the query in multi-head attention is set to the user attribute of the last month.
- w/o Memory: PA-LSTM without Memory, where the final score of classification is generated without dissatisfaction factor.
- w/o Attention: PA-LSTM without Multi-head Attention, where the hidden vector of LSTM in the last time step is directly regarded as the event sequence representation.
- w/o Memory and Attention: PA-LSTM without Memory and Multi-head Attention modules, where PA-LSTM degenerates to LSTM

The results showed in table 3 verify that all TCN, Memory, and Multi-head Attention can improve the model's performance. The TCN module's contribution is relatively small because the user attributes always remain stable in different months. The Memory module greatly improves the recall of the model. The Multi-head Attention with user attribute as a query can significantly improve the overall performance of the model.

Table 3. Effects of different modules in PA-LSTM

Method	ACC	Precision	Recall	F1
w/o TCN	0.951	0.828	0.901	0.863
w/o Memory	0.951	0.845	0.879	0.862
w/o Attention	0.944	0.815	0.872	0.842
w/o Memory and Attention	0.941	0.796	0.864	0.829

The above ablation experiments mainly focus on user attribute extraction and personalized prediction. Furthermore, we analyze the impact of the events feature extraction. Specifically, we study the impact of the number of LSTM layers. The experiment results are shown in Fig. 2. We can observe that the highest F1 score is obtained with 2 layers LSTM. The model performance will decline when the LSTM is too deep.

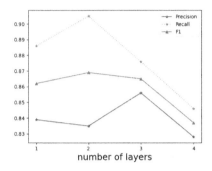

Fig. 2. The results of PA-LSTM with different LSTM layers

6   Conclusion

In the mobile service scenario, the prediction of dissatisfied users by analyzing event data is of great significance. In this paper, we focus on the prediction task and propose PA-LSTM model to distinguish the dissatisfied users, considering user event sequences, user attributes, and dissatisfaction tendencies of user groups. Our experiments on real-world dataset from CMCC demonstrate the effectiveness of PA-LSTM. In the future, we will further analyze specific reasons for dissatisfaction to assist CMCC in improving the quality of mobile services.

Acknowledgements. This research is supported by National Natural Science Foundation of China (No. 62077031, U1936206). We thank the reviewers for their constructive comments.

References

1. Arjovsky, M., Chintala, S., Bottou, L.: Wasserstein generative adversarial networks. In: International Conference on Machine Learning, pp. 214–223. PMLR (2017)
2. Babaev, D., et al.: Coles: contrastive learning for event sequences with self-supervision. In: Proceedings of the 2022 International Conference on Management of Data, pp. 1190–1199 (2022)
3. Babaev, D., Savchenko, M., Tuzhilin, A., Umerenkov, D.: ET-RNN: applying deep learning to credit loan applications. In: Proceedings of the 25th ACM SIGKDD International Conference on Knowledge Discovery & Data Mining, pp. 2183–2190 (2019)

4. Bai, S., Kolter, J.Z., Koltun, V.: An empirical evaluation of generic convolutional and recurrent networks for sequence modeling. arXiv preprint arXiv:1803.01271 (2018)

5. Choi, E., Bahadori, M.T., Sun, J., Kulas, J., Schuetz, A., Stewart, W.: Retain: an interpretable predictive model for healthcare using reverse time attention mechanism. In: Advances in Neural Information Processing Systems, vol. 29 (2016)

6. Dempster, A., Petitjean, F., Webb, G.I.: Rocket: exceptionally fast and accurate time series classification using random convolutional kernels. Data Min. Knowl. Disc. **34**(5), 1454–1495 (2020)

7. Faouzi, J.: Time series classification: a review of algorithms and implementations. Machine Learning (Emerging Trends and Applications) (2022)

8. Feng, W., Tang, J., Liu, T.X.: Understanding dropouts in MOOCS. In: Proceedings of the AAAI Conference on Artificial Intelligence, vol. 33, pp. 517–524 (2019)

9. Hochreiter, S., Schmidhuber, J.: Long short-term memory. Neural Comput. **9**(8), 1735–1780 (1997)

10. Ismail Fawaz, H., et al.: Inceptiontime: finding alexnet for time series classification. Data Min. Knowl. Disc. **34**(6), 1936–1962 (2020)

11. Karim, F., Majumdar, S., Darabi, H., Chen, S.: LSTM fully convolutional networks for time series classification. IEEE Access **6**, 1662–1669 (2017)

12. Lines, J., Taylor, S., Bagnall, A.: Hive-cote: the hierarchical vote collective of transformation-based ensembles for time series classification. In: 2016 IEEE 16th International Conference on Data Mining (ICDM), pp. 1041–1046. IEEE (2016)

13. Serrà, J., Pascual, S., Karatzoglou, A.: Towards a universal neural network encoder for time series. In: CCIA, pp. 120–129 (2018)

14. Vaswani, A., et al.: Attention is all you need. In: Advances in Neural Information Processing Systems, vol. 30 (2017)

15. Wang, Y., Shen, H., Liu, S., Gao, J., Cheng, X.: Cascade dynamics modeling with attention-based recurrent neural network. In: IJCAI, pp. 2985–2991 (2017)

16. Wang, Z., Yan, W., Oates, T.: Time series classification from scratch with deep neural networks: A strong baseline. In: 2017 International Joint Conference on Neural Networks (IJCNN), pp. 1578–1585. IEEE (2017)

17. Wu, W., Yan, J., Yang, X., Zha, H.: Discovering temporal patterns for event sequence clustering via policy mixture model. IEEE Trans. Knowl. Data Eng. **34**(2), 573–586 (2022)

18. Zhou, P., et al.: Attention-based bidirectional long short-term memory networks for relation classification. In: Proceedings of the 54th Annual Meeting of the Association for Computational Linguistics (volume 2: Short papers), pp. 207–212 (2016)

Customer Complaint Guided Fault Localization Based on Domain Knowledge Graph

Shuoshuo Sun[1], Zhihua Chai[1(✉)], Rui Wu[2], Jiawei Jin[2], Yonggeng Wang[2], Wenhao Xu[2], and Guilin Qi[1]

[1] Southeast University, Nanjing, China
{spespusliar,chaizhihua,gqi}@seu.edu.cn
[2] Ant Group, Hangzhou, China
{guli.wr,jinjiawei.jjw,yonggeng.wyg,hao.xuwh}@antfin.com

Abstract. Fault localization aims to identify where the faults occur, which is a critical task for online business systems. Currently, the work of localization is manually conducted. However, in complicated scenarios where thousands of applications are interrelated, it is difficult to quickly localize the fault even for experienced experts, which results in asset losses. The consequence urges the emergence of automatic fault localization which can assist emergency personnel. Existing automatic methods rely on learning from historical failures. However, faults rarely happen in mature systems of an enterprise, leading to the shortage of historical faulty data. To tackle this problem, we propose an Unsupervised Fault Localization (UFL) method. The proposed method utilizes customer complaints to guide localization from the perspective of semantics and leverages the domain knowledge graph to alleviate reliance on historical failures. The experimental results show that UFL outperforms existing methods for fault localization.

Keywords: Fault localization · Knowledge graph · Unsupervised learning · Language model

1 Introduction

Fault localization (FL) aims to identify the application in a system where the fault is occurring based on related information. Fault localization is difficult in a mature business system, which becomes very complex after years of development and upgrades and contains various applications with different functions. For example, Alipay is a payment system with more than nine hundred million users, including tens of thousands of applications, and there are complex relationships between applications. Once a fault occurs, it is time-consuming and labor-intensive to troubleshoot even for experienced experts. Therefore, an automated fault localization method is required to assist operation and maintenance personnel. The faults can be divided into several categories depending on the stage when they are detected, such as pre-intercepted faults, monitor-detected faults, and customer complaints faults. Customer complaint fault refers to a fault that is discovered at a late stage when a user's complaint, a paragraph of natural language text, is received. Among the faults of Alipay in 2021, customer complaints faults accounted for more than 53%.

Existing automated methods of FL are designed based on historical faulty patterns. For example, in [2], large numbers of error messages will be generated under an abnormal situation. The method here localizes the faults based on these messages, and there will be different error messages for each type of fault. Yang et al. [10] expand the scale of the model for FL to improve its ability, which strongly depends on labeled data. Labeled data can guide the model to learn faulty patterns through training. However, existing methods are not applicable to the localization of customer complaint faults, which severely lack historical failure data. Due to the perfection of pre-interception and monitoring technology in systems, customer complaint faults occur too infrequently to cover all faulty patterns through historical failures. At the same time, customer complaint faults lack deterministic information, such as error logs, because everything looks fine until customer complaints are received. To this end, an unsupervised FL method is urgently needed.

We propose a novel **U**nsupervised **F**ault **L**ocalization (UFL) method. UFL contains a **K**nowledge-aware domain **P**re-trained **L**anguage **M**odel (KPLM), semantic self-supervised learning of domain knowledge graph, and a scoring strategy. Firstly, due to the lack of related information on faults, we utilize customer complaints to guide localization by semantics matching. Pre-trained language models (PLM) are used because of their ability in natural language understanding. Furthermore, the descriptions of applications have strong domain characteristics, which reminds us to transfer the PLM to our field. Secondly, a domain knowledge graph (KG) is introduced to provide external knowledge, and we design a novel self-supervised learning method to extract the semantic information of the KG. Simultaneously, we inject knowledge into the language model through knowledge enhancement. Finally, we apply the above methods to score each application with a well-designed scoring strategy, and the K applications with the highest scores are the results of fault localization.

Experiments show that our UFL outperforms other methods and has great value in real scenarios. UFL has been deployed in the Alipay fault localization system. To sum up, the contributions of this paper include the following aspects:

- We introduce the domain knowledge graph to solve the problem of customer complaint faults localization, which can obviously improve the hitting ratio for FL according to the experiments.
- We propose a novel hierarchical self-supervised learning method to integrate the semantic and structural features of the knowledge graph.
- We conduct comprehensive experiments on a real-world dataset that contains customer complaints faults of the Alipay application in the last three years to demonstrate the effectiveness of our proposed method.

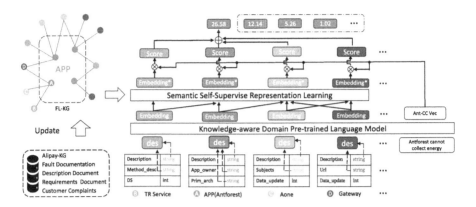

Fig. 1. The framework of our UFL. TR Service, Aone and Gateway are information related to applications. APPs are applications, and Antforest is an example.

2   Problem Formulation

Let $T_{cc} = \{w_1, \ldots, w_n\}$ represent a customer complaint, $\mathcal{A} = \{a_1, \ldots, a_N\}$ represents a collection of Alipay applications. T_{cc} is a sentence including n words, and w_i represents the i-th word in it. Customer complaint guided fault localization needs to identify the faulty application a_i according to T_{cc}. The method we propose focuses on semantic features. We use a PLM to extract the semantic feature, which is recorded as KPLM. In order to enrich the semantic features of application set $\mathcal{A}$, our domain knowledge graph $\mathcal{G} = (\mathcal{E}, \mathcal{R}, \mathcal{T})$ is introduced, where $\mathcal{E}, \mathcal{R}$ and $\mathcal{T}$ represent the collection of entities, relations and triples respectively. $\mathcal{D}$ represents the set of nodes descriptions in $\mathcal{G}$ and A is the adjacency matrix of it. After the introduction of the KG, the problem of customer complaint guided fault localization is transformed into localizing a faulty application node in the KG according to the customer complaint.

Based on the preliminaries above, we can formulate the problem of customer complaint guided fault localization. Given a customer complaint T_{cc} and the knowledge graph $\mathcal{G} = (\mathcal{E}, \mathcal{R}, \mathcal{T})$, the goal is to train a model which outputs the score set $\mathcal{S} = \{s_{(a_1)}, \ldots, s_{(a_N)}\}$ for each application. The K applications with the highest scores are the results of fault localization.

3   Methodology

3.1   Overview of Our Method

Figure 1 illustrates the framework of our method. The existing KG is Alipay's general KG, which provides support for various businesses rather than being specifically oriented to FL. Therefore, we perform semi-open information extraction [11] and keywords extraction [9] based on records of Alipay historical faults to update domain KG for FL. Then we tranfer PLM based on these corpora to map the customer complaints and application descriptions to the same semantic space. Also, we inject correlative knowledge into the PLM. Next, to match customer complaints with applications, we conduct semantic self-supervised learning to embed the KG. Finally, each application is assigned a score by our scoring strategy, and the applications with the top K scores are the results.

3.2 Update of Domain Knowledge Graph for Fault Localization

The existing domain KG is Alipay's general knowledge graph Alipay-KG. It is large in scale and contains 6,263,181 entities, including 10,190 applications as candidate localization results. Alipay-KG lacks professional knowledge related to faults, which leads to little effect on FL. Therefore, we first update Alipay-KG for FL.

In order to enrich the information of candidate applications and make them contain relevant knowledge of FL, we first adopt the USE [11] model to perform semi-open information extraction based on documents of historical faults. Closed-domain information extraction helps to build a high-precision vertical KG, while open-domain information extraction helps to build a rich and general KG. Semi-open information extraction moderately combines the two, which is suitable for the settings under the task of FL.

Furthermore, when a customer complaint reports faulty information, it is usually a short sentence, such as "W_1 can no longer W_2". If both W_1 and W_2 are keywords of an application, the probability of failure in this application should be high. So the keywords in it are significant. We use the KeyBERT [9] model to extract keywords. KeyBERT applies a PLM and is an end-to-end method that evaluates the importance of a word to the semantics of the entire sentence. Finally, we add the keywords extracted by KeyBERT to the KG. The KG updated is denoted as FL-KG.

3.3 Knowledge-aware Domain Pre-trained Language Model

The customer complaint guided fault localization mainly relies on semantics. In natural language understanding, the PLM has great ability of semantic extraction. However, the corpus used by PLM in the pre-training stage has different characteristics from the corpus in Alipay. Therefore, we transfer the model to our field. We collect corpus data of Alipay, with a total of 1,284,662 sentences, which are used to support the transfer of the PLM. At the same time, the task of customer complaint guided fault localization requires a certain knowledge of FL. In this paper, the knowledge is injected into the language model by combining the FL-KG in Sect. 3.2. The training is divided into two stages, domain transfer of pre-trained language model and knowledge enhancement.

Domain Transfer of Pre-trained Language Model. This stage aims to improve the model's understanding of the domain text. We use two tasks, including Masked Language Model (MLM) and classification, to transfer the PLM.

The method of MLM is consistent with the literature [1]. In order to make the model better adapt to our task of FL, that is, to match a customer complaint with the descriptive information of the application node in the FL-KG, we design a task of binary classification.

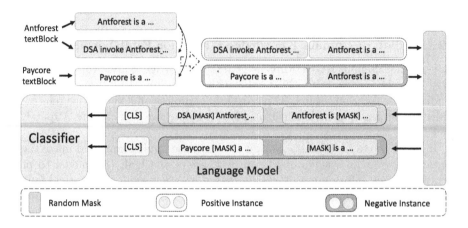

Fig. 2. Illustration of the classification task in pre-training.

The goal of the task is to make the model distinguish whether two sentences describe the same or different applications. As shown in Fig. 2, we use the descriptive texts of the nodes in FL-KG as the data source. The descriptive texts of an application are divided into text blocks of fixed length. Then we randomly select two text blocks from the same application as positive samples and two text blocks from different applications as negative samples. The block pairs separated by [SEP] is fed into the language model, and a multi-layer perceptron with a softmax function is used as the classifier. This task achieves the result that the model can discern common features from the same application and discrepancies from different applications.

Knowledge Enhancement. At this stage, we aim to inject the knowledge of FL-KG into the language model. Inspired by [12], we design a new task based on FL-KG to integrate the knowledge. Knowledge-aware pre-trained tasks in [12] use the triples in the KG and their corresponding encyclopedia texts, randomly masking the words in the text and the relationships in the triples, then making the model predict the parts masked to inject knowledge. We extract the triples of FL-KG and randomly mask head entities, tail entities, or relations. Similarly, we extract the descriptive text of the entities in triples, then randomly mask fifteen percent of the words (the descriptive texts of the two entities are processed in the same way as [1]). The two damaged descriptive texts are spliced after triples. Finally, we let the model predict these masked parts. The learning of this task is to allow the model to establish the corresponding relationship between descriptions and entities, and to predict the relationship between two entities through their description.

This module outputs a knowledge-aware domain pre-trained language model (KPLM). We utilize KPLM to encode the customer complaints and nodes in the FL-KG as Eq. (1) and (2).

$$X_{cc} = \text{KPLM}(T_{cc}), \tag{1}$$
$$X_{\mathcal{E}} = \text{KPLM}(\mathcal{D}), \tag{2}$$

where X_{cc} and $X_{\mathcal{E}}$ are the embeddings of customer complaints and $\mathcal{E}$.

3.4 Semantic Self-supervised Graph Representation Learning

In Sect. 3.3, we utilize KPLM to obtain the representations of customer complaints and application nodes. However, in FL-KG, application nodes contain not only their textual information but also structural information with other nodes. This kind of structural information enables other nodes to be a supplement to the nodes of applications. When the descriptive information of the target application node is missing or difficult to match with the customer complaints, neighbors' information could be helpful. We design a novel hierarchical self-supervised learning method and use the GraphSAGE [3] network to integrate the semantic and structural information. The advantage of GraphSAGE is its ability to handle dynamic changes in the graph that often occur in real-world scenarios, such as the addition or deletion of applications.

The mechanism of the GraphSAGE is its message passing when updating the node embedding, as Eq. (3),

$$X_{\mathcal{E}}^{(l+1)} = \text{UPDATE}(X_{\mathcal{E}}^{(l)}, \text{AGGREGATE}(X_{\mathcal{E}'}^{(l)}, \mathcal{E}' \in \mathcal{N}(\mathcal{E})\})) \tag{3}$$

where $X_{\mathcal{E}}^{(l)}$ represents the embeddings of nodes collection $\mathcal{E}$ in the l-th layer. $\mathcal{E}'$ represents the neighbors of nodes in $\mathcal{E}$. $\mathcal{N}(\mathcal{E})$ is the collection consisting of nodes which are neighbors of $\mathcal{E}$. AGGREGATE represents the aggregation method of neighbor nodes, UPDATE represents the update method of $X_{\mathcal{E}}$.

Inspired by BertGCN [6], we use $X_{\mathcal{E}}$ in Eq. (2) as the initialization of nodes in FL-KG. The initialization not only makes the embedding of the nodes and the customer complaints in the same semantic space, but also makes the subsequent use of GraphSAGE to update the node embedding completely in the semantic space instead of random space.

Hierarchical Self-supervised Learning Method In this section, we first construct a semantic graph, and then propose a hierarchical self-supervised learning method that can combine the semantic graph and the KG.

The process of constructing semantic graph is illustrated in Fig. 3. Semantic features of nodes $X_{\mathcal{E}}$ are from Eq. (2). Supposing our graph has $|\mathcal{E}|$ nodes and

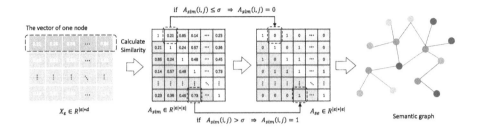

Fig. 3. Process of constructing the semantic graph.

the feature of each node is a d-dimensional vector, we have $X_{\mathcal{E}} \in R^{|\mathcal{E}|*d}$. Next, calculate the similarity between nodes and obtain the similarity matrix $A_{sim} \in R^{|\mathcal{E}|*|\mathcal{E}|}$. $A_{sim}(i, j)$ represents the similarity between i-th node and j-th node. σ denotes a similarity threshold. If $A_{sim}(i, j) > \sigma$, set $A_{sim}(i, j)$ to 1, otherwise, set it to 0, The updated matrix A_{se} is used as the adjacency matrix of the semantic graph.

We perform GraphSAGE on both the structural graph which means FL-KG and the semantic graph. The new embeddings of the two graphs are denoted as $X_{\mathcal{E}_{st}}$, $X_{\mathcal{E}_{se}}$. In order to perform deep extraction of nodes semantics, we proposes a hierarchical self-supervised learning method, which uses the constructed semantic graph as a supervision signal to enable GraphSAGE to map nodes to the deep semantic space. The self-supervised learning loss function is as Eq. (4).

$$\mathcal{L} = a\|X_{\mathcal{E}}^{st} - X_{\mathcal{E}}^{se}\|_2 + b\|X_{Local}^{st} - X_{Local}^{se}\|_2 + c\|X_{Global}^{st} - X_{Global}^{se}\|_2 \quad (4)$$

$$X_{Local} = \{\frac{1}{|\mathcal{C}_1|}\sum_i X_i^{\mathcal{C}_1}, \ldots, \frac{1}{|\mathcal{C}_m|}\sum_i X_i^{\mathcal{C}_m}\} \quad (5)$$

$$X_{Global} = \text{MeanPooling}(X_{\mathcal{E}}) \quad (6)$$

As shown in Fig. 4, the loss function includes three parts, namely node-level, local-level and global-level similarity loss. The loss at the node level is the L2 distance of $X_{\mathcal{E}}^{st}$, $X_{\mathcal{E}}^{se}$. Before calculating of the similarity loss at the local level, we use the K-means [4] algorithm to cluster FL-KG and semantic graph. The nodes belonging to the same class are viewed as a node, whose embeddings are from the mean of nodes embeddings in the class, as Eq. (5). $|\mathcal{C}_i|$ is the number of nodes in class $\mathcal{C}_i$. The L2 distance of the embedding matrixs X_{Local}^{se} and X_{Local}^{st} from the clustered graph is used as the loss at the local level. The global level embeddings X_{Global}^{st} and X_{Global}^{se} are directly obtained by using the mean-pooling algorithm as Eq. (6) and their L2 distance is used as the global similarity loss. Finally, the weighted summation of the three parts constitutes a hierarchical self-supervised loss, where a, b, and c are the weight coefficients. Semantic embeddings of nodes $X_{\mathcal{E}}'$ in this self-supervised learning setting are output for supporting FL.

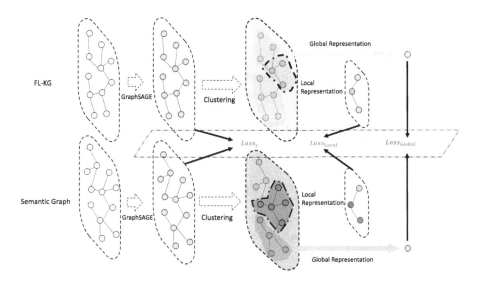

Fig. 4. Structure of semantic self-supervised learning.

3.5 Customer Complaint Fault Localization

In this section, we utilize a scoring strategy to conduct fault localization. Firstly, we calculate the similarity between the representation of the customer complaint and the embedding of applications respectively. Then a weighted scoring strategy is used to calculate the final score for each application. The higher the score is, the more likely the corresponding application match the fault. The applications with top K scores are the results of FL. Equation (7) illustrates the scoring strategy,

$$s_{a_i} = \alpha * Sim_{a_i} + \sum_{\text{type}} \beta_{\text{type}} * \sum_{j \in \mathcal{N}(a_i)} Sim_{a_j} \tag{7}$$

$$Sim_{a_i} = \cos(\mathrm{X_{cc}}, \mathrm{X}_i') \tag{8}$$

where s_{a_i} represents the score of node i, α and β_{type} is weight factor, Sim_{a_i} represents the similarity between node i itself and the customer complaint, while $\mathcal{N}(a_i)$ represents the set of neighbors of node i. The final fault localization result is selected based on this score.

4 Experiment

4.1 Dataset

Ant-cc: The dataset used in this paper is Alipay's customer complaint faults in the three years from October 2019 to October 2022. Each sample in the dataset includes a customer complaint and the fault localization result. A customer complaint is a short text, and the fault localization result is an application.

4.2   Compared Methods

We take the following commonly-used pre-trained models as baselines.

- **Word2vec** [8]: The Word2vec algorithm uses a neural network model to learn word embeddings from large text corpora, which can be pooled to represent sentence vectors.
- **Bert** [1]: A deep bidirectional transformer that achieves SOTA effect on multiple natural language understanding tasks.
- **ALBert** [5]: A lightweight version of the BERT model that reduces the size of the parameters.
- **Roberta** [7]: Similar to bert, but with more training corpus, dynamic MASK strategy and removing NSP task.
- **ERNIE** [12]: Inject knowledge into the pre-trained language model, making it capable of natural language understanding and generation at the same time.

Our Methods. We conduct ablation experiments by incrementally adding individual modules proposed in this paper and compare with baseline models.

- **UFL$_{\text{all}}$**: The complete framework proposed in this paper includes both a knowledge-aware domain pre-trained language model and a domain knowledge graph for fault localization.
- **UFL$_{+kg}$**: The only difference with UFL$_{all}$ is that the knowledge graph used is *Alipay-KG*.
- **UFL$_{+dm}$**: It does not use knowledge graph, but only uses knowledge-aware domain transfer pre-trained language model for semantic extraction.
- **UFL$_{base}$**: It only uses the knowledge-aware language model, and does not introduce knowledge graph and domain transfer.

Evaluation Metrics. In this paper, the evaluation metrics of customer complaint fault localization are HR@K, Mean Rank (MR) and Mean Reciprocal Ranking (MRR), where K is selected from 10, 50, 100. The calculation formulation of HR@K is $HR@K = \frac{1}{|S|} \sum_i \mathbb{I}(rank_i \leq K)$. $|S|$ represents the total number of samples, $rank_i$ is the rank of the result for customer complaint, and $\mathbb{I}$ is an indicator function. MR represents the predicted average rank, calculated as $MR = \frac{1}{|S|} \sum_i rank_i$ and $MRR = \frac{1}{|S|} \sum_i \frac{1}{rank_i}$.

4.3   Experimental Results

The experimental results are shown in Table 1. Word2vec performs better than the Bert series models without domain transfer because Word2vec is trained on the domain corpus and learns the language characteristics of the domain. This phenomenon shows the importance of domain language features and necessity of domain transfer. Bert series models outperform Word2vec models after transfer. Among baselines, the transferred Bert model is considered as the best, and the follow-up is completed under its support.

Table 1. The results of customer complaint fault localization. Methods marked with
† are models after transferring. Metrics marked with * stand for the corresponding
results on a filtered dataset.

Methods	HR@10	HR@50	HR@100	HR*@10	HR*@50	HR*@100	MR*	MRR*
Word2vec	15.79%	27.63%	39.47%	16.67%	30.56%	45.83%	<u>547.31</u>	0.0649
Bert	6.25%	15.62%	21.85%	6.67%	16.67%	23.33%	2321.73	0.0572
Bert†	17.19%	<u>35.94%</u>	<u>45.31%</u>	18.33%	<u>38.34%</u>	<u>48.35%</u>	1758.09	0.0959
Roberta	9.38%	17.19%	31.25%	10.18%	18.37%	33.34%	2351.09	0.0673
Roberta†	<u>20.31%</u>	31.25%	34.38%	<u>21.67%</u>	35.13%	38.33%	1977.75	0.0939
ERNIE	7.81%	20.31%	28.13%	8.33%	21.67%	30.01%	2369.01	0.0468
ERNIE†	16.75%	35.50%	41.75%	18.03%	38.17%	44.66%	1797.27	0.0947
ALBert	9.12%	18.75%	28.13%	10.07%	20.11%	30.20%	2712.65	0.0690
ALBert†	17.19%	28.13%	35.94%	18.35%	30.14%	38.33%	1458.25	<u>0.0974</u>
UFL$_{base}$	15.63%	32.81%	40.63%	16.76%	35.62%	43.31%	1838.09	0.0978
UFL$_{+dm}$	23.44%	35.94%	42.19%	25.14%	38.34%	45.43%	2009.79	0.0983
UFL$_{+kg}$	30.26%	59.21%	64.47%	34.33%	67.16%	73.13%	242.60	0.2011
UFL	**51.32%**	**84.21%**	**85.53%**	**58.21%**	**95.52%**	**97.01%**	**97.62**	**0.2710**

The last row of the table is the UFL proposed in this paper. Obviously, the
metrics of UFL have been greatly improved compared with the baselines. On the
HR@10 metric, our UFL achieved an effect of 51.32%, which means it could rank
the faulty application in the top 10 for more than half of the customer complaints.
This has a great practical effect on fault localization in reality, because emergency
personnel usually only check the results with higher rankings, like top ten, due
to time constraints. Once the emergency personnel find the faulty application
according to the results of UFL, they can immediately let the person in charge
of the application to solve the fault, thus greatly shortening the repair time and
reducing the asset loss for the enterprise.

In *Ant-cc*, a certain proportion of customer complaints can not match applications through semantics, which leads to the average MR are relatively high.
For better analysis, we re-stated the metrics (names of which with asterisks) by
removing these samples. On the HR*@100, our UFL reaches 97%, which means
that domain knowledge graph provides helpful domain knowledge and significantly improves the recall capability of UFL for the faulty applications.

4.4   Ablation Study

In order to illustrate the effectiveness and capability of each module, we further conduct ablation study, and the results are shown in Table 1. The results
show that, compared with the open-source pre-trained language model, all metrics in customer complaint fault localization are improved after transferring.
UFL$_{base}$ only does knowledge enhancement on open source pre-trained language
models. Its metrics are slightly lower than that of transferring, this is because

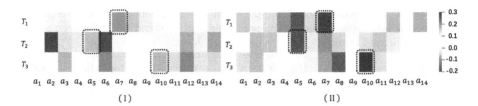

Fig. 5. Visualization of scores in real cases. (I) and (II) are generated by Bert[†] and UFL respectively. T_i indicate a customer complaint and a_i is an candidate application. The dotted boxes denote the ground truth for corresponding T_i.

the enhancement of knowledge only incorporates the knowledge of fault localization, and the ability of encoding the domain text has not been improved. UFL_{+dm} combines knowledge enhancement with domain transfer. From a macro perspective, domain transfer injects coarse-grained knowledge, while knowledge enhancement injects fine-grained knowledge. Therefore, the combination of the two will not cause conflict. The results of our experiment shows that after the combination of the two, all metrics are improved.

UFL_{+kg} introduces Alipay-KG and injects more domain-related knowledge, but this knowledge is general domain knowledge, not fault localization-oriented knowledge. The only difference between UFL_{all} and UFL_{+kg} is that the knowledge graph is fault localization-oriented or not. Table 1 shows that Hit@50 and Hit@100 are greatly improved by introducing the knowledge graph. After constructing the domain knowledge graph for fault localization, Alipay-KG is replaced by FL-KG, which further enriches the application information, and the metrics of customer complaint fault localization are improved again.

Overall, the continuous improvement of the metrics demonstrates the effectiveness of each module in UFL_{all} proposed by this paper.

4.5   Case Study

In order to demonstrate the effectiveness of the framework proposed in this paper, we will analyze several real Alipay fault localization cases in this section. We first select three customer complaints that recently happened. Then scoring results for applications generated by Bert[†] and UFL are visualized in Fig 5.

According to the customer complaint T_2, Bert[†] pays more attention to the application a_2, which is not the faulty one. Our UFL could precisely localize the a_5 by introducing our FL-KG and semantic self-supervised graph representation learning. For customer complaints T_1 and T_3, although Bert[†] have properly localized the faulty application, our UFL can still keep the correct result with a higher gap from unrelated applications. It is worth noting that a_8 is also assigned a high score given T_3. Through analysis, we find that a_8 and a_{10} are close in the domain knowledge graph, which means they have strong interaction in actual scenarios. This phenomenon certificates that our UFL can effectively utilize the knowledge graph to localize the faulty application.

5   Conclusion

This paper proposes an unsupervised fault localization method based on the domain knowledge graph, emphasizing the importance of semantic matching. Also, we transfer the pre-trained language model to the field of fault localization so that it has the ability to extract domain semantics. Further, the domain knowledge graph is introduced to inject the knowledge of customer complaint fault localization, and a semantic graph is constructed as a supervised signal to extract deep semantics and mine the knowledge of fault localization. The experimental results verify the effectiveness of our method. Our method has been deployed in the Alipay fault localization system[1].

References

1. Devlin, J., Chang, M.W., Lee, K., Toutanova, K.: Bert: pre-training of deep bidirectional transformers for language understanding. In: Proceedings of the 2019 Conference of the North American Chapter of the Association for Computational Linguistics: Human Language Technologies, Volume 1 (Long and Short Papers), pp. 4171–4186 (2019)
2. Elgamasy, M.M., Izzularab, M.A., Zhang, X.P.: Single-end based fault location method for VSC-HVDC transmission systems. IEEE Access **10**, 43129–43142 (2022)
3. Hamilton, W., Ying, Z., Leskovec, J.: Inductive representation learning on large graphs. In: Advances in Neural Information Processing Systems, vol. 30 (2017)
4. Krishna, K., Murty, M.N.: Genetic k-means algorithm. IEEE Trans. Syst. Man Cybern. Part B (Cybernetics) **29**(3), 433–439 (1999)
5. Lan, Z., Chen, M., Goodman, S., Gimpel, K., Sharma, P., Soricut, R.: Albert: a lite Bert for self-supervised learning of language representations. arXiv preprint arXiv:1909.11942 (2019)
6. Lin, Y., Meng, Y., Sun, X., Han, Q., Kuang, K., Li, J., Wu, F.: BertGCN: Transductive text classification by combining GNN and Bert. In: Findings of the Association for Computational Linguistics: ACL-IJCNLP 2021, pp. 1456–1462 (2021)
7. Liu, Y., et al.: Roberta: a robustly optimized Bert pretraining approach. arXiv preprint arXiv:1907.11692 (2019)
8. Mikolov, T., Chen, K., Corrado, G., Dean, J.: Efficient estimation of word representations in vector space. arXiv preprint arXiv:1301.3781 (2013)
9. Sharma, P., Li, Y.: Self-supervised contextual keyword and keyphrase retrieval with self-labelling (2019)
10. Yang, H., Zhao, X., Yao, Q., Yu, A., Zhang, J., Ji, Y.: Accurate fault location using deep neural evolution network in cloud data center interconnection. IEEE Trans. Cloud Comput. **10**, 1402–1412 (2020)
11. Yu, B., et al.: Semi-open information extraction. In: Proceedings of the Web Conference 2021, pp. 1661–1672 (2021)
12. Zhang, Z., Han, X., Liu, Z., Jiang, X., Sun, M., Liu, Q.: Ernie: enhanced language representation with informative entities. In: Proceedings of the 57th Annual Meeting of the Association for Computational Linguistics, pp. 1441–1451 (2019)

---

[1] This work was supported by Ant Group.

Self-Sampling Training and Evaluation for the Accuracy-Bias Tradeoff in Recommendation

Dugang Liu[1,2], Yang Qiao[3], Xing Tang[3], Liang Chen[3], Xiuqiang He[3],
Weike Pan[1(✉)], and Zhong Ming[1(✉)]

[1] College of Computer Science and Software Engineering, Shenzhen University,
Shenzhen, China
{panweike,mingz}@szu.edu.cn
[2] Guangdong Laboratory of Artificial Intelligence and Digital Economy (SZ),
Shenzhen, China
[3] FIT, Tencent, Shenzhen, China
{sunnyqiao,shawntang,leocchen,xiuqianghe}@tencent.com

Abstract. Research on debiased recommendation has shown promising results. However, some issues still need to be handled for its application in industrial recommendation. For example, most of the existing methods require some specific data, architectures and training methods. In this paper, we first argue through an online study that arbitrarily removing all the biases in industrial recommendation may not consistently yield a desired performance improvement. For the situation that a randomized dataset is not available, we propose a novel self-sampling training and evaluation (SSTE) framework to achieve the accuracy-bias tradeoff in recommendation, i.e., eliminate the harmful biases and preserve the beneficial ones. Specifically, SSTE uses a self-sampling module to generate some subsets with different degrees of bias from the original training and validation data. A self-training module infers the beneficial biases and learns better tradeoff based on these subsets, and a self-evaluation module aims to use these subsets to construct more plausible references to reflect the optimized model. Finally, we conduct extensive offline experiments on two datasets to verify the effectiveness of our SSTE. Moreover, we deploy our SSTE in homepage recommendation of a famous financial management product called Tencent Licaitong, and find very promising results in an online A/B test.

Keywords: Debiased recommendation · Self-sampling · Self-training · Self-evaluation

1 Introduction

A user will inevitably suffer from various biases during the interaction with a recommender system, which will lead to inherent variability in the feedback data. As a result, the collected data may not be able to reflect a user's true

X. Wang et al. (Eds.): DASFAA 2023, LNCS 13946, pp. 580–592, 2023.
https://doi.org/10.1007/978-3-031-30678-5_44

preferences [11]. Ignoring these biases will allow a recommendation model trained based on the feedback data to inherit and even amplify their influence, which is not conducive to the long-term and healthy development of a recommender system. Therefore, how to reasonably and effectively mitigate the bias problem in the feedback data is an important challenge.

Existing debiased recommendation methods can be mainly divided into two lines, including debiased recommendation with a randomized dataset and without a randomized dataset. A randomized dataset is collected with a specific uniform policy instead of a recommendation model, which can be regarded as a good unbiased proxy due to the random selection operation used for item assignment [2]. With the unbiased information contained in a randomized dataset, the first line focuses on designing different joint training modules to transfer them to a recommendation model trained on a biased feedback data [2,3,9]. In the case where a randomized dataset is unavailable, the second line mainly ensures the unbiasedness of an optimization objective and guides the design of the model architecture by introducing some theoretical framework [11,15,19]. Although the existing debiased recommendation methods have shown promising results, their application in an industrial recommendation is still lacking sufficient insight since most of them require some specific data, architectures, and training methods.

In particular, an important question that is rarely considered and answered is whether removing all the biases in an industrial recommendation is a desirable goal. To gain an initial insight into this problem, we first conduct a three-week online study in a real recommendation scenario, where an approximate uniform policy is deployed for comparison with the base model. This recommendation scenario comes from the homepage recommendation of Tencent Licaitong, which is one of the largest financial recommendation scenarios in China and its display homepage is shown on the left side of Fig. 1. Note that more information about this scenario and the evaluation metric COPM can be found in Sect. 4.4. From the right side of Fig. 1, we can find that the uniform policy will bring an expected performance improvement in the early stage, and this may be due to the unexpected recommendation brought by the random selection operation. However, in the later stage, the advantage of the uniform policy is not maintained, and degenerates to be similar to the base model. We argue that this may be due to the fact that the beneficial biases that improve the performance of the base model are removed in the uniform policy, e.g., a high-yield fund product should naturally receive more exposure in this recommendation scenario. Overall, this means that arbitrarily removing all the biases in an industrial recommendation may not consistently yield a desired performance improvement.

Therefore, in this paper, we propose to use an accuracy-bias tradeoff instead of removing all the biases. We then propose a simple but effective self-sampling training and evaluation (SSTE) framework to achieve this goal, which preserves the beneficial biases while removing the harmful ones. Specifically, our SSTE includes three customized modules: 1) a self-sampling module generates some corresponding auxiliary subsets with different degrees of bias from the original training set and validation set; 2) a self-training module combines the original training set and the auxiliary subsets for joint learning to infer the beneficial

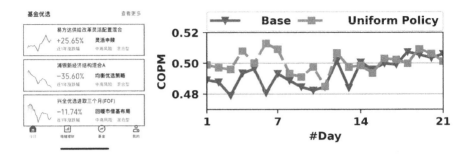

Fig. 1. A recommendation scenario on the homepage of Tencent Licaitong, and the results of a three-week online study conducted. The contents marked with the red boxes are the recommended items. Note that due to confidentiality, we have transformed the actual COPM values. (Color figure online)

biases and achieve a better accuracy-bias tradeoff; 3) a self-evaluation module combines the original validation set and the auxiliary subsets to construct a more reasonable reference to better reflect the optimized model offline. We conduct extensive offline experiments on a public dataset and a real product dataset to verify the effectiveness of our SSTE, including unbiased evaluation and compatibility analysis. In addition, we also show the strength of our SSTE in an online A/B test.

2   Related Work

In this section, we briefly review some related works on two research topics, including debiased recommendation and debiased evaluation.

Debiased Recommendation. According to the types of feedback data involved, existing debiased recommendation methods can be mainly divided into two categories, i.e., debiased recommendation with a randomized dataset and without a randomized dataset. The former aims to introduce a randomized dataset as an unbiased proxy, and then various ways of its joint training with the original biased feedback data can be designed to exploit the guidance of this unbiased information [2,3,9]. The latter considers mitigation of the bias problem in the case where a randomized dataset is not available. The main techniques include assuming and modeling the generation mechanism between a specific bias and some certain features [12,13], or introducing some theoretical frameworks to construct some corresponding unbiased estimators for this bias problem [11,15,18,19]. However, most of the existing methods require some specific data, architectures and training methods, which hinders the full exploration and sufficient insights for debiased recommendation in an industrial recommendation. Our SSTE aims to bridge the gap in this direction.

Debiased Evaluation. Due to the inherent biases in the feedback data, traditional evaluation metrics may not reflect the real performance of a recommendation model and will lead to a discrepancy between offline and online evaluations. To solve this problem, most previous works mainly consider from two aspects of measurement design and sample design. The former aims to design some corresponding unbiased versions for traditional metrics or propose some new unbiased evaluators [5,8,20], while the latter focuses on designing some methods that can construct an unbiased validation set [2,7]. However, most existing methods are inconvenient to application in an industrial recommendation due to the uncontrollable potential risks, i.e., evaluation errors. Different from them, our SSTE proposes a simple and effective self-evaluation method with the manageable potential risks.

3   The Proposed Framework

In this paper, we focus on alleviating the bias problem in implicit feedback without a randomized dataset. Suppose that the training set $\mathcal{D}_{tr} = \{(\boldsymbol{x}_i, y_i)\}_{i=1}^{m}$ with $x_i \in \mathcal{X}$ and $y_i \in \mathcal{Y}$ is drawn from a latent distribution $P(\boldsymbol{x}, y)$, where m is the number of training instances. $\mathcal{X} = \mathcal{X}_1 \times \cdots \times \mathcal{X}_d$ is a d-dimensional feature space, and $\mathcal{Y} = \{0, 1\}$ is a label space. And the validation set $\mathcal{D}_{val} = \{(\boldsymbol{x}_j, y_j)\}_{j=1}^{n}$ is drawn from a latent distribution $Q(\boldsymbol{x}, y)$, where n is the number of validation instances. Note that $y = 1$ and $y = 0$ indicate that a training or validation instance is a positive feedback and a negative feedback, respectively.

3.1   The Accuracy-Bias Tradeoff

Since the results in Fig. 1 suggest that arbitrarily removing all the biases in an industrial recommendation may not be an ideal choice, we propose a new accuracy-bias tradeoff goal to obtain a more desirable performance improvement, where the key idea is to treat all the biases in the feedback data as a combination of harmful and beneficial ones. To facilitate the understanding of the difference between our goal and the existing works, we give the causal diagrams of traditional recommendation, debiased recommendation and the proposed new goal in Fig. 2, respectively. We use U, V, M, C, A, and Y to denote the users, items, true matching preferences (i.e., U's specific preference for V), beneficial bias effects (i.e., the preference offset due to the beneficial biases such as high exposure bias for high-yield funds), harmful bias effects (i.e., the preference offset due to the harmful biases, such as position bias), and feedback labels, respectively. As shown in Figs. 2(a) and 2(b), traditional recommendation methods will encode the harmful bias effects, and debiased recommendation methods will remove both the beneficial and harmful bias effects. From Fig. 2(c) we can see that unlike them, our goal is to remove the harmful bias effects while retaining the beneficial bias effects.

3.2   Architecture

We propose a simple but effective self-sampling training and evaluation (SSTE) framework to achieve the desired accuracy-bias tradeoff, where its overall architecture is shown in Fig. 3. Given a training set $\mathcal{D}_{tr}$ and a validation set $\mathcal{D}_{val}$, a

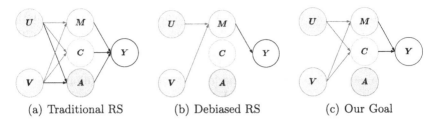

(a) Traditional RS   (b) Debiased RS   (c) Our Goal

Fig. 2. Causal diagrams w.r.t. (a) traditional recommendation, (b) debiased recommendation, and (c) the proposed solution with accuracy-bias tradeoff, where U, V, M, C, A, and Y denote the users, items, true matching preferences, beneficial bias effects, harmful bias effects, and feedback labels, respectively.

self-sampling module constructs a set of auxiliary subsets with different degrees of bias based on $\mathcal{D}_{tr}$ and $\mathcal{D}_{val}$, i.e., $\mathcal{A}_{tr} = \{\hat{\mathcal{D}}_{tr}^i\}_{i=1}^{T_1}$ and $\mathcal{A}_{val} = \{\hat{\mathcal{D}}_{val}^i\}_{i=1}^{T_2}$. $\hat{\mathcal{D}}_{tr}^i$ is an auxiliary subset sampled from $\mathcal{D}_{tr}$ based on a specific strategy, and T_1 is the number of auxiliary subsets equipped for $\mathcal{D}_{tr}$. $\hat{\mathcal{D}}_{val}^i$ and T_2 are similarly defined for $\mathcal{D}_{val}$. Then, a self-training module receives $\mathcal{D}_{tr}$ and $\mathcal{A}_{tr}$, and updates a recommendation model $\hat{\Theta}$ by jointly training with some shared parameters. The updated model $\hat{\Theta}$ is then passed to the self-evaluation module. And after combining $\mathcal{D}_{val}$ and $\mathcal{A}_{val}$, the defined new evaluation method is used to obtain the performance corresponding to the current training iteration. If the convergence condition is not met, the self-training module and the self-evaluation module continue to be executed alternately. And once it is met, the optimized recommendation model $\hat{\Theta}^*$ will be output.

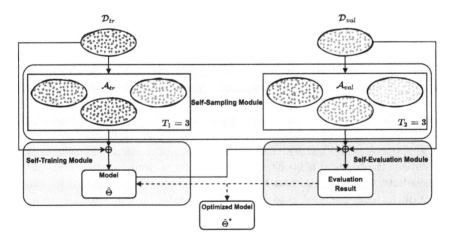

Fig. 3. The overall architecture of the proposed SSTE, where the core components are a self-sampling module, a self-training module and a self-evaluation module.

3.3   Training

Next, we describe each module in detail based on the training process.

The Self-Sampling Module. We propose an auxiliary subset sampling method based on truncated inverse propensity score (tIPS). Specifically, taking the training instances as an example, we first obtain the sampled probability of each instance $p(\boldsymbol{x}_i)$ by some existing IPS estimators in debiased recommendation, where a higher $p(\boldsymbol{x}_i)$ means this sample is more important for debiasing. We then choose a set of different truncation thresholds $\{\epsilon_{tr}^i\}_{i=1}^{T_1}$ and use each threshold separately to adjust the obtained $p(\boldsymbol{x}_i)$, i.e., keep $p(\boldsymbol{x}_i)$ when $p(\boldsymbol{x}_i) < \epsilon_{tr}^i$, otherwise modify $p(\boldsymbol{x}_i)$ to 1. This means that we control the level of debiasing by applying more protection to different proportions of important samples. Finally, based on the modified sampling probabilities, we can obtain a set of auxiliary subsets of $\mathcal{D}_{tr}$, i.e., $\mathcal{A}_{tr} = \{\hat{\mathcal{D}}_{tr}^i\}_{i=1}^{T_1}$, where each auxiliary subset corresponds to a truncation threshold. Similarly, by setting $\{\epsilon_{val}^i\}_{i=1}^{T_2}$ for the validation instances, we can obtain $\mathcal{A}_{val} = \{\hat{\mathcal{D}}_{val}^i\}_{i=1}^{T_2}$ corresponding to $\mathcal{D}_{val}$. The idea behind this sampling operation is to simulate the auxiliary subsets with different degrees of bias and different sets of biases based on the feedback data itself. Note that the self-sampling module can be executed only once as a preprocessing operation, or it can be re-executed after each round of training. We give an example of the sampling process in Fig. 4.

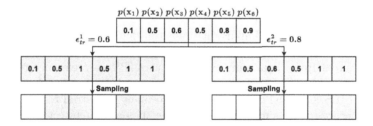

Fig. 4. The schematic diagram of a sampling process.

The Self-Training Module. In order to combine $\mathcal{D}_{tr}$ and $\mathcal{A}_{tr}$ to infer the beneficial bias information and constrain the model to achieve a better accuracy-bias tradeoff, we can use any model architecture with some shared parameters for training. Formally, let $\widetilde{\Theta} = \{\tilde{\theta}, \theta_s\}$ be the model parameters related to $\mathcal{D}_{tr}$, and $\hat{\Theta} = \{\hat{\theta}, \theta_s\}$ be the model parameters related to $\mathcal{A}_{tr}$, where θ_s is the shared parameter. The final optimization objective function of our SSTE can be expressed as follows,

$$\min_{\widetilde{\Theta}, \hat{\Theta}} \mathcal{L}_{SSTE} = \mathcal{L}_{\mathcal{D}_{tr}}\left(f\left(\widetilde{\Theta}, \boldsymbol{x}_i\right), y_i\right) + \mathcal{L}_{\mathcal{A}_{tr}}\left(g\left(\hat{\Theta}, \boldsymbol{x}_j\right), y_j\right) + \lambda\|\widetilde{\Theta}\| + \lambda\|\hat{\Theta}\|, \quad (1)$$

where $(\boldsymbol{x}_i, y_i) \in \mathcal{D}_{tr}$, $(\boldsymbol{x}_j, y_j) \in \mathcal{A}_{tr}$, $f(\cdot)$ and $g(\cdot)$ are the mapping functions, and λ and $\|\cdot\|$ are the tradeoff weight and the regularization terms, respectively.

Note that we will adopt the model $\hat{\Theta}$ during the inference phase. An intuitive interpretation for Eq. (1) is that joint training on data that simulates different environments forces the model to implicitly distinguish all the biases, where those biases that reach more consensus and encode more information are more likely to be beneficial, since the performance gains they bring are more robust across different environments.

The Self-Evaluation Module. To better capture the optimal model $\hat{\Theta}^*$ offline, we propose a bias-robust evaluation method. Specifically, let the performance of the model $\hat{\Theta}$ on $\mathcal{D}_{val}$ and $\hat{\mathcal{D}}^i_{val}$ be $e(\mathcal{D}_{val})$ and $e(\hat{\mathcal{D}}^i_{val})$, respectively, at the i-th training iteration, where $e(\cdot)$ denotes the main metric adopted. Combining $\mathcal{D}_{val}$ and $\mathcal{A}_{val} = \{\hat{\mathcal{D}}^i_{val}\}^{T_2}_{i=1}$, we can compute the maximum difference α in performance between any two sets,

$$\alpha = \max\{\max|e(\mathcal{D}_{val}) - e(\hat{\mathcal{D}}^i_{val})|, \max|e(\hat{\mathcal{D}}^i_{val}) - e(\hat{\mathcal{D}}^j_{val}|)\}, \qquad (2)$$

where $|\cdot|$ denotes an absolute value operation. Since $\mathcal{D}_{val}$ and $\mathcal{A}_{val}$ are simulated for different bias environments, and an ideal optimization model should have a stable performance in different environments, an intuitive idea is that the model with a smaller value of α is better. Finally, depending on whether $e(\cdot)$ is a higher-better metric, we use $e(\mathcal{D}_{val}) - \alpha$ or $e(\mathcal{D}_{val}) + \alpha$ as a modified performance result to better capture the optimal model $\hat{\Theta}^*$ offline with a manageable risk.

3.4 Remarks

IPS-based methods are an important branch in debiased recommendation, but the performance of these methods heavily depends on the estimation accuracy of IPS. Different from them, our SSTE only uses IPS as a reference to simulate different bias environments, and thus has a greater tolerance for the estimation accuracy of IPS. Our SSTE can be applied to many industrial recommendation scenarios because it does not depend on a specific data, architecture or training method, and is compatible with most industrial recommendation models. Furthermore, since the sampled auxiliary subset usually has a much smaller size than the original data, i.e., $|\hat{\mathcal{D}}^i_{tr}| \ll m$ and $|\hat{\mathcal{D}}^i_{val}| \ll n$, our SSTE does not increase the time and resource overhead too much.

4   Experiments

In this section, we first introduce the experimental setup, and then conduct extensive empirical studies and show the effectiveness of our SSTE.

4.1   Experimental Setup

Datasets. We use a very common benchmark dataset and a real product dataset in our experiments, i.e., Yahoo! R3 [13] and Product. Following the settings of most previous works [3,10], for Yahoo! R3, we first binarize each rating, where those greater than 3 are denoted as $y = 1$, and the rest as $y = 0$. For the biased feedback subset in Yahoo! R3, we randomly divide each user's feedback into training and validation sets with a ratio of 8 : 2. The randomized feedback subset in Yahoo! R3 is all used for unbiased evaluation, since it can be considered as the feedback generated by an unbiased scene. Product is a subset sampled from the log data collected in Tencent Licaitong's homepage recommendation business, involving about 2.8 million users, 560 items, and 9.6 million feedback. According to the different properties of the feedback data, we divide them into a biased feedback subset and a randomized feedback subset. After chronological ordering of the biased feedback subset, we obtain the training and validation sets in a ratio of 8 : 2. Similarly, the whole randomized feedback subset is used for unbiased evaluation. The statistics of the datasets are shown in Table 1.

Table 1. Statistics of the datasets. P/N represents the ratio between the numbers of positive and negative feedback.

	Yahoo! R3		Product	
	#Feedback	P/N (%)	#Feedback	P/N (%)
training set	254,713	67.02	7,133,519	5.21
validation set	56,991	67.00	1,633,716	5.20
test set	54,000	9.64	870,158	4.34

Evaluation Metrics. For all the experiments, we employ four evaluation metrics that are widely used in recommender systems, including the area under the ROC curve (AUC), precision (P@K), recall (R@K) and normalized discounted cumulative gain (nDCG). We report the results of P@K and R@K when K is 5 and 10, and the results of nDCG when K is 50. Since AUC is one of the most common metrics in an industrial recommendation, we choose it as our main evaluation metric, which is used to search for the best hyperparameters for all the candidate methods.

Baselines. To conduct a comprehensive evaluation of our SSTE, in the experiments, we select a set of representative debiased recommendation methods based on only a biased data (i.e., without a randomized dataset), including IPS [16], SNIPS [17], CVIB [18], AT [14], Rel [15] and DIB [11]. Furthermore, similar to most of the previous works, we adopt matrix factorization (MF) [6] and neural collaborative filtering (NCF) [4] as two backbone models. Therefore, each of these baselines has two corresponding versions based on different backbones. Note that we do not include CausE [2], KDCRec [9] and AutoDebias [3] because they require a randomized dataset for bias reduction.

Implementation Details. All the candidate methods have been implemented on TensorFlow. We set the optimizer to Adam, and use the hyperparameter search library *Optuna* [1] to speed up the hyperparameter search process with AUC as the target on the validation set. For our SSTE, we set both the number of self-samples T_1 and T_2 to 1, considering the tradeoff of performance and resource overhead. To avoid overfitting to the training set, we also employ an early stopping setting with a patience of 5. We tune the embedding dimension, the regularization weight, the batch size and the learning rate in the range of $\{5, 10, \cdots, 195, 200\}$, $\{1e^{-5}, 1e^{-4}, \cdots, 1e^{-1}, 1\}$, $\{2^7, 2^8, \cdots, 2^{13}, 2^{14}\}$ and $\{1e^{-4}, 5e^{-4} \cdots 5e^{-2}, 1e^{-1}\}$, respectively. Note that the source codes are available at https://github.com/dgliu/DASFAA23_SSTE.

4.2   Overall Results

The comparison results between our SSTE and the baselines are shown in Table 2. When using matrix factorization as the backbone model, as shown in the upper part of Table 2, our SSTE consistently outperforms all the baselines on all the metrics across the two datasets of Yahoo! R3 and Product. We can also observe that most debiasing methods can improve the unbiased performance of recommendation models to some extent, among which Rel and DIB are the two most competitive baselines. Compared with them, our SSTE can benefit from self-training and self-evaluation modules to achieve a better result. When using neural collaborative filtering as the backbone model, as shown in the lower part of Table 2, our SSTE outperforms all the baselines in most cases. We can find that although our SSTE is slightly weaker than DIB on P@10 and R@10 on Product, it still has a clear advantage on other metrics, especially the main metric AUC. This may be due to the impact caused by only considering AUC in parameter tuning. Overall, our SSTE has a better unbiased performance.

Table 2. Comparison results of unbiased evaluation, where the best results and the second best results are marked in bold and underlined, respectively. AUC is the main evaluation metric.

Method	Yahoo! R3						Product					
	AUC	nDCG	P@5	P@10	R@5	R@10	AUC	nDCG	P@5	P@10	R@5	R@10
MF	.7101	.0447	.0080	.0074	.0236	.0446	.8477	.0696	.0145	.0180	0687	.1707
IPS-MF	.7128	.0313	.0033	.0035	.0088	.0202	.8507	.0684	0139	.0180	.0663	.1717
SNIPS-MF	.7101	.0334	.0049	.0048	.0131	.0271	.8509	.0684	.0140	.0181	.0670	.1717
CVIB-MF	.7048	.0479	.0089	.0069	.0267	.0413	.8279	.0705	.0146	.0185	.0695	.1761
AT-MF	.7314	.0663	.0108	.0100	.0373	.0676	.8389	.0649	.0137	.0169	.0649	1595
Rel-MF	.7440	.0835	.0151	.0131	.0508	.0859	<u>.8519</u>	<u>.0785</u>	<u>.0177</u>	.0199	<u>.0838</u>	.1882
DIB-MF	<u>.7547</u>	<u>.0920</u>	<u>.0169</u>	<u>.0145</u>	<u>.0538</u>	<u>.0930</u>	.8510	.0781	.0159	<u>.0202</u>	.0764	<u>.1922</u>
SSTE-MF	**.7591**	**.0981**	**.0181**	**.0153**	**.0612**	**.0999**	**.8525**	0878	**.0222**	**.0211**	**.1062**	**.2010**
Method	Yahoo! R3						Product					
	AUC	nDCG	P@5	P@10	R@5	R@10	AUC	nDCG	P@5	P@10	R@5	R@10
NCF	.7244	.0294	.0022	.0026	.0068	.0147	.7930	.0795	.0192	.0195	.0907	.1845
IPS-NCF	.7229	.0291	.0031	.0036	.0084	.0199	.8024	.0802	.0187	.0188	.0886	.1781
SNIPS-NCF	.7224	.0314	.0034	.0034	.0091	.0183	.8026	.0869	.0195	.0206	.0921	.1945
CVIB-NCF	.7250	.0393	.0053	.0044	.0150	.0250	.8034	.0776	.0170	.0193	.0805	.1831
AT-NCF	.7167	.0351	.0056	.0048	.0157	.0271	.7967	.0741	.0181	.0179	.0856	.1688
Rel-NCF	.6895	.0505	.0083	.0073	.0246	.0451	.7970	.0827	.0183	.0198	.0869	.1880
DIB-NCF	<u>.7454</u>	<u>.0671</u>	<u>.0114</u>	<u>.0096</u>	<u>.0365</u>	<u>.0602</u>	**.8049**	<u>.1011</u>	<u>.0233</u>	**.0241**	**.1105**	**.2293**
SSTE-NCF	**.7561**	**.0696**	**.0115**	**.0101**	**.0370**	**.0638**	<u>.8061</u>	**.1033**	**.0247**	<u>.0228</u>	<u>.1175</u>	<u>.2171</u>

4.3   Compatibility Analysis

As described in Sect. 3, since our SSTE does not depend on a specific architecture or training method, it can be easily integrated with existing debiased recommendation methods and traditional recommendation methods. To verify the compatibility of our SSTE, in our experiments, we integrate it with all the baselines and compare it with the original baselines after re-searching for the best hyperparameters. We report the results on Yahoo! R3 in Fig. 5. We can find that our SSTE can bring a significant improvement over the unbiased performance of different baselines in most cases. In particular, we can observe that the positive effect of our SSTE will be weakened on the debiased recommendation method based on inverse propensity score (IPS). This may be due to the fact that the inaccurate estimation of IPS would bring an irreconcilable hurt to a recommendation model.

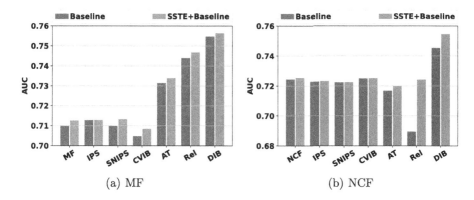

(a) MF   (b) NCF

Fig. 5. Recommendation results of our SSTE with different baselines on Yahoo! R3.

4.4   Online A/B Test

Finally, we deploy our SSTE in Tencent Licaitong, which is a large-scale internet financial platform dedicated to providing high-quality fund sales services for users. In this platform, there are tens of millions of active users every day, and a large number of feedback logs are generated and recorded. We conduct online A/B test for one month in the homepage recommendation scenario, which is the first page after user log in. In this recommendation scenario, a set of funds will be recommended to the user, and the user can perform some related operations, such as skip, click and purchase. The display page is shown in the left side of Fig. 1. The base model compared in the online test is a carefully tuned deep multi-task model in which clicks, conversions, and purchase amounts are predicted separately. We deploy SSTE on the same architecture, and both models are trained over the same training dataset, which contain more than 300 million logged feedback spanning two months. For online serving, 10% users are randomly selected as the experimental group and are served by SSTE, while another

30% users are in the control group for the base model. Different from the evaluation metrics adopted in offline experiments, we introduce three online evaluation metrics that are more concerned in financial recommendation, i.e., total clicks per mille (CLPM), total conversions per mille (COPM) and purchase amount per mille (PAPM). Specifically, CLPM, COPM and PAPM can be calculated by $\frac{Total\_Clicks}{Total\_Impressions} \times 1000$, $\frac{Total\_Conversions}{Total\_Impressions} \times 1000$ and $\frac{Purchase\_Amount}{Total\_Impressions} \times 1000$, respectively. The online A/B test results are shown in Fig. 6. We can find that our SSTE can bring a steady improvement on the three evaluation metrics. Overall, our SSTE can achieve an average improvement of 3.75%, 7.20% and 12.11% on CLPM, COPM and PAPM, respectively, in the whole online A/B test. This further demonstrates the effectiveness of our SSTE.

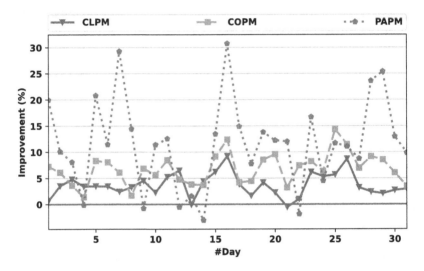

Fig. 6. The improvement of our SSTE compared with the base model in the online A/B test, including total clicks per mille (CLPM), total conversions per mille (COPM) and purchase amount per mille (PAPM).

5   Conclusions

In this paper, we first show through an online study that blindly removing all biases in an industrial recommendation application may not consistently yield a desired performance improvement. To achieve a better accuracy-bias tradeoff, we propose a simple yet effective self-sampling training and evaluation (SSTE) framework to preserve the beneficial biases while removing the harmful ones. Our SSTE contains three new modules, i.e., a self-sampling module constructs debiased subsets for training and validation, a self-training module aims to jointly learn the accuracy-bias tradeoff based on the original training data and debiased subset, and a self-evaluation module aims to capture an optimal model offline based on the original validation data and debiased subsets. We conduct extensive experiments on a public dataset and a real product dataset, and find that our SSTE can effectively improve the unbiased performance of the recommendation models, and is also of good compatibility. Finally, our SSTE demonstrates a steady improvement on core evaluation metrics in an online A/B test.

Acknowledgements. We thank the support of National Natural Science Foundation of China Nos. 61836005, 62272315 and 62172283.

References

1. Akiba, T., Sano, S., Yanase, T., Ohta, T., Koyama, M.: Optuna: a next-generation hyperparameter optimization framework. In: SIGKDD, pp. 2623–2631 (2019)
2. Bonner, S., Vasile, F.: Causal embeddings for recommendation. In: RecSys, pp. 104–112 (2018)
3. Chen, J., et al.: AutoDebias: Learning to debias for recommendation. In: SIGIR, pp. 21–30 (2021)
4. He, X., Liao, L., Zhang, H., Nie, L., Hu, X., Chua, T.S.: Neural collaborative filtering. In: TheWebConf, pp. 173–182 (2017)
5. Jadidinejad, A.H., Macdonald, C., Ounis, I.: The simpson's paradox in the offline evaluation of recommendation systems. ACM TOIS **40**(1), 1–22 (2021)
6. Koren, Y., Bell, R., Volinsky, C.: Matrix factorization techniques for recommender systems. Computer **8**, 30–37 (2009)
7. Liang, D., Charlin, L., Blei, D.M.: Causal inference for recommendation. In: Workshop on Causation: Foundation to Application co-located with the 32nd Conference on Uncertainty in Artificial Intelligence (2016)
8. Lim, D., McAuley, J., Lanckriet, G.: Top-N recommendation with missing implicit feedback. In: RecSys. pp. 309–312 (2015)
9. Liu, D., Cheng, P., Dong, Z., He, X., Pan, W., Ming, Z.: A general knowledge distillation framework for counterfactual recommendation via uniform data. In: SIGIR, pp. 831–840 (2020)
10. Liu, D., Cheng, P., Zhu, H., Dong, Z., He, X., Pan, W., Ming, Z.: Mitigating confounding bias in recommendation via information bottleneck. In: RecSys. pp. 351–360 (2021)
11. Liu, D., et al.: Debiased representation learning in recommendation via information bottleneck. In: ACM TORS (2022)
12. Liu, D., Lin, C., Zhang, Z., Xiao, Y., Tong, H.: Spiral of silence in recommender systems. In: WSDM, pp. 222–230 (2019)
13. Marlin, B.M., Zemel, R.S.: Collaborative prediction and ranking with non-random missing data. In: RecSys, pp. 5–12 (2009)
14. Saito, Y.: Asymmetric tri-training for debiasing missing-not-at-random explicit feedback. In: SIGIR, pp. 309–318 (2020)
15. Saito, Y., Yaginuma, S., Nishino, Y., Sakata, H., Nakata, K.: Unbiased recommender learning from missing-not-at-random implicit feedback. In: WSDM, pp. 501–509 (2020)
16. Schnabel, T., Swaminathan, A., Singh, A., Chandak, N., Joachims, T.: Recommendations as treatments: debiasing learning and evaluation. In: ICML, pp. 1670–1679 (2016)

17. Swaminathan, A., Joachims, T.: The self-normalized estimator for counterfactual learning. In: NeurIPS, pp. 3231–3239 (2015)
18. Wang, Z., Chen, X., Wen, R., Huang, S.L., Kuruoglu, E.E., Zheng, Y.: Information theoretic counterfactual learning from missing-not-at-random feedback. In: NeurIPS, pp. 1854–1864 (2020)
19. Wang, Z., He, Y., Liu, J., Zou, W., Yu, P.S., Cui, P.: Invariant preference learning for general debiasing in recommendation. In: SIGKDD, pp. 1969–1978 (2022)
20. Yang, L., Cui, Y., Xuan, Y., Wang, C., Belongie, S., Estrin, D.: Unbiased offline recommender evaluation for missing-not-at-random implicit feedback. In: RecSys, pp. 279–287 (2018)

Raven: Benchmarking Monetary Expense and Query Efficiency of OLAP Engines on the Cloud

Tongyu Wu[1], Rong Gu[1(✉)], Yang Li[2], Hongbin Ma[2], Yi Chen[1], Ying Zhu[1],
Xiaoxiang Yu[2], Tengting Xu[2], and Yihua Huang[1(✉)]

[1] State Key Laboratory for Novel Software Technology,
Nanjing University, Nanjing, China
{tongyu.wu,chenyi,201220068}@smail.nju.edu.cn, {gurong,yhuang}@nju.edu.cn
[2] Kyligence, Inc., Shanghai, China
{yang.li,hongbin.ma, xiaoxiang.yu,tengting.xu}@kyligence.io

Abstract. Nowadays, it is prevalent to build OLAP services on cloud platforms. Cloud OLAP adopters are eager to understand and characterize the performance of OLAP engines on the cloud. However, traditional OLAP benchmarks are usually designed for on-premise environments. When evaluating cloud OLAP engines, they have limitations on cloud environment adaption and cloud scenario benchmark execution. To address these issues, this paper proposes Raven, a cloud-oriented OLAP benchmark with flexible system architecture and diversified workloads. Raven supports cloud service deployment and various cloud OLAP engine integration. In addition, to simulate complex cloud query scenarios, we design a group of timeline-based and service-oriented workloads. We implement Raven on the Amazon AWS cloud platform and use it to evaluate typical types of widely-used OLAP engines, including Presto, SparkSQL, Kylin, and Athena. Experimental results show that Raven can effectively benchmark diversified OLAP engines. Besides, Raven can benchmark various configuration settings of an identical OLAP engine. We also explore an OLAP case study on the cloud using Raven.

Keywords: Cloud computing · OLAP engine · Query evaluation

1 Introduction

Online analytical processing (OLAP) is a widely-used data analysis technique to explore insights over large volumes of data [5,20]. These OLAP applications can promote decision support and business intelligence for organizations [22,32]. Nowadays, cloud computing provides a convenient and economic way to share computing resources over the public network [30]. Therefore, we observe that many cloud service companies set off to build OLAP services on their platforms [14]. Some design novel OLAP engines on cloud platforms, like Snowflake [6], Athena, Azure SQLDB, and AnalyticDB [34]. Others take on migrating existing OLAP engines to the cloud. Many on-cloud OLAP services are available in the Amazon AWS marketplace, such as Presto [21], SparkSQL [1], Hive [25], Impala [11], and

Druid [33]. Kyligence Inc. also presents Kyligence Cloud for deploying Apache Kylin [24] on the cloud. Users need to choose an appropriate OLAP engine and settings for real-world business scenarios on the cloud. Yet, it is challenging for users to evaluate OLAP performance precisely and economically on the cloud. Many researchers start to focus on benchmarking databases in the cloud environment [16]. They try to benchmark OLTP or interactive cloud serving systems [2,4,12]. The above benchmarks can adapt to cloud architectures. However, they cannot simulate specialized OLAP workloads. Therefore, users cannot migrate them directly for OLAP engine performance analysis.

Before cloud computing emerges, a variety of prior work also exists to address OLAP benchmarking on clusters [27] or virtual machines [28]. In addition, the Transaction Process Performance Council (TPC) benchmarks, such as TPC-H [26] and TPC-DS [19], are widely used for database performance evaluation. However, they have problems in benchmarking OLAP engines on the cloud. The problems are two-fold: On one hand, the benchmark needs to adapt to the cloud-based architecture. Unlike clusters or virtual machines, applications, storage, or monitoring are provided as a service on the cloud [15]. For example, due to computation-storage disaggregation, an OLAP benchmark on the cloud needs to interact with the storage service to fetch data. Yet, prior solutions cannot interact with these cloud services. On the other hand, the benchmark should accurately simulate diversified, complicated scenarios. Today's benchmarks require the simulation of a series of queries executed at different times with different characteristics. However, traditional benchmarks typically execute a set of queries sequentially. Thus, they can hardly simulate the complex OLAP scenarios on the cloud. These problems call for a cloud-adaptable benchmark to evaluate OLAP engines under complex workload scenarios on the cloud.

To address these problems, we propose *Raven*, a cloud environment-oriented OLAP engine performance benchmark platform. Raven supports cloud architecture adaptation and flexible workload execution. Raven is open-sourced at https://github.com/PasaLab/Raven/. The main contributions are as follows:

- **Cloud environment-oriented OLAP engine benchmark framework.** Raven is built over cloud service, allowing direct adoption of cloud services such as storage, monitoring, etc. Therefore, Raven supports convenient setup and OLAP performance evaluation. In addition, the design of Raven's Generic Engine Runtime (GER) allows multiple OLAP engines to plug in.
- **Stage and event-based workloads with timeline execution.** Raven defines workloads by *stage* and *event*. Stages and events build up the computing logic, consisting of operations like running a query, building indexes, and so on. The workload timeline controls the execution of the workload. Users can configure the timeline, stage, and events of the workload.
- **System implementation and evaluation.** We have implemented the above ideas and design in Raven, and adopted it to benchmark various OLAP engines on the AWS cloud platform. The experimental results show that Raven can effectively benchmark diversified OLAP engines on complex workloads. It can also benchmark various configuration settings of an identical OLAP engine. We also explore a cloud OLAP case study using Raven.

2   Benchmark Framework

2.1   System Architecture

Raven is a distributed benchmarking platform with four important components: Launcher, Configuration Handler, Workload Executor, and Performance Analyzer. The system architecture of Raven is illustrated in Fig. 1.

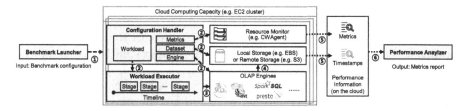

Fig. 1. System Architecture of Raven. The workflow contains three phases: initialization (dashed arrows), execution (solid arrows), and analysis (dotted arrows).

Launcher is the entry of Raven. It sets up the required cloud computing resource for OLAP performance benchmarking on the cloud. When users trigger Launcher, it follows these steps: First, Launcher creates a cloud computing cluster based on resource settings specified by users. Then, Launcher deploys Configuration Handler and Workload Executor on the cloud cluster. Finally, Launcher sends the benchmark specifications to Configuration Handler.

Configuration Handler parses the benchmark specification and configures the cloud cluster. Configuration Handler operates in the following three steps. First, it reads and parses the benchmark specification from Benchmark Launcher. Second, it distributes the parsed specifications to the four sub-modules shown in Fig. 1. Finally, the sub-modules configure the cloud cluster according to the specification, including the OLAP engine, workload, datasets, and metrics. Each sub-module handles one part of the specification.

Workload Executor generates and executes OLAP benchmarking workloads. Specifically, it reads the workload profile, generates and maintains the workload data structure, and executes operations or OLAP queries. A workload includes one or multiple stages on the execution timeline. A stage is a set of OLAP queries or non-query operations like building indexes. There exist a group of events in a stage. Events can be either a series of script commands or SQL queries. When an event is triggered, Workload Executor generates a thread and offloads an execution command to the OLAP engine. Then, the thread waits for the OLAP engine execution results. After that, Workload Executor persists the time-stamped logs to the remote storage for further analysis. More design details of the workload are in Sect. 3.

Performance Analyzer conducts in-depth performance analysis based on the metric data collected during the execution of the OLAP engine. First, Performance Analyzer collects performance-related information during execution. The information has two sources: local time-stamped logs and cloud monitoring service. Then, it calculates the information mentioned above and figures out metrics, including the QoS of the query and the expense on the cloud.

2.2   Raven Workflow

The workflow of Raven consists of three steps: benchmark startup, engine execution, and performance analysis. The arrows in Fig. 1 illustrate Raven's workflow
of benchmarking a specific OLAP engine.

Benchmark Startup. In this step, Raven launches a cloud cluster and pre-
pares the necessary environment for OLAP engine execution on the cluster.
First, Launcher reads the resource settings and creates the cloud computing
cluster (marked ①). For example, on Amazon AWS, Raven calls an AWS client
command to launch an EC2 cluster. Second, the user copies Raven and its configuration files to all instances on the cloud. Third, the user launches Configuration
Handler to set up the workload for query execution(marked ②).

Engine Execution. In this step, Raven generates a benchmark workload and
runs it on the targeted OLAP engine. Workload Executor generates a timeline
consisting of one or multiple stages. If the stage is made of non-query operations,
we call it an offline stage. In the offline stage, Workload Executor runs the
command as a shell script. If the stage consists of OLAP queries, we call it
an online stage. In the online stage, Workload Executor sends the SQL query
to the OLAP engine, and the engine executes the query (marked ③). Some
OLAP engines read the data schema from local storage, while others read data
from remote cloud storage before stage execution (marked ④). The stages are
executed one by one until all of them are finished.

Performance Analysis. In this step, Raven collects the performance information of the OLAP engine. While running the workload, Raven records the timestamps of each event. Also, cloud monitoring services emit all resource usage data
during workload execution. Raven sends the timestamps and data to the cloud
storage after the workload is done (marked ⑤). Then, Performance Analyzer
acquires the above information from the cloud storage (marked ⑥) and figures
out metrics. The metrics include the QoS of the query and the expense on the
cloud. If necessary, Performance Analyzer adopts a customized penalty function
to determine the overall cost based on both QoS and expense metrics.

Raven can evaluate different types of OLAP engines because of its generic
engine runtime (GER). GER presents the abstraction of OLAP engines on the
cloud. Raven designs the *Engine* abstract class to implement GER. This class
provides three major abstract functions. *setConf* function is for engine configuration setup. *launch* function is for the initialization of the OLAP engine. *query*
function is for the execution of OLAP queries. Each OLAP engine has its way
of configuring, launching, or querying. Therefore, they have different implementations of these functions. To sum up, Raven's design of generic engine runtime
provides the abstraction for various OLAP engines, which is of great help for
on-cloud engine performance evaluation.

3   Service-Oriented Query Benchmark Workload

In this section, we introduce the design of the timeline-based and service-oriented
benchmark workloads pattern. This design describes a set of workloads in sequential online or offline stages. On the one hand, the design of online stages helps to

describe OLAP queries into timestamped events. On the other hand, the design of offline stages enables the support of running scripts, allowing the implementation of pre-processing techniques, such as building indexes in advance. In conclusion, a service-oriented workload pattern is suitable for supporting various complex real-world OLAP scenarios on the cloud.

3.1 Workload Design

The design of our query workload pattern is shown in Fig. 2.

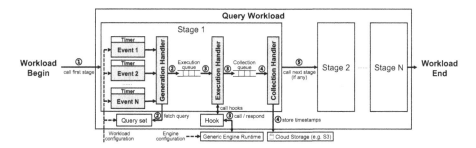

Fig. 2. Timeline-based, service-oriented query workload design. Dashed lines represents the preparation part, and solid lines represents the execution part.

Concepts. To introduce the workload pattern clearly, we first elaborate on the key concepts that are used in the workload pattern. *Event* is the abstraction of atomic operation in the benchmark workload. On the one hand, if the stage is online, events are OLAP queries that run given SQL statements on the OLAP engine. On the other hand, if the stage is offline, events are a series of commands. *Stage* is a set of one or multiple events, either online or offline. The former contains OLAP queries as events, while the latter consists of non-query operations, such as building indexes in advance. Running non-query commands before OLAP queries may help OLAP engines accelerate queries. *Workload timeline* is a sequence of multiple stages. The workload timeline controls the execution of workload stages in the benchmark. Benchmark users can customize and define the workload they want.

Procedure. Workload procedure has two parts: generation and execution. *Workload generation* includes the following steps: First, Workload Executor attaches a timestamp and a timer for each event. The timestamp on the event decides when to start. When the time of the timestamp is reached, the timer triggers the event. Second, Workload Executor builds and maintains an OLAP engine hook. This hook links to the generic engine runtime described in Sect. 2.2, and the generic engine runtime connects with the OLAP engine to benchmark. Third, Workload Executor loads the query set. The query set consists of all queries that may be used during workload execution. Finally, Workload Executor creates handlers and thread pools. There are three handlers in a stage: a generation handler, an execution handler, and a collection handler. Each handler is in charge of one thread pool. All the handlers communicate through synchronous message queues. After the four steps above, Workload Executor chains all stages up to a complete workload. *Workload execution* has the following steps: First, Workload Executor calls the first stage of the workload (marked ①). When one timer is triggered, Workload Executor sends the corresponding event to the generation handler. After the thread is ready, the generation handler pushes the event profile to the execution queue (marked ②). Second, the execution handler gets the event from the execution queue. If the stage is an online one, it calls the hook to link the OLAP engine via the generic engine runtime. If the stage is an offline one, it opens a command line for script execution. When the OLAP engine or the command line finishes and responds, the execution handler pushes the response to the collection queue (marked ③). Third, the collection handler persists the timestamps of the event to the remote storage on the cloud for further analysis (marked ④). Workload Executor records timestamps for each event's state changes by sending messages to the collection handler. Finally, when a stage is finished, Workload Executor finds the next stage to run (marked ⑤). The workload is over when all stages are finished.

3.2 Built-in Workloads

Raven allows customized workload configuration to generate diversified workloads. For convenience, Raven has several built-in workloads to simulate typical OLAP workloads. Their distribution histograms are shown in Fig. 3.

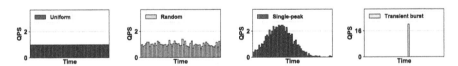

Fig. 3. Example histograms for built-in workloads.

Uniform input is common in updating stock market status or daily financial reports. In this workload, the same query is executed periodically. Raven triggers events according to the preset frequency, and each event runs the same query.

Random input can happen in biochemistry analysis, e.g. the analysis of cells when moving the microscope. The random input workload simulates irregular calls of the same query. Raven tags all events with a random timestamp to determine when to trigger it, and each event runs the same query.

Single-peak input usually happens in scenarios like customer behavior analysis in a dining hall. Throughout mealtime, the query frequency increases until reaching a peak, then the frequency decreases. The single-peak input workload simulates a set of queries following a normal distribution.

Transient burst input can simulate scenarios like uploading content analysis of social media when breaking news comes. The transient burst input simulates a huge number of queries coming at the same time.

Custom input is available if the models proposed above cannot meet the requirement of users. Users can specify the settings to define when to execute what type of commands or queries.

4 Evaluation

4.1 Experimental Setup

Hardware and Software. We use Amazon Web Services (AWS) to conduct all sets of evaluation experiments. Raven is implemented on the AWS framework, which includes the remote storage system S3, the computing platform EC2, and the metrics collection service CloudWatch. To send requests to and receive responses from AWS services, Raven uses the *boto3* Python package.

Unless otherwise specified, the experiments are conducted on an AWS cluster with 1 master instance of type $m5.xlarge$, and 4 core instances of $m5.4xlarge$. The CPUs of all $m5$ instances are Intel Xeon Platinum 8000 series. $m5.xlarge$ instances have 4 virtual CPUs and 16 GB memory space, and $m5.4xlarge$ instances have 16 virtual CPUs and 64 GB memory space.

As to the software, we use AWS EC2 instances with Hadoop 2.10.1. To monitor system metrics, we also install *CWAgent* and *collectd* on each EC2 instance. All instances require the following Python packages: numpy $\geq$ 1.21.1, boto3 $\geq$ 1.18.15, PyHive $\geq$ 0.6.4, requests $\geq$2.26.0, pyspark $\geq$ 3.1.2, and PyYAML $\geq$ 6.0.

OLAP Engines and Datasets. Raven can benchmark the performance of several OLAP engines on the cloud. In the experiments, we evaluate types of typical OLAP engines, including Spark SQL [1], Presto [21], Apache Kylin [24], and Amazon Athena. In our experiments, we use Hive 2.3.7, Spark SQL 2.4.7, Presto 0.245.1, Apache Kylin 3.1.2, and Amazon Athena 2. We use the datasets of Star Schema Benchmark (SSB) [18], at a scale factor of 100.

Workloads. As discussed in Sect. 3.2, Raven has several built-in workloads. For simplicity, we introduce the workloads to be used in later experiments with their abbreviation. *Uni(x)* workloads satisfy uniform distribution, where queries from the query set are randomly triggered every 10 s. The number in brackets stands for the workload duration in minutes. Workloads with a pattern like *Bur(x)* are transient burst workloads. The number in the brackets is the number of queries triggered simultaneously.

Metric Statistics. Raven collects information and calculates the quality of service (QoS) of queries and the expense on the cloud. *QoS of queries* is reported by the timestamps recorded for all events. Events in Raven include two time-related metrics: response time and query time. Response time is how long one event stays in the queue, whereas Query time is the time elapsed for event execution. *Expense on the cloud* is acquired from the AWS cost manager, including the money spent on AWS EC2 instances, Simple Storage Service, monitoring service CloudWatch, and other related services.

Table 1. QoS of query metrics

Metric	Engine	Avg time(s)	95% time(s)	Max time(s)
Response Time	Presto	1007.52	1833.41	1940.11
	Spark SQL	504.75	984.71	1021.47
	Apache Kylin	0.07	<0.01	4.51
	Athena	<0.01	<0.01	<0.01
Query Time	Presto	130.21	255.43	295.91
	Spark SQL	82.75	121.32	124.78
	Apache Kylin	2.30	5.46	14.52
	Athena	2.63	3.34	4.31

Table 2. Config sets to evaluate

Memory size parameters	Config set (GB)				
	M1	M2	M3	M4	M5
query.max-total-memory-per-node	10	20	30	40	50
query.max-total-memory	40	80	120	160	200
query.max-memory-per-node	8	16	24	32	40
query.max-memory	32	64	96	128	160
Thread concurrency parameters	**Config set**				
	TS	TM	TL		
task.max-worker-threads	16	32	64		
task.concurrency	16	32	64		

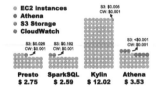

Fig. 4. Expense structure of different OLAP engines on the SSB dataset (SF=100), running a uniform input workload. Each circle stands for $0.1 of on-cloud expense.

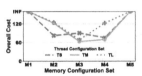

Fig. 5. Cost of various configuration of Presto on the SSB dataset (SF = 100), running a burst input workload. A lower cost is better. INF stands for runtime failure.

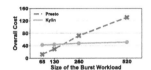

Fig. 6. Cost of Presto and Kylin on the SSB dataset (SF = 100), running burst input workloads of increasing size. A lower cost indicates better OLAP performance.

4.2 Benchmarking Different OLAP Engines Performance

In this subsection, we use Raven to evaluate different OLAP engines over the same workload. It verifies the effectiveness of Raven by running workloads on multiple engines. Attributed to the characteristics of Raven, users can make a proper choice among a number of OLAP engines.

This set of experiments evaluates Presto, Hive, Athena, and Spark SQL. The input data and query set is SSB, at a scale factor of 100. As to query plans, we use Uni(10) in these experiments. The benchmark collects metrics of on-cloud expense and QoS of queries. Figure 4 and Table 1 summarize the expense structure and QoS of query metrics in the experiments. These metrics help to decide which engine to choose. From the statistics, we find the following clues:

First, on average, running queries on Athena and Kylin is about 50× faster than that on Presto and Spark SQL. What's more, their response time is hundreds of thousands of times shorter. Athena provides fast, interactive query performance and automatically executes queries in parallel. Kylin runs queries faster because it builds cubes in advance to accelerate OLAP queries. Their response time is also lower due to their fast query execution. Since the OLAP engine can execute queries fast, the queries will not pile up in the waiting queue.

Second, the statistics show that the on-cloud expense is higher for the two faster engines, Athena and Kylin. Athena is more expensive because it runs as a service that needs extra expenses on AWS. The cloud expense of Kylin is 4 to 5 times more expensive because it has an offline stage to build cubes. The time spent on cube building reaches nearly 2 h in our evaluation. As a result, longer usage of EC2 instances leads to higher cloud expenses.

In addition, we also study Raven's overhead in this set of experiments. The overhead of Raven comes from two aspects. One is the extra time required by operations (e.g. parsing configuration), and the other is money spent by using monitoring services (e.g. CloudWatch). For the extra time, query execution logs show that the average time between the calls and the query is less than 4 milliseconds. For the monitoring services, the bill indicates that AWS CloudWatch spends only $0.001 on each workload, taking up less than 0.1% of the total expense. Therefore, the overhead is negligible compared to the workload execution.

This set of experiments reveals that Raven allows users to plug in various OLAP engines easily. Moreover, the overhead of the benchmark can be neglected.

4.3   Benchmarking Different Configurations of OLAP Engine

In this subsection, we use Raven to evaluate different configurations of the identical OLAP engine over the same workload. This set of experiments reveals how OLAP engine configurations affect its performance. Owing to the GER mentioned in Sect. 2.2, one OLAP engine can be easily configured and deployed.

This set of experiments takes Presto as a typical example to run with different configurations. The dataset and query set used here is SSB, at a scale factor of 100. The workload in this experiment is Bur(39). We use Eq. 1 to evaluate the OLAP engine's capability of running both short and long queries:

$$C = EoC \times \frac{q_{max} + 5q_{95\%} + 94q_{avg}}{100} \tag{1}$$

where C is the overall cost of the engine, EoC is the total expense on the cloud, and q_{max}, $q_{95\%}$, q_{avg} represents the maximum, 95% percentile, average query time. A lower cost indicates better OLAP engine performance. This set of experiments mainly focuses on two aspects of configuration: memory size and thread concurrency. The configuration sets we used in the experiments are listed in Table 2. Memory-related configuration sets are from $M1$ to $M5$, and concurrency-related ones are TS, TM, and TL. In practice, we run across 5 kinds of memory sizes and 3 kinds of thread concurrency configuration sets.

After the model's calculation, the results are illustrated in Fig. 5. From the results, we find the following clues: First, tuning the memory size too small or too large leads to a higher cost. When the memory size is too small, some queries fail because of running out of memory space. However, if the memory size goes too high, the memory allocated plus the heap memory exceeds the total available memory, leading to an illegal argument exception. Second, in a low-concurrency workload, increasing concurrency cannot reduce the cost. Specifying too many threads limits the memory space of each thread, slowing down OLAP queries.

This subsection proves that Raven is able to benchmark an identical OLAP engine with different configuration sets. Therefore, it helps users to tune for the best configuration for a given OLAP engine.

4.4   Case Study: What Are the Effects of Pre-processing?

This subsection studies the effects of pre-processing in Apache Kylin. Apache Kylin is a multi-dimensional OLAP (MOLAP) engine, and it builds *cubes* in advance on Hadoop and Spark for faster query speed. This case study explores the effectiveness of pre-processing with Raven quantitatively. Raven's service-oriented workload pattern enables the execution of pre-processing commands via RESTful APIs. What's more, Raven supports different sizes of workloads.

This set of experiments compares Apache Kylin (a MOLAP engine with pre-processing) with Presto (a typical OLAP engine without pre-processing). The experiments use an SSB dataset with a scale factor of 100. Several transient burst workloads are used to evaluate the quality of service. The evaluation begins with a Bur(65) workload, triggering 5 sets of 13 SSB queries simultaneously. Then, the number of sets of queries increases to 10, 20, and 40, naming Bur(130), Bur(260), and Bur(520), respectively. With Eq. 2, we evaluate whether the queries pushed into the execution queue stay long or not:

$$C = EoC \times \ln\left(r_{max}\right) \tag{2}$$

where C is the overall cost, EoC is the expense on the cloud, and r_{max} is the maximum response time. A lower cost indicates better OLAP performance.

For Apache Kylin, the computation includes an offline stage to pre-processing, e.g., building cubes. After that, Kylin runs the online stage. In the experiments, the expense of Apache Kylin is the sum of the two stages. Spark SQL only has an online stage, so the expense equals the online stage expense. In addition, both engines cache query results, which can interfere with the measurement of actual query time. To evaluate them fairly, caching of both engines is turned off.

The results are shown in Fig. 6. When the workload size is relatively small, Spark SQL has a lower overall cost. The number of queries is small, so running those queries is faster, making pre-processing seems time-consuming. However, when the workload size increases, the advantages of Kylin uncover. The online stage overhead is cut off by over 90% due to pre-processing techniques. The turning point of these two OLAP engines is between Bur(130), 10 sets, and Bur(260), 20 sets. After that point, Apache Kylin shows an overall cost advantage because pre-processing techniques accelerate queries.

5   Related Work

TPC-H [26], TPC-DS [19], and Star Schema Benchmark (SSB) [18] are popular OLAP benchmarks. These toolkits usually run all queries once at a time to evaluate their engine's performance. It is not suitable for cloud framework paradigms. BigDataBench [29] supports both offline and online big data query analysis. DIAMetrics [7,8] presents a novel framework for end-to-end benchmarking and performance observation of query engines developed in Google. Koalabench [17] benchmarks NoSQL databases, supporting models oriented from both columns and documents. SSB+ [3] improves the Star Schema Benchmark that dedicates to non-relational NoSQL models. These benchmarks generally focus on how to analyze big data querying. Similarly, existing research analyzes and further optimizes the performance of big data dataframe computation [10], file reading [9], and bloom filter indexing building [31] recently. However, they consider on-premise environments more than on-cloud ones.

Yahoo! Cloud service benchmark (YCSB) [4] and Kossmann's research [12] are common tools for performance comparisons of serving systems. However, evaluation benchmarks for analytical ones are relatively lacking. Kuschewski and Leis [13] propose a white-box OLAP performance modeling on the cloud for easier on-cloud instance configuration. Similar modeling analysis mainly considers computing servers [27]. Tan et al.'s paper [23] analyzes how to choose a cloud DBMS from an architectural view. However, there is still a blank spot to compare different query engines and tune a query engine for better performance.

6   Conclusion and Future Work

Effectively benchmarking OLAP on the cloud is essential. This paper proposes Raven, an OLAP benchmark on the cloud. Raven has a cloud environment-oriented OLAP engine benchmark framework and stage and event-based workloads with timeline execution. Experimental evaluation results demonstrate the effectiveness of Raven. In the future, we plan to explore how to evaluate on-cloud OLTP engines and how to benchmark engines over Serverless cloud architecture.

Acknowledgments. This work is funded in part by the China National Science Foundation (No. 62072230, U1811461), the Fundamental Research Funds for the Central Universities (No. 020214380089, 020214380098), Jiangsu Province Science and Technology Key Program (No. BE2021729), and the Collaborative Innovation Center of Novel Software Technology and Industrialization.

References

1. Armbrust, M., et al.: Spark SQL: relational data processing in Spark. In: SIGMOD Conference, pp. 1383–1394 (2015)
2. Battle, L., et al.: Database benchmarking for supporting real-time interactive querying of large data. In: SIGMOD Conference, pp. 1571–1587 (2020)
3. Chevalier, M., et al.: Benchmark for OLAP on nosql technologies comparing nosql multidimensional data warehousing solutions. In: RCIS, pp. 480–485 (2015)
4. Cooper, B.F., et al.: Benchmarking cloud serving systems with YCSB. In: SoCC, pp. 143–154 (2010)
5. Daase, B., et al.: Maximizing persistent memory bandwidth utilization for OLAP workloads. In: SIGMOD Conference, pp. 339–351 (2021)

6. Dageville, B., et al.: The snowflake elastic data warehouse. In: SIGMOD Conference, pp. 215–226 (2016)
7. Deep, S., et al.: DIAMetrics: benchmarking query engines at scale. SIGMOD Rec. **50**(1), 24–31 (2021)
8. Gruenheid, A., et al.: DIAMetrics: benchmarking query engines at scale. Proc. VLDB Endow. **13**(12), 3285–3298 (2020)
9. Gu, R., et al.: Improving in-memory file system reading performance by fine-grained user-space cache mechanisms. J. Syst. Archit. **1**(115), 1–15 (2021)
10. Gu, R., et al.: Octopus-DF: unified dataframe-based cross-platform data analytic system. Parallel Comput. **110**(2022), 1–12 (2022)
11. Kornacker, M., et al.: Impala: a modern, open-source SQL engine for hadoop. In: CIDR, pp. 1–10 (2015)
12. Kossmann, D., et al.: An evaluation of alternative architectures for transaction processing in the cloud. In: SIGMOD Conference, pp. 579–590 (2010)
13. Kuschewski, M., Leis, V.: White-box OLAP performance modeling for the cloud. In: CIDR, p. 1 (2021)
14. Lamb, A., et al.: The vertica analytic database: C-store 7 years later. Proc. VLDB Endow. **5**(12), 1790–1801 (2012)
15. Laszewski, T., Nauduri, P.: Chapter 1 - Migrating to the cloud. In: Migrating to the Cloud: Oracle Client/Server Modernization, pp. 1–19. Syngress, Boston (2012)
16. Li, C., et al.: The design and implementation of a scalable deep learning benchmarking platform. In: CLOUD, pp. 414–425 (2020)
17. Malki, M.E., et al.: Benchmarking big data OLAP nosql databases. In: UNet, pp. 82–94 (2018)
18. O'Neil, P.E., et al.: The star schema benchmark and augmented fact table indexing. In: TPCTC, pp. 237–252 (2009)
19. Pöss, M., et al.: TPC-DS, taking decision support benchmarking to the next level. In: SIGMOD Conference, pp. 582–587 (2002)
20. Queiroz-Sousa, P.O., Salgado, A.C.: A review on OLAP technologies applied to information networks. ACM Trans. Knowl. Discov. Data **14**(1), 8:1–8:25 (2020)
21. Sethi, R., et al.: Presto: SQL on everything. In: ICDE, pp. 1802–1813 (2019)
22. Steinmetz, N., et al.: Question answering on OLAP-like data sources. In: EDBT/ICDT Workshops, pp. 1–8 (2020)
23. Tan, J., et al.: Choosing a cloud DBMS: architectures and tradeoffs. Proc. VLDB Endow. **12**(12), 2170–2182 (2019)
24. The Apache Software Foundation: Apache Kylin — Analytical Data Warehouse for Big Data. http://kylin.apache.org/
25. Thusoo, A., et al.: Hive - a petabyte scale data warehouse using hadoop. In: ICDE, pp. 996–1005 (2010)
26. Transaction processing performance council: TPC-H homepage. http://www.tpc.org/tpch/
27. Varghese, B., et al.: Cloud benchmarking for performance. In: CloudCom, pp. 535–540 (2014)
28. Varghese, B., et al.: Container-based cloud virtual machine benchmarking. In: IC2E, pp. 192–201 (2016)
29. Wang, L., et al.: BigDataBench: a big data benchmark suite from internet services. In: HPCA, pp. 488–499 (2014)
30. Wu, Z., Li, K.: Vbtree: forward secure conjunctive queries over encrypted data for cloud computing. VLDB J. **28**(1), 25–46 (2019)
31. Xie, R., et al.: Hash adaptive bloom filter. In: IEEE ICDE Conference, pp. 636–647 (2021)

32. Xie, X., et al.: OLAP over probabilistic data cubes II: parallel materialization and extended aggregates. IEEE Trans. Knowl. Data Eng. **32**(10), 1966–1981 (2020)
33. Yang, F., et al.: Druid: a real-time analytical data store. In: SIGMOD Conference, pp. 157–168 (2014)
34. Zhan, C., et al.: AnalyticDB: real-time OLAP database system at Alibaba cloud. Proc. VLDB Endow. **12**(12), 2059–2070 (2019)

Dual Graph Multitask Framework for Imbalanced Delivery Time Estimation

Lei Zhang[1,2], Mingliang Wang[3], Xin Zhou[4], Xingyu Wu[3], Yiming Cao[1,2],
Yonghui Xu[2], Lizhen Cui[1,2(✉)], and Zhiqi Shen[4(✉)]

[1] School of Software, Shandong University, Jinan, China
clz@sdu.edu.cn
[2] Joint SDU-NTU Centre for Artificial Intelligence Research (C-FAIR),
Shandong University, Jinan, China
[3] Alibaba Group, Hangzhou, China
[4] School of Computer Science and Engineering, Nanyang Technological University,
Singapore, Singapore
zqshen@ntu.edu.sg

Abstract. Delivery Time Estimation (DTE) is a crucial component of
the e-commerce supply chain that predicts delivery time based on merchant information, sending address, receiving address, and payment time.
Accurate DTE can boost platform revenue and reduce customer complaints and refunds. However, the imbalanced nature of industrial data
impedes previous models from reaching satisfactory prediction performance. Although imbalanced regression methods can be applied to the
DTE task, we experimentally find that they improve the prediction performance of low-shot data samples at the sacrifice of overall performance.
To address the issue, we propose a novel Dual Graph Multitask framework for imbalanced Delivery Time Estimation (DGM-DTE). Our framework first classifies package delivery time as head and tail data. Then,
a dual graph-based model is utilized to learn representations of the two
categories of data. In particular, DGM-DTE re-weights the embedding
of tail data by estimating its kernel density. We fuse two graph-based
representations to capture both high- and low-shot data representations.
Experiments on real-world Taobao logistics datasets demonstrate the
superior performance of DGM-DTE compared to baselines.

Keywords: Delivery Time Estimation · Imbalanced Regression ·
Graph Neural Network

1 Introduction

As e-commerce proliferates, e-commerce logistics becomes a major industry
focus, and Delivery Time Estimation (DTE) is an important part of intelligent
e-commerce logistics. Accurate DTE can enhance the users' shopping experience
and increase the purchase rate to raises platform revenue [6].

In industrial e-commerce logistics scenarios, we focus on a category of Origin-
Destination (OD) DTE problems, where the delivery time of orders is predicted
based on known attributes, such as order merchant, sending address, receiving

© The Author(s), under exclusive license to Springer Nature Switzerland AG 2023
X. Wang et al. (Eds.): DASFAA 2023, LNCS 13946, pp. 606–618, 2023.
https://doi.org/10.1007/978-3-031-30678-5_46

Fig. 1. A DTE example on Taobao platform.

address, and payment time. Figure 1 presents a demonstration example of DTE when the user browses an item on the Taobao e-commerce platform.

Existing research formalizes the OD DTE problem as a regression problem, which uses end-to-end models such as Deep Neural Networks (DNNs) and representation learning [1,11,12] to predict the delivery time based on the order features. However, industrial e-commerce logistics data exhibits a skewed distribution of orders, *i.e.*, imbalanced data, as shown in Fig. 2(a). Most of the orders (about 90%) are delivered within 48–96 h (*i.e.*, the high-shot data region), with a portion of the data still in the medium-shot region (6.6%) and low-shot region (3.3%). As a result, models trained with such severely imbalanced data may have inferior performance on the medium- and low-shot data, as shown in Fig. 2(b). Besides, we find that the predicted values of these models are typically smaller than the real delivery time for orders within medium- and low-shot regions. Consequently, the platform may observe increasing user complaints and refund rates as orders cannot be received within the predicted time.

Dealing with imbalanced data in e-commerce logistics scenarios is a pressing challenge. There are two lines of research on imbalanced regression: synthesizing new samples for rare labeled data [3] and loss re-weighting [16,20]. Although these methods improve prediction performance for rare labeled data, they sacrifice prediction and representation performance for high-shot data, as shown in Table 3. Besides, current performance tests for imbalanced regressions are conducted on balanced test data, which does not make sense in practical industrial applications, where the test data is frequently also imbalanced.

To address the above challenges, this paper proposes a Dual Graph Multitask framework for imbalanced Delivery Time Estimation (DGM-DTE). Specifically, DGM-DTE first performs a classification task, which divides the orders into head and tail data according to the delivery time. Then, we leverage a dual graph-based representation module, one learns high-shot data representation in head data, and another re-weights the representations of tail data according to kernel density estimation of labels. For graph-based representation module, we build relation graphs from spatial, temporal, and merchant attributes of orders and use graph neural network (GNN) to capture both inter- and intra-correlations of attributes. Besides, we employ a simple but effective normalization for embedding de-biasing. The order representations learned from dual graph module are

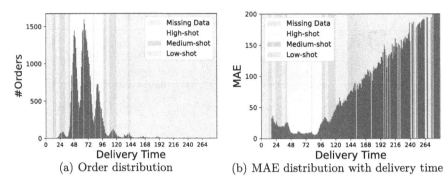

(a) Order distribution   (b) MAE distribution with delivery time

Fig. 2. Order distribution and MAE distribution with different delivery time.

then aggregated, so that the model can focus on both high-shot regional data and rare labeled data. Overall, we propose a multitask learning framework that predicts delivery time from two-view (classification and imbalanced regression). The main contributions of this paper are as follows.

- We focus on the imbalanced distribution of industrial e-commerce logistics data and propose a dual graph multitask model for imbalanced delivery time prediction.
- We design a GNN-based order representation module that can fully exploit the inter- and intra-correlation of order attributes.
- We conduct extensive experiments on real datasets from the Taobao platform to demonstrate the effectiveness of DGM-DTE in prediction performance. Various ablation studies validate the design of DGM-DTE is capable of improving the prediction accuracy of medium- and low-shot orders without compromising its performance on high-shot orders.

2   Related Work

2.1   Delivery Time Estimation

DTE is a category of estimated time of arrival problems, which is widely studied in transportation [8,10], logistics [1,13,21], and food delivery [9]. OD method is a line of research that predicts arrival time based on origin and destination without actual trajectories. [19] proposes a simple baseline model, which finds similar trips based on adjacent origin and destination. [2] introduces a simple static model with network optimization to predict travel time. [7,11] directly use DNN for end-to-end prediction. The above methods only use travel features for prediction without considering relations between attributes. MURAT [12] utilizes a multitask learning with GNNs to leverage the road network and spatio-temporal priors. Besides, BGE [13] is a bayesian graph model learning observed and unobserved attributes in logistics. However, existing methods ignore the imbalanced nature of data, resulting in unsatisfactory prediction performance.

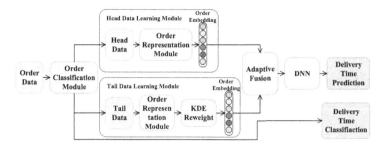

Fig. 3. The overall DGM-DTE framework.

2.2 Imbalanced Regression

The research of imbalanced regression is still in its initial stage, which can be divided into two streams: re-sampling and re-weighting. [3,4,17] introduce pre-processing strategies to re-sample and synthesize new samples for rare labeled data. Another line of research proposes the re-weight loss function to deal with imbalanced data. DenseWeight [16] weights the data based on the sparsity of the target value by kernel density estimation. [20] proposes two algorithms to smooth the distribution of labels and features. Besides, [15] designs a Balanced MSE loss function, which uses the training label distribution prior to recover the balanced prediction. However, existing methods compromise prediction performance over high-shot data and test the performance with balanced data, which is impractical in industrial applications.

3 The Proposed Model

The overall framework of the proposed DGM-DTE model is shown in Fig. 3. Firstly, we propose a classification module to divide orders into head and tail data. Then, we use the dual graph-based order representation module to learn the head and tail data embeddings separately. Finally, we aggregate the two parts of data embeddings for delivery time regression prediction.

3.1 Graph-based Order Representation Learning

The graph-based order representation module learns order embedding from the inter- and intra-correlation of order attributes, its structure is shown in Fig. 4.

We construct three graphs for three main attributes of orders, named spatial, temporal, and merchant relation graphs. For the weighted spatial relation graph $\mathcal{G}_S = (\mathcal{V}_{OD}, \mathcal{E}_S)$, where $\mathcal{V}_{OD}$ denotes the node set composed of OD pairs (*i.e.*, a group of sending and receiving addresses) of orders; $\mathcal{E}_S$ denotes a set of edges, which represents the relation of OD pairs in term of geographic location. For example, we define the weight of two OD pairs as the sum of the distance between their origins and the distance between their destinations. As for the unweighted temporal graph $\mathcal{G}_T = (\mathcal{V}_T, \mathcal{E}_T)$, which represents the periodicity of

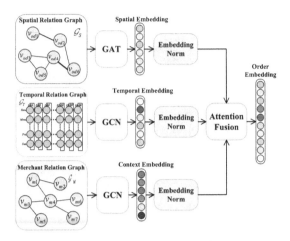

Fig. 4. Graph-based order representation model.

payment time in weeks and days. Each node in the temporal graph denotes the payment timestamp in hours, which is connected to adjacent hour nodes and its counterparts at the same hour of a week. Besides, the merchant graph $\mathcal{G}_M = (\mathcal{V}_M, \mathcal{E}_M)$ represents the similarity between merchants, which is manually defined based on historical orders.

To fully exploit the inter-correlation of attributes, GNN is used to learn node embeddings of attribute relation graphs. For the weighted spatial graph $\mathcal{G}_S$, we leverage Graph Attention Network (GAT) to learn OD node embedding,

$$
\boldsymbol{E}_{OD_i} = \sigma \left(\frac{1}{K} \sum_{k=1}^{K} \sum_{j \in \mathcal{N}\left(\mathcal{V}_{OD_i}\right)} \alpha_{ij} \boldsymbol{W} \boldsymbol{X}_{OD_j} \right),
\tag{1}
$$

where $\boldsymbol{X}_{OD_j} \in \mathbb{R}^{F_{OD}}$ denotes the initial feature of node $\mathcal{V}_{ODj}$, K denotes the number of head in multi-head attention. $\boldsymbol{W}$ is a trainable weight matrix, and $\mathcal{N}\left(\mathcal{V}_{OD_i}\right)$ is the neighbors of node $\mathcal{V}_{OD_i}$; and α_{ij} denotes the attention coefficient, which can be calculated as follows:

$$
\alpha_{ij} = \frac{\exp\left(\mathrm{ReLU}\left(f_a([\boldsymbol{W}\boldsymbol{X}_{OD_i} \mid \boldsymbol{W}\boldsymbol{X}_{OD_j}])\right)\right)}{\sum_{k \in \mathcal{N}\left(\mathcal{V}_{OD_i}\right)} \exp\left(\mathrm{ReLU}\left(f_a([\boldsymbol{W}\boldsymbol{X}_{OD_i} \mid \boldsymbol{W}\boldsymbol{X}_{OD_k}])\right)\right)},
\tag{2}
$$

where $f_a(\cdot)$ is a fully connected neural network. For the unweighted temporal graph $\mathcal{G}_T$, Graph Convolution Network (GCN) is utilized to learn the temporal node embedding.

$$
\boldsymbol{E}_T^l = \mathrm{GCN}\left(\boldsymbol{X}_T, \boldsymbol{A}_T\right) = \sigma \left(\widetilde{\boldsymbol{D}}_T^{-\frac{1}{2}} \widetilde{\boldsymbol{A}}_T \widetilde{\boldsymbol{D}}_T^{-\frac{1}{2}} \boldsymbol{E}_T^{l-1} \boldsymbol{W} \right),
\tag{3}
$$

where, $\widetilde{\boldsymbol{A}}_T = \boldsymbol{A}_T + \boldsymbol{I}$, $\widetilde{\boldsymbol{D}}$ denotes the degree matrix of $\widetilde{\boldsymbol{A}}_T$, $\boldsymbol{A}_T$ and $\boldsymbol{X}_T \in \mathbb{R}^{N_T \times F_T}$ are the adjacency matrix and initial node features of $\mathcal{G}_T$. Similarly, the merchant embedding can be learned via GCN as $\boldsymbol{E}_M = \mathrm{GCN}\left(\boldsymbol{X}_M, \boldsymbol{A}_M\right)$.

As representations of imbalanced data suffer from biased embedding, especially for attribute embedding. We further propose a simple but effective normalization method for embedding de-biasing on each attribute node embedding. Specifically, we use a per-dimension normalization to alleviate embedding bias, which calculates as $e_{mn}^{norm} = \frac{e_{mn}}{\|e_n\|}$.

Finally, we use an attention fusion to adaptively learn intra-correlation representation (*i.e.*, order embedding) according to spatial, temporal, and merchant embeddings. Due to varying contributions of different attributes to the order representation, we learn aggregation coefficients through a multi-head attention mechanism [18]. We use $Q = E_{OD}W^q$, $K = E_T W^k$, $V = E_M W^v$ as query, key, and value, respectively. $W^q \in \mathbb{R}^{d_{OD} \times d_O}$, $W^k \in \mathbb{R}^{d_T \times d_O}$, and $W^v \in \mathbb{R}^{d_m \times d_O}$ are weight matrices. The order embedding $E_O \in \mathbb{R}^{N \times d_O}$ can be represented as,

$$E_O = \text{CONCAT}\,(head_1, \ldots, head_h)\, W^O$$
$$head_i = \text{Attention}\left(QW_i^Q, KW_i^K, VW_i^V \right). \tag{4}$$

3.2 Delivery Time Classification

As Fig. 2 reveals, the imbalance of logistics data is mainly reflected in the large amount of data concentrated in the part of the header region. In contrast, the volume of data in the tail is small and spread over a wide area. Besides, it is also important to predict whether a package will arrive within a certain period (*e.g.*, three-day delivery) in practical e-commerce.

Therefore, we classify the delivery time as an auxiliary task of multitask learning. The division of binary classification is based on a defined time t_c, where orders with delivery time greater than t_c are in one category, *i.e.*, tail orders $z_{i,t}$, and vice versa are head orders $z_{i,h}$. In the order classification module, we first use a graph-based order representation to learn the order embedding E_O, then follow a MLP to obtain the classified output (*i.e.*, the prediction probability), $z_i = MLP(E_O) = [z_{i,h}, z_{i,t}]$, and $\widehat{y}_{ci} = argmax(z_i)$ is the classified prediction.

3.3 Dual Graph-based Order Representation

We propose a novel dual graph-based order representation learning for imbalanced regression learning, one for learning high-shot data representation in head data, and another for mining imbalanced tail data representation.

For the head data learning module, the input is the head data $O^{head} = O(\widehat{y}_{ci} = 0)$ predicted by the classification module, and the output (*i.e.*, order embedding) E_O^{head} is learned from the graph-based order representation module. However, it is inappropriate to directly use the same module for tail data $O^{tail} = O(\widehat{y}_{ci} = 1)$, as the tail data contains a wide range of delivery time and also shows an imbalanced distribution. So we propose an embedding re-weight strategy for the tail data learning module. A kernel density estimation is used to learn the imbalance property corresponding to continuous targets [20],

$$\widetilde{p}\,(y_t') \triangleq \int_{\mathcal{Y}} k\,(y_t, y_t')\, p(y_t) dy, \tag{5}$$

where k (y_t, y_t') is the Gaussian kernel function for tail label, and delivery time label are in the label space $\mathcal{Y}$, $i.e.$, $y_t, y_t' \in \mathcal{Y}$. Then, we can define the weight by square inverse of label density distribution, $\boldsymbol{w}_t = 1/\sqrt{\widetilde{p}\,(y_t')}$. The order embedding of tail data represented as $\boldsymbol{E}_O^{tail'} = \boldsymbol{w}_t \boldsymbol{E}_O^{tail}$.

3.4   Adaptive Delivery Time Prediction and Model Training

We leverage a DNN for regression prediction based on order representations for the main task delivery time prediction. First, an adaptive fusion is used to merge by index the head data embedding and re-weighted tail data embedding, The predicted delivery time is $\widehat{y}_r = \text{DNN}(merge(\boldsymbol{E}_O^{head}, \boldsymbol{E}_O^{tail'}))$.

Our proposed multitask framework has two tasks: delivery time classification as the auxiliary task and delivery time estimation as the main task. So the loss function contains two parts: Mean Absolute Error (MAE) for regression prediction and binary cross-entropy for classification,

$$\mathcal{L} = MAE(\boldsymbol{Y}_r, \widehat{\boldsymbol{Y}}_r) + BCE(\boldsymbol{Y}_c, \widehat{\boldsymbol{Y}}_c). \tag{6}$$

4   Experiments

4.1   Experimental Settings

Dataset. We collect two real e-commerce logistics datasets from Taobao, one of the world's largest e-commerce platforms. The first dataset, "D1", contains 452,917 orders received in Weihai, Shandong, China. The second dataset, named "D2", contains 1,048,575 orders with receiving addresses in Hangzhou, Zhejiang, China. We use time as the basis for dataset division since our goal is to predict the delivery time of future orders based on historical data. The details and division of the datasets are shown in Table 1.

To explore the imbalanced distribution of the dataset, we analyze the order distribution of two datasets, as shown in Fig. 5. Overall, both datasets exhibit seriously imbalanced distributions, with the more significant imbalance in dataset D1 manifesting as a wider distribution and a higher proportion of low-shot data. For D1, most orders are delivered within 48–96 h, but a few orders take an incredibly long time (>10 days). The delivery time of orders in D2 is generally shorter than D1. Most orders in D2 are delivered within 24–72 h, and the delivery time for low-shot data is typically greater than one week.

Evaluation Metrics. For the DTE task, we use the general regression metrics, $i.e.$, MAE, Mean Absolute Percentage Error (MAPE), and Window of Error (EW) [1] to evaluate the prediction performance. Where EW calculated as $p = \frac{1}{N}\sum_i^N H(EW - |y_r - \widehat{y}_r|)$, $H(\cdot)$ is the Heaviside step function, and p set as 90%, which measures the error window for 90% of orders.

Implementation Details. Our model is implemented on PyTorch and trained with a batch size of 2048. In the graph-based order representation module, the

Table 1. The Statistics of Experimental Datasets.

	#Order	#Merchant	#Sender	#Receiver	#Day	#OD pairs	#Train day	#Val. day	#Test day
D1	452,917	7,636	1,004	9	110	3,679	90	10	10
D2	1,048,575	9,954	1,152	57	51	10,689	37	7	7

(a) Order distribution in D1 (b) Order distribution in D2

Fig. 5. Order distribution in datasets.

initial node feature sizes of spatial, temporal, and merchant graphs are 207, 61, and 128, and node embedding dimensions after two layers GNN (*i.e.,* GAT, GCN) is 64 for all. The classification thresholds t_c for D1 and D2 datasets are 96 and 72 h, respectively. The order embedding size is 128, and the neuron numbers of DNN for regression prediction are 128, 64, and 32. We use Adam as the optimizer to train the model, and the learning rate is 0.0005.

Baselines. We compare our proposed DGM-DTE model with two types of baseline approaches for performance evaluation. The first type of baseline is the OD DTE models for the comparison of prediction performance, *e.g.,* **TEMP** [19] finding similar orders, **XGBoost** [5], **xDeepFM** [14], and **STNN** [11] directly using order features as the input, and graph-based model **MURAT** [12] and **BGE** [13]. Another type is the imbalanced models for validating the capability of the prediction for imbalanced data, such as **LDS** [20] and **BMSE** [15] for designing the re-weight loss function, and **SMOGN** [3] synthesizing data for pre-processing the imbalanced data.

4.2 Performance Comparison

We compare the proposed DGM-DTE model with existing OD DTE models and imbalanced regression models in terms of imbalanced DTE performance. The results are shown in Table 2. For a fair comparison with imbalanced regression models [3,15,20] that usually evaluated on balanced data, we also constructed a balanced test data for performance evaluation. We have the following findings by analyzing the experimental results: 1) Our DGM-DTE model outperforms existing models significantly on all datasets and evaluation metrics. The main

Table 2. DTE performance of different methods in terms of MAE, MAPE, and EW. The best results are **bold faces**, and the second best results are underlined.

Model	Types	D1			D2			Balanced Test		
		MAE	MAPE	EW	MAE	MAPE	EW	MAE	MAPE	EW
TEMP	OD	17.31	24.56%	30.05	11.22	23.56%	20.67	85.24	78.38%	178.38
XGBoost	DTE	16.15	24.43%	<u>28.68</u>	11.04	23.40%	<u>20.01</u>	88.62	60.12%	189.97
STNN		17.74	27.98%	33.73	12.57	24.46%	24.21	89.84	62.01%	186.97
BGE		15.61	25.36%	29.73	<u>10.26</u>	20.46%	20.67	87.02	61.19%	182.38
MURAT		17.74	29.52%	31.30	14.81	30.67%	25.08	86.28	59.96%	188.32
xDeepFM		16.76	39.49%	33.52	13.49	37.43%	29.64	86.11	69.98%	181.30
LDS	Imba-	17.15	25.65%	32.31	13.58	22.25%	22.08	86.46	<u>57.73%</u>	196.25
BMSE	lanced	<u>13.87</u>	<u>19.12%</u>	28.82	10.68	<u>19.26%</u>	21.02	<u>84.88</u>	59.16%	193.42
SMOGN	Models	20.75	28.91%	38.12	14.55	24.28%	24.37	85.09	60.24%	<u>173.77</u>
DGM-DTE	Ours	**11.97**	**16.81%**	**22.43**	**8.52**	**17.74%**	**19.77**	**83.30**	**57.43%**	**172.21**

Table 3. The MAE performance of different models in term of high-, medium-, and low-shot region data. The best results are in **bold faces**, and the second best results are underlined. The imrpov. is the improvement of ours vs. im-reg.

Shot	XGBoost	BGE	LDS	BMSE	SMOGN	im-reg	DGM-DTE	improve
High	13.5	12.04	22.46	12.18	35.78	<u>9.32</u>	**9.23**	1.0%
Medium	32.98	34.55	**30.35**	36.28	32.76	32.84	<u>31.88</u>	2.9%
Low	72.62	76.62	<u>62.27</u>	73.37	**60.57**	78.79	70.64	10.3%

reason is that our model considers the data imbalance and focuses on high-shot and rare labeled data, improving the performance on rare labeled data while maintaining high-shot data performance. For D1 dataset, the MAE of DGM-DTE outperforms existing models by 14% - 32%, and the MAE performance of our model improves by 17% - 42% on D2. For the balanced test, our model also outperforms existing imbalanced regression models, demonstrating that our model can be effectively used for imbalanced data prediction. 2) For OD DTE models, the performance of TEMP varies with the dataset, performing better on D2, because data in D1 is more substantial imbalanced than that in D2. Besides, the performance of XGBoost, STNN, and xDeepFM are unsatisfactory, indicating that only using order features is insufficient for DTE tasks. BGE achieves better performance since BGE introduces unobservable attributes of orders. 3) Some imbalanced regression models, such as LDS and SMOGN, perform well in the balanced test, but perform poorly on imbalanced real datasets. One plausible explanation is that such models focus on the rare labeled data and improve the predictive performance of rare labeled data, but at the expense of compromising the performance of high-shot data. 4) In general, D2 has better prediction performance than D1, especially on MAE and EW. The main reason is D1 suffers

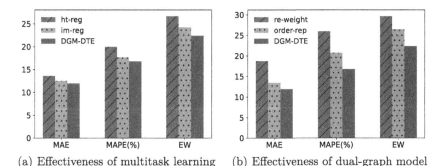

(a) Effectiveness of multitask learning   (b) Effectiveness of dual-graph model

Fig. 6. Ablation study of different parts of DGM-DTE model.

from severely imbalanced data, as shown in Fig. 5, D1 has a longer tail and a large number of data in the tails.

To better understand the effect of different methods on imbalanced data, we compare the performance of different methods on high-, medium- and low-shot data, which is displayed in Table 3. Our DGM-DTE outperforms the OD DTE models (*i.e.,* XGBoost and BGE) on all shots. The imbalanced regression models, LDS and SMOGN, achieved excellent performance in low-shot data, but inferior performance in high-shot, which suggests that these models enhance the performance of low-shot data at the expense of high-shot prediction. The im-reg is a variant of DGM-DTE, which directly uses imbalanced data as input of the dual graph module. The improvement shows that we can effectively improve the performance of low-shot data while ensuring high-shot performance by multitask learning with a dual graph module for the head and tail data separately.

4.3   Ablation Study

We design several variants and compare their prediction performance to verify the effectiveness of different components of DGM-DTE, as shown in Fig. 6.

To evaluate the effectiveness of multitask learning, we design two variants named ht-reg and im-reg. Both variants perform only regression task, where the input of ht-reg is manually divided into head and tail data, and im-reg with all imbalanced data as input for the dual graph model. The performance of ht-reg is worst since we cannot know whether the data is head or tail data in the test. Besides, im-reg model does not achieve better performance because it is insufficient to use re-weight to enable the model to focus more on low-shot data.

To investigate the effectiveness of the dual graph module, we design two variants: order representation model (short as order-rep, which only uses head data learning module with all data as input), and feature re-weight model (short as re-weight, which only uses tail data learning module with all data as input). The re-weight is a kind of imbalanced regression model, which pays more attention to rare labeled data, leading to poor performance on real imbalanced datasets.

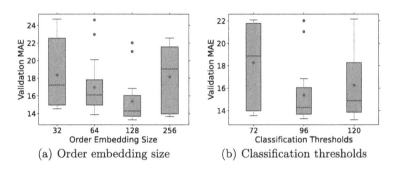

(a) Order embedding size   (b) Classification thresholds

Fig. 7. Performance of DGM-DTE with different parameters.

The order-rep does not consider the imbalance of the data and only uses GNN for order representation, which also fails to achieve satisfactory results.

4.4 Parameter Analysis

We analyze the sensitivity of the important parameters for prediction performance. The main parameters of DGM-DTE are order embedding size d_O and classification threshold t_c. To better observe the effect of different parameters on training, we analyze the MAE on the validation set, as shown in Fig. 7. For order embedding size, the MAE decreases and then increases significantly as d_O becomes larger, and shows optimal MAE performance at the size of 128. The MAE performs optimally at a classification threshold of 96 h, while the predictive performance is unstable at values of 72 and 120 h.

5   Conclusion

This paper presents a novel dual graph multitask framework for e-commerce delivery time estimation, which addresses the prevalent data imbalanced problem in the industry. We first classify the data into head and tail data depending on the delivery time, and then use a dual graph-based representation module to separately deal with the head and tail data from the classification, enabling the model to focus on both the high-shot data in the head and the rare labeled data in the tail. Finally, we aggregate two parts of data and estimate the delivery time. Experimental results on real datasets show that DGM-DTE can effectively improve the overall prediction performance, while also improving the predictive capability of rare labeled orders.

Acknowledgements. This work is partially supported by NSFC No.62202279; National Key R&D Program of China No. 2021YFF0900800; Shandong Provincial Key Research and Development Program (Major Scientific and Technological Innovation Project) (No. 2021CXGC010108); Shandong Provincial Natural Science Foundation (No. ZR202111180007); the Fundamental Research Funds of Shandong University.

This work is also supported, in part, by Alibaba Group through the Alibaba Innovative Research (AIR) Program and the Alibaba-NTU Singapore Joint Research Institute (AN-GC-2021-008-02); the State Scholarship Fund by the China Scholarship Council (CSC).

References

1. de Araujo, A.C., Etemad, A.: End-to-end prediction of parcel delivery time with deep learning for smart-city applications. IEEE IOTJ. **8**(23), 17043–17056 (2021)
2. Bertsimas, D., Delarue, A., Jaillet, P., Martin, S.: Travel time estimation in the age of big data. Oper. Res. **67**(2), 498–515 (2019)
3. Branco, P., Torgo, L., Ribeiro, R.P.: SMOGN: a pre-processing approach for imbalanced regression. In: Proceedings of LIDTA@PKDD/ECML 2017, vol. 74, pp. 36–50 (2017)
4. Branco, P., Torgo, L., Ribeiro, R.P.: REBAGG: resampled bagging for imbalanced regression. In: Proceedings of LIDTA@ECML/PKDD 2018, vol. 94, pp. 67–81. PMLR (2018)
5. Chen, T., Guestrin, C.: XGBoost: a scalable tree boosting system. In: Proceedings KDD 2016, pp. 785–794 (2016)
6. Cui, R., Lu, Z., Sun, T., Golden, J.: Sooner or later? Promising delivery speed in online retail. Promising Delivery Speed in Online Retail, 29 March 2020 (2020)
7. De Araujo, A.C., Etemad, A.: Deep neural networks for predicting vehicle travel times. In: 2019 IEEE SENSORS, pp. 1–4 (2019)
8. Fan, Y., Xu, J., Zhou, R., Liu, C.: Transportation-mode aware travel time estimation via meta-learning. In: Proceedings of DASFAA 2022, vol. 13246, pp. 472–488 (2022)
9. Gao, C., et al.: A deep learning method for route and time prediction in food delivery service. In: Proceedings of KDD 2021, pp. 2879–2889. ACM (2021)
10. Hong, H., Lin, Y., Yang, X., et al.: Heteta: heterogeneous information network embedding for estimating time of arrival. In: Proceedings of KDD 2020, pp. 2444–2454 (2020)
11. Jindal, I., Qin, T., Chen, X., Nokleby, M.S., Ye, J.: A unified neural network approach for estimating travel time and distance for a taxi trip. arXiv: 1710.04350 (2017)
12. Li, Y., Fu, K., Wang, Z., Shahabi, C., Ye, J., Liu, Y.: Multi-task representation learning for travel time estimation. In: Proceedings of KDD 2018, pp. 1695–1704 (2018)
13. Li, Y., et al.: Unsupervised categorical representation learning for package arrival time prediction. In: Proceedings of CIKM 2021, pp. 3935–3944 (2021)
14. Lian, J., Zhou, X., Zhang, F., Chen, Z., Xie, X., Sun, G.: XdeePFM: combining explicit and implicit feature interactions for recommender systems. In: Proceedings of KDD 2018, pp. 1754–1763 (2018)
15. Ren, J., Zhang, M., Yu, C., Liu, Z.: Balanced MSE for imbalanced visual regression. In: Proceedings of CVPR 2022, pp. 7916–7925. IEEE (2022)
16. Steininger, M., Kobs, K., Davidson, P., Krause, A., Hotho, A.: Density-based weighting for imbalanced regression. Mach. Learn. **110**(8), 2187–2211 (2021)
17. Torgo, L., Ribeiro, R.P., Pfahringer, B., Branco, P.: SMOTE for regression. In: Correia, L., Reis, L.P., Cascalho, J. (eds.) EPIA 2013. LNCS (LNAI), vol. 8154, pp. 378–389. Springer, Heidelberg (2013). https://doi.org/10.1007/978-3-642-40669-0_33

18. Vaswani, A., et al.: Attention is all you need. In: NeurIPS 2017, pp. 5998–6008 (2017)
19. Wang, H., Tang, X., Kuo, Y., Kifer, D., Li, Z.: A simple baseline for travel time estimation using large-scale trip data. ACM Trans. Intell. Syst. Technol. **10**(2), 19:1–19:22 (2019)
20. Yang, Y., Zha, K., Chen, Y., Wang, H., Katabi, D.: Delving into deep imbalanced regression. In: Proceedings of ICML 2021, vol. 139, pp. 11842–11851 (2021)
21. Zhou, X., Wang, J., Liu, Y., Wu, X., Shen, Z., Leung, C.: Inductive graph transformer for delivery time estimation. arXiv:2211.02863 (2022)

Real-Time Information Extraction for Phone Review in Car Loan Audit

Hongxuan Liu[1], Jie Wang[2(✉)], Yansong Wang[3], Shuling Yang[1],
Hanzhu Chen[1], and Binbin Fang[1]

[1] School of Data Science, University of Science and Technology of China, Hefei, China
{hxliu,slyang0916,chenhz,fangbinbin}@mail.ustc.edu.cn
[2] Department of Electronic Engineering and Information Science,
University of Science and Technology of China, Hefei, China
jiewangx@ustc.edu.cn
[3] Chery HuiYin Motor Finance Service Co., Ltd., Wuhu, China
wangyansong@cheryfs.cn

Abstract. Phone review is important in car loan audits, in which auditors contact applicants to make risk assessments by how applicants act to a sequence of questions. Due to the length of dialogues, auditors tend to miss important details, thus requiring an aiding system to record the dialogues in a compact form. Existing methods that utilize slot-value pairs to track the latest dialogue states fail to record the intermediate process which is critical for risk assessment. In this paper, we propose quadruples which consist of a dialogue act and a triple in a concept graph to represent the dialogue process, and model the dialogue recording task as a quadruple extraction problem for each utterance. To concisely construct quadruples, we convert slot-value pairs into a concept graph by disentangling domains from slots. In order to extract quadruples in real time, we design a model incorporating multi-head cross-attention mechanism and embedding sharing while considering parameter size and inference speed. Experiments on our real-world dialogue dataset show that our model achieves an accuracy of $\sim$82.7% which is similar to the best baseline with only $\sim$30 M parameters while performing real-time inference $\sim$3.6 times faster on an 8-core CPU with $\sim$90 ms per utterance.

Keywords: Information Extraction · Concept Graphs · Attention Mechanism

1 Introduction

Chery HuiYin Motor Finance Service Co., Ltd. is the leading company in China's car loan market. Car loans need to be audited before approval, where phone review plays an important role. According to how applicants act to a sequence of questions during phone review, auditors can assess the default and fraud risks of the loan. Dialogues for phone review are usually longer than those in

X. Wang et al. (Eds.): DASFAA 2023, LNCS 13946, pp. 619–630, 2023.
https://doi.org/10.1007/978-3-031-30678-5_47

widely researched dialogue datasets, which is shown in Table 1. During long conversations, auditors tend to forget questions they have asked as well as facts applicants have stated, which could lead to inaccurate risk assessments. Therefore, they require a real-time aiding system to record the dialogue process in a compact form to remind them of missed details.

Table 1. Comparison of our dataset to other dialogue datasets. EN stands for English and CN stands for Chinese, respectively. Dial. denotes dialogue.

Dataset	DSTC2	WOZ2.0	CrossWOZ	DuConv	KdConv	MovieChats	Ours
Language	EN	EN	CN	CN	CN	CN	CN
Avg. Turns/Dial	15.77	7.35	16.90	9.10	19.00	12.23	**45.84**
Avg. Tokens/Turn	8.47	11.27	16.25	10.60	13.50	14.77	**20.99**

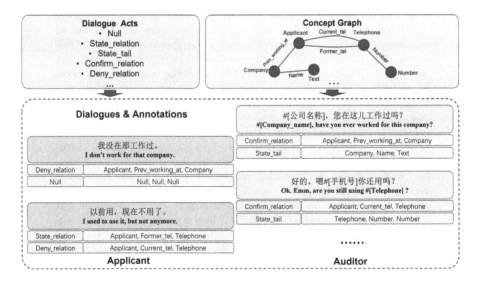

Fig. 1. An example of annotation. Quadruples are annotated below utterances. Red rectangles show dialogue acts. Green rectangles show triples. (Color figure online)

Existing works handle this problem by Dialogue State Tracking (DST) systems that track the latest dialogue state using slot-value pairs. However, these methods cannot record the dialogue process, which is important for auditors. For example, if an applicant's response changes for the same question, DST only retains the latest response, leading to the loss of previous information. Simply maintaining a value list for each slot is insufficient, as it does not capture the re-emphasis of unchanged values.

To handle these problems, we denote the dialogue process as a sequence of quadruples and model the dialogue recording task as a quadruple extraction problem for each utterance. An example is shown in Fig. 1. As shown in the example, each quadruple consists of a dialogue act and a triple of the subject, predicate, and object of a statement, and each utterance is denoted as a set of quadruples that each represents a question or a factual statement of the utterance.

For modeling triples concisely, we resort to concept graph, which is a kind of knowledge graph (KG) that contains nodes representing concepts and directed edges between concepts. However, constructing a concept graph from scratch is hard. To address this problem, we propose to build the concept graph and dialogue acts from existing slot-value pairs and dialogue acts. We map domains and values to concepts, and slots to relations, then decompose relations into existing concepts and new non-redundant relations to get the concept graph. Afterward, we adapt existing dialogue acts to the concept graph to improve precision.

We design a real-time quadruple extraction model with multi-head cross-attention and embedding sharing, which is composed of three modules stacked on top of each other. By selecting pre-trained language models carefully and incorporating embedding sharing, we achieved a significant reduction in the number of parameters. Additionally, by utilizing a single-pass setting that processes each utterance exactly once during inference, we can further improve the inference speed.

Evaluations of our proposed model and various baselines are performed on our dialogue dataset collected in a real-world scenario. These dialogues are produced by various auditors between different applicants to cover the richest possible dialogue content. Extensive experiments show that our proposed model achieves similar performance to best baseline models with fewer parameters and faster inference time while doing better on rare quadruples.

Overall, we make the following contributions:

- We denote the dialogue process dialogue as a sequence of quadruples and model the dialogue recording task as a quadruple extraction problem for each utterance.
- We propose to build concept graphs by converting existing multi-domain slot-value pairs into a concept graph and adapting dialogue acts to the concept graph.
- We design a model for real-time extraction of quadruples incorporating multi-head cross-attention mechanism and embedding sharing and employ multiple methods to reduce the parameter size and improve inference speed.
- Extensive experiments on our collected real-world dataset show that our proposed model can achieve similar results to the best baseline with only ∼30 M parameters while performing real-time inference on an 8-core CPU with ∼90 ms per utterance, which is ∼3.6 times faster than the best baseline.

2   Methods

In this section, we present the problem formulation of the dialogue recording task, the quadruple construction procedure, and the model design.

2.1   Problem Formulation

The goal of dialogue recording task is to extract a set of quadruples for the last utterance in each dialogue context. Given utterance u_i, its corresponding dialogue context $D_{i-\mu+1:i}$ is defined as

$$D_{i-\mu+1:i} = \{(u_{i-\mu+1}, r_{i-\mu+1}), (u_{i-\mu+2}, r_{i-\mu+2}), \ldots, (u_i, r_i)\} \qquad (1)$$

where μ is the context length, $u_j = \{u_{j,1}, u_{j,2}, \ldots, u_{j,m_j}\}$ is an utterance containing m_j tokens, and $r_j \in \{Auditor, Applicant\}$ denotes the speaker role of utterance u_j. As the final goal, quadruple set corresponding to u_i is defined as $Q_i = \{q_{i,1}, q_{i,2}, \ldots, q_{i,L}\}$ which contains L quadruples. Each $q_{i,k}$ is defined as $(a_{i,k}, s_{i,k}, p_{i,k}, o_{i,k}) \in Q_i$, $a_{i,k}$ denotes a dialogue act and $(s_{i,k}, p_{i,k}, o_{i,k})$ denotes a triple in the concept graph. Each $q_{i,k}$ can only be select from a finite candidate set $Q^{cand} = \{q_1^{cand}, q_2^{cand}, \ldots, q_N^{cand}\}$ of size N containing candidate quadruples. For utterance u_i , the dialogue recording task is defined as extracting Q_i from $D_{i-\mu+1:i}$. As we can see, extraction of each quadruple can be challenging, for the candidate set Q^{cand} contains all quadruples with valid combinations of a, s, p, and o. Figure 1 shows an annotation of $L = 2$, labeling the first utterance of the auditor as $q_{1,1} = (Confirm\_relation, Applicant, Prev\_working\_at, Company)$ and $q_{1,2} = (State\_tail, Company, Name, Text)$.

2.2   Quadruple Construction

Quadruples are built with dialogue acts and triples from a concept graph. In this section, we give a detailed description by first introducing the conversion from multi-domain slot-value pairs to a concept graph, then the method to adapt dialogue acts.

Converting multi-domain slot-value pairs to a concept graph takes two steps: mapping and decomposition.

- In the mapping step, we first categorize values into different concepts to increase reusability, then initialize the concept graph with previously constructed slot-value pairs, with domains and values to be concepts, and slots to be relations.
- In the decomposition step, by exploiting similar words and meanings between concepts and relations, we decompose relations into other concepts and relations without redundancy. For instance, $Owning\_company$ is decomposed into the existing domain $Company$ and a new relation $Owning$.

After concept graph construction, we find that dialogue acts are no longer precise, for values in slot-value pairs that are to be stated or asked are now decomposed into relations or concepts in a triple. For example, the dialogue act *Deny* cannot distinguish between *"I don't work for a company."* and *"I don't work for that company."* where the former applicant cannot work for any other company while the latter can. Therefore, we postfix 7 factually relevant dialogue acts which are always paired with slot-value pairs with _ *head*, _ *relation*, and _ *tail* to form 21 different dialogue acts. Finally, we get 44 dialogue acts, 79 concepts, and 163 relations from our multi-domain slot-value pairs, which are then constructed into 28,296 quadruples.

The first utterance of the auditor in Fig. 1 shows the power of combining dialogue acts and a concept graph: we can assign any position of (s, p, o) to be requested rather than just value position for slot-value pairs, which can remove redundancy between slots and domains.

2.3   Model Design

The proposed model consists of three parts: utterance encoder, context encoder, and quadruple decoder. An illustration of the model is shown in Fig. 2.

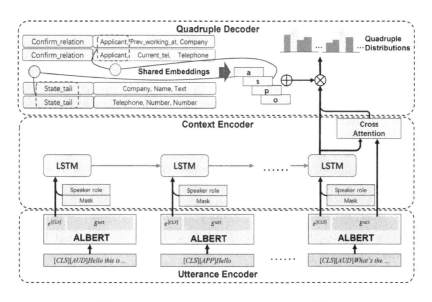

Fig. 2. Architecture of our proposed model.

Utterance Encoder. We use ALBERT [4], which is a lightweight language model for utterance encoding. Equipped with two parameter-reduction techniques, ALBERT is able to reduce parameter size significantly. In order to mitigate distribution shifts between the pre-training corpus and our dialogues, we incorporate extra fine-tuning on our dialogues with the masked language model (MLM) task and the sentence order prediction (SOP) task proposed by the original paper. During fine-tuning, we use $[AUD]$ and $[APP]$ as the auditor and applicant prefix of the utterance according to r_j. After fine-tuning, we format each (u_j, r_j) into the intra-utterance format $[CLS]\{[AUD], [APP]\}u_j$ and feed each utterance into ALBERT separately. We denote the output embedding of $[CLS]$ token as $e_j^{[CLS]} \in R^{d_{model}}$ and the embedding output of the whole utterance as $E_j^{utt} \in R^{m_j \times d_{model}}$ for u_j, where d_{model} is the hidden size of ALBERT.

Context Encoder. Utterance encoder only utilizes intra-utterance information and ignores inter-utterance information (or context information). However, context information serves as an important source for correctly extracting quadruples. To utilize contextual information, we use a Unidirectional Long Short-Term Memory Network (UniLSTM) [2] to encode inter-utterance information and multi-head cross-attention mechanism to refine intra-utterance representation. Moreover, UniLSTM can reduce the memory cost during inference by representing dialogue context with cell and hidden vectors of the last step.

For batched training which different context length, we combine $e_j^{[CLS]}$ with a mask bit $mask_j$ and a speaker role bit r_i to get

$$\tilde{e}_j^{[CLS]} = \text{Concat}(mask_j, r_i, e_j^{[CLS]}) \in R^{d_{model}+2} \tag{2}$$

UniLSTM takes $\tilde{E}_{i-\mu+1:i} = [\tilde{e}_{i-\mu+1}^{[CLS]}, \tilde{e}_{i-\mu+2}^{[CLS]}, \ldots, \tilde{e}_i^{[CLS]}]$ as input, and outputs inter-utterance embeddings $H_{i-\mu+1:i} = [h_{i-\mu+1}, \ldots, h_i] \in R^{\mu \times d_{hidden}}$ in the order of input, where d_{hidden} is the hidden size of UniLSTM.

We utilize multi-head cross-attention mechanism for intra-utterance representation refinement. For each attention head, hidden state h_i is mapped into query vector, and token embeddings E_i^{utt} are mapped into key vectors and value vectors. With n_{head} attention heads, multi-head cross-attention $MH_i \in R^{d_{hidden}}$ is computed as

$$MH_i = \text{Concat}(Head_{i,1}, Head_{i,2}, \ldots, Head_{i,n_{head}})W^O + h_i \tag{3}$$

where each attention head output $Head_{i,j} \in R^{d_{head}}$ is calculated by scaled dot-product attention [11].

Quadruple Decoder. The predicted score of each candidate quadruple q_j^{cand} for $q_{i,k}$ is calculated by

$$v_{i,j,k} = \frac{\exp(\hat{E}_i E_{j,k}^{cand^\top} + b_{j,k})}{\sum_{t=1}^N \exp(\hat{E}_i E_{t,k}^{cand^\top} + b_{t,k})} \tag{4}$$

where $\hat{E}_i = \text{Concat}(h_i, MH_i)$ is the estimated embedding, $E_{j,k}^{cand} \in R^{2d_{hidden}}$ and $b_{j,k} \in R$ are the learnable embeddings and biases of candidate quadruple

q_j^{cand} for $q_{i,k}$, respectively, $k \in \{1, 2, \ldots, L\}$. $E_{j,k}^{cand}$ and $b_{j,k}$ are shared across each utterance.

In order to improve the embeddings of rare quadruples with graph structure, we propose to share token embeddings across quadruples between the same tokens for each $q_{i,k}$. Quadruple embedding of q_j^{cand} for $q_{i,k}$ is define as

$$E_{j,k}^{cand} = a_{j,k}^{cand} + s_{j,k}^{cand} + p_{j,k}^{cand} + o_{j,k}^{cand} \tag{5}$$

where $a_{j,k}^{cand}, s_{j,k}^{cand}, p_{j,k}^{cand}, o_{j,k}^{cand}$ are shared for output $q_{i,k}$ among the same tokens across each utterance, and are trainable parameters with random initialization. Notice that embeddings are not shared between different ks.

Embedding sharing can also reduce parameter size. Let $T = |a| + |s| + |p| + |o|$ be the total number of dialogue act, subject, predicate, and object tokens. With embedding sharing, the quadruple embedding size can have a significant reduction from $O(N \times 2d_{hidden} \times L)$ to $O(T \times 2d_{hidden} \times L)$ if $N \gg T$, which is satisfied in our case with $N = 28,296$ and $T = 365$.

Loss Function. We use cross entropy loss for the predicted scores of $q_{i,k}$ as

$$Loss_{i,k} = -\frac{1}{N} mask_i \sum_{t=1}^{N} \log(v_{i,t,k}) y_{i,t,k} \tag{6}$$

where $y_{i,t,k}$ is the label, $mask_i = 0$ if utterance u_i is mask or 1 if not. Also, we add auxiliary losses calculated similarly for $u_{i-\mu+1}$ to u_{i-1} which are in the same context window as u_i. With the auxiliary losses, our model is encouraged to generate predictions for each utterance within the context window. The final loss is calculated as

$$Loss = -\frac{1}{\mu L} \sum_{j=i-\mu+1}^{i} \sum_{k=1}^{L} Loss_{j,k} \tag{7}$$

3   Experiments

3.1   Dataset Collection

Dataset collection is done in two steps: dialogue collection and dialogue annotation. We introduce these two steps in order, then present the dataset statistics.

Dialogue Collection. We collected the initial dataset of phone reviews from Chery HuiYin Motor Finance Service Co., Ltd. To improve data quality, we excluded dialogues whose corresponding audios are less than 3 min or less than 5 turns which indicates that dialogues were not proceeding as expected. Then we checked ASR text generated by our internal ASR system and excluded dialogues that are poorly transcribed because of poor recording quality. Finally, we corrected ASR texts manually to get the raw dataset.

Dialogue Annotation. We set the number of quadruples for each utterance $L = 2$ which is suitable for our dataset. In the case of containing only one or no statements in one utterance, the remaining quadruples are filled with null quadruples. Labels in the final dataset are highly unbalanced, where null quadruples take 23.3% of the samples in the first quadruple and 88.3% of the second.

Dataset Statistics. The final dialogue dataset we collect consists of 402 dialogues with 18,428 turns, where each turn contains 20.99 tokens on average. In the dataset, each dialogue is produced by a unique applicant. Preparing for 3-fold cross-validation, we randomly split the dataset into three folds in terms of dialogue. Then, each fold is split into a training set and a development set with train:dev=8:2. Statistics of the dataset are shown in Table 2.

Table 2. Dataset statistics for 3-fold cross-validation.

	Fold1		Fold2		Fold3		Total
	Train	Dev	Train	Dev	Train	Dev	
#Dialogues	108	26	108	26	108	26	402
#Turns	4,882	1,144	4,906	1,337	4,892	1,267	18,428
Avg. Turns/Dial	45.20	44.00	45.43	51.42	45.30	48.73	45.84
Avg. Tokens/Turn	21.01	21.18	20.90	19.85	21.18	21.61	20.99
#Auditors	47	20	42	22	50	20	66

3.2   Experiment Setups

Baselines. We compare our model to two types of baselines: generative baselines and discriminative baselines. The generative baselines are (1) GPT-2 [9], a general purpose Causal Language Model (CLM); (2) SimpleTOD [3], an approach designed for task-oriented dialogue with a single CLM. The discriminative baselines are (1) RoBERTa [5], a general purpose MLM; (2) TOD-BERT [14], a pre-trained MLM for task-oriented dialogues.

Implementation Details. We use a pre-trained ALBERT model which is trained with UER [19] on CLUE corpus [16]. We build our model with ALBERT model size $d_{model} = 1024$, UniLSTM hidden size $d_{hidden} = 512$, cross-attention heads $n_{head} = 2$,and head width $d_{head} = 256$. All three parts of the model are simultaneously trained with context length $\mu = 16$.

We use 3-fold cross-validation to evaluate model performance. For fold $n \in \{1, 2, 3\}$, we train models on training sets of fold $\{1, 2, 3\}\backslash\{n\}$ and do early stopping on development sets of the same folds, then perform evaluation on fold n.

We evaluate inference time on 8 cores of an Intel Xeon Silver 4214 CPU. During inference, we use a single-pass setting that set context length $\mu = \infty$ and collect outputs corresponding to each utterance as predictions. In the single-pass setting, each utterance gets processed exactly once per dialogue instead of μ times (once per context window), which leads to faster inference.

We use accuracy and accuracy* as metrics for our test results, where accuracy* denotes accuracy excluding null quadruples. We report results for all utterances, auditor's utterances, and applicant's utterances separately. Acc, Acc1, and Acc2 denote the accuracy of all, the first, and the second quadruple, respectively. Acc1* and Acc2* denote the accuracy* of the first and second quadruple, respectively. We also report models' parameter size and per-utterance inference time in milliseconds.

Table 3. Performance comparison of different methods. All the numbers are in percentage.

Model Name	Model Type	All					Auditor					Applicant				
		Acc	Acc1	Acc1*	Acc2	Acc2*	Acc	Acc1	Acc1*	Acc2	Acc2*	Acc	Acc1	Acc1*	Acc2	Acc2*
GPT-2	Generative	67.5	47.8	45.5	87.1	18.1	65.7	47.1	43.7	84.2	22.2	69.2	48.4	47.5	90.0	7.9
SimpleTOD	Generative	75.8	61.6	62.4	90.0	30.3	75.6	63.7	64.6	87.5	39.3	75.95	59.5	60.1	92.4	8.4
RoBERTa	Discriminative	79.0	67.1	64.6	90.9	*42.0*	79.4	69.8	67.7	88.9	*51.9*	78.6	64.4	61.5	92.8	*17.9*
TOD-BERT	Discriminative	**83.1**	**74.4**	**73.4**	*91.8*	41.2	*83.2*	**76.6**	**76.3**	*89.8*	51.2	**83.1**	**72.4**	**70.4**	**93.8**	16.8
Ours	Discriminative	*82.7*	*72.8*	*71.6*	**92.5**	**50.1**	**83.8**	*76.1*	*75.5*	**91.4**	**61.5**	*80.8*	*67.9*	*67.6*	*93.6*	**22.5**

Table 4. Per-utterance inference time and parameter size of each model. Per-utterance inference time is evaluated on an 8-core CPU.

Model Name	GPT-2	SimpleTOD	RoBERTa	TOB-BERT	Ours
Inference Time/ms	713.1	653.1	*302.8*	319.7	**89.6**
#Params	*97.3 M*	*97.3 M*	366.8 M	341.1 M	**30.1 M**

3.3 Main Results

Main experiment results are presented in Table 3. The results show that generative models GPT-2 and SimpleTOD have poorer performance than discriminative models, while the performance of our proposed model is on par with the discriminative model TOD-BERT with Acc. Though TOD-BERT performs better on first quadruples, our model performs much better than TOD-BERT with an absolute value of 8.9% on Acc2* evaluating over all speakers. Among all models, we observe a significant performance drop when using accuracy* as the evaluation metric compares to accuracy. Also, we notice that metrics for the second quadruple are always higher than the first one because the null quadruple counts for most correct cases in predictions. In addition, we observe that there is not always a positive correlation between accuracy* and accuracy.

Parameter sizes and per-utterance inference times of different models are shown in Table 4. Our proposed model achieves the best result in inference speed with the proposed model design and inference strategy, which is ~3.6 times faster than TOD-BERT and ~7.3 times faster than SimpleTOD while achieving a parameter size of ~30 M.

3.4 Ablation Study

Table 5 shows the results of ablation study. Starting from ALBERT, we add UniLSTM, multi-head cross-attention mechanism, and shared embeddings for candidate quadruples step-by-step. Results show that all structures we add give a performance boost to the Acc on all quadruples. UniLSTM gives a boost to all metrics. Multi-head cross-attention mechanism mainly improves the ability to identify null quadruples, while the benefit of shared embeddings is to extract rare quadruples with the same tokens more precisely.

Table 5. Ablation test for the proposed model. All the numbers are in percentage. "+UniLSTM" denotes adding UniLSTM, "+CrossAttn" denotes adding multi-head cross-attention mechanism, "+ShareEmb" denotes adding shared embeddings.

	All					Auditor					Applicant				
	Acc	Acc1	Acc1*	Acc2	Acc2*	Acc	Acc1	Acc1*	Acc2	Acc2*	Acc	Acc1	Acc1*	Acc2	Acc2*
ALBERT	76.7	61.3	56.1	92.1	42.4	81.5	72.1	69.7	90.8	55.4	72.1	50.9	42.3	93.3	10.5
+UniLSTM	81.6	70.7	69.6	92.4	47.7	82.5	73.7	73.2	91.2	59.1	80.7	67.7	65.9	93.6	19.8
+CrossAttn	82.4	72.2	70.4	92.5	46.5	83.4	75.6	74.2	91.2	58.4	81.3	68.8	66.5	93.7	17.6
+ShareEmb	82.7	72.8	71.6	92.5	50.1	83.8	76.1	75.5	91.4	61.5	80.8	67.9	67.6	93.6	22.5

4   Related Works

4.1   Dialogues with Knowledge Graphs

Recently, some dialogue datasets with KGs are proposed to integrate knowledge into dialogues [10,15,20]. MovieChats [10] collects chats in the movie domain by crowd-sourcing, then connects entities in a predefined KG to each utterance. KdConv [20] tags each utterance with a triplet from a KG covering three domains to record facts mentioned in utterances. Though they represent utterances with triples which is similar to ours, their dialogues are mainly generated by crowd-workers while ours are collected in real-world scenarios.

4.2   Utterance Tagging

Utterance tagging is a long-studied area. Researchers define different tags for utterances, like dialogue acts [7] and emotions [8]. Recently, unsupervised training and pre-trained models are introduced to benefit few-shot learning. Paranjape et al. [6] propose to combine pre-trained models to perform unsupervised learning on various dialogue datasets for low-resource utterance tagging. Zhang et al. [18] propose a two-step training method for few-shot intent detection task. While utilizing pre-training and pre-trained models, our work also employs embedding sharing to address the problem of rare quadruples.

4.3   Dialogue State Tracking

DST has long been researched in the task-oriented dialogue domain. DST aims at keeping dialogue states with slot-value pairs at each user utterance. Many DST datasets [12,13,21] and approaches [3,14] have been proposed, while some recent researchers also utilize relations between different slot-value pairs to enhance their model. SST [1] uses graph neural networks to fuse information between slots and utterances. STAR [17] incorporates self-attention to learn correlations between slots automatically. The graph of SST is built on both slots and tokens and the graph of STAR is built on slots, while we construct a concept graph by decomposing slots into concept level.

5   Conclusion

In this paper, we model the dialogue recording task as a quadruple extraction problem to extract each utterance into quadruples. To build quadruples, we propose to convert concept graphs from existing slot-value pairs and adapt existing dialogue acts. Since auditors require extraction results in real time, we design an extraction model incorporating multi-head cross-attention mechanism and shared token embeddings. Experiments demonstrate the effectiveness and efficiency of our model on the dialogue dataset collected in a real-world scenario.

Acknowledgements. This research was supported by Chery HuiYin Motor Finance Service Co., Ltd. and in part by National Nature Science Foundations of China grants U19B2026, 62021001, 61836011, and 61836006, and the Fundamental Research Funds for the Central Universities grant WK3490000004.

References

1. Chen, L., Lv, B., Wang, C., Zhu, S., Tan, B., Yu, K.: Schema-guided multi-domain dialogue state tracking with graph attention neural networks. In: Proceedings of the AAAI Conference on Artificial Intelligence, vol. 34, no. 05, pp. 7521–7528 (2020)
2. Hochreiter, S., Schmidhuber, J.: Long short-term memory. Neural Comput. **9**(8), 1735–1780 (1997)
3. Hosseini-Asl, E., McCann, B., Wu, C.S., Yavuz, S., Socher, R.: A simple language model for task-oriented dialogue. Adv. Neural Inf. Process. Syst. **33**, 20179–20191 (2020)
4. Lan, Z., Chen, M., Goodman, S., Gimpel, K., Sharma, P., Soricut, R.: ALBERT: a lite BERT for self-supervised learning of language representations. In: International Conference on Learning Representations (2020)

5. Liu, Y., et al.: RoBERTa: a robustly optimized BERT pretraining approach. arXiv preprint arXiv:1907.11692 (2019)
6. Paranjape, B., Neubig, G.: Contextualized representations for low-resource utterance tagging. In: Proceedings of the 20th Annual SIGdial Meeting on Discourse and Dialogue, pp. 68–74 (2019)
7. Pareti, S., Lando, T.: Dialog intent structure: a hierarchical schema of linked dialog acts. In: Proceedings of the Eleventh International Conference on Language Resources and Evaluation (LREC 2018) (2018)
8. Poria, S., Majumder, N., Mihalcea, R., Hovy, E.: Emotion recognition in conversation: research challenges, datasets, and recent advances. IEEE Access **7**, 100943–100953 (2019)
9. Radford, A., Wu, J., Child, R., Luan, D., Amodei, D., Sutskever, I.: Language models are unsupervised multitask learners (2019)
10. Su, H., et al.: MovieChats: chat like humans in a closed domain. In: Proceedings of the 2020 Conference on Empirical Methods in Natural Language Processing (EMNLP), pp. 6605–6619, November 2020
11. Vaswani, A., et al.: Attention is all you need. In: Advances in Neural Information Processing Systems, vol. 30 (2017)
12. Wen, T.H., et al.: A network-based end-to-end trainable task-oriented dialogue system. In: Proceedings of the 15th Conference of the European Chapter of the Association for Computational Linguistics: Volume 1, Long Papers, pp. 438–449 (2017)
13. Williams, J.D., Raux, A., Henderson, M.: The dialog state tracking challenge series: a review. Dialogue Discourse **7**(3), 4–33 (2016)
14. Wu, C.S., Hoi, S.C., Socher, R., Xiong, C.: TOD-BERT: pre-trained natural language understanding for task-oriented dialogue. In: Proceedings of the 2020 Conference on Empirical Methods in Natural Language Processing (EMNLP), pp. 917–929, November 2020
15. Wu, W., et al.: Proactive human-machine conversation with explicit conversation goal. In: Proceedings of the 57th Annual Meeting of the Association for Computational Linguistics, pp. 3794–3804 (2019)
16. Xu, L., Zhang, X., Dong, Q.: CLUECorpus2020: a large-scale chinese corpus for pre-training language model. arXiv preprint arXiv:2003.01355 (2020)
17. Ye, F., Manotumruksa, J., Zhang, Q., Li, S., Yilmaz, E.: Slot self-attentive dialogue state tracking. In: Proceedings of the Web Conference 2021, pp. 1598–1608 (2021)
18. Zhang, J., et al.: Few-shot intent detection via contrastive pre-training and fine-tuning. In: Proceedings of the 2021 Conference on Empirical Methods in Natural Language Processing, pp. 1906–1912, November 2021
19. Zhao, Z., et al.: UER: an open-source toolkit for pre-training models. EMNLP-IJCNLP **2019**, 241 (2019)
20. Zhou, H., Zheng, C., Huang, K., Huang, M., Zhu, X.: KdConv: a Chinese multi-domain dialogue dataset towards multi-turn knowledge-driven conversation. In: Proceedings of the 58th Annual Meeting of the Association for Computational Linguistics, pp. 7098–7108, July 2020
21. Zhu, Q., Huang, K., Zhang, Z., Zhu, X., Huang, M.: CrossWOZ: a large-scale Chinese cross-domain task-oriented dialogue dataset. Trans. Assoc. Comput. Linguist. **8**, 281–295 (2020)

HyperMatch: Knowledge Hypergraph Question Answering Based on Sequence Matching

Yongzhe Jia[1,2], Jianguo Wei[1(✉)], Zirui Chen[1], Dawei Xu[2], Lifan Han[1], and Yang Liu[1]

[1] College of Intelligence and Computing, Tianjin University, Tianjin, China
{jiayongzhe, jianguo, zrchen,hanlf,liruxru}@tju.edu.cn
[2] TechFantasy, Tianjin, China
xudawei@techfantasy.cn

Abstract. Question answering for automatic answer retrieval and knowledge hypergraphs for complex knowledge representations are currently two popular research areas. However, no knowledge hypergraph question answering method is available for answering complex questions. Secondly, current information retrieval-based methods cannot perform sequence ranking on a word basis when dealing with complex questions. They cannot add high weights to the relations of candidate entities related to the questions. We propose the HyperMatch method, which takes a single hyperedge as a unit and completes the extraction of candidate answer entities by sequence matching and multi-relation attention mechanism. Experiments show that HyperMatch can achieve a 9.44% improvement in the Hits@1 metrics. To the best of our knowledge, this is the first knowledge hypergraph question-answering method based on information retrieval.

Keywords: Knowledge hypergraph · Question answering · Sequence matching

1 Introduction

Question Answering over Knowledge Graphs (KGQA) has become an important research area over the last few years. Given a natural language question and a KG, the correct answer is derived based on the KG. Knowledge Hypergraphs (KH) is essentially a more expressive representation than knowledge graphs, in which the relation of each tuple is n-ary [17], allowing multi-hop information in the knowledge graph to be represented in a single hyperedge. As shown in Fig. 1, the knowledge that Marie Curie received the award needs to be represented by one knowledge hypergraph hyperedge or four knowledge graph triples. Therefore, using the knowledge hypergraph as the source of the QA system, the multi-hop question in the knowledge graph can be solved based on a single hyperedge.

However, existing KBQA methods based on information retrieval cannot complete Q&A based on the knowledge hypergraph structure. Most methods do not perform sequence matching at the word level nor aggregate matching

X. Wang et al. (Eds.): DASFAA 2023, LNCS 13946, pp. 631–642, 2023.
https://doi.org/10.1007/978-3-031-30678-5_48

results for decision-making to achieve the best matching between questions and candidate entities. Second, existing methods usually only consider entities and relations on the path from the topic entity to the candidate entity; other relations related to the candidate answer may contain helpful information about the candidate answer but are not considered.

In this paper, we propose HyperMatch, a KHQA method. The method adapts to the phenomenon that candidate entities contain multiple entities within a one-hop hyperedge and applies information retrieval to the knowledge hypergraph structure. Firstly, a matching model is applied to measure the similarity between questions and candidate entities, thus enabling word-level interactions through a bidirectional attention mechanism. Secondly, we incorporate contextual relationships with specific questions about candidate entities to enhance their representativeness. The method is evaluated on the knowledge hypergraph dataset WikiPeople-KH and the knowledge graph dataset WikiPeople-KG, constructed based on WikiPeople. The experimental results validate our hypothesis that the matching framework is better at finding the correct answer in KHQA and that including question-specific contextual relationships into the candidate representation can further improve the performance of KHQA.

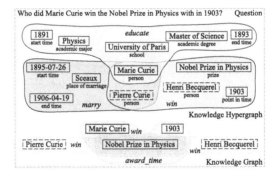

Fig. 1. An example of Marie Curie in knowledge hypergraph

The contributions of this paper are as follows:

- **Pipeline.** A knowledge hypergraph question answering pipeline, HyperMatch, is constructed to answer multi-hop complex questions in the knowledge graph based on a single hyperedge by exploiting the complex semantic properties of the knowledge hypergraph. The pipeline also solves the problem that the need for word-level alignment between two sequences and question-specific contextual relations to be given higher weights in traditional methods.
- **Benchmark.** Based on a single hyperedge of the knowledge hypergraph dataset WikiPeople, the Q&A dataset KHQuestions for knowledge hypergraphs is constructed. To evaluate the performance of the knowledge hypergraph and knowledge graph Q&A systems on KHQuestions, two knowledge

bases, WikiPeople-KH and WikiPeople-KG, are constructed. The system performance of both graph types can be evaluated on the same benchmark.

- **Experiments.** Our experiments have proven that the knowledge hypergraph-based Q&A system can perform better with the same data source, obtaining up to a 9.44% improvement in Hits@1 metrics. The superiority of completing Q&A based on the knowledge hypergraph structure is fully demonstrated.

The rest of this paper is organized as follows. Section 3 provides some preliminaries, including the knowledge hypergraph and the knowledge hypergraph question answering task. A detailed description of HyperMatch is provided in Sect. 4. Our performance evaluation of this matching method is reported in Sect. 5. Finally, we conclude this paper in Sect. 6.

2   Related Work

To the best of our knowledge, there is currently little related work in the KHQA field. Our approach is conceptually derived from two KGQA methods: semantic parsing and information retrieval. The latest progress in KGQA can be divided into the following categories:

2.1   Semantic Parsing

The SP-based methods follow a parse-then-execute procedure via a series of modules: question understanding, logical parsing, KB grounding, and KB execution. These modules will encounter different challenges for complex KBQA.

Firstly, question understanding becomes more difficult when the questions are complicated in both semantic and syntactic aspects. Many existing methods rely on syntactic parsings, such as dependencies [5] and Abstract Meaning Representation (AMR) [6], to provide better alignment between question constituents and logic form elements (e.g., entity, relation, entity types, and attributes). Secondly, logical parsing has to cover diverse query types of complex questions.

Moreover, a complex question involving more relations and subjects will dramatically increase the possible search space for parsing, which makes the parsing less effective. During parsing, traditional semantic parses (e.g., CCG [7]), which are developed without considering KB schemas, have shown their potential in parsing simple questions. Thirdly, the manual annotation of logic forms is expensive and labor-intensive, and it is challenging to train an SP-based method with weak supervision signals (i.e., question-answer pairs).

Based on the complexity and error propagation, we do not use the semantic parsing approach to handle the KHQA task.

2.2 Information Retrieval

The IR-based methods generate reasoning instructions by directly encoding questions as low-dimensional vectors through a neural network (e.g., LSTM) and extracting a question-specific graph from KBs.

Static reasoning instructions obtained through the above approaches cannot effectively represent the compositional semantics of complex questions. To comprehensively understand questions, recent work dynamically updated the reasoning instructions during the reasoning process. To focus on the currently unanalyzed part of the question [1], it proposed to update the reasoning instruction with information retrieved along the reasoning process.

Besides updating the instruction representation with the reasoned information, [2] proposed to focus on different parts of the question with a dynamic attention mechanism. This dynamic attention mechanism can promote the model to attend to other information conveyed by the question and provide proper guidance for subsequent reasoning steps. Instead of decomposing the semantics of questions, [3] proposed to augment the representation of the question with contextual information from the graph. After every reasoning step, they updated the reasoning instruction by aggregating information from the topic entity. Since simple questions only require one-hop reasoning on the neighborhood of the topic entity in KBs, IR-based methods are less likely to suffer from the inherent incompleteness of KBs [4]. It may be a severe problem for complex questions, where the correct reasoning path may need to be present in the question-specific graph.

There has yet to be a knowledge hypergraph-based information retrieval method. Compared to the current techniques, sequence matching at the word-level and relation weighting of candidate entities are achieved while obtaining multiple candidate entities based on a single hyperedge.

3 Preliminaries

In this section, the definition of the knowledge hypergraph and the question answering task performed on this structure are introduced.

3.1 Knowledge Hypergraph

A *knowledge hypergraph* is defined as $\mathcal{H} = (\mathcal{E}, \mathcal{R}, \mathcal{T})$, where $\mathcal{E}$, $\mathcal{R}$, and $\mathcal{T}$ is a finite set of entities, relations, and n-ary tuples, respectively. $t = r(\rho_1^r : e_1, \rho_2^r : e_2, ..., \rho_\alpha^r : e_\alpha)$ denotes a tuple where $r \in \mathcal{R}$ is a relation, each $e_i \in \mathcal{E}$ is an entity, each ρ_i^r is the corresponding role of relation r, and α is the non-negative integral arity of the relation r. While the triple of knowledge graph $r(e_1, e_2)$ is a special case of knowledge hypergraph in which the arity of any $r \in \mathcal{R}$ is two.

3.2   Knowledge Hypergraph Question Answering

A question is represented as a sequence of words $Q = (q_1, q_2, ..., q_m)$. Based on the knowledge hypergraph $\mathcal{H}$, the goal of the *knowledge hypergraph question answering* is to return a subset $\mathcal{A}$ of the entity set $\mathcal{E}$ as the answer.

4   Model

Like most previous work, our approach starts by identifying mentioned entities, i.e., entities mentioned in a given question. Then, these mentioned entities are used to identify candidate answer entities. In this work, the candidate entities are those entities that are directly associated with the mentioned entities through a single relation in a knowledge hypergraph.

A candidate sequence is constructed for each candidate entity using the entities and relations from the initially mentioned entity to the candidate entity. Then, using a neural network-based sequence matching model, all candidate sequences are matched against the question sequence to rank the candidate sequences and select the sequence with the first ranking as the answer. In the example shown in Fig. 1, Marie Curie, Nobel Prize in Physics, and 1903 are the mentioned entities. Considering only the one-hop relation, two entities can be derived from the mentioned entities: Pierre Curie and Henri Becquerel. An overview of HyperMatch is illustrated in Fig. 2.

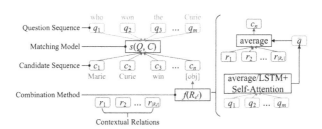

Fig. 2. An overview of HyperMatch

4.1   Candidate Entities

To identify a set of entities as candidate answers from the knowledge hypergraph, a set of mentioned entities is first identified from the given question. For example, given the question "Who won the Nobel Prize in Physics in 1902 at the same time as Marie Curie", Marie Curie, the Nobel Prize in Physics, and 1902 are the mentioned entities. The mentioned entities are obtained through the WikiData API. Use $E_Q^t \subset E$ to denote the mentioned entities found in question Q.

Next, for each mentioned entity $e^t \in E_Q^t$, all entities on the same hyperedge with e^t can be identified by tracking the hyperedges of the mentioned entities in the knowledge hypergraph. All entities on the same hyperedge as any topic entity are combined and considered candidate entities. Use $E_Q^c \subset E$ to refer to this candidate set.

The entities and relations in the hyperedges encompassing both the candidate entity and the corresponding topic entity are utilized to construct a sequence of candidates for a candidate entity. The sequences of words representing the entities and relations along the path are connected to form the candidate sequence.

4.2 Candidate Sequences

To enhance the candidate sequence of candidate entities e^c, further find other relations associated with e^c, where e^c can be any entities in the hyperedge. Let $R_{e^c} \subset R$ denote the set of relations connected to e^c, excluding those relations that connect e^c to a mentioned entity. This set of relations R_{e^c} provides some background information about the candidate entities and may be useful. For example, for the candidate entity Pierre Curie, the relation "marry" connected to it may help to match the candidate entity to the question better. Call R_{e^c} the contextual relation of the candidate entity e^c. The following section explains how to use the attention mechanism to give higher weight to these question-specific contextual relations.

4.3 Sequence Matching

First, each word is associated with an existing word embedding vector for the question sequence and the base candidate sequence. Then, for contextual relations in the augmented candidate sequences, each contextual relation is associated with a vector of randomly initialized relational embeddings. Both word embeddings and relation embeddings will be updated during the training process.

Let $\mathbf{Q} = (\mathbf{q}_1, \mathbf{q}_2, ..., \mathbf{q}_m)$ denotes the word embedding sequence of the question, $\mathbf{C} = (\mathbf{c}_1, \mathbf{c}_2, ..., \mathbf{c}_n)$ denotes the embedding vector sequence of augmented candidate sequences of candidate entities e^c. $(\mathbf{c}_1, \mathbf{c}_2, ..., \mathbf{c}_{n-1})$ is the word embedding in the candidate sequence of e^c, and the last embedding $\mathbf{c}_n$ is defined as the combination of R_{e^c} internal relational embeddings, i.e., the contextual relations associated with e^c, and it will be explained later how $\mathbf{c}_n$ is obtained from the relations of R_{e^c}.

Given two sequences $\mathbf{Q}$ and $\mathbf{C}$, get the matching scores between them. First, try to match $\mathbf{q}_i$ with $\mathbf{c}_j$ as follows:

$$e_{ij} = F(\mathbf{q}_i)^T F(\mathbf{c}_j) \tag{1}$$

where $F(\cdot)$ is a single nonlinear layer with ReLU as the activation function. e_{ij} essentially encodes the degree of match between $\mathbf{q}_i$ and $\mathbf{c}_j$.

The following normalized attention weights are then derived using e_{ij} as defined above.

$$a_{ij} = \frac{\exp(e_{ij})}{\sum_{i'=1}^{m} \exp(e_{i'j})} \tag{2}$$

where a_{ij} represents how $\mathbf{q}_i$ matches $\mathbf{q}_{i'}$ compared to the other tokens in the question, and $\mathbf{c}_j$. Similarly, another set of attention weights is also computed in the other direction:

$$b_{ij} = \frac{\exp(e_{ij})}{\sum_{j'=1}^{m} \exp(e_{ij'})} \tag{3}$$

where b_{ij} indicates how well $\mathbf{c}_j$ matches $\mathbf{c}_j$ compared to other tokens in the candidate sequence $\mathbf{c}_{j'}$.

Then, the weighted versions of $\mathbf{q}$ and $\mathbf{c}$ are defined as follows.

$$\overline{\mathbf{q}}_j = \sum_{i=1}^{m} a_{ij} \cdot \mathbf{q}_i \tag{4}$$

$$\overline{\mathbf{c}}_i = \sum_{j=1}^{n} b_{ij} \cdot \mathbf{c}_j \tag{5}$$

Here $\overline{\mathbf{q}}_j$ is a weighted sum of all $\mathbf{q}_i$ in the question sequence to match $\mathbf{c}_j$ in the candidate sequence. It follows the standard attention mechanism used in most previous work. The same idea applies to $\overline{\mathbf{c}}_i$.

Next, match $\mathbf{q}_i$ with $\overline{\mathbf{c}}_i$ and $\mathbf{c}_j$ with $\overline{\mathbf{q}}_j$ by defining the following two vectors.

$$\mathbf{v}_{1,i} = G\left(\begin{bmatrix} \mathbf{q}_i \odot \overline{\mathbf{c}}_i \\ (\mathbf{q}_i - \overline{\mathbf{c}}_i) \odot (\mathbf{q}_i - \overline{\mathbf{c}}_i) \end{bmatrix}\right) \tag{6}$$

$$\mathbf{v}_{2,j} = G\left(\begin{bmatrix} \mathbf{c}_j \odot \overline{\mathbf{q}}_j \\ (\mathbf{c}_j - \overline{\mathbf{q}}_j) \odot (\mathbf{c}_j - \overline{\mathbf{q}}_j) \end{bmatrix}\right) \tag{7}$$

where $\odot$ denotes element multiplication and $G(\cdot)$ is another feedforward neural network with ReLU activation. $\mathbf{v}_{1,i}$ measures the similarity between the question and its weighted version at the ith position. The same principle applies to $\mathbf{v}_{2,j}$.

Next, the sequence of $\mathbf{v}_{1,i}$ and $\mathbf{v}_{2,j}$ is aggregated using LSTM to extract two values from the vector generated through the maximal set:

$$\overline{\mathbf{V}}_1 = \mathrm{LSTM}([\mathbf{v}_{1,1}, \mathbf{v}_{1,2}, ..., \mathbf{v}_{1,m}]), \overline{\mathbf{v}}_1 = \max_i \overline{\mathbf{V}}_{1,i} \tag{8}$$

$$\overline{\mathbf{V}}_2 = \mathrm{LSTM}([\mathbf{v}_{2,1}, \mathbf{v}_{2,2}, ..., \mathbf{v}_{2,m}]), \overline{\mathbf{v}}_2 = \max_j \overline{\mathbf{V}}_{2,j} \tag{9}$$

Finally, the matching scores between the question sequence $\mathbf{Q}$ and the candidate sequence $\mathbf{C}$ are given by concatenating $\overline{\mathbf{v}}_1$ and $\overline{\mathbf{v}}_2$ and entering them to H:

$$s(\mathbf{Q}, \mathbf{C}) = H([\overline{\mathbf{v}}_1; \overline{\mathbf{v}}_2]) \tag{10}$$

A softmax layer is used on the matching scores of all candidate entities, and then the distribution of candidate entities can be derived.

4.4   Relation Attention

We now describe how c_n is derived from the contextual relations $\mathcal{R}_{e^c}$ for candidate e^c. The simplest way is to take the average of the relation embedding, which refer to as **A**:

$$c_n = \frac{1}{|\mathcal{R}_{e^c}|} \sum_{r \in \mathcal{R}_{e^c}} r \tag{11}$$

Then, apply an attention network to decide which relation is essential for the question. We define the β_r parameter, representing the importance of relation r concerning the question. We refer to this combination method as **RA**.

$$c_n = \sum_{r \in \mathcal{R}_{e^c}} r \cdot \frac{\exp(\mathbf{w}^T[\mathbf{r}; \mathbf{q}])}{\sum_{r' \in \mathcal{R}_{e^c}} \exp(\mathbf{w}^T[\mathbf{r}'; \mathbf{q}])} \tag{12}$$

The second term in the summation represents how important the token $\mathbf{q}_i$ is inside the question. This is the self-attention mechanism that has been widely used. This $\mathbf{q}'$ captures the more critical aspects of the question. Then we can use $\mathbf{q}'$ instead of $\mathbf{q}$ in Eq. (13) to obtain c_n. The method is denoted as **SA**.

$$\mathbf{q}' = \sum_{i=1}^{m} \mathbf{q}_i \cdot \frac{\exp(\mathbf{u}^T[\mathbf{q}_l; \mathbf{q}_i])}{\sum_{i'=1}^{m} \exp(\mathbf{u}^T[\mathbf{q}_l; \mathbf{q}_{i'}])} \tag{13}$$

In the training phase, the KL scatter between the actual and predicted distributions is used as the objective function to learn various model parameters. For prediction, the candidate entity with the highest probability is selected as the answer to the prediction.

5   Experiments

The performance of HyperMatch was tested on the knowledge hypergraph benchmark. Section 5.1 summarizes the experimental setups, such as datasets and baselines. All experiments in Sect. 5.2 were conducted to answer the question.

5.1   Experiment Settings

Datasets. The knowledge hypergraph WikiPeople-KH is constructed based on the WikiPeople dataset. We artificially generate different relation types to combine relation and attribute key-value pairs. For example, for the tuple containing the relation s̈pouseänd the remaining attribute s̈tart timeänd ënd time; the relation is uniformly named s̈pouse dUring the timeärtificially and stitched into the entity list of the tuple according to the original entity order. The detailed statistics of the datasets are summarized in Table 1.

Subsequently, one or more entities are randomly masked for different tuples as question answers. Then, different questions are generated based on the semantics of the relation and the remaining entities, resulting in the question dataset

KHQuestions. e.g. for the tuples {"award received_h": "Marie Curie", "award received_t": "Nobel Prize in Physics", "point in time": ["1903"], "together with": ["Henri Becquerel", "Pierre Curie"], "N": 5}, by overriding the two entities "Henri Becquerel" and "Pierre Curie" under the attribute "together with", the question "Who did Marie Curie win the Nobel Prize in Physics within 1903?" and the answer "Henri Becquerel" and "Pierre Curie" are generated.

To test the performance of the multi-hop knowledge graph baseline on KHQuestions, the star-to-clique decomposition framework is used to convert the constructed WikiPeople-KH into the knowledge graph WikiPeople-KG dataset.

Table 1. Dataset Statistics. The size of train, valid, and test represent the number of triples or tuples, respectively.

Dataset	#entities	#relations	#train	#valid	#test	#2-ary	#3-ary	#4-ary	#≥5-ary
WikiPeople-KH	47,765	707	305,725	38,223	38,281	337,914	25,820	15,188	3,307
WikiPeople-KG	85,229	184	242,754	80,919	80,918	404,591	0	0	0

Baselines. We compare our model with the Rce-KGQA [12], the EmbedKGQA [13], and the TransferNet [14] for WikiPeople-KG dataset. These methods can address the multi-hop question in knowledge graphs and open source.

- **Rce-KGQA** utilizes QA relational chain parsing to identify the semantics more accurately and leverages the structure information preserved in KG embedding to reason the implicit answer indirectly.
- **EmbedKGQA** utilizes the link prediction properties of KG embeddings to mitigate the KG incompleteness problem without using additional data. It trains the KG entity embeddings and uses them to learn question embeddings. During the evaluation, the scores (head entity, question) pair all entities again, and the highest-scoring entity is selected as an answer.
- **TransferNet** attends to different parts of the question, computes activated scores for relations, and then transfers the previous entity scores along activated relations differently.

Evaluation Metrics. Three evaluation metrics were employed to compare the performance of different question answering methods: Mean Reciprocal Rank (MRR), Hit@K, and Accuracy, where H@K is in %, and all results in Sect. 5.2 are rounded. The three metrics above are measured by ranking a test tuple t within a set of replaced tuples. For each tuple in the test set and each position i in the tuple, $|\mathcal{E}| - 1$ replaced tuples are generated by replacing the entity e_i with each entity in $\mathcal{E}\backslash\{e_i\}$.

Table 2. Results on the WikiPeople-KG and WikiPeople-KH Dataset.

Method	WikiPeople-KG/KH				
	MRR	Hit@1	Hit@3	Hit@10	Accuracy
Rce-KGQA	17.58	10.82	18.41	31.88	-
EmbedKGQA	19.47	12.15	21.01	34.25	-
TransferNet	18.62	11.76	19.43	32.99	10.05
HyperMatch	26.83	18.64	28.36	42.18	15.43
HyperMatch-EM	27.89	19.87	30.14	43.85	16.04
HyperMatch-EMA	28.49	20.63	31.46	45.30	16.43
HyperMatch-EMRA	28.90	21.22	32.65	47.73	16.87
HyperMatch-EMSA	**29.23**	**21.59**	**33.87**	**49.23**	**17.11**

5.2 Results

Our results are shown in Table 2. The first three rows are the experiment results of WikiPeople-KG for the baseline of multi-hop question answering over the knowledge graph, and the last five rows are the experiment results of Hyper-Match for WikiPeople-KH. All methods use KHQuestions, and since WikiPeople-KG is the knowledge graph obtained by the star-to-clique transformation of WikiPeople-KH, there is no data loss. The answer entities of the questions are available in the dataset. Ablation experiments are conducted on the HyperMatch model. EM, EMA, EMRA, and EMSA are the pipeline results with a matching framework combined with direct use of embedding, Average, Relation Attention, and Self-Attention, respectively.

The experiment results show that Average, Relational Attention, and Self-Attention effectiveness are sequentially improved. Even the HyperMatch model, which has the worst effect without the matching algorithm and any attention mechanism, yields better results than EmbedKGQA, a multi-hop Q&A method based on knowledge graphs. Compared with the multi-hop method, HyperMatch only takes the one-hop hyperedge of the target entity. As a result, its returns and does not use a complex multi-hop mechanism to obtain the answer to the question. The effectiveness of the knowledge hypergraph using n-ary relations to represent practical knowledge for answering complex questions is fully demonstrated. Secondly, by using the matching algorithm, the performance of Q&A can be further improved, and it is seen that word-level matching can obtain more accurate sequence-matching results. Then, by using the attention mechanism, the performance can be further significantly improved, specifically, SA outperforms RA. Judging the relevance of the association relations of candidate entities to the question helps to improve the model performance.

To check whether the contextual relations can encode helpful knowledge about the candidates, we extract the learned relation embeddings of some contextual relations and map them to a 2-dimensional space. We show these relations in Fig. 3. We can see that close relations tend to be associated with the same type of entities. For example, father and child are close to each other in Fig. 3,

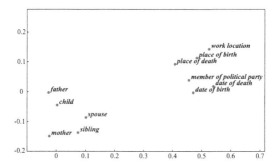

Fig. 3. Learned relation embeddings in 2-D space.

probably because these relations connect to entities which are people. We can also see that date of death and date of birth are also close, probably because they both connect to entities which are dates.

6   Conclusion

We propose the HyperMatch method, which takes a single hyperedge as a unit and completes the extraction of candidate answer entities by sequence matching and multi-relation attention mechanism. We conduct a comparative experiment on an extended version of the WikiPeople dataset. The HyperMatch can achieve a 9.44% improvement in the Hits@1 metrics. A future research direction is to extend single-hyperedge-based extraction to multi-hyperedge extraction.

Acknowledgements. Thanks to the project of Qinghai science and technology program (No. 2022-ZJ-T05), and the project of Tianjin science and technology program (No. 21JCZXJC00190).

References

1. Miller, A., Fisch, A., Dodge, J., Karimi, A.H., Bordes, A., Weston, J.: Key-value memory networks for directly reading documents (2016). arXiv preprint arXiv:1606.03126
2. He, G., Lan, Y., Jiang, J., Zhao, W.X., Wen, J.R.: Improving multi-hop knowledge base question answering by learning intermediate supervision signals. In: Proceedings of the 14th ACM International Conference on Web Search and Data Mining, pp. 553–561, March 2021
3. Sun, H., Dhingra, B., Zaheer, M., Mazaitis, K., Salakhutdinov, R., Cohen, W.W.: Open domain question answering using early fusion of knowledge bases and text (2018). arXiv preprint arXiv:1809.00782
4. Min, B., Grishman, R., Wan, L., Wang, C., Gondek, D.: Distant supervision for relation extraction with an incomplete knowledge base. In: Proceedings of the 2013 Conference of the North American Chapter of the Association for Computational Linguistics: Human Language Technologies, pp. 777–782, June 2013

5. Abujabal, A., Yahya, M., Riedewald, M., Weikum, G.: Automated template generation for question answering over knowledge graphs. In: Proceedings of the 26th International Conference on World Wide Web, pp. 1191–1200, April 2017

6. Kapanipathi, P., Abdelaziz, I., Ravishankar, S., Roukos, S., Gray, A., Astudillo, R., et al.: Question answering over knowledge bases by leveraging semantic parsing and neuro-symbolic reasoning. arXiv preprint arXiv:2012.01707 (2020)

7. Cai, Q., Yates, A.: Large-scale semantic parsing via schema matching and lexicon extension. In: Proceedings of the 51st Annual Meeting of the Association for Computational Linguistics (Volume 1: Long Papers), pp. 423–433, August 2013

8. Kwiatkowski, T., Choi, E., Artzi, Y., Zettlemoyer, L.: Scaling semantic parsers with on-the-fly ontology matching. In: Proceedings of the 2013 Conference on Empirical Methods in Natural Language Processing, pp. 1545–1556, October 2013

9. Reddy, S., Lapata, M., Steedman, M.: Large-scale semantic parsing without question-answer pairs. Trans. Assoc. Comput. Linguist. 2, 377–392 (2014)

10. Yih, W.T., Richardson, M., Meek, C., Chang, M.W., Suh, J.: The value of semantic parse labeling for knowledge base question answering. In: Proceedings of the 54th Annual Meeting of the Association for Computational Linguistics (Volume 2: Short Papers), pp. 201–206, August 2016

11. Yu, M., Yin, W., Hasan, K.S., Santos, C.D., Xiang, B., Zhou, B.: Improved neural relation detection for knowledge base question answering (2017). arXiv preprint arXiv:1704.06194

12. Jin, W., Yu, H., Tao, X., Yin, R.: Improving Embedded Knowledge Graph Multi-hop Question Answering by introducing Relational Chain Reasoning (2021). arXiv preprint arXiv:2110.12679

13. Saxena, A., Tripathi, A., Talukdar, P.: Improving multi-hop question answering over knowledge graphs using knowledge base embeddings. In Proceedings of the 58th Annual Meeting of the Association for Computational Linguistics, pp. 4498–4507, July 2020

14. Shi, J., Cao, S., Hou, L., Li, J., Zhang, H.: TransferNet: An effective and transparent framework for multi-hop question answering over relation graph (2021). arXiv preprint arXiv:2104.07302

15. Yu, H., Lu, J., Zhang, G.: An online robust support vector regression for data streams. IEEE Trans. Knowl. Data Eng. 34(1), 150–163 (2020)

16. Chen, Y., Subburathinam, A., Chen, C. H., Zaki, M.J.: Personalized food recommendation as constrained question answering over a large-scale food knowledge graph. In: Proceedings of the 14th ACM International Conference on Web Search and Data Mining, pp. 544–552, March 2021

17. Wen, J., Li, J., Mao, Y., Chen, S., Zhang, R.: On the representation and embedding of knowledge bases beyond binary relations (2016). arXiv preprint arXiv:1604.08642

18. Zhang, R., Li, J., Mei, J., Mao, Y.: Scalable instance reconstruction in knowledge bases via relatedness affiliated embedding. In: Proceedings of the 2018 World Wide Web Conference, pp. 1185–1194, April 2018

19. Fatemi, B., Taslakian, P., Vazquez, D., Poole, D.: Knowledge hypergraphs: Prediction beyond binary relations (2019). arXiv preprint arXiv:1906.00137

Demo Papers

IFGDS: An Interactive Fraud Groups Detection System for Medicare Claims Data

Jing Yu[1,2], Zhenyang Yu[1], Kaiming Zhan[2], Fan Wu[2], Bo Yin[2], and Lei Duan[1(✉)]

[1] School of Computer Science, Sichuan University, Chengdu, China
{yujing89,yuzhenyang}@stu.scu.edu.cn, leiduan@scu.edu.cn
[2] Jiuyuan Yinhai Software Co., Ltd., Chengdu, China
{zhankm,wufan,yinbo}@yinhai.com

Abstract. Safeguarding medicare fund security is a key concern in the medical and healthcare fields. Effectively detecting the fraud groups from massive amounts of medicare claims data is technically challenging. Therefore, we developed an interactive fraud groups detection system named IFGDS. It can screen out suspicious claims data from a large volume of medicare claims data and then detect fraud groups with "ganging up" medical visit behaviors and abnormal treatment characteristics. IFGDS provides abundant visual monitoring and analysis tools as well as a user-friendly interaction experience. It facilitates users to view patients-related information. Moreover, IFGDS has high cross-platform-portability to handle a comprehensive range of data resources. In practice, IFGDS has been applied to medicare regulatory scenarios, and has successfully recovered medicare fund among several cities.

Keywords: Fraud groups detection · Medicare claims · Big data

1 Introduction

Medicare fund fraud behaviors not only result in a waste of medical resources but also seriously threaten the security of medicare fund. Typical example is increasing the claim amount in various ways. Several characteristics of fraud claims data can reflect the above medicare fraud behaviors. If patients are aggregated by visit period, a group of patients would appear multiple times. Each time the claims/self-pay percentage corresponding to the group is high. The diseases treated corresponding to the group are different and irrelevant in multiple times. Figure 1 illustrates these characteristics graphically. It is noted that detecting fraud groups is a very important task.

However, traditional rule-based approaches rely on experts' domain knowledge seriously. The approaches are time-consuming, labor-intensive, and hardly meet the requirements of accurate monitoring. Existing machine learning-based

---

Supplementary Information The online version contains supplementary material available at https://doi.org/10.1007/978-3-031-30678-5_49.

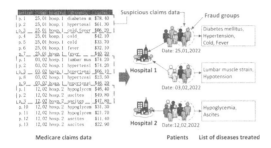 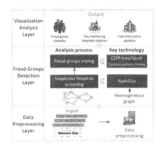

Fig. 1. An example of fraud groups detection   **Fig. 2.** Framework of IFGDS

approaches have achieved some success but cannot handle massive amounts of data efficiently in real-time. Therefore, we developed IFGDS, short for An Interactive Fraud Groups Detection System for Medicare Claims Data to find out the fraud claims data. It can identify suspicious claims data with the above-mentioned characteristics from a large amount of medicare data. Further, IFGDS could trace and target the corresponding patients which are fraud groups. The fraud groups found by IFGDS has loopholes in the medical treatment process or medical treatment records, such as a bunch of unnecessary checks during the medical treatment process. Relevant professionals can find out these loopholes based on experience and hold the claimants accountable. In this way, the medicare fund can be recovered. IFGDS combines heterogeneous graph cluster, association rules analysis, contrast pattern mining and distributed computing. It can effectively detect fraud groups. The main characteristics of IFGDS are summarized as follows.

- IFGDS utilizes big data techniques and can efficiently handle massive amounts of medicare data.
- IFGDS has a user-friendly interactive interface with abundant visual styles. It can display analysis results from multiple perspectives.
- IFGDS has been applied to a real-world scenarios. It has been successfully recovered medicare fund in some cities.

2   System Overview

Fig. 2 illustrates the overall framework of IFGDS. It consists of three layers: I) data preprocessing layer, II) fraud groups detection layer and III) visualization analysis layer.

Data preprocessing layer is necessary for the high-performance operations of IFGDS. It acquires the medicare data from the database, specifically including electronic medical records, medical costs, and medicare settlement information. This layer also provides various preprocessing ways including missing value processing, outlier processing, and column normalization according.

Fraud groups detection layer is the core of IFGDS. It contains the suspicious hospitals screening module and fraud groups mining module. The purpose

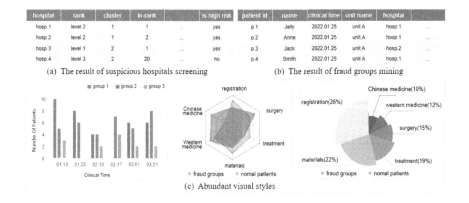

hospital	rank	cluster	in-rank		is high risk		patient id	name	clinical time	unit name	hospital	
hosp.1	level 2	1	1	...	yes		p.1	Jelly	2022.01.25	unit A	hosp.1	...
hosp.2	level 2	1	2	...	yes		p.2	Anne	2022.01.25	unit A	hosp.1	...
hosp.3	level 1	2	1	...	yes		p.3	Jack	2022.01.25	unit A	hosp.2	...
hosp.4	level 3	2	20	...	no		p.4	Smith	2022.01.25	unit A	hosp.1	...

(a) The result of suspicious hospitals screening   (b) The result of fraud groups mining

(c) Abundant visual styles

Fig. 3. Visualization analysis of IFGDS

of the suspicious hospitals screening module is to identify high-suspicious hospitals by clustering method. We built a directed heterogeneous bipartite network consisting of a "hospital-patient" network. The "patient-hospital" links indicate that different patients are admitted to different hospitals, while the "patient-patient" links indicate that different patients are seen together in the hospital at the same time. We use RankClus [2] to analyze the "hospital-patient" heterogeneous network. RankClus can cluster hospitals into different categories and rank hospitals in each category. Generally, hospitals with higher rank scores attract more patients, so that more patients may go to them at the same time. Figure 3(a) displays the clustering and ranking results. The fraud groups mining module aims to detect fraud groups. We use CSFP-tree [1] to discover the frequent itemsets of patients, i.e., to discover the patients going to the hospital together many times. Through the experiment, implementing CSFP-tree based on Spark [3] can take full advantage of Spark's parallel computing ability to obtain more comprehensive performance. We use the contrast pattern mining method to discover the characteristics of fraud groups. The characteristics to be analyzed include cost information, visit duration, visit age, etc. Figure 3(b) shows the mining results, which display information about the fraud groups in detail. Figure 3(c) shows partial results of the contrast pattern mining.

Visualization analysis layer provides rich visual styles. It is an interactive operation interface to display the visualization results from various dimensions. The main visualization contents include statistics on visits by fraud groups, statistics on the costs of fraud groups versus normal patients, and statistics on the cost composition of fraud groups. As illustrated in Fig. 3(c), these results can be displayed in the form of bar charts, radar charts, rose charts, and so on.

3   Demonstrations

In our demonstration, we utilize desensitized medicare claims data to introduce the core functions and interactive interfaces of IFGDS. To make it more visual,

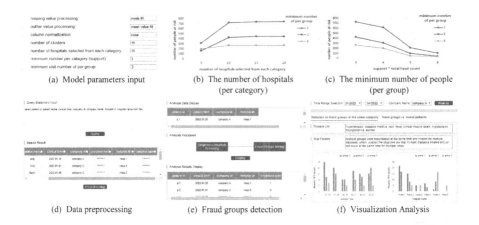

(a) Model parameters input

(b) The number of hospitals (per category)

(c) The minimum number of people (per group)

(d) Data preprocessing

(e) Fraud groups detection

(f) Visualization Analysis

Fig. 4. Demonstrations of IFGDS

we have posted a brief demo video and some sample data on GitHub[1] for access. The demonstration system realizes all functions of fraud groups identification. It is a simplified version of the monitoring system in the real world, not the real monitoring system itself. Fore intellectual property and privacy protection, the interface of the real monitoring system is not displayed.

IFGDS supports flexible data selection and processing so that users can set different model parameters to compare the mining results. It can be observed in Fig. 4(a), there are various adjustable parameters. In practical applications, users can tune the model parameters according to the actual situations to obtain optimal fraud groups detection performance. Here, we give some tuning results for two commonly used parameters. Figure 4(b) displays the results when setting different thresholds for the number of hospitals (per category). It can be found that when the threshold value is 10, IFGDS could achieve great results. Because further increasing cannot improve the performance significantly. This indicates that selecting the top ten suspicious hospitals is enough to discover fraud groups. Figure 4(c) presents the results when setting different thresholds for the minimum number of people (per group). It displays that the larger the threshold value is, the smaller the number of suspicious people screened out.

Figures 4(d)–4(f) shows the interactive interface of IFGDS. It includes three phases when detecting fraud groups from medicare claims data, corresponding to the hierarchical structure of IFGDS. Figure 4(d) represents the data preprocessing phase. The users first input SQL statements to obtain the relevant tables from the database, further choose some preprocessing operations to execute. Figure 4(e) presents the fraud groups detection phase. The users first set the number of clusters to screen out the suspicious hospitals set and then conduct the process of mining medicare fraud groups. Figure 4(f) is the interactive visualization analysis phase where the users can view the results from multiple

---

[1] https://github.com/scu-kdde/IFGDS-2023.

perspectives. In the interface, users can view a list of diseases used by fraud groups for reimbursement (upper part). The fraud factors are analyzed by our system (middle part). In addition, IFGDS provides abundant charts to visualize the analysis results (Fig. 3). For instance, Fig. 4(f) illustrates the statistics of the fraud groups (lower part), including the relationship between the number of clinical patients with clinal time or the hospitals.

Acknowledgement. This work was supported in part by the National Natural Science Foundation of China (61972268).

References

1. Jamsheela, J., Raju, G.: An improved frequent pattern tree: the child structured frequent pattern tree CSFP-tree. In: PAA, pp. 1–18 (2022)
2. Sun, Y., Han, J., Zhao, P., Yin, Z., Cheng, H., Wu, T.: Rankclus: integrating clustering with ranking for heterogeneous information network analysis. In: EDBT, pp. 565–576 (2009)
3. Zaharia, M., et al.: Resilient distributed datasets: A fault-tolerant abstraction for in-memory cluster computing. In: NSDI, pp. 15–28 (2012)

PFKMaster: A Knowledge-Driven Flow Control System for Large-Scale Power Grid

Huaiyuan Liu[1], Hongzhi Wang[1(✉)], Hekai Huang[2], Donghua Yang[1], Yong Tang[3], and Yanhao Huang[3]

[1] Harbin Institute of Technology, Harbin, China
wangzh@hit.edu.cn
[2] RHP Software Dept.I of ZTE Corporation, Shenzhen, China
[3] Laboratory of Power Grid Safety and Energy Conservation (China Electric Power Research Institute), Beijing, China

Abstract. Various stability analyses of the power system are based on the results of power flow calculation, which is not always convergent. In practice, large manual efforts are required to be repeated many times by electrical engineers to ensure the convergence of power flow calculation, which consumes much manpower and time cost. Motivated by this, we develop a novel knowledge-driven system that can automatically improve power flow convergence based on knowledge from experiences, called *PFKMaster*. Based on the features of power flow calculation, a *SAS triplet* mechanism is designed to better represent the experience knowledge of humans. In our system, a knowledge model is proposed to guide the power flow calculation automatically and universally. To achieve the goal, we build a specific knowledge base. In this system, new knowledge is discovered based on the method of representation learning and affair logic. To ensure knowledge quality, we also propose knowledge cleaning techniques in our system. We test *PFKMaster* on a real large-scale power grid, and the experimental results demonstrate that 94.8% of the non-convergent samples can be improved by 1.92 s per sample.

Keywords: Knowledge-driven · Power flow calculation · Knowledge graph

1 Introduction

The power flow calculation is the starting point for most power system analyses whose task is to determine the operating state of the power grid based on the given conditions. The convergence of power flow calculation means the difference between the solutions of two adjacent rounds in the process of iteratively solving equations is less than a certain threshold. Unfortunately, not all power flow

URL of demonstration video: https://youtu.be/sgfV1QUIjAs.

calculations are convergent. Thus, experienced power engineers have to adjust the output of some devices multiple times to ensure the convergence of the calculation, which is called power flow control.

The Need for PFKMaster . While the problem of non-convergence in power flow calculation, to a certain extent, can be solved by engineers, two major problems remain unresolved. On the one hand, humans being heavily involved in power flow control leads to much time consumption and labor cost, especially when the grid is complex and the number of parameters is large. On the other hand, the convergence of power flow calculation largely depends on expert knowledge, and even sometimes human experience is also insufficient to ensure the convergence of the calculation. The process of iterative optimization is as shown in Fig. 1.

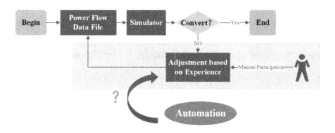

Fig. 1. Process of manual optimization

The knowledge can directly guide the actions of devices and then automatically optimize the parameters until the power flow calculation converges if these processes can be driven by a knowledge model and more new knowledge can be found to control power flow. In practice, it is challenging to build such a knowledge base due to the following major issues.

1) *Diversity:* The knowledge model needs to express both qualitative and quantitative relationships, logical and affair relationships between knowledge entities due to the diversity of knowledge.

2) *Complexity:* It is difficult to extract and distinguish the inherent laws of power grid data because of its high complexity.

3) *Correctness:* The new knowledge obtained by knowledge reasoning or discovery needs to be guaranteed usability. Unfortunately, there is no system that can solve these issues.

In order to overcome these challenges and fill this gap, we propose a novel system *PFKMaster*, a knowledge-driven flow control system for the large-scale power grid. The contributions of our work are as follows.

– *Pioneering:* In our knowledge, the proposed system is the first integrated system to control power flow based on the novel knowledge model.

– **Highly Automated:** We designed a new model to represent the complex knowledge of power flow control, and the process of manually optimizing parameters of power flow control is automated into a knowledge-driven pattern by this model.

– **Efficient:** The quality of knowledge determines the effectiveness of power flow control. Hence, we propose some knowledge enhancement techniques to ameliorate *PFKMaster*, including knowledge query, knowledge cleaning, and reasoning.

2   System Architecture and Key Techniques

Figure 2 shows the framework of *PFKMaster*. All formal knowledge is stored in the knowledge base. The users can interact with the modules of the system with a friendly interface to automatically control the power flow and manage the knowledge base. The power flow control mainly undergoes modules such as feature extraction, state matching, action query, and action execution. *Knowledge Enhancement Module* provides human-machine interaction, knowledge cleaning, knowledge reasoning, and query improvement for large knowledge graph.

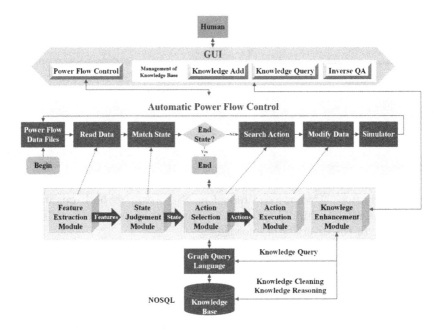

Fig. 2. The architecture of PFKMaster system

Initially, the system firstly extracts state features from the input files and then queries the matching state nodes in the knowledge base. These modules are typically time-saving based on specific knowledge expressions for power flow data whose type is varied. Next, the corresponding parameters in the data will be modified after matching the action nodes linked with the state nodes. Then, the system computes the new power flow based on the modified data and judges the convergence of the calculation. *PFKMaster* automatically optimizes the parameters until the power flow calculation is convergent, or all relevant knowledge has been tried but is still invalid.

The key techniques used in our system are described as follows.

Knowledge Model. In general, the convergence process of power flow calculation can be seen as a process that a state reaches a new state by some actions according to the characteristics of manual power flow control, called *SAS*. We design the corresponding knowledge form for the process and actions of the power flow control. The process of power flow control is described based on two triples since a simple relationship is insufficient to contain more information. The triples in the knowledge base include the following forms.

$< State > \rightarrow (Has\_action) \rightarrow < Action >$

$< Action > \rightarrow (Has\_state) \rightarrow < State >$

Thus, *Action* as an entity can express more information as follows. *Action* is a head node linked with *State*, and the tail node *Parameter* describes the parameter modified by this action. We can obtain a value from *Expression* composed of parameters, operational characters, and digits. The *Has_action* is an optional relationship which means the action has a child action.

$< Action > \rightarrow (Has\_object) \rightarrow < Parameter >$

$< Action > \rightarrow (Has\_value) \rightarrow < Expression >$

$< Action > \rightarrow (Has\_action) \rightarrow < Action >$

Besides, each parameter of the power grid has its *Location* in the input files. Then we can obtain the original value and the target location of action by the following triple.

$< Parameter > \rightarrow (Has\_location) \rightarrow < Location >$

Note that some actions do not need to be executed by humans or cannot be executed by humans, such as the calculation of power flow, which can be executed by some functions or programs. Such type of knowledge is formalized as follows.

$< Action > \rightarrow (Has\_function) \rightarrow < Function >$

Our system can automatically find the states and take action until reaching the termination state without manual participation based on such a triple mechanism. The schematic diagram of the application of knowledge and the automatic control of the power flow is given in Fig. 3.

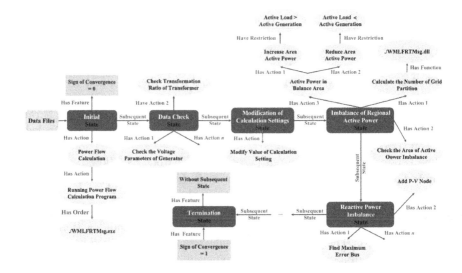

Fig. 3. An example of knowledge model

Knowledge Enhancement. As knowledge is crucial for the effectiveness of our system, we develop methods to improve the knowledge quality and speed up query processing.

Knowledge Query Processing. In consideration of the large scale of the knowledge graph, we design an approximate query algorithm. We use two sampling operators, MC-operator and Semi-RSF, to limit the usage of memory and establish the balance of resources [2]. The MC-operator uses the inverted index, called I-Index, to obtain a set of pairs (h, L_{uni}), where h is the keywords of the query Q, and L_{uni} is the unique numeric identifier of the entries related to h. Semi-RSF is based on a sophisticated index, called M-Index, recording the related entities for each vertex. Semi-RSF receives the output of the MC-operator, using M-Index to generate the final results. The application of the knowledge query algorithm ensures the rapidity and accuracy of our system when querying large-scale knowledge graphs.

Knowledge Cleaning. We propose a knowledge-cleaning method combined with semantics and embedding which judges the correctness of knowledge by judging the distance of vector and closure, and then some possibly correct knowledge will replace the wrong entity [1]. In addition, redundant knowledge will also be discovered and processed. The cleaning mechanism automatically runs in the system which improves the availability of knowledge in the knowledge base.

Knowledge Reasoning and Interactive Feedback Strategy. We present new knowledge generation methods based on knowledge reasoning technology [3,4]. And other reasoning methods are provided for selection, such as TransE, TransH,

AMIE+, etc. However, this new knowledge may be incorrect. Thus, we designed the component *Inverse-QA* which interactively inquires experts to figure out the unavailable ones and store the useful ones. At the same time, this new knowledge will also ensure their availability by knowledge cleaning mechanism.

3 Demonstration Scenarios

In this section, we demonstrate the core functions and successful use cases of our system.

Knowledge Management. We develop friendly interfaces to interact with users so that they can add, delete, modify, and query the knowledge of the knowledge base. In addition, our system provides batch operation, export, and generation of the knowledge graph, etc. At the same time, users can discover new knowledge by selecting different reasoning algorithms and can feed them back to experts for decision and operation. The related interfaces are shown in Fig. 4.

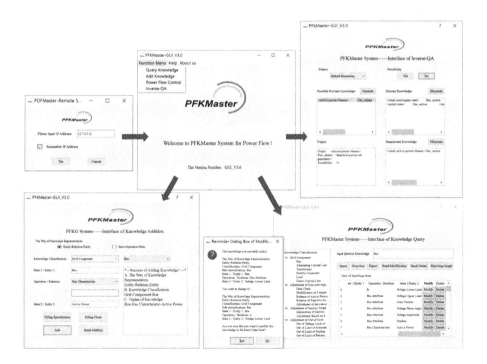

Fig. 4. Knowledge management

Automatic Power Flow Control. The system provides users with the functions of multi-step adjustment and single-step adjustment to control the power flow. That is, the users can automatically accomplish the whole process of power flow control according to the knowledge or select the specific operation of each adjustment step, respectively. The system has been successfully applied to the simulation analysis of the actual northeast power grid of China with 2 131 nodes and 1 748 edges, which can achieve a 94.8% of success rate in 271 min for 8 480 samples. Compared with manual control, it largely liberates human resources. The interface is depicted in Fig. 5.

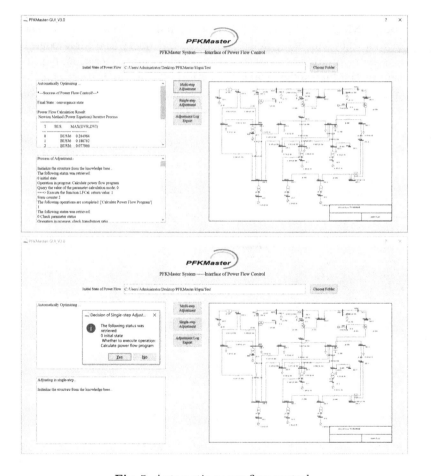

Fig. 5. Automatic power flow control

Acknowledgments. This paper was supported by NSFC grant (62232005, 62202126, U1866602) and Sichuan Science and Technology Program (2020YFSY0069).

References

1. Bordes, A., Usunier, N., Garcia-Duran, A., Weston, J., Yakhnenko, O.: Translating embeddings for modeling multi-relational data. Adv. Neural Inf. Process. Syst. **26**, 1–11 (2013)
2. Wan, X., Wang, H., Li, J.: LKAQ: large-scale knowledge graph approximate query algorithm. Inf. Sci. **505**, 306–324 (2019)
3. Wang, Y., Wang, H., He, J., Lu, W., Gao, S.: TAGAT: type-aware graph attention networks for reasoning over knowledge graphs. Knowl.-Based Syst. **233**, 107500 (2021)
4. Wang, Y., Wang, H., Lu, W., Yan, Y.: Metranse: manifold-like mechanism enhanced embedding for reasoning over knowledge graphs. Expert Syst. Appl. **209**, 118288 (2022)

PEG: A Partial Evaluation-based Distributed RDF Graph System

Shengyi Ji[1], Peng Peng[1(✉)], Jian Hu[1], Lei Zou[2], Zhen Huang[3], and Zheng Qin[1]

[1] Hunan University, Changsha, China
{jishengyi,hnu16pp,Prohujian,zqin}@hnu.edu.cn
[2] Peking University, Beijing, China
zoulei@pku.edu.cn
[3] National University of Defense Technology, Changsha, China
huangzhen@nudt.edu.cn

Abstract. In this demo, we implement a *Partial Evaluation*-based distributed RDF *Graph* system (PEG for short), which can implement partial evaluation without modifying the single-machine RDF graph system at each site. When a query is newly input, PEG rewrites it into a set of *local partial subqueries*. Then, each local partial match defined in the framework of partial evaluation can be also found by evaluating one of the local partial subqueries. To further prune some unnecessary results of local partial subqueries, PEG also associates each vertex with an attribute to specify whether it belongs to the partition at the site. In this demo, we demonstrate the main features of PEG and show how PEG loads and queries large RDF graphs in a distributed environment.

1 Introduction

RDF is a model widely used for representing knowledge. An RDF triple, the basic unit in this model, consists of a subject, a property, and an object, and describes two entities and a relationship between them. A set of such triples can be also represented as a graph, where subjects and objects are vertices and triples are edges with property names as edge labels. SPARQL is a query language to retrieve and manipulate RDF graphs, where the basic building block is the basic graph pattern (BGP). A BGP query can also be seen as a query graph, and answering a BGP query is equivalent to finding subgraph matches of the query graph over the RDF graph [5]. As the sizes of real RDF graphs increase, we should design a distributed RDF graph system to process queries. In this demo, we focus on designing a partial evaluation-based distributed RDF graph system.

Partial evaluation is a unifying framework that has been used for evaluating queries on distributed RDF graph systems [3,4]. There are two stages in the framework: the partial evaluation stage and the assembly stage. In the partial evaluation stage, each site first receives the full query and finds all *local partial matches*, where a local partial match actually is an overlapping part between a crossing match and partition stored at the site. Then, in the assembly stage, these local partial matches are assembled to form crossing matches. The partial evaluation-based distributed RDF graph systems can be independent of any specific partitioning strategies, but they need to modify the single-machine RDF graph system at each site since existing single-machine RDF graph systems can only support finding matches but not local partial matches.

In this demo, we present a new partial evaluation-based distributed RDF graph system, named PEG. PEG implements partial evaluation based on query rewriting and does not need to modify the single-machine RDF graph systems at each site. It provides some Web interfaces to visualize the procedures of loading and querying large RDF graphs.

2   System Overview

We now discuss more the main modules and workflow of PEG. Figure 1 shows the architecture of PEG. PEG is built on a cluster of sites, where each site installs a single-machine RDF graph system. There are two main modules in PEG: data loader and query processor. The data loader is to load RDF graphs in the cluster of sites, while the query processor is to handle the input query.

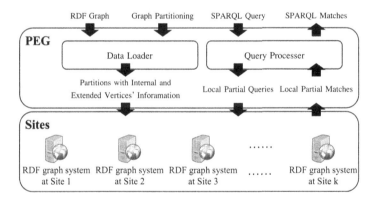

Fig. 1. System Architecture

2.1   Data Loader

In the module of the data loader, the RDF graph and its partitioning are input. It divides the RDF graph according to the partitioning and then distributes different partitions among sites. In PEG, an RDF graph is vertex-disjoint partitioned into several partitions, where each vertex is assigned to a single partition. Here, PEG also follows the principle of partial evaluation and is independent of any specific partitioning strategy, so it doesn't need the input to specify how to partition the RDF graph. The partitioning is formed as a file, where each line consists of two parts: the first part is the vertex and the second part is the identifier of the partition that the vertex belongs to.

In vertex-disjoint partitioning, some edges are cut across partitions. Here, we call them *crossing edges* and replicate them in the two partitions of their endpoints to guarantee completeness. Thus, for each partition, some vertices of other partitions are stored at it and the set of these vertices is called *extended vertices*. Meanwhile, the vertices of each partition are called *internal vertices* of the partition. Then, when the data loader partitions RDF graphs, it also associates each vertex with a value that denotes whether the vertex is extended or internal.

2.2   Query Processor

In the module of the query processor, the input query is first rewritten to a set of *local partial queries*. Then, these local partial queries are sent to all sites and evaluated there. The results of local partial queries are returned and joined together to form the final results. Here, we can show that any local partial match defined in the previous partial evaluation-based methods [3,4] is the match of one local partial query. Hence, PEG still follows the principle of partial evaluation.

In general, for an input query Q, a local partial query q is a query graph that meets the following constraints: 1) The set of query vertices $V(q) \subseteq V(Q)$ can be divided into two subsets $V^c(q)$ and $V^e(q)$, where the induced subgraph of $V^c(q)$ is weakly connected and any vertex in $V^e(q)$ is adjacent to at least one vertice in $V^c(q)$; 2) For any vertex v in $V^c(q)$, q contains a triple pattern (i.e. edge) which is v associated with a value of "Internal"; 2) For any vertex v in $V^e(q)$, q contains a triple pattern (i.e. edge) which is v associated with a value of "Extended"; 3) Any edge adjacent to one vertex in $V^c(q)$ is contained in q; 4) There does not exist an edge $u_1\vec{u}_2$ in q where $u_1 \in V^c(q) \wedge u_2 \in V^c(q)$.

In the query processor, the input query is rewritten to the set of all possible local partial queries. Then, we can prove the equivalence between local partial matches in [3,4] and matches of local partial queries. For each local partial match, we can formulate a query matching it and then prove that the query is a local partial query. Given a local partial match PM of a query Q with mapping function f, we use $f^{-1}(PM)$ to denote the subgraph (of Q) induced by a set of vertices, where for any vertex $v \in f^{-1}(PM)$ $f(v)$ is not NULL. We can formulate a query q^* by adding triple patterns in $f^{-1}(PM)$: if a vertex u in PM is internal, we add a triple pattern to associate $f^{-1}(u)$ with a value "Internal"; otherwise, we add a triple pattern to associate $f^{-1}(u)$ with a value "Extended". Then, we can find that q^* is a local partial query. This indicates that each local partial match can correspond to a match of one local partial query.

3   Demonstration

The PEG system can be downloaded at GitHub[1]. At each site, a partition is stored in the gStore RDF engine [5]. Implementations are in C++.

---

[1] https://bnu05pp.github.io/PEG/index.html.

3.1 How to Use PEG?

To help the users to use PEG, we use Vue.js to implement a Web interface of PEG, where Fig. 2 shows an example. Users can enter the SPARQL queries, select the corresponding database, and click the *execute* button to execute the query. Then, the results are listed below.

Fig. 2. Web Interfaces of PEG

3.2 Comparison with Existing Partial Evaluation-based Systems

In this experiment, we compare PEG with the previous partial evaluation-based system, gStoreD [3,4]. We use a well-known benchmark, LUBM [2], which includes seven benchmark queries $(LQ_1 - LQ_7)$ [1] . We generate a LUBM dataset of 100 million triples and divide it into 10 parts according to the university domain in LUBM. The experiment is conducted on a cluster of 10 machines in Alibaba Cloud running Linux, each of which has four CPUs, 32 GB memory, and 150 GB disk. Figure 3 shows the experiment results.

In general, PEG works better for all queries which need less chance for joining, because gStoreD proposes some built-in optimizations for queries when it modifies gStore at each site to implement the partial evaluation. These optimizations also take time, so for those queries which need less chance for joining it works worse. Because LQ_2 is a star query, the results $?x$ are in the same partition so it is not significant in LQ_2 compared with other benchmark queries.

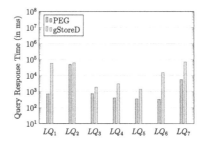

Fig. 3. Comparison with Existing Partial Evaluation-based Systems

4 Conclusion

In this demo, we implement a partial evaluation-based distributed RDF graph system named PEG. We show how to use PEG and compare it with the previous partial evaluation-based system, gStoreD.

Acknowledgment. This work was supported by the National Key R&D Projects (2022YFB3103500), Science and Technology Major Projects of Changsha City (No. kh2103003), NSFC under grants (U20A20174, 6177219), Natural Science Foundation of Hunan Province under grant (2022JJ30165), the Technology Projects of Hunan Province (2015TP1004) and Foundation of State Key Laboratory of Science and Technology on Parallel and Distributed Processing Laboratory (PDL) under grant (6142110200403).

References

1. Abdelaziz, I., Harbi, R., Khayyat, Z., Kalnis, P.: A survey and experimental comparison of distributed SPARQL engines for very large RDF data. PVLDB **10**(13), 2049–2060 (2017)
2. Guo, Y., Pan, Z., Heflin, J.: LUBM: a benchmark for OWL knowledge base systems. Web Semant. **3**(2–3), 158–182 (2005)
3. Peng, P., Zou, L., Guan, R.: Accelerating partial evaluation in distributed SPARQL query evaluation. In: ICDE, Macao, China, pp. 112–123. IEEE (2019)
4. Peng, P., Zou, L., Özsu, M.T., Chen, L., Zhao, D.: Processing SPARQL queries over distributed RDF graphs. VLDB J. **25**(2), 243–268 (2016)
5. Zou, L., Özsu, M.T., Chen, L., Shen, X., Huang, R., Zhao, D.: gStore: a graph-based SPARQL query engine. VLDB J. **23**(4), 565–590 (2014)

ODSS-RCPM: An Online Decision Support System Based on Regional Co-location Pattern Mining

Dongsheng Wang[1], Lizhen Wang[1,2(✉)], and Peizhong Yang[1]

[1] School of Information Science and Engineering, Yunnan University, Kunming 650091, China
Dswang1996@163.com, {lzhwang,ypz}@ynu.edu.cn
[2] Dianchi College of Yunnan University, Kunming 650228, China

Abstract. Regional co-location pattern mining (RCPM) is a branch of spatial co-location pattern mining, and is usually used to discover some sets of spatial features that do not often co-occur in large spatial scales, but co-occur in local regions. The decision-making effect can be maximized through the RCPs when regional decision-making is required. In this demonstration, an RCPM-based online decision support system, named ODSS-RCPM, is designed to improve the shortcomings of the regional division stage and the RCPM stage, respectively. First, the ODSS-RCPM divides the sub-regions through the distribution density of maximal cliques (MCs) of spatial instances, and mines RCPs based on MCs, which accelerates the RCPM process while solving the rationality of the divided sub-regions. Second, in ODSS-RCPM, users can select a sub-region to obtain the most suitable patterns within this sub-region or select an RCP to obtain the sub-regions of this RCP, when they need to make some decisions about regional planning.

Keywords: Decision Support System · Regional Co-location Pattern · Maximal Clique

1 Introduction

Spatial co-location pattern mining [1] is a spatial knowledge discovery technique that aims to discover spatial feature subsets with geographic correlation in spatial datasets. Regional co-location pattern mining (RCPM) [2] is a branch of spatial co-location pattern mining, which focuses on discovering pattern information implied in some sub-regions or finding sub-regions where some co-location patterns are likely to co-occur frequently. RCPM is more instructive for urban planning, land use and other scenarios that require regional planning. Two classical methods [3] of RCPM have been proposed, the division-based method and the cluster-based method. In these two existing types of methods, the division-based method breaks the connection between patterns, while the cluster-based method is influenced by the distribution of individual feature instances, and the pattern instance density of the divided sub-regions may not be high [3]. Deng et al. [4] proposed an adaptive spatial clustering algorithm based on Delaunay triangulation, which can generate different sub-regions adaptively, and use non-prevalent global co-location patterns as regional candidate patterns.

However, the impact of individual feature instance distribution on the mining results and the difficulty of identifying co-location patterns in low-density sub-regions are still not addressed in the existing literatures. The Delaunay triangle-based adaptive partitioning approaches divide the study area too thin and too fragmented, making the number of mined RCPs huge, and the divided sub-regions lack semantic information. In this demonstration, we first propose a multi-density sub-region partition strategy based on the MCs of spatial instances for the existing problems in the divided stage, which not only solves the influence of individual feature instances on the sub-region partition results but also endows these sub-regions with semantic information to represent different degrees of prevalent regions. In the mining stage, we designed a two-stage mining algorithm based on MCs for accelerating the RCPM process. Last, to make it easier for users to obtain valuable information from the mining results, and thus to provide decision support, a personalized display of the results is provided. Users can choose their interested sub-regions or patterns from the results, and the demonstration will be based on the user's selection, matching the most suitable RCPs in the selected sub-regions or the most suitable sub-regions of the selected patterns.

2   System Overview and Key Techniques

2.1   System Overview

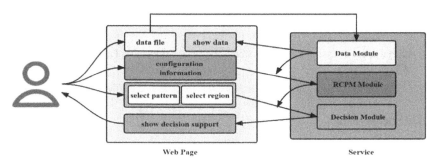

Fig. 1. The framework of ODSS-RCPM.

ODSS-RCPM is an online, friendly and interactive system based on the RCPM, designed to provide decision support for regional planning or commercial site selection, etc. Figure 1 shows the framework of ODSS-RCPM including the web page part and the service part, where the web page part is used to interact with users and display results, and the service part is used for data processing and RPCM.

Three main modules are included in ODSS-RCPM. **Data module** acquires the spatial dataset file uploaded by users and visually displays the distribution of dataset in the web page. **RCPM module** is the core of ODSS-RCPM, which performs the mining tasks of RCPs according to the uploaded dataset and the user-specified configuration information.

Different from the region partitioning stage of the existing RCPM methods, we use a novel partitioning approach based on MCs of spatial instances, which uses the density of MCs to measure the density of sub-regions and perform clustering operations. In the mining stage, we design a hash structure to store MCs and design a two-stage mining algorithm based on this hash structure to accelerate the RCPM process. **Decision module** provides decision support suggestions for users by letting users choose the regions or patterns of interest from the mining results.

2.2 Key Techniques

Within the **RCPM** module two main sections can be divided, the region partitioning section and the pattern mining section. In this demonstration system, we improved each of them to address the shortcomings of the traditional RCPM.

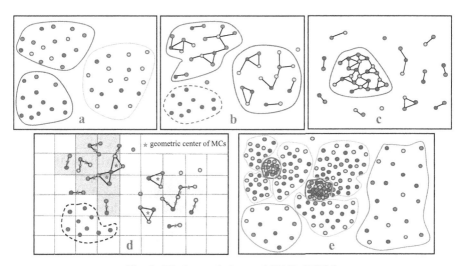

Fig. 2. The key techniques of ODSS-RCPM.

In the region division section, we propose using the MC as the clustering object for clustering operation. Unlike the case-based clustering methods, the MC-based approach is insensitive to individual feature instances. Figure 2(a) and Fig. 2(b) show the sub-regions obtained by clustering with spatial instances and MCs, respectively. The dashed region in Fig. 2(b) is an ignored interference area consisting of individual feature instances. In addition, the density of MCs can be used to determine the prosperity level of regions. As shown in Fig. 2(c), regions with high MC density not only have a high instance density but also a richer feature type.

To reduce the time consumed by the RCPM process, we use grid clustering method to cluster the MCs. Figure 2(d) depicts the core idea of the MCs-based partitioning approach in ODSS-RCPM. It extracts the geometric centers of MCs in cells as clustering objects and then divides cells that satisfy the current density and are adjacent to each other into the same sub-regions. Where, green and yellow regions represent two different sub-region division results, the dashed region will be ignored. Considering that the density of spatial data distribution is uneven in reality, we use multi-density clustering (A list of density threshold gradients) to fit the realistic scenario. Regions that do not meet the current density threshold will be used for the next level of clustering. Figure 2(e) shows the expected partitioning effect that can be achieved by multi-density clustering based on MCs in datasets with different density distributions.

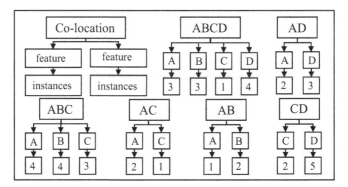

Fig. 3. The hash structure for storing MCs.

In the mining stage, we adopt a < co-location, < feature, instances > hash structure to store massive MCs, and design a two-stage mining algorithm based on this hash structure, which can accelerate the pattern mining process. Figure 3 shows the storage method of the hash structure. The first stage directly calculates the prevalence of co-location patterns included in the hash structure, and the second stage obtains the co-location patterns that are not identified in the first stage by decomposing the co-location patterns in the hash structure. For example, the pattern {B, D} can be obtained by decomposing {A, B, C, D}. The participating instances of the pattern {B, C} can be obtained by collecting {A, B, C, D} and {A, B, C} in the hash structure.

3   Demonstration Scenarios

ODSS-RCPM encapsulates a user-friendly online system through web services, where users can get appropriate decision support suggestions by interacting with ODSS-RCPM. In this demonstration, we use the Beijing vegetation distribution dataset with 12 spatial features and 25,000 spatial instances to show the ODSS-RCPM.

Figure 4 shows the main interface of ODSS-RCPM. First, the user needs to upload a spatial dataset to the system, and the system will return the spatial distribution of the dataset and display it on the web page, as shown in Fig. 4(a). Then, users enter the corresponding configuration parameters in Fig. 4(b). Figure 4(c) shows the region division results, and all discovered global and regional co-location patterns, but these results may be too many and too abstract for users to obtain valuable information, so we further provide an interactive interface for users to select patterns or sub-regions of interest. In Fig. 4(d), users can use the option tab on the left to obtain the most suitable spatial sub-regions by selecting the RCPs of interest, or use the option tab on the right to obtain the most suitable RCPs within the sub-region by choosing the sub-region of interest.

Overall, this demonstration greatly improves the readability of mining results, and users can obtain more intuitive decision support with simple interactive operations. Depending on ODSS-RCPM, users can apply RCPM to guide production and living practices. For example, in the mining results of the Beijing vegetation distribution dataset, selecting sub-region 3 can obtain the most suitable vegetation combination {A, C} and {A, H} in this sub-region, as Fig. 4(d) shows. Similarly, the most suitable living regions can also be obtained by choosing the vegetation combination {B, C}.

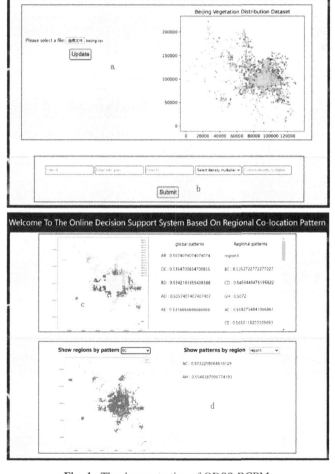

Fig. 4. The demonstration of ODSS-RCPM.

4 Conclusion

Regional co-location pattern mining (RCPM) reveals the implicit relationship between spatial features and geographic location, which provides us with spatial decision support. In this demonstration, we propose a sub-region partition method and RCPM mining method based on MCs of spatial instances, and design an online decision support system named ODSS-RCPM. Based on the RCPM results, the ODSS-RCPM can provide appropriate decision support suggestions according to the different choices of users, which can be widely used in urban planning, commercial site selection, agricultural production and living domains.

Acknowledgment. This work is supported by the National Natural Science Foundation of China (62276227, 61966036), the Project of Innovative Research Team of Yunnan Province (2018HC019), and the Yunnan Fundamental Research Projects (202201AS070015).

References

1. Wang, L., Fang, Y., Zhou, L.: Preference-Based Spatial Co-Location Pattern Mining. Springer, Singapore (2022). https://doi.org/10.1007/978-981-16-7566-9
2. Liu, Q., Liu, W., Deng, M., et al.: An adaptive detection of multilevel co-location patterns based on natural neighborhoods (2021)
3. Mohan, P., Shekhar, S., Shine, J. A., et al.: A neighborhood graph based approach to regional co-location pattern discovery: a summary of results (2011)
4. Deng, M., Cai, J., Liu, Q., et al.: Multi-level method for discovery of regional co-location patterns. Int. J. Geogr. Inf. Sci. **31**, 1846–1870 (2017)

PandaDB: An AI-Native Graph Database for Unified Managing Structured and Unstructured Data

Zihao Zhao[1,2], Zhihong Shen[1(✉)], Along Mao[1,2], Huajin Wang[1], and Chuan Hu[1,2]

[1] Computer Network Information Center, CAS, Beijing, China
{bluejoe,alongmao,wanghj,huchuan}@cnic.cn
[2] University of Chinese Academy of Sciences, Beijing, China

Abstract. In many applications, data are organized as graphs (e.g., social network and smart city). There could be unstructured data on such a graph, for example, the users' avatars and images included in a post. It is natural to think of these unstructured data as attributes of nodes or relationships. Then the users would tend to query the semantic information of unstructured data on the graph, namely hybrid queries. To meet the demand of hybrid queries, this paper introduces PandaDB, an AI-native graph database, and it has the following characteristics: (1) Unified management of unstructured data and graph data. (2) Online extracting and indexing semantic information of unstructured data. (3) Optimization of hybrid queries. The system and its concept have been verified by multiple applications based on it. Users could deploy PandaDB to support hybrid queries and data mining.

Keywords: Graph Database · AI · Unstructured Data

1 Introduction

In many applications, data are organized and managed as graphs, for example, social network [1] and the smart city [2]. An entity (e.g., a person or a university) is usually regarded as a node. The relationships (e.g., *studyAt* and *workFor*) between entities are regarded as edges. The attributes of entities and relationships are taken as properties (e.g., a person's name and birthday). In real applications, unstructured data (e.g., images, texts, audio, and videos) and structured data are often used to describe the properties. These unstructured data are properties of the entities. There are relationships between entities. Thus potential relationships exist among the multimedia data. For example, users who post similar content tend to construct a tight community (a.k.a, cluster). Two

This work was supported by the National Key R&D Program of China (Grant No. 2021YFF0704200) and Informatization Plan of Chinese Academy of Sciences (Grant No. CAS-WX2022GC-02).

X. Wang et al. (Eds.): DASFAA 2023, LNCS 13946, pp. 669–673, 2023.
https://doi.org/10.1007/978-3-031-30678-5_53

users who are thought unrelated are likely to know each other if they appear in the same photo. These two features would obviously help to recommend potential friends for the users. Thus, users tend to query the semantic information of multimedia data on the graph, namely hybrid queries. It implies the requirement of a database that could natively support hybrid queries.

Given the state-of-the-art works in both academic and industry communities, we are still facing the following challenges: (1) The graph databases can not manage and understand the semantic information of unstructured data. (2) The query language can not describe the statements of querying unstructured data on the graph [3]. (3) The newly introduced operations of unstructured data make it hard to optimize hybrid queries for databases. To solve these challenges, we propose an AI-native graph database, namely PandaDB[1] with the following characteristics: (1) Unified management of unstructured data and graph data. (2) Offering an extended query language to help users to query the unstructured data in a graph. (3) Online extracting and indexing of the semantic information of unstructured data, and optimizing the hybrid query plans according to the query features and data characteristics.

2   System Overview

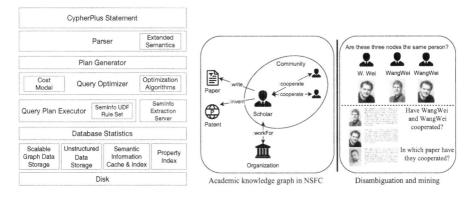

Fig. 1. Architecture of PandaDB   **Fig. 2.** Academic Graph Disambiguation

Figure 1 illustrates the architecture of the proposed system PandaDB. Compared with traditional databases, it is updated and enhanced on these modules and components: *Parser, Query Optimizer, Query Plan Executor, Unstructured Data Storage* and *Semantic Information Cache/Index*. The AI technology (e.g. AI models to understand unstructured data) is natively supported.

Parser: We proposed a new query language, namely CypherPlus, to facilitate the description of hybrid queries, with introducing new functions: *BLOB Functions, Semantic Information Extractor* and *Logical Comparison Symbols*. The parser are modified to support these newly introduced semantics.

---

[1] The project is open-sourced at: https://github.com/grapheco/pandadb-v0.1,.

Query Optimizer is to optimize the hybrid query plans. The cost model is designed to support estimating the cost of unstructured data operations in a hybrid query plan. The optimizer would observe the data statistics and distribution in real time, estimate each operation's cost by the cost model and re-order the operations by the optimization algorithm to get a query plan with minimal estimated cost. There are traditional optimization rules (e.g. predicate push down), the newly introduced operations of unstructured data extend the ordinary rules. For example, the combination of a *SORT* and *LIMIT* operation of unstructured data could be transferred to a kNN search operation of the semantic information in vector format. The latter would be much cheaper than the former. The optimization algorithm will take the new rules into consideration.

Query Plan Executor: The extraction and logical computation of semantic information are treated as UDFs (short for user defined functions) in PandaDB, the *SemInfo UDF Rule Set* defines how to execute these functions. PandaDB extracts the semantic information by AI models. A kind of semantic information corresponds to an AI model (one-to-one map). The AI models are maintained by the *SemInfo Extraction Server* and offer extraction service for the executor. Users are allowed to define the mapping between semantic information name and AI model in the *SemInfo UDF Rule Set*. Similarly, users could define the rules of semantic information logical computation. PandaDB offers a default implementation of *SemInfo UDF Rule Set*.

Unstructured Data Storage: PandaDB treats unstructured data as BLOB (Binary Large Object) and takes BLOB as the first-class citizen in the system. Traditional graph databases (e.g. Neo4j) , store unstructured data as *ByteArray* without the meta-data (e.g. the ID, MIME-TYPE and length). While in PandaDB, the query engine could obtain all the information of unstructured data, for example, the meta-data, unstructured contents, statistics and distribution information, and further, optimize the queries with reference to these information.

Semantic Information Cache and Index: PandaDB caches and indexes the semantic information to accelerate the query response. PandaDB offers two modes, namely, eager mode and lazy mode. Both of them would try to retrieve data from the cache when dealing with a query. The difference lies when they extract the semantic information. The eager mode would pre-extract all the semantic information and cache them. While the lazy mode does not extract the semantic information before it is queried for the first time.

3 Demonstration Scenarios

Academic Graph Disambiguation and Mining. NSFC (National Natural Science Foundation of China) stores and manages data about scholars, published papers, academic affiliations and scientific research funds details. Figure 2 shows the data overview in NSFC. There are about 1.5 TB of data, with 2 million scholars. Three example queries are shown in Fig. 2. All of them involve unstructured semantic information. About sixty different types of queries similar to these are carried on the system. PandaDB is deployed to support the ETL process of NSFC-KBMS [4], namely the Knowledge Base Management System of NSFC. We use AI technology to extract the author and scientific research organization information from the PDF files of the papers, then construct the corresponding association relationships between authors and their corresponding universities. This affiliation is used to build the connection between two graph nodes, as shown in Fig. 3.

Entertainment System. We extend the biggest movie comment and review dataset in China[2]. It contains more than 100 million movies, super stars, comments and users. We built a graph containing actors, movies, participation relationships and unstructured data (e.g. actors' photos and snapshots). PandaDB is evaluated to help users to find the super star in this graph. When the user submits a photo, PandaDB can find the superstar share the similar photo as the facial information of the input photo, then find the film in which the actor has played from the graph. This system is deployed and used in the production environment, and one demo video is in the link[3]. Figure 4 is the screenshot of the demo video, it shows how the GUI acts with PandaDB. The front-end application shows the content of the graph, which includes the images of actors. The user uploads an image and wants to query who is the man in the image he uploaded. The system should first extract the facial feature in the uploaded image, then compare with all the images in the database, and finally return the result. For this query, the front-end application needs only send a query to PandaDB, the statement is:

```
MATCH(n) WHERE n.image IS NOT NULL AND n.image <: IMAGE return n;
```

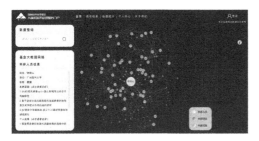

Fig. 3. Big Data Knowledge Service on NSFC   **Fig. 4.** Entertainment System

4 Conclusion

In this work we proposed an AI-native graph database system to meet the demands of hybrid queries. The plan optimizer and executor are designed to support hybrid queries optimization and execution, respectively. The system would then be able to understand the semantic information of unstructured data. Unstructured data could be efficiently managed by the storage layer. Finally, an new query language, namely CypherPlus is proposed to facilitate the hybrid query description.

References

1. Erling, O., et al.: The LDBC social network benchmark: interactive workload. In: Proceedings of the 2015 ACM SIGMOD International Conference on Management of Data (2015)
2. Usman, M., et al.: A survey on big multimedia data processing and management in smart cities. ACM Comput. Surv. (CSUR) 52(3), 1–29 (2019)
3. Francis, N., et al.: Cypher: an evolving query language for property graphs. In: Proceedings of the 2018 International Conference on Management of Data (2018)
4. Zhihong, S., Chang, Y., Hou Yanfei, W., Linhuan, L.Y.: Big linked data management: challenges, solutions and practices. Data Anal. Knowl. Disc. 2(1), 9–20 (2018)

Cleanits-MEDetect: Multiple Errors Detection for Time Series Data in Cleanits

Xiaoou Ding[iD], Yichen Song, Hongzhi Wang[(⊠)][iD], Donghua Yang, and Yida Liu

School of Computer Science and Technology, Harbin Institute of Technology, Harbin, China
{dingxiaoou,wangzh}@hit.edu.cn, 22S003013@stu.hit.edu.cn

Abstract. Data quality problems are seriously prevalent in time series data, and the data suffer from types of errors including single-point errors, continuous errors, and contextual errors. Since it is challenging to achieve high accuracy and efficiency in error detection tasks for time series data, we develop error detection system MEDetect in Cleanits, a data cleaning tool for multi-dimensional industrial time series. We propose an integration detection model for multiple errors, which holds the hierarchical variational automatic encoder as the main structure, and we propose a dimensionality reduction method for k-shape based clustering algorithm, which reduces the time costs of the detection process. MEDetect is designed to allow customized error detection, and users can choose detection and repairing algorithms on their demands.

Keywords: Time series data quality · Error detection · Hierarchical structure

1 Introduction

With the development of data collection techniques and the wide application of intelligent sensors, data is being accumulated at an unprecedented rate. As the high-quality data is the primary guarantee to accomplish valuable knowledge discovery of data assert, data quality problems are seriously prevalent in time series data [1,2], especially in IIoT fields. Considering the characteristics of time series data, it suffers from types of data errors, *e.g.*, the single errors [7], continuous errors [6], and contextual errors [4], etc.

It is challenging to detect various errors in real IIoT scenarios from the following aspects: (1) Various types of errors exist in time series, due to problems such as system delay and environmental interference, (2) errors are likely to have complicated dependencies or correlations, and (3) the labelled data are always limited, or even unavailable. These all pose great challenges to the accuracy and computational efficiency of error detection for time series data.

In this paper, we propose algorithms to detect multiple kinds of errors in time series data, and develop error detection subsystem *i.e.*, **MEDetect** for *Cleanits*,

X. Wang et al. (Eds.): DASFAA 2023, LNCS 13946, pp. 674–678, 2023.
https://doi.org/10.1007/978-3-031-30678-5_54

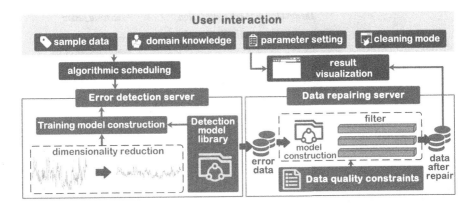

Fig. 1. Framework overview of MEDetect

a data *clean*ing system for multi-dimensional *i*ndustrial *t*ime *s*eries, based on our preliminary work [2]. The main **contributions** in this paper are summarized as follows.

(1) We develop **MEDetect** system to achieve effective detection of three types of errors, *i.e.*, single point errors, continuous errors, and contextual errors, in multivariate time series data. We propose an integration detection model for multiple errors, which holds the hierarchical variational automatic encoder as the main structure. It captures the features of types of errors from normal data, and achieves to learn the temporal and inter-sequence dependencies in time series.

(2) We propose a dimensionality reduction method for k-shape based clustering algorithm, which effectively retains the sequence information and obviously reduces the scale of the original data at the same time.

(3) We develop user-friendly interface in **Cleanits-MEDetect** and allow customized error detection with either supervised or unsupervised methods. Users can choose detection and repairing algorithms on their demands. **Our codes and demo video are available at Github**[1].

2   System Overview

Figure 1 outlines the framework of **MEDetect** module in Cleanits system combined with error repairing module and user interaction module. For user interaction, the system allows users to provide model parameters and translates these configurations into a script file. **MEDetect** server runs the script file and trains a deep model on the training set provided by users and stores the model.

For each *detection task*, the server calls the pre-trained model and labels errors based on a comparison of the L1 error between the reconstructed data and the original data. The main detection process is discussed in Sect. 3. After **MEDetect** obtains error labels, it parses the data and displays the identified

---

[1] https://github.com/Aries99C/MEDetect.

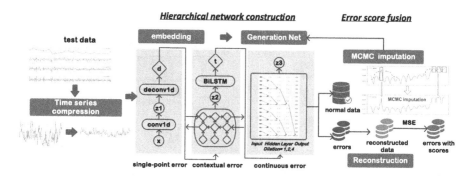

Fig. 2. Workflow of MEDetect

errors via the visualisation module. In the *repairing module*, it converts user-defined data quality constraints into functions that are recognised by the repair method. Then, the repairing server slices the original data and invokes the row and column constraint-based repairing method for error instances, which transforms all constraint functions into boundaries in the linear programming space and invokes the LP solver to directly obtain the repaired data values corresponding to the error data.

3   Implementations

Figure 2 shows the workflow of our solution for multiple-type error detection in time series data, which consists of the follow three steps.

(1) **Multidimensional data compression**. We first use PAA function to reduce the dimensionality of the data $S = \{S_1, ..., S_M\}$ in length to improve the computational efficiency of the unsupervised method [3]. We apply k-shape clustering method to efficiently obtain almost the same clustering results C on the compressed data as on the original data. Then, a representative sequence S_i^* is selected from each cluster C_i to form a sequence set $S' = \{S_1^*, ..., S_k^*\}$, where both the length and the dimension of S are significantly reduced.

(2) **Hierarchical network construction**. We design the detection model based on InterFusion [5] network. As shown in Fig. 2, we hierarchically use the characteristics of different network structures to learn the features of data in the cases of multiple types of errors. Concretely, we train a small-scale multi-layer convolutional neural network to learn the statistical characteristics of the input data, where one-dimensional convolution kernel is used to extract features of the data in the time dimension. We then use a deconvolution network to reconstruct the learned features so that the input to the next layer is the same as the original input. In the second layer, the bidirectional LSTM network is used to learn the context temporal dependence features in data, and such features are used to increase the reconstruction error of contextual errors. In the third layer, considering the high training complexity of the convolutional network $w.r.t$

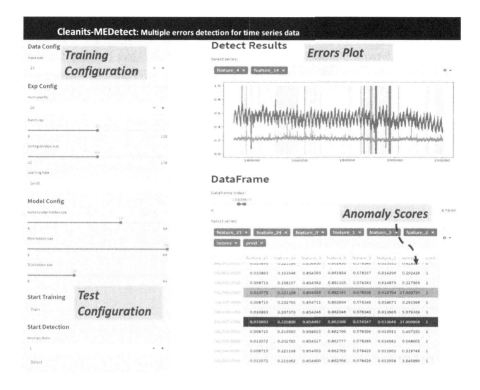

Fig. 3. Demo of **MEDetect** in **Cleanits**

continuous errors, we use the temporal convolutional neural network (TCN) to expand the receptive field of the feature layer in a hop-step manner to achieve high-level learning effectiveness on the long-subsequence features. After that, the generative network uses features learned by the inference network to reconstruct the original input data layer by layer. Thus, errors are identified according to the reconstruction error of the final reconstructed data.

(3) **Error score fusion**. After normalizing the error scores obtained by detection methods, we unify the scores and calculate the error classification according to the threshold values. In this step, we take the median of the error scores on each timestamp as the result.

4 Demonstration

We intend to demonstrate all the functions in **Cleanits-MEDetect** with a real-life IoT dataset from a large-scale fossil-fuel power plant. More than 60 attributes describing the water temperature control machine group are applied in the experiments.

Data is passed into the system by clicking the upload button on the home page. The system converts the user's configuration information into a script file

and transfer the data to the server's data set storage folder. By clicking the
"*Detection*" button on the left side of the page, users can execute the system's
default deep detection model for multiple anomaly types. After training the data
provided by users, the system obtains the error labels `dataset_pred_label`.npy
from the detection server, and displays them in the Plot area. Errors are marked
in red in Fig. 3. Note that **Cleanits** allows users to observe data with different
granularity and drag to adjust the length of the sliding window. In addition, it
is user-friendly to allow users to check the specific values of error instances in
table. User can input the legal start and end timestamps, and the table area will
automatically refresh in the page to present the data in such time interval.

Acknowledgements. This work was supported by the National Key Research and
Development Program of China (2021YFB3300502); National Natural Science Foun-
dation of China (NSFC) (62232005, 62202126, U1866602); China Postdoctoral Science
Foundation (2022M720957); Heilongjiang Postdoctoral Financial Assistance (LBH-
Z21137); and Sichuan Science and Technology Program (2020YFSY0069).

References

1. Ding, X., Wang, H., Li, G., Li, H., Li, Y., Liu, Y.: IoT data cleaning techniques:
   a survey. Intell. Conv. Netw. **3**(4), 325–339 (2022). https://doi.org/10.23919/ICN.
   2022.0026
2. Ding, X., Wang, H., Su, J., Li, Z., Li, J., Gao, H.: Cleanits: a data cleaning system
   for industrial time series. PVLDB **12**(12), 1786–1789 (2019)
3. Ding, X., Yu, S., Wang, M., Wang, H., Gao, H., Yang, D.: Anomaly detection
   on industrial time series based on correlation analysis. J. Softw. **31**(3), 22 (2020).
   https://doi.org/10.13328/j.cnki.jos.005907, (in Chinese)
4. Karkouch, A., Mousannif, H., Moatassime, H.A., Noël, T.: Data quality in internet
   of things: a state-of-the-art survey. J. Netw. Comput. Appl. **73**, 57–81 (2016)
5. Li, Z., et al.: Multivariate time series anomaly detection and interpretation using
   hierarchical inter-metric and temporal embedding. In: KDD 2021: The 27th ACM
   SIGKDD Conference on Knowledge Discovery and Data Mining, pp. 3220–3230.
   ACM (2021)
6. Song, S., Zhang, A.: IoT data quality. In: CIKM 2020: The 29th ACM International
   Conference on Information and Knowledge Management, Virtual Event, Ireland,
   19–23 October 2020, pp. 3517–3518. ACM (2020)
7. Wang, X., Wang, C.: Time series data cleaning: a survey. IEEE Access **8**, 1866–1881
   (2020)

TrajTrace: Tracing Moving Objects over Social Media

Zhihao Yang[1], Yunqi Zhang[1], Songda Li[1], Qinhui Chen[1], Hui Zhao[1,2(✉)],
Wei Cai[3], and Xi Lin[3]

[1] Software Engineering Institute, East China Normal University, Shanghai, China
`hzhao@sei.ecnu.edu.cn`
[2] Shanghai Key Laboratory of Trustworthy Computing, Shanghai, China
[3] Electronics Technology Group Corporation, The 51st Research Institue of China,
Beijing, China

Abstract. Online social media has lots of moving object information. Extracting movement information is a challenging work. It suffers from spatio-temporal information extraction, vague time, and vague location. Previous information extraction methods merely focus on the trajectory extraction. In this demonstration, we develop a web-based application, *TrajTrace*, to track moving objects. *TrajTrace* extracts triple <*object, movementState, location*> from social media text employing our proposed span-level joint entity and relation extraction model, *OMLer*. *OMLer* casts joint extraction as a token pair multi-categories classification task. It predicts the triple list corresponding to the input sequence. We employ BERT to encode the input sentence word by word. The self-attention mechanism and BiLSTM are applied to learn sequence features. Then, an order-first time matching algorithm is designed to solve the lacking temporal information problem in the extracted triples. Utilizing the proposed TF-IDF based clustering algorithm, we make the vague time accurate. The vague geographic location is converted to accurate latitude and longitude using the Bezier geodetic coordinate conversion algorithm. Toward aircraft and ships, besides the keyword search and the trajectories visualization, *TrajTrace* provides the historical activity area search of a specific object and the spatio-temporal distribution of moving objects at a given time or location.

Keywords: Moving objects · Trajectory · Joint entity and relation extraction · Attention mechanism · Token pair sequences

1 Introduction

People often share real-time moving object information on online social media, such as the takeoff time of an aircraft or the arrival location of a ship. Based on the movement information of moving objects on social media platforms, we can trace the moving object and mine the trajectories to find the movement behavior patterns.

X. Wang et al. (Eds.): DASFAA 2023, LNCS 13946, pp. 679–684, 2023.
https://doi.org/10.1007/978-3-031-30678-5_55

Movement information extraction is a challenging task compared with traditional information extraction. Traditional information extraction approaches seldom consider the movement action series of a moving object. Extracting moving object information from online social media faces the following challenges:

1. Traditional entity and relation extraction methods cannot extract the temporal information of relation triples. As for the sentence "Boeing 747 departed from Beijing on May 1", if only extracting the relation triple <*Boeing 747, /Plane/TakeOff/Location, Beijing*>, the temporal information "May 1" is omitted.
2. Overlapping relation triples are common. The movement information of a moving object often contains a series of actions. Thus a moving object has multiple movement statuses, which means the moving object is in multiple relation triples.
3. The expressions for moving objects are not formal, containing many colloquial expressions. Many expressions contain abbreviations and aliases mixed in Chinese and English. For example, the aircraft "Boeing 737" is commonly called "737", and the aircraft "Airbus A300" is called "A300".
4. Time and location expressions are vague, such as "recent days" and "about 60 nautical miles outside of East Wales". These unclear descriptions lead to inaccurate representations of the trajectories.

In this demonstration, we develop *TrajTrace*, a system to track moving objects, which extracts trajectories of aircraft and ships from open social media. Our implementations are summarized as follows:

1. We propose and train a span-level entity and relation joint extraction model, *OMLer*, to extract movement relation triples <*Object, MovementState, Location*>. We define four movement relations to describe the position movement of moving objects. The experimental results show that our model's precision is 89.4%, and outperforms CasRel by 3.4%.
2. We design an order-first time matching algorithm to match the time entities with their corresponding relation triples.
3. The vague time is converted to accurate expression by regular expression match and TF-IDF based clustering algorithm. The vague geographic location is made accurate by Bezier geodetic coordinate conversion algorithm.

2 System Overview and Key Techniques

This system consists of five modules as shown in Fig. 1: data module, NLP module, business module, gateway module, and visualization module.

OMLer. It is our proposed and trained span-level joint entity and relation extraction model. *OMLer* extracts relation triples <*object, movementState, location*>. The model includes a word embedding layer, a sequence labeling conversion layer, a hidden layer, and an output layer. Figure 2 illustrates the model architecture.

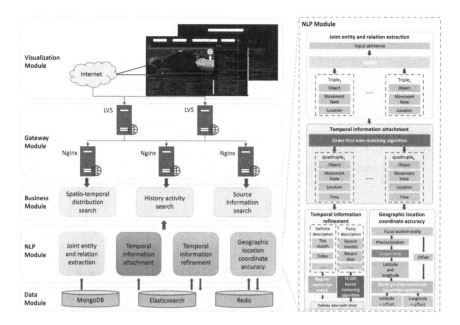

Fig. 1. System framework

In the word embedding layer, we re-implement the word segmentation function of BERT. Chinese words are segmented based on characters, and English words are segmented based on letters. Through the word segmentation function, the input sentence is converted into multiple tokens. The pre-trained model BERT [1] is leveraged to obtain the word embeddings of tokens. We define four movement relations: *TakeOff*, *Will-Appear*, *Appear*, and *Arrive*. TPLTS [4] encoder is applied to generate token pair sequences and tagged sequences of entities and each relation. There are three types tagged sequences: EH-ET (Entity Head-Entity Tail), SH-OH (Subject Head-Object Head), and ST-OT (Subject Tail-Object Tail). EH-ET indicates the tagged sequence of starting and ending positions of entities. Each relation type has SH-OH and ST-OT tagged sequences. SH-OH indicates the tagged sequence of the head positions of subjects and objects. ST-OT means the tagged sequence of the tail positions of subjects and objects.

In the hidden layer, the token pair sequence of the input sentence $hidden_{seq}$ is split into $h_{visible}$ and h_{repeat}. Through BiLSTM and self-attention mechanism [3], the model learns the sequence features of sentences and obtains h_{att}. Based on $h_{visible}$ and h_{repeat}, we perform conditional layer normalization (CLN) to obtain h_{mask}. Finally, we generate the output sequence h_{out} with h_{att} and h_{mask}. The output sequences of the entities and each relation are computed through independent network structures.

In the output layer, a fully connected layer is applied to predict the EH-ET sequence, SH-OH sequences and ST-OT sequences respectively. Based on the predicted sequences, the corresponding triples <*object, movementState, location*> are obtained through TPLTS decoder.

We compare our model with a baseline model for joint entity and relation extraction, CasRel [5], and a pipeline model that consists of BiLSTM-CRF [2] for named entity recognition and RNN for relation extraction. The experimental results show that our model reaches 89.4% in precision, 83.6% in recall, and 85.9% in F1 value. In precision, our model outperforms CasRel by 3.4% and the pipeline model by 21%. We also perform the ablation study. We utilize random inactivation to reduce model complexity and optimize the training process of the model. The ablation study result shows that the self-attention mechanism and random inactivation make an improvement by 0.3% in precision.

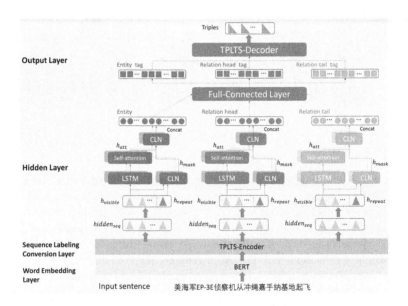

Fig. 2. *OMLer* architecture

Temporal Information Attachment. For movement relations, they are required to attach temporal information. The order-first time matching algorithm is designed to attach the temporal information to the predicted movement relation triples. If the sentence from which the triple is extracted contains time entities, the time entity nearest to the moving object is attached to the triple. Otherwise, the time of the blog posted is considered as the time entity. With the order-first time matching algorithm, a movement relation triple <*object, movementState, location*> is converted into a quadruple <*object, movementState, location, time*>.

Temporal Information Refinement. For the specific time description, such as "last month" and "yesterday", we convert the time to a standard format according to the microblog post time by regular expression match. For the unclear time representation, like "recent days", it is unable to determine the time by a single microblog. We first apply the TF-IDF based clustering algorithm to find the news describing the same event as the blog. Then, the found definite time entity among the clustered sentences is assigned to the triples which locate in the sentence with the unclear time representation.

Geographic Location Coordinate Accuracy. For the inaccurate location description, like the sentence "about 270 km east from Nantong, Jiangsu", we first employ Google Encoding Map API to query the coordinates of "Nantong, Jiangsu". Second, the direction "east" is converted to the azimuth. Then, the Bessel geodesic coordinate conversion algorithm is utilized to calculate the accurate latitude and longitude of the destination location.

3   System Demonstration

TrajTrace is a web-based system with an interactive map service interface. The system is designed to track aircraft and ships mentioned in online social media. Figure 3 displays the main scenarios of the demo.

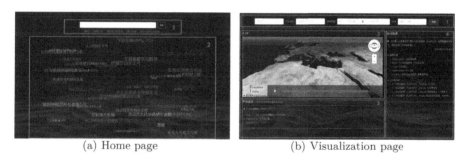

(a) Home page   (b) Visualization page

Fig. 3. System demonstration

Figure 3(a) displays the home page of our system. Users can input the name of a specific moving object Region 1 to query its movement information, including the location and the routine. Moreover, the ships or aircraft are displayed with high query frequency to users through the word cloud in Region 2. Users can also directly click the name to query.

Figure 3(b) shows the visualization page with the query results of the aircraft "F-22" as an example. First, users can query the movement information of some moving objects within a specific area or time in Region 1. Second, the spatio-temporal distribution of some ships or aircraft is visualized on the map in Region 2 and detailed in Region 3 to support further analysis. Besides, we provide a progress bar on the map. Users can click any time points on the progress bar to observe the changes in the spatio-temporal distribution. Third, The details of the original blogs and extracted triples are listed in Region 4.

4 Conclusion

In this demonstration, we develop *TrajTrace*, a system to track moving objects based on online social media. We propose and train *OMLer* to extract movement relation triples *<object, movementState, location>*. Using our designed order-first time matching algorithm, we convert extracted triples into quadruples *<object, movementState, location, time>*. We apply the TF-IDF based clustering algorithm to refine the time information. Vague geographic locations are formalized by Bezier geodetic coordinate conversion algorithm. Our demo provides users to search the movement trajectories and the spatio-temporal distribution of moving objects to further understand the movement pattern of moving objects.

Acknowledgements. This work is supported by National Key Research and Development Program (2019YFB2102600).

References

1. Devlin, J., Chang, M.W., Lee, K., Toutanova, K.: BERT: pre-training of deep bidirectional transformers for language understanding. In: ACL, pp. 4171–4186 (2019)
2. Lample, G., Ballesteros, M., Subramanian, S., Kawakami, K., Dyer, C.: Neural architectures for named entity recognition. In: NAACL, pp. 260–270 (2016)
3. Vaswani, A., et al.: Attention is all you need. In: NIPS, pp. 6000–6010 (2017)
4. Wang, Y., Yu, B., Zhang, Y., Liu, T., Zhu, H., Sun, L.: TPLinker: single-stage joint extraction of entities and relations through token pair linking. In: COLING, pp. 1572–1582 (2020)
5. Wei, Z., Su, J., Wang, Y., Tian, Y., Chang, Y.: A novel cascade binary tagging framework for relational triple extraction. In: ACL, pp. 1476–1488 (2020)

A Raft Variant for Permissioned Blockchain

Zheming Ye[1,2], Xing Tong[1,2], Wei Fan[1,2], Zhao Zhang[1,2(✉)], and Cheqing Jin[1,2]

[1] East China Normal University, Shanghai, China
{zhmye,xtong,wfan}@stu.ecnu.edu.cn, {zhzhang,cqjin}@dase.ecnu.edu.cn
[2] Engineering Research Center of Blockchain Data Management,
Ministry of Education, Beijing, China

Abstract. Many permissioned blockchain systems adopt Raft as one of the consensus protocol options, however, comparing with traditional distributed systems, blockchain system has more nodes which makes Raft faces challenges when applied to blockchain. In this study, we present a scalable Raft variant called SRaft for permissioned blockchain. SRaft decouples consensus into Replicating phase and Ordering phase, in Replicating Phase, SRaft adopts leaderless block replicating method and supports workload adaptive replicating pattern, so as to make full use of the bandwidth of all peers and greatly reduce the impact of unbalanced workload distribution. In Ordering phase, SRaft uses the features of the chain structure of blockchain to simplify the Ordering phase into a Sending-Receiving communication framework. In our demonstration, we use real-time dashboards containing visualization based on the process of SRaft and provide an interactive platform to give the attendees an in-depth understanding of what SRaft does.

1 Introduction

Benefiting from permissioned blockchain's strict entry mechanism, over 60% permissioned blockchain systems support crash fault tolerance(CFT) protocols [1]. As an efficient and understandable CFT consensus protocol, Raft [2] is widely used in distributed systems. However, blockchain systems have more peers which makes the leader become the bottleneck of Raft. The leader can be avoided becoming bottlenecks in some ways. For example, in related works, CRaft [3] proposes a Raft variant based on the Erasure code to reduce the network overhead of the leader. However, overloaded peers' uploading bandwidth may also limit the system's performance. To better solve the above problems, SRaft decouples Raft consensus into Replicating and Ordering phase. In Replicating phase, SRaft uses a leaderless replicating mechanism,i.e., all peers can replicate block, so as to avoid the bandwidth of the leader from limiting performance. In addition, in order to adapt to unbalanced workload distribution among peers, SRaft supports a workload adaptive replicating pattern. In Ordering phase, replicated blocks from Replicating phase are ordered by their hash value instead of the original blocks, so Ordering phase's leader will not become the bottleneck. In addition, SRaft uses the chain

X. Wang et al. (Eds.): DASFAA 2023, LNCS 13946, pp. 685–689, 2023.
https://doi.org/10.1007/978-3-031-30678-5_56

feature of blockchain to abstract Ordering phase into a Sending-Receiving communication framework, thus simplifying engineering implementation.

The purpose of our demonstration is to show a visualization result of the process of SRaft and provide a highly interactive platform for attendees. The attendees can have a more intuitive understanding of SRaft's work by interacting with our demonstration.

2   SRaft Overview

Comparing with original Raft, SRaft decouples the consensus process into Replicating and Ordering phases, thus SRaft can be well applied to permissioned blockchain. In Replicating phase, each peer can replicate a block, while in Ordering phase, there is only one leader which is responsible for ordering blocks. After receiving transactions sent by clients, in Replicating phase, the peer packs transactions into a block and replicates the block to other peers. Then, successfully replicated blocks' hash values are sent to the leader of Ordering phase. In Ordering phase, the leader orders the hash values of replicated blocks and drives this order to reach consensus among nodes. Next, we give a detailed description of the two phases mentioned above.

2.1   Replicating Phase

After receiving transactions, the peer constructs a block and replicates the block to other peers. SRaft allows each peer to replicate the block concurrently and independently. In order to maximize the utilization of bandwidth, SRaft allows each peer to allocate its available bandwidth to overloaded peers. Before replicating, the peer collects other peers' available bandwidth and constructs a block replicating route based on idle peers' available bandwidth, then, the block is replicated along this replicating route. For example, as Fig. 1 shows, P1 is ready to replicate a block and we suppose peers P2 and P3 have available bandwidth. If the bandwidth of P1 is sufficient for its current workload, the forwarding route constructed by P1 is one-to-all pattern. If P1 is overloaded, P1 constructs a forwarding route in tree pattern, specifically, P1 acts as tree's root node, and P2 and P3 act as branch nodes. When replicating the block, this forwarding route

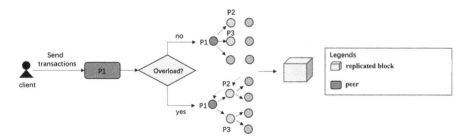

Fig. 1. The process of Replicating Phase

is also propagated to other peers. After replicating. After receiving the confirm messages from the majority of peers, corresponding block is considered successfully replicated. Finally, P1 sends successfully replicated block's hash value to the leader of Ordering phase.

2.2 Ordering Phase

After receiving the hash values of Replicating phase from other peers. The leader of Ordering phase orders these blocks. Then, the leader sends the ordering result to other peers and tries to make this result reach consensus among peers. In this process, SRaft uses same processing flow as Raft. First, the leader broadcasts the ordering result to other peers and waits for the majority confirm messages. After the majority confirm messages are received, the leader broadcasts the commit message and the blocks appeared in the ordering result are committed for execution. Comparing with Raft, blocks in the ordering result are represented by their hash values rather than original blocks, so as to ease network pressure of leader. Figure 2 shows the process of the Ordering Phase.

In addition, we leverage the chain feature of blockchain to pipeline the Ordering process and abstract it into a Sending-Receiving communication framework, so as to simplify engineering implementation.When the leader broadcasts the ordering result, it also collects newly replicated blocks' hash values from other peers. In the next round, the leader broadcasts the ordering result of these newly replicated blocks' hash values and commits the previous round's ordering result.

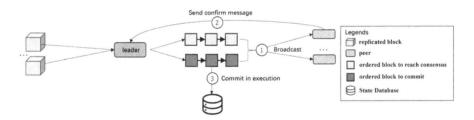

Fig. 2. The process of Ordering Phase

3 Demonstration

To give the attendees an in-depth understanding of what SRaft does, we simulate a system deployed SRaft, and provide an interactive front-end interface. Figure 3 shows a screenshot of our demonstration. We divided the main page of the demonstration into Panel A ~ D. Panel A visualizes the real-time status of each peer and the communication between each peer. Panel B provides a dashboard to directly display the current bandwidth and workload of each peer. Panel C gives the real-time overload ratio of each peer and shows it through an area chart. Panel D presents the latest information on blocks.

In our demonstration, each peer has a certain probability to be an initiator to replicate a block, then it contracts its forwarding route and replicates the

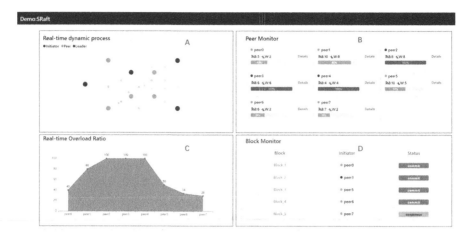

Fig. 3. The interface of Demonstration

block to other peers. At the same time, we set a leader to communicate with all peers at regular intervals. During the process, Panel A visualizes the process by coloring the peer in different statuses and different messages with different colors. Furthermore, the data in Panel B and the chart in Panel C change dynamically. The status of each peer is visually displayed by different color progress bar in Panel B, and Panel C puts the workload data of these peers in the same chart to better show the distribution of workload of each peer so that the attendees can intuitively understand the SRaft's work in Replicating Phase. Additionally, by clicking the "Details" button of each peer on Panel B, the attendees can view the detailed information of the corresponding peer. As shown in Fig. 4, Panel E visualizes the forwarding route and Panel F provides the current status of the blocks that packed locally by the current peer.

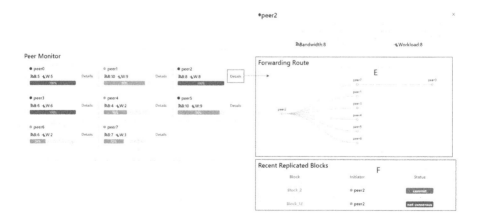

Fig. 4. The dashboard of peer's status

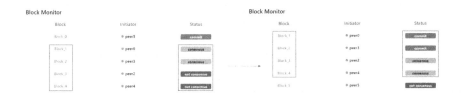

Fig. 5. Status transition of blocks

As the real-time status of the latest blocks is shown in Panel D, the attendees can see the transition of the block's status. For example, in Fig. 5, in the previous round, Block 1~ 2 has reached consensus among all peers but they haven't been committed yet. At the same time, Block 3 ~ 4 have just been sent to the leader for ordering. After the next round, attendees can observe the changes that Block 1 ~ 2 are committed and Block 3 ~ 4 reach consensus among all peers. Through the above, the attendees can further understand SRaft's Sending-Receiving communication framework in Ordering phase.

Acknowledgements. This work is supported by National Science Foundation of China (No. 61972152 and No.U1911203).

References

1. CAIFT.: Blockchain white paper (2021). (in Chinese)
2. Ongaro, D., Ousterhout, J.: In search of an understandable consensus algorithm. In: 2014 USENIX Annual Technical Conference, pp. 305–319 (2014)
3. Wang, Z., et al.: Craft: an erasure-coding-supported version of raft for reducing storage cost and network cost. In: FAST, pp. 297–308 (2020)

MOCache: A Cache Management Tool for Moving Object Databases

Bingya Wu, Xin Jin, and Jianqiu Xu$^{(\boxtimes)}$

College of Computer Science and Technology, Nanjing University of Aeronautics
and Astronautics, Nanjing, China
{wubingya,jinxin,jianqiu}@nuaa.edu.cn

Abstract. There is a lack of cache management in moving object
databases which is essentially important for system analysis. In addition,
users focus on queries in real scenarios, but cache replacement strategy is
not customized. To address these issues, we develop a cache management
tool called *MOCache* for moving object databases. The tool consists of
three components: data partition, cache strategy improvement and cache
status monitoring. First, we partition moving objects and 3D bbox to
establish mapping relationships between sub trips and cells. Then, we
propose a cache elimination strategy based on cell heat for the purpose
of exploiting spatio-temporal attributes. Finally, the tool visualizes cache
status and automatically generates warm-up commands. Demonstration
shows that the tool achieves hit ratio to 76.56% and greatly reduces execution time compared with alternative approaches, and helps researchers
evaluate factors affecting database cache.

Keywords: Moving object database · Cache replacement · Hot data

1 Introduction

The widespread use of GPS-enabled devices has made position data being easily
collected. In response to this, the research field of moving object data management has been extremely active in the past two decades. Several prototype
systems have been developed to process moving objects efficiently for retrieving all data in a particular spatio-temporal region via a limited number of I/O
accesses [1]. To optimize system performance, a number of techniques are proposed including storage method, index structure [2], and query optimization [3].
In particular, fully leveraging cache in existing moving object systems does not
receive adequate attention.

Cache is a fundamental but not transparent component in systems. Every
DBMS may implement its own caching layer tailored to customized requirements
and cache management generally includes: (i) what to cache, (ii) where to cache
and (iii) how to cache, which is closely related to data features and accessed
patterns [4]. Moving object databases (MODs) manage spatial objects that continuously change their locations over time. Foundational operations like range

© The Author(s), under exclusive license to Springer Nature Switzerland AG 2023
X. Wang et al. (Eds.): DASFAA 2023, LNCS 13946, pp. 690–695, 2023.
https://doi.org/10.1007/978-3-031-30678-5_57

queries and the nearest neighbour queries report objects with both temporal and spatial localities, which are consistent with cache locality principle. Data in MODs generally gets many query requirements and thus how to set up cache to hold data with high access frequency is a crucial issue. When the limited cache space overflows, determining which cached item to eliminate is challenging. In MODs, traditional replacement strategies [5] are utilized like *least recently used* (LRU), *two queues* (2Q), and *least frequently used* (LFU), which do not consider the characteristics of moving objects.

Therefore, in order to maximize cache performance in MODs, effective cache management is essentially required. We develop a tool called MOCache (Moving Objects Cache Management) to aid system developers to clearly understand data access patterns and make assessments about cache effectively. First, we propose a cache replacement strategy to calculate cell heat based on the partition of 3D bounding box. The method not only improves hit ratio and execution time, but also saves precious cache space by cutting moving objects into sub trips. Then, hot data in cells can automatically generate preheating commands for users to avoid cold start. Finally, we visualize cache fine-grained information after query operations for further system adjustment.

2   MOCache

The tool aims to help researchers enhance cache performance for MODs in aspects of improving cache hit ratio, reducing unnecessary cache space usage and visualizing cache status. The framework is composed of three main modules, as shown in Fig. 1: (i) User interface, (ii) Cache Function, (iii) Data storage.

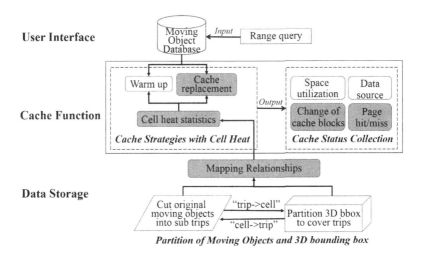

Fig. 1. The framework of MOCache.

2.1 Data Analysis and Partition

Generally, the historical movement of a moving object contains lots of consecutive points in a certain spatio-temporal range. For example, if we query *"Which taxis pass the railway station between 10:00 am and 10:30 am?"*, every targeting taxi with abundant points is read into cache to waste valuable space. Therefore, we partition original moving objects into sub trips according to movement attributes in real scenarios. Moving points collected in the same hour are partitioned into a new moving object. We calculate data distributions to obtain spatio-temporal range of the overall bounding box, then partition the three dimensions of longitude, latitude and time respectively to make a sub trip span three or more cells as little as possible. The purpose is to transfer the statistics of individual moving object to each cell to reduce overhead. Finally mapping relationships between cells and sub trips are established. When opening database, the system automatically creates a single instance to load the mapping relationships stored on disk originally into memory for counting heat values of cells when accessing data.

2.2 Cache Replacement Strategy

The classic LRU algorithm is utilized in cache replacement mechanism of MODs. However, when batch operation is performed, LRU will replace pages frequently to cause cache pollution. Data accessed frequently at certain periods or places should not be eliminated. To handle this problem, we propose a cache replacement strategy named LRU-CH (Cell Heat). We define a condition on LRU policy, which washes out a sub trip when it does not belong to any cell with high heat.

The specific steps are as follows. When executing every command, the mapping relationships can be employed to match sub trips to cells, and then the heat values of the accessed cells are updated. Those cells with high heat value are stored and maintained with a min heap of priority queue. The heat value at the top of the heap is the smallest. Therefore, we can directly compare the heat of victim found from the tail of the cache chain with the top of the min heap to judge whether to eliminate. In the cache blocks, if all cells where sub trips are located have high heat, we use a sentinel to point to the sub trip with the lowest cell heat and remove the victim. This approach not only preserves the fairness of LRU, but also takes advantage of the spatio-temporal property of moving objects. We also achieve a cumulative heat algorithm and heat values reflect previous database operations. Once the heat of one cell reaches the specified threshold, the heat values of all cells will be halved.

3 Demonstration

MOCache is implemented in an open-source and extensible moving object database system SECONDO [6], which is primarily for processing moving objects and spatio-temporal data. We provide the URL[1] to view the screencast video and source code.

---

[1] https://github.com/CamielCo/MOCacheDemo.git.

Data Preparation. We use the public Beijing taxi trajectories to conduct the demonstration. The dataset includes 919 moving objects and 1400000 temporal units. We perform an optimal partition over moving objects and achieve 119157 sub trips under the partition of 14900 cells such that 99.7% of sub trips are only located in one cell.

Query Generation. Range query is one of common queries in MODs. In the experimental evaluation, we use 1450 range queries to evaluate the performance of the cache management. Queries are produced as follows. We generate 100 spatio-temporal windows randomly that meet query specification, then repeat the first 10 entries 100 times, the 11th to 50th entries 10 times, and the 51st to 100th entries only once, resulting in 1450 queries. The template of the query expression is as follows:

let result1 = BJTaxisMBR_rtree BJTaxisMBR windowintersects [bbox([const upoint value (("2008-02-04-11:00:00" "2008-02-04-12:00:00" TRUE TRUE) (116.402884 39.05438 116.405651 39.913826))])] consume;

The *MBR* attribute representing spatio-temporal range of moving objects is created through the operator *bbox* in the system. The RTree index is built on *MBR* to ensure cache random access, then the operator *windowintersects* can be used to check whether two bounding boxes intersect each other.

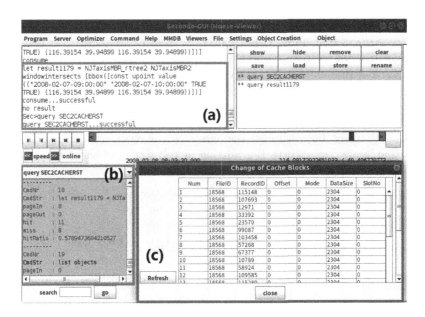

Fig. 2. Monitor cache status.

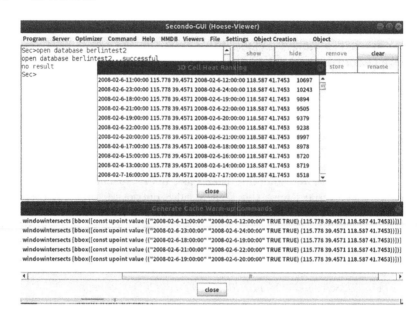

Fig. 3. Generate warm-up commands.

Demonstration. In order to well feedback cache internal information, collecting cache status after query operations is essential, which includes basic information and changes of cache chain. Users input query commands at Fig. 2(a). We design the system table *CACHERST* to record cache status. If executing the command "query CACHERST", the cache status of historical statements is tracked and presented at Fig. 2(b), such as hit ratio, changes of pages, especially page swapped out and in. Figure 2(c) monitors the changes of cache chain includes data source and the access order of cache blocks. For instance, the recently accessed moving object is placed at the head of the chain. Users can click the "Refresh" button to get the current cache chain information. Figure 3 shows that users can observe the ranking of cell heat and carry out warm-up commands automatically generated by hot data.

Table 1. The comparison of hit ratio.

	LRU	2Q	LFU	**LRU-CH**
Raw Data	15.65%	34.85%	37.5%	31.48%
Partitioned Data	35.29%	59.31%	60.44%	**76.56%**

Performance Evaluation. LRU-CH policy has a great impact on the performance of MOCache. By executing the above 1450 query commands, we evaluate the efficiency of LRU-CH from two aspects: cache hit ratio and execution time

Table 2. The comparison of execution time (min).

	LRU	2Q	LFU	LRU-CH
Raw Data	434.58 m	475.54 m	499.76 m	443.36 m
Partitioned Data	64.56 m	68.17 m	80.59 m	**47.16 m**

Table 3. The hit ratio of different cache size.

	16M	32M	64M
LRU-CH	23.82%	52.37%	76.56%

compared with other strategies in MODs. Table 1 and Table 2 show that LRU-CH speeds the hit ratio up to 76.56% and achieves the minimum execution time of 47.16 min after data partition. We also test LRU-CH by using the partitioned data and setting different cache size as shown in Table 3.

4 Conclusion

In this demonstration, a tool *MOCache* is proposed to do cache management for moving object databases. To help users observe and evaluate the impact of cache on database performance, *MOCache* monitors cache information and tracks hot data during query execution. We also achieve the LRU-CH policy to eliminate moving objects if cache overflows, which greatly improves hit ratio and execution time. In future, we plan to investigate more detailed methods for further improving our solutions.

Acknowledgements. This work is supported by National Natural Science Foundation of China under Grant No.61972198.

References

1. Cudre-Mauroux, P., Wu, E., Madden, S.: Trajstore: an adaptive storage system for very large trajectory data sets. In: ICDE, pp. 109–120 (2010)
2. Xu, J., Güting, R.H., Zheng, Y.: The tm-rtree: an index on generic moving objects for range queries. GeoInformatica **19**(3), 487–524 (2015)
3. Lan, H., Bao, Z., Peng, Y.: A survey on advancing the DBMS query optimizer: cardinality estimation, cost model, and plan enumeration. Data Sci. Eng. **6**(1), 86–101 (2021)
4. Durner, D., Chandramouli, B., Li, Y.: Crystal: a unified cache storage system for analytical databases. Proc. VLDB Endow. **14**(11), 2432–2444 (2021)
5. Liu, C., Ding, S., Ye, L., Chen, X., Zhu, W.: Cache replacement strategy based on user behaviour analysis for a massive small file storage system. In: ICCAE, pp. 178–183 (2022)
6. Güting, R.H., Behr, T., Düntgen, C., et al.: Secondo: a platform for moving objects database research and for publishing and integrating research implementations. Fernuniv., Fak. für Mathematik u. Informatik (2010)

Cnosdb: A Flexible Distributed Time-Series Database for Large-Scale Data

Yu Yan[1], Bo Zheng[2], Hongzhi Wang[1(✉)], Jinkai Zhang[1], and Yutong Wang[1]

[1] Harbin Institute of Technology, Harbin, China
{yuyan,wangzh}@hit.edu.cn, harbour.zheng@cnosdb.com
[2] Cnosdb Inc., Beijing, China

Abstract. With the development of the Internet of Things, the time series data generated by monitors, analyzers, and detection instruments in the industry has surged. The management of very large-scale time series data faces great challenges. However, the current distributed time series database is still poor in terms of data storage efficiency and data writing speed. In order to achieve the fast writing and high efficient storage of billions or even tens of billions of data points, we propose a cloud native distributed time series database, CnosDB. Our system integrates various data compression algorithms to achieve high compression rate in each data type. And we propose a three-layer storage policy to achieve fast writing under the premise of ensuring rapid time-based batch operations. In this paper, introduce the architecture and key techniques of CnosDB, and describe three key demo scenarios of our system.

1 Introduction

With the advent of big data, the scale of time series data surge in the industry, such as monitors, analyzers, and detection instruments in the electric power industry [10] and the chemical industry. Industrial data has three typical features: **Fast Generation Speed** [9]: Each monitoring point can generate large amount of data one second. **Unique Timestamp** [8]: Each piece of data has a dependent and unique timestamp. **Wide Collection Range** [13]: The conventional real-time monitoring system has thousands of monitoring points, which are generated data every second.

Faced with the real-time and large amount of time series data, traditional databases such as MySQL can no longer meet the requirements for massive data storage and management, and various types of time series databases have emerged.

Recently, in order to achieve efficient management of large-scale time series data, some time series database management systems have been developed. In the early days, researchers used other databases as the backend and developed a middleware for time series data management, such as TimescaleDB [3], [6], etc. Without own storage engines, middleware-based methods cannot effectively compress data and has weak aggregation capabilities. Later, the storage structure based on time series data gradually appeared [5] [1]. Graphite [5] designs the Whisper storage structure for time series data, which can store data at fixed time intervals and accelerate the writing speed of time series data. However, it does not support data shard. Time series data is stored in the file system which conducts bad scalability. To achieve high scalability, InfluxDB [7], FreeTSDB [4] and TDengine [2] have been proposed. They design more flexible shard architectures, which can realize the addition and deletion of data nodes. However, its writing speed and storage efficiency still cannot meet the demand of large-scale time series data. In fact, no management system that can efficiently handle very large-scale time series data access now has been proposed. The industry urgently needs a scalable database that can efficiently manage large-scale time series data.

Motivated by this, we analyze the features of time series data, and develop a scalable and efficient time series data management system. Considering the demands of large-scale time series data, we pay more attention to the efficiency of data writing policy and data compression. Our CnosDB has the following advantages:

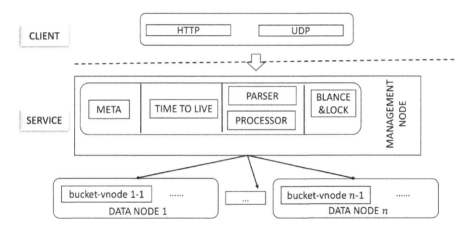

Fig. 1. System Architecture

- *High Compression Rate.* The system integrates compression algorithms for time series data, integers, floating point numbers and other data types, which can compress each type of data to the greatest extent.
- *Fast Data Writing.* We designed a new storage policy based on the features of time series data and adopted a hierarchical shard (called bucket-vnode), which reach higher performance by using disk batch sequential writing.
- *Friendly UI.* CnosDB provides friendly user interface. Our system not only supports command line interaction but also provides a friendly graphical interface.

2   System Architecture

In this section, we introduce the architecture of CnosDB as shown in Fig. 1. Our system is mainly constructed with Golang and uses the C/S structure which is flexible for extending nodes. Almost all distributed databases use the C/S architecture. In CLIENT, users could send the requests(such as create, insert, delete, update, query, etc.) by using HTTP or UDP. The SERVICE consists of a MANAGEMENT NODE and some DATA NODEs. The MANAGEMENT NODE is responsible for parsing requests and implementing queries. The META module is the core of the management node. It controls the meta data of the entire database, containing entity fields, verification rules, data types, etc. The TIME TO LIVE charges the data life cycle. In CnosDB, the TIME TO LIVE module will divide the series data into the bucket-vnode groups based on their timestamp. And then, TIME TO LIVE module could control data expiration by only operating bucket-vnode groups instead of filtering operations. Since time series data is usually expired on an entire group, using the entire set of expiration instead of partial expiration can greatly improve the efficiency of processing expired data. The combination of the PARSER and the PROCESSOR realizes the analysis and processing of the requests. The BLANCE&LOCK module charges the distributed nodes and achieves data balance and synchronization.

3   Key Technologies

In this section, we would introduce the key technologies of CnosDB, including data compression and storage structure.

3.1   Integrated Compactor

Since time series data is generated very fast and take up many storage resources, a reasonably designed compression algorithm can greatly improve storage efficiency. A variety of compression algorithms are embedded in our system for effectively compressing each type of data.

- *Timestamp* We develop the delta-of-delta algorithm to express the timestamp. The principle of delta-of-delta is that after calculating the difference between adjacent timestamps, the difference between adjacent timestamps is calculated again to achieve the smallest time span.
- *Integer* First we use zigzag encoding to encode integers. If all zigzag encoding values are less than simple8b.MaxValue, we use simple8b encoding; if any value is greater than simple8b.MaxValue, we use original value storage.
- *Float* The compression process is a byte stream (or bit stream) continuously appended according to a certain format [12] [11]. The first floating point number stored is the complete 8 bytes data, and the other numbers stored afterwards are the delta values obtained by XOR with the previous number.
- *Other Type* Strings are encoded using Snappy compression and Boolean values are encoded using a simple bit packing strategy.

Besides, to increase the compress ratio in total, we develop adaptive compression strategy to fit the data into periodic or non-periodic functions and store the data as these functions.

3.2 Storage Policy

Compared with other types of data, time series data is usually rarely queried, but requires fast writes. In addition, the life cycle of data in a time series database is also very important, and the database must be able to achieve rapid failure of the batch data. Taking into account the above two factors, we designed a three-layer storage strategy, including data layer, bucket layer and sequential write layer. Our policy could achieve fast writing while being able to perform timestamp-based batch operations.

Specifically, the first-level shard is based on the timestamp, and CnosDB automatically generates the shard interval according to the data retention policy set by the users. Based on this time interval, the inserted data will first be routed to the corresponding bucket-vnode groups. However, for avoiding hot write problems the time interval is usually set large, we designed a hashing-based algorithm to divide the groups within the same time range again. After the data is divided twice, it will reach one data node. On this data node, we build an LSM tree structure to implement sequential data writing.

4 Demonstration

We plan to demonstrate our CnosDB with three scenarios, containing management retention policy, database monitor, query implementation. More demonstrations and source code are available in github[1]. The video of the demonstration is shown in Youtube[2]. The functions to demo are shown as follows.

- *Manage Retention Policy.* Users can refresh, delete and switch retention policies easily through our interface.
- *Database Monitor.* Users can use the interface to view the current requests per second (read and write frequency) and transfer per second (the speed of write) to monitor the status of the database.
- *Query Implement.* Although the data in the time series database is not frequently queried, CnosDB provides a unified query interface for users to view the data.

We test our CnosDB with various data sets. Our system could write 20,000 points to the database with batch size 20,000 in seconds of time and process queries within milliseconds.

Acknowledgements. This paper was supported by NSFC grant (62232005, 62202126, U1866602) and Sichuan Science and Technology Program (2020YFSY0069).

---

[1] https://github.com/cnosdb/cnosdb.
[2] https://youtu.be/47T6cogP0aY.

References

1. Prometheus. https://prometheus.io/
2. Tdengine. https://www.taosdata.com/cn/
3. Timescaledb. https://docs.timescale.com/
4. Freetsdb-v0.1.1 (2021). https://github.com/freetsdb/freetsdb
5. Chris, D.: Graphite. https://db-engines.com/en/system/Graphite
6. Dan, H., Stroulia, E.: A three-dimensional data model in hbase for large time-series dataset analysis. In: Maintenance & Evolution of Service-oriented & Cloud-based Systems (2012)
7. Dix, P.: Influxdb - an open source distributed time series database (2017)
8. Garima, R.S.: Review on time series databases and recent research trends in time series mining. IEEE (2014)
9. Jensen, S.K., Pedersen, T.B., Thomsen, C.: Time series management systems: a survey. IEEE Trans. Knowl. Data Eng. **29**, 2581–2600 (2017)
10. Jia, N., Wang, J., Li, N.: Application of data mining in intelligent power consumption. In: IET Conference Proceedings, pp. 538–541 (2012). https://digital-library.theiet.org/content/conferences/10.1049/cp.2012.1035
11. Lindstrom, P., Isenburg, M.: Fast and efficient compression of floating-point data. IEEE Trans. Vis. Comput. Graph. **12**, 1245–1250 (2006)
12. Ratanaworabhan, P., Jian, K., Burtscher, M.: Fast lossless compression of scientific floating-point data. In: Data Compression Conference (2006)
13. Xu, L.: Telecom big data based user offloading self-optimisation in heterogeneous relay cellular systems. Int. J. Distrib. Syst. Technol. **8**(2), 27–46 (2017)

EduChain: A Blockchain-Based Privacy-Preserving Lifelong Education Platform

Xinzhe Huang[1], Yujue Wang[2], Hai Liang[1(✉)], Yong Ding[1,3], Qianhong Wu[4], Ziyi Zhang[5], and Qiang Qu[6]

[1] Guangxi Key Laboratory of Cryptography and Information Security,
School of Computer Science and Information Security,
Guilin University of Electronic Technology, Guilin, China
lianghai@guet.edu.cn
[2] Hangzhou Innovation Institute, Beihang University, Hangzhou, China
[3] Cyberspace Security Research Center, Pengcheng Laboratory, Shenzhen, China
[4] School of Cyber Science and Technology, Beihang University, Beijing, China
[5] PAAS Blockchain Service Team, Huawei Cloud Tech Co., Ltd, Beijing, China
[6] Blockchain Lab, Huawei Cloud Tech Co., Ltd, Shenzhen, China

Abstract. With the popularization of lifelong education, academic and vocational education run through the life of students. However, the traditional lifelong education platform has three problems, namely incoherent achievement data, poor privacy protection, and vulnerability to single-point attacks, which seriously hinder the development of the lifelong education industry. This paper proposes a lifelong education platform based on blockchain supporting privacy protection (EduChain). The access control on achievement data is realized based on proxy re-encryption, and anyone can trace achievement data to the issuing organization through digital signature technology. EduChain can realize the credible transfer and privacy protection of lifelong education archives between students and checkers. The experiments demonstrate the practicality of EduChain in education scenarios.

Keywords: Privacy protection · Lifelong education · Proxy re-encryption

1 Introduction

In the information era [1], the social demand for talent constantly develops in a diversified direction based on knowledge and technology, leading to an increasing importance of academic and vocational education [2]. In a lifelong education system, the achievement data of academic and vocational education constitute the student's lifelong education archives in different fields and stages [3]. The academic performance and practical training experience would enrich the students' lifelong education archives.

© The Author(s), under exclusive license to Springer Nature Switzerland AG 2023
X. Wang et al. (Eds.): DASFAA 2023, LNCS 13946, pp. 701–706, 2023.
https://doi.org/10.1007/978-3-031-30678-5_59

With existing technologies, the student achievement data of academic and vocational education are centrally stored in hardcopy or electronic form. Scattered achievement data would cause problems of data loss, forgery, much consumption of workforce and material resources, and heavy archives management. Moreover, the traditional centralized lifelong education platform takes full control of a large amount of student's private data, including identity and achievement data, which could enable these platforms to make profit from these sensitive data.

In order to solve the problems of incoherent achievement data, weak privacy protection, and vulnerability to single-point attacks, this paper proposes a privacy-preserving lifelong education platform based on blockchain (EduChain). EduChain uses smart contracts to support multi-party collaboration to formulate business specifications for managing lifelong education archives. Organizations can provide students with achievement data and upload them to the blockchain if the course requirements are satisfied. Students are able to generate verification codes for the lifelong education archives and delegate the decryption right to checkers. Any unauthorized checker cannot decrypt and view lifelong education archives even if he/she holds the verification code. Thus, EduChain solves many problems in the traditional lifelong platform and builds credible lifelong education archives for students. EduChain has the following advantages.

i) The cross-organization and cross-stage achievement data can be fully recorded.
ii) The privacy of students' lifelong education archives can be protected by proxy re-encryption [4] and the malicious/dishonest issuing organization can be traced by the ECDSA digital signature [5].
iii) The blockchain-based distributed system can reduce data loss from single-point attacks and prevent tampering.

2   System Model and Features

Fig. 1. EduChain's system model

As shown in Fig. 1, an EduChain system comprises four types of roles, namely, system manager, student, organization, and checker. EduChain consists of four phases, that is, initialization, accumulation, authorization, and inspection. In the initialization phase, the system manager is responsible for initializing EduChain, including deploying four smart contracts. In the accumulation phase, both academic and training organizations are able to record various achievement data, including course grades and internship experience, to the student. Various learning experiences finally form the student's lifelong education archives. In the authorization phase, the students can authorize some checker to decrypt the lifelong education archives through proxy re-encryption. In the inspection phase, the authorized checker is able to decrypt and view the student's lifelong education archives with his/her private key.

The data on EduChain include sensitive and public data. Sensitive data mainly contains students' identity and achievement data, which should be properly processed to guarantee privacy. The public data stored in a relational database are freely accessible to all users and can be updated by the owner. These public data are necessary in advancing the core business but do not need to be maintained in ciphertext format.

However, most of the data in lifelong education archives, such as education and internship experience, are sensitive. The proxy re-encryption and digital signature schemes are employed to protect the privacy, unforgeability, and traceability of achievement data. The organization signs the achievement data using its private key, which can only be confirmed and linked to the issuing organization using the corresponding public key. Also, the digital signature is utilized to authenticate the achievement data. Any modification of the achievement data will result in a failed verification. Proxy re-encryption is typically used to delegate decryption rights, so that EduChain can function as a proxy server between students and checkers, transmitting and processing keys and ciphertexts. Through the proxy server, the encrypted lifelong education archives can be converted to the ciphertexts that can be decrypted by the checker with his/her private key. In this way, the sharing of lifelong education archives can be achieved without revealing both parties' private key and plaintext data.

3   The EduChain System

To evaluate the effectiveness and practicality of our approach, we implemented the EduChain on a platform where the configuration contains a ubuntu 20.04 system with 2.2 GHz and 8 G RAM. FISCO BCOS (https://github.com/FISCO-BCOS/FISCO-BCOS) with four nodes was used as the underlying blockchain and the PBFT consensus was adopted to tolerate Byzantine errors such as deliberate deception by 1/3 of nodes. There are four solidity 0.4.25 smart contracts deployed in EduChain, namely the user management contract $SC_{PersonSol}$, the achievement template management contract $SC_{TemplateSol}$, the organization achievement management contract $SC_{EduRecordSol}$, and the lifelong education archives management contract $SC_{ArchiveSol}$. These smart contracts are used to regulate the blockchain-based execution of EduChain's transactions. Moreover, the system uses the relational database of Mysql 8.0 to maintain public data such as courses and majors to meet public access needs.

Fig. 2. EduChain data overview

Fig. 3. The person login page

Figure 2 demonstrates that the data on EduChain, including node operation and transaction and block generation, can be monitored over time. This means that any abnormal operating condition of EduChain can be quickly identified and addressed. Figure 3 shows that EduChain generates an on-chain account and certificate file for each student based on the ECDSA algorithm, where each certificate file represents a separate user. To protect the security of the student's

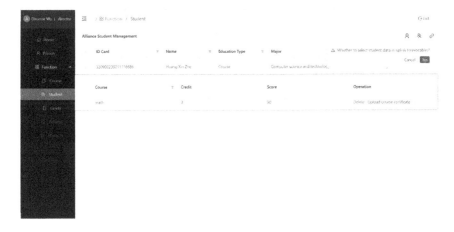

Fig. 4. The achievement data record page

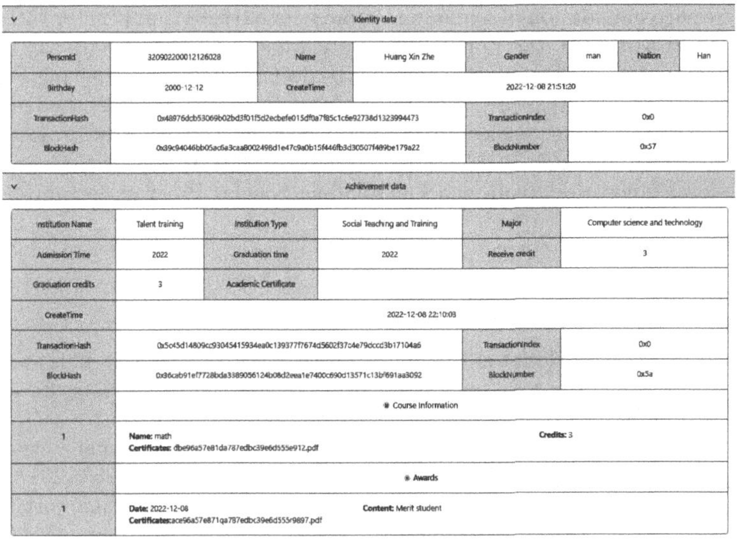

Fig. 5. The lifelong education archives

account, each student must not only enter the account password when logging in, but also submit the certificate file for supplementary verification. Figure 4 shows that EduChain invokes the $SC_{EduRecord}$ smart contract to store on the blockchain the addresses of organizations and students, course data, and digital signatures of organizations on achievement data. Concurrently, an immutable transaction is generated on the blockchain for all users to review. Figure 5 demonstrates that EduChain, serving as a proxy server, invokes the $SC_{ArchiveSol}$ contract to obtain the student's encrypted lifelong education archives and executes the re-encryption algorithm to convert them into ciphertexts that can be decrypted by the checker's private key. Then, the checker uses the private key to decrypt the ciphertext and review it, which includes identity and achievement data from various stages. The EduChain modules[1,2,3], source code[4] and video demonstration[5] are available online.

4 Conclusion

In this paper, we developed a blockchain-based privacy-preserving lifelong education platform, where the proxy re-encryption and digital signature were used to achieve privacy protection and traceability of achievement data. The experiments demonstrated the application of EduChain in the field of lifelong education, which can enable the credible transfer and privacy protection of lifelong education archives between students and checkers.

Acknowledgments. This article is supported in part by the National Key R&D Program of China under projects 2022YFB2702901 and 2020YFB1005600, the National Natural Science Foundation of China under projects 62162017, 62172119, U21A20467, 61932011, and 61972019, the Populus euphratica found CCF-HuaweiBC2021009, the Beijing Natural Science Foundation through project M21031, the Zhejiang Provincial Natural Science Foundation of China under Grant No. LZ23F020012, the Guangdong Key R&D Program under project 2020B0101090002, the Major Key Project of PCL under grants PCL2021A09, PCL2021A02, and PCL2022A03, and the special fund of the High-level Innovation Team and Outstanding Scholar Program for universities of Guangxi.

References

1. Webb, S., Hodge, S., Holford, J., Milana, M., Waller, R.: Aligning skills and lifelong learning for human-centred sustainable development. Int. J. Lifelong Educ. **41**(2), 127–132 (2022)
2. Hodge, S., Holford, J., Milana, M., Waller, R., Webb, S.: Vocational education and the field of lifelong education. Int. J. Lifelong Educ. **36**(3), 251–253 (2017)
3. Huang, X., et al.: A privacy-preserving credit bank supervision framework based on redactable blockchain. In: Svetinovic, D., Zhang, Y., Luo, X., Huang, X., Chen, X. (eds.) BlockSys 2022. Communications in Computer and Information Science, vol. 1679, pp. 18–30. Springer, Cham (2022). https://doi.org/10.1007/978-981-19-8043-5_2
4. Ateniese, G., Kevin, F., Green, M., Hohenberger, S.: Improved proxy re-encryption schemes with applications to secure distributed storage. ACM Trans. Inf. Syst. Secur. **9**(1), 1–30 (2006)
5. Johnson, D., Menezes, A., Vanstone, S.: The elliptic curve digital signature algorithm (ECDSA). Int. J. Inf. Secur. **1**(1), 36–63 (2001)

---

[1] https://check-ec.amsdba.com.
[2] https://person-ec.amsdba.com.
[3] https://organization-ec.amsdba.com.
[4] https://github.com/forguide/EduChain.
[5] https://youtu.be/jCkUirf3GBw.

ChainSync: A Real-Time Multi-chain ETL System for dApp Development

Shuyi Zhang[1]([✉])(iD), Xinyao Sun[1](iD), and Kyle Rehak[2](iD)

[1] White Matrix Inc., Nanjing, China
{tim,asun}@whitematrix.io
[2] Matrix Labs Inc., Vancouver, Canada
kyle@matrixlabs.org

Abstract. The increasing number of public blockchains has led to a need for infrastructures and tools for decentralized application (dApp) development. The blockchain Extract-Transform-Load (ETL) process, which involves extracting on-chain data, converting it into the desired format, and loading it into data stores, is essential for most dApp development. However, many blockchains lack blockchain ETL tools or frameworks, and building a blockchain ETL for each blockchain is complicated. To overcome this problem, we propose a multi-chain ETL system featuring (1) an ETL pipeline that can be easily extended to support multiple blockchains, (2) data stores optimized for dApp development, and (3) an event-based pub-sub system for dApps to consume real-time on-chain data. The proposed framework has been integrated with multiple blockchains and supports multiple dApps in production environments.

Keywords: Blockchain · ETL · Decentralized applications

1 Introduction

In recent years, more and more blockchains have been launched, leading to the massive growth of dApps, which are applications built with blockchains and smart contracts instead of centralized servers. When developing complex dApps such as blockchain games or non-fungible token (NFT) markets, it is difficult to store all of the information in the smart contract on the blockchain, so developers always need tools to extract core data from the blockchain and transform or enrich them into a format suitable for presentation on the front end [3]. Therefore, such ETL tools are needed to facilitate the development of dApps. For a prominent blockchain like Ethereum, the community is large enough to contribute tools such as EtherNet [2] and The Graph[1] for automating the ETL process for data from Ethereum. However, for smaller or newer blockchains, developers have to spend a considerable amount of time developing their own ETL pipeline, and this problem compounds when developing multi-chain dApps.

In this paper, we propose an extensible multi-chain ETL system named ChainSync. The contributions are summarized as follows:

---

[1] https://thegraph.com/

X. Wang et al. (Eds.): DASFAA 2023, LNCS 13946, pp. 707–711, 2023.
https://doi.org/10.1007/978-3-031-30678-5_60

– We propose a real-time ETL system for multiple blockchains, including Ethereum [1], BNB Chain[2], Polygon[3] and Flow[4], which can be easily extended to support more blockchains with minimal effort. It fills the gap for multi-chain ETL tools. Also, to the best of our knowledge, it is the first ETL tool to support all blockchain data on the Flow blockchain.
– We propose a multi-chain data store module that is optimized for fast queries on three types of blockchain data: transactions, events, and NFT information. The data stores and the query API empower many top-ranked dApps.
– Our system offers event pub/sub and handling to meet the needs of real-time dApps such as blockchain games and NFT marketplaces.

2   System Architecture

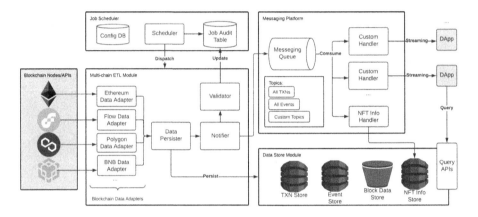

Fig. 1. The architecture of the ChainSync ETL system

The architecture of the system is shown in Fig. 1, which is divided into four parts: the job scheduler, the multi-chain ETL module, messaging platform, and the data stores.

Job Scheduler. The job scheduler continuously checks the latest block height of each integrated blockchain and the last successfully processed block. If the last successfully processed block is behind the latest block height, it dispatches jobs to the multi-chain ETL module to synchronize data from the designated blockchains.

---

[2] https://www.bnbchain.org/.
[3] https://polygon.technology/solutions/polygon-pos.
[4] https://flow.com/.

Multi-chain ETL Module. In the multi-chain ETL module, the blockchain data adapters fetch the blockchain data of a given block range from blockchain node APIs, then extract the transactions (TXNs), events, and block metadata and transform them into a unified format. The persister subsequently saves the TXN data and event data into data stores for further querying. Next, the notifier sends messages to notify downstream of processed events and TXNs. Finally, the validator will validate the checksum of processed data against the information in the block's metadata to ensure that all data in the block has been processed. The result of validation is updated in the ETL job audit table so the scheduler can pick up failed jobs to re-run.

The entire ETL process is idempotent and parallelizable, so it can be easily scaled up to catch up and synchronize to the latest block. Also, integrating with a new blockchain requires only minimal development of the blockchain data adapter instead of building a full ETL pipeline. We provide built-in implementations of data adapters for various blockchains, including Ethereum, Flow, BNB Chain, and Polygon.

Data Store Module. We propose a scalable multi-chain TXN store and event store design. We choose AWS DynamoDB [4] as the underlying database for our TXN store and event store for fast queries and great scalability to hold the growing data size from multiple blockchains. TXN store and event store are keyed by $\{blockchainId\}{:}\{txHash\}$ and $\{blockchainId\}{:}\{eventId\}$ respectively for low-latency key-value search. We also added indexes for *blockNumber* and *contractId* for query purposes. In addition, we also use AWS S3 to store the complete block data, whose paths are prefixed by $\{blockchainId\}{:}\{blockNumber\}$, for export and analytic purposes. All the stores come with HTTP query APIs for dApps to retrieve processed on-chain data more easily.

Messaging Platform. The messaging platform is a one-to-many pub/sub platform to dispatch the processed data to downstream handlers. We build it on top of Apache Kafka V3[5] for its scalability and throughput. Developers can write custom handlers that subscribe to specific topics for building real-time dApps. We also implement a built-in non-fungible token (NFT) data aggregation handler for each blockchain. This handler captures NFT transfer events and updates each token's metadata and ownership in the NFT token store, which is included in the data store module.

3   Demonstration Scenarios[6]

Multi-chain Data Integration. For demonstration, we create a web dashboard[7] ChainSync to provide better visibility. Figure 2(a) is the data adapter job dashboard that lists recent ETL jobs and their status, i.e., processing,

---

[5] https://kafka.apache.org/.
[6] Demonstration video: https://chainsyncdemo.info/videofor
[7] https://chainsyncdemo.info/.

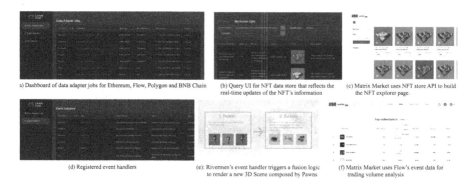

a) Dashboard of data adapter jobs for Ethereum, Flow, Polygon and BNB Chain   (b) Query UI for NFT data store that reflects the real-time updates of the NFT's information   (c) Matrix Market uses NFT store API to build the NFT explorer page.

(d) Registered event handlers   (e): Rivermen's event handler triggers a fusion logic to render a new 3D Scene composed by Pawns   (f) Matrix Market uses Flow's event data for trading volume analysis

Fig. 2. Interfaces of ChainSync dashboard and dApp showcases

success, and failure. The current version of ChainSync integrates with four popular blockchains, including Ethereum, Flow, BNB Chain, and Polygon.

Query the Latest Information from Data Stores. Figure 2(b) is the browser for the data stores of each supported blockchain. Users can query token information, including the token's ownership, metadata, and images, by contract address or user address via the Web UI. Aside from the UI, the query API is useful for dApp development. For instance, Matrix Market[8] and Phanta Bear's[9] NFT browser use the API to track and display the latest information for NFTs (Fig. 2(c)), eliminating their need to build a data back end.

Real-Time Event Subscription and Handling. Figure 2(d) shows the downstream applications and their event handlers registered to the topics. Matrix Market is an NFT marketplace that needs to scan all NFT transfer transactions, so it subscribes to the *All TXNs* and *All Events* topic of Flow blockchain to reflect real-time order updates. Rivermen[10], a blockchain puzzle game that allows users to fuse their 3D characters, its event handler listens to events emitted by its fusion smart contract deployed on Ethereum to trigger the rendering of the fused 3D models (Fig. 2(e)).

Data Export for Analytics. The data stored in ChainSync can be easily exported for analytical purposes. For example, Matrix Market loads Flow's event data from S3 to ClickHouse[11] to run analytic jobs to rank NFT collections by their trading volume over various time ranges (Fig. 2(f)).

Acknowledgment. This work is supported by the National Key Research and Development Program of China (No. 2021YFB2700900).

---

[8] https://matrixmarket.xyz/.
[9] https://ezek.io/products/.
[10] https://rivermen.io/.
[11] https://clickhouse.com/.

References

1. Buterin, V., et al.: A next-generation smart contract and decentralized application platform. White Paper **3**(37), 1–2 (2014)
2. Hou Su, V., Sen Gupta, S., Khan, A.: Automating ETL and mining of ethereum blockchain network. In: Proceedings of the 15th ACM International Conference on Web Search and Data Mining, pp. 1581–1584 (2022)
3. Khan, A.G., Zahid, A.H., Hussain, M., Farooq, M., Riaz, U., Alam, T.M.: A journey of web and blockchain towards the industry 4.0: an overview. In: 2019 International Conference on Innovative Computing (ICIC), pp. 1–7. IEEE (2019)
4. Sivasubramanian, S.: Amazon dynamodb: a seamlessly scalable non-relational database service. In: Proceedings of the 2012 ACM SIGMOD International Conference on Management of Data, pp. 729–730 (2012)

IoT-Assisted Blockchain-Based Car Rental System Supporting Traceability

Lipan Chen[1], Yujue Wang[2], Yong Ding[1,3]($\boxtimes$), Hai Liang[1], Changsong Yang[1], and Huiyong Wang[4]

[1] Guangxi Key Laboratory of Cryptography and Information Security,
School of Computer Science and Information Security,
Guilin University of Electronic Technology, Guilin, China
`stone_dingy@126.com`
[2] Hangzhou Innovation Institute, Beihang University, Hangzhou, China
[3] Cyberspace Security Research Center, Pengcheng Laboratory, Shenzhen, China
[4] School of Mathematics and Computing Science, Guilin University of Electronic
Technology, Guilin, China

Abstract. The majority of existing car rental systems rely on a centralized database to manage the data. Although these centralized architectures could make the system design and operation costs are reduced, there are many security problems in real world applications. If the information of the car obtained by the tenant is not completely credible, it will lead to security risks in the rental process. This paper proposes a blockchain-based car rental system with the Internet of Things (IoT), where the blockchain is engaged to store the use history of the car, so that the car condition can be traced to the source and the risk of using the car for the tenant can be reduced. The IoT technology is employed to enable renters to flexibly rent and unlock platform cars. Our system also provides visual operation interface. The evaluation results demonstrate the practicality of our system in the online car rental market with higher security and more flexible rental methods.

Keywords: Blockchain · Database · Internet of things · Car rental

1 Introduction

The online car rental system should provide convenience and security for renters. Nowadays, there are a lot of online car rental systems, which have greatly improved the efficiency of car rental compared with the traditional car rental mode, but all these online systems require the centralized database as the data storage platform [1]. This single data storage mechanism has obvious security problems, for example, the system administrators may arbitrarily operate the database, making the car information not trusted, and the tenant's car security cannot be guaranteed.

The code of this paper can be found in https://github.com/clp0321/srp-server.

© The Author(s), under exclusive license to Springer Nature Switzerland AG 2023
X. Wang et al. (Eds.): DASFAA 2023, LNCS 13946, pp. 712–718, 2023.
https://doi.org/10.1007/978-3-031-30678-5_61

Verma et al.'s [2] designed a blockchain-based secure car hiring system. The block chain was used to store the rental information of tenants, which improved the security of the system while considering convenience, but did not take into account the security of the information of the car itself. Osman et al. [3] developed an anonymous car rental system based on near field communication (NFC), which allows the renter to unlock the car conveniently through the NFC technology and protects the renter's privacy well to prevent information leakage. However, their system does not take into account the convenience of renting a car and the security of using the car. In Thakur's car rental system [4], renters are able to book cars online and make reasonable regional divisions of offline cars to realize convenient rental. However, same with these systems above, their system does not consider the security of using the car. If the platform rents out the car with potential dangers to the tenant, the tenant would be at a very disadvantages position once some problem occurs in the use of the car.

This paper proposes a blockchain-based car rental system with IoT technology. The blockchain is responsible for maintaining the use history of cars, making the car condition traceable to the source and reducing the risk of renters using cars. The IoT technology allows the tenants to flexibly rent and unlock platform cars. Moreover, tenant-friendly visual operation interfaces are provided to facilitate the usage for managers and tenants.

2   System Architecture

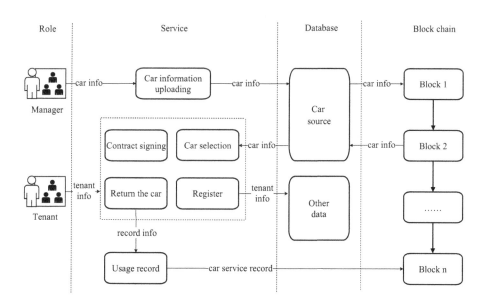

Fig. 1. System architecture

As shown in Fig. 1, the blockchain-based car rental system consists of two types of roles, namely, managers and tenants, and contains three components, that is, the car rental platform, database layer and blockchain layer.

Car Rental Platform: The car rental platform is composed of front-end service and back-end service, where the former displays the visual interface, and the latter provides the interface and data for the front-end page.

Database Layer: The database layer provides data for back-end functions and a portion of the space is used to store car source data downloaded from the blockchain via smart contracts, the other portion is employed to maintain other data such as user information. The car source data will be updated regularly according to the car source data on the blockchain.

Blockchain Layer: Blockchain layer provides security for car source data. With blockchain traceability, past use of the car source can be queried based on transaction hashes resulting from changes in car information.

3   System Implementation

We used HTML, CSS, and JavaScript to realize the front-end service, and Java language and spring boot framework to realize the back-end service of the car rental platform, respectively. In order to enable the data of the database to interact with the blockchain, the smart contract was designed for data upload and download. The main car use process of the system has the following three aspects:

Car Source Upload: The manager enters the new car source data into the database, and uploads it to the blockchain through the car source upload smart contract.

Car Rent:

- **Registration**: Tenants should first register on the platform through their mobile phone numbers. Any unregistered users can only view part information of the car source.
- **Car selection**: Tenants log in to the car rental platform to select cars and rental modes according to their preferences. The car source data displayed on the platform will be read from the database, and the database will update the car source data periodically from the blockchain, ensuring the efficiency of the system execution. Each record of the source has a corresponding transaction hash value, through which the authenticity of the data can be verified on the blockchain, thus ensuring the security of the system.
- **Contract signing**: Figure 2 shows the contract signing design. The tenant signs a lease contract on the platform. After the tenant initiates the contract signing request, the car source information in the database will be updated from the blockchain. After the tenant pays the deposit, the platform will automatically sign the contract and the tenant will obtain permission to unlock

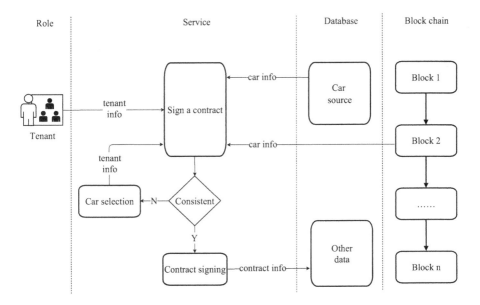

Fig. 2. Contract signing design

the car. If the information in the database is inconsistent with the data on the chain, the tenant's request will be rejected and the new car source data will be returned so that the tenant can sign the contract again. This mode of updating data from the blockchain reduces the number of blockchain data read requests and improves the performance of the blockchain system.

- **Car usage**: Cars operated by the platform are equipped with IoT-connected intelligent car locks, and unlocking and locking permissions for cars can be obtained online through the car rental platform. After that, tenants can use their mobile phones to open the locks. In this way, the convenience of renting cars can be realized.
- **Rental record upload**: After the tenant completes the rental order, the system will upload the usage record of the car to the blockchain.

After-Sales Service: If the tenant encounters various problems and disputes during the use of the car, the system supports contacting the platform for communication and mediation.

4   Demonstration

As shown in Fig. 3, we developed the functions of manager authority, including data statistics, car source data upload, car lock authorization, order overview, contract signing, dispute negotiation, appointment acceptance, and tenant authority management.

Data Presentation: The "Work" table page shows the overall operating status of the platform. The order status, reservation status and income and expenditure

Fig. 3. Management function

status of the last month and several months can be clearly displayed in the interface. In the upper right corner of the page, the design of the advertising column and a simple chart table. The detailed data list of "Order", "Contract", "Negotiation" and "Appointment" can be seen on the corresponding page in the left navigation bar.

Function: On the "Cars" page, only local administrators can upload local car source data, and the uploaded data will be stored in the database and blockchain. The manager can view the obtained auto lock unlocking authorization information on the "Equipment" page.

Authority Management: The authority of this system can be divided into two categories, namely managers and tenants, and there are two types of managers, namely super manager and local manager. The super manager is able to manage the authority of the entire system and supervise the operation of the system, which has all permissions but cannot upload car source data. The local manager can only upload the local car source data and deal with the rental problems of tenants.

Tenant-based functions include car selection, contract signing, on-chain record inquiry and after-sales negotiation. The on-chain record query page is shown in Fig. 4. By entering the hash value of some transaction in the interface, the tenant can view the corresponding on-chain car source information and lease history information.

Fig. 4. On-chain record query

5   Conclusion

In this paper, we designed a blockchain-based car rental system with IoT technology, that achieves convenient rental and offers security guarantees for car use. The IoT provides convenient for renters to hire cars, and the decentralized and distributed characteristics of blockchain provide security for renters to use cars. The storage structure of database + blockchain not only guarantees security but also takes into account the efficiency of the system.

Acknowledgments. This article is supported in part by the National Key R&D Program of China under project 2020YFB1006004, the Guangxi Natural Science Foundation under grant 2019GXNSFGA245004, the National Natural Science Foundation of China under projects 621621017, 61962012, and 62172119, the Zhejiang Provincial Natural Science Foundation of China under Grant No. LZ23F020012, the Guangdong Key R&D Program under project 2020B0101090002, the Major Key Project of PCL under grants PCL2021A09, PCL2021A02, and PCL2022A03, the Shenzhen Science and Technology R&D Fund under project JSGG20201102170000002, and the special fund of the High-level Innovation Team and Outstanding Scholar Program for universities of Guangxi.

References

1. Agrawal, A., Mathur, R.: Online vehicle rental system. Int. J. Sci. Res. Eng. Trends **6**(3), 1228–1230 (2020)
2. Sonakshi, V.S.: A blockchain-based secure car hiring system. In: Khanna, K., Estrela, V.V., Rodrigues, J.J.P.C. (eds.) Cyber Security and Digital Forensics . LNDECT, vol. 73, pp. 341–349. Springer, Singapore (2022). https://doi.org/10.1007/978-981-16-3961-6_29

3. Osman, M.N., Zain, N.Md., Paidi, Z., Sedek, K.A., Yusoff, M.N., Maghribi, M.: Online car rental system using web-based and SMS technology. Comput. Res. Innov. (CRINN). **2**, 277 (2017)
4. Thakur, A.: Car rental system. Int. J. Res. Appl. Sci. Eng. Technol. **9**(7), 402–412 (2021)

DiCausal: Exploiting Domain Knowledge for Interactive Causal Discovery

Wenbo Xu, Yueguo Chen$^{(\boxtimes)}$, Shengwei Huang, Xiongpai Qin, and Li Chong

DEKE Lab, Renmin University of China, Beijing, China
{xuwenbo_7777,chenyueguo,2021104388,qxp1990,chongli}@ruc.edu.cn

Abstract. We propose an interactive causal discovery system called *DiCausal*, which allows users to apply their simple domain knowledge of how variables are generated and interactively edit the graph during the causal discovery process without incurring too much burden. A novel form of domain knowledge representation and an adapted feature engineering method are introduced in *DiCausal*. Two existing causal discovery algorithms are adapted for verification. Experiment proves that such a way of incorporating domain knowledge into the discovery algorithms can achieve better results than pure data-driven methods.

Keywords: Causal Discovery · Domain Knowledge · Interaction

1 Introduction

Discovering causal relationships from a large number of variables is challenging due to the large amount of possible candidates among variables. Traditional trial-based solutions such as RCT (Randomised Controlled Trial) are not suitable because of the cost and ethical issues [2]. Data-driven causal discovery algorithms [2] are designed to model causal relationships as DAGs (Directed Acyclic Graphs), with a node representing a variable, and an edge representing a cause pointing to an effect. They aim to identify those edges with causal relationships. However, the performance of pure data-driven approaches is usually not good enough when the number of variables is large, which is very common in many scenarios. The lack of domain knowledge limits their performance on finding correct causal edges. Interactive causal discovery systems [5] are therefore introduced. However, they are typically designed for model tuning or knowledge discovery from the output graph. As far as we know, there is no interactive system that is able to optimize the causal discovery algorithms by intervening their execution process using domain knowledge.

We develop a domain-knowledge driven causal discovery system called *DiCausal*, which allows users to apply and revise their domain knowledge of how features are generated by interactively operating the DAGs during the causal discovery process. We will show that such a way of incorporating domain knowledge into the discovery algorithm can improve the performance of causal discovery. The main contributions are as follows:

© The Author(s), under exclusive license to Springer Nature Switzerland AG 2023
X. Wang et al. (Eds.): DASFAA 2023, LNCS 13946, pp. 719–723, 2023.
https://doi.org/10.1007/978-3-031-30678-5_62

1. We allow users to decompose DAGs into several parts (*sub-graphs*) according to the domain knowledge of how variables are physically generated.
2. *Sub-graphs* are cascade-coupled, which means causal effects can be propagated from upstream to downstream. A compatible feature engineering approach is proposed to facilitate the revision of domain knowledge.
3. Domain knowledge is integrated into two existing algorithms as heuristics and constraints. Such modification successfully introduces human guidance, reduces the search space, and therefore improves the performance.
4. The effect of the causal discovery system, *DiCausal*, is experimentally verified. Codes of *DiCausal* are available at: https://github.com/wswlfh/Dicausal.

2   System Overview and Key Techniques

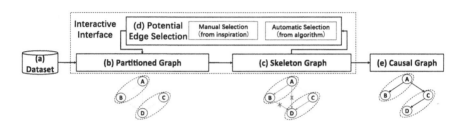

Fig. 1. The overall architecture of the *DiCausal* system.

System Architecture. Figure 1 illustrates the architecture of our system. Users can apply domain knowledge to divide variables into groups (*sub-graphs*) so that causal relationships can be categorized into inner-group ones and inter-group ones, to facilitate editing afterwards (b). The interactive interface shows the editable graph with information (e.g., name, type, weights) of edges and nodes, and enables users to edit the graph by setting an edge as a positive or negative candidate. For example, an edge $A \rightarrow D$ is labeled as a negative candidate while $A \rightarrow B$ is labeled as a positive one. We expect to derive a skeleton graph (c) as the input of the final causal discovery algorithm (e). Besides editing the graph from domain knowledge, the interactive interface will inspire users to adjust the edges. A cascading feature engineering approach is also proposed to predict potential edges automatically (d). So that, the loop (b,c,d) enables users to iteratively revise the skeleton graph using their domain knowledge. A causal discovery algorithm utilizes such skeleton graph as heuristics and constraints, and derives a causal graph accordingly (e). As we can see, *DiCausal* can leverage domain knowledge for causal discovery in an interactive manner.

Domain Knowledge. Traditional causal discovery algorithms start from a complete graph. However, most pairwise variables have no causal relationships, which can be easily captured using simple domain knowledge—all variables can be topologically grouped into several groups (*sub-graphs*) so that candidate edges across different groups can be significantly reduced as causal effects can only be propagated from upstream to downstream and not vice versa. Moreover, we allow users to set some variables as final effects and key effects. A key effect can be influenced by variables of the same group and the key effects of its upstream groups. It may further influence the key effects of its downstream groups. Final effects will not influence any other variables. Such settings require less domain knowledge compared with specifying certain edges [3,6] but help a lot to remove lots of unnecessary edges, and further to compute edge weights elegantly.

Cascading Feature Engineering. To effectively guide users on editing the skeleton graph, we need to compute the weights of edges. This is addressed by leveraging the cascading structure achieved by the key effects and final effects among groups. As mentioned above, only possible variables of an effect variable will be applied to the regression model. This allows us to conduct feature engineering (find candidate cause variables of an effect variable by regression contribution) in a cascading way, i.e., combining upstream key effects and variables of the same group as the predictors of an effect in each regression model.

Causal Discovery Incorporating Domain Knowledge. The framework can be applied to many existing causal discovery algorithms. We have adopted and improved two of them. PC Algorithm [2] generates DAGs by CI (conditional independence) test among nodes. However, different graph structures may have the same CI, which are called *equivalent* graphs. With domain knowledge, the performance of PC can be improved by removing false structures from the *equivalent* graphs. Zhu et al. [7] proposed a reinforcement learning (RL) model based on scoring the causal graphs and constraining the acyclicity. We modify their reward function (Eqn. 1) by adding $D(g, s)$ to measure how the estimated causal graph g fits domain knowledge represented by the skeleton graph s:

$$reward := [S(g) + \lambda_1 I(g \notin DAGs) + \lambda_2 h(g)] + \lambda_3 D(g, s) \qquad (1)$$

where $S(g)$ is score function to measure how causal graph g matches with given data; $I(\cdot)$ and $h(\cdot)$ constrains the acyclicity; λ_i are penalty parameters.

3   Demonstration

This section will demonstrate how *DiCausal* assists users with causal discovery through various interactive modules on the system interface. Figure 2 illustrates the interface of *DiCausal*. Module① supports the upload and overview of a tabular dataset. Module② enables users to construct *sub-graphs* with their domain knowledge. Users can create new *sub-graphs* by clicking "Add" button, and then partition variables and specify effects through checkbox. Thresholds of candidate edges can be set by scrollbar for cascading feature engineering afterwards. Settings can be saved for further usage. Module③ visualizes the editable skeleton graph and highlight the effect variables. Users can add or delete the edges by clicking and re-organize the nodes by dragging-and-dropping them into or out of *sub-graphs*. Information of a particular node or edge is shown in Module④ and Module⑤ respectively when hovering the mouse on it so that users can get hints to select potential causal edges. Module② ∼ ⑤ achieve the function of iterative adjustment i.e. loop (b, c, d) in Fig. 1. Module⑥ shows the causal graph derived from the algorithm assisted by domain knowledge. Similar to Module③④⑤, information on nodes and edges is also available.

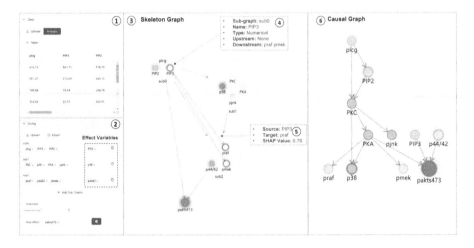

Fig. 2. The user interface of *DiCausal* system

4   Experiments and Conclusion

In our demo, we will use three datasets. Each dataset contains an observational tabular data and a DAG as the ground truth of causal structure. I.Sachs [2] dataset, a common benchmark for causal discovery. It uses the expression levels of biochemical indicators to model the causal structure among proteins and phospholipids. II.TE Process [4], a manufacturing dataset. It models the causal structure of consecutive measurements and manipulated variables in chemical engineering. III.A simulated dataset. A causal structure DAG is randomly generated and each edge between two nodes represents a generation function.

Considering two algorithms, PC [2] and RL [7], we separately compare their performance with and without domain knowledge applied (the suffix "-K" means methods with domain knowledge). We assume that positive candidates can be found during the interaction. *Sub-graphs* division in Sachs and simulated dataset are assumptions based on the ground truth DAG while *Sub-graphs* division in TE Process is according to the real plant structure. For Sachs dataset, 1 edge from true graph is randomly chosen as a positive candidate and 50 edges indirectly generated by dividing *Sub-graphs* are as negative ones. The counts of positive and negative candidates are 1 and 6 in simulated data; 5 and 378 in TE Process as it is large.

Table 1. Experimental results on different models and datasets

	Sachs Dataset				TE Process				Simulated Data			
	PC	PC-K	RL	RL-K	PC	PC-K	RL	RL-K	PC	PC-K	RL	RL-K
SHD	25	**21**	25	**20**	**67**	72	93	**91**	19	**16**	19	**16**
Precision	0.33	**0.45**	0.25	**0.32**	0.12	**0.15**	0.04	**0.05**	0.24	**0.33**	0.17	**0.27**
Recall	0.45	**0.65**	**0.35**	0.31	0.25	**0.39**	0.11	**0.14**	0.29	**0.43**	0.27	**0.35**
F1	0.38	**0.53**	0.29	**0.31**	0.16	**0.22**	0.06	**0.08**	0.26	**0.38**	0.22	**0.31**

We adopt 4 metrics to evaluate the output of causal discovery given a ground truth: structure hamming distance (SHD), consisting in computing the difference between the output graph and the ground truth by counting missing edges, redundant edges and edges with wrong direction; precision and recall, measuring the quality and quantity of derived edges; F1-Score, the harmonic mean of the precision and recall. Table 1 shows the performance. The results show that simple domain knowledge does help to improve the performance of causal discovery algorithms, although the performance improvement on the TE dataset is slightly low due to its large number of variables.

The experimental results prove the effectiveness of our method. It has huge potential for progress when more knowledge is input using *DiCausal*, which will be demonstrated by participants in the demo presentation. In future work, we will consider different forms of domain knowledge representation. In addition, how to improve the robustness of the model when there exists incorrect knowledge is also a problem that needs to be addressed.

Acknowledgements. This work is supported by National Key Research and Development Program (No. 2020YFB1710004) and the National Science Foundation of China under the grant 62272466.

References

1. Ge, X., Raghu, V.K., Chrysanthis, P.K., Benos, P.V.: CausalMGM: an interactive web-based causal discovery tool. Nucleic Acids Res. **48**, W597–W602 (2020)
2. Glymour, C., Zhang, K., Spirtes, P.: Review of causal discovery methods based on graphical models. Front. Genet. **10**, 524 (2019)
3. Kalainathan, D., Goudet, O., Dutta, R.: Uncovering causal relationships in python. JMLR **21**(37) (2020)
4. Menegozzo, G., Dall'Alba, D., Fiorini, P.: CIPCaD-Bench: continuous industrial process datasets for benchmarking causal discovery methods. In: CASE (2022)
5. Wang, J., Mueller, K.: Visual causality analysis made practical. In: VAST (2017)
6. Zhang, K., et al.: gCastle: a python toolbox for causal discovery. arXiv:2111.15155 (2021)
7. Zhu, S., Ng, I.: Causal discovery with reinforcement learning. In: ICLR (2019)

Flowris: Managing Data Analysis Workflows for Conversational Agent

Jiajia Sun, Juan Wang, Yueguo Chen[✉], and Xiongpai Qin

DEKE Lab, Renmin University of China, Beijing, China
{jiajiasun,2018202153,chenyueguo,qxp1990}@ruc.edu.cn

Abstract. Conversational agent has become a new way of conducting data analysis tasks, enabling people with different levels of analytic experience to interact with the system by providing natural language (NL) commands, choices, and parameters etc. However, flexibility becomes a challenge considering that users may want to integrate model services on the Web in such systems. To address it, we present Flowris, a prototype system that collects and manages provenance data for conversational agent. We will show how Flowris collects and manages provenance data, and how it supports the comparison and reproducibility of experiments.

Keywords: conversational agent · provenance · reproducibility

1 Introduction

With the prevalence of analytics-as-a-microservice, it is possible to integrate data analysis models on the Web, and conduct professional data analysis on demand by domain experts who may not have good programming skills. This brings one big challenge—how to provide usability and flexibility to domain users? Conversational agent for data analysis [3] is proposed to address the usability challenge. However, such systems provide less flexibility of data analysis considering that the applied workflows are typically predefined. In such applications, flexibility of the data analysis workflows is quite important because users may want to compare the performance of various models/algorithms/parameters. We therefore need a conversational data analysis system that is able to run multiple instances of a workflow and record the states/results of each instance so that comparison and reproducibility of experiments can be easy achieved by domain users.

Provenance can help solve this problem. It records the lineage of data, including its origin, the transformations that led to its current state, and the reasons for applying those transformations. The lineage of all data objects forms a provenance graph, i.e., a nested Directed Acyclic Graph (DAG). Each node captures either a computational unit (model/algorithm) or a data object. The edge between a data node and a computational node represents the input/output relationship. The provenance graph helps to interpret and reason about the results of complex workflows [4], and enables users to review the input and intermediate data. Comparison and reproducibility of experiments therefore can be achieved.

X. Wang et al. (Eds.): DASFAA 2023, LNCS 13946, pp. 724–728, 2023.
https://doi.org/10.1007/978-3-031-30678-5_63

In this paper, we introduce a conversational data analysis system called Flowris which supports provenance data management of complex workflows assembled from various analytical micro-services (algorithms). We apply a tool called Sematic to collect the provenance data and visualize provenance graphs relevant to the running data analysis pipeline. It is a lightweight development toolkit that automatically tracks the inputs and outputs of Sematic decorators embedded to the calling process of micro-services. The Sematic toolkit automatically obtains the provenance data and resolves the provenance graph. With provenance management, we show that Flowris supports the recommendation and trigger of workflows and computational units using NL commands, and therefore achieves good usability and flexibility for professional data analysis.

2   System Overview

The architecture of Flowris is illustrated in Fig. 1, consisting of three components: conversational agent, provenance collection, and provenance management.

Conversational agent for data analysis is built to facilitate domain experts to conduct data analysis. It allows users to freely express their analysis intents by using NL interaction. The blue part in Fig. 1 presents a high-level overview of the agent architecture. Each computational node/unit is associated with some NL commands that are able to trigger it. The entry point of the conversational agent is a natural language understanding module [2] that maps user's input onto a workflow or a particular computational unit. We design a state machine to keep track of which computational unit the user wants to execute, what arguments the user has provided so far in the conversation, and what arguments it still needs to request. Once enough information is ready, a computational unit will be executed, accompanied with the process of provenance data collection.

Provenance collection module is implemented using the Sematic toolkit. Sematic keeps track of those sematic functions (i.e., computational nodes in the provenance graphs) that are decorated with @sematic.func. Inputs and outputs of a sematic function are therefore serialized and tracked in the database. The NL commands that trigger workflows and computational nodes are tracked as well. With these information, Flowris visualizes a provenance graph according to the recorded workflow. As shown in Fig. 3-②, the provenance graph helps users to identify which part of the workflow contributes to the derived results. In this figure, white nodes represent computational nodes (with function calls), grey nodes represent data nodes (inputs or outputs of the computational nodes). Detailed information (e.g., function definitions, parameters, NL commands, dataset descriptions) of a node is accessible to users. As provenance data, these information will be used for the reuse and comparison of workflows.

3   Key Techniques

Provenance Management. In many data analysis scenarios, users often need to compare and try different algorithms and parameters to obtain the final

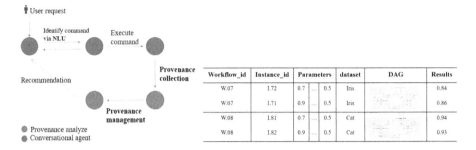

Fig. 1. System architecture   **Fig. 2.** Comparable table

desired experimental results. Meanwhile, the same workflow can be reused by different users, with different settings (e.g., datasets). How to utilize a large amount of provenance data about different execution instances of the same workflow becomes a critical problem. In this paper, we define the provenance data as a series of nested DAGs. Note that instances of a workflow may vary in many ways such as datasets, algorithms, parameters, or even the repetition of some nodes. A provenance graph of a workflow is actually the summary of all its instances of execution. As shown in Fig. 3-②, it can be identified by the sequential and parallel structures of the data/computational nodes, with difference in parameters are generally ignored. A new instance of execution may trigger the updates of its corresponding workflow, e.g., when it contains a new algorithmic node. An instance of a workflow execution is then modeled as a path instance from the source node to the target node, with parameters of each node recorded. With the models of workflow and its execution instances, comparisons among instances and workflows can be achieved using distance measures of sequences and graphs [1,5], where parameters of nodes need to be considered for matching. The provenance manager handles the evolution of experiments by comparing the execution instances. The provenance manager first traverses all the provenance graphs and calculates the distance between the graphs. When the user completes an execution of the workflow instance and enters the "compare" command, Flowris returns a comparable table. As shown in Fig. 2, the comparable table consists of two parts: other execution instances of the same workflow; workflow and its instances whose graph distance is closest to the current workflow graph. With this comparable table, users can easily compare the results of different instances to obtain insights of datasets/algorithms/parameters.

In addition to providing comparable instances, the provenance manager also provides users with model and parameter recommendations by analyzing historical provenance data. As users interact with the conversational agent, Flowris tracks their actions and maintains a contextual state that drives the system's recommendation. The contextual state is modeled as a combination of 1) provenance graph of the matched workflow, 2) nodes that have been activated/executed. Figure 3-② shows instances of the contextual state. The bold part indicates the activated nodes. The recommendation engine takes the contextual state and

historical provenance as inputs, returns an ordered list of NL command objects
(Fig. 3-③, which contain the recommended models or parameters. Given a contextual state, the system performs two steps to generate NL recommendations:
filtering and ranking. Flowris computes a ranking of the historical provenance
with respect to their graph distance to the existing execution path. According to
the activated/executed nodes, we can determine which nodes it should recommend. After shortlisting recommendation nodes, the system orders them based
on the interaction scores, which are computed by tracking the number of times
models or parameters are applied in the same context. We therefore can make
full use of historical experience to provide analysis guidance.

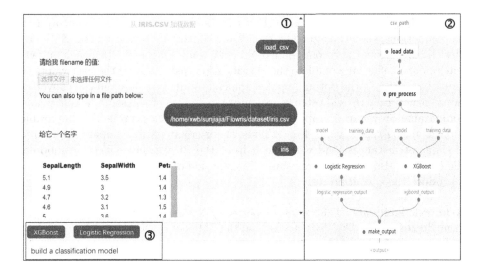

Fig. 3. The user interface of *Flowris* system

4   Demonstration

For user interface, the main workspace is divided into three areas: dialogue panel
(Fig. 3-①), provenance visualization panel (Fig. 3-②), and NL command recommendation panel (Fig. 3-③). In the dialogue panel, the system matches the NL
commands issued by users with the registered computational nodes and workflows. Once a matched workflow is detected, the system will guide users to complete the corresponding analysis task through a dialogue process, which may
include some choices (e.g., model selection or parameter settings). In the provenance visualization panel, the system presents the relevant historical provenance
graph in green and allows users to click provenance nodes to be visualized in
more details. The follow-up nodes recommended based on the bold subgraphs
and the historical experience also present in this panel. The recommendation
panel recommends a list of context-aware NL commands to guide users.

In our demonstration scenarios, users can try some classification and regression tasks. For example, a user issues a request to classify iris flowers base on their feature data. She can input "Train a model to classify irises", the conversational agent is able to recognize it as the classification analysis workflow, the provenance graph of classification analysis workflow will be visualized as a return in the provenance visualization panel. According to the connection between the workflow components, the user can first input "load csv" in the conversational panel. The conversational agent will trigger the data load function to accomplish this command. Then she can ask the agent to preprocess the dataset such as fill_na. The provenance graph of contextual state will be visualized in bold. Based on the context information and analysis experience, the provenance management module recommends models such as XGboost and Logistic Regression as the follow-up steps, as shown in Fig. 3-③. Looking through the follow-up recommendations, she is intrigued by the idea of continuing to construct a XGboost model, and selects the recommended NL command *XGboost*. With such analytical guidance, she can complete the classification task easily. If the user wants to figure out which is the optimal model, she can input "compare" in the conversational panel. Flowris will retrieve execution instances of the same workflow and comparable provenance graph from historical provenance, provides comparable table as return, as shown in Fig. 2. Thus, the user can easily compare the results of different instances, and with the help of the Provenance collection module, the user can also view the provenance information of any execution instance and reproduce the execution results.

Acknowledgements. This work is supported by National Key Research and Development Program (No. 2020YFB1710004) and the National Science Foundation of China under the grant 62272466.

References

1. Bergmann, R., Gil, Y.: Similarity assessment and efficient retrieval of semantic workflows. Inf. Syst. **40**, 115–127 (2014)
2. Chen, Q., Zhuo, Z., Wang, W.: Bert for joint intent classification and slot filling. arXiv preprint arXiv:1902.10909 (2019)
3. Fast, E., Chen, B., Mendelsohn, J., Bassen, J., Bernstein, M.S.: Iris: a conversational agent for complex tasks. In: CHI 2018, pp. 1–12 (2018)
4. Pimentel, J.F., Murta, L., Braganholo, V., Freire, J.: noWorkflow: a tool for collecting, analyzing, and managing provenance from python scripts. Proc. VLDB Endow. **10**(12) (2017)
5. Santos, E., Lins, L., Ahrens, J.P., Freire, J., Silva, C.T.: A first study on clustering collections of workflow graphs. In: Freire, J., Koop, D., Moreau, L. (eds.) IPAW 2008. LNCS, vol. 5272, pp. 160–173. Springer, Heidelberg (2008). https://doi.org/10.1007/978-3-540-89965-5_18

PhD Consortium

Fair and Privacy-Preserving Graph Neural Network

Xuemin Wang[1], Tianlong Gu[2(✉)], Xuguang Bao[1(✉)], and Liang Chang[1]

[1] Guangxi Key Laboratory of Trusted Software, Guilin University of Electronic Technology, Guilin 541004, China
bbaaooxx@163.com
[2] College of Cyber Security, Jinan University, Guangzhou 510632, China
gutianlong@jnu.edu.cn

Abstract. Graph neural networks (GNNs) have demonstrated superior performance in modeling graph-structured. They are vastly applied in various high-stakes scenarios such as financial analysis and social analysis. Among the fields, privacy issues and fairness issues have become the focus of attention. Currently, most studies focus on protecting data privacy or promoting the fairness of the model. However, how to both guarantee privacy and fairness is under-explored on GNN. To ensure GNNs behave in a socially responsible manner, it is necessary to protect the privacy and mitigate bias simultaneously. Therefore, our Ph.D. project aims at creating GNNs which can promote fairness and protect data privacy, preserving high utility performance.

Keywords: Fair · Privacy-preserving · Graph neural network

1 Problem and Motivation

Graph-structured data is pervasive in the real world, which includes social networks, bioinformatics networks, and trading networks. To gain deep insight from these graph data, lots of graph mining algorithms are proposed. In these efforts, graph neural networks (GNNs) have emerged as a powerful paradigm for learning graph representation. Besides, GNNs have demonstrated superior performance and made great progress in many high-stake areas such as credit scoring, drug discovery, and spam detection. In these risk-sensitive and security-critical scenarios, trustworthy issues such as unfair actions, lacking explainability, and ignoring privacy protection are becoming more critical. For example, in GNN-based credit scoring systems, the systems are unfair if they decline the loan application if the applicant is low-income; in GNN-based drug discovery systems, the process of finding new drugs is necessary to be explainable to humans; in GNN-based recommendation systems, when the trust third party is absent, the leakage of sensitive information in training data may happen. Recently, a trustworthy GNN [1] is proposed, aiming to solve trustworthiness issues from a technical viewpoint. It consists of four key dimensions: 1) respect for human autonomy, prevention of harm (Robustness & Privacy), fairness, and explainability. Currently, most studies only explore one aspect

© Springer International Publishing Switzerland 2023
X. Wang et al. (Eds.): DASFAA 2023, LNCS 13946, pp. 731–735, 2023.
https://doi.org/10.1007/978-3-031-30678-5_64

of trustworthy GNNs such as privacy-preserving GNNs, fair GNNs, and explainable GNNs. Actually, due to the diverse requirement, a human may expect GNNs to achieve more aspects of trustworthy GNNs, such as privacy and fairness issues.

Our doctor project aims to address privacy issues and fairness in GNN training, simultaneously, which consists of two main tasks. The first one is on centralized data. Concretely, fair GNNs need to get access to the sensitive information of users to determine whether he/she is in a protected group. At this time, when a trusted third party is absent, the individual privacy leakage of private users may happen. Similarly, without fairness consideration, privacy-preserving GNN may inherit bias from the data set and take unintended action for individuals or groups. The second one is on the distributed data. Federated GNN has been proposed to address the challenge of data-isolated islands. To improve efficiency, the federated model tends to overfit clients which possess more data. At this time, the federated model may possess worse performance on the client with fewer data. Without a fairness guarantee, the client gains the federated model with terrible performance, which decreases the trust of participants, and is unwilling to join the training. Finally, there are not enough users to provide data for federated training.

2   Current Development and Related Work

Graph Neural Network. Graphs are ubiquitous data structures, consisting of nodes and edges. They are widely applied to model complex scenarios in the real world such as social networks, traffic networks, and brain networks. Among the mining algorithms on these data, graph neural networks (GNNs) have demonstrated superior performance on analytical tasks, e.g., node classification, link prediction, graph classification, community detection, etc. Close on the heels of the successful adoption of GNNs, trustworthy issues have been arising, such as fairness and privacy. Currently, there are two directions for these two issues: fair GNN [2] and privacy-preserving GNN [3].

Fair Graph Neural Network. Fair graph neural network consists of three main categories. 1) *Group fairness*: Dai et al. [2] utilized adversarial learning to eliminate the bias of GNN by filtering out the information of sensitive attributes in learned node representation; 2) *Individual fairness*: Dong et al. [4] proposed a node ranking-based fairness, which is a novel notation of individual fairness. This individual fairness first finds the ranking list of each node in the input space and the output space based on the similarity between the node and its neighbors, and then requires these two ranking lists for each node to be consistent. 3) *Counterfactual fairness*: Agarwal et al. [5] considered both counterfactual fairness and stability. Their learned node representation is invariant to sensitive attribute values and perturbations of the graph structure.

Privacy-Preserving Graph Neural Network. Privacy-preserving graph neural network aims to protect privacy when training a GNN model. The first direction is to protect the privacy of data that is centralized in the third party. The privacy-preserving method consists of homomorphic encryption and different privacy, etc. [3]. The second direction is distributed training. Federated GNN [6] is a distributed collaborative graph learning paradigm, which can address the data isolation challenge. Although it may be vulnerable to inference attacks, it can preserve data privacy to an extent, when compared with centralized graph data to train the GNN model.

Fair and Privacy-Preserving Machine Learning. Fair and privacy-preserving machine learning requires the training framework to both protect data privacy and promote fairness. Dai et al. [7] considered an actual scenario where the sensitive attribute in graph data is limited to get access and protected by different privacy. Moreover, they proposed a training framework NT-FairGNN which can achiever group fairness and privacy on sensitive attributes, maintaining high node classification accuracy. Li et al. [8] proposed a novel optimization objective that gives higher relative weight to the device with higher loss, achieving a more uniform accuracy distribution of the federated model across each device.

3   Methodology

Model Fairness and Privacy-Preserving Centralized Data. Recently, various fairness promotion methods and privacy-preserving approaches on GNN have been proposed. To increase the trust in GNN, a training framework that can protect the privacy and promote fairness should be designed. Besides, the accuracy of GNN should be preserved. Hence, the GNN training model consists of three models: 1) the privacy-preserving module: aiming to protect the sensitive information in the user's terminal or trust the third party; 2) the fairness promotion module: focusing on adding fairness constraints or fairness loss into the objective function; 3) utility module: calculating the loss about the prediction. Fairness may improve or decrease when applying a privacy-preserving mechanism. Besides, in most attempts, accuracy may decrease when promoting fairness and privacy. To achieve the balance between accuracy, privacy, and fairness, three challenges are required to be addressed: (i) how to protect the privacy of the node; (ii) how to denoise on the server side to preserve model accuracy; (iii) how to promote fairness based on the noise data.

Fair Federated Graph Neural Network. To address the challenge of the data-isolated island in graph mining, a federated graph neural network is proposed. Most of the studies on federated GNN only consider how to learn a model with high utility. Furthermore, some personalized method has been applied in federated GNN, and privacy issues are also considered. However, federated GNNs also possess fairness issues due to the characteristics of distributed training such as performance fairness which requires the accuracy distribution of the federated model on each client data set to be uniform. The common method is providing a client with high aggregation weights when this client possesses high loss. In this way, it prevents the federated model to overfit a part of clients' data. There are two challenges: 1) how to set the aggregation weights, and 2) how to achieve performance fairness in federated GNNs setting concretely.

3.1   Validation and Exploitation of Results

Our Ph.D. project consists of two main directions: the first one focuses on the centralized data and the second one is on the distributed data. Firstly, the developed model will be evaluated and compared with the existing baseline models. Besides, the experiments are conducted on widely applied graph datasets such as Cora, credit, etc. On centralized data, to measure performance, we consider the privacy metric, fairness metric, and accuracy metric. When defending the inference attack, the accuracy of the attack model is adopted as privacy metrics. Moreover, the privacy budget is also used as the metric to measure privacy in different privacy. The fairness metrics can apply the already established metrics such as group fairness, individual fairness, and counterfactual fairness. The accuracy metrics are the accuracy that the backbone GNN model that can predict the target label. On distributed data, we first consider fairness metrics such as performance fairness and the federated model's utility. Furthermore, we also consider the privacy issues that existed in the federated GNN learning. After comparing with the baseline model, we conduct ablation studies to explore the effect of each parameter for each metric. Based on the above experiment results, we could validate the developed model and provide a corresponding conclusion.

4   Current Results and Future Work

Currently, we have achieved two GNN training frameworks that protect the privacy of the centralized data and promote the fairness of the model. Firstly, we focus on protecting data privacy when encountering curious third parties and promoting model fairness. The experiment results show that fairness and accuracy will decrease when increasing privacy protection. Besides, we consider avoiding inference attacks and promoting fairness. The experiment results show that our model promotes fairness when the privacy performance remains. Besides, the accuracy of the model decreased by less than 5%. In the future, we continue to provide a concrete method for federated GNN to improve fairness. Furthermore, we also consider privacy issues in the training process.

The main hypothesis for this project is that users require the GNN model to guarantee fairness and protect data privacy. Our project aims to provide a fair and privacy-preserving GNN approach that can promote fairness and protect the privacy of data while preserving competitive accuracy performance. The trustworthy GNN consists of four dimensions: respect for user autonomy, privacy, fairness, and explainable. In our project, to further achieve trustworthy GNN, we both consider privacy and fairness. Explainable is also an important aspect to be considered. Benefiting the explainable ability, we can acquire the causes of bias. After that, the explanation can help promote fairness more efficiently and protect privacy more flexibly. That is, in the future, the approach that can achieve privacy, fairness, and explainability for GNN can be provided.

Acknowledgement. This research was supported by grants (No. U22A2099, No. 62006057, No. 61966009, No. 62066010) from the National Natural Science Foundation of China.

References

1. Dai, E., et al.: A comprehensive survey on trustworthy graph neural networks: privacy, robustness, fairness, and explainability. arXiv preprint: arXiv:2204.08570 (2022)

2. Dai, E., Wang, S.: Say no to the discrimination: learning fair graph neural networks with limited sensitive attribute information. In: The 14th ACM International Conference on Web Search and Data Mining (WSDM), pp. 680–688 (2021)
3. Sajadmanesh, S., Gatica-Perez, D. Locally private graph neural networks. In: ACM SIGSAC Conference on Computer and Communications Security (CCS), pp. 2130–2145 (2021)
4. Dong, Y., Kang, J., Tong, H., Li, J.: Individual fairness for graph neural networks: a ranking based approach. In: 27th ACM SIGKDD Conference on Knowledge Discovery and Data Mining (KDD), pp. 300–210 (2021)
5. de Campos, C.P., Maathuis, M.H., Quaeghebeur, E.: Towards a unified framework for fair and stable graph representation learning. In: Proceedings of the Thirty-Seventh Conference on Uncertainty in Artificial Intelligence (UAI), pp. 2114–2124 (2021)
6. Liu, R., Yu, H.: Federated graph neural networks: overview, techniques and challenges. arXiv preprint: arXiv:2202.07256 (2022)
7. Dai, E., Wang, S.: Learning fair graph neural networks with limited and private sensitive attribute information. IEEE Trans. Knowl. Data Eng. (2022)
8. Sanjabi, M., Beirami, A., Smith, V.: Fair resource allocation in federated learning. In: 8th International Conference on Learning Representations (ICLR) (2020)

Improving Cross-Domain Named Entity Recognition from the Perspective of Representation

Jingyun Xu[1,2] and Yi Cai[1,2(✉)]

[1] School of Software Engineering, South China University of Technology,
Guangzhou 510650, China
`jingyun.x@qq.com, ycai@scut.edu.cn`
[2] The Key Laboratory of Big Data and Intelligent Robots, South China University
of Technology, and the Pazhou Lab, Guangzhou 510335, China

Abstract. Recently, cross-domain named entity recognition (cross-domain NER), which can reduce the high data annotation costs faced by fully-supervised methods, has drawn attention. Most competitive approaches mainly rely on pre-trained language models like BERT to represent words. As such, the original chaotic representations may bring challenges (*e.g.*, entity span detection errors and entity type misclassification) for them. Motivated by this, this proposal proposes to improve cross-domain NER by refining the original representations.

Keywords: Cross-domain Named Entity Recognition · Representation · Contrastive Learning

1 Introduction

Due to the capability of providing useful information for many NLP applications (*e.g.*, information retrieval [12,14,20,24,28], question answering [10,18] and so forth), Named Entity Recognition (NER) appeals to many researchers. With the development of deep learning techniques, fully supervised NER methods have achieved great success in NER. Unfortunately, on one hand, these approaches require large amounts of labeled data. On the other hand, in complex real-world scenarios, it is time-consuming and labor-sensitive to collect such annotated data. To address this shortcoming, cross-domain named entity recognition (cross-domain NER), which aims to transfer the knowledge learned from a high-resource source domain to a data-limited target domain, has gained increasing attention.

According to the tagging scheme, existing cross-domain NER works can be mainly classified into two categories. The first is to train models based on monolithic tags, where each token is labeled by a composition tag (*e.g.*, *B-person*) [16,29], we term them as compositional labeling-based methods. The second is to decompose the composition tag into two tags, where each token is labeled by an entity boundary tag (*e.g.*, *B*) and an entity type tag (*e.g.*, *person*) [26], we term them as modular learning-based approaches.

© The Author(s), under exclusive license to Springer Nature Switzerland AG 2023
X. Wang et al. (Eds.): DASFAA 2023, LNCS 13946, pp. 736–742, 2023.
https://doi.org/10.1007/978-3-031-30678-5_65

Though achieving promising performance, both types of methods mainly rely on pre-training language models (*i.e.*, the *query encoder* in the left part of Fig. 1) to represent words. As such, the original chaotic representations, which are shown in the right part of Fig. 1 (a), may bring two challenges. First, the mixed representations of entities belonging to different entity types are mixed (*e.g.*, the plus icons with different colors), which may lead to entity type misclassification. Second, the original chaotic representations do not distinguish between the entities and non-entities: the plus icons and the minus icons are mixed, which may lead to entity span detection errors.

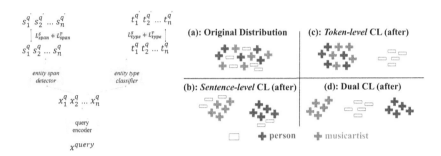

Fig. 1. Left: The baseline model. Right: The original representations and the refined representations after applying contrastive learning at different granularities. The plus icon indicates entities with different entity types (red: *musicartist*, blue: *person*), while the minus icon denotes the non-slot entities (*i.e.*, O). (Color figure online)

Recently, contrastive learning [3,8] has achieved remarkable success in representation learning. The key insight behind contrastive learning is that it firsts generates some positive keys and negative keys of the given queries. Then it could explicitly pull closer the distances between a query and its positive keys while pushing away negative keys to generate discriminative representations [8]. Inspired by this, this proposal attempts to utilize contrastive learning at different granularities to solve the above challenges in cross-domain NER from the perspective of representation: **1)** enhance the entity type classification by utilizing the sentence-level contrastive learning, as shown in Fig. 1 (b), **2)** enhance the entity span detection by utilizing the token-level contrastive learning, as shown in Fig. 1 (c), and **3)** simultaneously enhance entity span detection and entity type classification by utilizing the dual contrastive learning (*e.g.*, sentence-level contrastive learning and token-level contrastive learning), as shown in Fig. 1 (d). The structure of the next sections is listed as follows. Section 2 discusses various existing research approaches to cross-domain NER and contrastive learning. Section 3 describes the proposed method. Finally, a list of the near-future plans is drawn in Sect. 4.

2   Related Work

2.1   Cross-Domain Named Entity Recognition

Recently, many works have been explored for the cross-domain entity named entity recognition, which enables models to have better transferability by exploiting the correlation information between resource-rich and data-limited data [9,15]. In particular, a multi-cell compositional LSTM structure was proposed [11], which could capture the correlations between entity types separately and simultaneously consider which entity type a word belongs to. Meanwhile, [23] introduced a new architecture tailored to capture commonalities from multiple genres. Due to the lack of datasets that contains domain-specific entity types, it is hard to evaluate the performance of existing cross-domain NER methods. To this end, [16] introduces a new cross-domain NER dataset *crossNER*. Besides, they explore the power of pre-training language models for domain adaptation. Based on a machine reading comprehension framework, [21] further designs a new model [16]. Besides, to reduce the complexity of models and small the label space [26] explored a modular learning approach.

2.2   Contrastive Learning

In computer vision, various contrastive learning (CL) methods have been proposed, such as SimCLR [3], Moco [8], BYOL [7], CLIP [17], SwAV [2], SimSiam [4] and so on. Owing to its superior performance, some researchers also explore its power in the tasks of natural language understanding and pre-training [1,5,6,19,22,25,27].

To summarize, most existing cross-domain named entity recognition research works focus on building a transferable model by exploiting the correlation information between source and target domains. In contrast, from the perspective of representation, this proposal proposes to improve cross-domain NER by refining the original chaotic representations.

3   Model

Problem Statement. For the cross-domain NER, there are a number of labeled sentences $S = (S_1, S_2, \ldots, S_{N_s})$ from a source domain and a set of limited sentences $T = (T_1, T_2, \ldots, T_{N_t})$ from a target domain, where D_i denotes the ith sentence of the domain D, both N_s and N_t denote the sentence length. Given two sentences $S_i = (w_{i1}^S, w_{i2}^S, \ldots, w_{im}^S)$ and $T_i = (w_{i1}^T, w_{i2}^T, \ldots, w_{in}^T)$, one from each domain side, m and n denote the total number of words in the corresponding sentence. For the jth word in the sentence D_i, cross-domain NER aims to predict the label $y_{ij}^D \in \{$B-entity type, I-entity type, O$\}^1$ for each word w_{ij}^D.

Architecture. As shown in the right part of Fig. 2, given the sentence D_i (*i.e.*, S_i or T_i), we first construct it as "$[CLS]D_i[SEP]$", where [CLS] and [SEP] denote

---

[1] Here entity types are pre-defined, such as *location*, *organization*, and so on.

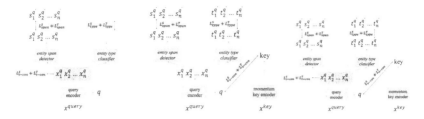

Fig. 2. The illustration of contrastive learning at different granularities: left: token-level contrastive learning, middle: sentence-level contrastive learning, right: dual contrastive learning (*i.e.*, the combination of token-level contrastive learning and sentence-level contrastive learning).

two special symbols [13]. Then we utilize BERT as the *query encoder* to extract the corresponding hidden representations, denoted as $z_i^S = (h_{iCLS}^S, h_{i1}^S, h_{i2}^S, \ldots, h_{im}^S, h_{iSEP}^S)$ of S_i and $z_i^T = (h_{iCLS}^T, h_{i1}^T, h_{i2}^T, \ldots, h_{in}^T, h_{iSEP}^T)$ of T_i. Meanwhile, two modules **token-level Contrastive Learning** and **Sentence-level Contrastive Learning** are introduced to refine original representations at different granularities. Finally, two modules *entity span detector* and the *entity type classifier* respectively take the refined representations after applying contrastive learning at different granularities (*i.e.*, token-level CL and sentence-level CL) to make predictions.

We briefly introduce the core part of our proposed method.

The Token-Level Contrastive Learning Module. To enhance entity span detection, we propose to apply token-level contrastive learning to cross-domain NER by minimizing the distance between representations of entities with the same type and maximizing the distance between representations of entities and non-entities, which is primarily shown in the left part of Fig. 2. As such, as shown in Fig. 1 (c), the representations of entities and non-entities are separated, solving the entity span detection errors,

The Sentence-Level Contrastive Learning Module. Based on data augmentation, we propose to apply contrastive learning to cross-domain NER by minimizing the distance between representations of entities with the same type and maximizing the distance between representations of entities belonging to different types in order to improve the applicability of the model in the target domain [8], which is primarily shown in the middle part of Fig. 2. As such, as shown in Fig. 1 (b), the representations of entities with different entity types are separated, solving the entity type misclassification.

4 Conclusion

This proposal outlined the current development of cross-domain NER and figured out the related challenges that should be considered. Previously, we have studied the feasibility of improving the performance of natural language processing tasks from the perspective of representation (*e.g.*, short text classification and aspect term extraction, sentiment analysis [20,24]). At the current stage, we are working on improving cross-domain NER from the perspective of representation. In particular, we propose to enhance entity span detection by utilizing token-level contrastive learning and enhance entity type classification by utilizing sentence-level contrastive learning, as shown in the right part of Fig. 2.

Acknowledgement. This work was supported in part by the National Natural Science Foundation of China under Grant 62076100, in part by the Fundamental Research Funds for the Central Universities, SCUT under Grant x2rjD2220050, in part by the Science and Technology Planning Project of Guangdong Province under Grant 2020B0101100002, in part by the Hong Kong Research Council under Grants PolyU 11204919 and C1031-18G, and in part by the Internal Research from the Hong Kong Polytechnic University under Grant 1.9B0V.

References

1. Barros, P., Sciutti, A.: All by myself: learning individualized competitive behavior with a contrastive reinforcement learning optimization. Neural Netw. **150**, 364–376 (2022)
2. Caron, M., Misra, I., Mairal, J., Goyal, P., Bojanowski, P., Joulin, A.: Unsupervised learning of visual features by contrasting cluster assignments. NIPS **33**, 9912–9924 (2020)
3. Chen, T., Kornblith, S., Norouzi, M., Hinton, G.: A simple framework for contrastive learning of visual representations. In: International Conference on Machine Learning, pp. 1597–1607 (2020)
4. Chen, X., He, K.: Exploring simple Siamese representation learning. In: Proceedings of the IEEE/CVF Conference on Computer Vision and Pattern Recognition, pp. 15750–15758 (2021)
5. Choi, S., Jeong, M., Han, H., Hwang, S.W.: C2l: causally contrastive learning for robust text classification. In: Proceedings of the AAAI Conference on Artificial Intelligence, pp. 10526–10534 (2022)
6. Giorgi, J., Nitski, O., Wang, B., Bader, G.: Declutr: deep contrastive learning for unsupervised textual representations. In: Proceedings of the 59th Annual Meeting of the Association for Computational Linguistics and the 11th International Joint Conference on Natural Language Processing, pp. 879–895 (2021)
7. Grill, J.B., et al.: Bootstrap your own latent-a new approach to self-supervised learning. Adv. Neural. Inf. Process. Syst. **33**, 21271–21284 (2020)
8. He, K., Fan, H., Wu, Y., Xie, S., Girshick, R.: Momentum contrast for unsupervised visual representation learning. In: Proceedings of the IEEE/CVF Conference on Computer Vision and Pattern Recognition, pp. 9729–9738 (2020)
9. Houlsby, N., et al.: Parameter-efficient transfer learning for NLP. In: International Conference on Machine Learning, pp. 2790–2799 (2019)
10. Huang, Q., et al.: Entity guided question generation with contextual structure and sequence information capturing. In: Proceedings of the AAAI Conference on Artificial Intelligence, vol. 35, pp. 13064–13072 (2021)

11. Jia, C., Zhang, Y.: Multi-cell compositional LSTM for NER domain adaptation. In: Proceedings of the 58th Annual Meeting of the Association for Computational Linguistics, pp. 5906–5917 (2020)

12. Jiang, D., Ren, H., Cai, Y., Xu, J., Liu, Y., Leung, H.F.: Candidate region aware nested named entity recognition. Neural Netw. **142**, 340–350 (2021)

13. Kenton, J.D.M.W.C., Toutanova, L.K.: Bert: pre-training of deep bidirectional transformers for language understanding. In: NAACL, pp. 4171–4186 (2019)

14. Lei, X., Cai, Y., Xu, J., Ren, D., Li, Q., Leung, H.: Incorporating task-oriented representation in text classification. In: Li, G., Yang, J., Gama, J., Natwichai, J., Tong, Y. (eds.) DASFAA 2019. LNCS, vol. 11447, pp. 401–415. Springer, Cham (2019). https://doi.org/10.1007/978-3-030-18579-4_24

15. Lin, B.Y., Lu, W.: Neural adaptation layers for cross-domain named entity recognition. In: Proceedings of the 2018 Conference on Empirical Methods in Natural Language Processing, pp. 2012–2022 (2018)

16. Liu, Z., et al.: CrossNER: evaluating cross-domain named entity recognition. In: Proceedings of the AAAI Conference on Artificial Intelligence, pp. 13452–13460 (2021)

17. Radford, A., et al.: Learning transferable visual models from natural language supervision. In: International Conference on Machine Learning, pp. 8748–8763 (2021)

18. Ren, D., Cai, Y., Lei, X., Xu, J., Li, Q., Leung, H.F.: A multi-encoder neural conversation model. Neurocomputing **358**, 344–354 (2019)

19. Shah, D., Gupta, R., Fayazi, A., Hakkani-Tur, D.: Robust zero-shot cross-domain slot filling with example values. In: Proceedings of the 57th Annual Meeting of the Association for Computational Linguistics, pp. 5484–5490 (2019)

20. Tan, X., Cai, Y., Xu, J., Leung, H.F., Chen, W., Li, Q.: Improving aspect-based sentiment analysis via aligning aspect embedding. Neurocomputing **383**, 336–347 (2020)

21. Tang, M., Zhang, P., He, Y., Xu, Y., Chao, C., Xu, H.: DoSEA: a domain-specific entity-aware framework for cross-domain named entity recogition. In: Proceedings of the 29th International Conference on Computational Linguistics, pp. 2147–2156 (2022)

22. Wang, L., et al.: Bridge to target domain by prototypical contrastive learning and label confusion: re-explore zero-shot learning for slot filling. In: Proceedings of the 2021 Conference on Empirical Methods in Natural Language Processing, pp. 9474–9480 (2021)

23. Wang, J., Kulkarni, M., Preoţiuc-Pietro, D.: Multi-domain named entity recognition with genre-aware and agnostic inference. In: ACL, pp. 8476–8488 (2020)

24. Xu, J., et al.: Incorporating context-relevant concepts into convolutional neural networks for short text classification. Neurocomputing **386**, 42–53 (2020)

25. Xu, S., Zhang, X., Wu, Y., Wei, F.: Sequence level contrastive learning for text summarization. In: Proceedings of the AAAI Conference on Artificial Intelligence, vol. 36, pp. 11556–11565 (2022)

26. Zhang, X., Yu, B., Wang, Y., Liu, T., Su, T., Xu, H.: Exploring modular task decomposition in cross-domain named entity recognition. In: SIGIR, pp. 301–311 (2022)

27. Zhang, Y., Zhang, R., Mensah, S., Liu, X., Mao, Y.: Unsupervised sentence representation via contrastive learning with mixing negatives. In: Proceedings of the AAAI Conference on Artificial Intelligence, pp. 11730–11738 (2022)

28. Zheng, C., Cai, Y., Xu, J., Leung, H., Xu, G.: A boundary-aware neural model for nested named entity recognition. In: Association for Computational Linguistics (EMNLP-IJCNLP) (2019)
29. Zheng, J., Chen, H., Ma, Q.: Cross-domain named entity recognition via graph matching. In: Findings of the Association for Computational Linguistics (ACL 2022), pp. 2670–2680 (2022)

Knowledge Hypergraph Reasoning Based on Representation Learning

Zhao Li[✉]

College of Intelligence and Computing, Tianjin University, Tianjin, China
lizh@tju.edu.cn

Abstract. The knowledge hypergraph, as a data carrier for describing real-world things and complex relationships, faces the challenge of incompleteness due to the proliferation of knowledge. It is an important research direction to use representation learning technology to reason knowledge hypergraphs and complete missing and unknown knowledge tuples. Most current methods extend directly from the binary relations of the knowledge graph to the n-ary relations without obtaining the position and role information of entities in each n-ary relation tuple, however, these semantic attribute information are crucial for knowledge hypergraph reasoning based on representation learning.

Keywords: Knowledge Hypergraph · Representation Learning · Knowledge Reasoning · Position and Role Information

1 Introduction

As an important cornerstone of the new generation of artificial intelligence, knowledge graph has been widely used in question answering system, personalized recommendation, search engine, and other fields. Knowledge graphs store facts of the form $r(h, t)$, where r is the binary relation, and h and t are the head and tail entities, respectively [13]. However, the real world is complex and coupled, and many relationships cannot be expressed using simple binary relationships. Take Freebase as an example, 61% of the relations in the knowledge base are n-ary relations, and more than a third of the entities in the entity set is involved in the composition of n-ary relations. These findings suggest that it is more general and necessary to use n-ary relations to describe things in the real world and their relationships to each other. The knowledge hypergraph, a large-scale semantic network that stores human knowledge in the form of a graph structure, can be seen as a generalization of the knowledge graph with greater expressive power by its formal use of n-ary relations to portray real-world things and their complex relationships.

Due to the large amount of knowledge in the real world, it is impractical to store all knowledge in the knowledge hypergraph, leading to the biggest challenge of the current knowledge hypergraph is its incompleteness, i.e. some of

PhD Supervisor: Prof. Xin Wang.

the links between entities are missing [9]. In the face of the incompleteness of the knowledge hypergraph, manually adding links between entities is very labor- and material-intensive, resulting in the need for automatic reasoning about missing links between entities. While representation learning-based knowledge graph reasoning techniques have proven to be an effective method for reasoning about binary relations, knowledge hypergraph reasoning remains a relatively unexplored area. Knowledge hypergraph reasoning aims to embed entities and relations as continuous low-dimensional vectors for efficient unknown tuple reasoning. By using these vector representations to efficiently represent the semantic associations between entities and relations, it is possible to infer whether missing tuples in the knowledge hypergraph are true, and thus complete the missing knowledge in the knowledge hypergraph. Knowledge hypergraph reasoning plays an important role in the construction and application of knowledge hypergraphs.

2   Related Work

Representation learning methods for knowledge hypergraph reasoning are mainly divided into three categories: translation models, tensor factorization models, and neural network models.

Translation Models. The translation-based model embeds entities and relations into the same vector space, and translates entity vectors based on relational embeddings so that the model learns the representations of entities and relations in the embedding space. The pioneering work m-TransH [15] proposes the idea of n-ary relations and is the first knowledge hypergraph representation learning model, which extends TransH [14] based on binary relations to n-ary relations. The RAE [11] builds on the m-TransH model and encourages modeling correlations between different attributes in order to achieve performance extensions to the m-TransH model.

Tensor Factorization Models. The tensor factorization-based models usually decompose a higher-order tensor into a sum of multiple lower-order tensors. The first work of this kind is the GETD [10] model, which is an extension of the knowledge graph TuckER [1] model with full expressiveness. The HSimplE [3] model based on SimplE [8] that can be applied to n-ary relations is proposed. Meanwhile, to consider the role of entity position semantic information in knowledge hypergraph representation learning, the HypE [3] model based on location to learn different embedding representations for different entities is proposed. Then, in order to address the problem of over-parameterization of models based on tensor factorization, the S2S [2] model achieves performance improvements for this class of models by sparsifying the core tensor to reduce the model parameters while retaining its expressive power using neural architecture search techniques. These models all treat each entity as equally important, but this is usually not the case. The RAM [11] model finds that current work has ignored the important semantic property of entity roles, and therefore proposes role-aware modeling at the role level, which encourages semantically related roles to have similar embedding representations.

Neural Network Models. The first work of this class of methods, NaLP [7], introduces the role-entity form of n-ary relations, using convolutional neural network (CNN) and fully connected network (FCN) modules to measure the compatibility of entities with their roles. And tNalp+ [6] further considers type information and optimizes negative sampling. HINGE [12] and NeuInfer [5] decompose the n-ary relation tuple into a knowledge graph triple and several role-entity pairs, relying on CNN and FCN models to complete the compatibility determination, respectively. These two models emphasize the importance of the principal triple, but the introduction of decomposition undermines the semantic and structural integrity of the n-ary relation fact. The StarE [4] brings significant performance benefits as a message-passing method that applies the CompGCN module to model decomposed knowledge graph triples.

3   Problem Definitions and Methodology

Currently, the existing knowledge hypergraph representation learning method is far from satisfactory results, which is a valuable and promising direction. The semantic information described by the knowledge hypergraph is very complex, and to better perform representation learning and knowledge reasoning tasks based on n-ary relations, we introduce two knowledge hypergraph semantic properties, entity position information and role information. As shown in Fig. 1, it is a real example of a knowledge hypergraph about a football star *Lionel Messi*. It can be seen that the entity position and role determine the different semantic information of the entity under the n-ary relation, i.e. the specific semantic information of the football star *Lionel Messi* in the tuple is determined by the position (3, 2, 1) and the corresponding role (father, football player, award winner) under the three different ternary relations respectively. Thus, the position and role information of entities is crucial for knowledge hypergraph representation learning and reasoning based on n-ary relations. Knowledge hypergraphs with entity location and role information are defined by us as

Definition 1. (Knowledge Hypergraph) *Given a finite set of entities $\mathcal{E}$, relations $\mathcal{R}$, and observed tuples $\mathcal{T}_O$, a knowledge hypergraph can be represented as $\mathcal{H} = (\mathcal{E}, \mathcal{R}, \mathcal{T}_O)$. The tuple t based on n-ary relations is denoted $r(\rho_1^r : e_1, \rho_2^r : e_2, ..., \rho_i^r : e_i, ..., \rho_\alpha^r : e_\alpha)$, where $r \in \mathcal{R}$ is a relation within the set of relations, $e_i \in \mathcal{E}$ is an entity within the set of entities, and i is the position index of the entity. ρ_i^r is the corresponding role of the relation r for an entity at index i, which can be explicitly defined or implicitly implied in the tuple semantics, and α is the arity of the relation r. Knowledge graphs based on the binary relation triple $r(e_1, e_2)$ are a special case of knowledge hypergraphs, where the arity of any relation $r \in \mathcal{R}$ in the set of relations is two.*

At present, the best-performing Transformer architecture will be used by us for knowledge hypergraph representation learning and reasoning, but the all-symmetric property of the Transformer results in its inability to capture position and role information, which is a major challenge. For entity role information encoding. Inspired by classical statistical methods in the field of information

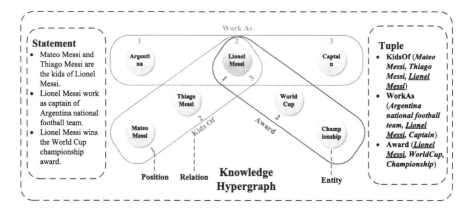

Fig. 1. A real-world example of knowledge hypergraph about a set of facts related to Lionel Messi, where each tuple is accompanied by different position information.

retrieval, we propose an entity role information encoding strategy *EF-IRF* based on prior information. Its main idea is that if an entity appears in the knowledge hypergraph with high frequency and contains few *n*-ary relations, the entity is considered to have good *n*-ary relation distinguishing ability and high role importance. According to the characteristics of knowledge hypergraph, logarithmic smoothing is adopted for both entity frequency index *EF* and inverse hyper-relation frequency index *IRF* to suppress the influence of noise data to a certain extent, and Laplacian smoothing is adopted for *IRF* to prevent the occurrence of uncomputable and negative probability anomalies. For position information encoding, there are generally only absolute position encoding and relative position encoding. In general, absolute position encoding has the advantages of simplicity and fast computation, while relative position encoding directly reflects the relative position signal, which is in line with the intuitive understanding and often has better practical performance. From the perspective of the complex number, we integrate absolute position encoding and relative position encoding, and propose a rotary position encoding strategy in complex number space for knowledge hypergraph, with the attention mechanism to achieve relative position encoding by absolute position encoding.

4   Conclusion

Knowledge hypergraph representation learning and reasoning is a challenging and unexplored area. In this paper, we plan to enhance the attention mechanism of the Transformer architecture combining a role information encoding and position information encoding strategy together to improve the representation learning and reasoning capability of the knowledge hypergraph.

Acknowledgments. This work is supported by the National Key R&D Program of China (2020AAA0108504), the National Natural Science Foundation of China (61972275).

References

1. Balažević, I., Allen, C., Hospedales, T.: Tucker: tensor factorization for knowledge graph completion. In: Proceedings of the 2019 Conference on Empirical Methods in Natural Language Processing and the 9th International Joint Conference on Natural Language Processing (EMNLP-IJCNLP), pp. 5185–5194 (2019)
2. Di, S., Yao, Q., Chen, L.: Searching to sparsify tensor decomposition for N-ary relational data. In: Proceedings of the Web Conference 2021, pp. 4043–4054 (2021)
3. Fatemi, B., Taslakian, P., Vazquez, D., Poole, D.: Knowledge hypergraphs: prediction beyond binary relations. In: Proceedings of the Twenty-Ninth International Conference on International Joint Conferences on Artificial Intelligence (IJCAI 2020), pp. 2191–2197 (2021)
4. Galkin, M., Trivedi, P., Maheshwari, G., Usbeck, R., Lehmann, J.: Message passing for hyper-relational knowledge graphs. In: Proceedings of the 2020 Conference on Empirical Methods in Natural Language Processing, pp. 7346–7359 (2020)
5. Guan, S., Jin, X., Guo, J., Wang, Y., Cheng, X.: NeuInfer: knowledge inference on N-ary facts. In: Proceedings of the 58th Annual Meeting of the Association for Computational Linguistics, pp. 6141–6151 (2020)
6. Guan, S., Jin, X., Guo, J., Wang, Y., Cheng, X.: Link prediction on N-ary relational data based on relatedness evaluation. IEEE Trans. Knowl. Data Eng. **35**(1), 672–685 (2021)
7. Guan, S., Jin, X., Wang, Y., Cheng, X.: Link prediction on N-ary relational data. In: The World Wide Web Conference, pp. 583–593 (2019)
8. Kazemi, S.M., Poole, D.: Simple embedding for link prediction in knowledge graphs. Adv. Neural Inf. Proc. Syst. **31** (2018)
9. Li, Z., Liu, X., Wang, X., Liu, P., Shen, Y.: TransO: a knowledge-driven representation learning method with ontology information constraints. World Wide Web **26**, 297–319 (2023)
10. Liu, Y., Yao, Q., Li, Y.: Generalizing tensor decomposition for N-ary relational knowledge bases. In: Proceedings of the Web Conference 2020, pp. 1104–1114 (2020)
11. Liu, Y., Yao, Q., Li, Y.: Role-aware modeling for N-ary relational knowledge bases. In: Proceedings of the Web Conference 2021, pp. 2660–2671 (2021)
12. Rosso, P., Yang, D., Cudré-Mauroux, P.: Beyond triplets: hyper-relational knowledge graph embedding for link prediction. In: Proceedings of the Web Conference 2020, pp. 1885–1896 (2020)
13. Shen, Y., Li, Z., Wang, X., Li, J., Zhang, X.: Datatype-aware knowledge graph representation learning in hyperbolic space. In: Proceedings of the 30th ACM International Conference on Information and Knowledge Management, pp. 1630–1639 (2021)
14. Wang, Z., Zhang, J., Feng, J., Chen, Z.: Knowledge graph embedding by translating on hyperplanes. In: Proceedings of the AAAI Conference on Artificial Intelligence, vol. 28 (2014)
15. Wen, J., Li, J., Mao, Y., Chen, S., Zhang, R.: On the representation and embedding of knowledge bases beyond binary relations. In: Proceedings of the Twenty-Fifth International Joint Conference on Artificial Intelligence, pp. 1300–1307 (2016)

Knowledge-Based Course Map Generation for Student Performance Prediction

Jianing Xia[(✉)]

School of Information Technology, Deakin University, Geelong, Australia
xiaji@deakin.edu.au

Abstract. Predicting students performance in online educational environment is a topic of considerable interest in recent years. While there has been some previous works endeavored to predict performance of students using some classification algorithms. Most of the work has focused on identifying vertex attributes that affect performance. Less work considers the effect of the explicit and implicit relationships between entities. Attributes contained in these relationships are also informative and can be used to significantly improve the accuracy of performance predictions. In this project, we present a comprehensive exploration on the topic of performance prediction based on heterogeneous course map. Specifically, we propose a novel definition of the Heterogeneous Course Map (HCM) to explore the explicit and implicit relationships between entities. We also develop robust, effective and efficient models and algorithms to fuse the structured data in HCM to predict the performance of students.

Keywords: Course map · Performance prediction · Prerequisite relations

1 Introduction

Online learning has received more and more attention, especially during the COVID-19 pandemic period. Flexibility, low cost, and convenience of online learning are the key factors to this anticipated boom. However, these factors also result in low completion rate and high drop-out rate associated with open online courses [1]. To address this phenomenon, e-learning often demands real-time and accurate assessment of students' performance. The result of student performance prediction can be used to build early warning systems and personalized recommendation systems to improve the students' learning experience, thereby creating opportunities to improve educational outcomes. There are multiple factors and attributes among different types of entities that affect students' performance and learning outcomes. Such as learning activities, specific concepts/knowledge points a student is targeting on, the courses the student has enrolled in and so on. These multi-typed entities and relationships between them

© The Author(s), under exclusive license to Springer Nature Switzerland AG 2023
X. Wang et al. (Eds.): DASFAA 2023, LNCS 13946, pp. 748–752, 2023.
https://doi.org/10.1007/978-3-031-30678-5_67

constitutes a heterogeneous information network (HIN). How to find the important factors or relations in HINs affecting the students performance the most is still an open problem. Course map generation is an essential task that helps educators and students visualize the structure of a course, the topics covered, and the relationships between them. By using data on student performance, course map generation can be enhanced to predict student performance and identify areas where students may struggle. The key challenge for constructing a course map is to handle heterogeneous student data and course data. So how to conduct a novel link prediction to find the explicit and implicit relations between heterogeneous entities, and identify the relationships which contribute the most to predicting student achievement has become the focus of this research.

2   Literature Review

In this chapter, we will briefly introduce the course map construction methods and review some of the existing popular prereqisite relation extraction techniques. At the same time, literature on students' performance prediction will also be reviewed.

2.1   Course Map Generation

Knowledge-based Course Map provides the Visualization of subject knowledge in educational environment. There are many efforts devoted to extract concept prerequisite relations and construct concept map in order to model the knowledge structure. These methods can be divided into recovery-based methods and learning-based methods. Recovery-based methods assume that prerequisite strength between two courses is a cumulative effect of the prerequisite strengths of all concept pairs. For instance, Yang et al. [2] created a concept graph by mapping courses to concepts and learning concept-level dependencies based on optimization method. Using the induced concept graph, they can predict unobserved precedence relationships among any courses. Similarly, Liang et al. [3] recovered a universally shared concept graph based on the observed course dependencies, course representation and a variant of soft-margin SVM algorithm. To demonstrate their superior performance, experiments had been conducted on both a synthetic data set and an actual university course data set. Another method called learning-based technique uses a number of manually created or automatically generated features to train prerequisite classifiers. Many works utilized the content of Wikipedia articles and their linkage structures as prerequisite evidence. Such as models of RefD from Liang et al. [4], Mooc-RF from Pan et al. [5], SEKEEO from Xiao et al. [6] and so on. In addition to Wikipedia resource, semantic, contextual and structural features based on course information are also included. However, gathering and processing data to select relevant and discriminating features takes time and is not trivial. Moreover, domain-dependent features extraction strongly rely on domain expertise which can't reach the fundamental generalisation, and non-optimal feature configurations could compromise classification accuracy.

2.2   Performance Prediction

Many works has been done to predict students' academic performance using different machine learning techniques, such as decision tree, artificial Neural networks based on Educational Data Mining [7,8]. It was observed that the Grade Point Average (GPA) and Cumulative GPA (CGPA) of students are the most common indicators used for evaluating and predicting students' academic achievement at the university level, other attributes such as quiz grades and attendance of students are also considered. However, these models only focused on general correlation analyses for making predictions, such as effect from GPA earning activities and other attributes of students. None fully consider the explicit and implicit relationships between entities in the heterogeneous information network. In this project, we are going to explore those relationships which help to predict students' performance more efficiently and accurately.

3   Methodology

In this work, we defined a Heterogeneous Course Map (HCM), in which the states of different kinds of entities and various relationships between them could be represented properly and explored. HCM is essentially an aggregation of groups of entities, where each group represents a particular type of entities known as the vertices, and the links between entities in the same group and/or across different groups are known as edges in the heterogeneous networks. Based on the HCM, we aim to explore the explicit and implicit relationships between entities using novel link prediction methods. For instance, the precedence dependencies between courses are already known from the curriculum design of universities. We have to explore the concepts covered by various courses and the prerequisite relationships among them. These relationships can be used in many educational applications. For instance, uncovered prerequisite relations between concepts are crucial for intelligent tutoring systems and curriculum planning, which can support instructors' educational offerings and enhance students' learning. Specifically, all the Wikipedia concepts are extracted from courses descriptions using Wikipedia API. Each concept node is initially represented using pretrained word embeddings. The concept representation is then updated using graph convolutional network (GCN) [9] by aggregating messages from its immediate neighbours. Concept representations are then fed into a Siamese network to predict the concept prerequisite relations. At the same time, concept pairwise features are extracted according to textual and structural information from course description and Wikipedia. All these features are combined to optimize the concept prerequisite relations in addition to course dependencies. Then we generate the concept map which integrates concept representation, courses dependencies and concept pairwise features. The concept prerequisite relationships also constitutes the important part of our HCM. As for predicting students' performance, machine Learning classification and regression algorithms have been demonstrated helpful. However, none of these methods consider the power of the interconnections between multi-typed entities. In this work, we make full use of the relationships obtained from our HCM to extract useful features and apply different models including Neural Network, Graph Embedding, and Random Forest to evaluate the learning performance. The challenge is to incorporate HCM and these classification methods effectively.

4   Current Results and Future Work

Current educational concept map analysis works do not have available public datasets to support related research. To solve this problem, researchers tend to generate private datasets by using concept and link extraction methods. However, the data collection is time-consuming and the private synthesis data is tendentious to their analysis result. So we built a comprehensive dataset named AuCM (Australian Courses Map data) for learning concept maps. Specifically, we collected 1292 undergraduate courses in Information technology (IT) and computer science (CS) from 14 Australian universities including information of course ID, category, prerequisite requirements among courses and so on. For comparative analysis, 7 universities were from "Group of Eight (G8)". To the best of our knowledge, this is the first dataset containing course information from Australian universities. Besides, we analyzed the semantic properties based on the concepts retrieved from the course description and visualized them to illustrate how our dataset could be used. Through the comparative analysis between universities from G8 and other non-G8 universities, our statistical analysis showed the superiority of the number of courses for a IT/CS bachelor degree offered by G8 universities. By semantic analysis of concepts/course descriptions, we found the variances in course design in terms of semantic richness and focus areas between G8 and non-G8 universities. We also constructed concept maps on AuCM to validate existing concept map generation baseline methods. In addition, we proposed a novel model to discover the prerequisite relationships between concepts which integrates the word representation, handcrafted features and courses dependencies using a neural network setting with better performance than baseline models. With obtained concept prerequisites, we could generate our HCM which consists of entities such as courses, concepts and students and the relationships among them. Future work would be to perform a more comprehensive model comparison and evaluation between our students' performance prediction model and other baselines based on HCM. Another direction would be to explore other educational applications using our HCM, such as optimal learning sequence generation and intelligent tutoring.

References

1. Lara, J.A., Aljawarneh, S., Pamplona, S.: Special issue on the current trends in E-learning Assessment. J. Comput. High. Educ. **32**, 1–8 (2020)
2. Yang, Y., Liu, H., Carbonell, J., Ma, W.: Concept graph learning from educational data. In: Proceedings of the Eighth ACM International Conference on Web Search and Data Mining (WSDM 2015), Shanghai, China, 2–6 February 2015, pp. 159–168 (2015)

3. Liang, C., Ye, J., Wu, Z., Pursel, B., Giles, C.L.: Recovering concept prerequisite relations from university course dependencies. In: Proceedings of the Thirty-First AAAI Conference on Artificial Intelligence, 4–9 February 2017, San Francisco, California, USA, pp. 4786–4791 (2017)
4. Liang, C., Wu, Z., Huang, W., Giles, C.L.: Measuring prerequisite relations among concepts. In: Proceedings of the 2015 Conference on Empirical Methods in Natural Language Processing (EMNLP 2015), Lisbon, Portugal, 17–21 September 2015, pp. 1668–1674. The Association for Computational Linguistics (2015)
5. Pan, L., Li, C., Li, J., Tang, J.: Prerequisite relation learning for concepts in MOOCs. In: Proceedings of the 55th Annual Meeting of the Association for Computational Linguistics, pp. 1447–1456. Association for Computational Linguistics (2017)
6. Xiao, K., Bai, Y., Wang, Z.: Extracting prerequisite relations among concepts from the course descriptions (SEKEEO-RN). Int. J. Softw. Eng. Knowl. Eng. **32**(4), 503–523 (2022)
7. Feng, G., Fan, M., Chen, Yu.: Analysis and prediction of students' academic performance based on educational data mining. IEEE Access **10**, 19558–19571 (2022)
8. Batool, S., Rashid, J., Nisar, M.W., Kim, J., Kwon, H.Y., Hussain, A.: Educational data mining to predict students' academic performance: a survey study. Educ. Inf. Technol. **28**(1), 905–971 (2023)
9. Kipf, T.N., Welling, M.: Semi-supervised classification with graph convolutional networks. In: 5th International Conference on Learning Representations (ICLR 2017), Toulon, France, 24–26 April 2017, Conference Track Proceedings (2017). https://openreview.net/

Author Index

Printed in the United States
by Baker & Taylor Publisher Services